Techniques and Concepts of High Energy Physics X

NATO Science Series

A Series presenting the results of activities sponsored by the NATO Science Committee. The Series is published by IOS Press and Kluwer Academic Publishers, in conjunction with the NATO Scientific Affairs Division.

A. **Life Sciences**	IOS Press
B. **Physics**	Kluwer Academic Publishers
C. **Mathematical and Physical Sciences**	Kluwer Academic Publishers
D. **Behavioural and Social Sciences**	Kluwer Academic Publishers
E. **Applied Sciences**	Kluwer Academic Publishers
F. **Computer and Systems Sciences**	IOS Press

1. **Disarmament Technologies**	Kluwer Academic Publishers
2. **Environmental Security**	Kluwer Academic Publishers
3. **High Technology**	Kluwer Academic Publishers
4. **Science and Technology Policy**	IOS Press
5. **Computer Networking**	IOS Press

NATO-PCO-DATA BASE

The NATO Science Series continues the series of books published formerly in the NATO ASI Series. An electronic index to the NATO ASI Series provides full bibliographical references (with keywords and/or abstracts) to more than 50000 contributions from internatonal scientists published in all sections of the NATO ASI Series.
Access to the NATO-PCO-DATA BASE is possible via CD-ROM "NATO-PCO-DATA BASE" with user-friendly retrieval software in English, French and German (WTV GmbH and DATAWARE Technologies Inc. 1989).

The CD-ROM of the NATO ASI Series can be ordered from: PCO, Overijse, Belgium

Series C: Mathematical and Physical Sciences – Volume 534

Techniques and Concepts of High Energy Physics X

edited by

T. Ferbel
University of Rochester,
Department of Physics and Astronomy,
Rochester, NY, U.S.A.

Kluwer Academic Publishers

Dordrecht / Boston / London

Published in cooperation with NATO Scientific Affairs Division

Proceedings of the NATO Advanced Study Institute on
Techniques and Concepts of High Energy Physics
St. Croix, US Virgin Islands,
June 18-29, 1998

A C.I.P. Catalogue record for this book is available from the Library of Congress.

ISBN 0-7923-5729-9 (HB)
ISBN 0-7923-5730-2 (PB)

Published by Kluwer Academic Publishers,
P.O. Box 17, 3300 AA Dordrecht, The Netherlands.

Sold and distributed in North, Central and South America
by Kluwer Academic Publishers,
101 Philip Drive, Norwell, MA 02061, U.S.A.

In all other countries, sold and distributed
by Kluwer Academic Publishers,
P.O. Box 322, 3300 AH Dordrecht, The Netherlands.

Printed on acid-free paper

Printed in the Netherlands.

This volume is dedicated to Bob and Jane Wilson,
for their friendship and support of the Institute.

TABLE OF CONTENTS

PREFACE

The Proceedings of the tenth Advanced Study Institute (ASI) on Techniques and Concepts of High Energy Physics are dedicated to Jane and Bob Wilson. Jane joined Bob at St. Croix for the first session of this Institute, after Bob had stepped down as director of Fermilab, and was scheming to build a modest charm factory in the parking lot of Columbia University's Nevis Laboratory. Through the years, Bob has been a great friend of the School, and much of its success and flavor can be attributed to his guidance and support.

The 1998 meeting was held once again at the Hotel on the Cay, and, as before, the work and the fun went on very enjoyably. We had a total of 76 participants from 23 countries, with the main financial support for the meeting provided by the Scientific Affairs Division of the North Atlantic Treaty Organization (NATO). The ASI was co-sponsored by the U.S. Department of Energy, by the Fermi National Accelerator Laboratory (Fermilab), by the U.S. National Science Foundation, and by the University of Rochester.

As in the case of the previous ASIs, the scientific program was designed for advanced graduate students and recent PhD recipients in experimental particle physics. The present volume of lectures should complement and update the material published (by Plenum) for the first nine ASIs and prove to be of value to a wider audience of physicists.

It is a pleasure to acknowledge the encouragement that I have continued to receive from students, colleagues and friends in organizing this meeting. I am indebted, as always, to Chris Quigg and to the other members of my Advisory Committee (listed in the back of the volume) for their patience and excellent advice. I am grateful to the enthusiastic lecturers for their participation in the ASI (both on the dance floor as well as in the lecture hall), and, of course, for their hard work in preparing their lectures and their manuscripts for the Proceedings. I thank Nicholas Savvas for organizing the student presentations, Amanda Weinstein for her brief contribution to the proceedings, and Zandy-Marie Hillis of the National Park Service and her colleagues for another fascinating description of the marine life and geology of St. Croix.

I thank P.K. Williams for support from the Department of Energy, and Marvin Goldberg for assistance from the National Science Foundation. I am grateful to John Peoples for providing the talents of Angela Gonzales for designing the poster for the School. At Rochester, I am indebted to Andrea Burne for help with budgeting issues and to Connie Jones for her exceptional organizational assistance.

I owe many thanks to Ann Downs and Marion Hazlewood for their and their staff's efficiency and hospitality at the Hotel on the Cay, and to Hurchell Greenaway and his staff at the Harbormaster for their cordiality and hard work in keeping us well fed and entertained. Finally, I thank Luis da Cunha and the NATO Division of Scientific Affairs for their cooperation and continued confidence.

T. Ferbel
Rochester, New York

LECTURERS

I. Abt	MPI, Munich, Germany
R. Aleksan	CE-Saclay, France
D. Bryman	TRIUMF, Vancouver, Canada
M. J. Herrero	Univ. Autonoma de Madrid, Spain
G. Ingelman	Uppsala University, Uppsala, Sweden
E. Kolb	University of Chicago, Chicago, IL, USA
K. Nakamura	KEK, Tsukuba, Japan
L. Nodulman	Argonne National Lab., Argonne, IL, USA
F. Pauss	ETH, Zurich, Switzerland
J. Stirling	University of Durham, Durham, UK
T. Virdee	CERN, Geneva, Switzerland
D. Whittum	SLAC, Stanford, CA, USA

ADVISORY COMMITTEE

B. Barish	Caltech, Pasadena, California
L. DiLella	CERN, Geneva, Switzerland
C. Fabjan	CERN, Geneva, Switzerland
H. Georgi	Harvard University, Cambridge, Massachusetts
C. Jarlskog	Stockholm University, Stockholm, Sweden
C. Quigg	Fermilab, Batavia, Illinois
P. Soding	DESY, Zeuthen, Federal Republic of Germany
M. Tigner	Cornell University, Ithaca, New York

DIRECTOR

T. Ferbel	University of Rochester, Rochester, New York

THE STANDARD MODEL

M. HERRERO
Departamento de Fisica Teorica
Facultad de Ciencias, C-XI
Universidad Autonoma de Madrid
Cantoblanco, 28049 Madrid, Spain

ABSTRACT

These lectures provide an introduction to the basic aspects of the Standard Model, $SU(3)_C \times SU(2)_L \times U(1)_Y$.

1. Introduction

All known particle physics phenomena are extremely well described within the Santard Model (SM) of elementary particles and their fundamental interactions. The SM provides a very elegant theoretical framework and it has succesfully passed very precise tests which at present are at the 0.1% level [1, 2, 3, 4].

We understand by elementary particles the point-like constituents of matter with no known substructure up to the present limits of $10^{-18} - 10^{-19} m$. These are of two types, the basic building blocks of matter themselves konwn as matter particles and the intermediate interaction particles. The first ones are fermions of spin $s = \frac{1}{2}$ and are classified into leptons and quarks. The known leptons are: the electron, e^-, the muon, μ^- and the τ^- with electric charge $Q = -1$ (all charges are given in units of the elementary charge e); and the corresponding neutrinos ν_e, ν_μ and ν_τ with $Q = 0$. The known quarks are of six different flavors: u, d, s, c, b and t and have fractional charge $Q = \frac{2}{3}, -\frac{1}{3}, -\frac{1}{3}, \frac{2}{3}, -\frac{1}{3}$ and $\frac{2}{3}$ respectively.

The quarks have an additional quantum number, the color, which for them can be of three types, generically denoted as q_i, $i = 1, 2, 3$. We know that color is not seen in Nature and therefore the elementary quarks must be confined into the experimentally observed matter particles, the hadrons. These colorless composite particles are classified into baryons and mesons.

T. Ferbel (ed.), Techniques and Concepts of High Energy Physics X, 1–59.

The baryons are fermions made of three quarks, qqq, as for instance the proton, $p \sim uud$, and the neutron, $n \sim ddu$. The mesons are bosons made of one quark and one antiquark as for instance the pions, $\pi^+ \sim u\bar{d}$ and $\pi^- \sim d\bar{u}$.

The second kind of elementary particles are the intermediate interaction particles. By leaving apart the gravitational interactions, all the relevant interactions in Particle Physics are known to be mediated by the exchange of an elementary particle that is a boson with spin $s = 1$. The photon, γ, is the exchanged particle in the electromagnetic interactions, the eight gluons g_α ; $\alpha = 1,..8$ mediate the strong interactions among quarks, and the three weak bosons, $W^\pm$, Z are the corresponding intermediate bosons of the weak interactions.

As for the theoretical aspects, the SM is a quantum field theory that is based on the gauge symmetry $SU(3)_C \times SU(2)_L \times U(1)_Y$. This gauge group includes the symmetry group of the strong interactions, $SU(3)_C$, and the symmetry group of the electroweak interactions, $SU(2)_L \times U(1)_Y$. The group symmetry of the electromagnetic interactions, $U(1)_{em}$, appears in the SM as a subgroup of $SU(2)_L \times U(1)_Y$ and it is in this sense that the weak and electromagnetic interactions are said to be unified.

The gauge sector of the SM is composed of eight gluons which are the gauge bosons of $SU(3)_C$ and the γ, $W^\pm$ and Z particles which are the four gauge bosons of $SU(2)_L \times U(1)_Y$. The main physical properties of these intermediate gauge bosons are as follows. The gluons are massless, electrically neutral and carry color quantum number. There are eight gluons since they come in eight different colors. The consequence of the gluons being colorful is that they interact not just with the quarks but also with themselves. The weak bosons, $W^\pm$ and Z are massive particles and also selfinteracting. The $W^\pm$ are charged with $Q = \pm 1$ respectively and the Z is electricaly neutral. The photon γ is massless, chargeless and non-selfinteracting.

Concerning the range of the various interactions, it is well known the infinite range of the electromagnetic interactions as it corresponds to an interaction mediated by a massless gauge boson, the short range of the weak interactions of about $10^{-16} cm$ correspondingly to the exchange of a massive gauge particle with a mass of the order of $M_V \sim 100\, GeV$ and, finally, the strong interactions whose range is not infinite, as it should correspond to the exchange of a massless gluon, but finite due to the extra physical property of confinement. In fact, the short range of the strong interactions of about $10^{-13} cm$ corresponds to the typical size of the ligthest hadrons.

As for the strength of the three interactions, the electromagnetic interactions are governed by the size of the electromagnetic coupling constant e or equivalently $\alpha = \frac{e^2}{4\pi}$ which at low energies is given by the fine structure constant, $\alpha(Q = m_e) = \frac{1}{137}$. The weak interactions at energies much lower than the exchanged gauge boson mass, M_V, have an effective (weak) strength given by the dimensionful Fermi constant $G_F = 1.167 \times 10^{-5}\, GeV^{-2}$. The name of strong interactions is due to their comparative stronger strength than the other interactions. This strength is governed by the size of the strong copling constant g_S or equivalently $\alpha_S = \frac{g_s^2}{4\pi}$ and is varies from large values to low energies, $\alpha_S(Q = m_{\mathrm{hadron}}) \sim 1$ up to the vanishing asymptotic limit $\alpha_S(Q \to \infty) \to 0$. This last limit indicates that the quarks behave as free particles when they are observed at infinitely large energies or, equivalently, inflinitely short distances and it is known as the property of asymptotic freedom.

Finally, regarding the present status of the matter particle content of the SM the situation is summarized as follows.

The fermionic sector of quarks and leptons are organized in three families with identical properties except for mass. The particle content in each family is:

$$1^{st}\text{ family: } \begin{pmatrix} \nu_e \\ e^- \end{pmatrix}_L, e_R^-, \begin{pmatrix} u \\ d \end{pmatrix}_L, u_R, d_R$$

$$2^{nd}\text{ family: } \begin{pmatrix} \nu_\mu \\ \mu^- \end{pmatrix}_L, \mu_R^-, \begin{pmatrix} c \\ s \end{pmatrix}_L, c_R, s_R$$

$$3^{rd}\text{ family: } \begin{pmatrix} \nu_\tau \\ \tau^- \end{pmatrix}_L, \tau_R^-, \begin{pmatrix} t \\ b \end{pmatrix}_L, t_R, b_R$$

and their corresponding antiparticles. The left-handed and right-handed fields are defined by means of the chirality operator γ_5 as usual,

$$e_L^- = \tfrac{1}{2}(1 - \gamma_5)e^-;\ e_R^- = \tfrac{1}{2}(1 + \gamma_5)e^-$$

and they transform as doublets and singlets of $SU(2)_L$ respectively.

The scalar sector of the SM is not experimentaly confirmed yet. The fact that the weak gauge bosons are massive particles, $M_W^\pm$, $M_Z \neq 0$, indicates that $SU(2)_L \times U(1)_Y$ is *NOT* a symmetry of the vacuum. In contrast, the photon being massless reflects that $U(1)_{em}$ is a good symmetry of the vacuum. Therefore, the Spontaneous Symmetry Breaking pattern in the SM must be:

$$SU(3)_C \times SU(2)_L \times U(1)_Y \to SU(3)_C \times U(1)_{em}$$

The above pattern is implemented in the SM by means of the so-called Higgs Mechanism which provides the proper masses to the $W^{\pm}$ and Z gauge bosons and to the fermions, and leaves as a consequence the prediction of a new particle: The Higgs boson particle. This must be scalar and electrically neutral. This particle has not been seen in the experiments so far [5].

These lectures provide an introduction to the basic aspects of the SM, $SU(3)_C \times SU(2)_L \times U(1)_Y$. They aim to be of pedagogical character and are especially addressed to non-expert particle physicists without much theoretical background in Quantum Field Theory. The lectures start with a review on some symmetry concepts that are relevant in particle physics, with particular emphasis in the concept of gauge symmetry. Quantum Electrodynamics (QED) is introduced next as a paradigmatic example of gauge theory. A short review on the most relevant precursors of Quantum Chromodynamics (QCD) and the Electroweak Theory are presented. Then, a brief introduction to the basics of QCD is presented. The central part of these lectures is devoted to the building of the Electroweak Theory and to review the Electroweak Symmetry Breaking in the SM. The concept of Spontaneous Symmetry Breaking and The Higgs Mechanism are explained. A short review on the present theoretical Higgs mass bounds is also included. The final part of these lectures is devoted to review the most relevant SM predictions.

There are many other important aspects of the SM that, due to the lack of space and time, are not covered here. Some of the complementary topics are covered by my fellow lecturers at this School. In particular, the lectures of J. Stirling cover QCD, those of R. Aleksan cover Quark Mixing and CP Violation and those of L. Nodulman cover the Experimental Tests of the SM.

2. Group Symmetries in Particle Physics

The existence of symmetries plays a crutial role in Particle Physics. We say that there is a symmetry S when the physical stystem under study has an invariance under the transformation given by S or, equivalently, when the Hamiltonian of this system H is invariant, i.e.,

$$SHS^{+} = H$$

Sometimes the set of independent symmetries of a system generates an algebraic structure of a group, in which case it is said there is a symmetry group [6].

2.1. TYPES OF SYMMETRIES

There are various ways of classifying the different symmetries. If we pay attention to the kind of parameters defining these symmetry transformations, they can be classified into:

1.- *Discrete Symmetries*
The parameters can take just discrete values. In Particle Physics there are several examples. Among the most relevant ones are the transformations of:

Parity P, Charge Conjugation C and Time Reversal T

On the other hand, by the CPT Theorem we know that all interactions must be invariant under the total transformation given by the three of them C, P and T, irrespectively of their order. It is also known that the elctromagnetic interactions and the strong interactions preserve in addition P, C and T separately, whereas the weak interactions can violate, P, C and PC.

2.- *Continous Symmetries*
The parameters take continous values. The typical examples are the rotations ,generically written as $R(\theta)$, where the rotation angle θ can take continous values. There are different kinds of continous symmetries. Here we mention two types:

1) *Space-Time symmetries*: Symmetries that act on the space-time. Typical examples are traslations, rotations, etc.

2) *Internal Symmetries*: Symmetries that act on the internal quantum numbers. Typical examples are $SU(2)$ Isospin symmetry, $U(1)_B$ baryon symmetry etc. Usually these symmetries are given by Lie groups.

2.2. IRREDUCIBLE REPRESENTATIONS OF A SYMMETRY GROUP

The classification of the particle spectra into irreducible representations of a given symmetry group is an important aspect of Particle Physics and helps in understanding their basic physical properties.

In the case of rotations, it is known that if a system given by a particle of spin j manifests $SO(3)$ rotational invariance, then it implies the existence of $(2j+1)$ degenerate energy levels which can be accommodated into the irreducible representation of $SO(3)$ of dimension $(2j+1)$. Let us see this in more detail.

Let $R(\theta)$ be a the rotation given by:

$$R(\theta) = e^{\,i\sum_{a=1}^{3}\theta_a J_a} \quad ; \; J_a = \text{angular momentum operators}$$

Invariance of the Hamiltonian under $R(\theta)$ implies the following sequence of statements,

$$RHR^{+} = H \Longrightarrow [H, J_a] = 0 \; (a = 1,2,3) \Longrightarrow \text{If } \{|n>\}/H|n>= E_n|n>$$
$$\text{Then } H(J_a|n>) = J_a(H|n>) = E_n(J_a|n>)$$

Thus, there are $(2j+1)$ states, the $J_a|n>$, that are degenerate and form the basis associated to angular momentum j. In addition, the angular momentum operators J_a are the generators of the symmetry group of rotations $SO(3)$ which is the set of all the 3×3 orthogonal matrices with unit determinant. By putting altogether, the conclusion is that the $SO(3)$ symmetry of H implies that each particle with angular momentum j has $(2j+1)$ degenerate levels which fit into the irreducible representation with dimension $(2j+1)$ of the $SO(3)$ group.

2.3. INTERNAL SYMMETRIES

The Internal Symmetries are transformations not on the space-time coordinates but on internal coordinates, and they transform one particle to another with different internal quantum numbers but having the same mass. In contrast to the case of space-time symmetries, the irreducible representations of internal symmetries are degenerate particle multiplets.

2.3.1. *SU(2) Isospin Symmetry*

The isospin symmetry is an illustrative example of internal symmetries. In this case the internal quantum number is isospin. Let us see that, in fact, invariance under isospin implies the existence of degenerate isospin multiplets. Let H_s be the Hamiltonian of strong interactions. Invariance of strong interactions under isospin rotations reads:

$$UH_sU^{+} = H_s$$

where, U is the isospin transformation and is given by

$$U = e^{i\sum_{a=1}^{3}\theta_a T_a}$$

with T_a $(a = 1,2,3)$ being the three generators of the $SU(2)$ group and θ_a the continous parameters of the transformation. The $SU(2)$ group is the set of 2×2 unitary matrices with unit determinant; and the $SU(2)$ algebra is defined by the conmutation relations of the generators:

$$[T_i, T_j] = \epsilon_{ijk}T_k; \; \epsilon_{ijk} = \text{structure constants of } SU(2)$$

As in the previous case of space-time transformations, one can show that invariance under the above U transformation implies,

$$[T_a, H_s] = 0; \; (a = 1,2,3)$$

and from this it is immediate to demonstrate the existence of degenerate isospin multiplets. Thus, for a given eigenstate of H_s one can always find, by application of the T_a generators, new eigenstates of H_s which are degenerate.

In terms of the physical states, the proton $|p>$, the neutron $|n>$, and the pions, $|\pi^+>$, $|\pi^->$ and $|\pi^0>$ the isospin rotations act as follows:

$$T_+|n>=|p>;\ T_-|p>=|n>$$
$$T_3|p>=\tfrac{1}{2}|p>;\ T_3|n>=-\tfrac{1}{2}|n>$$

$$T_+|\pi^->=\sqrt{2}|\pi^0>;\ T_+|\pi^0>=\sqrt{2}|\pi^+>$$
$$T_3|\pi^+>=|\pi^+>;\ T_3|\pi^0>=0;\ T_3|\pi^->=-|\pi^->$$

where, $T_\pm = T_1 \pm iT_2$.

Therefore the corresponding degenerate isospin multiplets are:

$$\begin{pmatrix} p \\ n \end{pmatrix} \text{ isospin doublet} \qquad \begin{pmatrix} \pi^+ \\ \pi^0 \\ \pi^- \end{pmatrix} \text{ isospin triplet}$$

Notice that neither the proton and neutron nor the three pions are exactly degenerate and therefore the isospin symmetry is not an exact symmetry of the strong interactions. In fact, the size of the mass-differences within a multiplet are indications of the size of the isospin breaking. Since the proton and neutron masses are pretty close and simmilarly for the masses of the three pions, it happens that the $SU(2)$ isospin symmetry is, indeed, an approximate symmetry of the strong interactions.

2.4. CLASSIFICATION OF INTERNAL SYMMETRIES AND RELEVANT THEOREMS

There are two distinct classes of internal symmetries:

1.- *Global symmetries*
The continous parameters of the transformation *DO NOT DEPEND* on the space-time coordinates. Some examples are: $SU(2)$ Isospin symmetry, $SU(3)$ flavor symmetry, $U(1)_B$ baryon symmetry, $U(1)_L$ lepton symmetry,...

2.- *Local (Gauge) symmetries*
The continous parameters of the transformation *DO DEPEND* on the space-time coordinates. Some examples are: $U(1)_{em}$ electromagnetic symmetry, $SU(2)_L$ weak isospin symmetry, $U(1)_Y$ weak hypercharge symmetry, $SU(3)_C$ color symmetry,...

There are two relevant theorems/principles that apply to the two cases above respectively and have important physical implications:

Noether's Theorem for Global Symmetries

If the Hamiltonian (or the Lagrangian) of a physical system has a global symmetry, there must be a current and the associated charge that are conserved.

Examples:

The $U(1)$ symmetries are global rotations by a given phase. For instance:

$$\Psi \to e^{i\alpha}\Psi$$

rotates the field Ψ by a phase α and it is the same for all space time points, i.e. it is a global phase. The $U(1)$ symmetry group is the set of complex numbers with unit modulus. We have already mentioned some examples in particle physics as the $U(1)_B$ and $U(1)_L$ global symmetries. The conserved currents are the barionic and leptonic currents respectively; and the associated conserved charges are the barionic and leptonic numbers respectively.

The Gauge Principle for Gauge Theories

Let Ψ be a physical system in Particle Physics whose dynamics is decribed by a Lagrangian $\mathcal{L}$ which is invariant under a global symmetry G. It turns out that, by promoting this global symmetry G from global to local, the originaly free theory transforms into an interacting theory. The procedure in order to get the theory invariant under local transformations is by introducing new vector boson fields, the so-called gauge fields, that interact with the Ψ field in a gauge invariant manner. The number of gauge fields and the particular form of these gauge invariant interactions depend on the particularities of the symmetry group G. More specifically, the number of associated gauge boson fields is equal to the number of generators of the symmetry group G.

Examples:

The local version of the previous example,

$$\Psi \to e^{i\alpha(x)}\Psi$$

with the phase α being a function of the space-time point $x \equiv x_\mu$, has one associated gauge boson field. This simplest case of $U(1)$ has just one generator and correspondingly one gauge field which is the exchanged boson particle and acts as the mediator of the corresponding interaction.

Other examples are: $SU(2)$ with three generators and the corresponding three gauge bosons and $SU(3)$ with eight generators and the corresponding eight gauge bosons. The generic case of $SU(N)$ has N^2-1 generators and correspondingly the same number of gauge bosons.

The above Gauge Principle is a very important aspect of Particle Physics and has played a crutial role in the building of the Standard Model.

The quantum field theories that are based on the existence of some gauge symmetry are called Gauge Theories. We have already mentioned the cases of $U(1)_{em}$, $SU(2)_L$, $U(1)_Y$ and $SU(3)_C$ gauge symmetries. The gauge theory based on $U(1)_{em}$ is Quantum Electrodynamics (QED), the gauge theory based on $SU(3)_C$ is Quantum Chromodynamics (QCD) and the corresponding one based on the composed group $SU(2)_L \times U(1)_Y$ is the so-called Electroweak Theory. The Standard Model is the gauge theory based on the total gauge symmetry of the fundamental interactions in particle physics, $SU(3)_C \times SU(2)_L \times U(1)_Y$.

3. QED: The paradigm of Gauge Theories

Quantum Electrodynamics is the most succesful Gauge Theory in Particle Physics and has been tested up to an extremely high level of precision. We show QED here as the paradigmatic example of the application of The Gauge Principle and its physical implications.

One starts with the following physical system: A free Dirac field Ψ with spin $s = \frac{1}{2}$, mass m and electric charge Qe.

The corresponding Lagrangian is:

$$\mathcal{L} = \overline{\Psi}(x)(i\partial\!\!\!/ - m)\Psi(x);\ \partial\!\!\!/ \equiv \partial_\mu \gamma^\mu$$

and the corresponding equation of motion is the Dirac equation:

$$(i\partial\!\!\!/ - m)\Psi(x) = 0$$

It is immediate to show the invariance of this Lagrangian under global $U(1)$ transformations which act on the fields and their derivatives as follows,

$$\Psi \to e^{iQ\theta}\Psi;\ \overline{\Psi} \to \overline{\Psi}e^{-iQ\theta};\ \partial_\mu\Psi \to e^{iQ\theta}\partial_\mu\Psi$$

here the global phase is $Q\theta$ and the continous papameter is θ.

By Noether's Theorem, this global $U(1)$ invariance of $\mathcal{L}$ implies the conservation of the electromagnetic current, J_μ, and the electromagnetic charge, eQ,

$$J_\mu = \overline{\Psi}\gamma_\mu eQ\Psi;\ \partial_\mu J^\mu = 0;\ eQ = \int d^3x J_0(x)$$

Now, if we promote the transformation from global to local, i.e, if the parameter θ is allowed to depend on the space-time point x, the corresponding transformations on the fields and their derivatives are,

$$\Psi \to e^{iQ\theta(x)}\Psi;\ \overline{\Psi} \to \overline{\Psi}e^{-iQ\theta(x)};\ \partial_\mu\Psi \to e^{iQ\theta(x)}\partial_\mu\Psi + iQ(\partial_\mu\theta(x))e^{iQ\theta(x)}\Psi$$

One can show that the Lagrangian in the form written above is not yet invariant under these local transformations. The solution to this question is provided by the Gauge Principle. One introduces one gauge vector boson field, the photon field $A_\mu(x)$ which interacts with the field Ψ and transforms properly under the $U(1)$ gauge transformations,

$$A_\mu \to A_\mu - \frac{1}{e}\partial_\mu \theta(x)$$

Here proper transformations means that it must compensate the extra terms introduced by $\partial_\mu \theta \neq 0$ such that the total Lagrangian be finally gauge invariant.

The most economical way of building this gauge invariant Lagrangian is by simply replacing the normal derivative, ∂_μ, by the so-called covariant derivative, D_μ,

$$D_\mu \Psi \equiv (\partial_\mu - ieQA_\mu)\Psi$$

which transforms covariantly, i.e. as the Ψ field itself,

$$D_\mu \Psi \to e^{iQ\theta(x)} D_\mu \Psi$$

Finally, in order to include the propagation of the photon field one adds the so-called kinetic term which must be also gauge invariant and is given in terms of the field strength tensor,

$$F_{\mu\nu} = \partial_\mu A_\nu - \partial_\nu A_\mu$$

The total Lagrangian is Lorentz and U(1) gauge invariant, and is the well known Lagrangian of QED,

$$\boxed{\mathcal{L}_{QED} = \overline{\Psi}(x)(i\not{D} - m)\Psi(x) - \frac{1}{4}F_{\mu\nu}(x)F^{\mu\nu}(x)}$$

Notice that it contains the wanted interactions within the $\overline{\Psi} i\not{D}\Psi$ term,

$$\overline{\Psi} eQA_\mu \gamma^\mu \Psi$$

Finally, the gauge group for electromagnetism is, correspondingly, $U(1)_{em}$ with one generator, Q and one parameter θ.

4. Strong Interactions before QCD: The Quark Model

The discovery of the Λ^0 and K^0 particles lead to the proposal of a new additive quantum number named 'strangeness' and denoted by S which is conserved by the strong interactions but is violated by the weak interactions.

The corresponding assignements are:

$$S(\Lambda^0) = -1,\ S(K^0) = +1,\ S(p) = S(n) = S(\pi) = 0.$$

Gell-Mann and independently Nishijima and Nakano by studing the properties of the hadrons noted in 1953 the linear relation among the three additive quantum numbers the strangeness S, the electromagnetic charge Q and the third component of the (strong) isospin T_3, given by the so-called *Gell-Mann-Nishijima relation* [7]:

$$\boxed{Q = T_3 + \frac{Y}{2}}$$

where the (strong) hypercharge Y, the baryon number B and the strangeness S are related by $Y = B + S$.

The existence of the new conserved quantum number S suggested to think of a larger symmetry than isospin $SU(2)$ for the strong interactions. Gell- Mann and Ne'eman in 1961 proposed the larger symmetry group $SU(3)$, named sometimes flavor symmetry, which in fact contains to $SU(2)$ [8]. They pointed out that all mesons and baryons with the same spin and parity can be grouped into irreducible representations of $SU(3)$. Thus, each particle is labelled by its (T_3, Y) quantum numbers and fits into one of the elements of these representations. Historically, it was named 'The Eightfold-Way' classification of hadrons since the first studied hadrons turned out to fit into representations of dimension eight, i.e. into octects of $SU(3)$. Later, higher dimensional representations, as decuplets etc., were needed to fit other hadrons. In Figure 1 some examples of $SU(3)$ octects are shown.

In 1964 Gell-Mann and Zweig [9] noted that the lowest dimensional irreducible representation, i.e. the triplet of $SU(3)$ with dimension equal to three, was not occupied by any known hadron and proposed the existence of new particles, named *quarks*, such that by fitting them into the elements of this fundamental representation and by making apropriate compositions with it, one could build up the whole spectra of hadrons. This brilliant idea, originaly based mainly on formal aspects of symmetries, led to the prediction of three new elementary particles, the three lightest quarks, distinguished by their flavors, the u (up), d (down) and s (strange) quarks. Correspondingly, their antiparticles, the antiquarks $\bar{u}$, $\bar{d}$ and $\bar{s}$ with the opposite quantum numbers, are fitted into the complex conjugate representation which is also a triplet. In Figure 1, these $SU(3)$ triplets are also shown.

The description of hadrons in terms of quarks by means of the $SU(3)$ irreducible representations and their properties is called the $SU(3)$ *Quark Model*. One uses group theory methods, for instance the Young Tableaux technique, to decompose products of irreducible representations into sums. Thus the mesons ($B = 0$) appear as composite states of a quark and an antiquark, and the baryons ($B = 1$) as composite states of three quarks:

$$\underline{\text{Mesons}}\text{: } q\bar{q} = 3 \otimes \bar{3} = 1 \oplus 8$$

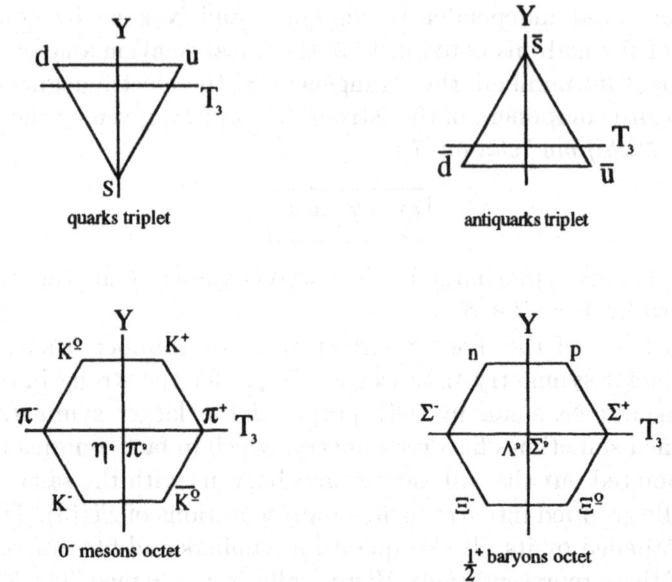

Figure 1. Examples of particle multiplets in the $SU(3)$ Quark Model

Here the 1 are the $SU(3)$ meson singlets as the $\eta' \sim (u\bar{u} + d\bar{d} + s\bar{s})$ with $J^P = 0^-$, the $\Phi \sim (u\bar{u} + d\bar{d} + s\bar{s})$ with $J^P = 1^-$... The 8 are the $SU(3)$ meson octets as the 0^- mesons, $\pi^+ \sim u\bar{d}$, $K^+ \sim u\bar{s}$ ...; the 1^- mesons, $\rho^+ \sim u\bar{d}$, $K^{*+} \sim u\bar{s}$..., and so on.

$$\underline{\text{Baryons:}}\ qqq = 3 \otimes 3 \otimes 3 = 1 \oplus 8 \oplus 8 \oplus 8 \oplus 10$$

Here the 1 are the $SU(3)$ baryon singlets; the 8 are the $SU(3)$ baryon octets as the $\frac{1}{2}^+$ baryons, $n \sim udd$, $p \sim uud$... ; the 10 are the $SU(3)$ decuplets as the $\frac{3}{2}^+$ baryons, $N^{*+} \sim uud$, $\Sigma^{*+} \sim suu$..., and so on.

Notice that the $SU(3)$ flavor symmetry is not an exact symmetry. As in the case of isospin symmetry, the mass differences within the members of a multiplet are signals of $SU(3)$ breaking. Similarly, the mass differences among the three quarks themselves are indications of this breaking as well. Although the breaking is certainly more sizeable in $SU(3)$ than in $SU(2)$, one can still deal with $SU(3)$ as an approximate symmetry.

Among the successful predictions of the Quark Model there are, for instance, the existence of some particles before their discovery as it is the case of the $\frac{3}{2}^+$ baryon $\Omega^- \sim sss$; and, in general, a large amount of the hadron properties are well described within this model. In particular, one may build up the hadron wave functions and compute some physical properties as, for instance, the hadron magnetic moments, in terms of the corresponding ones of the quarks components.

For example, the proton wave function with the spin up is:

$$|p\uparrow>=\sqrt{\tfrac{1}{18}}|u\uparrow u\downarrow d\uparrow +u\downarrow u\uparrow d\uparrow -2u\uparrow u\uparrow d\downarrow +perm.>$$

The predicted proton and neutron magnetic moments are:

$$\mu_p=\tfrac{4}{3}\mu_u-\tfrac{1}{3}\mu_d;\ \mu_n=\tfrac{4}{3}\mu_d-\tfrac{1}{3}\mu_u;\ \text{with, } \mu_q=\tfrac{Q_q e}{2m_q}$$

and their ratio is, therefore,

$$\boxed{\tfrac{\mu_n}{\mu_p}=-\tfrac{2}{3}}$$

which is rather close to its experimental measurement,

$$\boxed{\left.\tfrac{\mu_n}{\mu_p}\right|_{exp}=-0.68497945\pm 0.00000058}$$

The Quark Model has some noticeable failures which were the reason to abandon it as the proper model for strong interactions. The most famous one, from the historial point of view, is the so-called *paradox of the* Δ^{++}. Its wave function for the case of spin up is given by,

$$|\Delta^{++}\uparrow>\sim|u\uparrow u\uparrow u\uparrow>$$

which is apparently totally symmetric since it is symmetric in space, flavor and spin. However, by Fermi-Dirac statistics it should be antisymmetric as it corresponds to an state with identical fermions. This apparent paradox, was solved by Gell-Mann with the proposal of the quarks carrying a new quantum number, *the color* and, in consequence, being non-identical from each other. Correspondingly, a quark q can have three different colors, generically, q_i $i=1,2,3$. Since, this property of color is not seen in Nature, the colors of the quarks must be combined such that they produce colorless hadrons. In the group theory language it is got by requiring the hadrons to be in the singlet representations of the color group, $SU(3)_C$. Since the singlet representation is allways antisymmetric, by including this color wave function one gets finally the expected antisymmetry of the Δ^{++} total wave function.

5. QCD: The Gauge Theory of Strong Interactions

Quantum Chromodynamics is the gauge theory for strong interactions and has provided plenty of successful predictions so far [3].

It is based on the gauge symmetry of strong interactions, i.e. the local color transformations which leave its Hamiltonian (or Lagrangian) invariant. The gauge symmetry group that is generated by these color transformations is the non-abelian Lie group $SU(3)_C$. Here C refers to colors and

3 refers to the three posible color states of the quarks which are assumed to be in the fundamental representation of the group having dimension three. The gluons are the gauge boson particles associated to this gauge symmetry and are eight of them as it corresponds to the number of $SU(3)$ generators. The gluons are the mediators of the strong interactions among quarks. Genericaly, the quarks and gluons are denoted by:

$$\underline{\text{quarks:}}\ q_i,\ i = 1,2,3; \qquad \underline{\text{gluons:}}\ g_\alpha,\ \alpha = 1,...,8$$

The building of the QCD invariant Lagrangian is done by following the same steps as in the QED case. In particular, one applies the gauge principle as well with the particularities of the non-abelian group $SU(3)$ taken into account. Thus, the global symmetry $SU(3)$ of the Lagrangian for the strong interactions is promoted to local by replacing the derivative of the quark by its covariant derivative which in the QCD case is,

$$D_\mu q \equiv \left(\partial_\mu - ig_s(\tfrac{\lambda_\alpha}{2})A_\mu^\alpha\right) q$$

where,

$$q = \begin{pmatrix} q_1 \\ q_2 \\ q_3 \end{pmatrix}$$

$$\begin{aligned} q_i &= \text{quark fields; } i = 1,2,3 \\ g_s &= \text{strong coupling constant} \\ \tfrac{\lambda_\alpha}{2} &= SU(3) \text{ generators} \\ A_\mu^\alpha &= \text{gluon fields; } \alpha = 1,...,8 \end{aligned}$$

The QCD Lagrangian is then written in terms of the quarks and their covariant derivatives and contains in addition the kinetic term for the gluon fields,

$$\boxed{\mathcal{L}_{QCD} = \sum_q \overline{q}(x)(i\not{D} - m_q)q(x) - \frac{1}{4}F_{\mu\nu}^\alpha(x)F_\alpha^{\mu\nu}(x)}$$

The gluon field strength is,

$$F_{\mu\nu}^\alpha(x) = \partial_\mu A_\nu^\alpha(x) - \partial_\nu A_\mu^\alpha(x) + g_s f^{\alpha\beta\gamma} A_{\mu\beta} A_{\nu\gamma}$$

and contains a bilinear term in the gluon fields as it corresponds to a non-abelian gauge theory with structure constants $f^{\alpha\beta\gamma}$ $(\alpha, \beta, \gamma = 1, ..., 8)$.

It can be shown that the above Lagrangian is invariant under the following $SU(3)$ gauge transformations,

$$\begin{cases} q(x) & \rightarrow\ e^{i\theta_\alpha(x)\frac{\lambda^\alpha}{2}} q(x) \\ D_\mu q(x) & \rightarrow\ e^{i\theta_\alpha(x)\frac{\lambda^\alpha}{2}} D_\mu q(x) \\ A_\mu^\alpha(x) & \rightarrow\ A_\mu^\alpha(x) - \frac{1}{g_s}\partial_\mu \theta^\alpha(x) + f^{\alpha\beta\gamma}\theta_\beta(x) A_{\mu\gamma}(x) \end{cases}$$

where $\theta_\alpha(x)$ $\alpha = 1,...,8$ are the parameters of the transformation.

Similarly to the QED case, the gauge interactions among the quarks and gluons are contained in the $\bar{q}i\not{D}q$ term,

$$\bar{q}g_s\frac{\lambda^\alpha}{2}A^\alpha_\mu\gamma^\mu q$$

There is, however, an important difference with the QED case. The gluon kinetic term $F^\alpha_{\mu\nu}F^{\mu\nu}_\alpha$ contains a three gluons term and a four gluons term. These are precisely the selfinteraction gluon vertices which are genuine of a non-abelian theory.

5.1. $SU(3)$ GROUP PROPERTIES

In this section we list the basic properties of the $SU(3)$ group which are relevant for QCD and in particular for the computation of color factors in processes mediated by strong interactions.

$SU(3)$ is the set of 3×3 unitary matrices with unit determinant. Any element of $SU(3)$, U, can be written in terms of its 8 generators, $\frac{\lambda_\alpha}{2}$ and a set of 8 real parameters θ_α as,

$$U = e^{i\theta^\alpha\frac{\lambda_\alpha}{2}};\ \alpha = 1,...,8$$

The generators are 3×3 traceless hermitian matrices $\frac{\lambda_\alpha}{2}$ and are given in terms of the so-called Gell-Mann matrices, λ_α,

$$\lambda_1 = \begin{pmatrix} 0 & 1 & 0 \\ 1 & 0 & 0 \\ 0 & 0 & 0 \end{pmatrix},\ \lambda_2 = \begin{pmatrix} 0 & -i & 0 \\ i & 0 & 0 \\ 0 & 0 & 0 \end{pmatrix},\ \lambda_3 = \begin{pmatrix} 1 & 0 & 0 \\ 0 & -1 & 0 \\ 0 & 0 & 0 \end{pmatrix}$$

$$\lambda_4 = \begin{pmatrix} 0 & 0 & 1 \\ 0 & 0 & 0 \\ 1 & 0 & 0 \end{pmatrix},\ \lambda_5 = \begin{pmatrix} 0 & 0 & -i \\ 0 & 0 & 0 \\ i & 0 & 0 \end{pmatrix}$$

$$\lambda_6 = \begin{pmatrix} 0 & 0 & 0 \\ 0 & 0 & 1 \\ 0 & 1 & 0 \end{pmatrix},\ \lambda_7 = \begin{pmatrix} 0 & 0 & 0 \\ 0 & 0 & -i \\ 0 & i & 0 \end{pmatrix},\ \lambda_8 = \frac{1}{\sqrt{3}}\begin{pmatrix} 1 & 0 & 0 \\ 0 & 1 & 0 \\ 0 & 0 & -2 \end{pmatrix}$$

Some basic properties of the $SU(3)$ generators are:

$$\boxed{\left[\frac{\lambda_\alpha}{2}, \frac{\lambda_\beta}{2}\right] = if_{\alpha\beta\gamma}\frac{\lambda_\gamma}{2}} \qquad \boxed{Tr\left(\frac{\lambda_\alpha}{2}\frac{\lambda_\beta}{2}\right) = \frac{1}{2}\delta_{\alpha\beta}}$$

The tensor $f_{\alpha\beta\gamma}$ is totally antisymmetric and its elements are the structure constants of $SU(3)$. The non-vanishing elements are,

$$f_{123} = 1,\ f_{147} = \tfrac{1}{2},\ f_{156} = -\tfrac{1}{2},\ f_{246} = \tfrac{1}{2},\ f_{257} = \tfrac{1}{2},$$
$$f_{345} = \tfrac{1}{2},\ f_{367} = -\tfrac{1}{2},\ f_{458} = \sqrt{\tfrac{3}{2}}\ f_{678} = \sqrt{\tfrac{3}{2}}$$

Some useful relations for practical computations and relevant group factors are,

$$\delta_{\alpha\beta} C_A = \sum_{\gamma\delta} f_{\alpha\gamma\delta} f_{\beta\gamma\delta} \ ; \ C_A = 3$$

$$\delta_{ik} C_F = \sum_{\alpha l} \frac{\lambda^{\alpha}_{il}}{2} \frac{\lambda^{\alpha}_{lk}}{2} \ ; \ C_F = \frac{4}{3}$$

$$\delta_{\alpha\beta} T_F = \sum_{ki} \frac{\lambda^{\alpha}_{ik}}{2} \frac{\lambda^{\beta}_{ki}}{2} \ ; \ T_F = \frac{1}{2}$$

$$(i, k = 1, 2, 3 \ ; \ \alpha, \beta, \gamma, \delta = 1, ..., 8)$$

5.2. COMPUTING COLOR FACTORS IN QCD

For practical computations, sometimes it is convenient to define color factors associated to a physical proccess in QCD. These color factors are genuine of QCD and can be computed apart by using the $SU(3)$ group relations.

For illustrative purposes, we present here one particular example: the computation of the color factor associated to the scattering proccess of two different quarks, $qq' \to qq'$.

There is just one Feynman diagram contributing to the scattering amplitude of this process which is the diagram with one gluon exchanged in the t-channel. Notice that this diagram is similar to the one contributing to the QED proccess $e^-\mu^- \to e^-\mu^-$ where one photon is exchanged in the t-channel.

One can use the known result from QED for the spin-averaged squared amplitude of the proccess $e^-\mu^- \to e^-\mu^-$ in terms of the Mandelstam variables, s, t, and u,

$$\overline{|F|}^2 = 2e^4 \left(\frac{s^2 + u^2}{t^2} \right)$$

and by simply replacing the electromagnetic coupling constant e by the strong coupling constant g_s and by adding a color factor F_C one obtains the corresponding squared amplitude of the proccess $qq' \to qq'$, at tree level in QCD,

$$\overline{|F|}^2 = F_C 2 g_s^4 \left(\frac{s^2 + u^2}{t^2} \right)$$

The color factor F_C can be now computed apart. From the QCD Feynman rules for the quark-gluon-quark vertex and for the gluon propagator; and by averaging in initial colors and summing in final colors one gets,

$$F_C = \frac{1}{9} \sum_{ijlm\alpha\beta\alpha'\beta'} \left(\frac{\lambda_{ij}^{\alpha}}{2} \right) \delta_{\alpha\beta} \left(\frac{\lambda_{lm}^{\beta}}{2} \right) \left(\frac{\lambda_{ij}^{\alpha'}}{2} \right)^* \delta_{\alpha'\beta'} \left(\frac{\lambda_{lm}^{\beta'}}{2} \right)^*$$

The above expression can be simplified by using the properties of the $SU(3)$ generators,

$$\begin{aligned} F_C &= \tfrac{1}{9} \textstyle\sum_{\alpha\alpha'} \left[\sum_{ij} \left(\frac{\lambda_{ij}^{\alpha}}{2} \right) \left(\frac{\lambda_{ji}^{\alpha'}}{2} \right) \right] \left[\sum_{lm} \left(\frac{\lambda_{lm}^{\alpha}}{2} \right) \left(\frac{\lambda_{ml}^{\alpha'}}{2} \right) \right] \\ &= \tfrac{1}{9} \textstyle\sum_{\alpha\alpha'} \left[\delta_{\alpha\alpha'} T_F \right] \left[\delta_{\alpha\alpha'} T_F \right] \\ &= \tfrac{2}{9} \end{aligned}$$

6. Weak Interactions before The Electroweak Theory

The existence of new interactions of weak strength were proposed to explain the experimental data indicating long lifetimes in the decays of known particles, as for instance,

$$n \to p e^- \overline{\nu}_e \; ; \; \tau_n = 920 \; sec$$

$$\pi^- \to \mu^- \overline{\nu}_\mu \; ; \; \tau_{\pi^-} = 2.6 \times 10^{-8} \; sec$$

$$\mu^- \to e^- \overline{\nu}_e \nu_\mu \; ; \; \tau_\mu = 2.2 \times 10^{-6} \; sec$$

These are much longer lifetimes than the typical decays mediated by strong interactions as,

$$\Delta \to p\pi \; ; \; \tau_\Delta = 10^{-23} \; sec$$

and by electromagnetic interactions as,

$$\pi^0 \to \gamma\gamma \; ; \; \tau_{\pi^0} = 10^{-16} \; sec$$

The history of weak interactions before the formulation of the Standard $SU(2)_L \times U(1)_Y$ Theory is an interesting example of the relevant interplay between theory and experiment. There were a sequence of proposed models which were confronted systematically with the abundant experimental data and which needed to be either refined or rejected in order to be compatible with the observations. All this relevant phenomenology of weak interactions together with the advent of the gauge theories in particle physics led finally to the formulation of the Electroweak Theory, i.e. the gauge theory of electroweak interactions. Among the most relevant predecessor theories of electroweak interactions are the following: Fermi Theory, V-A Theory of Feynman and Gell-Mann and the IVB theory of Lee, Yang and Glashow.

6.1. FERMI THEORY OF WEAK INTERACTIONS

In 1934 Fermi proposed the four-fermion interactions theory [10] in order to describe the neutron β-decay $n \to pe^-\overline{\nu}_e$,

$$\boxed{\mathcal{L}_F = -\frac{G_F}{\sqrt{2}}\left[\overline{p}(x)\gamma_\lambda n(x)\right]\left[\overline{e}(x)\gamma^\lambda \nu_e(x)\right] + h.c.}$$

where the fermion field operators are denoted by their particle names and,

$$G_F = 1.167 \times 10^{-5}\ GeV^{-2}$$

is the so-called Fermi constant which provides the effective dimensionful coupling of the weak interactions.

The Fermi Lagrangian above assumes a vector structure, as in the electromagnetic case, for both the hadronic current, $J_\lambda^{(h)}(x) = \overline{p}(x)\gamma_\lambda n(x)$, and the leptonic current, $J_\lambda^{(l)}(x) = \overline{\nu}_e(x)\gamma_\lambda e(x)$; and postulates a local character for the four fermion interactions, namely, the two currents are contracted at the same space-time point x.

Due precisely to the above vector structure of the weak currents, the Fermi Lagrangian does not explain the observed parity violation in weak interactions.

6.2. PARITY VIOLATION AND THE V-A FORM OF CHARGED WEAK INTERACTIONS

The observation of Kaon decays in two different final states with opposite parities,

$$K^+ \to \pi^+\pi^0 \text{ and } K^+ \to \pi^+\pi^+\pi^-$$

led Lee and Yang in 1956 to suggest the non-conservation of parity in the weak interactions responsible for these decays [11]. Parity violation was discovered by Wu and collaborators in 1957 [12] by analizing the decays of Co nuclei

$$^{60}Co \to ^{60}Ni^*\ e^-\ \overline{\nu}_e$$

which proceed via neutron decay $n \to pe^-\overline{\nu}_e$.

The nuclei are polarized by the action of an external magnetic field such that the angular momenta for Co and Ni are $J = 5$ and $J = 4$ respectively, both aligned in the direction of the external field. By conservation of the total angular momentum, the angular momentum of the combined system electron-antineutrino is inferred to be $J(e^- \, \overline{\nu}_e) = 1$ and must be aligned with the other angular momenta. Therefore both the electron and the antineutrino must have their spins polarized in this same direction. The electron from the decay is seeing always moving in the opposite direction to the external field. By total momentum conservation, the undetected antineutrino is, in consequence, assumed to be moving in the opposite direction to the electron. Altogether leads to the conclusion that the produced electron has negative helicity and the antineutrino has positive helicity. Therefore, the charged weak currents responsible for these decays always produce left-handed electrons and right-handed antineutrinos. The non-observation of left-handed antineutrinos nor right-handed neutrinos in processes mediated by charged weak interactions is a signal of parity violation since the parity transformation changes a left-handed fermion into the corresponding right-handed fermion and viceversa. In fact, it is an indication of *maximal parity violation* which implies that the charged weak current must be neccessarily of the vector minus axial vector form,

$$\boxed{J_\mu \sim V_\mu - A_\mu}$$

Let us see this in more detail. The vector and axial vector currents transform under parity as follows,

$$V^\mu = \overline{\Psi}\gamma^\mu\Psi \xrightarrow{P} \begin{cases} +\overline{\Psi}\gamma^0\Psi \\ -\overline{\Psi}\gamma^k\Psi \; ; \; k = 1,2,3 \end{cases}$$

$$A^\mu = \overline{\Psi}\gamma^\mu\gamma^5\Psi \xrightarrow{P} \begin{cases} -\overline{\Psi}\gamma^0\gamma^5\Psi \\ +\overline{\Psi}\gamma^k\gamma^5\Psi \; ; \; k = 1,2,3 \end{cases}$$

Therefore the various products transform as,

$$V_\mu V^\mu \xrightarrow{P} V_\mu V^\mu$$

$$A_\mu A^\mu \xrightarrow{P} A_\mu A^\mu$$

$$A_\mu V^\mu \xrightarrow{P} -A_\mu V^\mu$$

Any combination of vector and axial vector currents as $J_\mu \sim \alpha V_\mu + \beta A_\mu$ will generate parity violation in the Lagrangian, $\mathcal{L} \sim J_\mu J^{\mu+}$. But maximal parity violation is only reached if $J_\mu \sim V_\mu - A_\mu$, since

$$J_\mu J^{\mu+} \sim (V_\mu - A_\mu)(V^\mu - A^\mu) \xrightarrow{P} (V_\mu + A_\mu)(V^\mu + A^\mu)$$

which translates into that charged weak interactions only couple to left-handed fermions or right-handed antifermions. This can be seen simply by rewritting the current J_μ in terms of the field components. For instance, the leptonic current can be rewritten in terms of the left-handed fields as,

$$J_\mu \sim V_\mu - A_\mu = \overline{\nu}_e \gamma_\mu (1-\gamma_5) e = 2\overline{(\nu_e)}_L \gamma_\mu e_L$$

6.3. V-A THEORY OF CHARGED WEAK INTERACTIONS

After the discovery of parity violation in weak interactions, Feynman and Gell-Mann in 1958 proposed the V-A Theory [13] which incorporated the success of the Fermi Theory and solved the question of parity non-conservation by postulating instead a V-A form for the charged weak current. The current-current interactions are, like in the Fermi Theory, of local character, being contracted at the same space-time point. The effective weak copling is, as in the Fermi Theory, given by the Fermi constant, G_F.

The Lagrangian of the V-A Theory for the two first fermion generations is as follows,

$$\boxed{\mathcal{L}_{V-A} = -\frac{G_F}{\sqrt{2}} J^{CC}_\mu(x) J^{\mu CC+}(x)}$$

$$\boxed{J^{CC}_\mu = \overline{\nu}_e \gamma_\mu (1-\gamma_5) e + \overline{\nu}_\mu \gamma_\mu (1-\gamma_5) \mu + \overline{u} \gamma_\mu (1-\gamma_5) d'}$$

Notice that the d-quark field appearing in this Lagrangian, denoted by d', is the weak interactions d-quark eigenstate which is different than the d-quark mass eigenstate, denoted in these lectures by d. They are related by a rotation of the so-called Cabibbo angle θ_c,

$$d' = cos\theta_c d + sin\theta_c s$$

The idea of the rotated d-quark states was proposed by Cabibbo in 1963 [14] to account for weak decays of 'strange' particles and, in particular, to explain the suppression factor of the kaon decay rate as compared to the pion decay rate which experimentally was found to be about $\frac{1}{20}$. By comparing the theoretical prediction from the V-A Theory with the experimental data, the numerical value of the θ_c angle is inferred,

$$\frac{\Gamma(K^- \to \mu^- \overline{\nu}_\mu)}{\Gamma(\pi^- \to \mu^- \overline{\nu}_\mu)} \sim \frac{sin^2\theta_c}{cos^2\theta_c} \sim \frac{1}{20} \quad \Longrightarrow \quad \boxed{\theta_c \simeq 13^\circ}$$

The value of the effective coupling of the weak interactions, G_F is deduced from the meassurement of the μ lifetime,

$$\tau_\mu^{\rm exp} = 2.2 \times 10^{-6} \; sec$$

The prediction in the V-A Theory to tree level and by neglecting the electron mass is,

$$\frac{1}{\tau_\mu} = \Gamma(\mu^- \to e^- \overline{\nu}_e \nu_\mu) = \frac{G_F^2 m_\mu^5}{192\pi^3}$$

and from this,

$$\boxed{G_F = 1.167 \times 10^{-5} \; GeV^{-2}}$$

The V-A Theory described reasonably well the phenomenology of weak interactions until the discovery of the neutral currents in 1973 [15]. Notice that the neutral currents were not included in the formulation of the V-A Theory. Besides, The V-A Theory presented some non-appealing properties from the point of view of the consistency of the theory itself. In particular, the V-A Theory violates unitarity and it is a non-renormalizable theory. The unitarity violation property can be seen, for instance, by comparing the prediction in the V-A Theory of the cross section for elastic scattering of electron and neutrino,

$$\sigma_{V-A}(\nu e^- \to \nu e^-) = \frac{G_F^2}{6\pi} s$$

with the unitarity bound for the total cross section which is obtained from the general requirement of unitarity of the scattering S-matrix,

$$SS^+ = S^+ S = I \implies |a_J(s)|^2 \leq 1 \, \forall J \implies$$
$$\sigma(s)_{\text{tot}} = \frac{16\pi}{s} \sum_J (2J+1)|a_J(s)|^2 \leq \frac{16\pi}{s} \sum_J (2J+1)$$

It is clear that for high energies the prediction from the V-A Theory surpasses the unitarity bound and, therefore, it should not be trusted. It happens roughly at $\sqrt{s} \sim 300 \; GeV$.

The non-renormalizability of the V-A Theory can be seen, for instance, by computing loop contributions to the cross section and realizing that there appear quadratic divergences which cannot be absorbed into redefinitions of the parameters of this theory. As in the previous discussion on unitarity, it is due to the 'bad' behaviour of the V-A Theory at high energies. The V-A Theory is said to be non-predictive at high energies and it should only be used as an effective theory at low enough energies.

6.4. INTERMEDIATE VECTOR BOSON THEORY

The Intermediate Vector Boson (IVB) Theory of weak interactions assumed that these are mediated by the exchange of massive vector bosons with spin,

$s = 1$. First, it was proposed the existence of intermediate charged vector bosons $W^{\pm}$ for the charged weak interactions [16] and later the intermediate neutral vector boson Z for the neutral weak interactions [17]. Notice that these bosons were not true gauge bosons yet.

The interaction Lagrangian of the IVB Theory, including both the charged (CC) and the neutral (NC) currents, is given by,

$$\boxed{\begin{array}{c} \mathcal{L}_{IVB} = \mathcal{L}_{CC} + \mathcal{L}_{NC} \\ \mathcal{L}_{CC} = \frac{g}{\sqrt{2}}(J_\mu W^{\mu +} + J_\mu^+ W^{-\mu}) \\ \mathcal{L}_{NC} = \frac{g}{cos\theta_w} J_\mu^{NC} Z^\mu \end{array}}$$

where,

$$\boxed{J_\mu = \sum_l \overline{\nu}_l \gamma_\mu \left(\frac{1-\gamma_5}{2}\right) l + \sum_q \overline{q} \gamma_\mu \left(\frac{1-\gamma_5}{2}\right) q'}$$

$$\boxed{J_\mu^{NC} = \sum_{f=l,q} g_L^f \overline{f} \gamma_\mu \left(\frac{1-\gamma_5}{2}\right) f + \sum_{f \neq \nu} g_R^f \overline{f} \gamma_\mu \left(\frac{1+\gamma_5}{2}\right) f}$$

Here the $W_\mu^\pm$ and Z_μ are the charged and neutral intermediate vector bosons respectively and g is the dimensionless weak coupling. The weak angle, θ_w, defines the rotation in the neutral sector from the weak eigenstates to the physical mass eigenstates, and relates the weak coupling to the electromagnetic coupling, $g = \frac{e}{sin\theta_w}$.

Notice that the current-current interactions are non-local, in contrast to the V-A Theory, due to the propagation of the intermediate bosons. Besides, the new proposed neutral currents have both V-A and V+A components, although experimentaly it is known that the V-A component dominates.

The prediction of neutral currents in 1961 [17] was corroborated experimentally 12 years later! in neutrino-hadron scattering by the Gargamelle collaboration at CERN [15]. It was a great success of the IVB Theory which was incorporated later into the construction of the SM.

The relation between the parameters of the IVB Theory and the V-A Theory, which is also incorporated in the construction of the SM, can be obtained by comparison of the predictions from the two theories for $e\nu \to e\nu$ scattering at low energies ($\sqrt{s} << M_W$),

$$\boxed{\frac{G_F}{\sqrt{2}} = \frac{g^2}{8M_W^2}}$$

Finally, the IVB Theory is not free of problems either. It shares with the V-A Theory the problems of non-renormalizability and violation of unitarity at high energies. At low energies, say below the M_W threshold, the IVB Theory is a well behaved effective theory of the weak interactions, but above it the theory behaves badly. The problem of non-renormalizability can be seen for instance by studing the $e^+e^- \to e^+e^-$ scattering proccess at one loop. There are one-loop diagramms with W bosons propagating in the internal lines that diverge quadratically at high energies due to the bad behaviour of the W boson propagator in the IVB Theory,

$$(-i\Delta_W)_{IVB} \overset{k^2>>M_W^2}{\longrightarrow} \frac{1}{M_W^2}$$

This should be compared with the well behaved W boson propagator in the Electroweak Gauge Theory,

$$(-i\Delta_W)_{\text{gauge}} \overset{k^2>>M_W^2}{\longrightarrow} \frac{1}{k^2}$$

The violation of the unitarity bound occurs at slightly higher energies that in the V-A Theory case. For instance, the cross-section for the production of two longitudinal gauge bosons from neutrinos in the IVB Theory at tree level is,

$$\sigma_{IVB}(\nu\bar{\nu} \to W_L^+ W_L^-) \sim \frac{g^4}{M_W^4} s$$

which surpasses the unitarity bound at approximately, $\sqrt{s} \sim 500\ GeV$. Notice that there is just one contributing diagramm, the one with an electron in the t-channel. Notice also that the IVB Theory does not include the vector bosons self-interactions which are generic of non-abelian gauge theories. These are precisely the 'repairing' interactions ocurring in the Electroweak Theory . The prediction from the SM for the previous scattering proccess includes the contribution from an extra diagramm with a Z boson exchanged in the s-channel which couples to the final $W_L^+ W_L^-$ pair with a typical non-abelian Yang Mills coupling. This new diagramm cancels the bad high energy behaviour of the previous one. This dramatic cancellation also occurs in many other proccesses. See, for instance, the meassurement of the cross-section for $e^+e^- \to W^+W^-$ at LEP presented in Nodulman's lectures, where these cancellations are shown.

7. Building The Electroweak Theory

7.1. SOME NOTES ON HISTORY

The proposal of the symmetry group for the Electroweak Theory, $SU(2)_L \times U(1)_Y$, was done by Glashow in 1961 [17]. His motivation was rather to unify weak and electromagnetic interactions into a symmetry group that contained $U(1)_{em}$. The predictions included the existence of four physical vector boson eigenstates, $W^\pm$, Z, and γ, obtained from rotations of the weak eigenstates. In particular, the rotation by the weak angle θ_w which defines the Z weak boson was introduced already in this work. The massive weak bosons $W^{\pm}$ and Z were considered as the exchanged bosons in the weak interactions, but they were not considered yet as gauge bosons. The vector boson masses M_W and M_Z were parameters introduced by hand and the interaction Lagrangian was that of the IVB Theory.

Another key ingredient for the building of the Electroweak Theory is provided by the Goldstone Theorem which was initiated by Nambu in 1960 and proved and studied with generality by Goldstone in 1961 and by Goldstone, Salam and Weinberg in 1962 [18]. This theorem states the existence of massless spinless particles as an implication of spontaneous symmetry breaking of global symmetries.

The spontaneous symmetry breaking of local (gauge) symmetries, needed for the breaking of the electroweak symmetry $SU(2)_L \times U(1)_Y$, was studied by P. Higgs, F.Englert and R.Brout, Guralnik, Hagen and Kibble in 1964 and later [19]. These works were inspired in previous studies within the context of condensed-matter physics as those by Nambu and Jona-Lasinio on BCS Theory of superconductivity and works by Schwinger in 1962 and by Anderson in 1963 [20]. The procedure for this spontaneous breakdown of gauge symmetries is referred to as the Higgs Mechanism.

The Electroweak Theory as it is known nowadays was formulated by Weinberg in 1967 and by Salam in 1968 who incorporated the idea of unification of Glashow [1]. This Theory, commonly called Glashow-Weinberg-Salam Model or SM, was built with the help of the gauge principle and the knowledgde of gauge theories and incorporated all the good phenomelogical properties of the pregauge theories of the weak interactions, and in particular those of the IVB theory. The SM is indeed a gauge theory based on the gauge symmetry of the electroweak interactions $SU(2)_L \times U(1)_Y$ and the intermediate vector bosons, γ, $W^{\pm}$ and Z are the four associated gauge bosons. The gauge boson masses, M_W and M_Z, are generated by the Higgs Mechanism in the Electroweak Theory and, as a consequence, it respects unitarity at all energies and is renormalizable.

The important proof of renormalizability of gauge theories with and without spontaneous symmetry breaking was provided by 't Hooft in 1971 [21].

The first firm indication that the Sandard Model was the correct theory of electroweak interactions was probably the discovery of Neutral Currents in 1973 [15] which included the first meassurement of $sin^2\theta_w$. By using this experimental input for θ_w and the values of the electromagnetic coupling and G_F, the SM provided the first estimates for M_W and M_Z at that time which were already very close to the present values.

Another important ingredients of the SM are: fermion family replication, quark mixing and CP violation. After the proposal of the $d-s$ quark mixing given by the Cabibbo angle [14], the charm quark was postulated [22] as the companion of the s quark in the charged weak interactions. Futhermore, Glashow, Iliopoulos and Maiani showed in 1970 [23] that any sensible weak interaction theory must have this extra associated hadronic current in order to suppress to an acceptable level the induced strangeness-changing-neutral current effects. This suppression mechanism of flavour-changing-neutral currents (FCNC), usualy called GIM Mechanism, although invented before the general acceptance of gauge theories, can best be explained in that context. The existence of the c quark was confirmed in 1974 [24] with the discovery of the $J-\Psi$ particle which is interpreted as a $c\bar{c}$ bound state. With the discovery of the τ and ν_τ leptons [25] and the b quark [26], the fermion scenario with three families was set in. Finally, the discovery of the top quark in 1994 (17 years later!) [27] has completed this scenario. The quark mixing in the three generations case is given by the so-called Cabibbo-Kobayashi-Maskawa matrix [28] which incorporates the needed phase for CP violation in the SM.

The gold success of the SM was clearly the discovery of the gauge bosons $W^\pm$ and Z at the SpS collider at CERN in 1983 [30]. Since then there have been plenty of succesfull tests of the SM.

7.2. CHOICE OF THE GROUP $SU(2)_L \times U(1)_Y$

In order to follow the argument for the choice of the relevant group in the Electroweak Theory, $SU(2)_L \times U(1)_Y$, it is sufficient to consider the $e^-\nu_e$ component of the charged weak current that we write now in the form,

$$J_\mu = \overline{\nu}\gamma_\mu \left(\frac{1-\gamma_5}{2}\right) e = \overline{\nu_L}\gamma_\mu e_L = \overline{l_L}\gamma_\mu \sigma_+ l_L$$

$$J_\mu^+ = \overline{e}\gamma_\mu \left(\frac{1-\gamma_5}{2}\right) \nu = \overline{e_L}\gamma_\mu \nu_L = \overline{l_L}\gamma_\mu \sigma_- l_L$$

and we have introduced the lepton doublet notation and the σ_i $(i = 1, 2, 3)$ Pauli matrices,

$$l_L = \begin{pmatrix} \nu_L \\ e_L \end{pmatrix}, \ \overline{l_L} = \begin{pmatrix} \overline{\nu_L} & \overline{e_L} \end{pmatrix}, \ \sigma_\pm = \frac{1}{2}(\sigma_1 \pm i\sigma_2)$$

$$\sigma_1 = \begin{pmatrix} 0 & 1 \\ 1 & 0 \end{pmatrix}, \ \sigma_2 = \begin{pmatrix} 0 & -i \\ i & 0 \end{pmatrix}, \ \sigma_3 = \begin{pmatrix} 1 & 0 \\ 0 & -1 \end{pmatrix}$$

The 2×2 matrices $T_i = \frac{\sigma_i}{2}$ $i = 1, 2, 3$ are the three generators of $SU(2)$.

Notice that in the charged currents there are just two generators T_1 and T_2. A third generator T_3 is needed in order to close the $SU(2)$ algebra. This implies the formulation of the third current that is relevant for electroweak interactions,

$$J^3_\mu = \overline{l_L}\gamma_\mu \frac{\sigma_3}{2} l_L = \frac{1}{2}(\overline{\nu_L}\gamma_\mu \nu_L - \overline{e_L}\gamma_\mu e_L)$$

The weak isospin group is the $SU(2)$ group that is generated by these three generators and is usually denoted by $SU(2)_L$, where the subscript L refers to the left-handed character of the three weak currents. The weak isospin algebra is correspondingly,

$$\left[\frac{\sigma_i}{2}, \frac{\sigma_j}{2}\right] = i\epsilon_{ijk}\frac{\sigma_k}{2}$$

where the $SU(2)$ structure constants are the completely antisymmetric Levi-Civita symbols ϵ_{ijk}.

By Noether's Theorem there are three associated conserved weak charges,

$$T^i = \int d^3x J^i_0(x)\ ,\ i = 1, 2, 3$$

It is interesting to notice that the above introduced neutral weak current is none of the two physical known neutral currents, J^{em}_μ and J^{NC}_μ. Futhermore, none of these two currents have definite properties under $SU(2)_L$ transformations, whereas J^3_μ does. With the motivation of unifying the electromagnetic and weak interactions, Glashow proposed to include the electromagnetic current by adding to $SU(2)_L$ a new $U(1)$ group which should be different than $U(1)_{em}$ in order to the get the proper conmutation relations among the $U(1)$ and $SU(2)_L$ generators. The new proposed group is the weak hypercharge group $U(1)_Y$ with one generator $\frac{Y}{2}$ which indeed, as it must be, conmutes with the three $SU(2)_L$ generators. The associated neutral current is the weak hypercharge current, J^Y_μ, and the conserved charge is the weak hypercharge Y. Within this formalism there is some sort of electromagnetic and weak interactions unification since the $U(1)_{em}$ group appears as a subgroup of the total electroweak group,

$$U(1)_{em} \subset SU(2)_L \times U(1)_Y$$

The relation among the charges associated to the three neutral currents, J^{em}_μ, J^3_μ and J^Y_μ, is a replica of the Gell-Mann Nishijima relation,

$$\boxed{Q = T_3 + \tfrac{Y}{2}}$$

where now,

$$Q = \text{electric charge}\ ,\ T_3 = \text{weak isospin}\ ,\ Y = \text{weak hypercharge};$$

and the corresponding relation among the currents is,

$$J_\mu^{em} = J_\mu^3 + J_\mu^Y$$

Therefore, if the following are used as inputs

$$J_\mu^{em} = (-1)\overline{e_L}\gamma_\mu e_L + (-1)\overline{e_R}\gamma_\mu e_R$$

$$J_\mu^3 = \left(-\frac{1}{2}\right)\overline{e_L}\gamma_\mu e_L + \left(\frac{1}{2}\right)\overline{\nu_L}\gamma_\mu \nu_L$$

one can get J_μ^Y and the orthogonal combination J_μ^{NC} as outputs,

$$J_\mu^Y = 2(J_\mu^{em} - J_\mu^3) = (-1)\overline{e_L}\gamma_\mu e_L + (-2)\overline{e_R}\gamma_\mu e_R + (-1)\overline{\nu_L}\gamma_\mu \nu_L$$

$$J_\mu^{NC} \perp J_\mu^{em} \Rightarrow$$

$$J_\mu^{NC} = c_w^2 J_\mu^3 - s_w^2 \frac{J_\mu^Y}{2} = (-\frac{1}{2} + s_w^2)\overline{e_L}\gamma_\mu e_L + (s_w^2)\overline{e_R}\gamma_\mu e_R + (\frac{1}{2})\overline{\nu_L}\gamma_\mu \nu_L$$

where,

$$c_w = cos\theta_w \; ; \; s_w = sin\theta_w \; ; \; \theta_w = \text{weak angle}$$

Notice that the currents that couple to the physical neutral bosons A_μ and Z_μ are J_μ^{em} and J_μ^{NC} respectively, and it is this last one, J_μ^{NC}, that inherits the generic name of neutral current.

From the above expressions for the neutral currents, one can also extract the values of the corresponding charges and couplings. For instance, from J_μ^{NC} one gets the relevant factors in the weak neutral couplings to electrons and neutrinos , $g_L^e = -\frac{1}{2} + s_w^2$, $g_R^e = s_w^2$, $g_L^\nu = \frac{1}{2}$ and $g_R^\nu = 0$.

If one includes the contributions from all the quarks and leptons of the three families, the neutral currents are written generically as:

$$J_\mu^Y = \sum_f Y_{f_L}\overline{f_L}\gamma_\mu f_L + \sum_f Y_{f_R}\overline{f_R}\gamma_\mu f_R$$

$$J_\mu^{NC} = \sum_f g_L^f\overline{f_L}\gamma_\mu f_L + \sum_{f\neq\nu} g_R^f\overline{f_R}\gamma_\mu f_R$$

$$J_\mu^{em} = \sum_f Q_f\overline{f_L}\gamma_\mu f_L + \sum_f Q_f\overline{f_R}\gamma_\mu f_R$$

$$J_\mu^3 = \sum_f T_3^f\overline{f_L}\gamma_\mu f_L$$

where,

$$g_L^f = T_3^f - Q_f s_w^2 \; ; \; g_R^f = -Q_f s_w^2$$

TABLE 1. Lepton quantum numbers

Lepton	T	T_3	Q	Y
ν_L	$\frac{1}{2}$	$\frac{1}{2}$	0	−1
e_L	$\frac{1}{2}$	$-\frac{1}{2}$	−1	−1
e_R	0	0	−1	−2

TABLE 2. Quark quantum numbers

Quark	T	T_3	Q	Y
u_L	$\frac{1}{2}$	$\frac{1}{2}$	$\frac{2}{3}$	$\frac{1}{3}$
d_L	$\frac{1}{2}$	$-\frac{1}{2}$	$-\frac{1}{3}$	$\frac{1}{3}$
u_R	0	0	$\frac{2}{3}$	$\frac{4}{3}$
d_R	0	0	$-\frac{1}{3}$	$-\frac{2}{3}$

The corresponding quantum numbers for the fermions of the first family are collected in Tables 1 and 2. The fermions of the second and third family have the same quantum numbers as the corresponding fermions of the first one.

Similarly, the charged current is written generically as,

$$J_\mu = \sum_f \overline{f_L} \gamma_\mu \sigma_+ f_L$$

Finally, the electroweak interaction Lagrangian is written in terms of the currents and the physical fields as,

$$\mathcal{L}_{\text{int}} = \mathcal{L}_{CC} + \mathcal{L}_{NC} + \mathcal{L}_{em}$$

where,

$$\mathcal{L}_{CC} = \frac{g}{\sqrt{2}}(J_\mu W^{\mu\,+} + J_\mu^+ W^{\mu\,-})$$

$$\mathcal{L}_{NC} = \frac{g}{c_w} J_\mu^{NC} Z^\mu$$

$$\mathcal{L}_{em} = e J_\mu^{em} A^\mu$$

Notice that $\mathcal{L}_{CC}$ and $\mathcal{L}_{NC}$ are the same Lagrangians as in the IVB Theory.

8. SM: The Gauge Theory of Electroweak Interactions

The SM is the gauge theory for electroweak interactions and has provided plenty of successful predictions with an impressive level of precision.

It is based on the gauge symmetry of electroweak interactions, namely, the symmetry $SU(2)_L \times U(1)_Y$ previously introduced which is required to be a local symmetry of the electroweak Lagrangian. As before, $SU(2)_L$ is the weak isospin group which acts just on left-handed fermions and $U(1)_Y$ is the weak hypercharge group. The $SU(2)_L \times U(1)_Y$ group has four generators, three of which are the $SU(2)_L$ generators, $T_i = \frac{\sigma_i}{2}$ with $i = 1, 2, 3$, and the fourth one is the $U(1)_Y$ generator, $\frac{Y}{2}$. The conmutation relations for the total group are:

$$[T_i, T_j] = i\epsilon_{ijk}T_k \; ; \; [T_i, Y] = 0 \; ; \; i, j, k = 1, 2, 3$$

The left-handed fermions transform as doublets under $SU(2)_L$,

$$f_L \to e^{i\vec{T}\vec{\theta}} f_L \; ; \; f_L = \begin{pmatrix} \nu_L \\ e_L \end{pmatrix}, \begin{pmatrix} u_L \\ d_L \end{pmatrix}, \ldots$$

whereas the right-handed fermions transform as singlets,

$$f_R \to f_R \, ; \; f_R = e_R \, , \, u_R \, , \, d_R, \ldots$$

The fermion quantum numbers are as in Tables 1 and 2 and the relation

$$Q = T_3 + \frac{Y}{2}$$

is also incorporated in the SM.

The number of associated gauge bosons, being equal to the number of generators, is four:

<u>gauge bosons</u>

W^i_μ , $i = 1, 2, 3$. These are the weak bosons of $SU(2)_L$

B_μ. This is the hypercharge boson of $U(1)_Y$

The building of the SM Lagrangian is done by following the same steps as in any gauge theory. In particular, the $SU(2)_L \times U(1)_Y$ symmetry is promoted from global to local by replacing the derivatives of the fields by the corresponding covariant derivatives. For a generic fermion field f, its covariant derivative corresponding to the $SU(2)_L \times U(1)_Y$ gauge symmetry is,

$$D_\mu f = \left(\partial_\mu - ig\vec{T}.\vec{W}_\mu - ig'\frac{Y}{2}B_\mu\right) f$$

where,

$g=$ coupling constant corresponding to $SU(2)_L$
$g'=$ coupling constant corresponding to $U(1)_Y$

For example, the covariant derivatives for a left-handed and a right-handed electron are respectively,

$$D_\mu e_L = \left(\partial_\mu - ig\frac{\vec{\sigma}}{2}.\vec{W}_\mu + ig'\frac{1}{2}B_\mu\right)e_L \; ; \; D_\mu e_R = \left(\partial_\mu + ig'B_\mu\right)e_R$$

As in the previous cases of QED and QCD, the gauge invariant electroweak interactions are generated from the $\overline{f}i\not{D}f$ term. After replacing the covariant derivative above, and by rotating the weak bosons to the physical basis, one can check that the interaction terms obtained for the electroweak bosons with the quarks and leptons are the same as those in the interaction Lagrangian $\mathcal{L}_{\text{int}}$ given in the previous section.

9. Lagrangian of The Electroweak Theory I

In order to get the total Lagrangian of the Electroweak Theory one must add to the previous fermion terms containing the kinetic and fermion interaction terms, the gauge boson kinetic terms and the gauge boson self-interaction terms. The SM total Lagrangian can be written as,

$$\boxed{\mathcal{L}_{SM} = \mathcal{L}_f + \mathcal{L}_G + \mathcal{L}_{SBS} + \mathcal{L}_{YW}}$$

where, the fermion Lagrangian is,

$$\boxed{\mathcal{L}_f = \textstyle\sum_{f=l,q} \overline{f}i\not{D}f}$$

and the Lagrangian for the gauge fields is,

$$\boxed{\mathcal{L}_G = -\tfrac{1}{4}W^i_{\mu\nu}W^{\mu\nu}_i - \tfrac{1}{4}B_{\mu\nu}B^{\mu\nu} + \mathcal{L}_{GF} + \mathcal{L}_{FP}}$$

which is written in terms of the field strength tensors,

$$W^i_{\mu\nu} = \partial_\mu W^i_\nu - \partial_\nu W^i_\mu + g\epsilon^{ijk}W^j_\mu W^k_\nu$$

$$B_{\mu\nu} = \partial_\mu B_\nu - \partial_\nu B_\mu$$

$\mathcal{L}_{GF}$ and $\mathcal{L}_{FP}$ are the gauge fixing and Faddeev Popov Lagrangians respectively that are needed in any gauge theory. We omit to write them here for brevity. These have also been omited in the cases of QCD and QED.

Notice that this gauge Lagrangian contains the wanted self-interaction terms among the three $W^i_\mu, i = 1, 2, 3$ gauge bosons, as it corresponds to a non-abelian $SU(2)_L$ group.

The last two terms, $\mathcal{L}_{SBS}$ and $\mathcal{L}_{YW}$ are the Symmetry Breaking Sector Lagrangian and the Yukawa Lagrangian respectively. As will be discussed in the forthcomming sections, these terms are needed in order to provide the wanted M_W and M_Z gauge boson masses and m_f fermion masses.

One can show that $\mathcal{L}_{SM}$ is indeed invariant under the following $SU(2)_L \times U(1)_Y$ gauge transformations:

$$\begin{array}{lcl} f_L & \to & e^{i\vec{T}\vec{\theta}(x)} f_L \\ f_R & \to & f_R \\ f & \to & e^{i\frac{Y}{2}\alpha(x)} f \\ W^i_\mu & \to & W^i_\mu - \frac{1}{g}\partial_\mu \theta^i(x) + \epsilon^{ijk}\theta^j W^k_\mu \\ B_\mu & \to & B_\mu - \frac{1}{g'}\partial_\mu \alpha(x) \end{array}$$

The physical gauge bosons $W^\pm_\mu$, Z_μ and A_μ are obtained from the electroweak interaction eigenstates by the following expressions,

$$\boxed{\begin{array}{c} W^\pm_\mu = \frac{1}{\sqrt{2}}(W^1_\mu \mp iW^2_\mu) \\ Z_\mu = c_w W^3_\mu - s_w B_\mu \\ A_\mu = s_w W^3_\mu + c_w B_\mu \end{array}}$$

where, θ_w defines the rotation in the neutral sector. The relations among the various couplings are obtained by identifying the interactions terms with those of $\mathcal{L}_{\text{int}}$. Thus one gets,

$$\boxed{g = \frac{e}{s_w}} \quad \boxed{g' = \frac{e}{c_w}}$$

Finally, note that mass terms as $M^2_W W_\mu W^\mu$, $\frac{1}{2}M^2_Z Z_\mu Z^\mu$ and $m_f \bar{f} f$ are forbidden by $SU(2)_L \times U(1)_Y$ gauge invariance. This is a new situation which is not found in QED or QCD. The needed gauge boson masses must be generated in a gauge invariant way. The spontaneous breaking of the $SU(2)_L \times U(1)_Y$ symmetry and the Higgs Mechanism provide indeed this mass generation. To this subject we come next.

10. The Concept of Spontaneous Symmetry Breaking and The Higgs Mechanism

One of the key ingredients of the SM of electroweak interactions is the concept of Spontaneous Symmetry Breaking (SSB), giving rise to Goldstone-excitations [18] which in turn can be related to gauge boson mass terms [5]. When this SSB refers to a gauge symmetry instead of a global symmetry, then the Higgs Mechanism operates [19]. This procedure is needed in order to describe the short ranged weak interactions by a gauge theory without spoiling gauge invariance. The discovery of the $W^\pm$ and Z gauge bosons at CERN in 1983 [30] may be considered as the first experimental evidence of the SSB phenomenon in electroweak interactions. In present and future experiments one hopes to get insight into the nature of this Symmetry Breaking Sector (SBS) and this is one of the main motivations for constructing the next generation of accelerators. In particular, it is the most exiciting challenge for the LHC collider being built at CERN.

In the SM, the symmetry breaking is realized linearly by a scalar field which acquires a non-zero vacuum expectation value. The resulting physical spectrum contains not only the massive intermediate vector bosons and fermionic matter fields but also the Higgs particle, a neutral scalar field which has escaped experimental detection until now. The main advantage of the SM picture of symmetry breaking lies in the fact that an explicit and consistent formulation exists, and any observable can be calculated perturbatively in the Higgs self-coupling constant. However, the fact that one can compute in a model doesn't mean at all that this is the right one.

The concept of spontaneous electroweak symmetry breaking is more general than the way it is usually implemented in the SM. Any alternative SBS has a chance to replace the standard Higgs sector, provided it meets the following basic requirements: 1) Electromagnetism remains unbroken; 2) The full symmetry contains the electroweak gauge symmetry; 3) The symmetry breaking occurs at about the energy scale $v = (\sqrt{2}G_F)^{-\frac{1}{2}} = 246\ GeV$ with G_F being the Fermi coupling constant.

In the following it is reviewed the basic ingredients of the symmetry breaking phenomenom in the Electroweak Theory. Some relevant topics related with this breaking are also discussed.

10.1. THE PHENOMENON OF SPONTANEOUS SYMMETRY BREAKING

A simple definition of the phenomenon of SSB is as follows:

A physical system has a symmetry that is spontaneously broken if the interactions governing the dynamics of the system possess such a symmetry but the ground state of this system does not.

An illustrative example of this phenomenon is the infinitely extended ferromagnet. For this purpouse, let us consider the system near the Curie temperature T_C. It is described by an infinite set of elementary spins whose interactions are rotationally invariant, but its ground state presents two different situations depending on the value of the temperature T.

Situation I: $T > T_C$

The spins of the system are randomly oriented and as a consequence the average magnetization vanishes: $\vec{M}_{\text{average}} = 0$. The ground state with these disoriented spins is clearly rotationally invariant.

Situation II: $T < T_C$

The spins of the system are all oriented parallely to some particular but arbitrary direction and the average magnetization gets a non-zero value: $\vec{M}_{\text{average}} \neq 0$ (*Spontaneous Magnetization*). Since the directions of the spins are arbitrary, there are infinite possible ground states, each one corresponding to one possible direction and all having the same (minimal) energy. Futhermore, none of these states are rotationally invariant since there is a privileged direction. This is, therefore, a clear example of SSB since the interactions among the spins are rotationally invariant but the ground state is not. More specifically, it is the fact that the system 'chooses' one among the infinite possible non-invariant ground states what produces the phenomenon of SSB.

On the theoretical side, and irrespectively of what could be the origen of such a physical phenomenon at a more fundamental level, one can parametrize this behaviour by means of a symple mathematical model. In the case of the infinitely extended ferromagnet one of these models is provided by the Theory of Ginzburg and Landau [29]. We present in the following the basic ingredients of this model.

For T near T_C, $\vec{M}$ is small and the free energy density $u(\vec{M})$ can be approached by (here higher powers of $\vec{M}$ are neglected):

$$
\begin{aligned}
u(\vec{M}) &= (\partial_i \vec{M})(\partial_i \vec{M}) + V(\vec{M}) \; ; \; i = 1,2,3 \\
V(\vec{M}) &= \alpha_1 (T - T_C)(\vec{M}.\vec{M}) + \alpha_2 (\vec{M}.\vec{M})^2 \; ; \; \alpha_1, \alpha_2 > 0
\end{aligned}
$$

The magnetization of the ground state is obtained from the condition of extremum:

$$
\frac{\delta V(\vec{M})}{\delta M_i} = 0 \Rightarrow \vec{M}.\left[\alpha_1 (T - T_C) + 2\alpha_2 (\vec{M}.\vec{M})\right] = 0
$$

There are two solutions for $\vec{M}$, depending on the value of T:

Solution I:

$$
\text{If}\, T > T_C \Rightarrow \left[\alpha_1 (T - T_C) + 2\alpha_2 (\vec{M}.\vec{M})\right] > 0 \Rightarrow \vec{M} = 0
$$

The solution for $\vec{M}$ is the trivial one and corresponds to the situation I described before where the ground state is rotational invariant. The potential $V(\vec{M})$ has a symmetric shape with a unique minimum at the origen $\vec{M} = 0$ where $V(0) = 0$. This is represented in Fig.2a for the simplified bidimensional case, $\vec{M} = (M_X, M_Y)$.

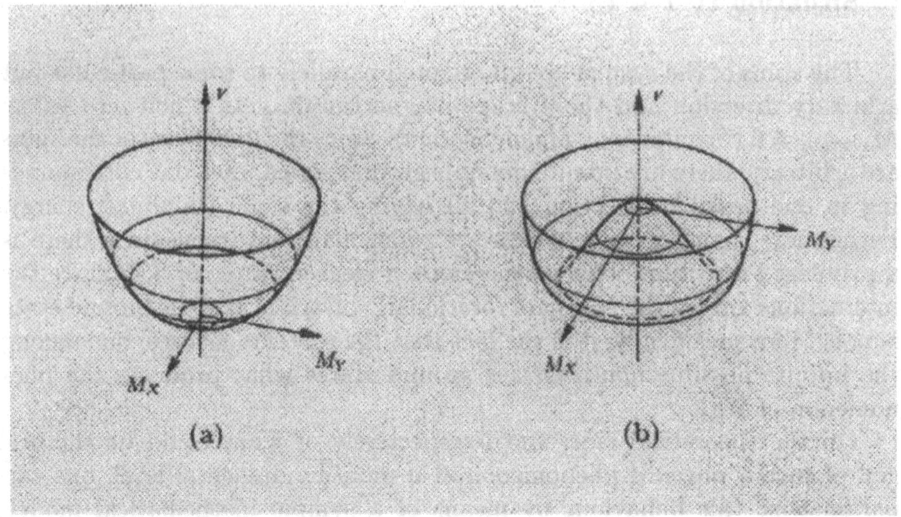

Figure 2. The potential $V(\vec{M})$ in the symmetric phase (a) and in the spontaneously broken phase (b)

Solution II:

If $T < T_C \Rightarrow \vec{M} = 0$ is a local maximum and the condition of minimun requires:

$$\alpha_1(T - T_C) + 2\alpha_2(\vec{M}.\vec{M}) = 0 \Rightarrow |\vec{M}| = \sqrt{\frac{\alpha_1(T_C - T)}{2\alpha_2}}$$

Namely, there are infinite absolute minima having all the same $|\vec{M}|$ above, but different direction of $\vec{M}$. This corresponds to the situation II where the system has infinite possible degenerate ground states which are not rotationally invariant. The potential $V(\vec{M})$ has a 'mexican hat shape' as represented in Fig.2b for the bidimensional case.

Notice that it is the choice of the particular ground state what produces, for $T < T_C$, the spontaneous breaking of the rotational symmetry.

10.2. SPONTANEOUS SYMMETRY BREAKING IN QUANTUM FIELD THEORY: QCD AS AN EXAMPLE

In the language of Quantum Field Theory, *a system is said to possess a symmetry that is spontaneously broken if the Lagrangian describing the dynamics of the system is invariant under these symmetry transformations, but the vacuum of the theory is not.* Here the vacuum $|0>$ is the state where the Hamiltonian expectation value $<0|H|0>$ is minimum.

For illustrative purposes we present in the following the particular case of QCD where, besides the color gauge symmetry, there is an extra global symmetry, named chiral symmetry, which turns out to be spontaneously broken . For simplicity let us consider QCD with just two flavours. The Lagrangian is that in section 5 with just the two ligthest quarks, u and d.

One can check that for $m_{u,d} = 0$, $\mathcal{L}_{QCD}$ has the chiral symmetry $SU(2)_L \times SU(2)_R$ that is defined by the following transformations:

$$\Psi_L \to \Psi'_L = U_L \Psi_L$$

$$\Psi_R \to \Psi'_R = U_R \Psi_R$$

where,

$$\Psi = \begin{pmatrix} u \\ d \end{pmatrix} \; ; \; \Psi_L = \frac{1}{2}(1-\gamma_5)\Psi \; ; \; \Psi_R = \frac{1}{2}(1+\gamma_5)\Psi$$

$$U_L \in SU(2)_L \; ; \; U_R \in SU(2)_R$$

U_L and U_R can be written in terms of the 2x2 matrices T_L^a and T_R^a $(a = 1, 2, 3)$ corresponding to the generators Q_L^a and Q_R^a of $SU(2)_L$ and $SU(2)_R$ respectively:

$$U_L = \exp(-i\alpha_L^a T_L^a) \; ; \; U_R = \exp(-i\alpha_R^a T_R^a)$$

It turns out that the physical vacuum of QCD is not invariant under the full chiral $SU(2)_L \times SU(2)_R$ group but just under the subgroup $SU(2)_V = SU(2)_{L+R}$ which is precisely the already introduced isospin group. The transformations given by the axial subgroup, $SU(2)_A$, do not leave the QCD vacuum invariant. Therefore, QCD with $m_{u,d} = 0$ has a chiral symmetry which is spontaneously broken down to the isospin symmetry:

$$SU(2)_L \times SU(2)_R \to SU(2)_V$$

The fact that in Nature $m_{u,d} \neq 0$ introduces an extra explicit breaking of this chiral symmetry. Since the fermion masses are small this explicit breaking is soft. The chiral symmetry is not an exact but approximate symmetry of QCD.

One important question is still to be clarified. How do we know from experiment that, in fact, the QCD vacuum is not $SU(2)_L \times SU(2)_R$ symmetric?. Let us assume for the moment that it is chiral invariant. We will see that this assumption leads to a contradiction with experiment.

If $|0>$ is chiral invariant $\Rightarrow$

$$U_L|0> = |0> \; ; \; U_R|0> = |0> \Rightarrow$$

$$T_L^a|0>=0; T_R^a|0>=0 \Rightarrow Q_L^a|0>=0\,;\, Q_R^a|0>=0$$

In addition, if $|\Psi>$ is an eigenstate of the Hamiltonian and parity operator such that:

$$H|\Psi>=E|\Psi>\,;\; P|\Psi>=|\Psi>$$

then,

$$\exists|\Psi'>=\frac{1}{\sqrt{2}}(Q_R^a-Q_L^a)|\Psi>\;/\;H|\Psi'>=E|\Psi'>\,;\; P|\Psi'>=-|\Psi'>$$

In summary, if the QCD vacuum is chiral invariant there must exist pairs of degenerate states in the spectrum, the so-called parity doublets as $|\Psi>$ and $|\Psi'>$, which are related by a chiral transformation and have opposite parities. The absence of such parity doublets in the hadronic spectrum indicates that the chiral symmetry must be spontaneously broken. Namely, there must exist some generators Q^a of the chiral group such that $Q^a|0>\neq 0$. More specifically, it can be shown that these generators are the three Q_5^a $(a=1,2,3)$ of the axial group, $SU(2)_A$. In conclusion, the chiral symmetry breaking pattern in QCD is $SU(2)_L\times SU(2)_R\to SU(2)_V$ as announced.

10.3. GOLDSTONE THEOREM

One of the physical implications of the SSB phenomenom is the appearance of massless modes. For instance, in the case of the infinitely extended ferromagnet and below the Curie temperature there appear modes connecting the different possible ground states, the so-called spin waves.

The general situation in Quantum Fied Theory is described by the Goldstone Theorem [18]:

If a Theory has a global symmetry of the Lagrangian which is not a symmetry of the vacuum then there must exist one massless boson, scalar or pseudoscalar, associated to each generator which does not annihilate the vacuum and having its same quantum numbers. These modes are referred to as Nambu-Goldstone bosons or simply as Goldstone bosons.

Let us return to the example of QCD. The breaking of the chiral symmetry is characterized by $Q_5^a|0>\neq 0$ $(a=1,2,3)$. Therefore, according to Goldstone Theorem, there must exist three massless Goldstone bosons, $\pi^a(x)$ $a=1,2,3$, which are pseudoscalars. These bosons are identified with the three physical pions.

The fact that pions have $m_\pi\neq 0$ is a consequence of the soft explicit breaking in $\mathcal{L}_{QCD}$ given by $m_q\neq 0$. The fact that m_π is small and that there is a large gap between this mass and the rest of the hadron masses can be seen as another manifestation of the spontaneous chiral symmetry breaking with the pions being the pseudo-Goldstone bosons of this breaking.

10.4. DYNAMICAL SYMMETRY BREAKING

In the previous sections we have seen the equivalence between the condition $Q^a|0>\neq 0$ and the non-invariance of the vacuum under the symmetry transformations generated by the Q^a generators:

$$U|0>\neq|0> \; ; \; U=\exp(i\epsilon^a Q^a)$$

In Quantum Field Theory, it can be shown that an alternative way of characterizing the phenomenom of SSB is by certain field operators that have non-vanishing vacuum expectation values (v.e.v.).

$$\text{SSB} \Longleftrightarrow \exists \Phi_j / <0|\Phi_j|0>\neq 0$$

This non-vanishing v.e.v. plays the role of the order parameter signaling the existence of a phase where the symmetry of the vacuum is broken.

There are several possibilities for the nature of this field operator. In particular, when it is a composite operator which represents a composite state being produced from a strong underlying dynamics, the corresponding SSB is said to be a dynamical symmetry breaking. The chiral symmetry breaking in QCD is one example of this type of breaking. The non-vanishing chiral condensate made up of a quark and an anti-quark is the order paremeter in this case:

$$<0|\bar{q}q|0>\neq 0 \Rightarrow SU(2)_L\times SU(2)_R \rightarrow SU(2)_V$$

The strong interactions of $SU(3)_C$ are the responsible for creating these $\bar{q}q$ pairs from the vacuum and, therefore, the $<0|\bar{q}q|0>$ should, in principle, be calculable from QCD.

It is interesting to mention that this type of symmetry breaking can happen similarly in more general $SU(N)$ gauge theories. The corresponding gauge couplings become sufficiently strong at large distances and allow for spontaneous breaking of their additional chiral-like symmetries. The corresponding order paremeter is also a chiral condensate: $<0|\overline{\Psi}\Psi|0>\neq 0$.

10.5. THE HIGGS MECHANISM

The Goldstone Theorem is for theories with spontaneously broken global symmetries but does not hold for gauge theories. When a spontaneous symmetry breaking takes place in a gauge theory the so-called Higgs Mechanism operates [19]:

The would-be Goldstone bosons associated to the global symmetry breaking do not manifest explicitely in the physical spectrum but instead they 'combine' with the massless gauge bosons and as result, once the spectrum of the theory is built up on the asymmetrical vacuum, there appear massive vector particles. The number of vector bosons that acquire a mass is precisely equal to the number of these would-be-Goldstone bosons.

There are three important properties of the Higgs Mechanism for 'mass generation' that are worth mentioning:

1.- It respects the gauge symmetry of the Lagrangian.
2.- It preserves the total number of polarization degrees.
3.- It does not spoil the good high energy properties nor the renormalizability of the massless gauge theories.

We now turn to the case of the SM of Electroweak Interactions. We will see in the following how the Higgs Mechanism is implemented in the $SU(2)_L \times U(1)_Y$ Gauge Theory in order to generate a mass for the weak gauge bosons, $W^{\pm}$ and Z.

The following facts must be considered:

1.- The Lagrangian of the SM is gauge $SU(2)_L \times U(1)_Y$ symmetric. Therefore, anything we wish to add must preserve this symmetry.
2.- We wish to generate masses for the three gauge bosons $W^{\pm}$ and Z but not for the photon, γ. Therefore, we need three would-be-Goldstone bosons, ϕ^+, ϕ^- and χ, which will combine with the three massless gauge bosons of the $SU(2)_L \times U(1)_Y$ symmetry.
3.- Since $U(1)_{em}$ is a symmetry of the physical spectrum, it must be a symmetry of the vacuum of the Electroweak Theory.

From the above considerations we conclude that in order to implement the Higgs Mechanism in the Electroweak Theory we need to introduce 'ad hoc' an additional system that interacts with the gauge sector in a $SU(2)_L \times U(1)_Y$ gauge invariant manner and whose self-interactions, being also introduced 'ad hoc', must produce the wanted breaking, $SU(2)_L \times U(1)_Y \rightarrow U(1)_{em}$, with the three associated would-be-Goldstone bosons ϕ^+, ϕ^- and χ. This sytem is the so-called SBS of the Electroweak Theory.

11. The Symmetry Breaking Sector of the Electroweak Theory

In this section we introduce and justify the simplest choice for the SBS of the Electroweak Theory.

Let Φ be the additional system providing the $SU(2)_L \times U(1)_Y \rightarrow U(1)_{em}$ breaking. Φ must fulfil the following conditions:

1.- It must be a scalar field so that the above breaking preserves Lorentz invariance.

2.- It must be a complex field so that the Hamiltonian is hermitian.

3.- It must have non-vanishing weak isospin and hypercharge in order to break $SU(2)_L$ and $U(1)_Y$. The assignment of quantum numbers and the choice of representation of Φ can be done in many ways. Some possibilities are:

- Choice of a non-linear representation: Φ transforms non-linearly under $SU(2)_L \times U(1)_Y$.

- Choice of a linear representation: Φ transforms linearly under $SU(2)_L \times U(1)_Y$. The simplest linear representation is a complex doublet. Alternative choices are: complex triplets, more than one doublet, etc. In particular, one may choose two complex doublets H_1 and H_2 as in the Minimal Supersymmetric SM.

4.- Only the neutral components of Φ are allowed to acquire a non-vanishing v.e.v. in order to preserve the $U(1)_{em}$ symmetry of the vacuum.

5.- The interactions of Φ with the gauge and fermionic sectors must be introduced in a gauge invariant way.

6.- The self-interactions of Φ given by the potential $V(\Phi)$ must produce the wanted breaking which is characterized in this case by $< 0|\Phi|0 > \neq 0$. Φ can be, in principle, a fundamental or a composite field.

7.- If we want to be predictive from low energies up to very high energies the interactions in $V(\Phi)$ must be renormalizable.

By taking into account the above seven points one is led to the following simplest choice for the system Φ and the Lagrangian of the SBS of the Electroweak Theory:

$$\begin{aligned}
\mathcal{L}_{SBS} &= (D_\mu\Phi)^\dagger(D^\mu\Phi) - V(\Phi) \\
V(\Phi) &= -\mu^2\Phi^\dagger\Phi + \lambda(\Phi^\dagger\Phi)^2 \,;\; \lambda > 0
\end{aligned}$$

where,

$$\begin{aligned}
\Phi &= \begin{pmatrix} \phi^+ \\ \phi_0 \end{pmatrix} \\
D_\mu\Phi &= (\partial_\mu - \frac{1}{2}ig\vec{\sigma}\cdot\vec{W}_\mu - \frac{1}{2}ig'B_\mu)\Phi
\end{aligned}$$

Here Φ is a fundamental complex doublet with hypercharge $Y(\Phi) = 1$ and $V(\Phi)$ is the simplest renormalizable potential. $\vec{W}_\mu$ and B_μ are the gauge fields of $SU(2)_L$ and $U(1)_Y$ respectively and g and g' are the corresponding gauge couplings.

It is interesting to notice the similarities with the Ginzburg-Landau Theory. Depending on the sign of the mass parameter $(-\mu^2)$, there are two possibilities for the v.e.v. $< 0|\Phi|0 >$ that minimizes the potential $V(\Phi)$,

1) $(-\mu^2) > 0$: The minimum is at:

$$<0|\Phi|0>=0$$

The vacuum is $SU(2)_L \times U(1)_Y$ symmetric and therefore no symmetry breaking occurs.

2) $(-\mu^2) < 0$: The minimum is at:

$$|<0|\Phi|0>| = \begin{pmatrix} 0 \\ \frac{v}{\sqrt{2}} \end{pmatrix} \; ; \; \text{arbitrary } arg\,\Phi \; ; \; v \equiv \sqrt{\frac{\mu^2}{\lambda}}$$

Therefore, there are infinite degenerate vacua corresponding to infinite posssible values of $arg\,\Phi$. Either of these vacua is $SU(2)_L \times U(1)_Y$ non-symmetric and $U(1)_{em}$ symmetric. The breaking $SU(2)_L \times U(1)_Y \to U(1)_{em}$ occurs once a particular vacuum is chosen. As usual, the simplest choice is taken:

$$|<0|\Phi|0>| = \begin{pmatrix} 0 \\ \frac{v}{\sqrt{2}} \end{pmatrix} \; ; \; arg\,\Phi \equiv 0 \; ; \; v \equiv \sqrt{\frac{\mu^2}{\lambda}}$$

The two above symmetric and non-symmetric phases of the Electroweak Theory are clearly similar to the two phases of the ferromagnet that we have described within the Ginzburg Landau Theory context. In the SM, the field Φ replaces the magnetization $\vec{M}$ and the potential $V(\Phi)$ replaces $V(\vec{M})$. The SM order papameter is, consequently, $<0|\Phi|0>$. In the symmetric phase, $V(\Phi)$ is as in Fig.2a, whereas in the non-symmetric phase, it is as in Fig.2b.

Another interesting aspect of the Higgs Mechanism, as we have already mentioned, is that it preserves the total number of polarization degrees. Let us make the counting in detail:

1) **Before SSB**
4 massless gauge bosons: $W^{\mu}_{1,2,3}, B^{\mu}$
4 massless scalars: The 4 real components of Φ: $\phi_1, \phi_2, \phi_3, \phi_4$

Total number of polarization degrees $= 4 \times 2 + 4 = 12$

2) **After SSB**
3 massive gauge bosons: $W^{\pm}, Z$
1 massless gauge boson: γ
1 massive scalar: H

Total number of polarization degrees: $3 \times 3 + 1 \times 2 + 1 = 12$

Furthermore, it is important to realize that one more degree than needed is introduced into the theory from the beginning. Three of the real components of Φ, or similarly $\phi^{\pm} \equiv \frac{1}{\sqrt{2}}(\phi_1 \mp i\phi_2)$ and $\chi = \phi_3$, are the needed would-be Goldstone bosons and the fourth one ϕ_4 is introduced just to complete the complex doublet. After the symmetry breaking, this extra degree translates into the apparition in the spectrum of an extra massive scalar particle , the Higgs boson particle H.

12. Lagrangian of The Electroweak Theory II

In order to get the particle spectra and the particle masses we first rewrite the full SM Lagrangian which is $SU(2)_L \times U(1)_Y$ gauge invariant:

$$\mathcal{L}_{SM} = \mathcal{L}_f + \mathcal{L}_G + \mathcal{L}_{SBS} + \mathcal{L}_{YW}$$

where, $\mathcal{L}_f$, and $\mathcal{L}_G$ have been given previously and $\mathcal{L}_{SBS}$ and $\mathcal{L}_{YW}$ are the SBS and the Yukawa Lagrangians respectively,

$$\boxed{\mathcal{L}_{SBS} = (D_\mu\Phi)^\dagger(D^\mu\Phi) + \mu^2\Phi^\dagger\Phi - \lambda(\Phi^\dagger\Phi)^2}$$

$$\boxed{\mathcal{L}_{YW} = \lambda_e \bar{l}_L \Phi e_R + \lambda_u \bar{q}_L \widetilde{\Phi} u_R + \lambda_d \bar{q}_L \Phi d_R + h.c. + 2^{nd} \text{ and } 3^{rd} \text{ families}}$$

Here,

$$\begin{aligned} l_L &= \begin{pmatrix} \nu_L \\ e_L \end{pmatrix} ; \; q_L = \begin{pmatrix} u_L \\ d_L \end{pmatrix} \\ \Phi &= \begin{pmatrix} \phi^+ \\ \phi_0 \end{pmatrix} ; \; \widetilde{\Phi} = i\sigma_2\Phi^* = \begin{pmatrix} \phi_0^* \\ -\phi^- \end{pmatrix} \end{aligned}$$

Notice that $\mathcal{L}_{SBS}$ is needed to provide the M_W and M_Z masses and $\mathcal{L}_{YW}$ is needed to provide the m_f masses.

The following steps summarize the procedure to get the spectrum from $\mathcal{L}_{SM}$:

1.- A non-symmetric vacuum must be fixed. Let us choose, for instance,

$$< 0|\Phi|0 >= \begin{pmatrix} 0 \\ \frac{v}{\sqrt{2}} \end{pmatrix}$$

2.- The physical spectrum is built by performing 'small oscillations' around this vacuum. These are parametrized by,

$$\Phi(x) = \exp\left(i\frac{\vec{\xi}(x)\vec{\sigma}}{v}\right)\begin{pmatrix} 0 \\ \frac{v+H(x)}{\sqrt{2}} \end{pmatrix}$$

where $\vec{\xi}(x)$ and $H(x)$ are 'small' fields.

3.- In order to eliminate the unphysical fields $\vec{\xi}(x)$ we make the following gauge transformations:

$$\begin{aligned}
\Phi' &= U(\xi)\Phi = \begin{pmatrix} 0 \\ \frac{v+H}{\sqrt{2}} \end{pmatrix} ;\ U(\xi) = \exp\left(-i\frac{\vec{\xi}\vec{\sigma}}{v}\right) \\
l'_L &= U(\xi)l_L ;\ e'_R = e_R ;\ q'_L = U(\xi)q_L ;\ u'_R = u_R ;\ d'_R = d_R \\
\left(\frac{\vec{\sigma}\cdot\vec{W}'_\mu}{2}\right) &= U(\xi)\left(\frac{\vec{\sigma}\cdot\vec{W}_\mu}{2}\right)U^{-1}(\xi) - \frac{i}{g}(\partial_\mu U(\xi))U^{-1}(\xi) ;\ B'_\mu = B_\mu
\end{aligned}$$

4.- Finally, the weak eigenstates are rotated to the mass eigenstates which define the physical gauge boson fields:

$$\begin{aligned}
W_\mu^\pm &= \frac{W'^1_\mu \mp iW'^2_\mu}{\sqrt{2}}, \\
Z_\mu &= c_w W'^3_\mu - s_w B'_\mu, \\
A_\mu &= s_w W'^3_\mu + c_w B'_\mu,
\end{aligned}$$

It is now straightforward to read the masses from the following terms of $\mathcal{L}_{SM}$:

$$\begin{aligned}
(D_\mu\Phi')^\dagger(D^\mu\Phi') &= \left(\frac{g^2v^2}{4}\right)W_\mu^+W^{\mu-} + \frac{1}{2}\left(\frac{(g^2+g'^2)v^2}{4}\right)Z_\mu Z^\mu + \ldots \\
V(\Phi') &= \frac{1}{2}(2\mu^2)H^2 + \ldots \\
\mathcal{L}_{YW} &= \left(\lambda_e\frac{v}{\sqrt{2}}\right)\bar{e}'_L e'_R + \left(\lambda_u\frac{v}{\sqrt{2}}\right)\bar{u}'_L u'_R + \left(\lambda_d\frac{v}{\sqrt{2}}\right)\bar{d}'_L d'_R + \ldots
\end{aligned}$$

and get finally the tree level predictions:

$$\begin{aligned}
M_W &= \frac{gv}{2} ;\ M_Z = \frac{\sqrt{g^2+g'^2}v}{2} \\
M_H &= \sqrt{2}\mu \\
m_e &= \lambda_e\frac{v}{\sqrt{2}} ;\ m_u = \lambda_u\frac{v}{\sqrt{2}} ;\ m_d = \lambda_d\frac{v}{\sqrt{2}} ;\ldots
\end{aligned}$$

where,

$$v = \sqrt{\frac{\mu^2}{\lambda}}$$

Finally one can rewrite $\mathcal{L}_{SBS}$ and $\mathcal{L}_{YW}$, after the application of the Higgs Mechanism, in terms of the physical scalar fields, and get not just the mass terms but also the kinetic and interaction terms for the Higgs sector,

$$\mathcal{L}_{SBS} + \mathcal{L}_{YW} \to \mathcal{L}_H^{\text{free}} + \mathcal{L}_H^{\text{int}} + \ldots$$

where,

$$\boxed{\mathcal{L}_H^{\rm free} = \tfrac{1}{2}\partial_\mu H \partial^\mu H - \tfrac{1}{2}M_H^2 H^2}$$

and,

$$\boxed{\begin{array}{c}\mathcal{L}_H^{\rm int} = -\frac{M_H^2}{2v}H^3 - \frac{M_H^2}{8v^2}H^4 \\ -\frac{m_f}{v}\bar{f}Hf \\ +M_W^2 W_\mu^+ W^{\mu -}\left(1 + \frac{2}{v}H + \frac{1}{v^2}H^2\right) \\ +\frac{1}{2}M_Z^2 Z_\mu Z^\mu \left(1 + \frac{2}{v}H + \frac{1}{v^2}H^2\right)\end{array}}$$

Some comments are in order.

- All masses are given in terms of a unique mass parameter v and the couplings g, g', λ, λ_e, etc..
- The interactions of H with fermions and with gauge bosons are proportional to the gauge couplings and to the corresponding particle masses:

$$f\bar{f}H \;:\; -i\frac{g}{2}\frac{m_f}{M_W}\;;\quad W_\mu^+ W_\nu^- H \;:\; igM_W g_{\mu\nu}\;;\quad Z_\mu Z_\nu H \;:\; \frac{ig}{c_w}M_Z g_{\mu\nu}$$

- The v.e.v. v is determined experimentally form μ-decay. By identifying the predictions of the partial width $\Gamma(\mu \to \nu_\mu \bar{\nu}_e e)$ in the SM to low energies ($q^2 << M_W^2$) and in the V-A Theory one gets,

$$\frac{G_F}{\sqrt{2}} = \frac{g^2}{8M_W^2} = \frac{1}{2v^2}$$

  And from here,

$$v = (\sqrt{2}G_F)^{-\frac{1}{2}} = 246\ GeV$$

- The values of M_W and M_Z were anticipated successfully quite before they were measured in experiment. The input parameters were θ_w, the fine structure constant α and G_F. Before LEP these were the best measured electroweak parameters.
- In contrast to the gauge boson sector, the Higgs boson mass M_H and the Higgs self-coupling λ are completely undetermined in the SM. They are related at tree level by, $\lambda = \frac{M_H^2}{2v^2}$.
- The hierarchy in the fermion masses is also completely undetermined in the SM.

13. Theoretical Bounds on M_H

In this section we summarize the present bounds on M_H from the requirement of consistency of the theory.

13.1. UPPER BOUND ON M_H FROM UNITARITY

Unitarity of the scattering matrix together with the elastic approximation for the total cross-section and the Optical Theorem imply certain elastic unitarity conditions for the partial wave amplitudes. These, in turn, when applied in the SM to scattering processes involving the Higgs particle, imply an upper limit on the Higgs mass. Let us see this in more detail for the simplest case of scattering of massless scalar particles: $1+2 \to 1+2$.

The decomposition of the amplitude in terms of partial waves is given by:

$$T(s, \cos\theta) = 16\pi \sum_{J=0}^{\infty} (2J+1) a_J(s) P_J(\cos\theta)$$

where P_J are the Legendre polynomials.

The corresponding differential cross-section is given by:

$$\frac{d\sigma}{d\Omega} = \frac{1}{64\pi^2 s} |T|^2$$

Thus, the elastic cross-section is written in terms of partial waves as:

$$\sigma_{\text{el}} = \frac{16\pi}{s} \sum_{J=0}^{\infty} (2J+1) |a_J(s)|^2$$

On the other hand, the Optical Theorem relates the total cross-section with the forward elastic scattering amplitude:

$$\sigma_{\text{tot}}(1+2 \to \text{anything}) = \frac{1}{s} Im\, T(s, \cos\theta = 1)$$

In the elastic approximation for σ_{tot} one gets $\sigma_{\text{tot}} \approx \sigma_{\text{el}}$. From this one finally finds,

$$Im\, a_J(s) = |a_J(s)|^2 \; ; \forall J$$

This is called the elastic unitariry condition for partial wave amplitudes. It is easy to get from this the following inequalities:

$$|a_J|^2 \leq 1 \; ; \; 0 \leq Im\, a_J \leq 1 \; ; \; |Re\, a_J| \leq \frac{1}{2} \; ; \; \forall J$$

These are necessary but not sufficient conditions for elastic unitarity. It implies that if any of them are not fulfiled then the elastic unitarity condition also fails, in which case the unitarity of the theory is said to be violated.

Let us now study the particular case of $W_L^+W_L^-$ scattering in the SM and find its unitarity conditions. The $J = 0$ partial wave can be computed from:

$$a_0(W_L^+W_L^- \to W_L^+W_L^-) = \frac{1}{32\pi}\int_{-1}^{1} T(s, \cos\theta)d(\cos\theta)$$

where the amplitude $T(s, \cos\theta)$ is given by,

$$\begin{aligned} T(W_L^+W_L^- \to W_L^+W_L^-) = & -\frac{1}{v^2}\{-s - t + \frac{s^2}{s - M_H^2} + \frac{t^2}{t - M_H^2} + 2M_Z^2 + \\ & \frac{2M_Z^2 s}{t - M_Z^2} + \frac{2t}{s}(M_Z^2 - 4M_W^2) - \frac{8s_W^2 M_W^2 M_Z^2 s}{t(t - M_Z^2)}\} \end{aligned}$$

By studying the large energy limit of a_0 one finds,

$$|a_0| \overset{s>>M_H^2, M_V^2}{\longrightarrow} \frac{M_H^2}{8\pi v^2}$$

Finally, by requiring the unitarity condition $|Re\ a_0| \leq \frac{1}{2}$ one gets the following upper bound on the Higgs mass:

$$M_H < 860\ GeV$$

One can repeat the same reasoning for different channels and find similar or even tighter bounds than this one [31, 32].

At this point, it should be mentioned that these upper bounds based on perturbative unitarity do not mean that the Higgs particle cannot be heavier than these values. The conclusion should be, instead, that for those large M_H values a perturbative approach is not valid and non-perturbative techniques are required. In that case, the Higgs self-interactions governed by the coupling λ become strong and new physics phenomena may appear in the $O(1\ TeV)$ range. In particular, the scattering of longitudinal gauge bosons may also become strong in that range [31] and behave similarly to what happens in $\pi\pi$ scattering in the $O(1\ GeV)$ range. Namely, there could appear new resonances, as it occurs typically in a theory with strong interactions. This new interesting phenomena could be studied at the next hadron collider, LHC [31, 33].

13.2. UPPER BOUND ON M_H FROM TRIVIALITY

Triviality in $\lambda\Phi^4$ theories [34] (as, for instance, the scalar sector of the SM) means that the particular value of the renormalized coupling of $\lambda_R = 0$ is the unique fixed point of the theory. A theory with $\lambda_R = 0$ contains non-interacting particles and therefore it is trivial. This behaviour can already be seen in the renormalized coupling at one-loop level:

$$\lambda_R(Q) = \frac{\lambda_0}{1 - \frac{3}{2\pi^2}\lambda_0 \log(\frac{Q}{\Lambda})} \; ; \; \lambda_0 \equiv \lambda_R(Q = \Lambda)$$

As we attempt to remove the cut-off Λ by taking the limit $\Lambda \to \infty$ while λ_0 is kept fixed to an arbitrary but finite value, we find out that $\lambda_R(Q) \to 0$ at any finite energy value Q. This, on the other hand, can be seen as a consequence of the existence of the well known Landau pole of $\lambda\Phi^4$ theories.

The trivilaty of the SBS of the SM is cumbersome since we need a self-interacting scalar system to generate M_W and M_Z by the Higgs Mechanism. The way out from this apparent problem is to assume that the Higgs potential $V(\Phi)$ is valid just below certain 'physical' cut-off Λ_{phys}. Then, $V(\Phi)$ describes an effective low energy theory which emerges from some (so far unknown) fundamental physics with Λ_{phys} being its characteristics energy scale. We are going to see next that this assumption implies an upper bound on M_H [35].

Let us assume some concrete renormalization of the SM parameters. The conclusion does not depend on this particular choice. Let us define, for instance, the renormalized Higgs mass parameter as:

$$M_H^2 = 2\lambda_R(v)v^2$$

where,

$$\lambda_R(v) = \frac{\lambda_0}{1 - \frac{3}{2\pi^2}\lambda_0 \log(\frac{v}{\Lambda_{\text{phys}}})}$$

Now, if we want $V(\Phi)$ to be a sensible effective theory, we must keep all the renormalized masses below the cut-off and, in particular, $M_H < \Lambda_{\text{phys}}$. However, one can see that for arbitrary values of Λ_{phys} it is not always possible. By increasing the value of Λ_{phys}, M_H decreases and the other way around, by lowering Λ_{phys}, M_H grows. There is a crossing point where $M_H \approx \Lambda_{\text{phys}}$ which happens to be around an energy scale of approximately $1\,TeV$. Since we want to keep the Higgs mass below the physical cut-off, it implies finally the announced upper bound,

$$M_H^{1-\text{loop}} < 1\,TeV$$

Of course, this should be taken just as a perturbative estimate of the true triviality bound. A more realistic limit must come from a non-perturbative treatment. In particular, the analyses performed on the lattice [36] confirm this behaviour and place even tighter limits. The following bound is found,

$$M_H^{\text{Lattice}} < 640\ GeV$$

Finally, a different but related perturbative upper limit on M_H can be found by analysing the renormalization group equations in the SM to one-loop. Here one includes, the scalar sector, the gauge boson sector and restricts the fermionic sector to the third generation. By requiring the theory to be perturbative (i.e. all the couplings be sufficiently small) at all the energy scales below some fixed high energy, one finds a maximum allowed M_H value [37]. For instance, by fixing this energy scale to $10^{16}\ GeV$ and for $m_t = 170\ GeV$ one gets:

$$M_H^{\text{RGE}} < 170\ GeV$$

Of course to believe in perturbativity up to very high energies could be just a theoretical prejudice. The existence of a non-perturbative regime for the scalar sector of the SM is still a possibility and one should be open to new proposals in this concern.

13.3. LOWER BOUND ON M_H FROM VACUUM STABILITY

Once the asymmetric vacuum of the $SU(2)_L \times U(1)_Y$ theory has been fixed, one must require this vacuum to be stable under quantum corrections. In principle, quantum corrections could destabilize the asymmetric vacuum and change it to the symmetric one where the SSB does not take place. This phenomenom can be better explained in terms of the effective potential with quantum corrections included. Let us take, for instance, the effective potential of the Electroweak Theory to one loop in the small λ limit:

$$V_{\text{eff}}^{1-\text{loop}}(\Phi) \simeq -\mu^2\Phi^\dagger\Phi + \lambda_R(Q_0)(\Phi^\dagger\Phi)^2 + \beta_\lambda(\Phi^+\Phi)^2\log\left(\frac{\Phi^\dagger\Phi}{Q_0^2}\right)$$

where, $\beta_\lambda \equiv \frac{d\lambda}{dt} \simeq \frac{1}{16\pi^2}\left[-3\lambda_t^4 + \frac{3}{16}(2g^4 + (g^2 + g'^2)^2)\right]$.

The condition of extremum is:

$$\frac{\delta V_{\text{eff}}^{1-\text{loop}}}{\delta\Phi} = 0$$

which leads to two possible solutions: a) The trivial vacuum with $\Phi = 0$; and b) The non-trivial vacuum with $\Phi = \Phi_{\text{vac}} \neq 0$. If we want the true vacuum to be the non-trivial one we must have:

$$V_{\text{eff}}^{1-\text{loop}}(\Phi_{\text{vac}}) < V_{\text{eff}}^{1-\text{loop}}(0)$$

However, the value of the potential at the minimum depends on the size of its second derivative:

$$M_H^2 \equiv \frac{1}{2}\{\frac{\delta^2 V}{\delta\Phi^2}\}_{\Phi=\Phi_{\rm vac}}$$

and, it turns out that for too low values of M_H^2 the condition above turns over. That is, $V(0) < V(\Phi_{\rm vac})$ and the true vacuum changes to the trivial one. The condition for vacuum stability then implies a lower bound on M_H [38]. More precisely,

$$M_H^2 \quad > \quad \frac{3}{16\pi^2 v^2}(2M_W^4 + M_Z^4 - 4m_t^4)$$

Surprisingly, for $m_t > 78\ GeV$ this bound dissapears and, moreover, $V_{\rm eff}^{1-\rm loop}$ becomes unbounded from below!. Apparently it seems a disaster since the top mass is known at present and is certainly larger than this value. The solution to this problem relies in the fact that for such input values, the 1-loop approach becomes unrealistic and a 2-loop analysis of the effective potential is needed. Recent studies indicate that by requiring vacuum stability at 2-loop level and up to very large energies of the order of $10^{16}\ GeV$, the following lower bound is found [39]:

$$M_H^{\rm v.stab.} \quad > \quad 132\ GeV$$

This is for $m_t = 170\ GeV$ and $\alpha_s = 0.117$ and there is an uncertainty in this bound of 5 to 10 GeV from the uncertainty in the m_t and α_s values.

14. SM predictions

In the following we present the tree level predictions from the SM and compare them with the present experimental values. The experimental values presented here (unless explicitely stated otherwise) have been borrowed from the talk by D.Karlen given at the ICHEP'98 Vancouver Conference [4]. For a more detailed discussion on the experimental tests of the SM see the lectures of L. Nodulman.

14.1. GAUGE BOSON MASSES

Before the discovery of the $W^\pm$, Z gauge bosons, the best known SM parameters were α, G_F and $sin^2\theta_w$. The present values are highly precise:

$$\alpha_{exp}^{-1} = 137.0359895 \pm 0.0000061$$

from atomic, molecular and nuclear data, and

$$G_F^{exp} = (1.16639 \pm 0.000022) \times 10^{-5}\ GeV^{-2}$$

from $\mu-$decay.
$sin^2\theta_w$ was measured firstly in the seventies in νN scattering experiments. The ratio of the cross-sections for neutral currents and charged currents as predicted in the SM is a known function of $sin\theta_w$,

$$\frac{\sigma_{NC}(\nu q \to \nu q)}{\sigma_{CC}(\nu q \to l q')} \sim f(sin^2\theta_w)$$

It is the measurement of this ratio what provides a measurement of $sin^2\theta_w$. The present experimental value is,

$$sin^2\theta_w|_{\rm exp} = 0.2255 \pm 0.0021$$

The SM does not predict a numerical value for M_W and M_Z but provides some relations among the relevant parameters. These relations are different to tree level than to, for instance, one-loop level. In particular, the following relations hold to tree level in the SM,

$$M_W = \frac{gv}{2}\ ;\ \frac{G_F}{\sqrt{2}} = \frac{g^2}{8M_W^2} = \frac{1}{2v^2}\ ;\ g = \frac{e}{s_w}\ ;\ \rho_{SM}^{\rm tree} \equiv \frac{M_W^2}{M_Z^2 c_w^2} = 1$$

From these expressions it is inmediate to derive the two following tree level relations,

$$M_W = \left(\frac{\pi\alpha}{G_F\sqrt{2}}\right)^{\frac{1}{2}} \frac{1}{sin\theta_w}$$

$$M_Z = \left(\frac{\pi\alpha}{G_F\sqrt{2}}\right)^{\frac{1}{2}} \frac{1}{sin\theta_w cos\theta_w}$$

Finally, by inserting the experimental values of α, G_F and θ_w into these expressions one gets the tree level values for the gauge boson masses,

$$M_W^{\rm tree} = 78\ GeV\ ;\ M_Z^{\rm tree} = 89\ GeV$$

The discovery of the $W^\pm$ and Z gauge bosons in 1983 at the CERN SpS collider [30] lead to the definitive confirmation of the validity of the SM. Notice that the measured masses were surprisingly close to the SM tree level predictions,

$$M_W^{\rm SpS} = (81 \pm 2)\ GeV\ ;\ M_Z^{\rm SpS} = (93 \pm 3)\ GeV$$

The present experimental values are very precise,

$$\begin{aligned} M_W^{\rm exp} &= (80.41 \pm 0.09)\ GeV\ (p\bar{p}) \\ &\quad\ (80.37 \pm 0.09)\ GeV\ (\text{LEP}) \end{aligned}$$

$$M_Z^{exp} = (91.1867 \pm 0.0021)\ GeV\ (\mathrm{LEP})$$

14.2. GAUGE BOSON DECAYS

The $W^\pm$ and Z gauge bosons can decay either in quarks or in leptons within the SM. The dominant decays are clearly into quarks due to the extra color factor, N_C, which is not present in the leptonic decays. The tree level predictions for the partial widths in the approximation of neglecting the fermion masses are the following,

$$\Gamma(W^+ \to e^+\nu_e) = \frac{g^2}{48\pi}M_W = \frac{G_F M_W^3}{6\sqrt{2}\pi} = 0.232\ GeV$$

$$\Gamma(W^+ \to \mu^+\nu_\mu) = \Gamma(W^+ \to \tau^+\nu_\tau) = \Gamma(W^+ \to e^+\nu_e)$$

$$\Gamma(W^+ \to u_i\bar{d}_j) = N_C|U_{ij}|^2\frac{G_F M_W^3}{6\sqrt{2}\pi} = 0.232\ N_C|U_{ij}|^2\ GeV$$

$$\Gamma(Z \to f\bar{f}) = \kappa_f\frac{G_F M_Z^3}{6\sqrt{2}\pi}(g_{Vf}^2 + g_{Af}^2) = 0.3318\kappa_f(g_{Vf}^2 + g_{Af}^2)\ GeV$$

where, U_{ij} are the CKM matrix elements and,

$$\kappa_f = 1\ ,\ f = l, \nu\ ;\ \kappa_f = N_C\ ,\ f = q\ ;\ g_{Vf} = T_3^f - 2Q^f s_w^2\ ;\ g_{Af} = T_3^f$$

In Table 3 it is shown the tree level predictions in the SM, for the total Z and $W^\pm$ widths, Γ_Z and Γ_W, and the ratios:

$$R_e = \frac{\Gamma(Z \to \mathrm{hadrons})}{\Gamma_e}\ ;\ R_b = \frac{\Gamma(Z \to b\bar{b})}{\Gamma(Z \to \mathrm{hadrons})}\ ;\ R_c = \frac{\Gamma(Z \to c\bar{c})}{\Gamma(Z \to \mathrm{hadrons})}$$

The numerical predictions shown here use as input the present experimental values for M_Z, G_F, α and $sin^2\theta_w$, and the M_W value that one gets with those experimental values put into the previous SM tree level relation. That is, $M_W = 80.94\ GeV$.

By comparing the tree level SM results with the present experimental values it is clear that they provide reasonable good predictions. However, due to the high level of precision of the present measurements one can also conclude from this table that some of the tree level predictions are already not compatible with data. In fact in some observables they are out by several standard deviations. This is a clear indication that the SM radiative corrections must be included in the theoretical predictions [40]. The present experimental analysis of the SM parameters, in fact, do include these radiative corrections. The summary of measurements included in the

TABLE 3. Confronting tree level SM predictions with data

Parameter	Tree Level SM	Exp. value
$\Gamma_Z(GeV)$	2.474	2.4948 ± 0.0025
$\Gamma_W(GeV)$	2.09	2.06 ± 0.06
R_e	20.29	20.765 ± 0.026
R_b	0.219	0.21656 ± 0.00074
R_c	0.172	0.1733 ± 0.0044

combined analysis of SM parameters from LEP and SLC can be found in Nodulman lectures.

14.3. TOP QUARK PHYSICS

As has been shown before, the top mass m_t is not predicted in the SM. Instead, the SM provides, via The Higgs Mechanism, the tree level relation,

$$m_t = \lambda_t \frac{v}{\sqrt{2}} = \lambda_t \left(\frac{1}{2\sqrt{2}G_F}\right)^{\frac{1}{2}}$$

which gives m_t in terms of the top Yukawa coupling λ_t. But, λ_t as the other fermion Yukawa couplings are unknown parameters in the SM.

The existence of the top quark, however, was never questioned seriously, since there were several strong arguments supporting the need of this third generation fermion. On one hand, the top quark was needed to avoid unwanted flavour changing neutral currents (FCNC). On the other hand the top quark was needed to avoid unwanted $SU(2)_L \times U(1)_Y$ anomalies. Thus, The top quark was expected for a long time. Its discovery finally occured in 1994 at the $p\overline{p}$ TeVatron collider at Fermilab [27]. The present experimental value of the top mass as provided by the two Tevatron experiments is,

$$m_t^{\rm exp} = (173.8 \pm 5.0)\ GeV\ (\mathrm{CDF} + \mathrm{D0})$$

It is remarkable this much larger mass value than the rest of the fermion masses. In fact, for this top mass value one can extract the corresponding Yukawa coupling and get $\lambda_t \sim 1$ which is a rather large value, although it can still be considered as a perturbative coupling. To the question, why the top quark is so heavy?, there is no answer within the SM.

Concerning the top quark decays, the dominant one is by far the decay into a W^+ gauge boson and a bottom quark. The SM tree level prediction

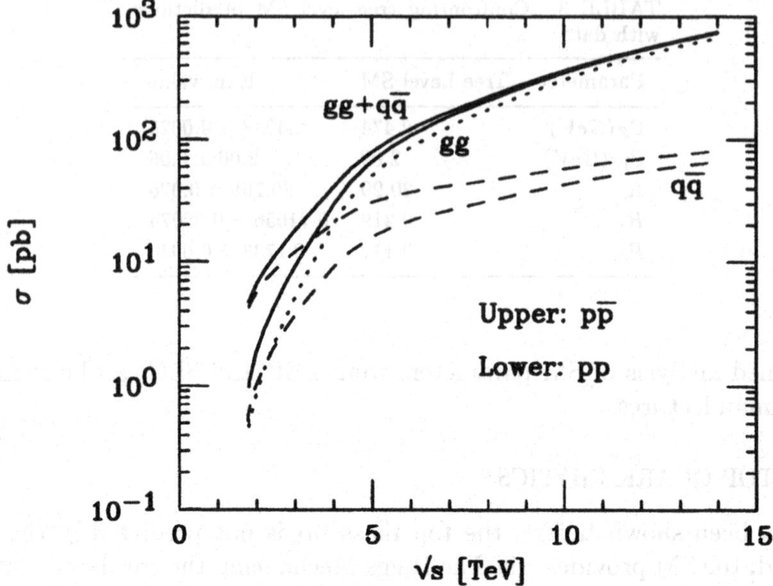

Figure 3. $t\bar{t}$ production at hadron colliders

for the partial width, in the approximation of neglecting m_b, is

$$\Gamma(t \to W^+ b) = \frac{G_F m_t^3}{8\pi\sqrt{2}} |U_{tb}|^2 \left(1 - \frac{M_W^2}{m_t^2}\right)\left(1 + 2\frac{M_W^2}{m_t^2}\right) \sim 2\ GeV$$

There is not experimental meassurement of the total or partial width yet.

Regarding the top production, it is obvious that, given the large mass value, it can only be produced at present in the TeVatron collider. The future hadron collider LHC at CERN will provide additional interesting information on top quark physics. In Figure 3 it is shown the cross-section for $t\bar{t}$ production at TeVatron and LHC from the various possible channels. It is clear from the figure that $q\bar{q} \to t\bar{t}$ is the dominant process at TeVatron, whereas $gg \to t\bar{t}$ will dominate at LHC [41].

14.4. HIGGS PHYSICS

The Higgs mass M_H is not predicted in the SM either. The Higgs Mechanism provides M_H as a function of the Higgs self-coupling λ and $v = 246\ GeV$,

$$M_H = \sqrt{2}\mu = \sqrt{2v^2\lambda}$$

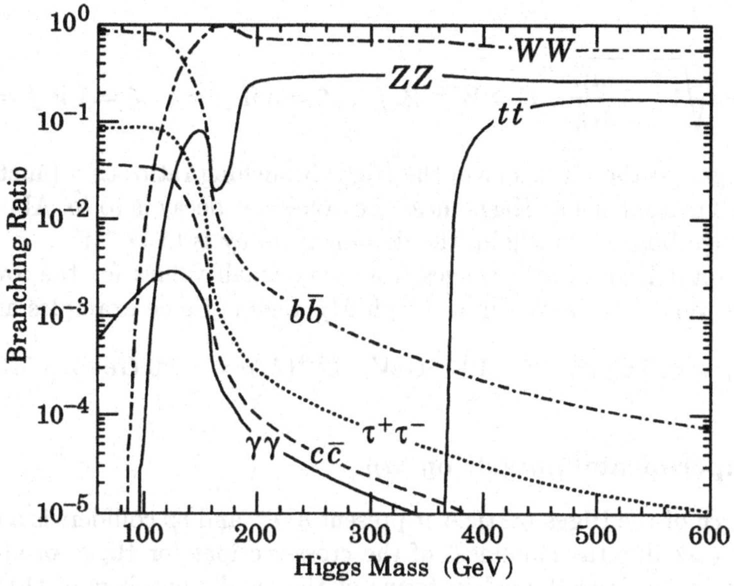

Figure 4. **Branching ratios for Higgs decays as a function of M_H**

but, λ is also unknown. Therefore, M_H can take any value in the SM. As we have seen, the unique restrictions on M_H come from the consistency of the theory, that is from unitarity, triviality, and vacuum stability arguments.

Contrary to the top quark case, there are no strong theoretical arguments (anomalies, etc) supporting the need of this scalar elementary particle, H.

Regarding the Higgs decays, the tree level SM predictions for the partial widths are the following,

$$\Gamma(H \to W^+W^-) = \frac{G_F M_H^3}{8\sqrt{2}\pi}\beta_W\left(\beta_W^2 + \frac{12M_W^4}{M_H^4}\right)$$

$$\Gamma(H \to ZZ) = \frac{G_F M_H^3}{16\sqrt{2}\pi}\beta_Z\left(\beta_Z^2 + \frac{12M_Z^4}{M_H^4}\right)$$

$$\Gamma(H \to f\bar{f}) = \frac{G_F m_f^2 M_H}{4\sqrt{2}\pi}\beta_f^3\xi$$

where,

$$\beta_P = \sqrt{1 - 4\frac{m_P^2}{M_H^2}}\ ,\ P = W^{\pm}, Z, f\ ;\ \xi = 3 \text{ if } f = q\ ,\ \xi = 1 \text{ if } f = l$$

In Figure 4 there are shown the Higgs branching ratios as a function of the Higgs mass. For low Higgs mass the dominant decay is to $b\bar{b}$. Above the weak gauge bosons threshold, the dominant decay is to W^+W^-.

The total Higgs width ranges from very small values for the low M_H region to very large values in the high M_H region. Some examples are,

$$\Gamma_H^{\text{tot}}(M_H = 100\ GeV) \sim 5 \times 10^{-3}\ GeV\ ;\ \Gamma_H^{\text{tot}}(M_H = 1000\ GeV) \sim 570\ GeV$$

15. Experimental bounds on M_H

The search of the Higgs particle at present e^+e^- and $\bar{p}p$ colliders is a rather difficult task due the smallness of the cross-sections for Higgs production which, in turn, is explained in terms of the small couplings of the Higgs particle to light fermions: $Hf\bar{f} \leftrightarrow -i\frac{g}{2}\frac{m_f}{M_W}$. On the other hand, at present available energies, the dominant decay channel is $H \to b\bar{b}$ which is not easy to study due to the complexity of the final state and the presence of large backgrounds [42].

15.1. HIGGS SEARCH AT E^+E^- COLLIDERS (LEP, SLC)

The Higgs search during the first period of LEP (LEPI) and SLC was done mainly by analysing the process:

$$e^+e^- \to Z \to Z^*H$$

with the virtual Z^* decaying as $Z^* \to l^+l^-$, $\nu\bar{\nu}$, $q\bar{q}$ and the Higgs particle decaying as $H \to b\bar{b}$.

At LEPI with a center-of-mass-energy adjusted to the Z mass, $\sqrt{s} \sim M_Z$, a very high statistic was reached and a systematic search of the Higgs particle for all kinematically allowed M_H values was possible. The absence of any experimental signal from the Higgs particle implied a lower bound on M_H. The last reported bound from LEPI was:

$$M_H > 66\ GeV\ (95\% C.L.)\ (LEPI)$$

In the second phase of LEP, LEPII, a center-of-mass-energy of up to $\sqrt{s} \sim 189 GeV$ is at present reached. The relevant process for Higgs searches is:

$$e^+e^- \to Z^* \to ZH$$

where now the intermediate Z boson is virtual and the final Z is on its mass shell. The analyses of the various relevant Z and H decays at LEPII give at present the following averaged lower Higgs mass bound:

$$M_H > 89.8\ GeV\ (95\%C.L)\ (LEPII)$$

In addition to the direct bounds on M_H from LEP and SLC data, a great effort is being done also in the search of indirect Higgs signals from its contribution to electroweak quantum corrections. In fact, there are already interesting upper experimental bounds on M_H from the meassurement of observables as $\Delta\rho$, Δr and other related ones whose prediction in the SM is well known. It is interesting to mention that neither the Higgs particle nor the top quark decouple from these low energy observables. It means that the quantum effects of a virtual H or t do not vanish in the asymptotic limit of infinitely large M_H or m_t respectively. For instance, the leading corrections to $\Delta\rho$, and Δr in the large m_t and large M_H limits are respectively:

$$\begin{aligned}
(\Delta\rho)_t &= \frac{\sqrt{2}G_F3}{16\pi^2}m_t^2 + ... \\
(\Delta\rho)_H &= -\frac{\sqrt{2}G_F M_W^2}{16\pi^2}3\frac{s_w^2}{c_w^2}\left(\log\frac{M_H^2}{M_W^2} - \frac{5}{6}\right) + ... \\
(\Delta r)_t &= -\frac{c_w^2}{s_w^2}\frac{\sqrt{2}G_F3}{16\pi^2}m_t^2 + ... \\
(\Delta r)_H &= \frac{\sqrt{2}G_F M_W^2}{16\pi^2}\frac{11}{3}\left(\log\frac{M_H^2}{M_W^2} - \frac{5}{6}\right) + ...
\end{aligned}$$

Whereas the top corrections grow with the mass as m_t^2, the Higgs corrections are milder growing as $\log M_H^2$. It means that the top non-decoupling effects at LEP are important. In fact they have been crucial in the search of the top quark and have provided one of the first indirect indications of the 'preference of data' for large m_t values. This helped in the search and final discovery of the top quark at TeVatron.

The fact that the Higgs non-decoupling effects are mild was announced a long time ago by T.Veltman in the so-called *Screening Theorem* [32]. This theorem states that, *at one-loop, the dominant quantum corrections from a heavy Higgs particle to electroweak observables grow, at most, as* $\log M_H$. *The Higgs corrections are of the generic form:*

$$g^2(\log\frac{M_H^2}{M_W^2} + g^2\frac{M_H^2}{M_W^2} + ...)$$

and the potentially large effects proportional to M_H^2 *are 'screened' by additional small* g^2 *factors.*

The present global analysis of SM parameters at LEP and SLC with the radiative corrections included, gives the following upper Higgs mass bound:

$$M_H < 280\ GeV\ (95\%\ C.L.)\ \ (LEP + SLC)$$

15.2. HIGGS SEARCH AT HADRONIC COLLIDERS

The relevant subprocesses for Higgs production at hadronic pp and $p\bar{p}$ colliders are: gluon-gluon fusion ($gg \to H$), WW and ZZ fusion ($qq \to qqH$), $t\bar{t}$ fusion ($gg \to t\bar{t}H$) and Z (W) bremsstrahlung ($q\bar{q} \to Z(W)H$).

At present available energies the dominant subprocess is gg-fusion. This is the case of TeVatron with a center-of-mass-energy of $\sqrt{s} = 1.8\ TeV$ and an integrated luminosity of $L = 100\ pb^{-1}$ per experiment. However, due to the cleanest signature of the Z and W bremsstrahlung subprocesses and the fact that this has less background, this channel is the most studied one at TeVatron. The Higgs searches at TeVatron have not provided yet any lower Higgs mass bound. The sensitivity of the present search [43] is limited by statistics to a cross section approximately two orders of magnitude larger than the predicted cross section for SM Higgs production. For the next TeVatron run there will be a twenty-fold increase in the total integrated luminosity per experiment. However, it is still insufficient to reach say a 120 GeV Higgs mass, unless the total detection efficiency be improved by one order of magnitude. The viability of this improvement is at present under study.

The Higgs search at the LHC collider being built at CERN, is very promising. In particular, it will cover the whole Higgs mass range and hopefully will be able to distinguish between the various possibilities for the SBS of the Electroweak Theory. For a review on Higgs searches at LHC see F. Pauss lectures.

16. Acknowledgements

I thank Tom Ferbel for the invitation to this interesting and enjoyable school and for the perfect organization. I am grateful to all the students and lecturers for making many interesting questions and comments during my lectures and for keeping a very pleasant atmosphere for fruitful discussions. I have also profited from discussions with J.Trocóniz. I wish to thank C.Glasman, P.Seoane and G.Yepes for their help with the figures. This work was supported in part by the Spanish CICYT under project AEN97-1678.

References

1. S.L.Glashow, Nucl.Phys.22(1961),579;
   S.Weinberg, Phys.Rev.Lett.19(1967),1264;
   A.Salam in Elementary Particle Physics (Nobel Symp. N.8), Ed. N.Svartholm, Almquist and Wiksells, Stockholm (1968), p.367
2. For a pedagogical introduction to The Santard Model see the text books,
   'Gauge Theory of Elementary Particle Physics', T.P.Cheng and L.F.Li, Oxford Univ. Press, 1991 (reprinted);
   'Quarks and Leptons: An introductory course in Modern Particle Physics, F.Halzen and A.D.Martin , Ed. John Wiley: New York, 1984;
   'Gauge Theories in Particle Physics', I.J.R.Aitchison and A.J.G.Hey, Second Edition, Inst. of Phys. Publishing, Bristol and Philadelphia, 1993 (reprinted). Graduate Student Series in Physics;
   'Dynamics of The Standard Model', J.Donoghue, E.Golowich and B.R.Holstein, Cambridge Monographs on Particle Physics, Nuclear Physics and Cosmology, CUP, 1994 (reprinted).
3. For an introduction to QCD and related phenomenology see,
   'QCD and collider physics', R.K.Ellis, W.J.Stirling and B.R.Webber, cambridge University Press (1996);
   'Foundations of Quantum Chromodynamics', T.Muta, World Scientific (1987), Singapore;
   'The Theory of Quarks and Gluon Interactions', F.J.Yndurain, Springer-Verlag 2nd. Edition (1993), (New Edition in press, 1999);
   'QCD: Renormalization for the Practicioner, R.Pascual and R. Tarrach, Springer-Verlag (1984).
4. For a recent review on SM tests see,
   D.Karlen, Plenary talk 'Experimental Status of the Standard Model' at the International Conference on High Energy Physics, ICHEP98, Vancouver, July 1998. To be published in the Proceedings.
5. For an introduction to the Symmetry Breaking Sector of the Electroweak Theory see, 'Introduction to the Symmetry Breaking Sector', M. Herrero, Proceedings of the XXIII International Meeting on Fundamental Physics, Comillas, Santander, Spain, May 1995, Eds. T.Rodrigo and A.Ruiz. World Sci.Pub.Co(1996), p.87.
6. A summary on Group Theory in the context of Particle Physics is given in,
   'Gauge Theory of Elementary Particle Physics', T.P.Cheng and L.F.Li, Oxford Univ. Press, 1991 (reprinted).
7. M.Gell-Mann, Phys.Rev.92(1953)833;
   K.Nishijima and T.Nakano Prog.Theor.Phys.10(1953)581.
8. M.Gell-Mann, California Institute of Technology Synchroton Laboratory Report No. CTSL-20 (1961);
   Y.Ne'eman, Nucl.Phys.26(1961)222.
9. M.Gell-Mann, Phys.Lett.8(1964)214;
   G.Zweig, CERN Report No. 8182/TH 401, 1964.
10. E.Fermi, Nuovo Cimento 11(1934)1; Z.Phys.88(1934)161.
11. T.D.Lee and C.N.Yang, Phys. Rev.104(1956)254.
12. C.S.Wu et al., Phys.Rev.105(1957)1413.
13. R.Feynman and M.Gell-Mann, Phys.Rev.109(1958)193.
14. N.Cabibbo, Phys.Rev.Lett.10(1963)531.
15. F.J.Hasert, et al., Phys.Lett.**B46**(1973)138.
16. T.D.Lee, M.Rosenbluth and C.N.Yang, Phys.Rev.75(1949)9905;
    T.D.Lee, CERN Report 61-30(1961).
17. S.L.Glashow, Nucl.Phys.22(1961)579.
18. Y.Nambu, Phys.Rev.Lett.4(1960),380;
    J.Goldstone, Nuovo Cimento 19 (1961),154;

J.Goldstone, A.Salam, S.Weinberg, Phys.Rev.127(1962),965.
19. P.W.Higgs, Phys.Lett.12(1964),132;
F.Englert and R.Brout, Phys.Rev.Lett.(1964),321;
G.S.Guralnik, C.R.Hagen and T.W.B.Kibble, Phys.Rev.Lett.13(1964)585;
P.W.Higgs, Phys.Rev.145(1966),1156;
T.W.B.Kibble, Phys.Rev.155(1967),1554.
20. Y.Nambu and G.Jona-Lasinio, Phys.Rev.122(1961),345; Phys.Rev.124(1961), 246;
J.Schwinger, Phys.Rev.125(1962)397;
P.Anderson, Phys.Rev.130(1963),439.
21. G. 't Hooft, Nucl.Phys.B33(1971),173; B35(1971)167.
22. J.D.Bjorken and S.L.Glashow, Phys.Lett.11(1964)255.
23. S.L.Glashow, J.Iliopoulos and L.Maiani, Phys.Rev.D2(1970)1285.
24. J.J.Aubert et al., Phys.Rev.Lett.33(1974)1404;
J.E.Augustin et al., Phys.Rev.Lett.33(1974)1406.
25. M.L.Perl et al., Phys.Rev.Lett.35(1975)1489; Phys.Lett.B70(1977)487
26. S.W.Herb et al., Phys.Rev.Lett.39(1977)252;
L.M.Lederman, In Proc. 19th Int. Conf. on High Energy Phys., Tokyo 1978, Ed. G.Takeda, Physical Society of Japan, Tokyo.
27. CDF Collaboration: F. Abe et al., Phys. Rev. Lett.73(1994), 225; Phys. Rev.D50 (1994), 2966;
CDF Collaboration: F.Abe et al., Phys.Rev.Lett.74(1995)2626;
D0 Collaboration: S.Abachi et al., Phys.Rev.Lett.74(1995)2632
28. M.Kobayashi and K.Maskawa, Prog.Theor.Phys.49(1973)652.
29. V.L.Ginzburg and L.D.Landau, J.Expl.Theoret.Phys.USSR 20 (1950),1064.
30. UA1 Collaboration, G.Arnison et al., Phys.Lett.B122(1983)103;
UA2 Collaboration, M.Banner et al., Phys.Lett.B122(1983)476
31. B.Lee, C.Quigg and H.Thacker, Phys.Rev.D16(1977), 1519;
M.Chanowitz and M.K.Gaillard, Nucl.Phys.B261(1985),379.
32. M.Veltman, Act.Phys.Pol.B8(1977),475; Nucl.Phys.B123(1977),89.
33. A.Dobado, M.Herrero, J.Terron, Z.Phys.C50(1991),205; Z.Phys.C50(1991)465.
34. K.Wilson, Phys.rev.B4(1971),3184;
K.Wilson and J.Kogut, Phys.Rep.12C(1974),75.
35. R.Dashen and H.Neuberger, Phys.Rev.Lett.50(1983),1897;
A.Hasenfratz and P.Hasenfratz, Phys.Rev.D34(1986),3160.
36. P.Hasenfratz and J.Nager, Z.Phys.C37(1988);
A.Hasenfratz and T.Neuhaus, Nucl.Phys.B297(1988),205;
J.Kuti, L.Lin and Y.Shen, Phys.Rev.Lett.61(1988),678;
M.Luscher and P.Weisz, Phys.Lett.B212(1988),472;
A.Hasenfratz in Quantum Fields on The Computer, Ed. M.Creutz, World Sci. Singapore, 1992, p.125.
37. N.Cabibbo et al., Nucl.Phys.B158(1979),295.
38. M.Lindner, Z.Phys.31(1986), 295;
M.Sher, Phys.Rep.179(1989),273;
M.Lindner, M.Sher and H.W.Zaglauer, Phys.Lett.B228(1989),139.
39. M.Sher, Phys.Lett.B331(1994),448;
G.Altarelli and G.Isidori, Phys.Lett.B357(1994),141;
J.A.Casas, J.R.Espinosa and M.Quiros, Phys.Lett.B342(1995),171.
40. For an introduction to electroweak radiative corrections in the SM, see for instance, 'Electroweak Theory', W.Hollik, hep-ph/9602380.
41. For a review on top physics see, 'Top Quark Physics: Overview', S.Parke, presented at the International Symposium on 'QCD Corrections and New Physics', Hiroshima, Japan, October 1997. To be published in the proceedings, hep-ph/9712512.
42. For a general overview on Higgs searches see 'The Higgs Hunters Guide', J.Gunion et al, Frontier in Physics, Addison-Wesley, Menlo Park, 1990.

43. CDF Collaboration:F.Abe et al., Report No. Fermilab-PUB-98-252-E,1998. To appear in Phys.Rev.Lett.

TOPICS IN PERTURBATIVE QCD

W.J. STIRLING
Departments of Mathematical Sciences and Physics,
University of Durham,
South Road, Durham DH1 3LE, UK

1. Introduction

Quantum Chromodynamics (QCD), the gauge field theory which describes the interactions of coloured quarks and gluons, is one of the components of the SU(3)×SU(2)×U(1) Standard Model. At short distances, equivalently high energies, the effective coupling is small and the theory can be studied using perturbative techniques. Nowadays detailed tests of perturbative QCD are performed at all the high-energy colliders, and in the production and decay of heavy quark systems. Some of the most direct information comes from high-energy processes involving leptons and photons. The colour neutrality of these particles, together with the relative ease with which they can be accelerated and detected in experiments, allows for particularly precise theoretical calculations and experimental measurements. The paradigm process is the investigation of the short-distance structure of hadrons using virtual electroweak gauge boson probes ($\gamma, W^{\pm}, Z^0$) emitted from high-energy beams of charged leptons or neutrinos — deep inelastic scattering. Here we see the asymptotic property directly, as the gauge bosons scatter incoherently off the weakly interacting quarks and gluons, for example $\gamma^* q \to q$ and $\gamma^* g \to q\bar{q}$. From such experiments we learn how the partons (i.e. the quarks and gluons) share the momentum and quantum numbers of the hadron. By studying how the 'structure functions' vary with the momentum transferred by the probe, precision measurements of the short-distance coupling can be made. The information obtained in this way is a vital input to signal and background cross–section calculations at the LHC.

Another fundamental QCD process is the production of hadrons in electron–positron annihilation, $e^+e^- \to q\bar{q}$ at lowest order. The importance of this process is that it allows a detailed study of how quarks 'shower' into

T. Ferbel (ed.), Techniques and Concepts of High Energy Physics X, 61–109.

multiparton states, and how these materialize into jets of hadrons. The quark and gluon spins, the non–Abelian vertices of the theory and the short-distance coupling can be measured from these final states.

In these lectures we will discuss these and other high-energy processes, with the common theme of providing detailed phenomenological tests of perturbative QCD. Much of the basic theoretical framework is covered in the lectures by Maria Jose Herrero, but to make the discussion relatively self–contained, we first of all (Section 2) review some of the fundamental properties of the theory which are particularly relevant for the processes under consideration. Of particular importance in this context is the definition of the 'running' coupling α_s, the fundamental parameter of the theory, and the colour algebra identities which are necessary for performing calculations. In Section 2 we review the application of perturbative QCD to high-energy electron-positron annihilation, focusing in particular on the total cross section and multijet final states. In Section 3 we introduce the parton model of short–distance hadron structure and discuss how the basic 'scaling' property is modified by perturbative QCD corrections, and how parton distributions can be determined from experiment. Some applications to high–energy hadron collider processes are also discussed.

Lack of space precludes an in–depth treatment of most of these issues, but further information can of course be found in the literature. In particular, Reference [1] covers all the topics discussed in these lectures in significantly greater detail.

2. Basics of Perturbative QCD

2.1. THE QCD LAGRANGIAN

The QCD Lagrangian is, up to gauge–fixing terms,

$$\begin{aligned}
\mathcal{L}_{QCD} &= -\frac{1}{4} F^{(a)}_{\mu\nu} F^{(a)\mu\nu} + \sum_q \bar{\psi}^q_i \left(i\gamma^\mu (D_\mu)_{ij} - m_q \delta_{ij} \right) \psi^q_j \\
F^{(a)}_{\mu\nu} &= \partial_\mu A^a_\nu - \partial_\nu A^a_\mu + g_s f_{abc} A^b_\mu A^c_\nu \\
(D_\mu)_{ij} &= \delta_{ij} \partial_\mu - i g_s T^a_{ij} A^a_\mu
\end{aligned} \qquad (1)$$

where g_s is the QCD coupling constant, T^a_{ij} and f_{abc} are the SU(3) colour matrices and structure constants respectively, the $\psi^q_i(x)$ are the 4–component Dirac spinors associated with each quark field of colour i and flavour q, and the $A^a_\mu(x)$ are the eight Yang-Mills gluon fields. From this Lagrangian, the Feynman rules can be derived in the usual way, see the Table on the next page.

A, α p B, β

$$\delta^{AB}\left[-g^{\alpha\beta}+(1-\lambda)\frac{p^{\alpha}p^{\beta}}{p^{2}+i\varepsilon}\right]\frac{i}{p^{2}+i\varepsilon}$$

A p B

$$\delta^{AB}\frac{i}{p^{2}+i\varepsilon}$$

a, i p b, j

$$\delta^{ab}\frac{i}{(\hat{p}-m+i\varepsilon)_{ji}}$$

B, β q p r A, α C, γ

$$-gf^{ABC}\left[g^{\alpha\beta}(p-q)^{\gamma}+g^{\beta\gamma}(q-r)^{\alpha}+g^{\gamma\alpha}(r-p)^{\beta}\right]$$

(all momenta incoming)

A, α B, β C, γ D, δ

$$-ig^{2}f^{XAC}f^{XBD}\left(g_{\alpha\beta}g_{\gamma\delta}-g_{\alpha\delta}g_{\beta\gamma}\right)$$
$$-ig^{2}f^{XAD}f^{XBC}\left(g_{\alpha\beta}g_{\gamma\delta}-g_{\alpha\gamma}g_{\beta\delta}\right)$$
$$-ig^{2}f^{XAB}f^{XCD}\left(g_{\alpha\gamma}g_{\beta\delta}-g_{\alpha\delta}g_{\beta\gamma}\right)$$

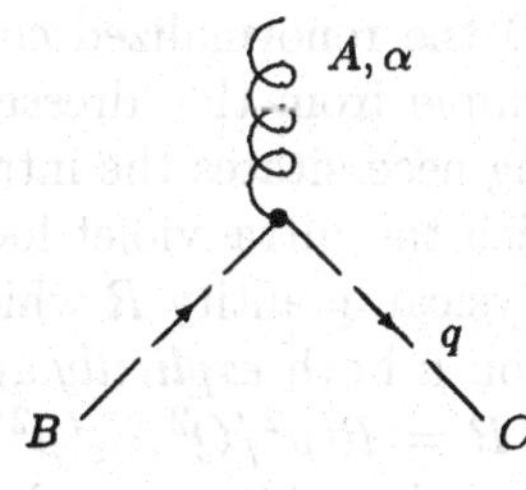

$$gf^{ABC}q^{\alpha}$$

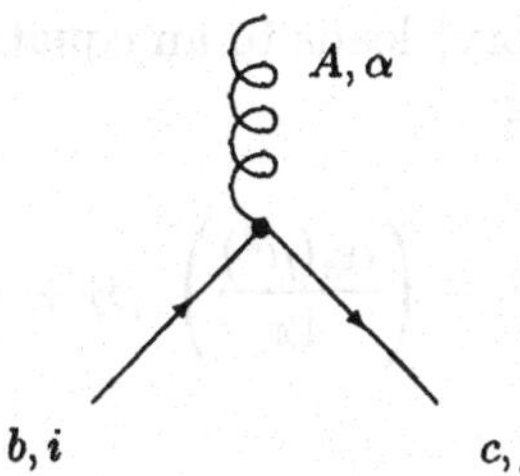

$$-ig\left(T^{A}\right)_{cb}\left(\gamma^{\alpha}\right)_{ji}$$

TABLE 1.

Explicit forms for the SU(3) colour matrices and structure constants can be found, for example, in Ref. [1]. The following are some useful identities:

$$\begin{aligned}
[T^a, T^b] &= i f^{abc} T^c \\
\{T^a, T^b\} &= d^{abc} T^c + \frac{1}{3}\delta^{ab} \\
f^{acd} f^{bcd} &= C_A \delta^{ab} \\
(T^a T^a)_{ij} = T^a_{ik} T^a_{kj} &= C_F \delta_{ij} \\
\mathrm{Tr}(T^a T^b) = T^a_{ij} T^b_{ji} &= T_F \delta^{ab} \\
C_A &= N_c = 3 \\
C_F &= \frac{N_c^2 - 1}{2N_c} = \frac{4}{3} \\
T_F &= \frac{1}{2} \\
\mathrm{Tr}(T^a T^b T^c) &= \frac{i}{4} f^{abc} + \frac{1}{4} d^{abc} \\
f^{abc} f^{abc} &= 24 \\
d^{abc} d^{abc} &= \frac{40}{3}
\end{aligned} \tag{2}$$

where summation over repeated indices is understood.

2.2. THE QCD COUPLING CONSTANT

Quantum Chromodynamics is an asymptotically free gauge field theory, that is, the strength of the interaction between the quarks and gluons becomes weaker in the short–distance limit. In QCD the renormalized coupling can be defined in a variety of ways, for example from the 'dressed' qqg or ggg vertices. Renormalization of the coupling necessitates the introduction of a scale μ — effectively the scale at which the ultra–violet loop divergences are subtracted off. A dimensionless physical quantity R which depends on some energy scale Q will depend also on μ both *explicitly* and *implicitly* through the renormalized coupling, i.e. $R = R(\mu^2/Q^2, \alpha_s(\mu^2))$. The fact that such a quantity should not depend on the arbitrary scale μ (when calculated to all orders in perturbation theory) leads to an equation for the μ dependence of the renormalized coupling:

$$\begin{aligned}
\frac{\mu^2}{\alpha_s(\mu^2)} \frac{\partial \alpha_s(\mu^2)}{\partial \mu^2} &= -\frac{\alpha_s(\mu^2)}{4\pi}\beta_0 - \left(\frac{\alpha_s(\mu^2)}{4\pi}\right)^2 \beta_1 - \left(\frac{\alpha_s(\mu^2)}{4\pi}\right)^3 \beta_2 + \ldots \\
\beta_0 &= 11 - \frac{2}{3} n_f
\end{aligned}$$

$$\beta_1 = 102 - \frac{38}{3} n_f$$
$$\beta_2(\overline{\rm MS}) = \frac{2857}{2} - \frac{5033}{18} n_f + \frac{325}{54} n_f^2, \qquad (3)$$

for n_f massless quark flavours. Note that the coefficients in the above perturbative expansion depend, in general, on the renormalization scheme (RS), although for massless quarks the first two coefficients, β_0 and β_1, are RS independent. In essentially all phenomenological applications the $\overline{\rm MS}$ RS is used.

At leading order, i.e. retaining only the coefficient β_0, Eq. (3) can be solved for α_s to give

$$\alpha_s(\mu^2) = \frac{\alpha_s(\mu_0^2)}{1 + \alpha_s(\mu_0^2)\, b \ln(\mu^2/\mu_0^2)} \qquad (4)$$

or

$$\alpha_s(\mu^2) = \frac{1}{b \ln(\mu^2/\Lambda^2)}, \qquad (5)$$

where $b = \beta_0/4\pi = (33 - 2n_f)/(12\pi)$.

These two expressions are entirely equivalent — they differ only in the choice of boundary condition for the differential equation, $\alpha_s(\mu_0^2)$ in the first case and the dimensionful parameter Λ in the second. In fact nowadays Λ is disfavoured as the fundamental parameter of QCD, since its definition is not unique beyond leading order (see below), and its value depends on the number of 'active' quark flavours. Instead, it has become conventional to use the value of α_s in the $\overline{\rm MS}$ scheme at $\mu^2 = M_Z^2$ as the fundamental parameter. The advantage of using M_Z as the reference scale is that it is (a) very precisely measured [2], (b) safely in the perturbative regime, i.e. $\alpha_s(M_Z^2) \ll 1$, and (c) far from quark thresholds, i.e. $m_b \ll M_Z \ll m_t$.

The parameter Λ is, however, sometimes still used as a book-keeping device. At next-to-leading order there are two definitions of Λ which are widely used in the literature:

$$\text{definition 1}: \qquad b \ln \frac{Q^2}{\Lambda^2} = \frac{1}{\alpha_s(\mu^2)} + b' \ln\left(\frac{b'\alpha_s(\mu^2)}{1 + b'\alpha_s(\mu^2)}\right), \qquad (6)$$

$$\text{definition 2}: \qquad \alpha_s(\mu^2) = \frac{1}{b \ln(\mu^2/\Lambda^2)}\left[1 - \frac{b'}{b}\frac{\ln\ln(\mu^2/\Lambda^2)}{\ln(\mu^2/\Lambda^2)}\right], \qquad (7)$$

where $b' = \beta_1/4\pi\beta_0 = (153 - 19n_f)/(2\pi(33 - 2n_f))$. The first of these solves Eq. (3) exactly when β_2 and higher coefficients are neglected, while the second (the 'PDG' definition [2]) provides an explicit expression for $\alpha_s(\mu^2)$ in terms of μ^2/Λ^2 and is a solution of Eq. (3) up to terms of order

$1/\ln^3(\mu^2/\Lambda^2)$.[1] Note that these two Λ parameters are *different* for the *same* value of $\alpha_s(M_Z^2)$, the difference being about one quarter the size of the current measurement uncertainty:

$$\Lambda_1^{(5)} - \Lambda_2^{(5)} \simeq 15 \text{ MeV} \simeq \frac{1}{4}\delta_{\text{exp}}\Lambda^{(5)}. \tag{8}$$

A second difficulty with the above definitions is that Λ depends on the number of active flavours. Values of Λ for different numbers of flavours can be defined by imposing the continuity of α_s at the scale $\mu = m$, where m is the mass of the heavy quark. For example, for the b-quark threshold: $\alpha_s(m_b^2, 4) = \alpha_s(m_b^2, 5)$. Using the next–to–leading order form (6) for $\alpha_s(\mu^2)$ one can show that

$$\Lambda(4) \approx \Lambda(5)\left(\frac{m_b}{\Lambda(5)}\right)^{\frac{2}{25}}\left[\ln\left(\frac{m_b^2}{\Lambda(5)^2}\right)\right]^{\frac{963}{14375}}. \tag{9}$$

Since in practice most higher order QCD corrections are carried out using the $\overline{\text{MS}}$ regularization scheme, one uses either of the above results for $\alpha_s(\mu^2)$ with $\Lambda \equiv \Lambda_{\overline{\text{MS}}}$. Table 2 gives the conversion between $\Lambda_{\overline{\text{MS}}}^{(5)}$ and $\alpha_s(M_Z^2)$ using definition 1 in (6).

In these lectures we will be mainly concerned with QCD physics at e^+e^- colliders and in deep inelastic scattering (DIS). Both processes offer several essentially independent measurements of α_s, summarized in Table 3. Note that all of these use the $q\bar{q}g$ vertex to measure α_s, with the high Q^2 scale provided by an electroweak gauge boson, for example a highly virtual γ^* in DIS or an on–shell Z^0 boson at LEP1 and SLC. There are two main theoretical issues which affect these determinations. The first is the effect of unknown higher-order (next–to–next–to–leading order (NNLO) in most cases) perturbative corrections, which leads to a non–negligible renormalisation scheme dependence uncertainty in the extracted α_s values. This is particularly true for the 'event shape' measurements at e^+e^- colliders (see later). The exceptions here are the total e^+e^- hadronic cross section (equivalently, the Z^0 hadronic decay width) and the DIS sum rules, which are known to NNLO. The second issue concerns the residual impact of $\mathcal{O}(1/Q^n)$ power corrections. For some processes it can be shown that the leading corrections are $\mathcal{O}(1/Q)$ (for example $\mathcal{O}(1/M_Z)$ for the corrections to event shapes at LEP1 and SLC) which can easily be comparable in magnitude to the NLO perturbative contributions. In deep inelastic scattering, the higher–twist power corrections to structure functions $F_i(x, Q^2)$ are $\mathcal{O}(1/Q^2(1-x))$ and must be included in scaling violation fits especially at large x. Such power corrections (and their uncertainties) must

[1]The expressions for α_s can be generalized to include also the β_2 term [3].

$\Lambda_{\overline{\rm MS}}^{(5)}$ (MeV)	$\alpha_s(M_Z^2)$
50	0.0970
100	0.1060
150	0.1122
200	0.1170
250	0.1210
300	0.1245
350	0.1277
400	0.1305
450	0.1332
500	0.1356
550	0.1379
600	0.1401

TABLE 2. $\alpha_s(M_Z^2)$ for various $\Lambda_{\overline{\rm MS}}^{(5)}$.

be taken into account in α_s determinations, either using phenomenological parametrizations or theoretical models.

Figure 1, which updates Table 12.1 of Ref. [1], summarizes the $\alpha_s(M_Z^2)$ measurements from some of the most accurate recent determinations. For experiments performed at energy scales different from M_Z, the α_s values measured at $\mu^2 = Q^2_{\rm exp}$ are converted to $\alpha_s(M_Z^2)$ using the above expressions. The consistency of the various measurements is remarkable — α_s is indeed a universal parameter. Defining a 'world average' value presents a technical difficulty, however. Since the errors of most of the measurements are largely theoretical — often based on estimates of unknown higher-order corrections or non–perturbative effects — and neither gaussian nor completely independent, the overall error on the combined value of $\alpha_s(M_Z^2)$ cannot be obtained from standard statistical techniques. The average value[2] of the measurements presented in Fig. 1 is

$$\text{WORLD AVERAGE:} \qquad \alpha_s(M_Z^2) = 0.118 \pm 0.004\,. \qquad (10)$$

Following Ref. [1], the error here is defined as 'the uncertainty equal to that of a typical measurement by a reliable method'.[3] In view of the consistency

[2]obtained by χ^2 minimisation, as described in Ref. [1].

[3]In view of the recent improvements in the lattice, Z^0 hadronic width, and DIS (νN)

	quantity	perturbation series
e^+e^-	R_{ee}, R_Z, R_τ	$R = R_0[1+\alpha_s/\pi+\ldots]$
	event shapes, f_3, ...	$1/\sigma d\sigma/dX = A\alpha_s + B\alpha_s^2 + \ldots$
	$D^h(z,Q^2)$	$\partial D^h/\partial \ln Q^2 = \alpha_s D^h \otimes P + \ldots$
ℓN DIS	$F_i(x,Q^2)$	$\partial F_i/\partial \ln Q^2 = \alpha_s F_i \otimes P + \ldots$
		$\int dx F_i(x,Q^2) = A + B\alpha_s + \ldots$
	$\sigma(2+1\,\text{jet})$	$\sigma = A\alpha_s + B\alpha_s^2 + \ldots$

TABLE 3. Summary of the most important processes for α_s determinations in e^+e^- collisions and in deep inelastic lepton–hadron scattering.

of all the measurements, and in particular of those with the smallest uncertainties, it seems unlikely that future 'world average' values of α_s will deviate significantly, if at all, from the current value given in (10).

3. QCD in High Energy e^+e^- Collisions

Many of the basic ideas and properties of perturbative QCD can be illustrated by considering the process $e^+e^- \to$ hadrons. We begin by discussing how the order α_s corrections to the total hadronic cross section are calculated, and how renormalization scheme dependence enters at order α_s^2. This cross section also provides one of the most precise measurements of the strong coupling, see Fig. 1.

Perturbative QCD also predicts a rich 'jet' structure for the final state hadrons. We show how jet cross sections can be defined, and how some of the predictions compare with experiment.

3.1. THE TOTAL CROSS SECTION FOR $E^+E^- \to$ HADRONS

One of the theoretically most straightforward predictions of perturbative QCD is for $R^{e^+e^-}$, the ratio of the total e^+e^- hadronic cross section to the

determinations, it seems appropriate to decrease the uncertainty of ±0.005 in Ref. [1] to ±0.004.

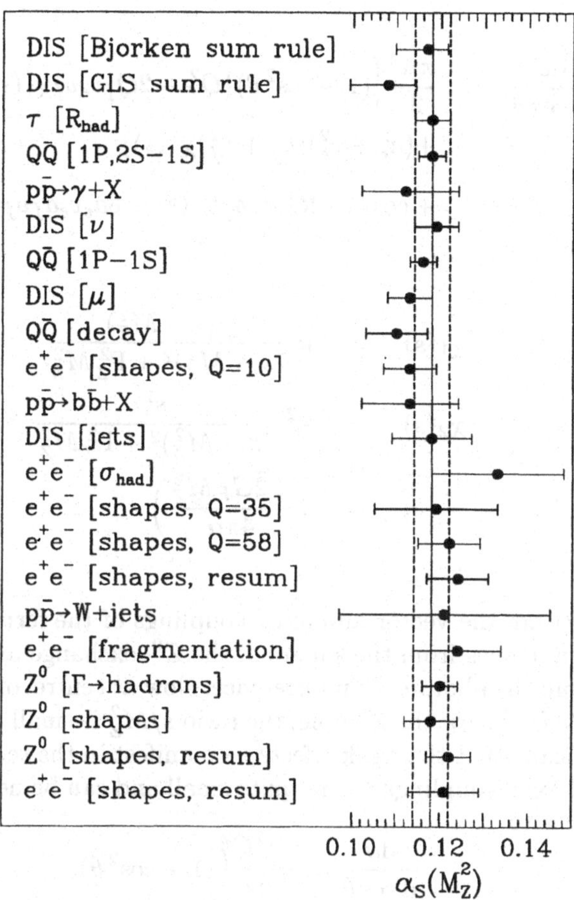

Figure 1. Measurements of $\alpha_s(M_Z^2)$, in the $\overline{\rm MS}$ renormalisation scheme, updated from Ref. [1].

muon pair production cross section. On the Z^0 pole, as for example at LEP and SLC, the analogous quantity is the ratio of the partial decay widths of the Z^0 to hadrons and to muon pairs.

We begin by considering the high–energy $2 \to 2$ process $e^+e^- \to f\bar{f}$ with f a light charged fermion, $f \neq e$. In lowest order, the process is mediated by either a virtual photon or a Z^0 in the s–channel. With θ the centre–of–mass scattering angle of the final state fermion pair, the differential cross section is:

$$\frac{d\sigma}{d\cos\theta} = \frac{\pi\alpha^2}{2s}\Big[(1+\cos^2\theta)\{Q_f^2 - 2Q_f v_e v_f \chi_1(s) + (a_e^2+v_e^2)(a_f^2+v_f^2)\chi_2(s)\} + \cos\theta(-4Q_f a_e a_f \chi_1(s) + 8a_e v_e a_f v_f \chi_2(s))\Big] \quad (11)$$

where

$$\chi_1(s) = \kappa\frac{s(s-M_Z^2)}{(s-M_Z^2)^2+\Gamma_Z^2 M_Z^2}$$

$$\chi_2(s) = \kappa^2\frac{s^2}{(s-M_Z^2)^2+\Gamma_Z^2 M_Z^2}$$

$$\kappa = \left(\frac{\sqrt{2}G_F M_Z^2}{4\pi\alpha}\right) \quad (12)$$

$$(13)$$

and (v_f, a_f) are the vector and axial couplings of the fermions to the Z^0.[4] The χ_2 term comes from the square of the Z^0–exchange amplitude and the χ_1 term from the photon–Z^0 interference. Now at centre–of–mass scattering energies $\sqrt{s}$ far below the Z^0 pole, the ratio s/M_Z^2 is small and so $1 \gg \chi_1 \gg \chi_2$. This means that the weak effects — manifest in the terms involving the vector and axial couplings — are quite small and can be neglected. Eq. (11) then reduces to

$$\frac{d\sigma}{d\cos\theta} = \frac{\pi\alpha^2 Q_f^2}{2s}(1+\cos^2\theta). \quad (14)$$

Integrating over θ and setting $Q_f = -1$ gives the total cross section for $e^+e^- \to \mu^+\mu^-$:

$$\sigma_0 = \frac{4\pi\alpha^2}{3s}, \quad (15)$$

where $\sqrt{s}$ is the total centre-of-mass energy. On the Z^0 pole, $\sqrt{s} = M_Z$, the χ_2 term in (11) dominates and the (peak) cross section is

$$\sigma_0 = \frac{4\pi\alpha^2\kappa^2}{3\Gamma_Z^2}(a_e^2+v_e^2)^2. \quad (16)$$

When an electron and a positron annihilate they can also produce hadrons in the final state. The formation of the observed final state hadrons is not governed by perturbation theory. Why then would one expect perturbation theory to give an accurate description of the hadronic production

[4] i.e. $v_e = (-1+4\sin^2\theta_W)/2$, $a_e = -1/2$ etc.

cross section? The answer can be understood by visualizing the event in space–time. The electron and positron form a photon of virtuality $Q = \sqrt{s}$ which fluctuates into a quark and an antiquark. By the uncertainty principle, this fluctuation occurs in a space time volume $1/Q$, and if Q is large the production rate should be predictable in perturbation theory. Subsequently the quarks and gluons form themselves into hadrons. This happens at a later time scale characterized by the scale $1/\Lambda$, where Λ is the typical mass scale of the strong interactions. The interactions which change quarks and gluons into hadrons modify the outgoing state, but they occur too late to modify the probability for an event to happen.

In leading–order perturbation theory, therefore, the total hadronic cross section is obtained by simply summing over all kinematically accessible flavours and colours of quarks. Ignoring the Z^0 exchange contributions (i.e. assuming $\sqrt{s} \ll M_Z$) we have

$$R^{QPM} = \frac{\sum_q \sigma(e^+e^- \to q\bar{q})}{\sigma(e^+e^- \to \mu^+\mu^-)} = 3\sum_q Q_q^2 \,. \tag{17}$$

With $q = u, ..., b$ we obtain $R^{QPM} = 11/3 = 3.67$. At $\sqrt{s} = 34$ GeV the measured value is about 3.9 (see for example Ref. [2]). Even allowing for the Z^0 contribution ($\Delta R_Z \simeq 0.05$), this result is some 5% higher than the lowest–order prediction. It turns out that the difference is due to higher–order QCD corrections, and in fact the comparison between theory and experiment gives one of the most precise determinations of the strong coupling constant.

The $O(\alpha_s)$ corrections to the total hadronic cross section are calculated from both real and virtual one–gluon emission diagrams. For the real gluon contributions, it is convenient to write the three-body phase space integration as

$$\mathrm{d}\Phi_3 \sim \mathrm{d}\alpha\, \mathrm{d}\beta\, \mathrm{d}\gamma\, \mathrm{d}x_1\, \mathrm{d}x_2, \tag{18}$$

where α, β, γ are Euler angles, and $x_1 = 2E_q/\sqrt{s}$ and $x_2 = 2E_{\bar{q}}/\sqrt{s}$ are the energy fractions of the final state quark and antiquark. Integrating out the Euler angles gives a matrix element which depends only on x_1 and x_2 and the contribution to the total cross section is

$$\sigma^{q\bar{q}g} = \sigma_0\, 3\sum_q Q_q^2 \int \mathrm{d}x_1 \mathrm{d}x_2\, \frac{2\alpha_s}{3\pi}\, \frac{x_1^2 + x_2^2}{(1-x_1)(1-x_2)} \tag{19}$$

where the integration region is defined by $0 \leq x_1, x_2 \leq 1$, $x_1 + x_2 \geq 1$. Unfortunately, the integrals are divergent at $x_i = 1$! These singularities come from regions of phase space where the gluon is *collinear* with either quark, $\theta_{qg} \to 0$, or where the gluon is *soft*, $E_g \to 0$. Evidently we require some sort of regularization procedure — to render the integrals finite — before the calculation can be completed. A variety of methods are suitable. One can give the gluon a small mass, or take the final state quark and antiquark off–mass–shell by a small amount. In each case the singularities are then manifest as logarithms of the regulating mass.

A more elegant procedure is to use dimensional regularization, with the number of space–time dimensions > 4. With the three–body phase space integrals now cast in n dimensions, the soft and collinear singularities appear as *poles* at $n = 4$. Details of how the calculation proceeds can be found for example in [1]. The result is that the cross section of Eq. (19) becomes

$$\sigma^{q\bar{q}g} = \sigma_0 \, 3 \sum_q Q_q^2 \, \frac{2\alpha_s}{3\pi} \, H(\epsilon) \left[\frac{2}{\epsilon^2} - \frac{3}{\epsilon} + \frac{19}{2} + O(\epsilon) \right], \qquad (20)$$

where $\epsilon = (n-4)/2$ and $H(\epsilon) = 1 + O(\epsilon)$.

The virtual gluon contribution can be calculated in a similar fashion, with dimensional regularization again used to render finite the infra–red divergences in the loops. The result is

$$\sigma^{q\bar{q}(g)} = \sigma_0 \, 3 \sum_q Q_q^2 \, \frac{2\alpha_s}{3\pi} \, H(\epsilon) \left[-\frac{2}{\epsilon^2} + \frac{3}{\epsilon} - 8 + O(\epsilon) \right]. \qquad (21)$$

When the two contributions (20) and (21) are added together the poles exactly cancel and the result is *finite* in the limit $\epsilon \to 0$:

$$R^{e^+e^-} = 3 \sum_q Q_q^2 \left\{ 1 + \frac{\alpha_s}{\pi} + O(\alpha_s^2) \right\}. \qquad (22)$$

Note that the next-to-leading order correction is positive, and with a value for α_s of about 0.15, can accommodate the experimental measurement at $\sqrt{s} = 34$ GeV.[5]

The cancellation of the soft and collinear singularities between the real and virtual gluon diagrams is not accidental. Indeed, there are theorems — the Bloch, Nordsieck [4] and Kinoshita, Lee, Nauenberg [5] theorems — which state that suitably defined inclusive quantities will be free of singularities in the massless limit. The total hadronic cross section is an example of such a quantity, whereas the cross section for the exclusive $q\bar{q}$ final state, i.e. $\sigma(e^+e^- \to q\bar{q})$, is not.

The $O(\alpha_s^2)$ and $O(\alpha_s^3)$ corrections to $R^{e^+e^-}$ are also known. At these higher orders we encounter the ultra–violet divergences associated with the

[5]In contrast, the corresponding correction is negative for a scalar gluon.

renormalization of the strong coupling. Writing

$$\begin{aligned} \sigma_{tot} &= \frac{4\pi\alpha^2}{3s} R, \\ R &= K_{QCD}\, 3 \sum_q Q_q^2, \\ K_{QCD} &= 1 + \sum_{n\geq 1} C_n \left(\frac{\alpha_s}{\pi}\right)^n , \end{aligned} \tag{23}$$

the coefficients C_1, C_2 and C_3 are (in the $\overline{\text{MS}}$ scheme with the renormalization scale choice $\mu = \sqrt{s}$):

$$\begin{aligned} C_1 &= 1 \\ C_2 &= \left(\frac{2}{3}\zeta(3) - \frac{11}{12}\right) n_f + \left(\frac{365}{24} - 11\zeta(3)\right) \\ &\simeq 1.986 - 0.115 n_f \\ C_3 &= \left(\frac{87029}{288} - \frac{1103}{4}\zeta(3) + \frac{275}{6}\zeta(5)\right) \\ &\quad - \left(\frac{7847}{216} - \frac{262}{9}\zeta(3) + \frac{25}{9}\zeta(5)\right) n_f \\ &\quad + \left(\frac{151}{162} - \frac{19}{27}\zeta(3)\right) n_f^2 \\ &\quad - \frac{\pi^2}{432}(33 - 2n_f)^2 + \eta\left(\frac{55}{72} - \frac{5}{3}\zeta(3)\right) \\ &\simeq -6.637 - 1.200 n_f - 0.005 n_f^2 - 1.240\eta , \end{aligned} \tag{24}$$

where $\eta = (\sum_q Q_q)^2/(3\sum_q Q_q^2)$ and the sum extends over the (n_f) quarks which are effectively massless at the energy scale $\sqrt{s}$. The result for C_3 is from Ref. [6]. Apart from the η term, the QCD corrections in K are the same for the ratio of hadronic to leptonic Z^0 decay widths: $R_Z = \Gamma_h/\Gamma_\mu$. In practice, quark masses (particularly m_b and m_t) have a small but non-negligible effect [7] and must be taken into account in precision fits to data.

Experiments at LEP and SLC are able to measure R_Z very accurately. From such measurements, a very precise value of $\alpha_s(M_Z^2)$ can be obtained. In practice, α_s is measured simultaneously with other parameters in a global electroweak fit, and the value obtained is correlated to some extent with m_t, M_H, etc. A recent fit of this kind [8] gives

$$\alpha_s(M_Z^2) = 0.120 \pm 0.003(\text{fit}) \pm 0.002(\text{theory}), \tag{25}$$

as displayed in Fig. 1.

Through $\mathcal{O}(\alpha_s^3)$, the explicit μ–dependence of the perturbation series for R is restored by the replacements:

$$\begin{aligned} \alpha_s &\rightarrow \alpha_s(\mu^2) \\ C_2 &\rightarrow C_2 - C_1\frac{\beta_0}{4}\log\frac{s}{\mu^2} \\ C_3 &\rightarrow C_3 + C_1\left(\frac{\beta_0}{4}\right)^2\log^2\frac{s}{\mu^2} - (C_1\frac{\beta_1}{16} + C_2\frac{\beta_0}{2})\log\frac{s}{\mu^2}. \end{aligned} \quad (26)$$

where β_0 and β_1 have been defined in Section 2 above. Note that the μ^2–dependence of the second order coefficient is exactly as specified by the renormalization group equation, i.e. the coefficient of $\log(s/\mu^2)$ is proportional to the β function coefficient defined in (3).

In general the coefficients of any QCD perturbative expansion depend on the choice made for the renormalization scale μ in such a way that as μ is varied, the change in the coefficients exactly compensates the change in the coupling $\alpha_s(\mu^2)$. However this μ–independence breaks down whenever the series is *truncated.* One can show in fact that changing the scale in a physical quantity such as $R^{e^+e^-}$ — which has been calculated to $O(\alpha_s^n)$ — induces changes of $O(\alpha_s^{n+1})$. This is illustrated in Fig. 2, taken from Ref. [1], which shows $K_{QCD} = 1 + \delta$ for R_Z as a function of μ, as the higher–order terms are added in. As expected, the inclusion of higher–order terms leads to a more definite prediction. In the absence of higher–order corrections, one can try to guess the 'best' choice of scale, defined as the scale which makes the truncated and all–orders predictions equal. In the literature, two such choices have been advocated in particular. In the *fastest apparent convergence* approach [9], one chooses the scale $\mu = \mu_{\rm FAC}$, where

$$R^{(1)}(\mu_{\rm FAC}) = R^{(2)}(\mu_{\rm FAC}). \quad (27)$$

On the other hand, the *principle of minimal sensitivity* [10] suggests a scale choice $\mu = \mu_{\rm PMS}$, where

$$\mu\frac{\mathrm{d}}{\mathrm{d}\mu}R^{(2)}(\mu)\Big|_{\mu_{\rm PMS}} = 0. \quad (28)$$

These two special scales can be identified in Fig. 2. It is, however, important to remember that there are no theorems that prove that any of these schemes are correct. All one can say is that the theoretical error on a quantity calculated to $O(\alpha_s^n)$ is $O(\alpha_s^{n+1})$. Varying the scale is simply one way of quantifying this uncertainty.

3.2. JET CROSS SECTIONS

The expression given for the total hadronic cross section in the previous section is very concise, but it tells us nothing about the kinematic *distribution*

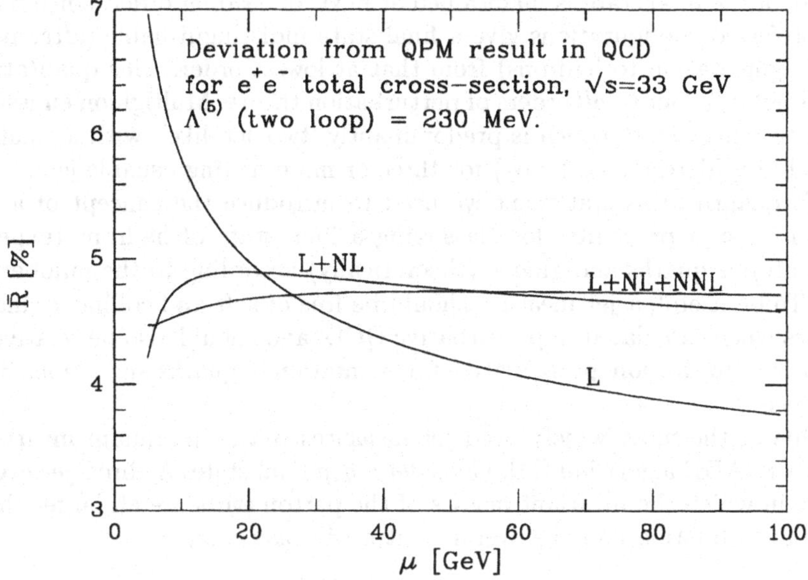

Figure 2. The effect of higher order QCD corrections to R_Z, as a function of the renormalization scale μ, from Ref. [1].

of hadrons in the final state. If the hadronic fragments of a fast moving quark have limited transverse momentum relative to the quark momentum, then the lowest order contribution — $e^+e^- \to q\bar{q}$ — can naively be interpreted as the production of two back–to–back jets. In this section we investigate how higher–order perturbative corrections modify this picture.

Consider first the next–to–leading process $e^+e^- \to q\bar{q}g$. From the previous section (Eq. (19)), we have

$$\frac{1}{\sigma}\frac{\mathrm{d}^2\sigma}{\mathrm{d}x_1\mathrm{d}x_2} = \frac{2\alpha_s}{3\pi}\,\frac{x_1^2+x_2^2}{(1-x_1)(1-x_2)}\,. \tag{29}$$

Recall that the cross section becomes infinitely large when either (a) the gluon is collinear with one of the outgoing quarks, or (b) the gluon momentum goes to zero. This corresponds to (a) only one and (b) both of the x_i approaching 1 respectively. In other words the gluon prefers to be soft and/or collinear with the quarks. If the gluon is *required* to be well–separated in phase space from the quarks — a configuration corresponding to a 'three jet event' — then the cross section is suppressed relative to lowest order by one power of α_s. It would appear, therefore, that the two–jet nature of the final state is maintained at next–to–leading order, since both the preferred configurations give a final state indistinguishable (after parton fragmentation to hadrons) from that at lowest order. This qualitative result holds in fact to *all* orders of perturbation theory. Multigluon emission leads to a final state which is predominantly 'two–jet–like', with a smaller probability (determined by α_s) for three or more distinguishable jets.

To quantify this statement we need to introduce the concept of a *jet measure*, i.e. a procedure for classifying a final state of hadrons (experimentally) or quarks and gluons (theoretically) according to the number of jets. To be useful, a jet measure should be free of soft and collinear singularities when calculated in perturbative QCD, and should also be relatively insensitive to the non–perturbative fragmentation of quarks and gluons into hadrons.

One of the most widely used jet measures is the 'minimum invariant mass' or JADE algorithm [11]. Consider a $q\bar{q}g$ final state. A three–jet event is one in which the invariant masses of the parton pairs are all larger than some fixed fraction y of the overall centre–of–mass energy:

$$(p_i + p_j)^2 > ys, \qquad i, j = q, \bar{q}, g. \tag{30}$$

It is immediately clear that this region of phase space avoids the soft and collinear singularities of the matrix element. In fact in terms of the energy fractions, Eq. (30) is equivalent to

$$0 < x_1, x_2 < 1 - y, \qquad x_1 + x_2 > 1 + y. \tag{31}$$

If we define f_2 and f_3 to be the two– and three–jet fractions defined in this way, then to $O(\alpha_s)$ we obtain

$$\begin{aligned} f_3 &= \frac{2\alpha_s}{3\pi}\left[(3-6y)\log\left(\frac{y}{1-2y}\right) + 2\log^2\left(\frac{y}{1-y}\right) + \frac{5}{2} - 6y - \frac{9}{2}y^2 \right. \\ &\quad \left. + 4\,\mathrm{Li}_2\left(\frac{y}{1-y}\right) - \frac{\pi^2}{3}\right], \quad \mathrm{Li}_2(y) = -\int_0^y \frac{\mathrm{d}z}{1-z}\log z\,, \\ f_2 &= 1 - f_3\,. \end{aligned} \tag{32}$$

Note that the soft and collinear singularities reappear as large logarithms in the limit $y \to 0$. Clearly the result only makes sense for y values large enough such that $f_2 \gg f_3$, so that the $O(\alpha_s)$ correction to f_2 is perturbatively small.

The generalization to multijet fractions is straightforward. Starting from an n–parton final state, identify the pair with the minimum invariant mass squared. If this is greater then ys then the number of jets is n. If not, combine the minimum pair into a single 'cluster'. Then repeat for the $(n -$

1)–parton/cluster final state, and so on until all parton/clusters have a relative invariant mass squared greater than ys. The number of clusters remaining is then the number of jets in the final state. Note that an n–parton final state can give any number of jets between n (all partons well-separated) and 2 (for example, two hard quarks accompanied by soft and collinear gluons).

Since a soft or collinear gluon emitted from a quark line does not change the multiplicity of jets, the cancellation of soft and collinear singularities that was evident in the total cross section calculation can still take place, and the jet fractions defined this way are free of such singularities to all orders in perturbation theory.

Now in general we have

$$\begin{aligned} f_{n+2}(\sqrt{s},y) &= \left(\frac{\alpha_s(s)}{\pi}\right)^n \sum_{j=0}^{\infty} C_{nj}(y)\left(\frac{\alpha_s(s)}{\pi}\right)^j, \quad n \geq 0, \\ \sum_{n=2}^{\infty} f_n &= 1\,. \end{aligned} \tag{33}$$

Since the jet–defining parameter y is dimensionless, all the energy dependence of the jet fractions is contained in the coupling $\alpha_s(s)$. One can therefore exhibit the *running* of the strong coupling by measuring a decrease in f_3 as $\sqrt{s}$ increases, see Fig. 3. Note that experimentally the algorithm is applied to final state *hadrons* rather than *partons*. However studies using parton shower/fragmentation Monte Carlos have shown that — at least at very high energy — the fragmentation corrections are small ($\mathcal{O}(1/Q)$) and therefore the QCD parton–level predictions can be reliably compared with the experimental data. A quantitative discussion can be found in Ref. [15], for example.

The next–to–leading order corrections to f_3 have been calculated [12]. Because the hadronization corrections to f_3 are relatively small, the three–jet rate provides one of the most precise measurements of α_s at LEP and SLC. A typical fit is shown in Fig. 4.

While the above definition is well suited to experimental jet measurements, it is not quite optimum from a theoretical point of view. The reason is that when y becomes small (as happens in practice), the large logarithms of y explicit in (32) begin to dominate the theoretical predictions. It is straightforward to show that higher–order corrections to jet fractions such as f_2 and f_3 will contain terms like $\alpha_s^n \log^{2n} y$. When y is small enough that $\alpha_s \log^2 y \sim 1$, these terms must be resummed to obtain a reliable prediction. Unfortunately, the JADE algorithm is not well-suited to this type of resummation [16], and so a variant — the 'Durham' or k_T algorithm — has been proposed [17]. In this modified algorithm, the invariant mass

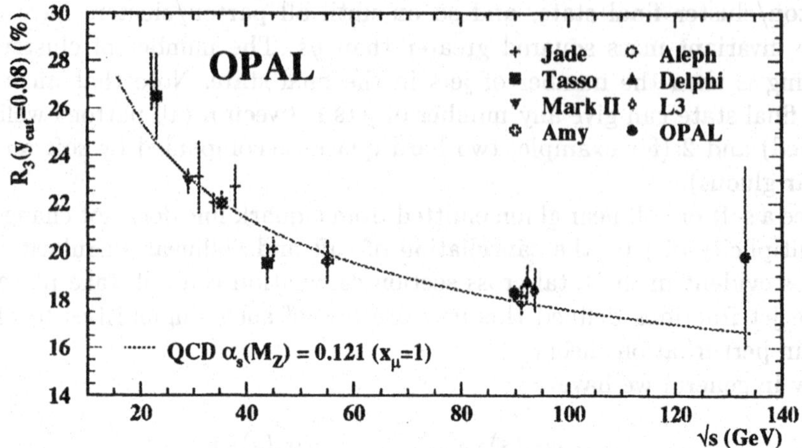

Figure 3. A compilation of three-jet fractions (R_3) at different e^+e^- annihilation energies, from the OPAL collaboration [13].

measure of two partons (hadrons) given in (30) is replaced by the minimum of the relative transverse momenta:

$$\min k_{Tij}^2 = \min(E_i^2, E_j^2) \sin\theta_{ij}^2 > ys, \quad i,j = q, \bar{q}, g\,, \tag{34}$$

in the e^+e^- centre–of–mass frame with massless quarks and gluons. With this new definition of the jet measure, resummation of all large logarithms can be performed [17]. Finite, next–to–leading order corrections have also been calculated [15] and comparisons of theory and experiment have been performed, as for the JADE algorithm. It is interesting that the (presumably) more reliable α_s values from resummed perturbative jet measures tend to be slightly larger than those obtained without resummation, see for example Fig. 1.

3.3. EVENT SHAPE VARIABLES

The other high–precision determination of α_s at LEP and SLC comes from *event shape* variables, quantities that characterize the 'shape' of an event, for example whether the distribution of hadrons is pencil–like, planar, spherical etc. This is more general than the jet cross section approach discussed above, since a jet-finding algorithm will *always* find jets in a hadronic final state even when none existed in the first place, for example in the limiting case when the hadronic energy is distributed uniformly over the 4π solid angle. The procedure is to define a quantity X which measures some

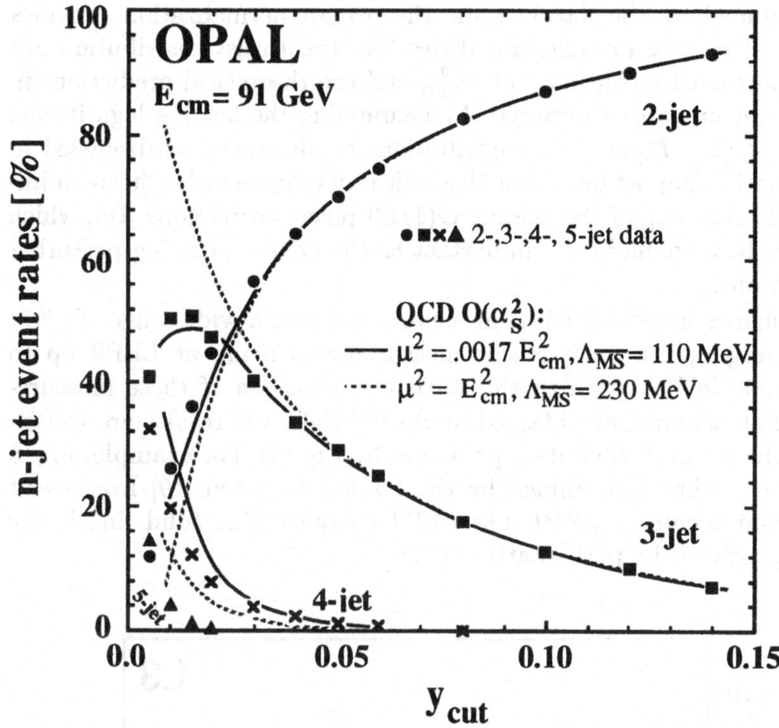

Figure 4. Multijet rates measured by the OPAL collaboration [14], with perturbative QCD fits.

particular aspect of the shape of the hadronic final states. The distribution $\mathrm{d}\sigma/\mathrm{d}X$ can be measured and compared with the theoretical prediction. For the latter to be calculable in perturbation theory, the variable should be infra–red safe, i.e. insensitive to the emission of soft or collinear gluons. A typical example is the thrust variable:

$$T = \max_{\mathbf{n}} \frac{\sum_i |\mathbf{p_i} \cdot \mathbf{n}|}{\sum_i |\mathbf{p_i}|}. \tag{35}$$

Thus a pure two–particle final state (e.g. $q\bar{q}$) has $T = 1$, while for $T < 1$ the leading order (parton) contribution to the thrust distribution comes from the $q\bar{q}g$ final state:

$$\frac{1}{\sigma}\frac{d\sigma}{dT} = \alpha_s A_1(T) + \alpha_s^2 A_2(T) + \ldots + \mathcal{O}\left(\frac{1}{E_{\rm cm}}\right), \tag{36}$$

where the coefficient functions $A_1(T)$ and $A_2(T)$ have been calculated. Thus the *shape* of the distribution tests, via $A_1(T)$, the basic QCD $q\bar{q}g$ interaction vertex (scalar gluons would, for example, give a different shape and can be excluded by the data), while the overall normalization provides a measure of α_s. At present, quantities like the thrust distribution are known in perturbation theory to $\mathcal{O}(\alpha_s^2)$, and the theoretical predictions in the $T \to 1$ region can be improved by resumming the leading logarithmic $A_n \sim \ln^{(2n-1)}(1-T)/(1-T)$ contributions to all orders, as discussed in Ref. [1]. Another important recent theoretical development has been an improved understanding of the leading $\mathcal{O}(1/E)$ power corrections [18], which at LEP can be as numerically important as the next–to–leading perturbative corrections.

Event shapes have yielded α_s measurements over a wide range of e^+e^- collision energies, the most recent measurements being at LEP2 up to $\sqrt{s} = \mathcal{O}(190 \text{ GeV})$. Although the statistical precision of these measurements cannot match that obtained at the Z^0 pole, the results are consistent with the Q^2 evolution of α_s predicted by Eq. (3). For example, Fig. 5 shows the α_s values determined by the L3 collaboration [19] from event shape measurements at LEP1 and LEP2 energies. The solid line is the evolution predicted by perturbative QCD.

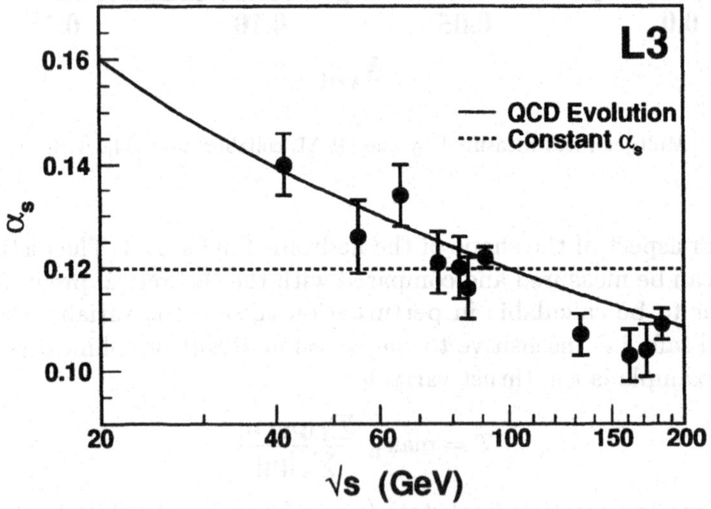

Figure 5. Measurements of α_s from event shapes at LEP1 and LEP2 from the L3 collaboration [19]. The errors correspond to experimental uncertainties only.

There are many other detailed tests of QCD which can be performed at high–energy e^+e^- colliders, but lack of space precludes a detailed discussion here. Some of most important are: (i) using four–jet events to test the non–Abelian strucure of QCD via the $e^+e^- \to q\bar{q}gg$ process, (ii) studying the detailed structure of (light and heavy) quark jets and gluon jets, (iii) comparing measured particle multiplicities with leading-logarithm ('MLLA') QCD predictions. Further information can be found in Ref. [1].

4. Deep Inelastic Scattering

The original, and still one of the the most powerful, test of perturbative QCD is the breaking of Bjorken scaling in the structure functions measured in deep inelastic lepton–hadron scattering. Nowadays, structure function analyses not only provide some of the most precise tests of the theory but also determine the momentum distributions of partons in hadrons for use as input in predicting cross sections in high energy hadron–hadron collisions. In this section we first describe the basic features of the parton model in deep inelastic scattering and then discuss how the picture is modified by perturbative corrections. Comprehensive reviews of deep inelastic scattering, the parton model and QCD can be found in Refs. [1] and [20], for example.

4.1. THE PARTON MODEL

Consider the deep inelastic lepton–proton scattering process $lp \to lX$. Label the incoming and outgoing lepton four-momenta by k^μ and k'^μ respectively, the incoming proton momentum by p^μ $(p^2 = M^2)$ and the momentum transfer by $q^\mu = k^\mu - k'^\mu$. The standard deep inelastic variables are defined by:

$$\begin{aligned}
Q^2 &= -q^2 \qquad p^2 = M^2 \\
x &= \frac{Q^2}{2p\cdot q} = \frac{Q^2}{2M(E-E')} \\
y &= \frac{q\cdot p}{k\cdot p} = 1 - E'/E \\
s &= (k+p)^2 = M^2 + \frac{Q^2}{xy},
\end{aligned} \tag{37}$$

where the energies are defined in the rest frame of the target. Analogous expressions can be derived for lepton–hadron *colliders*, such as HERA. The hadronic structure functions $F_i(x, Q^2)$ are then defined in terms of the inclusive lepton scattering cross sections. For example, for charged lepton (neutral current) scattering via virtual photon exchange, $lp \to lX$,

$$\frac{d^2\sigma^{em}}{dxdy} = \frac{4\pi\alpha^2(s-M^2)}{Q^4}\left[\left(\frac{1+(1-y)^2}{2}\right)2xF_1^{em}\right.$$

$$+(1-y)(F_2^{em}-2xF_1^{em})-\frac{M^2}{s-M^2}xyF_2^{em}\Bigg], \quad (38)$$

and for neutrino or antineutrino (charged current) scattering via virtual W exchange, $\nu(\bar{\nu})p \to lX$,

$$\frac{d^2\sigma^{\nu(\bar{\nu})}}{dxdy} = \frac{G_F^2(s-M^2)}{2\pi}\left[\left(1-y-\frac{M^2}{s-M^2}xy\right)F_2^{\nu(\bar{\nu})}\right.$$
$$\left.+y^2xF_1^{\nu(\bar{\nu})}+(-)y(1-y/2)xF_3^{\nu(\bar{\nu})}\right]. \quad (39)$$

In the quark–parton model, these structure functions are related to the quark 'distribution functions' or 'densities' $q(x,\mu^2)$, where $q(x,\mu^2)dx$ is the probability that a parton carries a momentum fraction x of the target nucleon's momentum when probed (by a gauge boson γ^*, W or Z) at momentum transfer scale μ. In deep inelastic scattering the relevant scale is the virtuality of the gauge boson probe, i.e. $\mu^2 = Q^2$. Thus, assuming four approximately massless quark flavours,

$$\begin{aligned}
F_2^{\nu} &= 2x[d+s+\bar{u}+\bar{c}] \\
xF_3^{\nu} &= 2x[d+s-\bar{u}-\bar{c}] \\
F_2^{\bar{\nu}} &= 2x[u+c+\bar{d}+\bar{s}] \\
xF_3^{\bar{\nu}} &= 2x[u+c-\bar{d}-\bar{s}] \\
F_2^{em} &= x\left[\frac{4}{9}(u+u+c+\bar{c})+\frac{1}{9}(d+\bar{d}+s+\bar{s})\right] \\
F_L \equiv F_2-2xF_1 &= 0.
\end{aligned} \quad (40)$$

This last result, the vanishing of the structure function for longitudinal virtual photon scattering, is called the Callan–Gross relation and follows from the spin$-1/2$ property of the quarks. Note that when the nature of the target is unambiguous, the notation $q(x,\mu^2)$ and $g(x,\mu^2)$ for the quark and gluon densities can be used, otherwise a general notation is $f_{a/A}(x,\mu^2)$, where a = u, d, ... g and A = p, n, Fe, Cu, etc. In the 'naive' parton model the structure functions *scale*, i.e. $F(x,Q^2) \to F(x)$ in the asymptotic (Bjorken) limit: $Q^2 \to \infty$, x fixed. In fact, it was the observation of scaling in the original SLAC experiments that provided the first evidence of pointlike parton structure in the hadron. To a first approximation, therefore, one can take the parton distributions to be functions of x only: $q(x,\mu^2) \to q(x)$. We shall see below how perturbative QCD induces logarithmic deviations from scaling, exactly in line with more recent high–precision experimental measurements.

Individual quark distributions can be determined from measurements of the various structure functions in (40). A picture emerges in which a proton consists of three valence quarks (uud) and a 'sea' of $q\bar{q}$ pairs and gluons. In the most simple version of this parton model the sea would be (three) flavour symmetric and hence the net quark distributions would be given by $u = u_V + S$, $d = d_V + S$ and $s = \bar{s} = \bar{u} = \bar{d} = S$, with the sum rules

$$\int_0^1 u_V(x)dx = 2\int_0^1 d_V(x)dx \quad = \quad 2\,, \tag{41}$$

$$\int_0^1 x[u_V(x) + d_V(x) + 6S(x)]dx \quad = \quad 1\,. \tag{42}$$

These represent conservation of proton quantum numbers and total momentum fraction respectively.

Figure 6 shows a typical set of 'modern' quark and gluon distributions $xf_i(x,\mu^2)$ [21] in the proton extracted from fits to deep inelastic and other data, when probed at a momentum scale $\mu^2 = 20$ GeV^2, a typical value for fixed–target deep inelastic experiments. Notice that the sea is definitely *not* SU(3) flavour symmetric, rather the strange quark distribution is roughly a factor of 2 smaller than the light sea quarks, and there is even a significant asymmetry between the $\bar{u}$ and $\bar{d}$ quarks in the sea. Neither of these features is quantitatively understood at present. Qualitatively, one would expect smaller distributions for heavier sea quarks, i.e. $\bar{u},\bar{d} > s > c > b > \ldots$, and some sort of Fermi exclusion principle ($u > d \Rightarrow \bar{u} < \bar{d}$) might explain the asymmetry between $\bar{u}$ and $\bar{d}$. Notice also that at this scale a small charm quark component *is* observed, consistent with the expectation that the virtual photon should be able to resolve $c\bar{c}$ pairs in the quark sea when $Q^2 > O(m_c^2)$. The sum rule (41) is experimentally well verified, but the net momentum fraction (42) carried by the quarks alone is found to be only about 50%, with the gluons (not directly measured in leading–order deep inelastic scattering, see below) accounting for the other 50%.

4.2. SCALING VIOLATIONS – THE DGLAP EVOLUTION EQUATIONS

In QCD, Bjorken scaling is broken by logarithms of Q^2. Physically, a quark in the proton can emit a gluon, with probability determined by α_s, and lose momentum as a result. Since the higher the Q^2 the more phase space is available for gluon emission, the expectation is that parton distributions should shrink to small x as Q^2 increases, with the rate of shrinkage being controlled by α_s.

In describing quantitatively the way in which scaling is violated it is convenient to define singlet and non–singlet quark distributions:

$$F^{NS}(x,Q^2) \quad = \quad q_i(x,Q^2) - q_j(x,Q^2)\,,$$

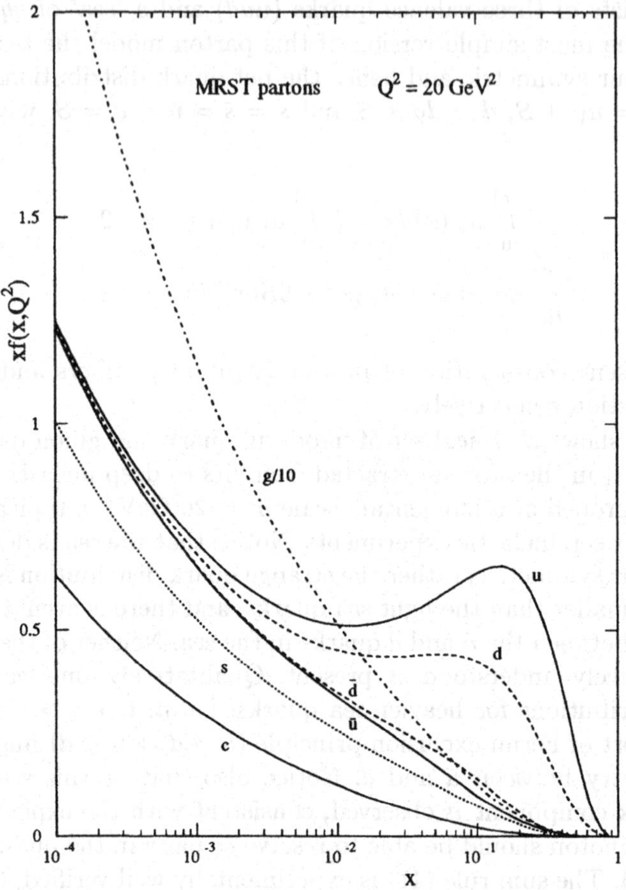

Figure 6. **Quark and gluon distribution functions at $\mu^2 = 20$ GeV^2, from Ref. [21].**

$$F^S(x,Q^2) = \sum_i \left[q_i(x,Q^2) + \bar{q}_i(x,Q^2)\right] , \tag{43}$$

where we have restored the explicit Q^2 dependence. The non–singlet structure functions have non–zero values of flavour quantum numbers such as isospin or baryon number. The variation with Q^2 of these functions is described by the Dokshitzer–Gribov–Lipatov–Altarelli–Parisi (DGLAP) equations [22]:

$$\begin{aligned}
Q^2\frac{\partial F^{NS}}{\partial Q^2} &= \frac{\alpha_s(Q^2)}{2\pi}P^{qq}*F^{NS}\\
Q^2\frac{\partial F^{S}}{\partial Q^2} &= \frac{\alpha_s(Q^2)}{2\pi}\left(P^{qq}*F^{S}+2n_fP^{qg}*g\right)\\
Q^2\frac{\partial g}{\partial Q^2} &= \frac{\alpha_s(Q^2)}{2\pi}\left(P^{gq}*F^{S}+P^{gg}*g\right),
\end{aligned}\tag{44}$$

where $*$ denotes a convolution integral:

$$f*g=\int_x^1\frac{dy}{y}f(y)g\left(\frac{x}{y}\right). \tag{45}$$

In leading order the DGLAP kernels (or 'splitting functions') are

$$\begin{aligned}
P^{qq} &= \frac{4}{3}\left(\frac{1+x^2}{1-x}\right)_+\\
P^{qg} &= \frac{1}{2}\left[x^2+(1-x)^2\right]\\
P^{gq} &= \frac{4}{3}\left[\frac{1+(1-x)^2}{x}\right]\\
P^{gg} &= 6\left[\frac{1-x}{x}+x(1-x)+\left(\frac{x}{1-x}\right)_+\right]\\
&\quad -\left[\frac{1}{2}+\frac{n_f}{3}\right]\delta(1-x).
\end{aligned}\tag{46}$$

Note the 'plus prescription' for those functions which are singular as $x\to 1$:

$$\int_0^1 dx f(x)\left(g(x)\right)_+ = \int_0^1 dx[f(x)-f(1)]g(x). \tag{47}$$

The DGLAP equations can be solved analytically by defining *moments* (formally, Mellin transforms) of the structure functions, $M_n^{NS}=\langle F^{NS}\rangle_n\equiv\int_0^1 dx x^{n-1}F^{NS}$ etc. The convolution integral then becomes a simple product. Introducing the leading–order expression for the QCD coupling constant derived in Section 2.2,

$$\alpha_s(Q^2)=\frac{4\pi}{\beta_0\ln(Q^2/\Lambda^2)}, \tag{48}$$

one obtains, for the non–singlet solution,

$$M_n^{NS}(Q^2)=M_n^{NS}(Q_0^2)\left(\frac{\alpha_s(Q^2)}{\alpha_s(Q_0^2)}\right)^{-d_n}, \tag{49}$$

where $d_n = 2\langle P^{qq}\rangle_n/\beta_0$. Note that $d_1 = 0$ and that $d_n < 0$ for $n \geq 2$, which implies that the parton distributions decrease and increase with increasing Q^2 at large and small x respectively, as argued on physical grounds above. Solutions for the singlet and gluon moments can be found in a similar way, by first diagonalizing the coupled equations. In practice, it is often more convenient to solve the DGLAP equations numerically by iterating small steps in $\log Q^2$, starting from a set of 'input' parton distributions $f_i(x, Q_0^2)$. Figure 7 shows the same set of parton distribution functions as in Fig. 6, but now DGLAP–evolved to the much higher scale $\mu^2 = 10^4$ GeV^2, typical of measurements at the HERA ep collider. By comparing the two figures one can clearly see the shrinkage to small x as Q^2 increases.

The precision of contemporary deep inelastic data demands that the QCD predictions are calculated beyond leading order. This amounts to the replacements (shown schematically):

$$\begin{aligned} P(x) \quad &\rightarrow \quad P(x,Q^2) = P^{(0)}(x) + \frac{\alpha_s(Q^2)}{2\pi} P^{(1)}(x) + \ldots \\ F = x\sum_q e_q^2 q \quad &\rightarrow \quad F = x\sum_q (C_q * q + C_g * g) \end{aligned}$$

$$C_q = e_q^2\delta(1-x) + O(\alpha_s(Q^2)), \qquad C_g = O(\alpha_s(Q^2)). \tag{50}$$

The functions C_q and C_g are called coefficient functions. Beyond leading order, the definition of parton distributions (like the definition of α_s) becomes (factorization) scheme dependent (see Ref. [1] for a more detailed discussion). Different schemes have different coefficient and higher–order splitting functions, and correspondingly different parton distributions to render the (physical) structure functions scheme independent. In the 'DIS' scheme, for example, $C_q(x) = e_q^2\delta(1-x)$, $C_g = 0$. It is conventional nowadays to work in the $\overline{\rm MS}$ scheme, where for example

$$\begin{aligned} x^{-1}F_2^{\mu p}(x,Q^2) \;=\; & \sum_q e_q^2\, q^{\overline{\rm MS}}(x,Q^2) \\ & + \frac{\alpha_s^{\overline{\rm MS}}(Q^2)}{2\pi} \sum_q \int_x^1 \frac{dz}{z}; c_{2,q}(z)\, q^{\overline{\rm MS}}\left(\frac{x}{z},Q^2\right) \\ & + \left(\sum_q e_q^2\right) \frac{\alpha_s^{\overline{\rm MS}}(Q^2)}{2\pi} \int_x^1 \frac{dz}{z}\, c_{2,g}(z)\, g^{\overline{\rm MS}}\left(\frac{x}{z},Q^2\right) \\ & + O(\alpha_s^2)\,. \end{aligned} \tag{51}$$

The scaling violations predicted by perturbative QCD are clearly visible in the data. Figure 8 shows high precision data on the structure functions $F_2^{\nu N}$ and $xF_3^{\nu N}$ from the CCFR collaboration [23]. As expected, the slopes

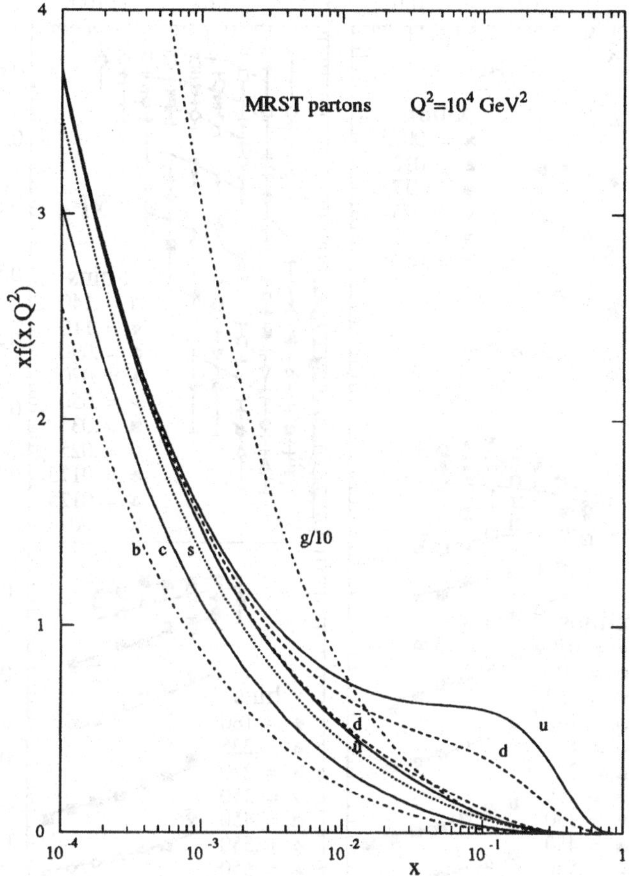

Figure 7. The same parton distribution functions as in Fig. 6, but now at $\mu^2 = 10^4$ GeV^2, from Ref. [21].

$\partial F_{2,3}/\partial \ln Q^2$ are negative at large x and positive at small x respectively. From data such as those shown in Fig. 8, the predictions of perturbative QCD for scaling violations (44) can be tested, and a precise measurement [23] of the strong coupling $\alpha_s(Q^2)$ can be made:

$$\mathrm{CCFR}(F_{2,3}) \ : \quad \alpha_s(M_Z^2) = 0.119 \pm 0.002(\mathrm{exp.}) \pm 0.001(\mathrm{HT}) \pm 0.004(\mathrm{scale}). \tag{52}$$

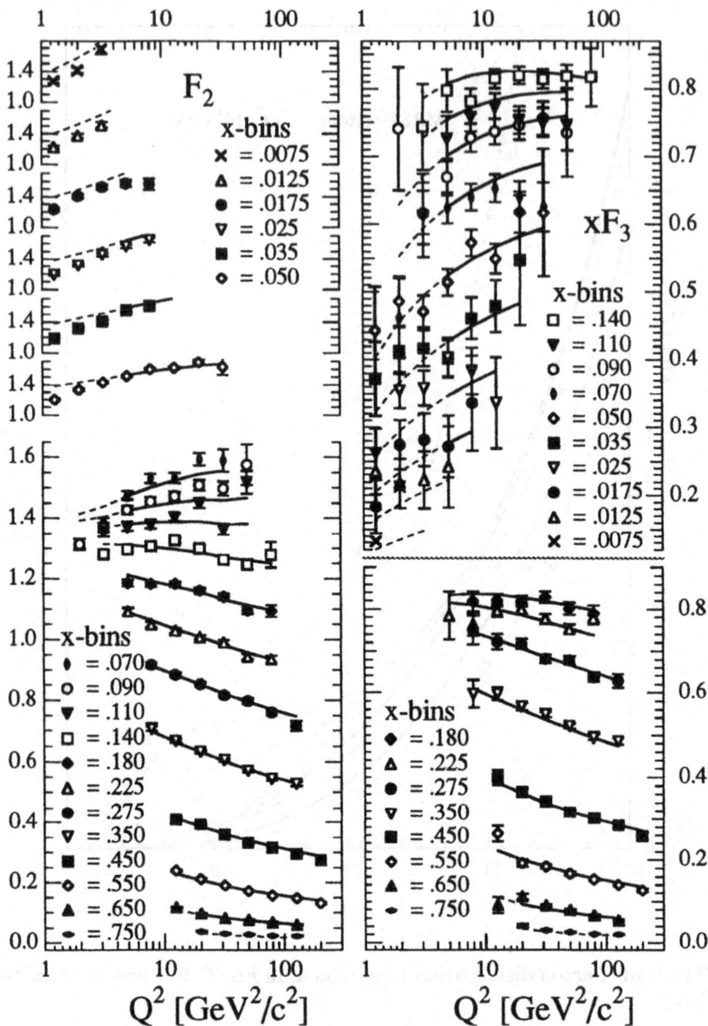

Figure 8. Measurements of the structure function $F_2^{\nu N}$ from the CCFR collaboration together with a NLO QCD fit, from Ref. [23].

The second error is from an estimate of the higher–twist contribution:[6]

$$F(x,Q^2) \;=\; F^{(2)}(x,Q^2) \;+\; \frac{F^{(4)}(x,Q^2)}{Q^2} \;+\; \dots\,, \tag{53}$$

[6]The superscripts on the right–hand side of (53) refer to the 'twist' = (dimension – spin) of the contributing operators.

using the model of Ref. [24], and the third is the scale dependence uncertainty. Notice that, except at large x, the Q^2 variation of F_2 is sensitive to the *a priori* unknown gluon distribution and there is potentially a strong α_s–gluon correlation. Non–singlet structure functions such as F_3 do not suffer from the gluon correlation problem (see Eq. (44)), but these are only measurable experimentally by constructing differences between cross sections, e.g. $\sigma^{\nu N} - \sigma^{\bar{\nu} N}$. This inevitably introduces additional systematic and statistical uncertainties. The α_s value (52) is one of the most precise determinations, see Fig. 1. It agrees perfectly with the values measured in e^+e^- annihilation, showing that α_s is indeed a universal parameter, independent of whether the short distance process is spacelike (DIS) or timelike (e^+e^-).

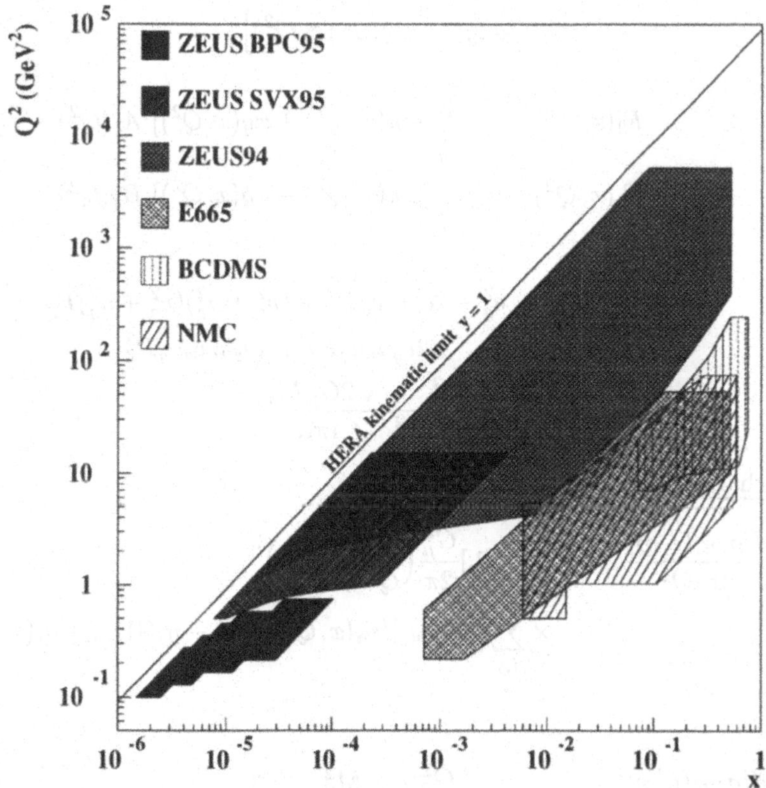

Figure 9. Regions in the $x - Q^2$ plane covered by various deep inelastic scattering experiments.

Deep inelastic fixed–target experiments measure quark distributions very accurately over a broad range in x ($\sim 0.01 - 0.8$) up to scales of order $\mu^2 \sim 200\ \mathrm{GeV}^2$. The HERA high-energy $e^{\pm}p$ collider, with $\sqrt{s} \sim 300$ GeV, is able to extend the x range down to very small values, and the Q^2 range up to very high values, see Fig. 9 [29]. At the high $Q^2 > \mathcal{O}(10^4\ \mathrm{GeV}^2)$ values measured at HERA, the W and Z contributions to the ep cross sections cannot be neglected. The neutral current cross section (38) must be modified to include Z exchange, and a corresponding charged current cross section (for $ep \to \nu X$) introduced. Ignoring the proton mass, the expressions are:

- neutral current

$$\begin{aligned} \frac{d^2\sigma_{NC}(e^{\pm}p)}{dxdQ^2} &= \frac{2\pi\alpha^2}{xQ^4}\Big[[1+(1-y)^2]F_2(x,Q^2) - y^2 F_L(x,Q^2) \\ &\quad \mp 2y(1-y)xF_3(x,Q^2)\Big] \end{aligned} \tag{54}$$

$$\begin{aligned} F_2(x,Q^2) &= \sum_q [xq(x,Q^2) + x\bar{q}(x,Q^2)]\ A_q(Q^2) \\ xF_3(x,Q^2) &= \sum_q [xq(x,Q^2) - x\bar{q}(x,Q^2)]\ B_q(Q^2) \end{aligned} \tag{55}$$

$$\begin{aligned} A_q(Q^2) &= e_q^2 - 2e_q v_e v_q P_Z + (v_e^2 + a_e^2)(v_q^2 + a_q^2)P_Z^2 \\ B_q(Q^2) &= -2e_q a_e a_q P_Z + 4v_e a_e v_q a_q P_Z^2 \\ P_Z &= \frac{Q^2}{Q^2 + M_Z^2}\ \frac{\sqrt{2}G_\mu M_Z^2}{4\pi\alpha} \end{aligned} \tag{56}$$

- charged current

$$\begin{aligned} \frac{d^2\sigma_{CC}(e^-p)}{dxdQ^2} &= [1-\mathcal{P}_e]\frac{G_\mu^2}{2\pi}\Big(\frac{M_W^2}{Q^2+M_W^2}\Big)^2 \\ &\quad \times \sum_{i,j}\Big[|V_{u_i d_j}|^2 u_i(x,Q^2) + (1-y)^2 |V_{u_j d_i}|^2 \bar{d}_i(x,Q^2)\Big] \end{aligned} \tag{57}$$

$$\begin{aligned} \frac{d^2\sigma_{CC}(e^+p)}{dxdQ^2} &= [1+\mathcal{P}_e]\frac{G_\mu^2}{2\pi}\Big(\frac{M_W^2}{Q^2+M_W^2}\Big)^2 \\ &\quad \times \sum_{i,j}\Big[|V_{u_i d_j}|^2 \bar{u}_i(x,Q^2) + (1-y)^2 |V_{u_j d_i}|^2 d_i(x,Q^2)\Big] \end{aligned} \tag{58}$$

From these expressions we see that (i) the charged current cross section is suppressed by $\mathcal{O}(Q^4)$ at small Q^2 where the neutral current cross section is dominated by photon exchange, and (ii) at very high $Q^2 \gg \mathcal{O}(M_V^2)$, the charged and neutral cross sections are of the same order. The HERA data confirm this behaviour: Fig. 10 shows the neutral and charged current cross sections for e^+p scattering at high Q^2 measured by ZEUS [30], together with the Standard Model predictions.

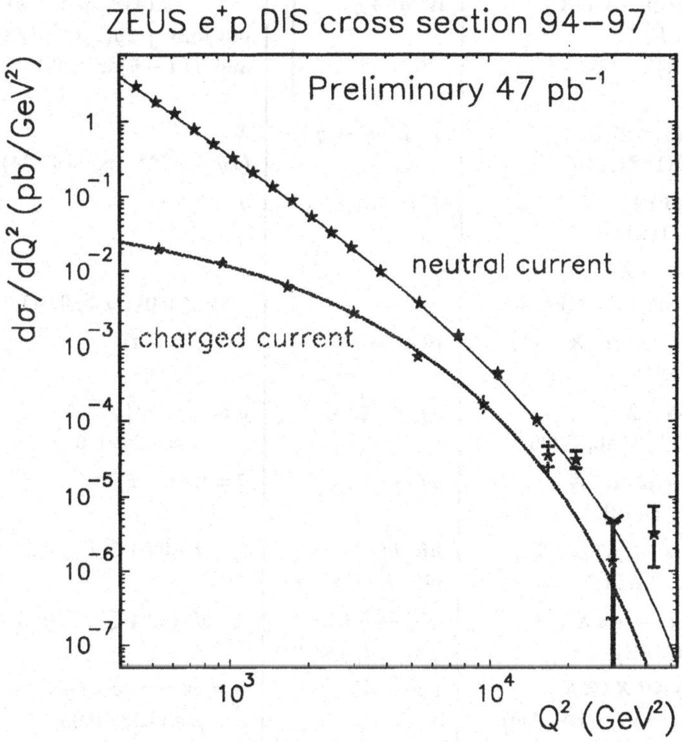

Figure 10. **Charged and neutral current DIS cross sections at high Q^2, as measured by the ZEUS collaborations in e^+p scattering at HERA.**

Parton distributions at some starting scale μ_0^2 are a byproduct of DGLAP fits to DIS data. These can then be evolved to higher μ^2 and used for hadron collider phenomenology. Instead of laboriously integrating the DGLAP equations each time a parton distribution is required, it is useful to have an analytic approximation, valid to a sufficient accuracy over a prescribed (x, μ^2) range. Such parametrizations are discussed in the following section.

Process/ Experiment	Leading order subprocess	Parton behaviour probed
DIS ($\mu N \to \mu X$) $F_2^{\mu p}, F_2^{\mu d}, F_2^{\mu n}/F_2^{\mu p}$ (SLAC, BCDMS, NMC, E665)*	$\gamma^* q \to q$	Four structure functions → $u + \bar{u}$ $d + \bar{d}$ $\bar{u} + \bar{d}$ s (assumed $= \bar{s}$), but only $\int xg(x, Q_0^2)dx \simeq 0.35$ and $\int(\bar{d} - \bar{u})dx \simeq 0.1$
DIS ($\nu N \to \mu X$) $F_2^{\nu N}, xF_3^{\nu N}$ (CCFR)*	$W^* q \to q'$	
DIS (small x) F_2^{ep} (H1, ZEUS)*	$\gamma^*(Z^*)q \to q$	λ ($x\bar{q} \sim x^{-\lambda_S}$, $xg \sim x^{-\lambda_g}$)
DIS (F_L) NMC, HERA	$\gamma^* g \to q\bar{q}$	g
$\ell N \to c\bar{c}X$ F_2^c (EMC; H1, ZEUS)*	$\gamma^* c \to c$	c ($x \gtrsim 0.01$; $x \lesssim 0.01$)
$\nu N \to \mu^+\mu^- X$ (CCFR)*	$W^* s \to c$ $\hookrightarrow \mu^+$	$s \approx \frac{1}{4}(\bar{u} + \bar{d})$
$pN \to \gamma X$ (WA70*, UA6, E706, ...)	$qg \to \gamma q$	g at $x \simeq 2p_T^\gamma/\sqrt{s} \to$ $x \approx 0.2 - 0.6$
$pN \to \mu^+\mu^- X$ (E605, E772)*	$q\bar{q} \to \gamma^*$	$\bar{q} = ...(1 - x)^{\eta_S}$
$pp, pn \to \mu^+\mu^- X$ (E866, NA51)*	$u\bar{u}, d\bar{d} \to \gamma^*$ $u\bar{d}, d\bar{u} \to \gamma^*$	$\bar{u} - \bar{d}$ ($0.04 \lesssim x \lesssim 0.3$)
$ep, en \to e\pi X$ (HERMES)	$\gamma^* q \to q$ with $q = u, d, \bar{u}, \bar{d}$	$\bar{u} - \bar{d}$ ($0.04 \lesssim x \lesssim 0.2$)
$p\bar{p} \to WX(ZX)$ (UA1, UA2; CDF, D0) **→ $\ell^\pm$ asym** (CDF)*	$ud \to W$	u, d at $x \simeq M_W/\sqrt{s} \to$ $x \approx 0.13$; 0.05 slope of u/d at $x \approx 0.05 - 0.1$
$p\bar{p} \to t\bar{t}X$ (CDF, D0)	$q\bar{q}, gg \to t\bar{t}$	q, g at $x \gtrsim 2m_t/\sqrt{s} \simeq 0.2$
$p\bar{p} \to$jet $+ X$ (CDF, D0)	$gg, qg, qq \to 2j$	q, g at $x \simeq 2E_T/\sqrt{s} \to$ $x \approx 0.05 - 0.5$

TABLE 4. Processes studied in the global MRST analysis [21] (* indicates data fitted).

4.3. PARTON DISTRIBUTIONS IN HADRONS

As we have seen in the previous sections, the distributions of quarks (and, indirectly via the DGLAP equations, gluons) in the proton are determined by values of the structure functions $F_i(x, Q^2)$ measured in the various deep inelastic scattering experiments. It is relevant to ask why we should devote so much effort to the study of the distributions of partons in the proton. There are two main reasons, one experimental and one theoretical. First, a detailed knowledge of parton distribution functions (pdfs) is an essential ingredient in all 'hard' interactions involving protons, and so they are needed to estimate the production rates of the various hard processes that may occur at present and future colliders. Second, the parton structure, as encoded in the f_i, is interesting in its own right. In particular, novel perturbative QCD effects are expected to become apparent at small x. The reason is that at small x the sum over soft gluons emitted off the incoming parton leads to a power series in $\alpha_s \ln(1/x)$, which on resummation, via the Lipatov (or BFKL) equation [25], suggests that the gluon and quark singlet distributions behave as

$$xg,\ xq_S \ \sim\ x^{-\lambda} \tag{59}$$

as $x \to 0$, with $\lambda = 12 \ln 2\alpha_s/\pi$ predicted to be about 0.5. Such an increase in $xg(x, \mu^2)$ and $xq_S(x, \mu^2)$ as x decreases cannot go on indefinitely. If the density of gluons becomes too large they can no longer be treated as free partons, and the effects of recombination or shadowing must be included. The 'naive' BFKL predictions (59) for the small x behaviour of the parton distributions is valid only for *asymptotically small* x values. It is far from clear whether the values attainable at HERA (see Fig. 9) are in fact small enough for the leading behaviour to be observable. Indeed, standard NLO DGLAP evolution provides a satisfactory explanation of the observed small–x behaviour, with approximately flat (i.e. $xf_i \sim x^0$) distributions at a starting scale $\mu_0^2 \sim 1\ \mathrm{GeV}^2$, see the discussion in Ref. [21] for example.

In fact, approximately flat starting distributions are in line with long-standing non–perturbative Regge arguments for structure functions in the $x \to 0$ limit. According to Regge theory, the high–energy behaviour of an elastic hadron scattering amplitude is controlled by a sequence of Regge trajectories corresponding to the exchange of families of particles with different spin, see for example Ref. [31]. In the small–x limit, the quark–proton amplitude $\mathcal{A}_{qp}$ in deep inelastic scattering is probed at high energy, $s \sim Q^2/x$, for which we would expect

$$\mathcal{A}_{qp} \sim \beta_{\mathcal{P}} s^{\alpha_{\mathcal{P}}-1} + \beta_{\mathcal{R}} s^{\alpha_{\mathcal{R}}-1} + \dots\ , \tag{60}$$

where the leading trajectories are

$$\alpha_{\mathcal{P}} \ \simeq\ 1 \qquad \text{pomeron } \mathcal{P}$$

$$\alpha_{\mathcal{R}} \simeq \frac{1}{2} \qquad \rho, \omega, a_2, f_2, \ldots \tag{61}$$

Inserting this behaviour into the parton–model calculation of the F_2 structure function gives the leading small–x behaviour

$$F_2(x) \sim \beta_{\mathcal{P}} x^{1-\alpha_{\mathcal{P}}} + \beta_{\mathcal{R}} x^{1-\alpha_{\mathcal{R}}} \,. \tag{62}$$

We may interpret the two terms in (62) as the contributions to the structure function from the flavour–singlet quark sea, with behaviour determined by the leading 'pomeron' trajectory $\alpha_{\mathcal{P}}$, and from the flavour–non–singlet valence quarks, with behaviour controlled by the 'Reggeon' trajectory $\alpha_{\mathcal{R}}$. We would likewise expect that the behaviour of the gluon distribution at small x is also determined by the pomeron trajectory, yielding the predictions

$$q_S,\, g \sim x^{-1}, \qquad q_V \sim x^{-\frac{1}{2}} \tag{63}$$

in the $x \to 0$ limit, or equivalently

$$F_2^p \sim x^0, \qquad F_2^p - F_2^n \sim x^{\frac{1}{2}} \,. \tag{64}$$

A detailed analysis of small x, modest Q^2 structure function measurements at HERA collider and fixed–target energies shows that they are indeed approximately consistent with the predictions of Regge theory, see for example the recent ZEUS measurements [29] of F_2 at small x in Fig. 11. However the data also show an apparent steepening of the behaviour at small x as Q^2 increases, exactly as expected from perturbative QCD DGLAP evolution as described above. Therefore, although much has been written about the theoretical 'BFKL' behaviour of the small–x parton distributions,[7] there is as yet no compelling experimental evidence and so we shall not discuss this further here.

There are currently three collaborations producing sets of parton distributions which are widely used in high-energy collider phenomenology: MRS (Martin–Roberts–Stirling), CTEQ (Collaboration for Theoretical and Experimental Studies in Quantum Chromodynamics) and GRV (Glück–Reya–Vogt) (see, respectively, [21, 26, 27] and references therein). The first two of these use the concept of 'global fits' to determine each parton distribution as accurately as possible from high-precision data on deep inelastic structure functions and other hard scattering processes. The GRV analysis is in the context of the 'dynamical parton model' [28] in which the partons evolve from valence–like distributions at a low Q^2 scale. These starting distributions are tuned to fit the data at higher Q^2.

[7] For a review of small–x physics, and a list of references, see for example Ref. [1].

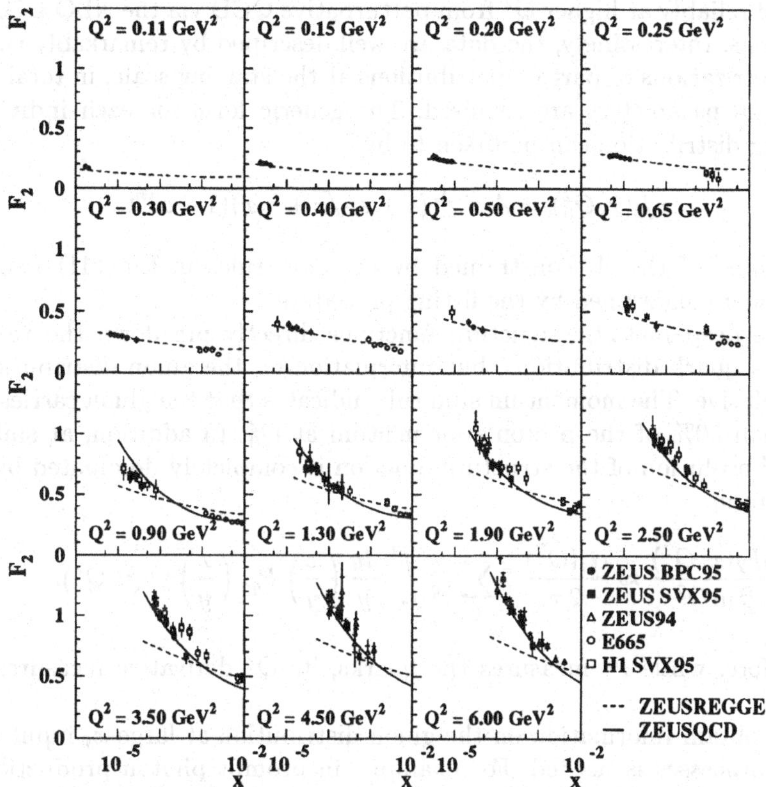

Figure 11. Low-Q^2 F_2 data for different Q^2 bins together with a Regge fit (dashed curves) to the ZEUS BPC95 data [29]. Also shown at larger values of Q^2 is the ZEUS NLO DGLAP fit (full curves).

The last few years have seen a spectacular improvement in the precision and in the kinematic range of the experimental measurements of deep inelastic and related hard scattering processes. As a consequence the pdfs are much better known, with tight constraints on the gluon and the quark sea for Bjorken x as low as 10^{-4}. In what follows we will summarize the recent MRST pdf analysis of Ref. [21]. This is the most recent of the global analyses, and takes into account all the new information as well as incorporating new theoretical developments in the heavy quark sector.

Table 4 illustrates the variety of data used in the recent MRST analysis [21]. The basic procedure is to parametrize the f_i at a sufficiently large 'starting scale' ($Q_0^2 = 1$ GeV^2 for MRST) so that the $f_i(x, Q^2)$ can be calculated reliably at higher Q^2 from perturbative QCD via the NLO DGLAP equations. Interestingly, the data are well described by remarkably simple parametrizations of parton distributions at the starting scale; in total only about 20 parameters are required. The generic form for each individual starting distribution can be taken to be

$$xf_i(x, Q_0^2) = A_i x^{-\lambda_i}(1 + \epsilon_i\sqrt{x} + \gamma_i x)(1 - x)^{\eta_i}, \tag{65}$$

with some of the A_i constrained by the sum rules in Eq. (41) and the remainder constrained by the fitting procedure.

The deep-inelastic structure functions directly pin down the valence and sea quark distributions, but information on the gluon distribution is more elusive. The momentum sum rule indicates that the gluon carries just less than 50% of the proton's momentum at Q_0^2. In addition, at small x the Q^2 evolution of the structure function is completely dominated by the gluon term:

$$\frac{\partial F_2(x, Q^2)}{\partial \ln Q^2} \approx \frac{\alpha_s(Q^2)}{2\pi} 2\sum_q e_q^2 \int_x^1 \frac{dy}{y} \left(\frac{x}{y}\right) P_{qg}\left(\frac{x}{y}\right) yg(y, Q^2). \tag{66}$$

Therefore, while F_2 measures the quarks, its Q^2 derivative measures the gluon.

To obtain information on the gluon distribution at large x, input from other processes is needed. For example, in prompt photon production in hadron-hadron (pN) collisions the gluon enters at leading order via the QCD subprocess $gq \to \gamma q$, in contrast to $p\bar{p} \to \gamma X$ where the annihilation process $q\bar{q} \to \gamma g$ is much more important. The relevant data are from the WA70 and E706 collaborations [32, 33] which determine the gluon in the region $x \sim 0.2 - 0.5$. Combined with the momentum sum rule constraint, this gives a reasonable measurement of the gluon at large x, see Fig. 12, although additional assumptions are needed concerning the 'intrinsic transverse momentum' distribution of the partons in the proton, see for example the discussion in [21]. Data on the Drell–Yan $pN \to \mu^+\mu^- X$ process, which is mediated at LO by $q_{\text{val}}\bar{q}_{\text{sea}} \to \gamma^*$, constrain the large-$x$ $(1 - x)^{\eta_s}$ behaviour of the sea quark distributions, see Section 5. Finally data on W and Z production at $\bar{p}p$ colliders impose tight constraints on the u and d distributions, particularly when the accurate measurements of $F_2^{\mu n}/F_2^{\mu p}$ have to be fitted simultaneously.

As mentioned already, a feature of recent parton determinations is the marked difference between the $\bar{u}$ and $\bar{d}$ pdfs, see Fig. 6, motivated by new precise experimental measurements. The DIS structure function measurements (of $F_2^{\mu p}, F_2^{\mu n}, F_2^{\nu N}$ and $xF_3^{\nu N}$) determine $(\bar{u} + \bar{d})$, but not $(\bar{u} - \bar{d})$.

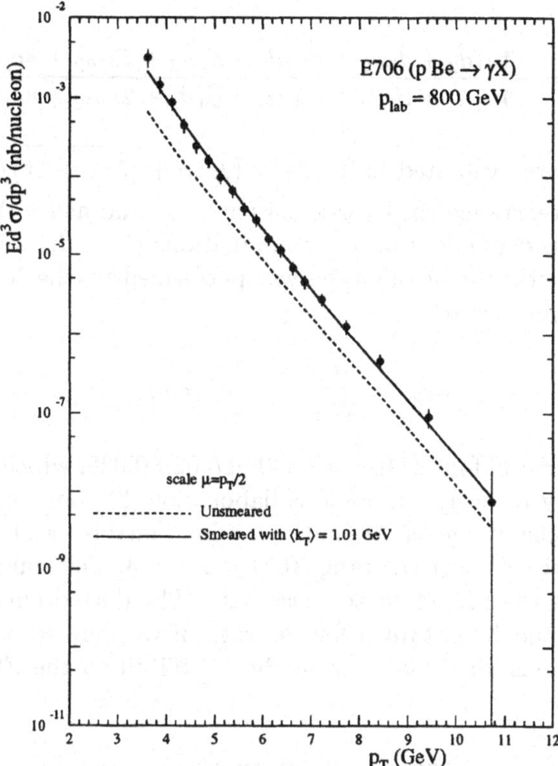

Figure 12. Comparison of the E706 prompt photon data [33] data at 800 GeV with the MRST parton set [21]. The scale is chosen to be $p_T/2$ and the effect of including parton transverse momentum is shown. These data are used to constrain the large-x gluon.

Historically the first indication of the $\bar{u} \neq \bar{d}$ flavour asymmetry of the sea came from the evaluation of the Gottfried sum

$$I_{\rm GS} \equiv \int_0^1 \frac{dx}{x} \left(F_2^{\mu p} - F_2^{\mu n}\right) \tag{67}$$

by NMC [34]. This gives information on the integral of $\bar{u} - \bar{d}$ and indicates that, on average, $\bar{d}$ is greater than $\bar{u}$.

For a direct determination of $\bar{u} - \bar{d}$ consider, for example, the asymmetry of Drell–Yan production in pp and pn collisions [35]

$$A_{\rm DY} \equiv \frac{\sigma_{pp} - \sigma_{pn}}{\sigma_{pp} + \sigma_{pn}} = \frac{1-r}{1+r}, \tag{68}$$

where $r = \sigma_{pn}/\sigma_{pp}$ and where $\sigma \equiv d^2\sigma/dMdx_F$ with M and x_F being the invariant mass and the Feynman x of the produced lepton pair. At leading order we have

$$r \equiv \frac{\sigma_{pn}}{\sigma_{pp}} = \frac{(4u_1\bar{d}_2 + d_1\bar{u}_2 + 4\bar{u}_1 d_2 + \bar{d}_1 u_2 + 2s_1 s_2 + 8c_1 c_2)}{(4u_1\bar{u}_2 + d_1\bar{d}_2 + 4\bar{u}_1 u_2 + \bar{d}_1 d_2 + 2s_1 s_2 + 8c_1 c_2)} \tag{69}$$

where the pdfs are evaluated at $x_1, x_2 = (\pm x_F + \sqrt{x_F^2 + 4\tau})/2$, with $\tau = M^2/s$. We may rearrange the expression for $1 - r$, and hence that for $A_{\rm DY}$, to show that it is dependent on the combinations $(\bar{u}_1 - \bar{d}_1)$ and $(\bar{u}_2 - \bar{d}_2)$.

The first experiment of this type was performed by the NA51 collaboration [36] who measured

$$R_{dp} \equiv \frac{\sigma_{pd}}{2\sigma_{pp}} = \frac{1}{2}(1 + r) \tag{70}$$

at $x_1 = x_2 = 0.18$ and found $A_{\rm DY} = -0.09 \pm 0.02 \pm 0.025$, which corresponds to $\bar{d}/\bar{u} \simeq 2$. Very recently the E866 collaboration [37] have measured R_{dp} over a much wider range of M and x_F, which enables a study of the x dependence of $(\bar{u} - \bar{d})$ over the range $0.04 < x < 0.3$. The continuous curve in Fig. 13 shows the MRST fit to these data. The dotted curve shows the values which would be obtained for the ratio if we were to set $\bar{u}$ equal to $\bar{d}$. The implications for $\bar{d}$ and $\bar{u}$ from the MRST fit to the E866 data are shown in Fig. 6.

5. Hard Processes in Hadronic Collisions

5.1. INTRODUCTION

It was first pointed out by Drell and Yan [38] that parton model ideas developed for deep inelastic scattering could be extended to certain processes in hadron-hadron collisions. The paradigm process was the production of a massive lepton pair by quark-antiquark annihilation — the Drell–Yan process — and the hadronic cross section σ was to be obtained by weighting the subprocess cross section $\hat{\sigma}$ for $q\bar{q} \to \mu^+\mu^-$ with the parton distribution functions $f_{q/A}(x)$ extracted from deep inelastic scattering:

$$\sigma_{AB} = \int dx_a dx_b \; f_{a/A}(x_a) f_{b/B}(x_b) \, \hat{\sigma}_{ab \to X} \; , \tag{71}$$

where for the Drell–Yan process, $X = l^+l^-$ and $ab = q\bar{q},\ \bar{q}q$. The domain of validity is the asymptotic 'scaling' limit (the analogue of the Bjorken scaling limit in deep inelastic scattering) $M_X \equiv M^2_{l^+l^-}, s \to \infty$, $\tau = M^2_{l^+l^-}/s$ fixed. The good agreement between theoretical predictions and the measured cross sections provided confirmation of the parton model formalism, and allowed for the first time a rigorous, quantitative treatment of hadronic

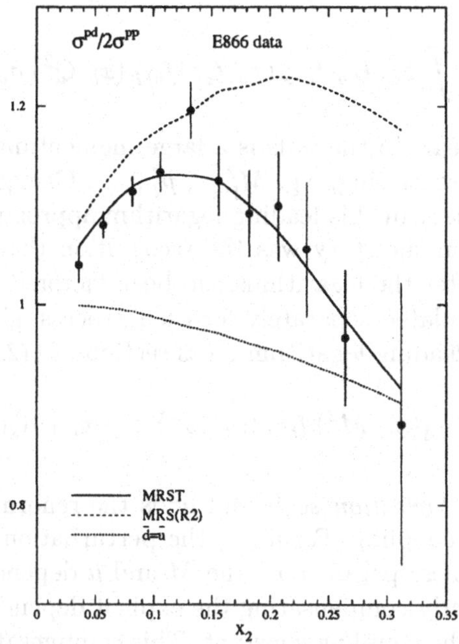

Figure 13. The continuous curve is the MRST description of the E866 [37] data for the ratio of the cross sections for hadroproduction of dileptons for proton and deuterium targets versus x_2, the fractional momentum of the parton in the target. The other curves are for comparison only.

cross sections. Studies were extended to other 'hard scattering' processes, for example the production of hadrons and photons with large transverse momentum, with equally successful results. Problems, however, appeared to arise when perturbative corrections from real and virtual gluon emission were calculated. Large logarithms from gluons emitted collinear with the incoming quarks appeared to spoil the convergence of the perturbative expansion. It was subsequently realised that these logarithms were the same as those which arise in deep inelastic scattering structure function calculations (see Section 4), and could therefore be absorbed, via the DGLAP equations, in the definition of the parton distributions, giving rise to logarithmic violations of scaling. The key point was that *all* logarithms appearing in the Drell–Yan corrections could be factored into renormalized parton distributions in this way, and *factorization theorems* which showed that this was a general feature of hard scattering processes were derived [39]. Taking into account the leading logarithm corrections, Eq. (71) simply becomes:

$$\sigma_{AB} = \int dx_a dx_b \, f_{a/A}(x_a, Q^2) f_{b/B}(x_b, Q^2) \, \hat{\sigma}_{ab \to X} \, . \tag{72}$$

The Q^2 which appears in the pdfs is a large momentum scale which characterizes the hard scattering, e.g. $M^2_{l^+l^-}$, p^2_T, ... Changes to the Q^2 scale of $\mathcal{O}(1)$ are equivalent in this leading logarithm approximation.

The final step in the story was the recognition that the *finite* corrections left behind after the logarithms had been factored were not universal and had to be calculated separately for each process, giving rise to $\mathcal{O}(\alpha_s)$ corrections to the leading logarithm cross section of (72). Schematically

$$\sigma_{AB} = \int dx_a dx_b \, f_{a/A}(x_a, M^2) \, f_{b/B}(x_b, M^2) \, \times \, [\hat{\sigma}_0 + \alpha_s(\mu^2) \, \hat{\sigma}_1 + ...]_{ab \to X} \, . \tag{73}$$

Here M^2 is the *factorization scale* and μ^2 is the renormalization scale for the QCD running coupling. Formally, the perturbation series is invariant under changes in these parameters, the M and μ dependence of the coefficients, e.g. $\hat{\sigma}_1$, exactly compensating the explicit dependence of the parton distributions and the coupling constant. This compensation becomes more exact as more terms are included in the perturbation series. To avoid unnaturally large logarithms reappearing in the perturbation series it is sensible to choose M and μ values of the order of the typical momentum scales of the hard scattering process, and $M = \mu$ is also often assumed.

In general, all the important hadronic processes have now been calculated to next–to–leading order (NLO), i.e. up to and including the $\hat{\sigma}_1$ terms. One process — the Drell–Yan process — is even calculated to one order higher (see below). This allows a very high degree of precision in a wide variety of processes. In many cases, the residual renormalization and factorization scale dependence is weak, and the precision of the theoretical prediction is limited only by uncertainties in the knowledge of the parton distributions.

What, then, are the most important applications of this formalism? One can, for example, attempt to measure α_s, particularly from those processes involving α_s at leading order, i.e. in $\hat{\sigma}_0$, and also study final–state QCD jets in parton scattering processes. One can also obtain information on parton distributions, particularly the gluon and sea quark distributions, complementary to that from deep inelastic scattering, as described in the previous section. However, perhaps the most important application is the prediction of various Standard Model and New Physics cross sections at high energy colliders such as the Tevatron ($p\bar{p}$) and LHC (pp). There are many examples of situations where the ability to detect a signal for new particle production depends crucially on the accuracy of the Standard Model background estimate. For reference, we show in Fig. 14 the predictions for some important Standard Model cross sections at $p\bar{p}$ and pp colliders, calculated at next-to-leading order in QCD perturbation theory using the latest MRST pdfs [21].[8]

We have already mentioned that the Drell–Yan process is the paradigm hadron–collider hard scattering process, and so we will discuss this in some detail in what follows. Many of the remarks apply also to other processes, in particular those shown in Fig. 14, although of course the higher–order corrections and the initial–state parton combinations are process dependent.

5.2. THE DRELL–YAN PROCESS

The Drell–Yan process is the production of a lepton pair (e^+e^- or $\mu^+\mu^-$ in practice) of large invariant mass M in hadron-hadron collisions by the mechanism of quark–antiquark annihilation [38]. In the basic Drell–Yan mechanism, a quark and antiquark annihilate to produce a virtual photon, $q\bar{q} \to \gamma^* \to l^+l^-$. At high energy colliders, such as the Tevatron and LHC, there is of course sufficient centre–of–mass energy for the production of on–shell W and Z bosons as well, see below. The cross section for quark-antiquark annihilation to a lepton pair via an intermediate massive photon is easily obtained from the fundamental QED $e^+e^- \to \mu^+\mu^-$ cross section, with the addition of the appropriate colour and charge factors.

$$\sigma(q\bar{q} \to e^+e^-) = \frac{4\pi\alpha^2}{3\hat{s}}\frac{1}{N}Q_q^2, \tag{74}$$

where Q_q is the quark charge: $Q_u = +2/3$, $Q_d = -1/3$ *etc.* The overall colour factor of $1/N = 1/3$ is due to the fact that only when the colour of the quark matches with the colour of the antiquark can annihilation into a colour–singlet final state take place.

In general, the incoming quark and antiquark will have a spectrum of centre–of–mass energies $\sqrt{\hat{s}}$, and so it is more appropriate to consider the differential mass distribution:

$$\frac{d\hat{\sigma}}{dM^2} = \frac{\hat{\sigma}_0}{N}Q_q^2\delta(\hat{s} - M^2), \quad \hat{\sigma}_0 = \frac{4\pi\alpha^2}{3M^2}, \tag{75}$$

where M is the mass of the lepton pair. In the centre–of–mass frame of the two hadrons, the components of momenta of the incoming partons may be written as

$$p_1^\mu \;=\; \frac{\sqrt{s}}{2}(x_1, 0, 0, x_1)$$

[8] Also shown, for comparison, is the *total* cross section calculated using a (non–perturbative) Regge–based model [2].

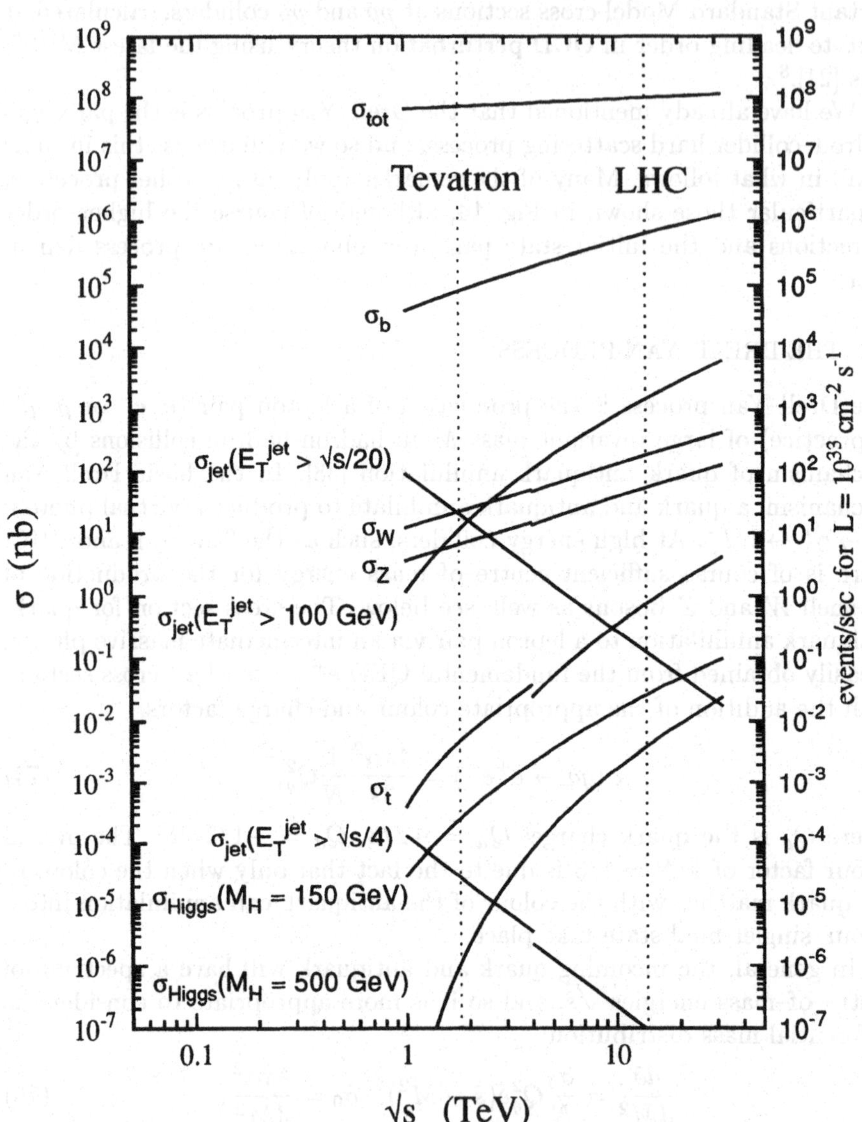

Figure 14. **Standard Model cross sections at the Tevatron and LHC colliders, calculated using MRST partons.**

$$p_2^\mu = \frac{\sqrt{s}}{2}(x_2, 0, 0, -x_2) . \tag{76}$$

The square of the parton centre–of–mass energy $\hat{s}$ is related to the corresponding hadronic quantity by $\hat{s} = x_1 x_2 s$. Folding in the momentum distribution functions for the initial state quarks and antiquarks in the beam and target gives the hadronic cross section:

$$\begin{aligned} \frac{d\sigma}{dM^2} &= \frac{\hat{\sigma}_0}{N} \int_0^1 dx_1 dx_2 \delta(x_1 x_2 s - M^2) \\ &\times \Big[\sum_k Q_k^2 (q_k(x_1, M^2) \bar{q}_k(x_2, M^2) + [1 \leftrightarrow 2]) \Big] . \end{aligned} \tag{77}$$

Note that the virtual photon is a *timelike* ($Q^2 > 0$) probe of the hadronic structure.

Apart from the mild logarithmic M^2 dependence in the distribution functions, the lepton–pair cross section exhibits *scaling* in the variable $\tau = M^2/s$:

$$\begin{aligned} M^3 \frac{d\sigma}{dM} &= \frac{8\pi\alpha^2\tau}{3N} \int_0^1 dx_1 dx_2 \delta(x_1 x_2 - \tau) \\ &\times \Big[\sum_k Q_k^2 (q_k(x_1, M^2) \bar{q}_k(x_2, M^2) + [1 \leftrightarrow 2]) \Big] \\ &= F(\tau, M^2) . \end{aligned} \tag{78}$$

From (76), the rapidity of the produced lepton pair is found to be $y = 1/2 \ln(x_1/x_2)$, and hence

$$x_1 = \sqrt{\tau}\, e^y , \qquad x_2 = \sqrt{\tau}\, e^{-y}. \tag{79}$$

The double–differential cross section is therefore

$$\frac{d\sigma}{dM^2 dy} = \frac{\hat{\sigma}_0}{Ns} \Big[\sum_k Q_k^2 (q_k(x_1, M^2) \bar{q}_k(x_2, M^2) + [1 \leftrightarrow 2]) \Big]. \tag{80}$$

with x_1 and x_2 given by (79). By measuring the distribution in the rapidity and mass of the lepton pair one can in principle directly measure the quark distribution functions of the colliding hadrons, see below. This is particularly important for pion distributions, which are not accessible from deep inelastic scattering.

Another variable which is sometimes used is the longitudinal momentum fraction of the lepton pair $x = 2p_L/\sqrt{s}$. In the parton model, it follows from (76) that

$$x = x_1 - x_2 , \tag{81}$$

which leads to (*cf.* (79))

$$x_1 = \frac{1}{2}\left(x + \sqrt{x^2 - 4\tau}\right), \qquad x_2 = \frac{1}{2}\left(x - \sqrt{x^2 - 4\tau}\right). \qquad (82)$$

Both the cross sections $d\sigma/dM^2dy$ and $d\sigma/dM^2dx$ can therefore be used to probe the parton distributions. Note also that the ranges of the variables y and x are obtained by requiring $x_1, x_2 \leq 1$:

$$-\frac{1}{2}\log\frac{1}{\tau} \leq y \leq \frac{1}{2}\log\frac{1}{\tau}, \qquad -1+\tau \leq x \leq 1-\tau. \qquad (83)$$

As mentioned in the introduction to this section, in QCD there exists a systematic procedure for calculating the perturbative corrections to all orders. The next-to-leading order corrections are obtained from one-gluon real and virtual emission diagrams:

$$\begin{aligned}\frac{d\sigma}{dM^2} = \; & \frac{\sigma_0}{Ns}\int_0^1 dx_1 dx_2 dz \delta(x_1 x_2 z - \tau) \\ & \Big\{\Big[\sum_k Q_k^2(q_k(x_1,\mu^2)\bar{q}_k(x_2,\mu^2) + [1 \leftrightarrow 2])\Big] \\ & \times\Big[\delta(1-z) + \frac{\alpha_s(\mu^2)}{2\pi} f_q(z)\Big] \\ & +\Big[\sum_k Q_k^2(g(x_1,\mu^2)(q_k(x_2,\mu^2) + \bar{q}_k(x_2,\mu^2)) \\ & +[1 \leftrightarrow 2])\Big]\Big[\frac{\alpha_s(\mu^2)}{2\pi} f_g(z)\Big]\Big\}, \end{aligned} \qquad (84)$$

where μ is the (arbitrary) factorization/renormalization scale. Explicit expressions for the f_q and f_g correction terms [40] can be found, for example, in Ref. [1]. The $O(\alpha_s^2)$ corrections to $d\sigma/dM^2$ have also been calculated [41], but the expressions are again too cumbersome to be presented here.

The size of the perturbative corrections depends on the lepton-pair mass and on the overall centre-of-mass energy. At fixed-target energies and masses the correction is generally large and positive, of order 50% or more. In this regime of relatively large τ, the (negative) contribution from the quark-gluon scattering terms in (84) is quite small. However at $p\bar{p}$ and pp collider energies, where τ is much smaller, the f_g term is more important and the overall correction is smaller.

Several important pieces of information can be obtained from Drell–Yan data. Low-mass lepton-pair production in high energy hadron collisions is, at least in principle, sensitive to the small x behaviour of the parton distributions. In pp or pN collisions the cross section is proportional to the sea–quark distribution, $\bar{q}(x, Q^2)$. This provides complementary information to deep inelastic scattering, and in fact Drell–Yan data can be used to constrain the sea-quark distributions in global parton distribution fits.

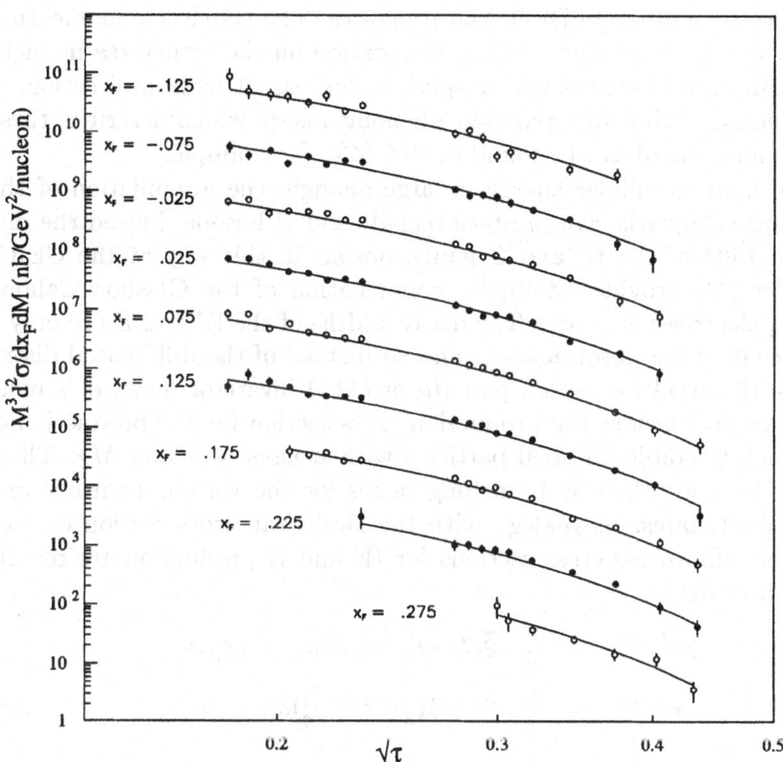

Figure 15. Hadroproduction of dileptons computed from the MRST parton set compared with the E605 data [42]. The theory curves include an additional K' factor of 0.9. No correction for the heavy target has been made. The scale on the left–hand axis is appropriate for the theory and data at $x_F = -0.125$. For display purposes the normalization is then decreased by a factor of ten for each step upwards in x_F.

As an example, Fig. 15, from Ref. [21], shows a comparison of data from the E605 collaboration [42] on the cross section $M^3 d^2\sigma/dM dx_F$ for $p\,\mathrm{Cu} \to \mu^+\mu^-$ at $p_{\mathrm{lab}} = 800$ GeV/c ($\sqrt{s} = 38.8$ GeV) with theoretical (NLO QCD) predictions calculated at next–to–leading order. The data are used in the global MRST fit to constrain the sea quarks in the interval $0.15 \lesssim x \lesssim 0.4$. The factorization and renormalization scales are here set equal to the invariant mass M of the lepton pair, and an overall phenomenological normalization parameter, which allows for possible higher-order effects, is included.

Other important information can be obtained from Drell–Yan cross section measurements. The distributions of quarks in *pions* can be extracted from data in πp and πN collisions. The 'EMC effect (see Ref. [20]) — the apparent difference between quark distributions in light and heavy nuclei — can also be studied in Drell–Yan processes. The transverse momentum of the lepton pair also gives direct information on the transverse momentum distribution of quarks with respect to the parent hadron direction. A comprehensive review of Drell–Yan phenomenology which describes these issues in more detail can be found in Ref. [43], for example.

If the hadron collider energy is large enough, the annihilation of the quarks and antiquarks can produce real W and Z bosons. Indeed the discovery in 1983 of the W and Z gauge bosons in this way at the CERN $p\bar{p}$ collider [44] provided dramatic confirmation of the Glashow–Salam–Weinberg electroweak model. The decay widths of the W and Z are only a few per cent of the boson masses, and so instead of the differential distribution in the resulting lepton pair ($l\nu$ or l^+l^-) invariant mass, it is more appropriate to consider the production cross section for the production of approximately stable on–shell particles with masses M_W and M_Z. These can then be multiplied by branching ratios for the various hadronic and leptonic final states. In analogy with the Drell–Yan cross section derived above, the subprocess cross sections for W and Z production are readily calculated to be

$$\begin{aligned} \hat{\sigma}^{q\bar{q}'\to W} &= \frac{\pi}{3}\sqrt{2}G_F M_W^2 |V_{qq'}|^2 \delta(\hat{s} - M_W^2) \\ \hat{\sigma}^{q\bar{q}\to Z} &= \frac{\pi}{3}\sqrt{2}G_F M_Z^2 (v_q^2 + a_q^2) \delta(\hat{s} - M_Z^2), \end{aligned} \tag{85}$$

where $V_{qq'}$ is the appropriate Cabibbo–Kobayashi–Maskawa matrix element, and v_q (a_q) is the vector (axial vector) coupling of the Z to the quarks. The $O(\alpha_s)$ perturbative QCD correction to the W and Z cross sections is the same as the Drell–Yan correction (for a photon of the same mass) discussed in the previous section — the gluon is 'flavour blind' and couples in the same way to the annihilating quark and antiquark.

As already noted, these cross sections have now been calculated to next–to–next–to–leading order (i.e. $O(\alpha_s^2)$) [41]. Figure 16 shows the cross sections for $W^\pm$ and Z^0 production as a function of the collider energy $\sqrt{s}$. The curves are calculated using the results of Ref. [41] (in the $\overline{\rm MS}$ scheme) with MRST parton distributions [21] and the renormalization and factorization scales $M = \mu = M_W, M_Z$. The data points are from UA1 [45], UA2 [46], CDF [47] and D0 [48] at $\sqrt{s} = 630$ GeV and 1.8 TeV. The net effect of

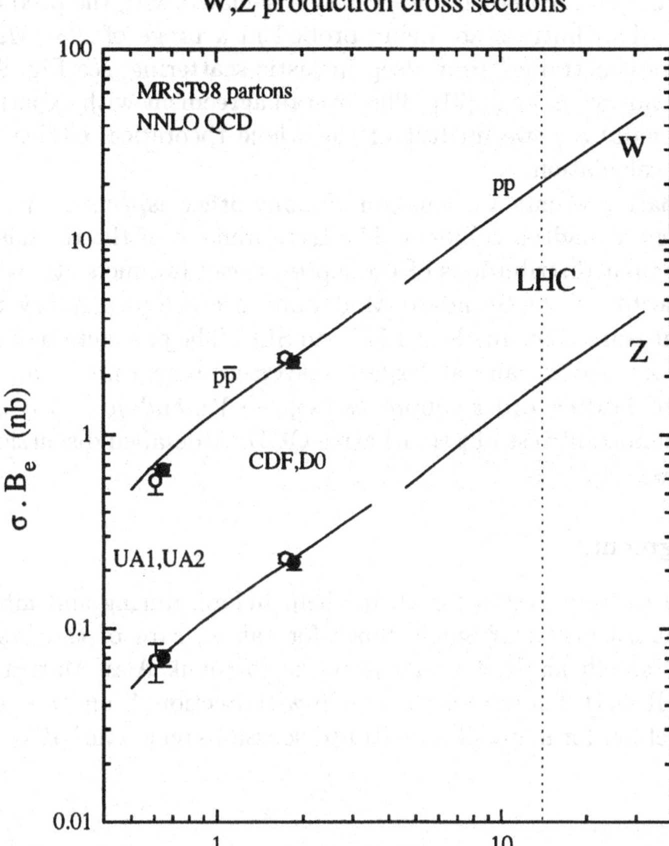

Figure 16. Theoretical (NNLO QCD) predictions for the $W^{\pm}$ and Z^0 total production cross sections in pp and $p\bar{p}$ collisions, as a function of $\sqrt{s}$, with data from UA1 [45], UA2 [46], CDF [47] and D0 [48].

the NLO and NNLO corrections is to increase the lowest order cross section by about 30%. The NLLO correction is significantly smaller than the NLO correction, due to a partial cancellation between the positive second order corrections involving the $q\bar{q}$ initial state and the negative corrections from the qg initial state. Perhaps the most important point to note from Fig. 16 is that, aside from unknown (and presumably small) $O(\alpha_s^3)$ corrections, there is virtually no theoretical uncertainty associated with the predictions — the parton distributions are being probed in a range of $x \sim M_W/\sqrt{s}$ where they are constrained from deep inelastic scattering, see Fig. 9, and the scale dependence is weak [21]. This overall agreement with experiment, therefore, provides a powerful test of the whole theoretical edifice which goes into the calculation.

Lack of space prevents a discussion of many other aspects of W and Z phenomenology at hadron colliders. The measurement of the W mass and width, the angular distributions of the lepton decay products etc. test the electroweak sector of the Standard Model and are complementary to the precision Z measurements made at LEP and SLC. The production of Drell–Yan (γ^*, W, Z) lepton pairs at large transverse momentum — mediated by the next–to–leading–order subprocesses $q\bar{q} \to Vg$ and $qg \to Vq$ — also provides an important test of perturbative QCD. A detailed discussion can be found in Ref. [1].

Acknowledgments

I am grateful to Tom Ferbel for all his help before, during and after the Institute. Thanks also to Connie Jones for taking care of the practical arrangements so efficiently. I would also like to thank Alan Martin, Dick Roberts and Robert Thorne for their help with Section 4, and Keith Ellis and Bryan Webber for many illuminating discussions on perturbative QCD.

References

1. *QCD and Collider Physics*, R.K. Ellis, W.J. Stirling and B.R. Webber, Cambridge University Press (1996).
2. *Review of Particle Properties*, Phys. Rev. **D54** (1996) 77.
3. W.J. Marciano, Phys. Rev. **D29** (1984) 580.
4. F. Bloch and A. Nordsieck, Phys. Rev. **52** (1937) 54.
5. T. Kinoshita, J. Math. Phys. **3** (1962) 650.
   T.D. Lee and M. Nauenberg, Phys. Rev. **B133** (1964) 1549.
6. M.A. Samuel and L.R. Surguladze, Phys. Rev. Lett. **66** (1991) 560; **66** (1991) 2416(e).
   S.G. Gorishny, A.L. Kataev and S.A. Larin, Phys. Lett. **259B** (1991) 144.
7. B.A. Kniehl and J.H. Kuhn, Nucl. Phys. **B329** (1990) 547.
8. LEP Electroweak Working Group report CERN-PPE/96-183 (1996).
9. G. Grunberg, Phys. Lett. **95B** (1980) 70.
10. P.M. Stevenson, Nucl. Phys. **B150** (1979) 357.
11. JADE collaboration: S. Bethke *et al.*, Phys. Lett. **213B** (1988) 235.
12. G. Kramer and B. Lampe, Fortschr. Phys. **37** (1989) 161.
13. OPAL collaboration: G. Alexander *et al.*, Z. Phys. **C72** (1996) 191.
14. OPAL collaboration: M.Z. Akrawy *et al.*, Phys. Lett. **B235** (1990) 389.
15. S. Bethke, Z. Kunszt, D.E. Soper and W.J. Stirling, Nucl. Phys. **B370** (1992) 310.
16. N. Brown and W.J. Stirling, Phys. Lett. **252B** (1990) 657.
17. S. Catani, Yu.L. Dokshitzer, M. Olsson, G. Turnock and B.R. Webber, Phys. Lett. **269B** (1991) 432.

18. B.R. Webber, Phys. Lett. **B339** (1994) 148.
Yu.L. Dokshitzer and B.R. Webber, Phys.Lett. **B404** (1997) 321; Phys. Lett., **B352** (1995) 451.
Yu.L. Dokshitzer, G. Marchesini and B.R. Webber, Nucl. Phys. **B469** (1996) 93.
G.P. Korchemsky and G. Sterman, preprint hep-ph/9505391; Nucl. Phys., **B437** (1995) 415.
P. Nason and M.H. Seymour, Nucl. Phys. **B454** (1995) 291.
R. Akhoury and V.I. Zakharov, Phys. Lett. **B357** (1995) 646; Nucl. Phys. **B465** (1996) 295.
19. L3 collaboration: M. Acciarri *et al.*, preprint CERN-EP/98-148, 1998.
20. *The Structure of the Proton*, R.G. Roberts, Cambridge University Press (1990).
21. A.D. Martin, R.G. Roberts, W.J. Stirling and R. Thorne, Eur. Phys. J. C4 (1998) 463.
22. L.N. Lipatov, Sov. J. Nucl. Phys. **20** (1975) 95.
V.N. Gribov and L.N. Lipatov, Sov. J. Nucl. Phys. **15** (1972) 438.
G. Altarelli and G. Parisi, Nucl. Phys.**B126** (1977) 298.
Yu.L. Dokshitzer, Sov. Phys. JETP **46** (1977) 641.
23. CCFR collaboration: W.G. Seligman *et al.*, preprint hep-ex/9701017 (1997).
24. M. Dasgupta and B.R. Webber, preprint hep-ph/9704297 (1997).
25. E.A. Kuraev, L.N. Lipatov and V.S. Fadin, Phys. Lett. **B60** (1975) 50; Sov. Phys. JETP **44** (1976) 433 and **45** (1977) 199.
Ya.Ya. Balitsky and L.N. Lipatov, Sov. J. Nucl. Phys. **28** (1978) 822.
26. CTEQ collaboration: H.-L. Lai *et al.*, Phys. Rev. **D55** (1997) 1280.
27. M. Glück, E. Reya and A. Vogt, Zeit. Phys. **C67** (1995) 433.
28. G. Altarelli, N. Cabibbo, L. Maiani and R. Petronzio, Nucl. Phys. **B69** (1974) 531.
M. Glück and E. Reya, Nucl. Phys. **B130** (1977) 76.
29. ZEUS collaboration: J. Breitweg *et al.*, preprint DESY 98-121 (August 1998).
30. ZEUS collaboration: presented by A. Quadt at DESY, March 1998.
31. P.D.B. Collins, *Introduction to Regge Theory*, Cambridge University Press (1977).
32. WA70 collaboration: M. Bonesini *et al.*, Z. Phys. **C38** (1988) 371.
33. E706 collaboration: L. Apanasevich *et al.*, hep-ex/9711017 (1997).
34. NMC collaboration: P. Amaudruz *et al.*, Phys. Rev. Lett. **66** (1991) 2712.
35. S.D. Ellis and W.J. Stirling, Phys. Lett. **256B** (1991) 258.
36. NA51 collaboration: A. Baldit *et al.*, Phys. Lett. **B332** (1994) 244.
37. E866 collaboration: E.A. Hawker *et al.*, hep-ex/9803011, to be published in Phys. Rev. Lett.
38. S.D. Drell and T.M. Yan, Ann. Phys. **66** (1971) 578.
39. See for example: J.C. Collins and D.E. Soper, Ann. Rev. Nucl. Part. Sci. **37** (1987) 383, and references therein.
40. G. Altarelli, R.K. Ellis and G. Martinelli, Nucl. Phys. **B143** (1978) 521; **B146** (1978) 544(e); **B147** (1979) 461.
J. Kubar-Andre and F.E. Paige, Phys. Rev. **D19** (1979) 221.
J. Kubar-Andre, M. LeBellac, J.L. Meunier and G. Plaut, Nucl. Phys. **B175** (1980) 251.
41. R. Hamberg, T. Matsuura and W.L. van Neerven, Nucl. Phys. **B345** (1990) 331; Nucl. Phys. **B359** (1991) 343.
W.L. van Neerven and E.B. Zijlstra, Nucl. Phys. **B382** (1992) 11.
42. E605 collaboration: G. Moreno *et al.*, Phys. Rev. **D43** (1991) 2815.
43. K. Feudenreich, Int. J. Mod. Phys **A19** (1990) 3643.
44. UA1 collaboration: G. Arnison *et al.*, Phys. Lett. **B122** (1983) 103.
UA2 collaboration: G. Banner *et al.*, Phys. Lett. **B122** (1983) 476.
45. UA1 collaboration: C. Albajar *et al.*, Z. Phys. **C44** (1989) 115.
46. UA2 collaboration: J. Alitti *et al.*, Phys. Lett. **276B** (1992) 365.
47. CDF collaboration: F. Abe *et al.*, Phys. Rev. Lett. **76** (1996) 3070.
48. D0 collaboration: S. Abachi *et al.*, Phys. Rev. Lett. **75** (1995) 1456.

RARE AND EXOTIC DECAYS

DOUGLAS BRYMAN[0]
*TRIUMF, 4004 Wesbrook Mall, Vancouver, B.C.
Canada V6T 2A3*

Abstract. These lectures deal with current measurements of kaon and muon decays which address issues related to the presence of three generations (or flavors) of quarks and leptons, i.e. the generation puzzle. Advanced searches for evidence of direct CP violation in $K \to \pi\pi$ decays are expected to yield new results shortly. The Standard Model prediction for the flavor-changing neutral current reaction $K^+ \to \pi^+ \nu\overline{\nu}$ is being challenged at the 10^{-10} level. New experiments are being developed to study the especially attractive CP-violating channel $K_L^0 \to \pi^0 \nu\overline{\nu}$. Very sensitive searches for lepton flavor violation and other non-Standard Model effects are reaching unprecedented levels of sensitivity. The experiments are treated in the context of their theoretical motivations.

1. Introduction

Although the present state of affairs in particle physics is dominated by the agreement of virtually all measurements with the predictions of the Standard Model (SM), there remain many unanswered questions – particularly related to the apparently redundant multiplicity of the "generations" (or "families", "flavors", etc.) of quarks and leptons. In the quark sector, there are three generations of quark doublets which are related by the Cabibbo-Kobayashi-Maskawa (CKM) mixing matrix. $K^0 - \overline{K}^0$ oscillations ($d\overline{s} \to \overline{d}s$), for example, represent extraordinary evidence that quark flavor is not conserved. Unitarity of the CKM matrix with at least three generations of quarks allows for the "technical" accommodation of a complex CP-violating phase parameter, possibly providing a clue to the generation puzzle. Assuming that the product CPT is invariant, as required by general

[0]E-mail: doug@triumf.ca

T. Ferbel (ed.), Techniques and Concepts of High Energy Physics X, 111–145.

principles of relativistic quantum field theory, CP violation also implies T violation. This may be of more than passing importance since understanding CP (T) violation may provide insight into phenomena as basic as the dominance of matter over anti-matter in the universe.

In the lepton sector, the situation is simpler, but no less puzzling. Three generations of "electrons" (e, μ, τ) and their associated neutrinos appear to be isolated replications of each other, except for mass. The weak coupling strengths of the three generations to the gauge bosons are identical to within considerable experimental precision.[1] No confirmed mixing of leptons has been observed (although there is recent "action" in the neutrino oscillation camp![2]) at minute levels of experimental sensitivity. One can speculate that some global or accidental symmetry is responsible for the apparent conservation of lepton flavor or, as in many alternate theories, that flavor-violating interactions are suppressed at very high mass scales. In addition, there seem to be obvious, but unexplained symmetries between the quarks and leptons (including the equality of the charge of the electron and proton, the matching pairs of doublets, etc.)

K decays have long been a rich and surprising source of information on fundamental questions in particle physics and have played a crucial role in the development of the successful SM. Startling quantum mechanical phenomena like $K^0 - \overline{K}^0$ oscillations and regeneration, the first hint of parity violation in the weak interactions, the absence of flavor changing neutral currents which led to the prediction of charm, and the only observed evidence of CP violation are among the historical credits belonging to the remarkable K meson system. Rare and ultra-rare K and μ decays continue to play an important role in the study of SM phenomena and in the search for information on the generation puzzle.

CP violation in the neutral K system is conventionally discussed by describing the short lived component K_S ($\tau_S = 0.9 \times 10^{-10}s$) and the long lived K_L ($\tau_L = 5.2 \times 10^{-8}s$) as mixtures of CP eigenstates K_1 and K_2

$$K_S = \frac{(K_1 + \bar{\epsilon} K_2)}{\sqrt{1+\bar{\epsilon}^2}} \tag{1}$$

and

$$K_L = \frac{(K_2 + \bar{\epsilon} K_1)}{\sqrt{1+\bar{\epsilon}^2}} \tag{2}$$

where $\bar{\epsilon}$ is a measure of the mixing. In the SM, CP violation also occurs "directly" in the decay amplitudes as a consequence of the phase parameter in the CKM matrix. An alternate explanation involves the hypothesis of a superweak force[3] which would require only mixing to explain the CP violation phenomena observed in the K system. Currently, the elucidation of CP violation is focused on experiments studying $K \to \pi\pi$ decays which are aimed at revealing evidence for direct CP violation. A previous round of experiments did not allow an unambiguous interpretation of the existence of direct CP violation, so additional efforts have been mounted. In the future, new experiments in the K and B systems may provide precise information on CP violation as predicted in the SM or find evidence for a deviation from those expectations.

Since flavor changing neutral currents (FCNC) can only be realized in the SM by a two step process involving successive exchange of gauge bosons, higher order K decays are potentially fertile testing grounds because nearly the entire known spectrum of quarks, leptons, and bosons must interact in the internal loops which lead to the final observed states. In contrast to most meson decays, the ultra-rare "dynamic duo" $K^+ \to \pi^+\nu\overline{\nu}$ and $K_L^0 \to \pi^0\nu\overline{\nu}$ are unusually attractive candidates for study since uncertainties from hadronic effects are minimal and very precise calculations have been made. For this reason, $K^+ \to \pi^+\nu\overline{\nu}$ and $K_L^0 \to \pi^0\nu\overline{\nu}$ may be among the best sources of unambiguous information on SM parameters such as top quark mixing and the CP-violating CKM phase.

Lepton flavor-violating (LFV) reactions have continued to be pursued vigorously on many fronts in spite of 50 years of null experimental searches.[1] In most extensions of the SM, these reactions occur naturally. Actually, they are often difficult to eliminate without inventing seemingly artificial mechanisms. Progress on LFV experiments has spanned many orders of magnitude recently and there is no indication that the limits of achievable sensitivity have been exhausted. Since the LFV rates are supposedly suppressed by heavy force carriers, the virtual mass scales accessed by these experiments have been pushed to extremely high levels reaching $m_h \sim$TeV$/c^2$.

In the following, I will discuss the context and status of three groups of topical experiments involving rare K and μ decays which bear on the questions of CP violation, quark mixing and the quest for new physics to solve the generation puzzle. Section 2 deals with the search for direct CP violation, particularly measurements of ϵ'/ϵ in $K \to \pi\pi$ decay. Section 3 covers measurements of the charged and neutral $K \to \pi\nu\overline{\nu}$ decays. Section 4 describes experiments sensitive to forbidden (non-SM) effects including lepton flavor violation. For more details and references on rare kaon decay experiments there are a number of excellent reviews available.[5, 6, 7]

[1]The first search for $\mu \to e\gamma$ was done in 1948 by Hincks and Pontecorvo.[4]

2. The Search for Direct CP violation in K Decays

2.1. ϵ/ϵ' IN $K \to \pi\pi$ DECAYS

Since the weak interactions do not conserve strangeness, if CP invariance held, the weak eigenstates of the neutral kaon would be

$$K_1 = \frac{(K^0 + \overline{K^0})}{\sqrt{2}} \tag{3}$$

and

$$K_2 = \frac{(K^0 - \overline{K^0})}{\sqrt{2}} \tag{4}$$

where K^0 and $\overline{K^0}$ are the CP conjugate strong interaction states with strangeness -1 and 1. K_1 and K_2 are then CP eigenstates with CP=1 and CP=-1, respectively.

The 1964 discovery[8] that the K_L which normally decays via the CP-odd three pion channel (as well as the semi-leptonic modes $\pi e\nu$ and $\pi\mu\nu$) also occasionally decays to the CP-even two pion states $\pi^+\pi^-$ and $\pi^0\pi^0$ led to the conclusion that CP was violated in the weak interactions. As indicated in eqns. 1 and 2, K_S which decays primarily to the CP-even two pion states is now identified as predominantly K_1 with a small admixture of K_2, and K_L is predominantly K_2 with a small admixture of K_1.

The SM with three generations accommodates CP violation via inclusion of a single complex phase parameter in the unitary 3×3 CKM quark mixing matrix. Charged current interactions of quarks are parameterized as

$$J_\mu = (\overline{u}, \overline{c}, \overline{t})\gamma_\mu \begin{pmatrix} V_{ud} & V_{us} & V_{ub} \\ V_{cd} & V_{cs} & V_{cb} \\ V_{td} & V_{ts} & V_{tb} \end{pmatrix} \begin{pmatrix} d \\ s \\ b \end{pmatrix}.$$

The Wolfenstein representation[9] of the CKM matrix conveniently expresses it in terms of four independent parameters ($\lambda = V_{us} = 0.22, A, \rho$ and η)

$$\begin{pmatrix} 1-\frac{\lambda^2}{2} & \lambda & A\lambda^3(\rho - i\eta) \\ -\lambda & 1-\frac{\lambda^2}{2} & A\lambda^2 \\ A\lambda^3(1-\rho-i\eta) & -A\lambda^2 & 1 \end{pmatrix}.$$

The parameter η is then the embodiment of CP violation in the SM.

The experimental observables describing CP violation in $K^0 \to \pi\pi$ decay are conventionally characterized in terms of the decay amplitudes

A($K^0_{L(S)} \to \pi\pi$) as

$$\eta_{+-} = \frac{A(K^0_L \to \pi^+\pi^-)}{A(K^0_S \to \pi^+\pi^-)} = |\eta_{+-}|e^{i\phi_{+-}} \tag{5}$$

and

$$\eta_{00} = \frac{A(K^0_L \to \pi^0\pi^0)}{A(K^0_S \to \pi^0\pi^0)} = |\eta_{00}|e^{i\phi_{00}}. \tag{6}$$

In addition, there is one parameter associated with CP violation in the semi-leptonic decays

$$\delta = \frac{\Gamma(K_L \to \pi^- l^+ \nu) - \Gamma(K_L \to \pi^+ l^- \overline{\nu})}{\Gamma(K_L \to \pi^- l^+ \nu) + \Gamma(K_L \to \pi^+ l^- \overline{\nu})}. \tag{7}$$

The current values for the CP violation observables are $\eta_{+-} = 2.275 \pm 0.019 \times 10^{-3}$, $\phi_{+-} = 43.5 \pm 0.6°$, $\eta_{00} = 2.285 \pm 0.019 \times 10^{-3}$, $\phi_{00} = 43.4 \pm 1.0°$, and $\delta = 3.27 \pm 0.012 \times 10^{-3}$.[10]

CPT invariance, which holds under general assumptions for a local field theory with Lorenz invariance and normal spin-statistics requirements, implies that if CP is violated so is time-reversal (T) invariance. Recently, the CPLEAR group has presented explicit evidence for T violation[11] derived from the time-dependent asymmetry of semi-leptonic K decays

$$A(\tau) = \frac{\Gamma_{\overline{K^0} \to \pi^- e^+ \nu}(\tau) - \Gamma_{K^0 \to \pi^+ e^- \nu}(\tau)}{\Gamma_{\overline{K^0} \to \pi^- e^+ \nu}(\tau) + \Gamma_{K^0 \to \pi^+ e^- \nu}(\tau)}.$$

They found a time averaged value for the asymmetry of

$$< A(\tau) >= (6.6 \pm 1.3_{\text{stat}} \pm 1.0_{\text{syst}}) \times 10^{-3}.$$

which indicates a clear departure from T invariance.

Assuming CPT invariance and using the symmetry of the wave functions for bosons, the decay amplitudes A($K^0 \to \pi\pi$) can be written as

$$A(K^0 \to \pi\pi) = A_0 e^{i\delta_0} + A_2 e^{i\delta_2} \tag{8}$$

and

$$A(\overline{K^0} \to \pi\pi) = A_0^* e^{i\delta_0} + A_2^* e^{i\delta_2} \tag{9}$$

where the terms A_I (I=0,2) represent the allowed isospin (I) amplitudes of the $\pi\pi$ system and δ_I is the associated $\pi\pi$ strong interaction phase shift. The observation that the I=0 amplitudes dominate is known as the $\Delta I = 1/2$ rule. If CP symmetry were invariant, the amplitudes A_I would be real.

Performing an isospin decomposition of the charged and neutral two pion states, one obtains

$$A(K^0 \to \pi^+\pi^-) = A_0 e^{i\delta_0} + \frac{1}{\sqrt{2}} A_2 e^{i\delta_2} \tag{10}$$

and

$$A(K^0 \to \pi^0\pi^0) = A_0 e^{i\delta_0} - \sqrt{2} A_2 e^{i\delta_2}. \tag{11}$$

Finally, employing eqns. 1-6, 10 and 11 results in

$$\eta_{+-} = \epsilon + \epsilon' \tag{12}$$

and

$$\eta_{00} = \epsilon - 2\epsilon' \tag{13}$$

where

$$\epsilon = \overline{\epsilon} + i\frac{ImA_0}{ReA_0}, \tag{14}$$

$$\epsilon' = \frac{1}{\sqrt{2}} e^{i(\delta_2 - \delta_0 + \pi/2)} \left(\frac{ImA_2}{ReA_2} - \frac{ImA_0}{ReA_0}\right)\omega, \tag{15}$$

and $\omega = \frac{ReA_2}{ReA_0}$. In the SM, direct CP violation in $K \to 2\pi$ decays is expected to be a small effect compared to the component due to mixing so $\epsilon' << \epsilon$.

2.2. CALCULATIONS OF ϵ AND ϵ'.

In the SM, the parameter ϵ has its main component due to mixing ($\overline{\epsilon}$) as well as a small contribution due to direct CP violation in the amplitude (ImA_0) as indicated in eqn. 14. [2] ϵ can be extracted from the ratio of the imaginary part over the real part of the box diagram for $K - \overline{K}$ mixing, shown in fig. 1. As detailed by Buras and Fleischer,[12] the main uncertainty in the calculation of ϵ comes from the non-perturbative dynamics, namely, the estimate for the matrix element

$$< \overline{K^0} | (\overline{s}\gamma_\mu(1-\gamma_5)d)(\overline{s}\gamma^\mu(1-\gamma_5)d) | K^0 > \equiv \frac{8}{3} B_K f_K^2 m_K^2. \tag{16}$$

B_K is a factor estimated using non-perturbative techniques like lattice calculations and f_K is the K decay constant. Due to the large uncertainty in B_K it is difficult to obtain a precise constraint on ρ and η from the measurements of ϵ.

The value of ϵ' in eqn. 15 results from the difference in CP violation in the $I = 0$ and $I = 2$ channels of $K \to 2\pi$ decays. A non-zero value for ϵ'

[2] In other models it is possible to choose a phase convention where ImA_0 is zero. See the discussion in ref. [6].

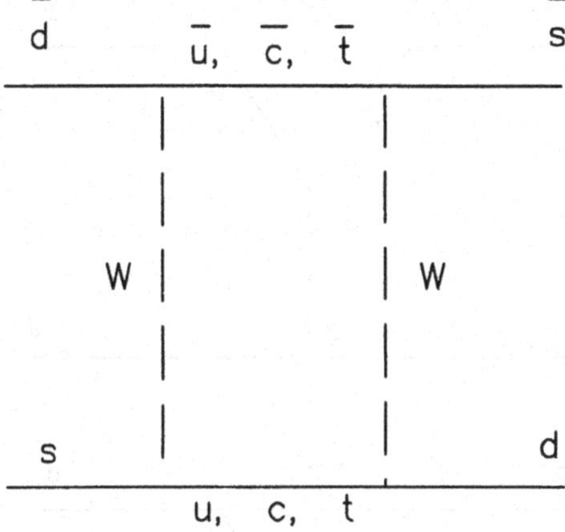

Figure 1. Box diagram for $K - \overline{K}$ mixing.

would therefore indicate the presence of direct CP violation due to one or both of the I$= 0, 2$ channels (ImA_0,ImA_2) and rule out the pure superweak interpretation of CP violation in the K system. Nevertheless, ϵ' could still be extremely small if the two terms in eqn. 15 nearly cancel.

The primary diagrams contributing to the calculation of ϵ' are shown in fig. 2. The so-called gluonic penguin graph fig. 2a introduced by Gilman and Wise[13] gives rise to the term $\frac{ImA_0}{ReA_0}$ in eqn. 15 whereas the electromagnetic penguin graph of fig. 2b results in the $\frac{ImA_2}{ReA_2}$ term. These terms tend to cancel for large m_t[14] and for present values the theoretical estimate[12] is

$$\frac{\epsilon'}{\epsilon} = (3.6 \pm 3.4) \times 10^{-4} \tag{17}$$

where the main uncertainty arises from the scalar operators between hadronic states. Thus, for the present, the theoretical uncertainties due to hadronic effects are large for both ϵ' and ϵ.

2.3. MEASUREMENTS OF ϵ'/ϵ

Regardless of detailed theoretical questions, a confirmed non-zero measurement of ϵ'/ϵ would establish the consistency of the SM origin of CP violation by providing evidence for direct CP violation. To determine ϵ'/ϵ accurately, it is useful to measure the double ratio of decay rates

$$R = \frac{\frac{\Gamma(K_L \to \pi^0\pi^0)}{\Gamma(K_S \to \pi^0\pi^0)}}{\frac{\Gamma(K_L \to \pi^+\pi^-)}{\Gamma(K_S \to \pi^+\pi^-)}} = \left|\frac{\eta_{00}}{\eta_{+-}}\right|^2 \tag{18}$$

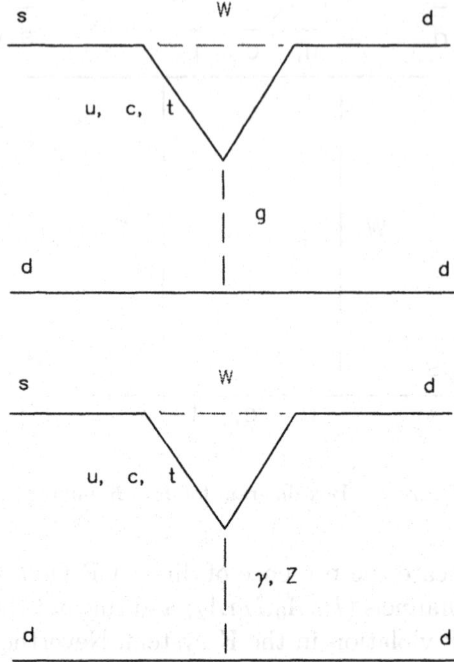

Figure 2. Gluonic (a) and electomagnetic (b) penguin diagrams relevant to the calculation of ϵ'/ϵ.

since it can be shown that

$$R \approx 1 - 6\frac{\epsilon'}{\epsilon}\,. \tag{19}$$

The most recent values for ϵ'/ϵ based on measurements of R were reported several years ago:

$\epsilon'/\epsilon = (23 \pm 3.5 \pm 6.0) \times 10^{-4}$ (NA31) [15]

and

$\epsilon'/\epsilon = (7.4 \pm 5.2 \pm 2.9) \times 10^{-4}$ (FNAL E731)[16].

The apparent non-zero value for ϵ'/ϵ found by CERN experiment NA31 (which is 3 σ from zero) was not confirmed by E731 at Fermilab, although the difference between the results represents only a 2 σ "discrepancy". Now, a new round of major experimental efforts has been mounted at Fermilab

(KTEV-E832), CERN(NA48) and DAΦNE(KLOE) to measure ϵ'/ϵ with greater precision.

2.3.1. *NA48 AND E832*

Determining ϵ'/ϵ via eqn. 18 allows for good suppression of many potentially biasing effects related to acceptances, rates and backgrounds which nearly cancel in the measurement of the double ratio. The main issue in such a measurement is the accurate counting of both the neutral and charged decays modes which are generally detected using different experimental techniques.

The new experiments E832 and NA48 are aiming for an improvement in precision of about a factor of 5 over previous efforts, possibly approaching the 10^{-4} level for ϵ'/ϵ. This means that statistics of the $K_L \rightarrow 2\pi$ decays must reach at least several $\times 10^6$ observed decays for each channel. Even larger numbers of events are required for studies of systematic effects. The major sources of potential systematic uncertainty in measurements of R are background subtraction, energy calibration, differences in acceptance for K_L and K_S decays, and accidental coincidences. Backgounds can arise from common decay modes like $K \rightarrow 3\pi$ when one pion is missed or from semi-leptonic decays $K \rightarrow \pi l \nu$ when the lepton l (e or μ) is misidentified. Energy calibration uncertainties can develop in the transformation from the laboratory frame to the center-of-momentum frame, particularly for the neutral modes. Here, calorimeter performance is crucial. Since K_S and K_L have very different lifetimes, there can also be uncertainties related to their decay vertex distributions. Finally, the rate dependent effects of accidental coincidences and beam flux variations on R can be minimized if all four modes are taken simultaneously.

The E832 experiment at Fermilab[17] and NA48[18] at CERN represent significant refinements and improvements of the techniques previously employed in E731 and NA31, respectively, and each employs state-of-the-art beams and detector technologies. In both E832 and NA48, high energy beams of neutral kaons (which also contain prodigious contamination by neutrons) decay in evacuated decay regions. In the forward direction, magnetic spectrometers with drift chamber trackers are used for measuring the momenta of charged particles from $K_{L,S} \rightarrow \pi^+\pi^-$ decays. The four photons from $K_{L,S} \rightarrow \pi^0\pi^0$ decays are detected by non-sampling segmented electromagnetic calorimeters of the highest quality (liquid Kr for NA48 and pure CsI crystals for E832) used to measure the photon energies and positions. In both new experiments, all four decay modes are collected simultaneously. Sophisticated triggers use hit and cluster counting, pattern recognition, and energy in the calorimeter to select two pion modes for data acquisition. Photon veto detectors surrounding the decay volumes eliminate events with extra particles (e.g. γs from $K \to 3\pi$ decays) which may escape the main detector. Muons are identified by their detection after massive filters. The techniques of E832 and NA48 differ mainly through the production of the beams of K_S and K_L and the methods for determining the acceptances.

In E832 shown in fig. 3 two identical K_L beams with energies up to about 140 GeV are generated from a single production target by 5×10^{12} 800 GeV protons/spill ($\sim$ 20 s). An active-detector regenerator switches between the beams on a pulse-by-pulse cycle to create the K_S beam via coherent regeneration from one of the K_L beams. Detector acceptance differences for the vacuum and regenerated beams are calculated in detailed Monte Carlo simulations. The use of a regenerator insures that the momentum distributions of the K_L and K_S are quite similar although there are potentially additional backgrounds and biases from scattering in the regenerator and misidentification of the beam of origin for a decay event due to incoherent regeneration. Initial running[17] has demonstrated γ energy resolution of 0.65% at 20 GeV, more than five times better than in E731 and momentum resolution of $\frac{\Delta P}{P} = 0.25\%$ (twice as good as E731). The resultant 2π invariant mass resolutions have been improved by factors of 2-3 to $\Delta m \approx 1.7$ MeV/c^2. In addition, the photon veto capabilities have been substantially upgraded. Already, 4×10^6 $K_L \to \pi^0\pi^0$ decays have been recorded and first results with improved precision are expected soon.

The NA48 detector is shown in fig. 4. K_S and K_L beams are generated simultaneously from separate production targets, and decays of K_L or K_S are identified by reconstruction in one or the other beam. A 450 GeV proton beam of intensity 1.5×10^{12}/spill is used to produce the K_L beam. A small fraction of the beam 3×10^7/spill is channeled using a bent crystal and redirected back onto the K beam line. These protons hit a K_S production target shown in fig. 5. The protons are individually detected with timing resolution $\sigma < 200$ ps to tag the K_S decay events. Kaons in the energy range of $70 < E_K < 170$ GeV are used. The energy resolution of the liquid Kr calorimeter has been determined from initial running to be about 0.7% at 60 GeV. The $\pi^0\pi^0$ and $\pi^+\pi^-$ invariant mass resolutions have been measured to be 3.5 MeV/c^2 and 2.5 MeV/c^2, respectively.

In NA48, the correction for acceptance is made by binning the decays according to the K_L and K_S lifetimes. This technique minimizes the potential uncertainties from simulation on knowledge of the acceptance, although at the expense of statistics. More than 6.5×10^5 $K_L \to \pi^0\pi^0$ decays were collected in an initial run[18] and additional data is presently being acquired. Systematic effects are estimated to be at the desired level of precision (i.e. $<$ several $\times 10^{-4}$).

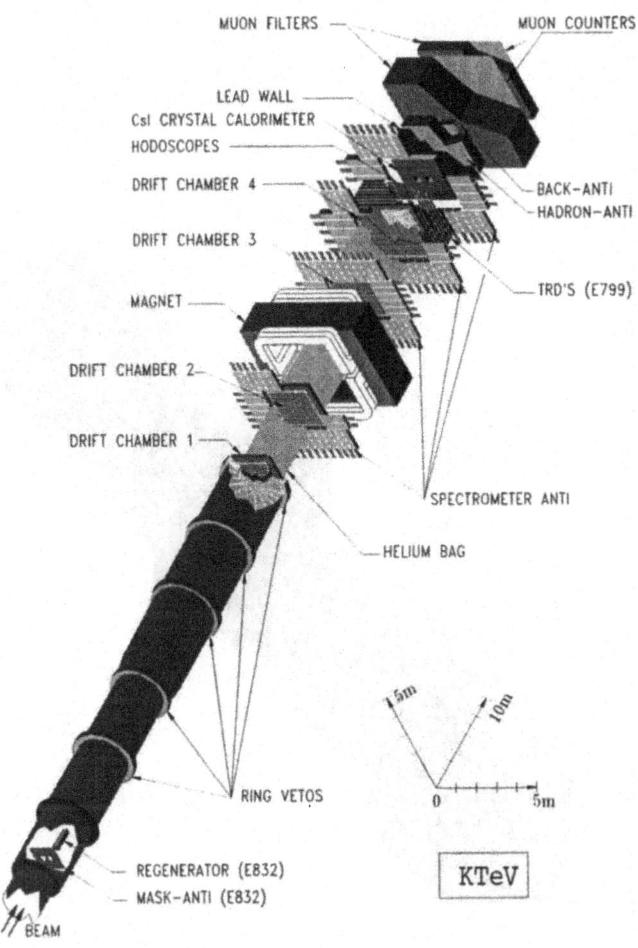

Figure 3. The KTEV detector used to measure ϵ'/ϵ and rare decays at Fermilab.

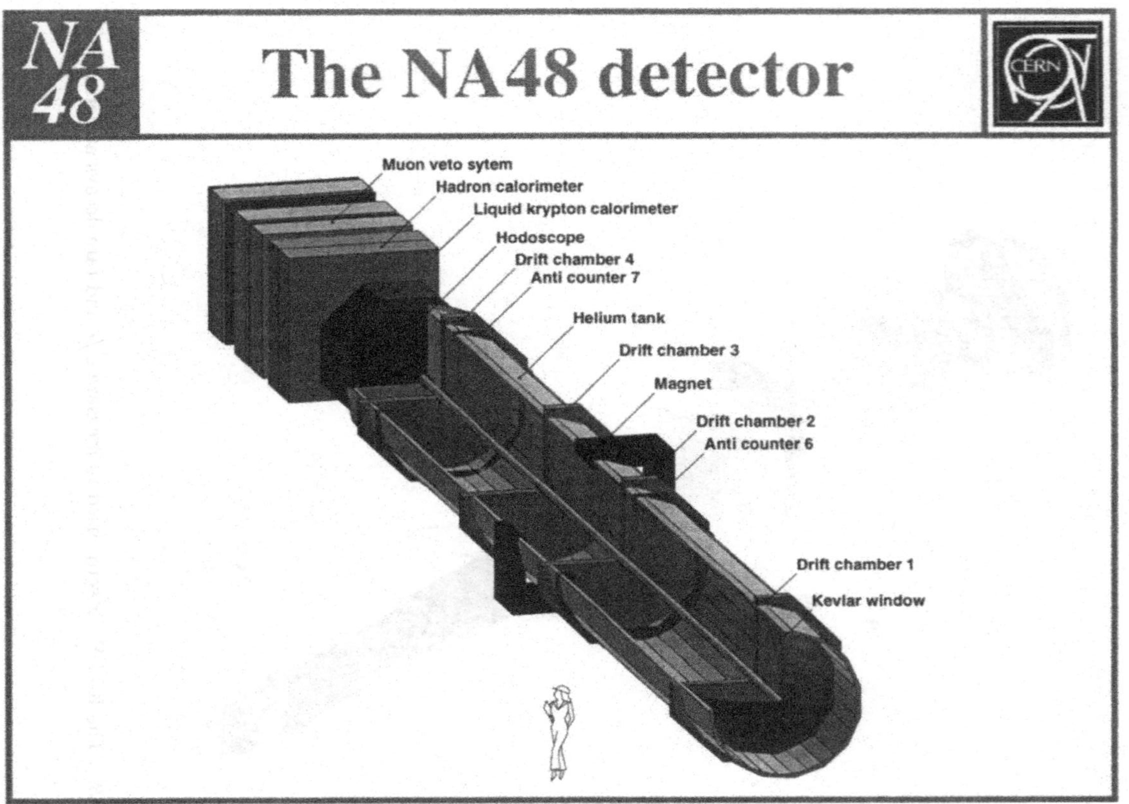

Fig. 4. The NA48 detector at CERN used to measure ϵ'/ϵ and rare decays.

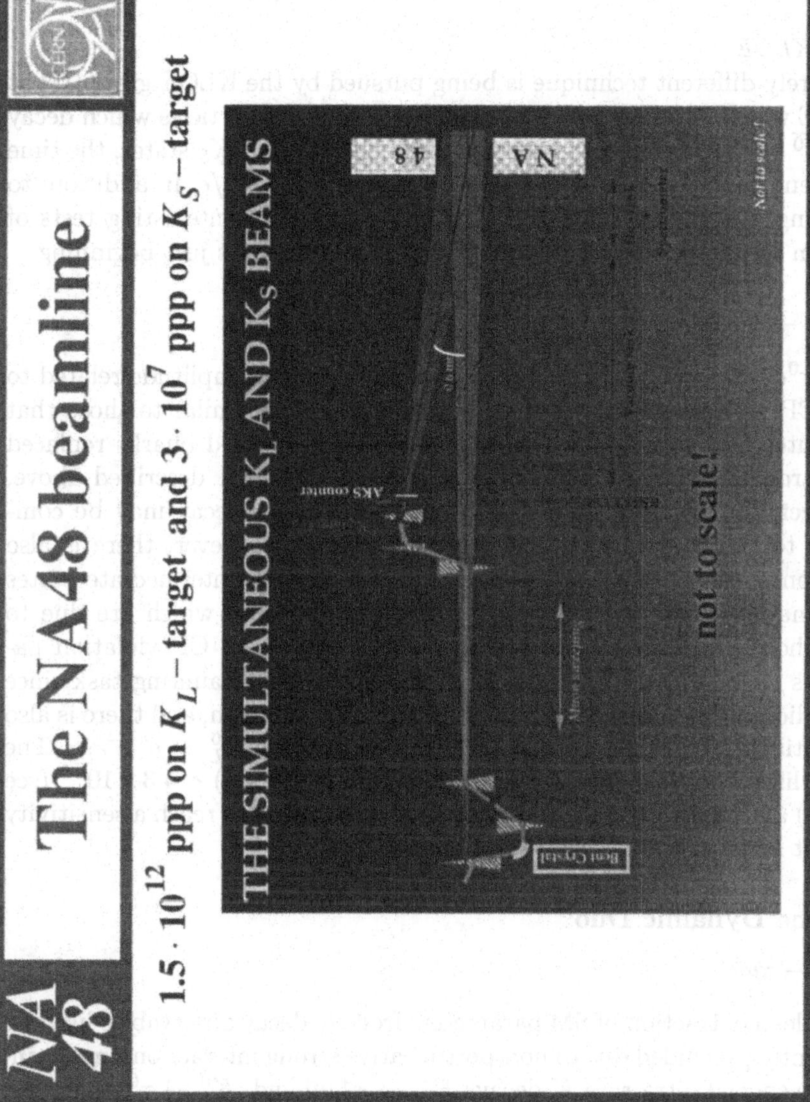

Fig. 5. The NA48 beam arrangement described in the text.

Both experiments have taken initial data, but at the time of writing, no new results on ϵ'/ϵ were available.

2.3.2. *KLOE*

An entirely different technique is being pursued by the KLOE group[19] at DAΦNE where e^+ and e^- beams collide to produce ϕ particles which decay to $K^0\overline{K^0}$ pairs. Since the ϕ is a superposition of K_L and K_S states, the time dependent decay rate for $K_L, K_S \to 2\pi$ is sensitive to ϵ'/ϵ. In addition, to measuring ϵ'/ϵ the 10^{-4} level, the KLOE group plans interesting tests of quantum mechanics and CPT. The KLOE experiment is just beginning.

2.4. $K_L^0 \to \pi^0 e^+ e^-$

$K_L^0 \to \pi^0 e^+ e^-$ is another process expected to have an amplitude related to direct CP violation. It arises from penguin diagrams similar to those that contribute to ϵ'/ϵ shown in fig 2b (but with the lower d quarks replaced by electrons). In comparison with the case of $\epsilon'/\epsilon << 1$ described above, the direct CP violation amplitude in $K_L^0 \to \pi^0 e^+ e^-$ decay may be comparable to the component due to $K - \overline{K}$ mixing. However, there is also a CP-conserving interaction arising from two-photon intermediate states which may compete with the CP-violating amplitudes which are due to single-photon intermediate states.[20] Extracting direct CP violation parameters from $K_L^0 \to \pi^0 e^+ e^-$ measurements will be a daunting task since the predicted overall branching ratio is in the 10^{-11} region, and there is also a potentially serious background arising from decays $K_L^0 \to e^+ e^- \gamma\gamma$. The current limit[3] on the branching ratio is $\mathrm{B}(K_L^0 \to \pi^0 e^+ e^-) < 4.3 \times 10^{-9}$ (see ref. [21]) and the KTEV group is presently attempting to reach a sensitivity of about 10^{-11}.[22]

3. "The Dynamic Duo"

3.1. $K \to \pi\nu\overline{\nu}$

While clean extraction of SM parameters from K decay observables like ϵ'/ϵ is presently precluded due to non-perturbative strong interaction uncertainties, the special ultra-rare K decays $K^+ \to \pi^+\nu\overline{\nu}$ and $K_L^0 \to \pi^0\nu\overline{\nu}$ do not suffer from these maladies. These reactions may eventually prove to be excellent sources of information on SM parameters such as top quark mixing and CP-violating phases. In $K^+ \to \pi^+\nu\overline{\nu}$ and $K_L^0 \to \pi^0\nu\overline{\nu}$, hadronic effects are well known from measurements of the similar decays K$\to \pi e \nu$ which are related by isospin. In addition, the simple final states allow unusually precise calculations to be made.[12] Furthermore, long distance effects (e.g.

[3] All limits presented will be at the 90 % confidence level.

those due to internal photon or meson exchange) are negligible in comparison with short distance effects (i.e. those related to quark, lepton and gauge boson exchange).

Only a few other possible SM observables (e.g. comparison of B_s and B_d mixing, and CP asymmetries in $B \to \Psi K_s$ decays) provide similarly "golden" opportunities for unambiguously revealing SM effects. Any deviations of SM quark mixing (CKM) parameters derived from the cleanest observables in the K and B systems would signify the presence of new physics.

$K^+ \to \pi^+\nu\overline{\nu}$ presents a unique opportunity for evaluating the detailed predictions of higher order weak interactions in the SM because only second order weak effects like those shown in fig. 6 are significant. There are also diagrams where the $\bar{t}$ quark is replaced by $\overline{u}$ and $\overline{c}$ quarks and all three lepton generations contribute. In fig. 6, the u quark spectator is explicitly shown. $K^+ \to \pi^+\nu\overline{\nu}$ has comparable contributions in the SM due to top and charm quark exchange but is free of long-distance contributions down to the 10^{-13} level.[23] Thus, SM or new physics effects, can be cleanly extracted by studying $K^+ \to \pi^+\nu\overline{\nu}$.

In the Standard Model, $K^+ \to \pi^+\nu\overline{\nu}$ is sensitive to important (but poorly known) parameters, in particular, the magnitude of the coupling of the top quark to the down quark, $|V_{td}| = |A\lambda^3(1-\rho-i\eta)|$. The branching ratio can be expressed in simplified form as

$$B(K \to \pi^+\nu\overline{\nu}) \approx 4 \times 10^{-11} A^4 X_0^2(x_t)(\eta^2 + (\rho_0 - \rho)^2), \tag{20}$$

where

$$X_0(x) = \frac{x}{8}\left[\frac{x+2}{x-1} + \frac{3x-6}{(x-1)^2}\ln x\right], \tag{21}$$

$x_t = \frac{m_t^2}{m_W^2}$ and ρ_0 represents the charm quark contribution. Detailed calculations by Buras and collaborators[12, 24, 25] beyond leading logarithms (using a two-loop renormalization group analysis for the charm contribution and O(α_s) for top) indicate theoretical uncertainties no greater than 5% for typical SM parameters. The present range of branching ratios calculated in the SM is $(0.6 - 1.5) \times 10^{-10}$.

$K_L \to \pi^0\nu\overline{\nu}$ is also a FCNC process that is induced through the loop effects of fig. 6 but in this case with a spectator d quark. The most striking feature here is that $K_L \to \pi^0\nu\overline{\nu}$ is totally dominated by direct CP violation because only the imaginary part of V_{td} survives in the amplitude.[26] This is in stark contrast to all other potential approaches to direct CP violation in the K system.

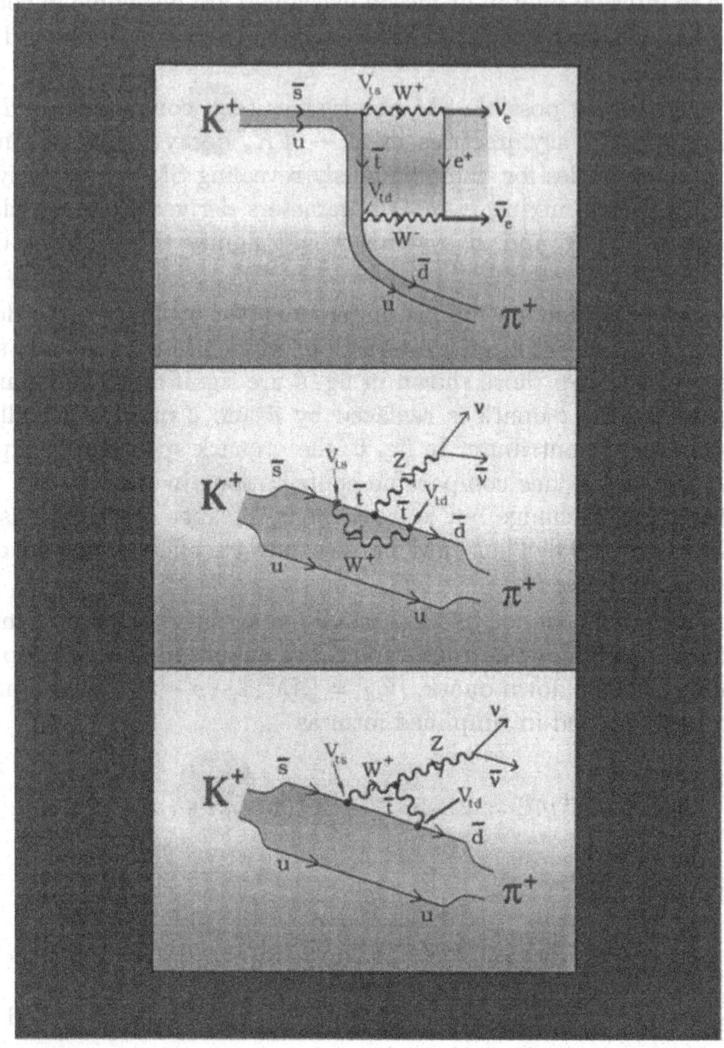

Figure 6. **Examples of second order weak Feynman diagrams leading to $K^+ \to \pi^+ \nu\overline{\nu}$ decay. In addition to the $\overline{t}$ quark exchange shown, diagrams with $\overline{u}$ and $\overline{c}$ quarks also contribute. Furthermore, diagrams with each of the three lepton generations have comparable weight.**

The expression for the $K_L \to \pi^0 \nu\overline{\nu}$ branching ratio can be written as[24]

$$B(K_L \to \pi^0 \nu\overline{\nu}) = B(K^+ \to \pi^0 e^+ \nu) \frac{\tau(K_L)}{\tau(K^+)} \frac{3\alpha^2 r}{2\pi^2 \sin^4 \Theta_W} A^4 \lambda^8 X^2(x_t) \eta^2 \tag{22}$$

where r=0.923 is the product of the leading isospin-breaking correction factor (0.944)[27] and the QCD correction factor (0.985). The full expression for $X^2(x_t)$ is given in ref. [24]. As indicated by eqn. 22, B($K_L^0 \rightarrow \pi^0\nu\bar{\nu}$) is proportional to η^2 with few sources of theoretical uncertainty. The estimated error from purely theoretical sources is in the range of only 1%. Using present estimates for SM parameters, the branching ratio for $K_L^0 \rightarrow \pi^0\nu\bar{\nu}$ is expected to be B($K_L^0 \rightarrow \pi^0\nu\bar{\nu}$)= $(3 \pm 2) \cdot 10^{-11}$. Although measuring such a tiny branching ratio is a great experimental challenge, recent efforts in this field are promising as will be described below.

Taken together $K^+ \rightarrow \pi^+\nu\bar{\nu}$ and $K_L^0 \rightarrow \pi^0\nu\bar{\nu}$ can be used to determine the unitarity triangle illustrated in fig. 7. [4] A clean measure of the height of the unitarity triangle is provided by the $K_L \rightarrow \pi^0\nu\bar{\nu}$ branching ratio. V_{td} (essentially the side from (ρ, η) to (1,0)) can be accurately determined by measuring $K^+ \rightarrow \pi^+\nu\bar{\nu}$ after a correction is applied for a charm quark contribution (which is indicated by the solid line extension to the right of (1,0) in fig. 7). This may be one of the best methods for specifying the unitary triangle. The angles (α, β and γ) will be sought by studying various observables in B decays.[28] Determining $\sin 2\beta$ from asymmetries in $B \rightarrow \Psi K_S$ decays is theoretically clean whereas extraction of $\sin 2\alpha$ from $B \rightarrow \pi\pi$ is complicated by the presence of penguin contributions as in the case of ϵ'/ϵ discussed above. Eventually, results from the CP violation experiments on B physics, ϵ'/ϵ and $K \rightarrow \pi\nu\bar{\nu}$ measurements may be combined to achieve the best picture of CP violation and the highest precision on quark mixing parameters as illustrated in fig. 8 due to Buras.[24]

3.2. $K^+ \rightarrow \pi^+\nu\bar{\nu}$ EXPERIMENT: E787

The E787 group working at the 30 GeV Alternating Gradient Synchrotron (AGS) of Brookhaven National Laboratory has reported evidence for the reaction $K^+ \rightarrow \pi^+\nu\bar{\nu}$ based on the observation of a single clean event.[29] The experiment employs an advanced design low energy kaon beam and a sophisticated detection apparatus shown in fig. 9 which is located inside a 1 T, 3 m diameter solenoidal magnet. The beam line was specially constructed to reduce the fraction of contaminating particles (primarily π's) to less than 25% of the total. The detector has a variety of technically in-

[4] The unitarity of the CKM matrix results in the relation

$$V_{ud}V_{ub}^* + V_{cd}V_{cb}^* + V_{td}V_{tb}^* = 0 \qquad (23)$$

or

$$1 + \frac{V_{td}V_{tb}^*}{V_{cd}V_{cb}^*} = -\frac{V_{ud}V_{ub}^*}{V_{cd}V_{cb}^*} \equiv \rho + i\eta \qquad (24)$$

which determines a triangle in the (ρ, η) plane.

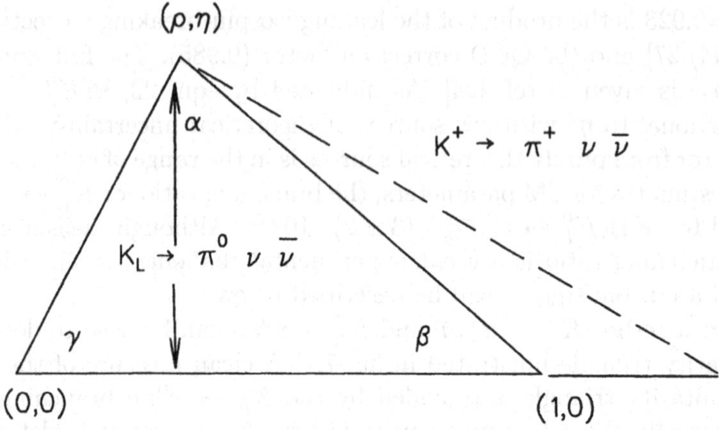

Figure 7. The unitarity triangle.

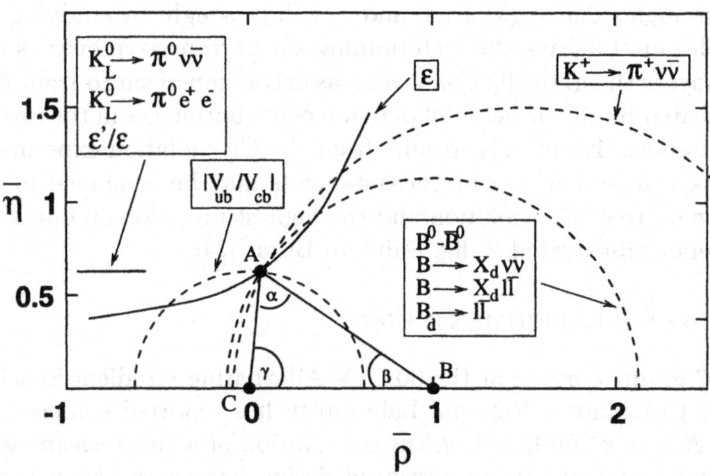

Figure 8. Constraints on the Wolfenstein parameters from K and B physics as depicted by Buras. [24] $\overline{\rho}$ and $\overline{\eta}$ are higher order generalizations of the basic parameters.

teresting features including a scintillating fiber target read out by 500 MHz GaAs CCD transient digitizers, a low mass central tracking drift chamber using inflated cathode-strip foils for the z-position measurement, and highly efficient calorimeters employing pure CsI crystals and Pb/scintillator sandwich detectors. Pions are distinguished from muons by kinematics and by observing the $\pi \rightarrow \mu \rightarrow e$ decay sequence at rest using 500 MHz flash-ADC transient digitizers. The nearly 4π sr photon veto calorimeter is one element

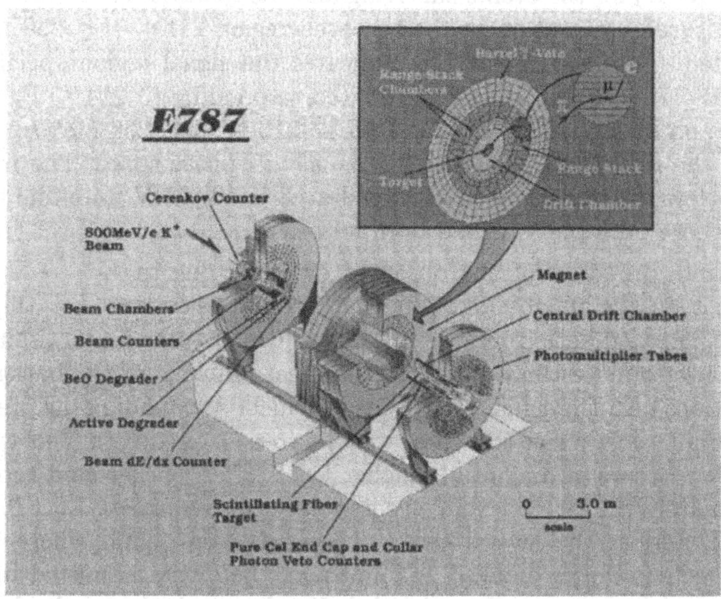

Figure 9. The E787 detector at BNL used to measure $K^+ \to \pi^+\nu\overline{\nu}$ and other rare decays. The inset shows a cross section of the central region where the pion kinematics and decay sequence are measured.

of an elaborate scheme to suppress backgrounds from processes which, although similar to $K^+ \to \pi^+\nu\overline{\nu}$, occur at rates of 10^8 to 10^9 higher.

Attempting to detect a three body decay at the 10^{-10} level in which only one daughter particle is visible represents an experimental challenge. In E787, the signature for $K^+ \to \pi^+\nu\overline{\nu}$ is a K^+ decay at rest to a π^+ of momentum $P < 227$ MeV/c and no other observable product. Definitive observation of this signal requires suppression of all backgrounds to well below the sensitivity for the signal and reliable estimates of the residual background levels. Major background sources include the copious two-body decays $K^+ \to \mu^+\nu_\mu$ ($K_{\mu 2}$) with a 64% branching ratio and $P = 236$ MeV/c and $K^+ \to \pi^+\pi^0$ ($K_{\pi 2}$) with a 21% branching ratio and $P = 205$ MeV/c. The only other important background sources are scattering of pions in the beam and K^+ charge exchange (CEX) reactions resulting in decays $K_L^0 \to \pi^+ l^- \overline{\nu}$, where $l = e$ or μ. To suppress the backgrounds, techniques were employed that incorporated redundant kinematic and particle identification measurements and efficient elimination of events with additional particles.

The observation of the first $K^+ \to \pi^+\nu\overline{\nu}$ candidate event was made in data taken in 1995 with an upgraded E787 detector. A reconstruction of the event is shown in fig. 10. Figure 11(a) shows the range in scintillator (R)

vs. Energy (E) for the events surviving all other analysis cuts. Only events with measured momentum in the accepted region $211 \leq P \leq 230$ MeV/c are plotted. The rectangular box indicates the signal region specified as range $34 \leq R \leq 40$ cm of scintillator (corresponding to $214 \leq P_\pi \leq 231$ MeV/c) and energy $115 \leq E \leq 135$ MeV ($213 \leq P_\pi \leq 236$ MeV/c) which encloses the upper 16.2% of the $K^+ \rightarrow \pi^+\nu\bar{\nu}$ phase space. The residual events below the signal region clustered at $E = 108$ MeV were due to $K_{\pi 2}$ decays where both photons had been missed.

The interpretation of the observed event as due to $K^+ \rightarrow \pi^+\nu\bar{\nu}$ is dependent on the reliability of the background estimates. For this data set, the estimated background level was $b = 0.08 \pm 0.03$ events. Confidence in the background estimates and in the measurements of the background distributions near the signal region was provided by extending the predicted background levels when the cuts were relaxed in predetermined ways so as to allow orders of magnitude higher levels of all background types. At approximately the $20 \times b$ level 2 events were observed where 1.6 ± 0.6 were expected, and at the level $150 \times b$, 15 events were found where 12 ± 5 were expected. Under detailed examination, the events admitted into the enlarged regions were consistent with being due to the known background sources. The event of fig. 10 also satisfied the most demanding criteria designed in advance for candidate evaluation. This put it in a region with an additional background rejection factor of 10. In this region, $b' = 0.008 \pm 0.005$ events would be expected from known background sources while 55% of the acceptance for $K^+ \rightarrow \pi^+\nu\bar{\nu}$ is retained. Since the explanation of the observed event as background was highly improbable, it was concluded that a kaon decay $K^+ \rightarrow \pi^+\nu\bar{\nu}$ was likely observed. If the observed event was due to $K^+ \rightarrow \pi^+\nu\bar{\nu}$, the branching ratio is $B(K^+ \rightarrow \pi^+\nu\bar{\nu}) = 4.2^{+9.7}_{-3.5} \times 10^{-10}$.

The observation of an event with the signature of $K^+ \rightarrow \pi^+\nu\bar{\nu}$ is consistent with the expectations of the SM, although the central value is well above it. Based on the result for $B(K^+ \rightarrow \pi^+\nu\bar{\nu})$ and the relations given in ref. [25], $|V_{td}|$ lies in the range $0.006 < |V_{td}| < 0.06$. By the end of the 1998 run, E787 expects to have obtained a single event sensitivity of $< 7 \times 10^{-11}$, about a factor of 7 below the sensitivity of the 1995 data set which produced the observed event.

3.3. $K_L^0 \rightarrow \pi^0\nu\bar{\nu}$ EXPERIMENTS

Using the apparatus shown in fig. 3, the KTEV group (E799-II) recently reported[30] a new preliminary result on the search for $K_L^0 \rightarrow \pi^0\nu\bar{\nu}$ using detection of the Dalitz decay $\pi^0 \rightarrow e^+e^-\gamma$. The disadvantage of the small branching ratio of the Dalitz decay mode (1.2%) was overcome by superior event reconstruction using the charged particles (compared with the pri-

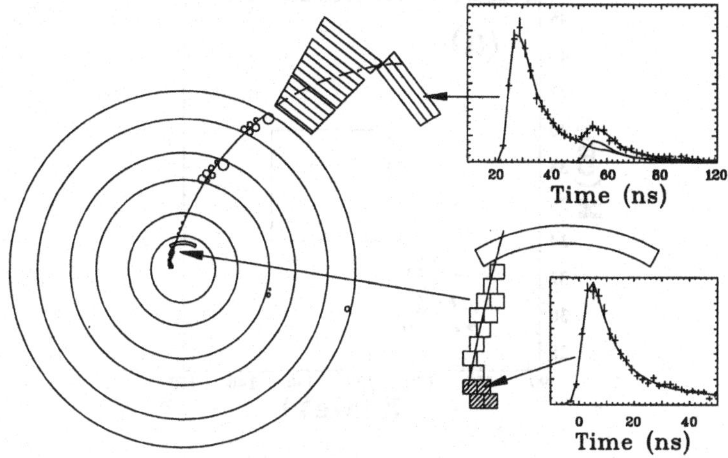

Figure 10. Reconstruction of the E787 $K^+ \rightarrow \pi^+\nu\overline{\nu}$ candidate event. On the left is the end view of the detector showing the track in the target, drift chamber (indicated by drift-time circles), and range stack (indicated by the layers that were hit). At the lower right is a blowup of the target region where the hatched boxes are kaon hits, the open boxes are pion hits, and the inner trigger counter hit is also shown. The pulse data sampled every 2 ns (crosses), in one of the target fibers hit by the stopped kaon is displayed along with a fit (curve) to the expected pulse shape. At the upper right of the figure is the $\pi \rightarrow \mu$ decay signal in the range stack scintillator layer where the pion stopped. The crosses are the pulse data sampled every 2 ns, and the curves are fits for the first, second and combined pulses.

mary $\pi^0 \rightarrow \gamma\gamma$ decay) which helps suppress potentially serious backgrounds from $K_L \rightarrow \pi^0\pi^0$, $\Xi \rightarrow \Lambda\pi^0$, $\Lambda \rightarrow \pi^0 n$, and neutron interactions which produce pions from the residual gas in the vacuum pipe. The result based on all the KTEV data taken so far is B($K_L^0 \rightarrow \pi^0\nu\overline{\nu}$)$< 5.9 \times 10^{-7}$. This result is about two orders of magnitude better than the limit reported previously by this group in 1994.[31] Future progress on the search is anticipated by the KTEV group using the $\pi^0 \rightarrow \gamma\gamma$ decay.

Another $K_L^0 \rightarrow \pi^0\nu\overline{\nu}$ experiment was recently proposed at KEK[32] seeking a single event sensitivity of 2×10^{-11} employing a highly collimated "pencil" beam. The Fermilab KTEV/KAMI group is also exploring the possibility of a new search for $K_L^0 \rightarrow \pi^0\nu\overline{\nu}$ using a K beam produced by the new 120 GeV main injector synchrotron.[33]

Although definitively measuring $K_L^0 \rightarrow \pi^0\nu\overline{\nu}$ at levels comparable to the lowest predictions of the SM presents many difficulties, a concept for achieving this goal was recently proposed at BNL. The method to be used in BNL E926 (see ref. [34]) is based on determining the K_L momentum by time-of-flight so that the detected π^0 can be reconstructed in the kaon center-of-mass system. A low energy (1 GeV) neutral beam produced at

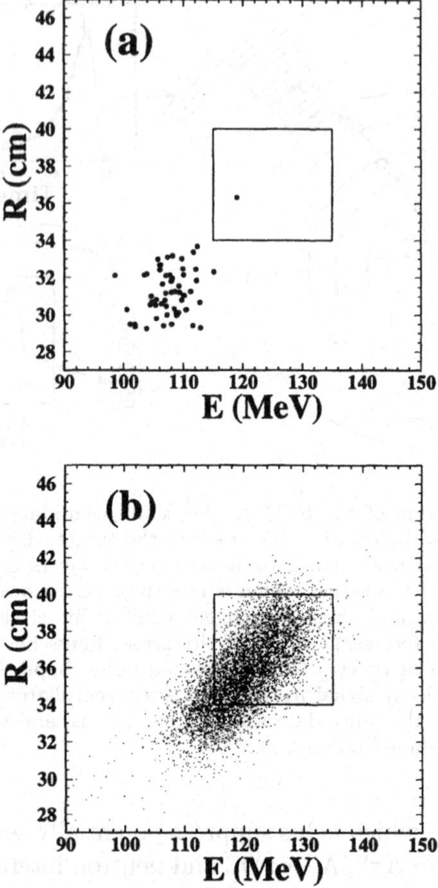

Figure 11. (a) Range (R) vs. energy (E) distribution for the $K^+ \to \pi^+ \nu\bar{\nu}$ data set with the final cuts applied. The box enclosing the signal region contains a single candidate event. (b) The Monte Carlo simulation of $K^+ \to \pi^+ \nu\bar{\nu}$ with the same cuts applied.

large angle to the 30 GeV protons is used to facilitate the time-of-flight measurement. With the proposed detector sketched in fig. 12 all possible kinematic measurements on the photons from $\pi^0 \to \gamma\gamma$ decay are made. Energy, time, conversion point and angle of the individual photons are determined and a high efficiency veto system insures that no other particles are present in coincidence. Based on photon detection efficiency measurements made in E787 and using the extra information available from the kaon time-of-flight, it is expected that the inefficiency for detecting π^0s can be reduced to 10^{-8}, thereby suppressing the most dangerous background

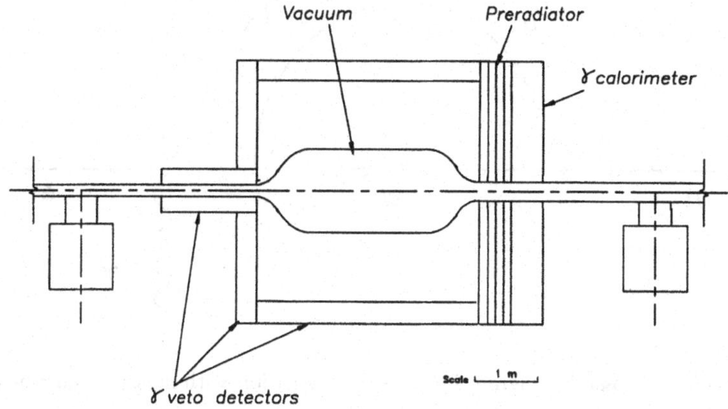

Figure 12. Proposed E926 $K_L^0 \rightarrow \pi^0\nu\bar{\nu}$ detector. Neutral kaons at approximately 1 GeV enter the detector from the left. In the forward detection region the photon detector system consists of a fine grained preradiator in which the photons are converted and the first e^+e^- pair is tracked followed by an 18 radiation length calorimeter. The entire decay path is surrounded by γ veto detectors.

from $K \rightarrow \pi^0\pi^0$ decays (when two of the photons are not detected) at a level that is an order of magnitude below the expected signal. Using these and other special techniques, the goal of a $K_L^0 \rightarrow \pi^0\nu\bar{\nu}$ signal in excess of 50 events may be reached in the presence of what otherwise may be insurmountable backgrounds.

4. Searches for Non-Standard Model processes.

4.1. $K \rightarrow \pi$ +"NOTHING"

Non-SM contributions to the experimental signature $K^+ \rightarrow \pi^+$ 'nothing' may include $K^+ \rightarrow \pi^+\nu\bar{\nu}$ with exotic intermediate states (e.g. leptoquarks or supersymmetric particles), lepton flavor violation via $K^+ \rightarrow \pi^+\nu\bar{\nu}'$, $K^+ \rightarrow \pi^+X_0$, and $K^+ \rightarrow \pi^+X_0 + X_0'$, where X_0 represents some new weakly interacting neutral particle. One possibility for X_0 in the decay $K^+ \rightarrow \pi^+X_0$ is Wilczek's familon,[35] a Goldstone boson which arises in theories of spontaneously broken family symmetry. An unexpected kinematic signature for the π (e.g. a peak away from the two-body K$\rightarrow \pi\pi$ momentum) or a decay rate in conflict with the SM predictions could signal the existence of new physics.

The likelihood that the E787 candidate event mentioned above was due to $K^+ \rightarrow \pi^+X_0$ ($m_{X_0} = 0$) is small, however. Based on the measured resolutions, the χ^2 CL for consistency with this hypothesis was 0.8% and the upper limit $B(K^+ \rightarrow \pi^+X_0) < 3.0 \times 10^{-10}$ was derived. New E787 data

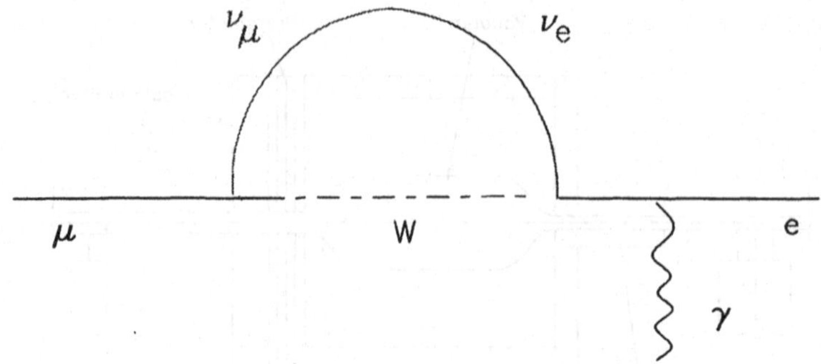

Figure 13. Possible diagram for $\mu \to e\gamma$ transition with massive neutrinos.

should eventually improve the sensitivity by another order of magnitude.

4.2. LEPTON FLAVOR VIOLATION

Lepton flavor violation (LFV) is absent in the SM with massless neutrinos. A minimal extension that includes small neutrino masses might allow LFV reactions such as $\mu \to e\gamma$ via diagrams like that shown in fig. 13. Even if the masses of the three known neutrinos were at their present experimental limits the branching ratio for $\mu \to e\gamma$ would be tiny:[36]

$$B(\mu \to e\gamma) = \frac{\Gamma(\mu \to e\gamma)}{\Gamma(\mu \to e\nu\bar{\nu})} = 10^{-39}\left(\frac{m_{\nu_1}^2 c^4 - m_{\nu_2}^2 c^4}{400\ \mathrm{eV}^2}\right)^2. \qquad (25)$$

Many more exotic extensions of the SM have natural LFV interactions but heavy LFV force carriers suppress the branching ratios by factors of m_{H}^{-4}. With present limits on LFV reactions like $\mu \to e\gamma$, high mass scales are being probed as in some supersymmetric grand unified theories. Stimulated by recent suggestions of atmospheric neutrino oscillations[2], Hisano, Nomura and Yanigida[37] presented an SU(5) model incorporating right-handed neutrinos at the GUT scale and $\Delta m^2 = m_{\nu_\tau}^2 - m_{\nu_\mu}^2 = 10^{-3} - 10^{-4}$ eV^2/c^4 in which the $\mu \to e\gamma$ branching ratio would have a *lower* bound of 10^{-14}, only a few orders of magnitude below the current limit and possibly within reach experimentally. In another model, Barbieri and Hall[38] found that a large top-quark Yukawa coupling led to rates of lepton flavor violating processes which were one or two orders of magnitude below present experimental bounds. The uncertainties for predicting the rates of LFV processes in terms of the masses of supersymmetric partner particles like "selectrons" ($e_{\tilde{R}}$) may be much less model dependent than for competing processes like proton decay (the former standard for GUTs) which would also require baryon number non-conservation. $\mu \to e\gamma$ appears to be the most sensitive process in these models for a given branching ratio, but nuclear $\mu \to e$ conversion discussed below is also competitive since the experimental levels reached so far (and perhaps, in the future) are significantly lower.

Lepton flavor violating reactions have continued to be pursued vigorously on many fronts including muon, tau and kaon decays and in ep collisions at HERA.[39] Table 1 gives some current results and goals for experiments in progress.

TABLE 1. Limits on flavor violating and other exotic K and μ decays.

Reaction	Present limit (90% c.l.)	Reference	In-progress goal	Lab.
$\mu \to e\gamma$	4.9×10^{-11}	[40]	7×10^{-13}	LAMPF
$\mu \to 3e$	1×10^{-12}	[41]		
$\mu^- Ti \to e^- Ti$	6.1×10^{-13}	[42]	few $\times 10^{-14}$	PSI
$\mu^+ e^- \to \mu^- e^+$	8.2×10^{-11}	[43]		
$K_L \to \mu e$	5×10^{-12}	[48]		
$K^+ \to \pi^+ \mu^+ e^-$	2.1×10^{-10}	[46]	3×10^{-12}	AGS
$K_L \to \pi^0 \mu e$	3.2×10^{-9}	[47]	3×10^{-11}	FNAL
$K^+ \to \pi^+ X$	3×10^{-10}	[29]	10^{-11}	AGS

4.3. MUONIUM-ANTIMUONIUM CONVERSION

Processes like the spontaneous conversion of muonium "atoms" to antimuonium ($\mu^+ e^- \to \mu^- e^+$) appear in models containing Majorana neutrinos which are natural consequences of the "see-saw mechanism", a candidate approach to account for the tiny (or "zero") masses of neutrinos.

A clever technique designed to have high sensitivity to muonium-antimuonium transitions has been used in a PSI experiment.[43] The $\mu^- \to e^- \nu\nu$ decay is detected in a large solid angle magnetic spectrometer. The residual positron from the antimuonium is accelerated electrostatically and then observed using the 511 keV annihilation radiation. The spectrometer is shown in fig. 14. Fig. 15 shows the data in which 1 event was observed in the signal region where 1.7 ± 0.2 events were expected from backgrounds. Based on this result the probability for muonium-antimuonium conversion was set at $P_{M\overline{M}} < 8.2 \times 10^{-11}$ and the hypothetical coupling constant for such transitions limited to $G_{M\overline{M}} < 3.3 \times 10^{-3} G_F$. This represents a three order of magnitude improvement in the limit on $P_{M\overline{M}}$.

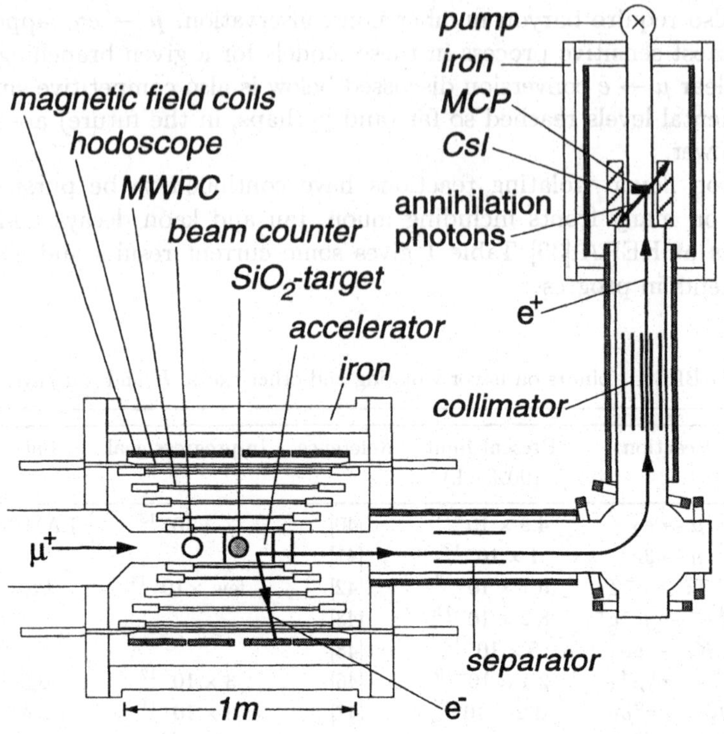

Figure 14. The spectrometer used to search for muonium-antimuonium conversion at PSI. [43]

4.4. $\mu \to e\gamma$

Following several generations of "meson factory" experiments at LAMPF, SIN/PSI and TRIUMF, a major thrust at LAMPF during the past decade has been to gain another order of magnitude or more sensitivity on the search for $\mu \to e\gamma$ using the MEGA detector.[44] Data acquisition has been completed for the MEGA experiment which employed a large solid angle detection system suitable for operation at very high instantaneous rates using extremely thin cylindrical MWPCs and high precision pair spectrometers.

Future improvements in sensitivity on $\mu \to e\gamma$ may depend on suppression of accidental backgrounds (which occur when the positron from one muon decay is detected in coincidence with a photon from a distinct radiative muon decay $\mu^+ \to e^+\nu\nu\gamma$) and on improvements in detector resolutions including energy, timing and photon position and angular measurements. At LAMPF, the 7% duty factor (ratio of the beam on time to beam off time) was a limiting factor in suppressing accidentals. Operation of future

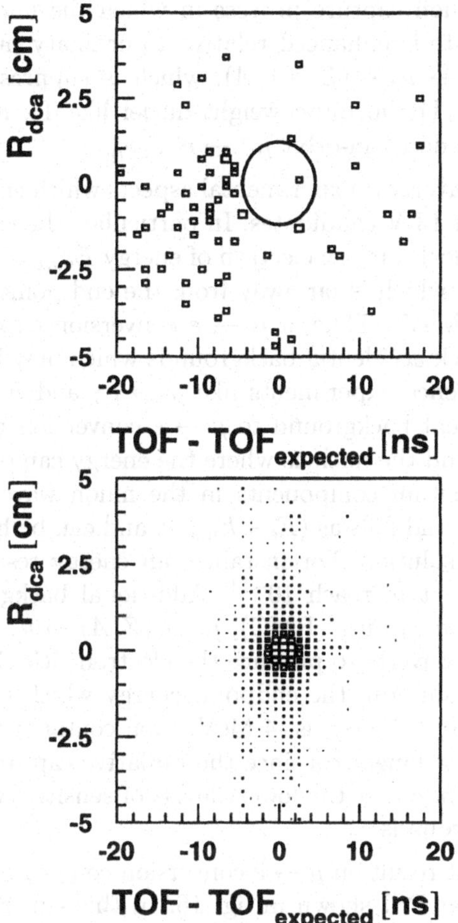

Figure 15. The difference in traced back positions (R_{dca}) for the $\mu \rightarrow e$ decay and the positron vs. the time-of-flight difference for data (upper plot) and simulation (lower plot) in the search for muonium-antimuonium transitions [43].

$\mu \rightarrow e\gamma$ experiments in the environment of a high average intensity, (possibly 10^8 μ^+/s beam) where the duty factor is an order of magnitude greater (at PSI and TRIUMF the duty factor is 100%) may be needed to reach levels near 10^{-14} which currently appear theoretically interesting.

4.5. $\mu \rightarrow e$ conversion

Nuclear $\mu \rightarrow e$ conversion from a muon bound in atomic orbit, $\mu^- + (Z, A) \rightarrow e^- + (Z, A)$, is another process being pursued vigorously. The coherent $\mu \rightarrow e$ muon capture process in which the nucleus is left in its original ground state is enhanced relative to ordinary nuclear muon capture, $\mu^- + (Z, A) \rightarrow \nu_\mu + (Z-1, A)$, which is suppressed by the Pauli exclusion principle. For medium weight nuclei like Ti, nuclear μ capture dominates over the decay-in-orbit $\mu^- \rightarrow e^- \overline{\nu}_e \nu_\mu$.

There are also favorable experimental aspects which enhance $\mu \rightarrow e$ conversion above other LFV candidates. In particular, the signature of $\mu \rightarrow e$ conversion is a monochromatic electron of energy $E_{\text{max}} = m_\mu c^2 - E_{\text{binding}} - E_{\text{recoil}} \approx 104$ MeV, which is far away from the end point of the free μ decay spectrum (53 MeV). Thus, a $\mu \rightarrow e$ conversion experiment does not suffer from two-track accidental backgrounds which may limit the sensitivity of LFV coincidence experiments like $\mu \rightarrow e\gamma$ and $K \rightarrow \mu e$. However, the principal inherent background to $\mu \rightarrow e$ conversion comes from muon decay-in-orbit around the nucleus where the energy can extend up to E_{max} due to high momentum components in the muon wave function. Fortunately, this background falls as $(E_e - E_{\text{max}})^5$ and can be handled by having excellent energy resolution. For instance, an energy resolution of 2 MeV (FWHM) is sufficient to reach 10^{-14}. Additional backgrounds can come from radiative muon capture (RMC), $\mu^- + (Z, A) \rightarrow \nu_\mu + (Z-1, A) + \gamma$, where the photon converts to produce the electron. RMC processes can be suppressed due to shifts in the endpoint energy which for the case of the reaction $\mu + \text{Ti} \rightarrow \nu_\mu + \text{Sc} + \gamma$ is -4 MeV. Pion contamination in the muon beam is particularly dangerous since the radiative capture photons extend up to energy $E = m_\pi c^2$. At the lowest levels of sensitivity, cosmic rays can also present backgrounds.

The most recent result on $\mu \rightarrow e$ conversion comes from the SINDRUM II experiment[42] at PSI shown in fig. 16. In this experiment, an intense μ^- beam at the rate of about 5×10^6/s with momentum P= 88 MeV/c entered along the axis of the 1.2 Tesla super-conducting solenoid through collimators and beam counters to stop in a Ti target. Outgoing electrons with P> 70 MeV/c were detected in a large solid angle spectrometer. Since most of the decay-in-orbit electrons have transverse momenta well below 70 MeV/c, relatively slow drift chambers could be used for tracking. Thus, good resolution $\frac{\Delta P}{P} \approx 1.5\%$ was obtained by measuring at least one turn of the helical trajectories in radial drift chambers with CO_2- and He-based gases. The use of light gases limited the contributions to the resolution from multiple Coulomb scattering. Induced charge readout on cathode strips at the outer cathodes of these chambers allowed for 3-dimensional trajectory points to be obtained with good resolution ~ 400 μm (FWHM). Čerenkov detectors and scintillators were used for fast triggering of candidate electron events. Since the μ^- capture lifetime in Ti is about 329 ns, electrons from the target were sought in the time interval from a few ns after the arrival of the incident beam particle (to reduce prompt backgrounds due to π capture) to a few hundred ns.

A new preliminary upper limit[42] on the branching ratio for $\mu \to e$ conversion in Ti has been obtained based on no observed events. The data is shown in fig. 17 at various stages of analysis along with a simulated signal. The result was

$$B\,(\mu^- + \mathrm{Ti} \to e^- + \mathrm{Ti}) \;=\; \frac{\Gamma(\mu^- + \mathrm{Ti} \to e^- + \mathrm{Ti})}{\Gamma(\mu^- + \mathrm{Ti} \to \nu_\mu + (Z,A)')} \qquad (26)$$
$$< \frac{2.3}{N_\mu f_{\mathrm{capture}} \epsilon} = 6.1 \times 10^{-13}$$

where (Z,A)$'$ represents all nuclear final states from muon capture, $N_\mu = 3 \times 10^{13}$ μ^- stopped in the target, $f_{\mathrm{capture}} = 0.853$ is the nuclear capture probability in Ti and $\epsilon = 0.146$ is the efficiency for detecting 104 MeV electrons. The factor 2.3 is the Poisson statistics value for a 90% c.l. limit when 0 events are observed.[5] The result in eqn. 26 is the lowest of the LFV limits and is about a factor 7 below the previous search done a decade ago with the TRIUMF time projection chamber.[45]

The SINDRUM II group is awaiting the development of a new higher intensity muon beam to attempt to reach the level of a few $\times 10^{-14}$. The new beam is expected to have a stopping intensity of $> 4 \times 10^7$ μ^-/s with <1% pion contamination.

4.6. LFV IN KAON DECAYS

Also prominent among LFV reactions are $K_L^0 \to \mu e$ which is potentially sensitive to new axial-vector and pseudoscalar interactions and $K^+ \to \pi^+\mu^+e^-$ which can occur via vector or scalar currents. These reactions are especially attractive in non-SM approaches which have overall "generation" numbers conserved considering both the quark and lepton sectors, and they are extremely sensitive to the presence of high mass scales, possibly reaching into TeV territory.

Two major experiments E871 and E865 were mounted at BNL in recent years to search for the LFV reactions $K_L^0 \to \mu e$ and $K^+ \to \pi^+\mu^+e^-$. These improved searches have been made possible in part by the substantial recent upgrading of the beam intensity at the AGS which has now exceeded 6×10^{13} protons/1.6 s spill. E865 at BNL is a fine-tuned upgrade of the experiment performed by the group which obtained a limit of B($K^+ \to \pi^+\mu^+e^-$)$< 2.1 \times 10^{-10}$ (see ref. [46]). The present goal is 3×10^{-12} which is being pursued using a more intense 6 GeV beam, a higher acceptance

[5]See the discussion of Poisson statistics in [10].

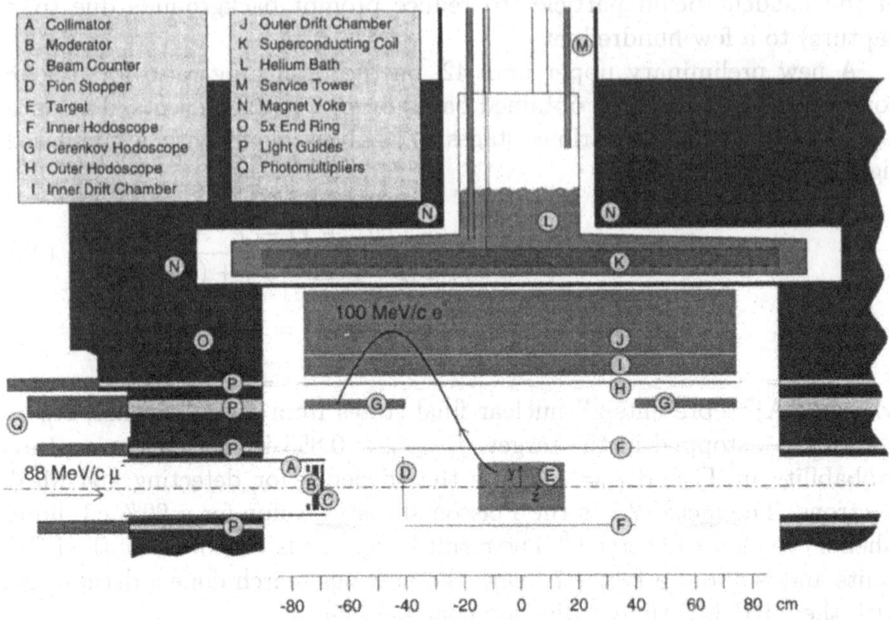

Figure 16. The $\mu - e$ conversion experiment apparatus SINDRUM II at PSI.

detector with improved resolutions and greater redundancy in the tracking and particle identification systems. The new experiment took data in 1996 and is continuing. Another search dealing with the related process $K_L^0 \to \pi^0\mu^\pm e^\mp$ is being made at Fermilab[47].

Recently, the results of a new search for $K_L^0 \to \mu e$ were reported[48] by the BNL E871 group. They used a dual bend magnetic spectrometer shown in fig. 18 which incorporated a high rate tracking apparatus employing small diameter (5 mm) straw tubes. In addition, an unusual approach of placing a beam plug in the high rate neutral beam upstream of the main apparatus was employed successfully to reduce counting rates in the detectors. The $K_L^0 \to \mu e$ signature is a reconstructed $(\mu^\pm e^\mp)$ pair with no missing transverse momentum and invariant mass $m_{\mu e} = m_K$. Accidental coincidences of the decay products from $K \to \pi e\nu$ to produce the electron and either another $K \to \pi\mu\nu$ decay or a $\pi \to \mu\nu$ decay to produce the μ can result in the searched-for μe final state signature. However, such background events are suppressed by kinematics and the V-A matrix element for $K \to \pi e\nu$ decay resulting in sparse population near the electron's end point energy. Another possible source of background arises when the electron or π from a $K \to \pi e\nu$ decay scatters in the chamber materials to confuse the reconstruction and the signature is completed by $\pi \to \mu\nu$ decay.

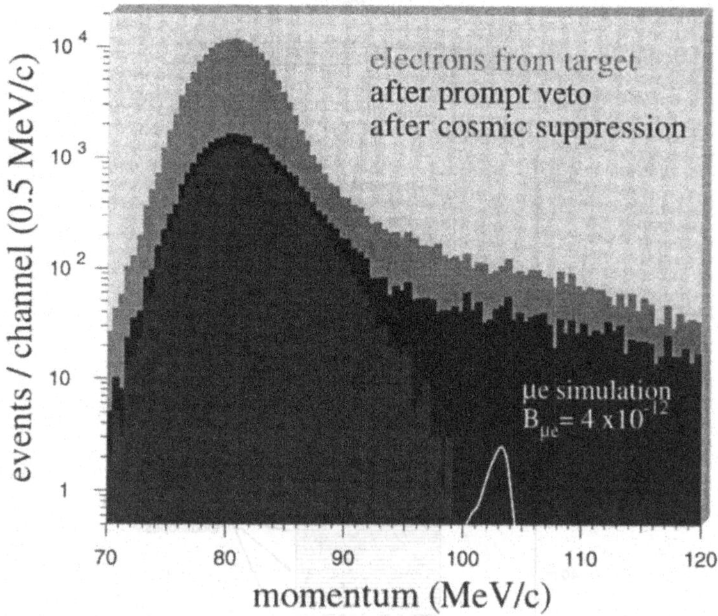

Figure 17. The $\mu \rightarrow e$ conversion data (lower histogram) observed by the SINDRUM II group at PSI. The upper curves represent various stages of analysis and the peak is from a simulation of a signal at the 4×10^{-12} level. [42]

A preliminary analysis indicates that backgrounds due to $K \rightarrow \pi e \nu$ decay may come in at the 10^{-13} level and limit future searches for $K_L^0 \rightarrow \mu e$. In the recent experiment, no events were observed in the signal region where the background level expected was 0.11 events resulting in a preliminary upper limit of $B(K_L^0 \rightarrow \mu e) < 5 \times 10^{-12}$.

4.7. FUTURE SEARCHES FOR LFV

The long term future of LFV searches may be focused on muon reactions like $\mu \rightarrow e\gamma$ and $\mu \rightarrow e$ conversion (see table 1) where the main impediment to progress appears to be beam intensity rather than backgrounds. New proposals are under consideration at several laboratories. For example, a new proposal to study $\mu \rightarrow e$ conversion aiming at $< 10^{-16}$ has been presented at BNL.[50] Major improvements in beam intensity and quality along with technical advances in detectors are required. High duty factor is especially important for these rare decay experiments, some of which also require special pulsing time structure and possibly polarized beams. Innovative beam approaches involving solenoid collection schemes and muon cooling[51] may prove to be the keys to major advances and discoveries.

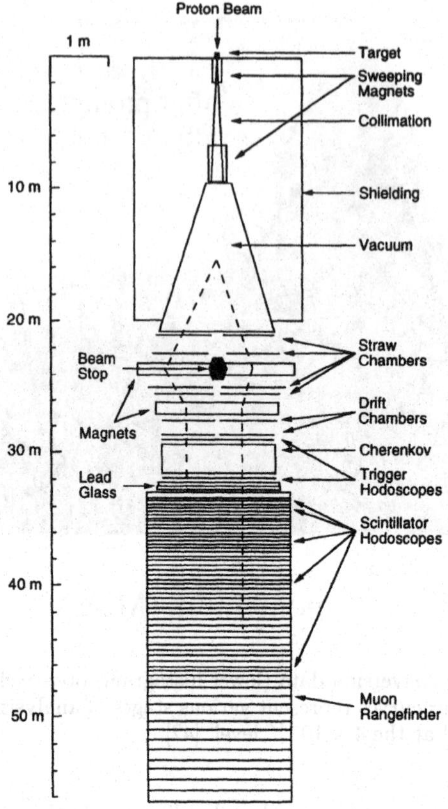

Figure 18. The E871 detector[49] used in the search for $K_L^0 \to \mu e$ and other decays.

4.8. T-violation IN $K_{\mu 3}$ DECAY

In $K^+ \to \pi^0 \mu^+ \nu$ decay, transverse muon polarization (P_μ^T, out of the decay plane), defined by the triple product

$$P_\mu^T = \frac{\vec{s} \cdot (\vec{p_\mu} \times \vec{p_\pi})}{|\vec{p_\mu} \times \vec{p_\pi}|}, \tag{27}$$

is odd under time reversal. Since in the SM, P_μ^T is expected to be near zero ($\sim 10^{-6}$), a measurement of a much larger effect could signal the presence of a new source of T violation. For instance, in some multi-Higgs models, a diagram involving exchange of a charged Higgs particle could interfere with the SM W boson exchange process leading to effects as large as $P_\mu^T \sim 10^{-3}$. While previous experiments have produced only null results at a higher level $P_\mu^T = 0.00185 \pm 0.0036$ (see ref. [52]), a current experiment is operating at KEK (E246).[53] In this effort aiming to substantially improve the sensitivity to P_μ^T, a technique involving stopped kaons (K^+), a toroidal spectrometer and an efficient muon polarimeter is being employed. New results are expected soon. Other new efforts to search for non-SM T-violation in $K_{\mu 3}$ decay have also been proposed for BNL.[54, 55]

5. Conclusion

Considerable progress is being made in measurements of rare kaon and muon decays. New results on ϵ'/ϵ at the 10^{-4} level of precision from studies of $K \to 2\pi$ decays are expected soon from NA48 and E832. These experiments may discover the first clear evidence for direct CP violation, providing consistency with the Standard Model picture for the origin of CP violation.

In order to obtain precise information on poorly known CKM parameters from the K system, it will likely be necessary to measure the ultra-rare decays $K \to \pi\nu\bar{\nu}$. Extraordinary theoretical precision allows the measurements of $K^+ \to \pi^+\nu\bar{\nu}$ and $K_L^0 \to \pi^0\nu\bar{\nu}$ to unambiguously test the SM origin of CP violation and, perhaps, to eventually yield the most accurate determinations of the CKM CP violating phase η and $|V_{td}|$. Although the branching ratios are small, these rare decays provide exceptional opportunities for making progress in our understanding of flavor-dynamics and CP violation. Since some non-SM approaches predict that different values of CKM parameters may be found in the K and B systems, whereas, in the SM, the same values must obtain, having measurements in both systems is essential.

At BNL, E787, having observed a candidate $K^+ \to \pi^+\nu\bar{\nu}$ event at the 10^{-10} level, is now approaching the sensitivity necessary to closely examine the SM prediction. A new experiment has been proposed at BNL with the aim of making a definitive measurement of $K_L^0 \to \pi^0\nu\bar{\nu}$. Advanced efforts to search for $K_L^0 \to \pi^0\nu\bar{\nu}$ are also planned at KEK and Fermilab. Absence of $K^+ \to \pi^+\nu\bar{\nu}$ or $K_L^0 \to \pi^0\nu\bar{\nu}$ within the expected ranges or conflicts with other measures of quark mixing parameters would indicate new physics.

Rare forbidden kaon and muon reactions such as muonium-antimuonium conversion, nuclear $\mu \to e$ conversion and $K_L^0 \to \mu e$ which do not conserve lepton flavor are being sought by current and planned experiments of unprecedented sensitivity. Recent experiments have pushed the limits down by several orders of magnitude; for instance, the branching ratio limit for $\mu \to e$ conversion was set at $< 6.1 \times 10^{-13}$ in an experiment at PSI. New experiments are in the planning stages to make large further gains in sensitivity. Confirmed non-zero findings in any of them would be revolutionary.

There is also a growing list of improved results and first time observations of rare radiative decays like $K_L \to e^+e^-$, $K \to \gamma\gamma$, $K_L \to \mu\mu$, $K \to \pi\gamma\gamma$, $K \to \pi\mu\mu$ and $K^+ \to \mu\nu\gamma$ which have been produced in conjunction with the current round of experiments at BNL, KEK, Fermilab and CERN. These are of considerable interest in light of theoretical developments in low energy QCD, in particular chiral perturbation theory.[20] In addition, there is great promise for improved tests of CP and CPT violation at the new Φ factory DAΦNE.

References

1. See, for example, Bryman, D.A., Comm. Nucl. Part. Phys. **21**, 101 (1993); A. Pich, *Proc. Int. Worksp. on Phys. Beyond the Standard Model: From Theory to Experiment* (Valencia 97), Valencia, Spain, hep-ph/9802257 (1997).
2. See the lecture notes by Nakamura, K., in these proceedings.
3. Wolfenstein, L., Phys. Rev. Lett. **13**, 562 (1964); see also Wolfenstein, L., Comm. Nucl. Part. Phys. **21**, 275 (1994).
4. Hincks, E.P. and Pontecorvo, B., Phys. Rev. **74**, 697 (1948).
5. Ritchie, J.L. and Wojcicki, S.G., Rev. Mod. Phys. **65**, 1149 (1993).
6. Winstein, B. and Wolfenstein, L., Rev. Mod. Phys. **65**, 1113 (1993).
7. Littenberg, L. and Valencia, G. Ann. Rev. Nucl. Part. Sci. **43**, 729 (1993).
8. Christenson, J., Cronin, J., Fitch, V., and Turlay, R., Phys. Rev. Lett. **13**, 138 (1964).
9. Wolfenstein, L., Phys. Rev. Lett. **51**, 1945 (1983).
10. Particle Data Group, Eur. Phys. J. **C3**, 1 (1998).
11. Kokkas, P., *et al.*, Proc. Int. Conf. H.E.P., ed. A. Astbury, Vancouver (1998), to be published.
12. Buras, A.J. and Fleischer, R., Proc. Heavy Flavours II, World Scientific (1997), eds. Buras, A.J., and Linder, M., (to be published) (hep-ph/9704376).
13. Gilman, F., and Wise, M., Phys. Lett. **B83**, 83 (1979).
14. Flynn, J.M. and Randall, L., Phys. Lett. **B224**, 221 (1989); erratum *ibid.*Phys. Lett. **B235**, 412 (1990).
15. Gibbons, L.K., *et al.*, Phys. Rev. **D55**, 6625 (1997).
16. Barr, G.D., *et al.*, Phys. Lett. **B317**, 233 (1993).
17. Cheu, E., *et al.*, *Proc. Int. Conf. H.E.P.*, *op.cit.*
18. Bluemer, H., *et al.*, *Proc. Int. Conf. H.E.P.*, *op.cit.*
19. See Patera, V., *Proc. Workshop on K Physics*, ed. L. Iconomidou-Fayard, Editions Frontieres, p.99 (1997).
20. See Pich, A., *Proc. Workshop on K Physics*, *op. cit.*, p.353.
21. Harris, D.A., *et al.*, Phys. Rev. Lett. **71**, 3918 (1993).
22. O'Dell, V., *Proc. Workshop on K Physics*, *op. cit.*, p.87.
23. Rein, D., and Sehgal, L.M., Phys. Rev. **D39**, (1989); Hagelin, J.S., and Littenberg, L.S., Prog. Part. Nucl. Phys. **23**, 1 (1989); Lu, M., and Wise, M.B., Phys. Lett. **B324**, 461 (1994).
24. Buras, A.J., *Probing the the Standard Model Interactions*, eds. David, F., and Gupta, R., Elsevier, (to be published) (1998).
25. Buchalla, G., and Buras, A.J., Nucl. Phys. **B412**, 106 (1994).
26. Littenberg, L.S., Phys. Rev. **D39**, 3322 (1989).
27. Marciano, W., and Parsa, A., Phys. Rev. **D53**, R1 (1996).
28. See the lecture notes by Aleksan, R., in these proceedings.
29. Adler, S., *et al.* Phys. Rev. Lett. **79**, 2204 (1997).
30. Barker, A., *et al.*, *Proc. Intl. Conf. H.E.P.*, *op.cit.*
31. Weaver, M., *et al.*, Phys. Rev. Lett. **72**, 3758(1994).

32. see Inagaki, T., *Proc. Symp. on Flavor Changing Neutral Currents*, ed. D. Cline, World Sci., p. 109 (1997).
33. Cheu, E., *et al.*, FERMILAB-PUB-97-321-E.
34. Chiang, I.H., *et al.*, BNL AGS proposal E926.
35. Wilczek, F., Phys. Rev. Lett. **49**, 6782(1982).
36. Gvozdev, A.A., *et al.*,Phys. Lett. **B345**, 490 (1995).
37. Hisano, J., Nomura, D., Yanigida, T., KEK-TH-548, hep-ph/9711348 (1997).
38. Barbieri, R., and Hall, L., Phys. Lett. **B338**, 212 (1994).
39. Sirois, Y., *Proc. WEIN Symp.*, Santa Fe (1998), eds. Hoffman, C., and Herczeg, P., to be published.
40. Bolton, R., *et al.*, Phys. Rev. **D38**, 2121 (1988).
41. Belgardt, U., *et al.*, Nucl. Phys.**B299**, 1 (1988).
42. Wintz, P., *Proc. Int. Conf. H.E.P.*, *op.cit.*
43. Willmann, L., *et al.*, hep-ex/9807011 (1998).
44. Stanislaus, S., *et al.*, *Proc. Symp. on Flavor Changing Neutral Currents*, *op. cit.*, p. 143.
45. Ahmed, S., *et al.* Phys. Rev. **D38**, 2102 (1988).
46. Lee, A.M., *et al.*, Phys. Rev. Lett. **64** 165 (1990).
47. Arisaka, K., *et al.*, Phys. Lett. **B432**, 230 (1998).
48. Bachman, M., *et al. Proc. Int. Conf. H.E.P.*, *op.cit.*
49. Belz, J., *et al.* UTEXAS-HEP-98-13 (1998); hep-ex/9808037.
50. see Molzon, W., *Proc. Lepton-Baryon '98*, ed. Klapdor, H., to be published.
51. Palmer, R., *et al.* BNL-65627, (1998).
52. Blatt, S.R., *et al.*, Phys. Rev. **D27**, 1056 (1983).
53. see Kuno, Y., *et al.*, *Proc. Int. Conf. H.E.P.*, *op.cit.*
54. Imazato, J., *et al.*, BNL proposal E923 (1996).
55. Diwan, M., *et al.*, BNL E923 proposal (1996).

EXPERIMENTAL TESTS OF THE STANDARD MODEL

L. NODULMAN
Argonne National Laboratory
High Energy Physics Division
Argonne IL 60439 USA

1. Introduction

The title implies an impossibly broad field, as the Standard Model includes the fermion matter states, as well as the forces and fields of $SU(3) \times SU(2) \times U(1)$. For practical purposes, I will confine myself to electroweak unification, as discussed in the lectures of M. Herrero. Quarks and mixing were discussed in the lectures of R. Aleksan, and leptons and mixing were discussed in the lectures of K. Nakamura. I will essentially assume universality, that is flavor independence, rather than discussing tests of it.

I will not pursue tests of QED beyond noting the consistency and precision of measurements of α_{EM} in various processes including the Lamb shift, the anomalous magnetic moment (g-2) of the electron, and the quantum Hall effect. The fantastic precision and agreement of these predictions and measurements is something that convinces people that there may be something to this science enterprise.

Also impressive is the success of the "Universal Fermi Interaction" description of beta decay processes, or in more modern parlance, weak charged current interactions. With one coupling constant G_F, most precisely determined in muon decay, a huge number of nuclear instabilities are described. The slightly slow rate for neutron beta decay was one of the initial pieces of evidence for Cabbibo mixing, now generalized so that all charged current decays of any flavor are covered.

QCD has also evolved an impressive ability to predict a wide range of measurements with a universal coupling, α_S. Tests of QCD were covered in the lectures of J. Stirling. Clearly the issues of associating final state jets with quarks and gluons, and of analyzing proton structure in terms of quarks and gluons will be important in many experimental tests of electroweak unification.

T. Ferbel (ed.), Techniques and Concepts of High Energy Physics X, 147–192.

The lack of renormalizability of the Fermi theory of charge current weak interactions, that is the inability to calculate radiative corrections, and thus bad behavior in the high energy limit, was the motivation for models of unification. One of several such models, which, in the early 70's went under the name "Weinberg-Salam," [1] has become "Standard." A simple-minded picture of this model is that by combining an isosinglet and a isotriplet of gauge bosons, one mixes up the γ for QED, heavy $W^{\pm}$ bosons for the weak charged current, and a heavy Z^0 which predicted the weak neutral current interaction. A third parameter describing the neutral weak interaction can be taken as some definition of the weak neutral triplet/singlet mixing angle, Θ_W, or more practically, as the rather precisely measured mass of the Z^0 boson. A fourth parameter is needed, associated with consistency in heavy gauge boson masses; this may be taken as the Standard Model Higgs mass, which is largely decoupled from observables.

For practical purposes, I will consider the Standard Model to include the simplest Higgs mechanism, one complex doublet, with one residual Higgs particle with a mostly unpredicted mass. Implementation of the Higgs mechanism in terms of fundamental scalar multiplets may well be just a mathematical trick; one certainly hopes that nature could not really be so unimaginative. But since the simplest scheme is still viable, I will take it as standard. I note that in terms of multiple Higgs states, infinite variety is possible, although there are some constraints from measurements of the ρ parameter, as M. Herrero discussed in terms of "custodial symmetry."

The unified electroweak theory does allow calculation of radiative corrections. These corrections give terms involving squares of fermion masses, and logarithms of scalar masses, for example, in predicting the weak mixing angle and/or the W boson mass. That is, heavier masses imply larger effects. These would include particles we may not know about, as well as particles regarded as standard. So if everything hangs together in terms of the top quark mass and the Higgs mass, we obtain constraints on what else could be out there.

Precision measurements to challenge these predictions have been an essential feature of e^+e^- collider studies culminating in the LEP and SLC programs. The predictions serve as a motivator for high energy hadron collider programs, from the $Sp\bar{p}S$ collider at CERN, to the current Fermilab Tevatron Collider, and the LHC being built at CERN. Most processes involve "oblique" or propagator corrections; these involve top mass squared and Higgs mass logarithmic terms. Notably the Z^0 decay rate to b pairs, R_b, depends on vertex corrections which depend essentially on the top mass. One of the claims of success of this precision measurement program has been the consistency of the indirectly implied top mass with the hadron collider top mass limits and, eventually, top mass measurements. Thus, the measurement of the top mass is an essential part of the program, which has moved on to constraining the Higgs.

I will review several measurements to illustrate this program: The new muon (g-2) experiment at Brookhaven illustrates both weak and QED measurement. The NuTeV experiment at Fermilab illustrates the historically important and currently still competitive contribution of neutrino physics. The LEP precision Z^0 lineshape and Z^0 decay asymmetry measurements are clearly the core of this program. The ultimate Z^0 asymmetry measurement comes, of course, from SLC. The increasingly precise W boson mass measurements at the Tevatron have been joined by measurements at LEP2. The top mass measurements complete the indirect Higgs picture, but the direct Higgs search at LEP2 has a significant impact as well.

2. BNL E815 MUON (g-2)

2.1. GOALS OF THE MEASUREMENT

The anomalous magnetic moment of the muon, defined as

$$a_\mu \equiv \mu_\mu/(e\hbar/2m_\mu) - 1 = (g_\mu - 2)/2,$$

is a less favorable QED test than the electron magnetic moment. The heavier muon mass makes radiative corrections involving hadronic states relatively important. The heavier muon mass also makes electroweak radiative corrections relatively important. Previous measurements at CERN[2] were precise enough to establish the presence of hadronic corrections. The goal at Brookhaven is to measure a_μ well enough to get a handle on the electroweak radiative corrections; a requirement for this is a precise enough prediction of hadronic corrections to allow the EWK corrections to be isolated.[3]

TABLE 1. Values and corrections to a_μ in units of 10^{-11}.

Quantity	Value	Error
QED prediction	116584706	2
EWK correction	151	4
HAD correction	6771	77
Overall prediction	116591628	77
CERN measurement	116592300	840

The magnetic moment numbers are given in Table 1. The QED prediction is calculated to α^5[4] and the EWK correction is calculated to two

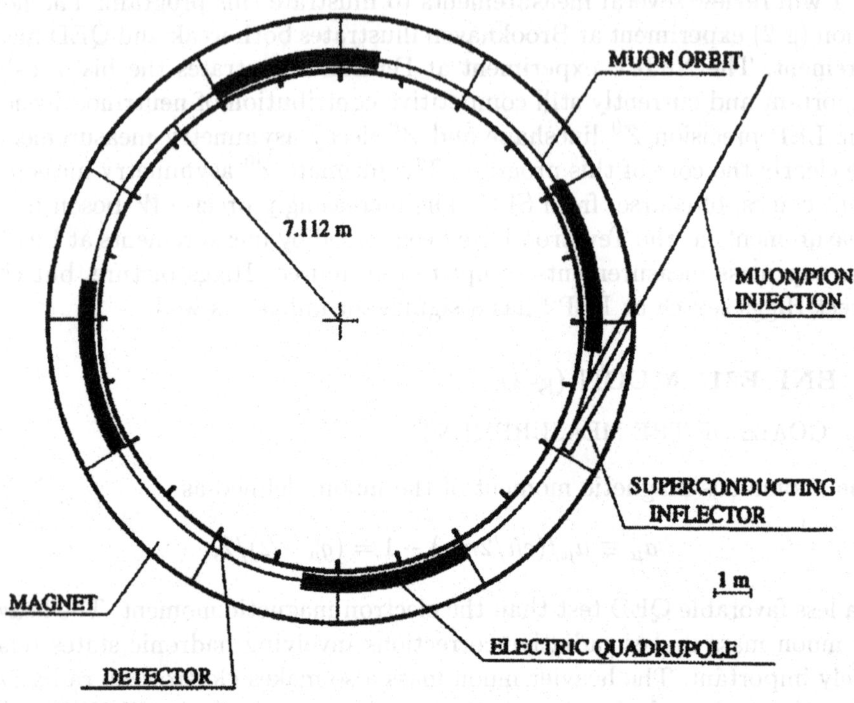

Figure 1. The Brookhaven g-2 muon storage ring.

loops.[4, 5] The leading order hadronic correction, a vacuum polarization loop with hadrons, must be determined with dispersion relations, using the low energy $e^+e^- \to hadrons$ measurements.[6] These measurements are being improved at CMD2 in Novosibirsk, BES, and in τ decay studies. Higher order hadronic corrections, and light-by-light or diagrams with four photons converging on a loop[7] are relatively under control.

While it is clear that the current level of prediction creates a market for improving on the CERN measurement, the predicted level of EWK corrections at 151×10^{-11} is not all that much bigger than the error of 77×10^{-11} in the prediction, which is dominated by the hadronic uncertainty. Improving the low energy e^+e^- measurements is a clear concurrent goal. The Brookhaven goal for measurement precision is $\pm 35 \times 10^{-11}$ for both μ^+ and μ^-.

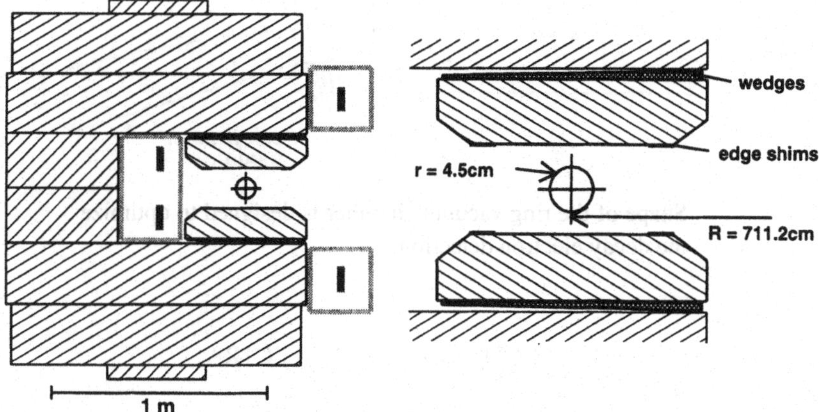

Figure 2. The profile of the g-2 muon storage ring magnet. The boxes are the vessels for superconducting cable. The detail on the right shows allowance for field adjustment.

2.2. METHOD OF MEASUREMENT

The charged pion decays by the weak charged current to $e\bar{\nu}_e$ or $\mu\bar{\nu}_\mu$. The V-A form implies that the zero (or negligible) mass of the neutrino forces helicity selection. To conserve angular momentum in the decay of the zero spin pion, the massive charged lepton must be in the helicity state disfavored by V-A. Since the muon is much heavier than the electron, the electron channel is much more suppressed by helicity, $\sim 10^{-4}$. The muons are longitudinally polarized. So if you have a beam of pions, muons from forward decay have the least momentum change and will most likely remain in the beam. Thus, one can readily obtain beams of longitudinally polarized muons. The basic method of the experiment is to put polarized muons in a magnetic field and measure their spin precession. The highest momentum electrons in the decay $\mu \to e\bar{\nu}_e\nu_\mu$ analyze the muon spin.

The experiment is done by putting muons into a storage ring. Lots of physics results come from various storage rings, but the g-2 ring, shown in Fig. 1, is rather different. For a precision measurement, you need as well known and as uniform a magnetic field as possible. Thus, the ring is a continuous bending magnet. The injection needs to be carefully done to avoid messing that up. The time scale for storage is short due to the muon lifetime, so imbedded electrostatic quadrupoles are sufficient to capture the beam. Decay electrons are detected in stations at the array of windows.

Gedanken Problem. The g-2 ring uses an iron dominated superconducting magnet, which would not be much stronger than typical conventional storage ring bending magnets. The design momentum is 3.1 GeV/c. The

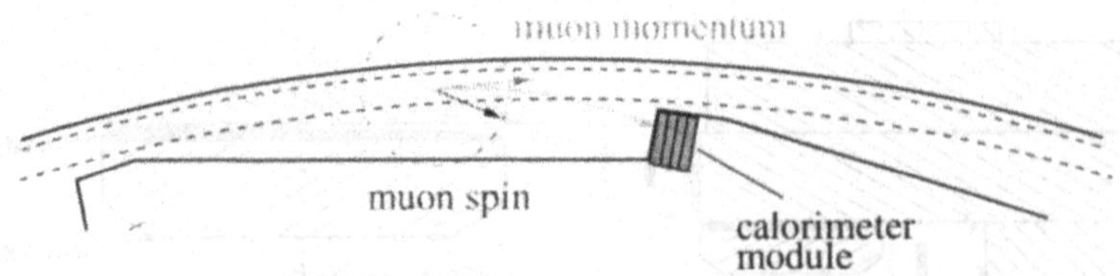

Shape of the ring vacuum chamber is designed to optimize the decay electron detection.

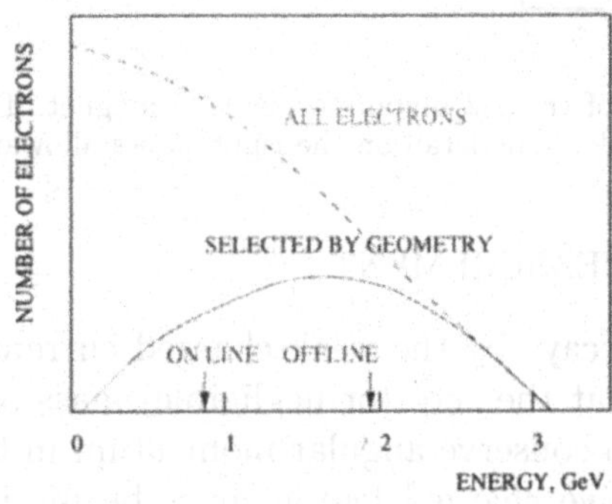

High energy electrons provide most information about spin rotation signal.

Figure 3. Top: the decay electron window in the vacuum. Bottom: the energy spectrum of detectable electrons with online and offline thresholds marked.

SPEAR storage ring at SLAC, original home of Ψs, χs, Ds, τs etc., has a design momentum of 4 GeV/c, yet it is something like five times larger than the g-2 ring. Why?

Injection is clearly an important problem. So far, they have used pion decay. at the appropriate fortunate time, to give a slight momentum kick for injection. This is rather inefficient, and the pions which do not decay run into things, creating a blast of stuff in the detectors, so that measurements must wait till the detectors are stable. A fast kicker is in the works so that a muon beam can be directly injected.

Magnet quality is an ongoing project. The ring magnet profile is shown in Fig. 2. Field mapping and orbit studies feed back to shimming the iron pole tips. The profile and absolute value are both important; an NMR

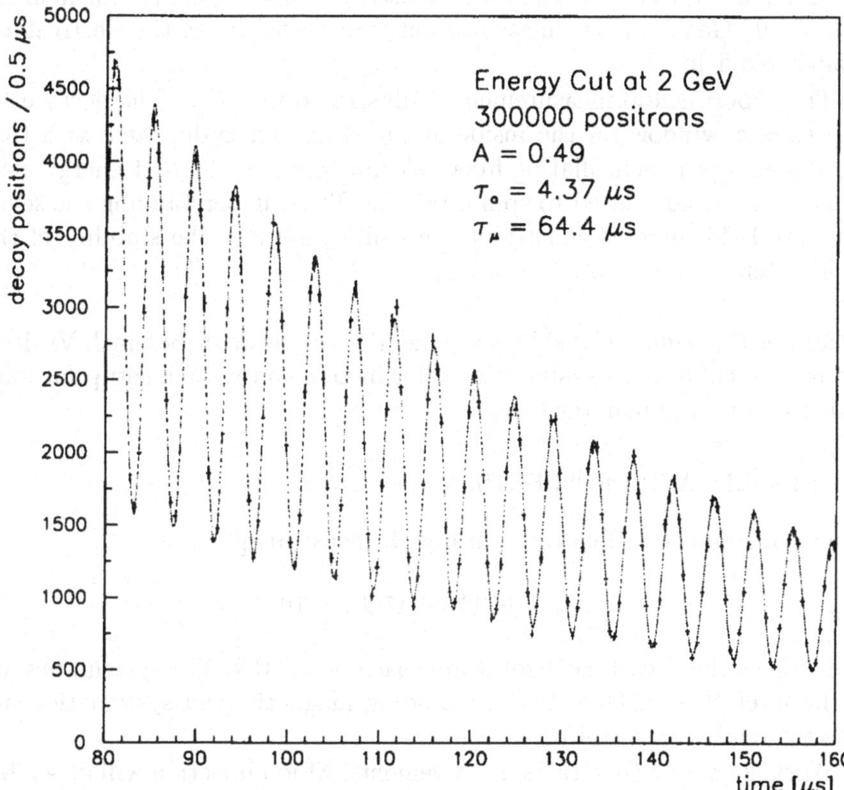

Figure 4. **The time spectrum of detected electrons. The period gives a_μ. They may eventually improve G_F.**

probe provides the absolute field value by reference to the proton magnetic moment.

Gedanken Problem. The iron dominated superconducting magnet provides good opportunity to control field quality. For the beyond LHC generation hadron collider, VLHC or Eloiseatron, what are the tradeoffs between high field (coil dominated) and iron dominated bending magnets?

What you want to measure is the difference in the cyclotron frequency and the spin precession frequency given by

$$\omega_a = \frac{e}{mc}[a_\mu B - (a_\mu - 1/(\gamma^2 - 1))(B \times E)].$$

At Brookhaven, as at CERN, they choose the "magic gamma" for momentum of 3.09 GeV/c, so the measurement is not sensitive to the electrostatic quadrupole field.

The experimental measurement is illustrated in Fig. 3. The decay electron exits a window on the inside of the ring, and is detected with timing and energy in scintillating fiber calorimeters. The highest energy electrons are most correlated to spin direction. Thus, understanding the selection threshold imposes constraints on calibration and the stability of the calorimeter.

Gedanken Problem. Consider the general cases, as described in J. Virdee's lectures, of calibration systems for calorimeters contrasting setup for data taking and *in situ* maintenance.

2.3. RESULTS AND PROSPECTS

The measurement is illustrated in Fig. 4, measuring[8]

$$a_{\mu^+} = 116592500(1500) \times 10^{-11}.$$

The level of the statistical error is approaching CERN. The systematics are at the level of $\sim \pm 300 \times 10^{-11}$, including magnetic field systematics and detector related uncertainty.

Work is going into various improvements. Muon injection will give a big statistical boost. Field improvements with shims and trim coils will improve systematics, as will improved detector understanding and tracking of the electrons to verify where the muon beam is. And, of course, the hadronic correction prediction is also getting attention.

In 1998, they expect to get to the level of $\pm 115 \times 10^{-11}$ for μ^+. Full design precision for both muon charges should be realized in 2002.

3. NuTeV

3.1. GOALS OF THE MEASUREMENT

NuTeV is the latest incarnation of the CCFR (Chicago, Columbia, Fermilab, Rochester) neutrino experiment at Fermilab. They were off looking for wrong-sign heavy leptons when Gargamelle observed neutral currents. Their confirmation of neutral currents at the 1974 London Rochester conference silenced the many vocal skeptics, thus contributing to the recognition of electroweak unification.

The basic measurement then and now is to measure the relative rate of neutral current (NC) and charged current (CC) interactions, measuring

$$\sin^2\theta_W^{\text{on-shell}} \equiv 1 - \frac{{M_W}^2}{{M_Z}^2} = 0.2268 \pm 0.0037$$

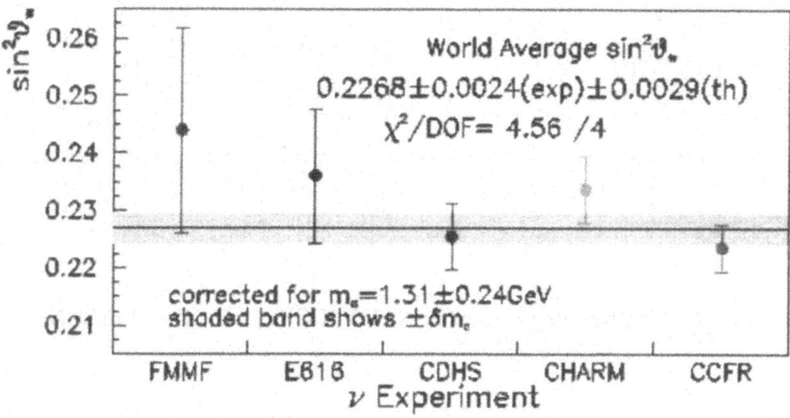

All other experiments are corrected to
CCFR & CDHS average m_c and to large M_{top} ($M_{top} > M_W$)

Figure 5. The evolution of neutrino weak mixing measurements showing the charm systematic limit.

the on-shell weak mixing angle

$$sin^2\Theta_W^{on-shell} \equiv 1 - \frac{m(W)^2}{m(Z)^2}.$$

The experiment uses muon neutrinos and distinguishes NC from CC by the absence of penetrating muons in the final state. Over several generations of such experiments, systematics such as detailed understanding of the boundaries of the detectors, and more importantly, the charm quark mass, have limited the measurement, as seen in Fig. 5. Neutrinos can interact with a strange sea quark, producing charm; the charm decay can give a muon, complicating the measurement. The charm mass is needed to predict the CC rate. Two muon events can be used to get some measure, but this seems to have plateaued at the level, in the appropriate QCD order, of $m_c \sim 1.3 \pm 0.3$ GeV/c^2. Another important systematic effect is the level and understanding of ν_e in the beam, as electron CC look like muon NC. Electron neutrinos from decays of K_L^0 from the production target are a problem due to significant uncertainties in the K_L^0 production rate.

The basic goal of the NuTeV measurement is to escape these systematic limits. By creating a beam with very little neutrino/antineutrino cross talk, it can use the Paschos-Wolfenstein relation[9]

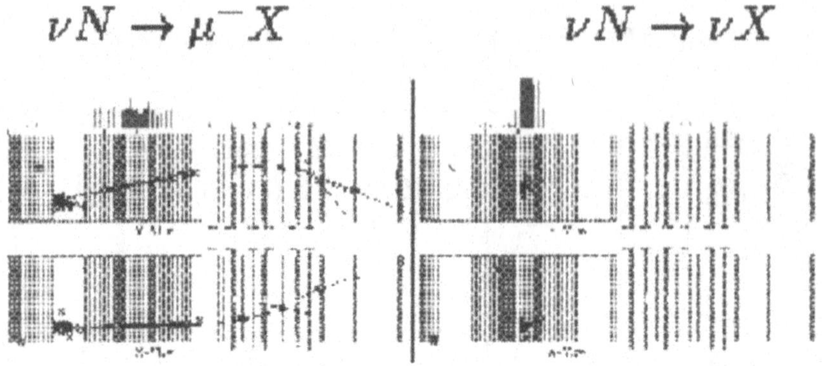

Figure 6. Neutrino interactions in NuTeV/CCFR, left CC and right NC.

$$(\sigma_{NC} - \sigma_{\bar{N}C})/(\sigma_{CC} - \sigma_{\bar{C}C}) = \rho^2(1/2 - sin^2\Theta_W),$$

obtaining weak mixing from the difference of neutrino and antineutrino cross sections. Thus, the sea cancels and the remaining difference in charm production, due to valence d quark, is Cabbibo suppressed. The goal is to make a competitive inference of the W mass.

3.2. METHOD OF MEASUREMENT

The CCFR detector consists of 690 tons of tracking calorimeter with transverse square planes of iron, liquid scintillator, and drift chambers. The central 390 tons are considered fiducial. This is followed by an extensive system of drift chambers and iron toroids to measure muon momentum. The basic measurements are illustrated with event pictures, see Fig. 6.

The most significant change for NuTeV, compared to CCFR, is the beam. Neutrino experiments typically use zero degree production to optimize yield, using magnetic horn focusing for broad-band beams, or beam line (quadrupole) focusing for narrow-band beams. The NuTeV beam, shown in Fig. 7, looks away from zero degrees, avoiding sensitivity to upstream scraping, reducing the contribution from K_L^0 decays, and thoroughly removing wrong sign particles. The usual decay region is followed by muon shielding, then the detector. Thus the beam should be well understood, as is illustrated in Fig. 8. Note the two peaks corresponding, left-to-right, to neutrinos from pion and kaon decays. Residual beam systematics are dominated by the $K^\pm$ e3 branching ratio uncertainty. The flux can be studied with quasielastic CC events, $\nu p \to \mu n$ with perhaps nuclear breakup, but not pion production.

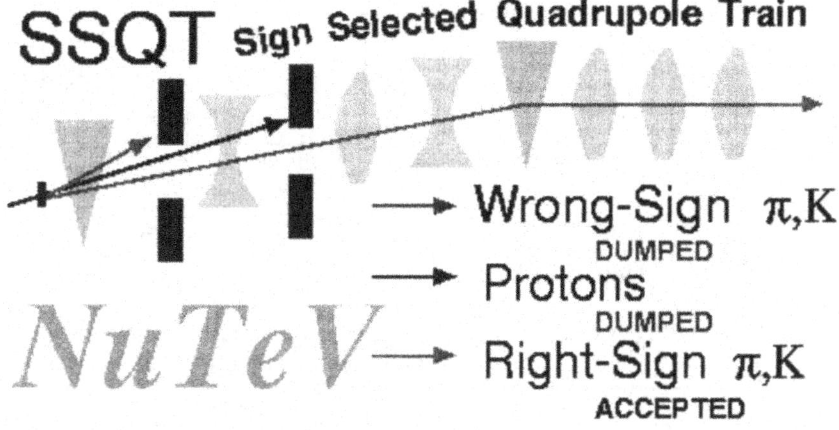

Figure 7. NuTeV production target and beam line into the decay region.

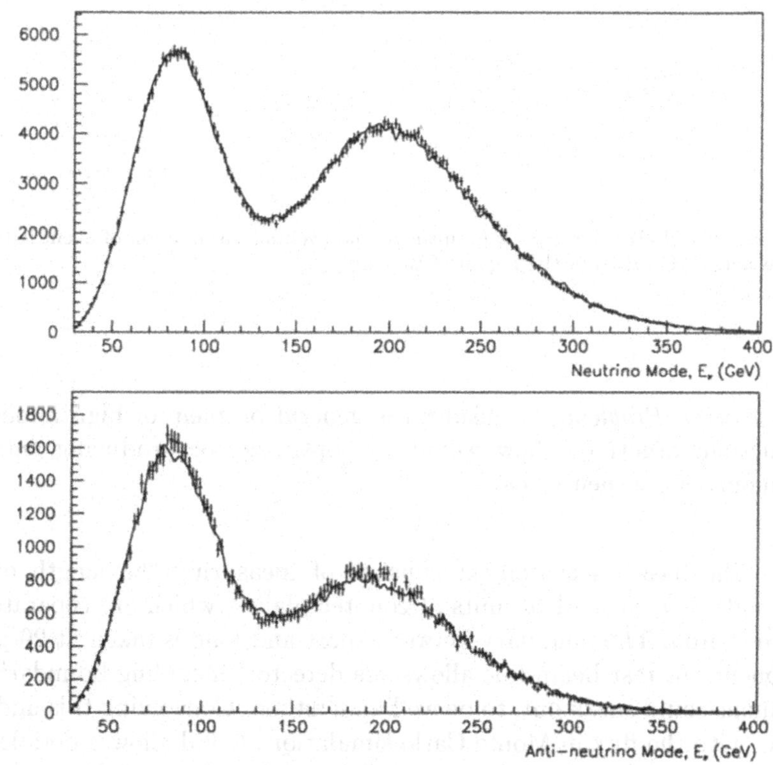

Figure 8. NuTeV neutrino and antineutrino flux, detected and MC.

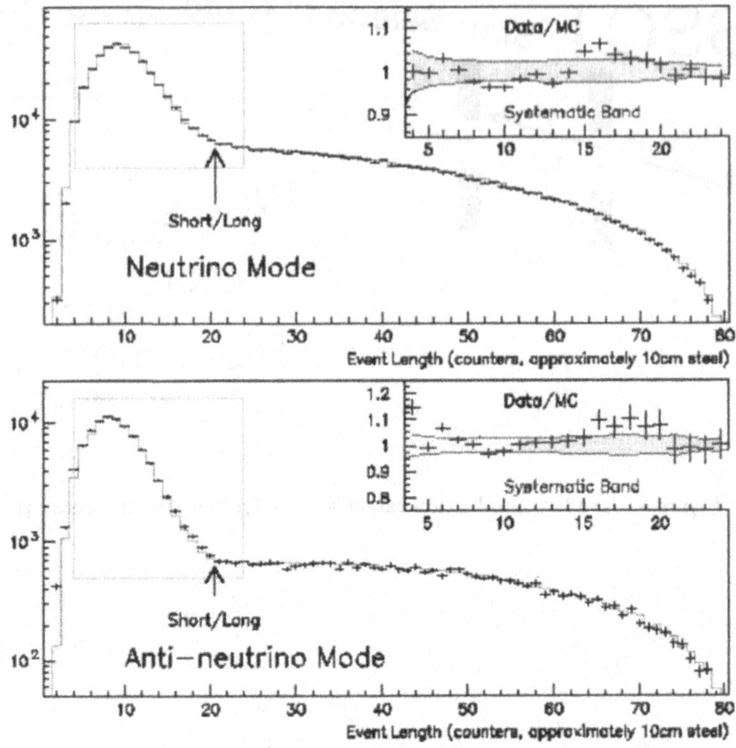

Figure 9. NuTeV neutrino and antineutrino events as a function of event length. Insets are data/MC ratios in the region of the cut.

Gedanken Problem. Consider the general problem of high intensity production targeting - how would you optimize for producing antiprotons, muons, or tau neutrinos?

The basic measurement consists of measuring the length of events. Length is measured in units of counter planes, which are separated by 10 cm of iron. The boundary between short and long is taken at 20 planes. A concurrent test beam line allows the detector, including boundaries where muons could sneak out, to be well understood. Convolving this understanding with the flux in Monte Carlo simulation should allow a detailed understanding of length distributions as illustrated in Fig. 9. The region of the cut is reasonably well described, as shown in the insets, and the ratio short/long can be unfolded to give NC/CC.

3.3. RESULTS AND PROSPECTS

The NuTeV run was completed in 1997, and preliminary results have been given.[10] The statistical precision for $sin^2\Theta_W$ is ± 0.00190, comparable to the CCFR result. The physics model systematics level is ± 0.00070, with detector systematics at ± 0.00075. Unlike CCFR, the NuTeV measurement is statistically dominated. Using the measured top mass, the NC/CC ratio gives $sin^2\Theta_W = 0.2253 \pm 0.0022$, which corresponds to $m(W) = 80.26 \pm 0.11$ GeV/c^2. Combining NuTeV with the CCFR result gives

$$sin^2\Theta_W^{on-shell} = 0.2255 \pm 0.0021.$$

Their goals have been largely achieved. Some improvement in experimental systematics may come with further analysis. No further data taking is planned.

4. Digression on Collider Detectors

The experiments remaining are at either e^+e^- or $\bar{p}p$ colliders. Almost all such collider detectors, as well as those at HERA, are quite similar. Good examples, ATLAS and CMS for LHC, were thoroughly described in the lectures of F. Pauss. A tracking volume around the interaction is defined inside a solenoid. These days drift chambers for magnetic tracking are complemented, as discussed in the lectures of I. Abt, by silicon detectors. There may also be particle ID. The tracking volume is surrounded by calorimeters, and a muon identification system surrounds that. This is illustrated by SLD, shown in Fig. 10.

Electrons are identified by matching a track to suitable calorimeter measurements. Muons match a track inside to a track or track stub outside. To identify τ leptons, one demands one or three tracks corresponding to a relatively narrow calorimeter energy cluster, isolated from other activity. In the case of hadron colliders, the detectors must make ID information available for fast triggers.

The silicon detectors are used to identify b quarks by observing secondary vertices. This procedure is usually marginal for charm, and c quarks tend to be identified by reconstructing exclusive D decay modes or using the soft pion from $D^* \to \pi^\pm D$ with partial reconstruction of the D to show the low Q for the pion. Clearly, b and c identification are correlated; charm decay is only three times faster, and bottom decays produces charm.

Gedanken Problem. How would you trigger, at a hadron collider, on τ leptons? Has this worked?

The first detector of this form was Mark I at SPEAR in the early 70s. CDF at the Tevatron was the first such for hadron colliders. Notable ex-

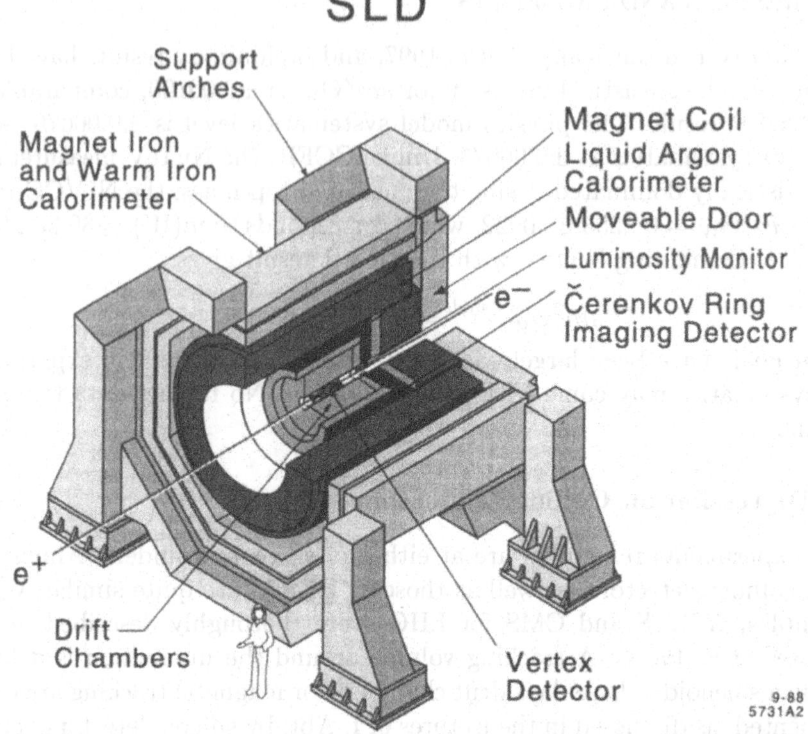

Figure 10. SLD as an example of a generic collider detector. The warm iron calorimeter doubles as a muon detector.

ceptions with no central field are the Crystal Ball, UA2, and pre-upgrade DØ. UA1 had a dipole magnet.

5. LEP1 *Z* Studies

5.1. GOALS OF THE MEASUREMENT

The basic goal of the LEP1 program was a comprehensive study of the Z boson. The Z is observed as an s channel resonance in e^+e^- collisions. The important issues here are the Z lineshape, basic aspects of decays, and in particular the charge and τ polarization asymmetries which result from parity violation. The presence of both vector and axial vector coupling gives asymmetries which measure weak mixing. A definition of the weak mixing angle $sin^2\Theta_{Weff}$ is used which avoids loss of precision to uncertain top and Higgs mass dependence. For leptons, this is simply

$$sin^2\Theta_{eff}^{lept} \equiv 1/4(1 - g_{V\ell}/g_{A\ell}).$$

Vacuum Chamber Current Correlation

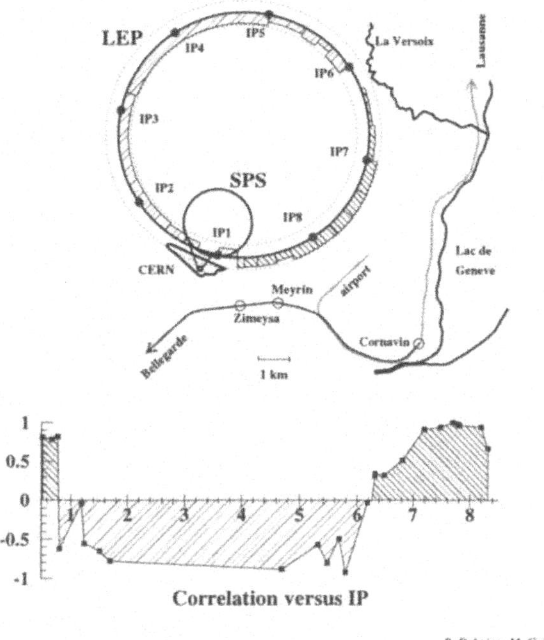

B. Dehning, M. Geitz

Figure 11. The LEP ring showing proximity to Lake Geneva and electric rail lines. Also shown is ground current in the beam pipe. L3, Aleph, Opal and Delphi are at IPs 2, 4, 6 and 8. The circumference is 27 km.

Checking the electroweak radiative corrections was one of the central original motivations for the LEP program.

5.2. METHOD OF MEASUREMENT

Four detectors, Aleph, Delphi, L3, and Opal, use the LEP collider, shown in Fig. 11. The detectors need to identify the interactions with definable efficiencies, and tell the $\pm e^+$ direction. One concern is to measure the absolute luminosity; small high precision calorimeters are used to count small angle Bhabha (e^+e^- elastic) scattering. The experimental method has evolved sufficiently that the luminosity uncertainty is dominated by the QED calculation.[11]

A greater concern is the absolute beam energy calibration, which determines the precision of the Z mass measurement. Survey and field mapping, as for g-2, is not good enough by itself. The basic measurement uses res-

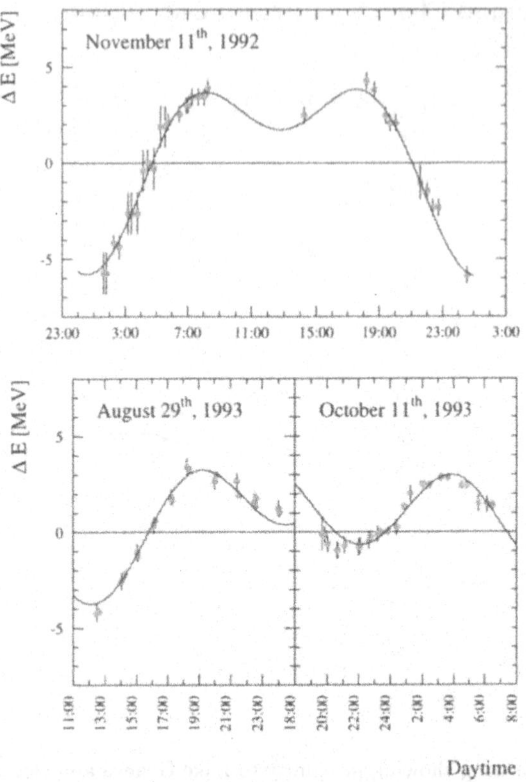

Figure 12. The LEP beam energy correlation to the moon, revealing the effect of tides.

onant depolarization.[12] With favorable accelerator parameters avoiding resonances, synchrotron radiation will tend to polarize the beams. Applying transverse RF, with precisely the correct frequency, causes the polarization to be lost. The intrinsic width for this procedure corresponds to 200 KeV. One corrects for RF energy gain and synchrotron radiation loss as appropriate from the measurement to each interaction point.

In practice, during energy scans, this measurement is made at the end of stores. The measurements show a 5 MeV spread in beam energy predicted from field and orbit measurements minus resonant depolarization values. If the depolarization measurement times are representative, one can combine them statistically to get absolute energy. But precision and confidence are improved if one can understand and model the time dependence which causes the spread.

The LEP program is clearly big science. The first correlation found was with the phase of the moon, shown in Fig. 12. The tidal expansion and contraction of the LEP ring is noticeable. Another influence on the

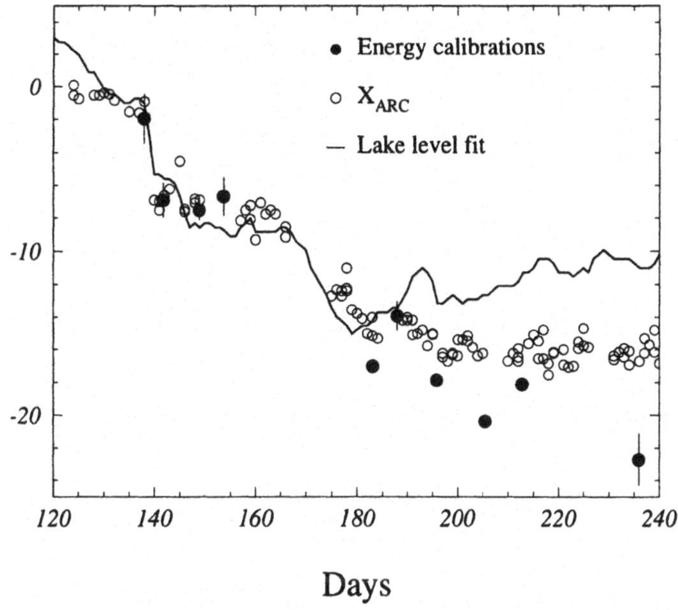

Figure 13. The LEP beam energy correlation to the water level of Lake Geneva.

size of the LEP ring is the amount of water in the ground; this is fairly well modelled by the water level of Lake Geneva, as see in Fig. 13. These correlations gave reasonable confidence in measurements through 1994.

A major energy scan was undertaken in 1995. During the run, unexpected time dependences were found in NMR readings. The patterns were found to be reasonably regular with daytime. Eventually these were found to be correlated with running of the TGV trains, as seen in Fig. 14. Apparently the TGV actually uses the ground to return current, and significant levels of current were found in the LEP beam pipe, shown in Fig. 11. A time dependent correction was determined, and a retroactive correction applied to previous data.

Charge asymmetries are straightforward to measure, although physics interpretation for hadrons can involve considerable sophistication in QCD and in relating jets to quarks and understanding charge correlations and efficiencies. The polarization of τ leptons is analyzed in the decay modes $\pi\nu$ and $\rho\nu$, as well as $a_1\nu$ and leptonic decays. The polarization angular

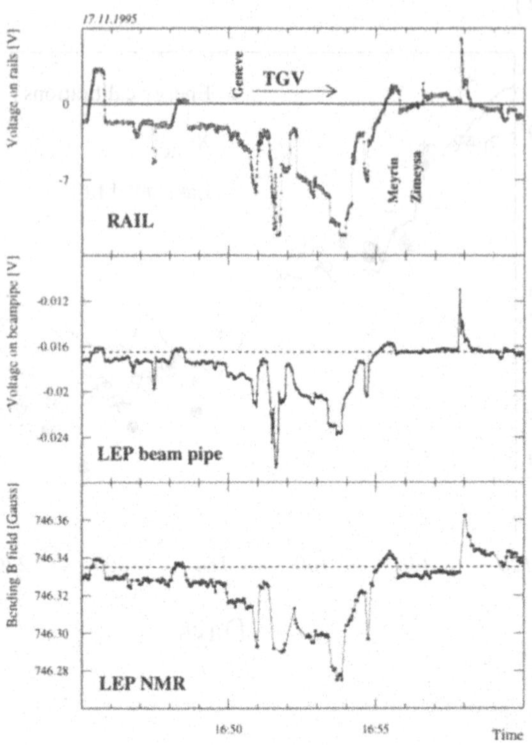

Figure 14. **The LEP beam energy (NMR) correlation to current in the beampipe and power to the TGV.**

distribution varies as

$$P_\tau(cos\theta) = (A_\tau(1 + cos^2\theta) + 2A_e cos\theta)/(1 + cos^2\theta + 2A_\tau A_e cos\theta),$$

where $A_f \equiv 2g_V g_A/(g_V^2 + g_A^2)$ for Zf coupling.

5.3. RESULTS AND PROSPECTS

The L3 version of the Z scan is shown in Fig. 15. I will quote combined results from Vancouver 98.[13] The lineshape is well described by

$$m_Z = 91.1867(21)\ \mathrm{GeV/c^2}$$
$$\Gamma_Z = 2.4939(24)\ \mathrm{GeV}.$$

With efficiencies and luminosity normalization, one obtains a peak cross section of 41.491 (58) nb and an average leptonic decay width of 83.91 (10)

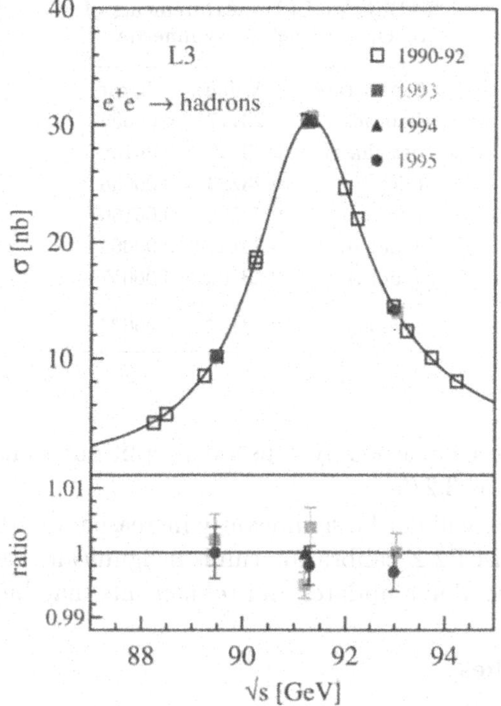

Figure 15. The LEP Z lineshape as seen by L3.

MeV. So there are three flavors of neutrinos, and no significant rate to any channel we don't know about.

The Z asymmetry measurements are summarized in Table 2. The leptonic measurements are a combination of $\mu^+\mu^-$ (most precise), with e^+e^- (diluted by t channel exchange), and $\tau^+\tau^-$ (lower efficiency). Of hadronic final states, the best measurement is for b quarks, as the tag tends to tell you where the quark really is. Identified charm and jet charge, which represents light quarks, have similar precision. Clearly, there are correlations among the hadronic measurements.

Gedanken Problem. Using several samples to make a given measurement, where there is cross talk between the samples, is a general problem. How do you deal with it?

The τ pair decay polarization angular distribution terms, which depend on electron and on tau coupling, are listed separately. All the asymme-

TABLE 2. LEP measurements of $sin^2\Theta_{Weff}$ from Z asymmetries.

Final state	Value	Error
Leptons	0.23117	0.00054
Jet charge	0.23210	0.00100
b jet	0.23223	0.00038
c jet	0.23200	0.00100
τ pol A_e	0.23141	0.00065
τ pol A_τ	0.23202	0.00057
Average	0.23187	0.00024

try measurements are reasonably consistent; split out as listed, the χ^2 per degree of freedom is 3.2/5.

The LEP1 data will not be significantly increased; the statistical impact of the occasional LEP2 Z calibration run is insignificant. Most analyses are reasonably mature. Some updates and refinements may be expected.

6. SLC Z Studies

6.1. GOALS OF THE MEASUREMENT

The goal of SLC is to use initial state electron polarization to give the most precise possible measurement of the Z asymmetry, and thus weak mixing. An overriding goal, not relevant here, is to demonstrate the feasibility of linear colliders.

6.2. METHOD OF MEASUREMENT

To measure the Z asymmetry A_{LR}, one simply measures the cross section difference for left- and right-handed electrons at the Z pole. All the detector needs to do is count events $(N_L - N_R)/(N_L + N_R)$, then divide out the beam polarization. The result depends on $sin^2\Theta_{Weff}$ as

$$A_{LR} = 2(1 - 4sin^2\Theta_W)/(1 + (1 - 4sin^2\Theta_W)^2).$$

The tricky part is obtaining and understanding the beam polarization.

SLC is shown in Fig. 16.[14] Positrons are made by accelerating electrons down most of the length of the linac, hitting a target from which positive charges are collected and returned to the start of the linac. After some acceleration, they are put into a damping ring.

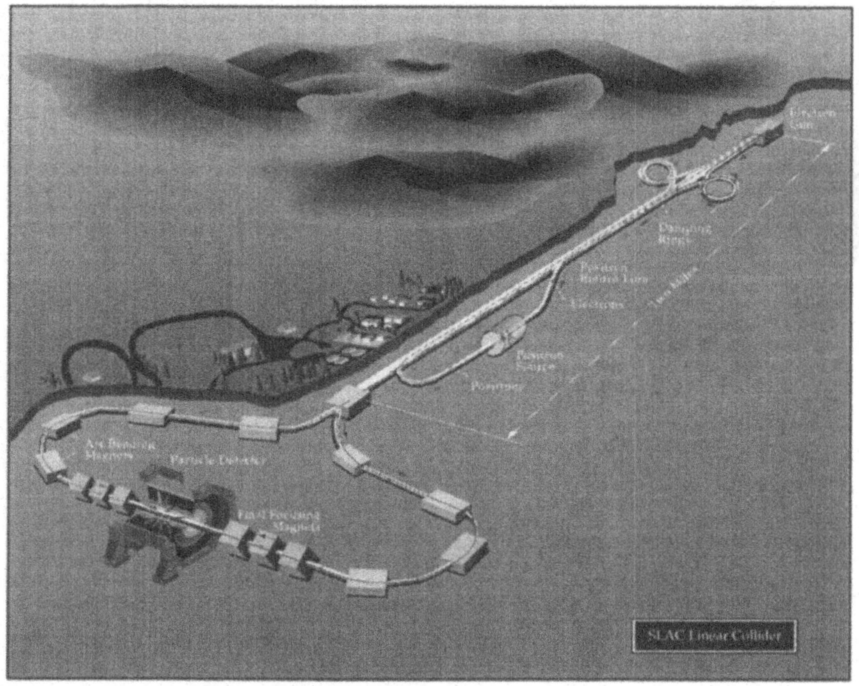

Figure 16. The linear collider at SLAC. The linac is 3.2 km long.

The polarized electron source was a great technical triumph. It produces right- or left-handed electrons based on a random number. These are accelerated a bit, then transferred, with spin rotated to vertical, to a damping ring. The damping rings are synchrotron radiation cooling rings, compressing the phase space of the electrons and positrons to allow a small beamspot and high luminosity.

For a pulse of the collider, the bunch of electrons, spin rotated back to longitudinal, are followed by the positron bunch, accelerated down the linac. At the end of the linac the electron polarization is rotated back to vertical for the arc, and checked with a Moller polarimeter. Preserving the polarization through the arc is also tricky, requiring delicate alignments. The polarization is rotated back to longitudinal going into the final focus and IP. The electron polarization is measured downstream of the IP with a Compton polarimeter, shown in Fig. 17; a Cerenkov detector array at several angles measures the asymmetry for recoil electrons which scatter off circularly polarized laser light.

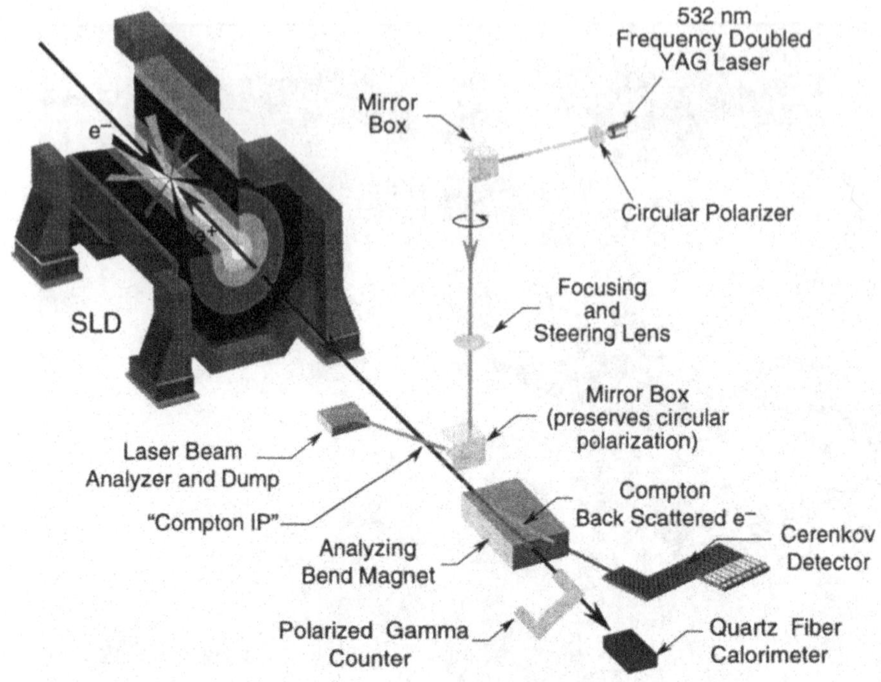

Figure 17. The Compton polarimeter.

A litany of systematic effects has been investigated, including laser polarization, noise, beam optics at the IP, bunch tails, and possible positron polarization. Everything seems to be under control. Average polarization is $77.25 \pm 0.52\%$.

6.3. RESULTS AND PROSPECTS

The result, as of Vancouver 1998,[15] is $A_{LR} = 0.1510 \pm 0.0025$; where the error includes $\sim \pm 0.0010$ systematics in quadrature. This gives $sin^2\Theta_{Weff} = 0.23101(31)$. This is indeed the most precise single measurement, and it continues the historic trend of being noticeably lower than other determinations. In combination with the LEP measurements one gets

$$sin^2\Theta_{Weff} = 0.23155(19).$$

With LEP itemized as listed in Table 2, this corresponds to χ^2 per degree of freedom of 8.1/6. So, in principle, a PDG S^* factor for combining measurements should be used to increase the combined error by $\sim \times 1.15$. In previous years the consistancy has been considerably worse, and in practice even the PDG struggles dealing with it.[16]

SLC/SLD has just completed running. They have some data not included above, and some systematic studies continue. A final update may come during 1999.

7. The Tevatron W Mass

7.1. GOALS OF THE MEASUREMENT

For a while, the only way to study W and Z bosons was at the $Sp\bar{p}S$ collider at CERN. Parts of that program have continued at the Tevatron collider at Fermilab, remaining competitive in the LEP2 era. These include rare W decay searches, the W width, and of note here, the W mass. Each of the Tevatron collider experiments, CDF and DØ, would like to measure the W mass, with existing data, as well as any one LEP experiment will, $\sim \pm 100$ MeV/c^2. Eventually after running with the ongoing upgrades complete, each hopes to improve on the final LEP2 precision.

7.2. METHOD OF MEASUREMENT

When detecting e^+e^- collisions at the Z pole, if you see particles which vaguely correspond in energy to twice the beam energy, that is a Z. Vector bosons are readily produced in hadron collisions, but come in association with soft particles, which I will call "X" as in $\bar{p}p \rightarrow W + X$. Generally, Z and W bosons are observed at hadron colliders in leptonic decays.

X is made of low transverse momentum (few hundred MeV/c^2) particles which evenly occupy longitudinal phase space, pseudorapidity $\eta = -ln(tan(\theta/2))$. For high luminosity, the inelastic cross section is large enough that extra "minimum bias" inelastic events overlap the events of interest. These events consist of soft particles distributed evenly in η. For the Tevatron running six bunches, 3.5 microseconds between bunches, an average of one overlap event corresponds to a luminosity of $\sim 6 \times 10^{30} cm^{-2} sec^{-1}$. The most recent Tevatron data had an average luminosity of almost twice that. As luminosity goes up, so does X. For the high luminosity upgrade, the luminosity will be spread out in more bunches, which will keep X down. At LHC, the bunches are 25 ns apart, but X gets rather large for luminosity anywhere near design.

W and Z production is described, as discussed by J. Stirling, by PDFs giving probabilities for finding partons in the proton and antiproton, and the Drell-Yan quark annihilation process.[17] QCD resummation[18] can be used to define parameters to describe $p_T(W)$ in the low $p_T(W)$ region relevant to measuring the W mass. These parameters are determined or

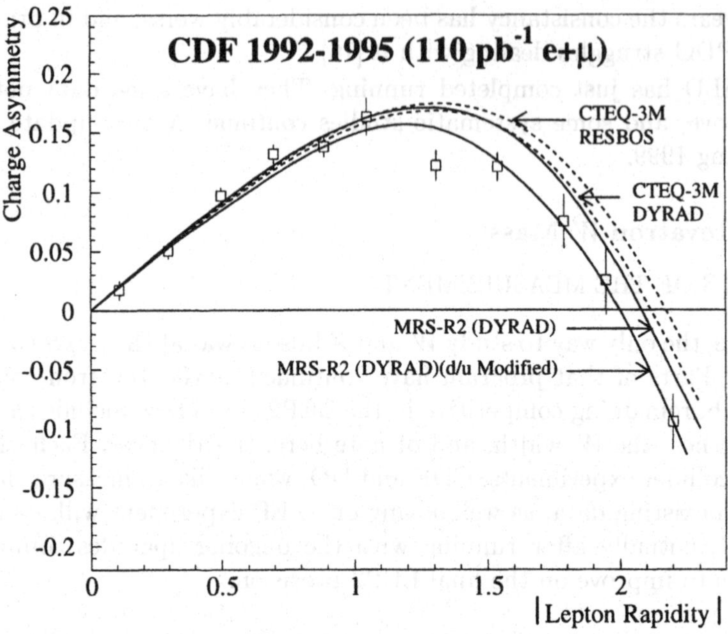

Figure 18. The CDF measurement of W lepton charge asymmetry. CTEQ and MRS are collaborations which do general fits for PDFs. DYRAD and RESBOS are NLO and resummed Monte Carlo generators respectively.

checked by measuring $p_T(Z)$. If $p_T(W)$ were perfectly predicted, then the p_T of the decay charged lepton would be used directly to estimate the W mass, minimizing the effect of X.

As pointed out in Stirling's discussion of PDFs, the momentum fraction distributions for u and d in the proton are different in shape. As Ws are produced $u\bar{d} \to W^+$ and charge conjugate, there is a net W charge correlation, with W^+ tending to be produced with net longitudinal momentum along the proton direction. For central values of η, this charge correlation applies to the decay lepton, as seen in Fig. 18.[19] For higher absolute η, parity violation in the decay reverses the correlation seen in the decay leptons. Measuring this charge asymmetry gives a PDF constraint that is quite useful for measuring the W mass.

While the net longitudinal energy flow is essentially unmeasured, due to the finite longitudinal acceptance, the net transverse energy is reasonably well measured for modest X. The net calorimeter energy flow, other than the lepton, is the sum of response to the hadronic recoil against $p_T(W)$ and X. The neutrino is inferred from the net calorimeter imbalance. Typical W

event selection requires lepton and missing E_T above 20 GeV.

The optimal strategy for the W mass uses transverse mass,

$$m_T \equiv \sqrt{(E_T(\ell) + E_T(\nu))^2 - (\vec{p}_T(\ell) + \vec{p}_T(\nu))^2},$$

as the mass estimator; this minimizes the uncertainty resulting from variability of the $p_T(W)$ distribution.

Gedanken Problem. Under what circumstances is lepton transverse energy a better mass estimator than transverse mass?

In general, the Z sample can be used to calibrate all responses for measuring the W mass, but the cross section times leptonic branching ratio for Zs is an order of magnitude smaller than for Ws.

The measurement depends crucially on calibrating the lepton energy scale. For the CDF magnetic detector this involves using $\psi \to \mu\mu$ to calibrate tracking, and understanding dE/dx and tracking systematics to extrapolate to the higher momenta of W and Z muon decays. The calibration is transferred to electron measurement by understanding the tracking material radiation length, and matching predicted E/p for W electrons. So far, for the most recent large data sample, CDF analysis is ongoing and only a preliminary muon result has been quoted.[20]

The DØ detector had no magnet for the data taken so far. They use the $Z \to ee$ mass with angular correlation, $\psi \to ee$, the π^0 mass, and the test beam measured linearity of their calorimeter to set the electron energy scale.[21] Taking $E(measured) = \alpha_{EM} \times E(true) + \delta_{EM}$, the constraints are shown in Fig. 19. For the final uncertainty, the deviation from linearity allowed by the test beam data is accounted, expanding the allowed region.

In reconstructing the recoil to $p_T(W)$, the DØ calorimeter reconstructs more than 80% of the net transverse energy. By contrast, CDF reconstructs slightly less than 60%. Most of the difference is due to the absence of a magnet in DØ; DØ has a statistical advantage due to the resulting better transverse mass resolution. The leptonic Z samples are used to calibrate this response.

Gedanken Problem. Do you know of any instance when making a compensating calorimeter (DØ as opposed to CDF, ZEUS as opposed to H1), as discussed by J. Virdee, created an actual physics measurement advantage?

There is no analytic form to describe the transverse mass distribution, shown for the DØ 94-95 data in Fig. 20. A fast Monte Carlo generator including all that is known about W production and detector response is used to make transverse mass templates as a function of assumed W mass.

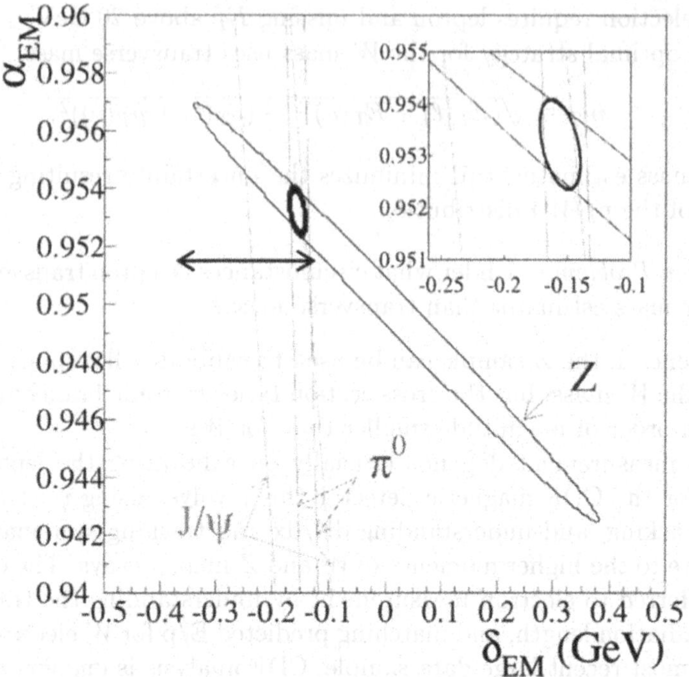

Figure 19. **The DØ electron energy scale determination. α is the scale and δ is the effective pedestal offset.**

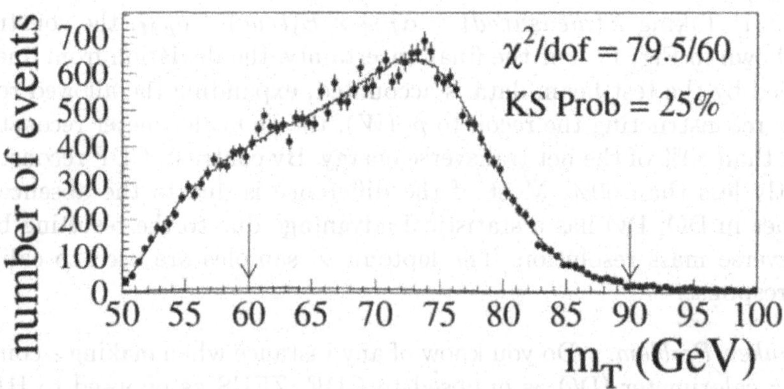

Figure 20. **The DØ electron W transverse mass distribution. The fitting region is marked, as is the small background contribution.**

The templates are then used in a likelihood fit. We will see this procedure of fitting data to Monte Carlo templates several times.

TABLE 3. Measurement errors in MeV/c^2 for the 94-95 DØ W mass determination.

Statistics	95
Lepton systematics	35
Recoil measurement	40
QCD model (PDF, resum)	25
QED radiative cor.	15
Background	10
Other	15
ALL systematics	70
ALL	120

DØ has completed and published their analysis.[21] The errors are listed in Table 3. Note that the 95 MeV statistical error includes, in quadrature, 65 MeV for Z statistics in setting the electron energy scale. The overall modelling (theory) error is at the level of $\sim \pm 30$ MeV which bodes well for the future.

They measure m(W) = 80.44 (12) GeV/c^2 for 94-95 data, and 80.43 (11) when combined with their earlier data. The preliminary 94-95 CDF muon measurement gives 80.43 (16), and combined with previous measurements gives 80.38 (12). Combining DØ, CDF, and UA2, and accounting common uncertainties, gives m(W)= 80.41 $\pm$ 0.09 GeV/c^2.

DØ has reached its goal, while CDF is still working on theirs. Both detectors are undergoing major upgrades; "run 2" data taking should begin in 2000. A solenoid and magnetic tracking is part of the DØ upgrade. The accelerator upgrade includes the Main Injector and much improved antiproton collection. More bunches will be used to spread out the increased luminosity, the six bunches for existing data ("run 1") will increase to 36 and eventually to $\sim$ 100 bunches.

The run 1 existing datasets correspond to $\sim$ 110 pb^{-1}. By 2003 each experiment hopes to collect 2-4 fb^{-1}. Further improvements could produce samples of perhaps 20 fb^{-1} by the end of 2005. Such samples should allow the final LEP2 W mass precision of 35-40 MeV/c^2 to be seriously challenged.

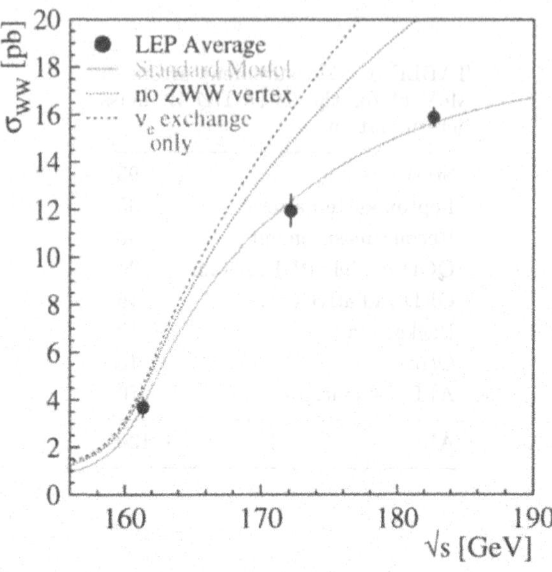

Figure 21. The W pair cross section at LEP.

8. The LEP2 W Mass

8.1. GOALS OF THE MEASUREMENT

The twin goals of the LEP2 program are the study of W bosons and searches. Part of the former is a goal for an eventual combined W mass determination to 40 MeV/c^2 or better. My discussion of progress toward this goal will follow the recent review by Glenzinski.[22]

8.2. METHOD OF MEASUREMENT

With increased RF available, the energy of the LEP ring has risen sufficiently to allow $e^+e^- \to W^+W^-$. The initial W mass measurement came from running just over threshold and measuring the cross section, just as the τ mass was precisely measured at BES.[23] The left-most point in Fig. 21 best determines how much the solid curve can slide horizontally. This threshold analysis is statistically limited, but the drive to search for new things raised the energy and changed the strategy to direct reconstruction.

The most important channels for direct reconstruction are both Ws decaying hadronically, "$qqqq$," or else one decaying hadronically and the other producing an electron or muon, "$\ell\nu qq$." Some collaborations include $\tau\nu qq$, but I will ignore that complication. Lepton measurements and net energy flow have been discussed previously. The identification of jets of energy flow with final state partons was discussed by Stirling. The LEP collaborations typically use the JADE[24] and DURHAM[25] algorithms for this analysis.

The analyses use kinematic constrained fits. These apply energy and momentum conservation to a set of measurements that can be pulled within their measurement errors, defining a χ^2. For the $qqqq$ case, energy conservation means the energy of each W is given by the beam energy, neglecting initial state radiation (ISR). The electron or positron is likely to radiate a photon down the beampipe before interacting, calculable in QED. Each component of momentum should sum to zero; in the z or beam direction this again implies neglecting ISR. So one assigns the energy flow to four jets, hypothesizes two pairs as making up each of the 2 Ws (there are three choices), and calculates a χ^2 for the hypothesis.

In addition to the four constraints (4C) above, one may constrain the two separate W masses to be the same within an error which includes the effect of $\Gamma(W)$, 5C. When good enough solutions are found, either the lowest χ^2 solution, or a weighted average of acceptable solutions, is used to give the mass measurement for that event. The distribution of measurements is fit to Monte Carlo templates varying the assumed W mass; the Monte Carlo includes the ISR neglected in fitting. Four jet W pair candidates are typically selected with an efficiency of $\sim 85\%$ and a purity of $\sim 80\%$. This is illustrated by the ALEPH $qqqq$ data at 183 GeV shown in Fig. 22.[26]

The event selection for $\ell\nu qq$ is typically $\sim 87\%$ efficient and 95% pure. Since net momentum measurement must be used for the neutrino, we are left with one constraint, or two for equal W masses. This channel is illustrated for ALEPH $e\nu qq$ data at 183 GeV in Fig. 23.[26] Note that in both cases, the use of the beam energy makes the result less sensitive to absolute jet energy measurement systematics.

Gedanken Problem. Why is there a tail on the high side in Fig. 23? Does raising the LEP energy help the W mass measurement by direct reconstruction?

The DELPHI collaboration uses a somewhat different approach, as discussed in the talk by Martijn Mulders. They attempt to measure ISR event by event, and fit reconstructed mass to an analytic form.[27]

LEP runs at the Z from time to time to help monitor detector responses.

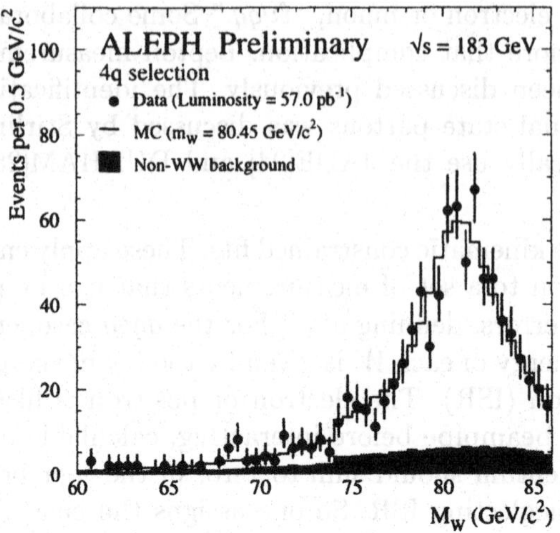

Figure 22. **The ALEPH W mass reconstruction at 183 GeV in $qqqq$.**

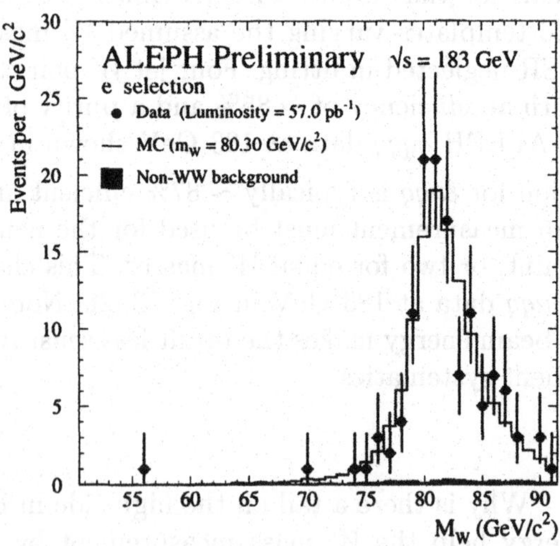

Figure 23. **The ALEPH W mass reconstruction at 183 GeV in $e\nu qq$.**

TABLE 4. ALEPH 183 GeV W mass systematics in MeV/c^2. Those marked * are correlated in all experiments.

SOURCE	$\ell\nu qq$	$qqqq$
ISR*	5	10
Hadronize*	25	35
Detector	22	24
Fit	15	14
Beam energy*	22	22
CR/BE*	-	56

8.3. RESULTS AND PROSPECTS

ALEPH preliminary systematic errors for 183 GeV measurements are given in Table 4.[26] "Hadronize" refers to the systematics of associating measured energy with final state partons, including soft gluons. Detector effects include boundaries and linearity. The fit error includes selection biases, background uncertainties, and Monte Carlo statistics. "CR/BE" refers to QCD final state correlations: color reconnection, as the Ws decay so quickly that decay parton color fields from both Ws overlap, and Bose-Einstein correlations for final state pions from both Ws. Such final state correlations are being looked for, and can be measured or limited.

The 183 GeV measurements, quoted at Vancouver for the combined LEP experiments[13] are m(W) = 80.28 (12) for $\ell\nu qq$ and 80.34 (14) for $qqqq$. The overall direct reconstruction result is 80.36 (9). Combining that with the threshold measurement of 80.40 (20) gives $m(W) = 80.37(9)$ for the overall LEP2 direct measurement. Combining this with hadron collider measurements one obtains

$$m(W) = 80.39 \pm 0.06 \text{ GeV/c}^2$$

as the overall direct measurement, with if anything, a shortage of χ^2 in combining results.

The LEP2 run is just getting started. Progress is being made on systematics so that by the end of the run in 2000, the desired precision looks possible.

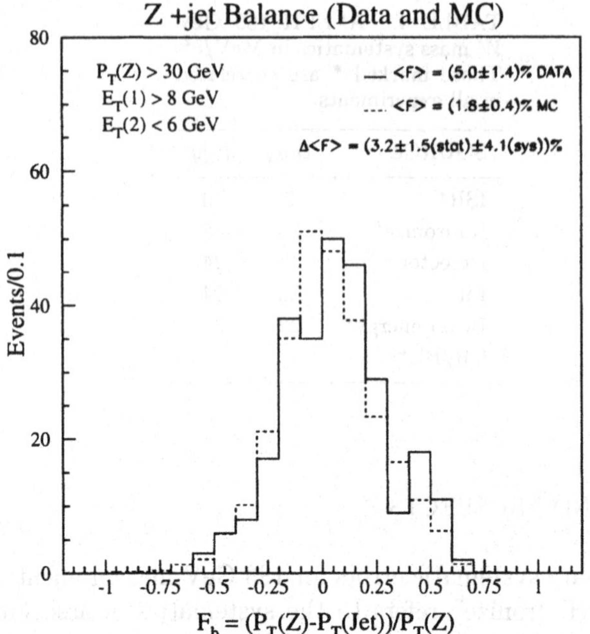

Figure 24. The Z jet balance: lepton p_T minus jet p_T over lepton p_T for CDF $Z + jet$ events, data and Monte Carlo.

9. The Tevatron Top Mass

9.1. GOALS OF THE MEASUREMENT

Having established the presence of the top pair signal, the goals are to confirm the characteristics of that signal in all possible ways and relevant here, to measure the top mass as accurately as possible.

9.2. METHOD OF MEASUREMENT

At the Tevatron, top is predominantly pair produced in the subreaction $\bar{q}q \to g \to \bar{t}t$. Top decays are essentially 100% to Wb, so the final state for top pairs contains two Ws like LEP2, a b and $\bar{b}$, and of course, X. Like LEP2, the channels are characterized by the W decays, again taking leptons as e or μ, all hadronic "$qqqqbb$," lepton plus jets "$\ell\nu qqbb$," and dilepton "$\ell\nu\ell\nu bb$." The all hadronic channel has relatively poor signal-to-noise. The dilepton channel has poor statistics, and is underconstrained due to the two neutrinos, and thus is tricky. I will discuss the lepton plus jets channel, which gives most of the precision.

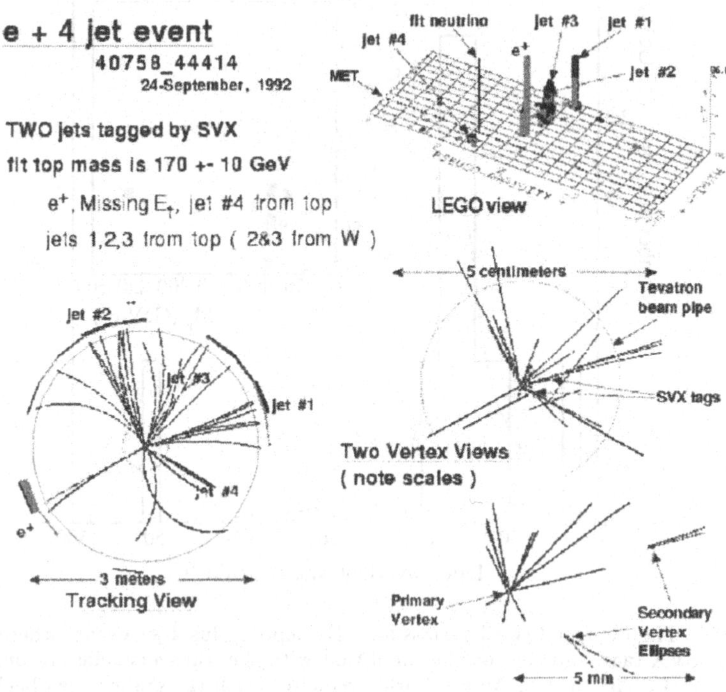

Figure 25. A reasonably convincing CDF electron plus four jet top candidate. The LEGO view shows calorimeter energy in $\eta \times \phi$; the other views are in ϕ.

This measurement is based on jet energies, without the beam energy calibration possible at LEP. Jets are defined as calorimeter clusters of energy in a circle ("cone") in $\eta \times \phi$ space of radius typically ~ 0.4. We have seen, in the W mass discussion, the EM calorimeter calibration. For CDF, the hadronic calibration starts from the test beam, then uses jet fragmentation, with the nonlinearity of calorimeter response measured from test beam and *in situ* isolated particle measurements. There are corrections for final state hadrons coming from the relevant parton falling outside the cone, and for X getting into the cone. As luminosity varies, X varies, which can be corrected on average for a given sample; there is also a jet threshold bias. These effects are studied by combining top events, real and Monte Carlo, with one or more minimum bias events.

The detailed jet response study is done with the help of tracking, that is for calorimeters covering central rapidity. The scale is transferred to the rest of the calorimeters using dijet balance. The process can be checked

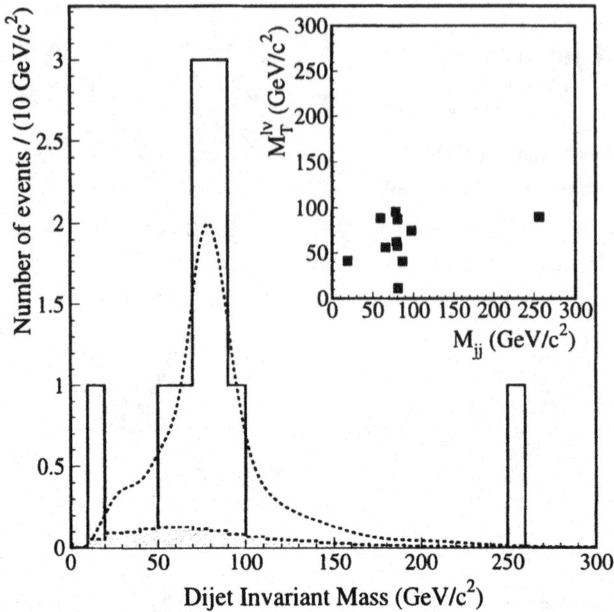

Figure 26. **The untagged two jet mass for CDF lepton plus 4 jet events where two jets have loose SVX tags, and the correlation of that with the transverse mass reconstructing the leptonic W. The curve is Monte Carlo prediction with the shaded area background.**

using photon/jet event balance, but one needs to worry that selecting or triggering on a photon puts an initial state intrinsic transverse momentum "k_T" bias. The initial partons tend to be moving in the photon direction. Top specific jet corrections account mostly for muons in b decays. The final check that this all makes sense is to look at the transverse energy balance in $Zjet$ events. This is shown in Fig. 24 for CDF Z events, where the lepton pair p_T is above 30 GeV/c. This is balanced essentially by one jet, with no other jet above 6 GeV E_T. The agreement is well within the expected systematic error.

DØ employs an equivalent procedure, starting from the electromagnetic scale and studying photon jet balance etc. Although with no magnet, they have fewer handles but their corrections are smaller.

For obtaining a mass estimate, kinematic fitting is again used. One starts with a sample of events with a lepton plus four jets and missing E_T. Only overall transverse momentum balance is available, and that is used to define the neutrino p_T. The longitudinal momentum of the neutrino is unmeasured. There are two W mass constraints, one on a jet pair and the other on the lepton and the inferred neutrino. There are usually two viable constraint solutions for the neutrino longitudinal momentum ("ambiguity"). A further constraint comes from demanding that the two top masses be consistent. The net constraint is 2C.

The fit is well illustrated by considering an event, shown in Fig. 25, where two jets are identified as *b*s by silicon vertex tags ("SVX"). Jets 2 and 3 seem together in the calorimeter display, but are clearly separated in the tracking view. The secondary vertices at a few mm identify jets 1 and 4 as *b*s. Thus jets 2 and 3 should be a W; the invariant mass of the pair, not fit, is 79 GeV/c^2. The only ambiguities are: which b goes with which W, and which neutrino solution to use. The best χ^2 assigns jet 4 to the leptonic W, and gives an event top mass of 170 ± 10 GeV/c^2. The *W*s observed for all candidates events, when there are two loose SVX tags, are shown in Fig. 26. With enough statistics the W peak may become a calibration.

If you don't have the b tag information, you simply try all the possibilities. Sometimes the best χ^2 is the wrong combination; the top mass resolution degrades as the number of possible combinations grows. Both CDF and DØ tag b jets using associated e or μ from heavy quark decay, called soft lepton tag, SLT. The statistics, signal and background levels, and resolution are illustrated for CDF in Fig. 27. In defense of the SLT sample, it should be noted that SLT tagged events that are also SVX tags are removed from the SLT sample and kept as SVX, just as all tagged events have been removed from the no tag sample.[28]

DØ defines four variables, with minimal mass bias, that discriminate the top from the background. The background is predominantly $W + jets$ with some fake leptons. The variables are missing E_T, acoplanarity, the centrality of the non-leading jets, and a measure of the smallest 2 jet separation. They do a joint fit to a discriminant or a neural net (NN) output constructed from the four variables, and top mass, as illustrated in Fig. 28. The discriminant and neural net analyses are combined, accounting correlations, to give their result.[29] The combination uses the technique of pseudoexperiments, analyzing many Monte Carlo samples of the same size as the data to understand measurements and correlations. Such exercises allow you to understand whether a given fit result makes sense.

9.3. RESULTS AND PROSPECTS

The several top mass determinations are listed in Table 5. The predominant systematic error comes from the jet energy scale. This is $\sim \pm 5$ GeV/c^2 in lepton plus jets. It includes jet systematics, as discussed, as well as variation with different assumptions about how much gluon radiation goes where. With enough data, even the gluon variation can be constrained. The

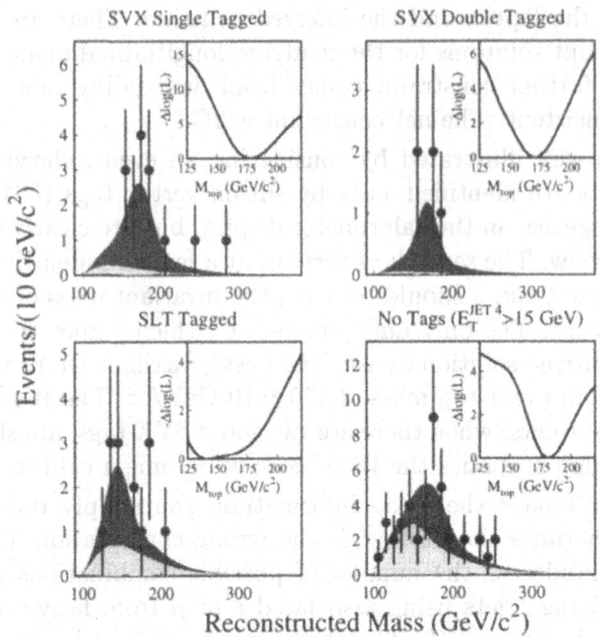

Figure 27. CDF top mass distributions and fits for the SVX 1 tag, SVX 2 tag, SLT tag and no tag samples. The insets are likelihood results. The shaded areas are the fit results with dark signal and light background.

TABLE 5. Top mass measurements, in GeV/c^2.

Channel	Experiment	Value
$\ell\nu qqbb$	CDF	175.9 ± 6.9
$\ell\nu qqbb$	DØ	173.3 ± 8.4
$\ell\nu\ell\nu bb$	DØ	168.4 ± 12.8
$\ell\nu\ell\nu bb$	CDF	167.4 ± 11.4
$qqqqbb$	CDF	186 ± 13

several results have been combined,[30] accounting correlations, to give

$$m(top) = 173.8 \pm 5.0 \text{ GeV/c}^2$$

Only minor refinements to the analysis of existing data may be expected.

It is difficult to predict how much improvement can be achieved with the Tevatron upgrades, given the jet energy scale systematic level. But our

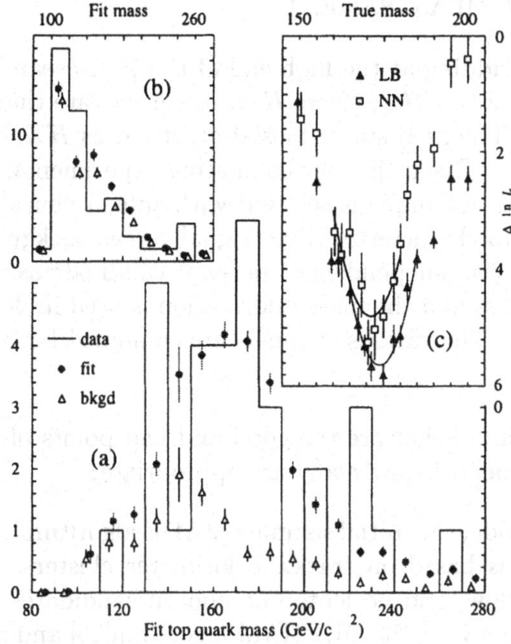

Figure 28. DØ top mass distribution and fit for (a) predominantly signal events and (b) predominantly background events. The likelihoods for the two discriminants are shown in (c).

measurements with $\sim 110\ \text{pb}^{-1}$ are almost as precise as we predicted a few years ago for the upgrade $2\ \text{fb}^{-1}$ samples.[31] A factor of two improvement seems reasonably safe, for twenty times the data with improved detectors.

10. The LEP2 Higgs Search

10.1. GOALS OF THE MEASUREMENT

One of the accomplishments of the LEP program has been to search for the Standard Model Higgs over the complete kinematically allowed range of the Higgs mass. So far nothing has been found. In the LEP2 era, as the energy rises the search is extended, typically to twice the beam energy minus the Z mass. As an illustration, I will describe the L3 analysis of the 183 GeV data.[32]

For incrementally adding at the high end of the Higgs search, the relevant process is $e^+e^- \rightarrow Z^* \rightarrow HZ$, where $H \rightarrow \bar{b}b$ since bs are the heaviest available decay mode. The final states looked at, listed as HZ, are $bbqq$, $bb\nu\nu$, $bb\ell\ell$, $\tau\tau qq$ and $bb\tau\tau$. Given the four competing experiments, the analysis is fairly sophisticated. A sample is selected with cuts, a neural net (NN) discriminant is used to characterize Higgs signal versus background, as in the D0 top mass analysis, independent of mass. A candidate sample is checked as a function of mass and the mass information is used in defining a purity used to set a limit. The various channels are combined for a final result.

Gedanken Problem. What are the good and bad points of having several competing experiments? How many are appropriate?

For the $bbqq$ mode, the initial sample is JADE algorithm[24] 4 jet events. Further selection is based on tracks, calorimeter clusters, visible energy, small net energy flow, and no lepton or photon candidates. Energy is apportioned to the jets with the DURHAM algorithm[25] and a 4C fit defines a kinematic χ^2_{fit}. How well an appropriate jet pair gives a Z mass determines the selection on a χ^2_{mass}. The 321 events selected compare to 315 predicted for background, mainly W pairs and $qq\gamma$. For the NN, tracks, clusters, event shape, b tag and χ^2_{fit} are used. The results are shown in Fig. 29. Twenty events have NN$>$ 0.5, but there is no sign of any signal.

For the $bb\nu\nu$ mode, events are selected on tracks and clusters. Two jet events (DURHAM) with a recoil mass between 40 and 115 are consistent with $Z \rightarrow \nu\nu$. Net energy flow and b tag probability are required. The 56 events found compare to 50 predicted as background. The NN uses b tag, angles, recoil mass, jet masses, net E_T and χ^2_{fit}. The results are shown in Fig. 30. Again there is no sign of a signal.

For the $bb\ell\ell$ mode, where ℓ as usual is e or μ, the Z is observed as the lepton pair; 6 ee candidates and 2 $\mu\mu$ candidates are found, again consistent with background. The NN uses b tag, angles, m(Z), jet masses, and χ^2_{fit}. Only one event has NN$>$ 0.1; it gives a mass less than 70.

For the two $\tau\tau$ modes, τs are selected as two isolated 1/3 prongs of opposite charge. A cut discriminant uses jets, angles, masses, b tag and χ^2_{fit}. One event is selected with a predicted background of 2.4.

There is no sign of a signal and systematics are included in a pseudoexperiment exercise for combining channels. Systematics include luminosity (0.3%), detector efficiency (4%) and background uncertainty, taken as correlated between channels (10%).

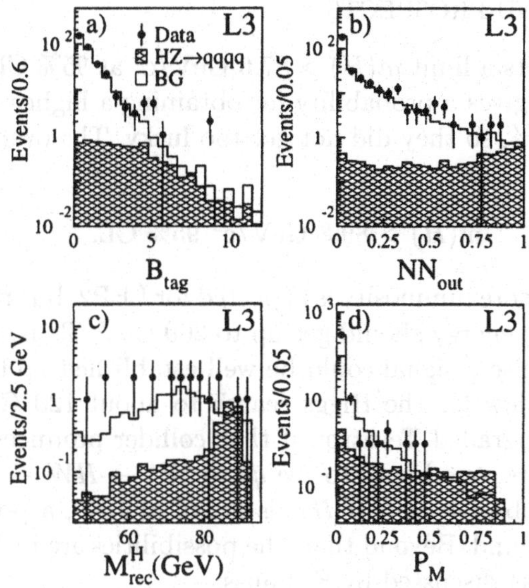

Figure 29. For *bbqq*; the *b* tag discriminant (a), NN output (b), *H* mass for the events with NN> 0.5 (c) and purity (d). The open histogram is background, and the shaded one is a nominal 87 GeV Higgs signal.

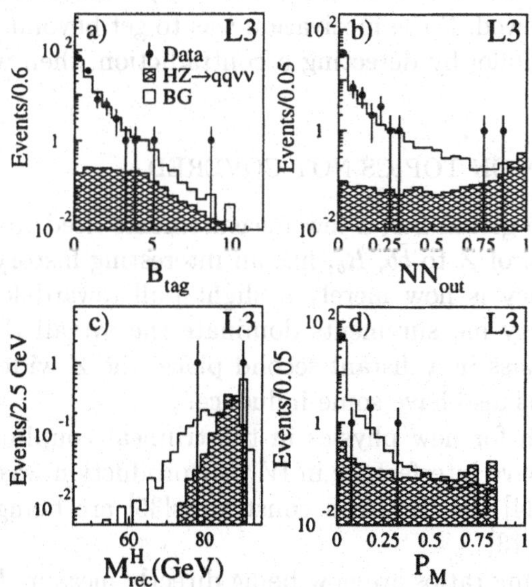

Figure 30. For $bb\nu\nu$; the *b* tag discriminant (a), NN output (b), *H* mass for the two events with NN> 0.5, (c) and purity (d). The open histogram is background, and the shaded one is a nominal 87 GeV Higgs signal.

10.3. RESULTS AND PROSPECTS

The L3 analysis sets a limit $m(H) > 87.6$ GeV/c^2 at 95% CL. The pseudo-experiment study gives a probability for obtaining a higher limit given the data sample of 35%, so they did not get too lucky. The overall LEP2 limit is[33]

$$\mathrm{m}(H) > 89.8 \text{ GeV/c}^2 \text{ 95\% CL.}$$

Considerably more luminosity is expected for LEP2, but more important for this search, the energy should get up to 200 GeV. Thus, the limit could get up to 109, or else a signal could be well established up to 99 GeV/c^2.

A further window for the Higgs search to about 120 GeV/c^2, will be available at the upgraded Tevatron if that collider produces data samples of ~ 20 fb^{-1}. There, one hopes to see $\bar{q}q \to W^* \to HW \to bb\ell\nu$. It would be interesting to observe HW if HZ is found at LEP, a possibility which should be kept in mind. Beyond that, the possibilities are rather thoroughly covered by LHC, as discussed by F. Pauss.

11. Conclusions

The program of precision electroweak measurements is a great success since all the measurements are consistent. The simple-minded minimal Higgs scenario is still allowed. If the motivation was to get beyond the limitations of the Standard Model by detecting a contradiction, then we must report failure.

11.1. DIGRESSION ON TOPICS NOT COVERED

Before we get into global fits, a few measurements need to be mentioned. The decay fraction of Z to $\bar{b}b$, R_b, has an interesting history; the formerly exciting discrepancy is now merely a slight pull toward lower top mass. While Z asymmetry measurements dominate the overall Higgs mass constraint, with W mass in a distant second place, the Z width and leptonic branching fractions also have some influence.

One can search for new physics in the trilinear couplings, *eg.* ZWW. While the destructive interference in W pair production was demonstrated at the Tevatron,[34] the Tevatron constraints[35] are being overtaken by the LEP2 studies.[13]

The W branching ratios are now being directly measured at LEP2. The leptonic fractions will soon be more precise than the Tevatron constraints from the leptonic Z/W cross section ratio. The Tevatron still has much larger samples, given a trigger signature, and continues to have the more accurate direct W width measurement, and better reach for rare W decays.

Vancouver 1998

	Measurement	Pull
m_Z [GeV]	91.1867 ± 0.0021	.08
Γ_Z [GeV]	2.4939 ± 0.0024	-.80
σ^0_{hadr} [nb]	41.491 ± 0.058	.31
R_e	20.765 ± 0.026	.66
$A^{0,e}_{fb}$	0.01683 ± 0.00096	.72
A_e	0.1479 ± 0.0051	.24
A_τ	0.1431 ± 0.0045	-.80
$\sin^2\theta^{lept}_{eff}$	0.2321 ± 0.0010	.54
m_W [GeV]	80.370 ± 0.090	.01
R_b	0.21656 ± 0.00074	.90
R_c	0.1733 ± 0.0044	.24
$A^{0,b}_{fb}$	0.0991 ± 0.0021	-1.78
$A^{0,c}_{fb}$	0.0714 ± 0.0044	-.47
A_b	0.856 ± 0.036	-2.18
A_c	0.638 ± 0.040	-.74
$\sin^2\theta^{lept}_{eff}$	0.23101 ± 0.00031	-1.78
$\sin^2\theta_W$	0.2255 ± 0.0021	1.06
m_W [GeV]	80.410 ± 0.090	.45
m_t [GeV]	173.8 ± 5.0	.50
$1/\alpha$	128.896 ± 0.090	-.04

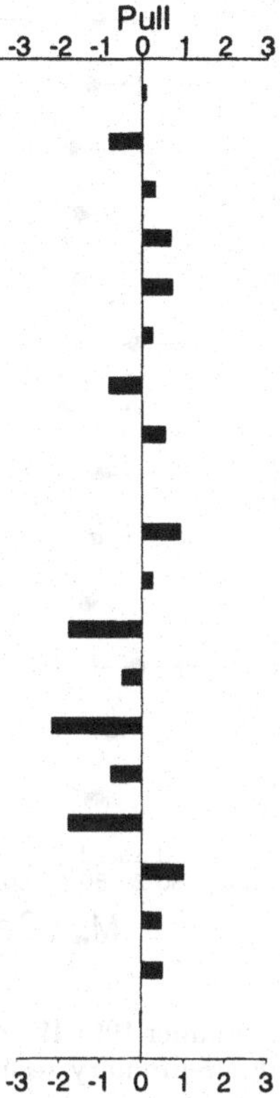

Figure 31. The summer 1998 LEPEWWG global fit pull distribution. The top 15 measurements are combined LEP results, then SLD, NuTeV, two from the Tevatron Collider, and calculated α. The fit also gives an α_S in agreement with other measurements.

I also note that in many cases flavor universality is tested as well as assumed. Axial and vector couplings can be measured separately, and many assumptions can be relaxed and checked.

11.2. GLOBAL FITS

Although the g-2 experiment may not soon provide input to the global electroweak fits, all the other measurements discussed do. I will use the LEPEWWG version as quoted at Vancouver;[13] PDG gets similar results.[16] That it all hangs together is shown in the pulls plot, that is how much each measurement deviates from the fit value, Fig. 31. The usual suspects are off a bit, but none have either the statistical significance or the connection

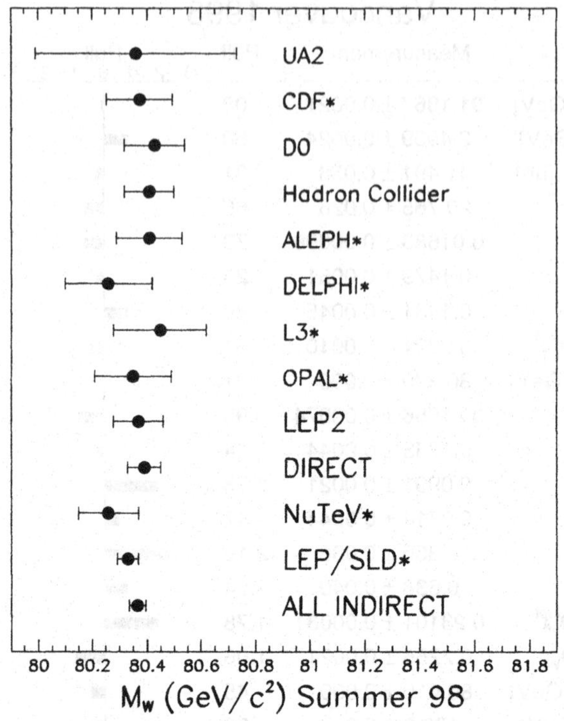

Figure 32. The summer 1998 W mass determinations, direct and indirect. Measurements marked * include preliminary results.

to a popular SUSY scenario to make them noteworthy.

Let us check if the W mass, calculated indirectly through radiative corrections, agrees with the direct measurements. This is shown in Fig. 32. The direct measurements agree among themselves, perhaps too well. The NuTeV analysis inputs the measured top mass, while LEP/SLC does not; they also indirectly infer a top mass. For the global indirect W mass, the direct top mass measurement is included, moving the result up.

If utility is defined as providing the greatest constraint on the Standard Model Higgs mass, then given the incredibly precise measurements of α, G_F and m(Z), the Z asymmetries are most useful. This constraint is illustrated in Fig. 33. While a couple of individual measurements would prefer the Higgs to have been found some time ago, the general trend can accomodate a Higgs mass such that there is no need for new physics to the Plank scale.

The correlation of the W mass and the top mass is shown, in the NuTeV world view, in Fig. 34. The Z asymmetry dominates the width of the region

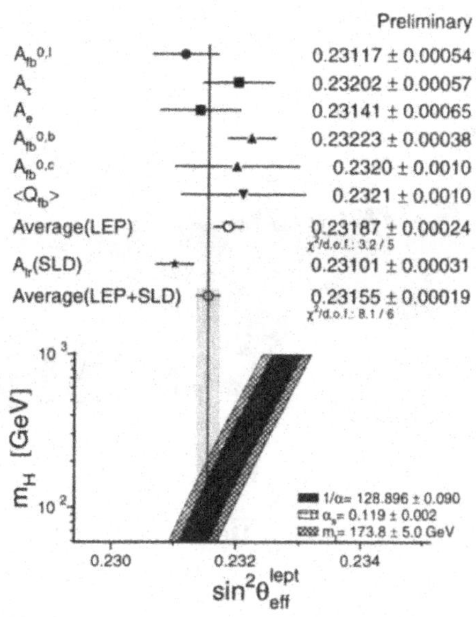

Figure 33. The summer 1998 LEPEWWG/SLD Z asymmetry measurements.

allowed by LEP/SLC indirect measurements as m(H) changes. A factor of two improvement on both direct measurements will help a lot.

The overall Higgs constraint is shown in Fig. 35. This is quoted as giving a one-sided 95% CL upper limit on the Higgs mass, increased since Moriond 1998, of 280 GeV/c^2.[36] But one should not really ignore the fact that the left side of the plot has been ruled out. Even in the most unfavorable MSSM scenario, SUSY Higgs below 70 GeV/c^2 are ruled out.[37] It may be more appropriate to call the limit $\sim$ 90% CL; 5% of what is left on the right side of the plot corresponds to a rather higher mass limit. No allowance has been made for measurement discrepancies; the limit depends strongly on the SLD A_{LR} result. So the SUSY establishment, hoping that the presence of a low mass Higgs will be established, needs to remain patient.

The simplest Standard Model Higgs scenario remains viable. Except for those SUSY scenarios which imitate the Standard Model, more complicated Higgs scenarios generally imply that there is no constraint.

Gedanken Problem. Using all the information about Fig. 35, and avoiding religious prejudice, what would you quote for a Higgs mass upper limit?

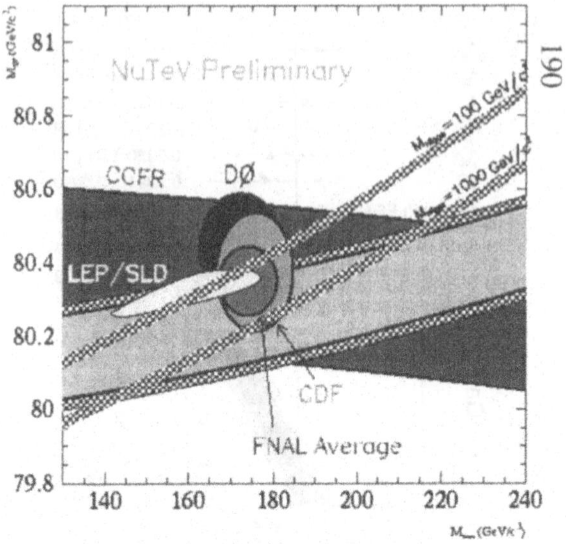

Figure 34. The parameter space of W mass and top mass, with bands shown for Higgs mass values and contours showing measurement constraints.

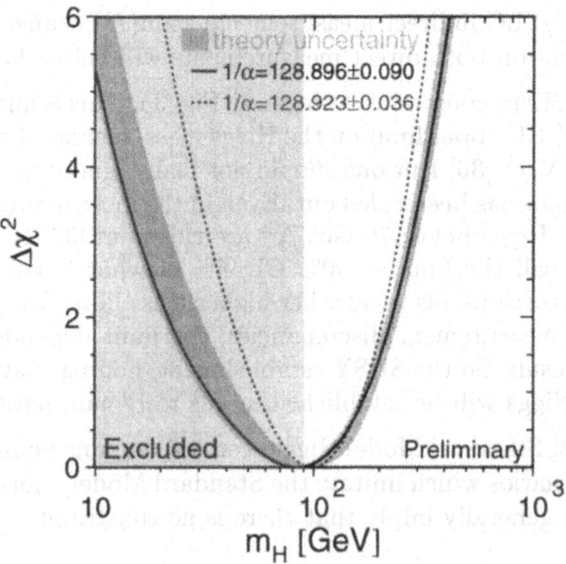

Figure 35. The global electroweak fit χ^2 versus Standard Model Higgs mass value.

11.3. PROSPECTS

The BNL g-2 experiment is just getting started. LEP1 and SLC have finished running, with some analysis updates pending. LEP2 is just getting going and one can anticipate W mass precision improvements, as well as an extension of the Higgs search in what seems to be a promising region.

The Tevatron is on hiatus, upgrading. Some updates, particularly on the W mass, are pending. Once Tevatron running resumes, substantial improvements may be expected in top mass and W mass precision, giving these measurements comparable electroweak precision to the Z asymmetries. There is even some window of opportunity to search for $H \to \bar{b}b$, slightly extending the LEP2 range.

If we are persistent and patient, LHC results must eventually clarify the picture. I certainly hope that we learn something more than a value for the Standard Model Higgs mass.

12. Acknowledgements

I thank Tom Ferbel for being our perfect host. I am grateful to the students, and to my fellow lecturers for keeping things so interesting. I would like to thank Cosmas Zachos, Lee Roberts, Doug Glenzinski, Jae Yu, Randy Keup, Darien Wood, Alain Blondel and Ruth Hill for their help. This work was supported in part by the U. S. Department of Energy, contract W-31-109-ENG-38.

References

1. S. Weinberg, Phys. Rev. Lett. **19**, 1264 (1967); A. Salam in*Elementary Particle Theory*, ed. N. Svartholm (Almquist and Wiksells, Stockholm 1969) p. 367.
2. J. Bailey *et al.*, Nucl. Phys. **B150**, 1 (1979).
3. L. Roberts, Proc. ICHEP96 (Warsaw), ed. Z. Ajduk and A. K. Wroblewski (World Sci., 1997) pp 1035-1039.
4. A. Czarnecki, B. Krause and W. Marciano, Phys. Rev. Lett. **76** 3267 (1996).
5. A. Czarnecki, B. Krause and W. J. Marciano, Phys. Rev. **D52**, R2619 (1995).
6. R. Alemeny, M. Davier and A. Höcker, Eur. Phys. Jour. **C2**, 123 (1998), M. Davier and A. Höcker, Phys. Lett **B419**, 419 (1998).
7. M. Hayakawa and T. Kinoshita, Phys. Rev **D57**, 465 (1998).
8. C. Timmermans, Talk at ICHEP98 (Vancouver), see http://ichep98.triumf.ca/info/sessions/list.asp.
9. F. A. Paschos and L. Wolfwenstein, Phys. Rev. **D7**, 91 (1973).
10. T. Bolton, Talk at ICHEP98 (Vancouver), *loc. cit.*, see also K. S. McFarland *et al.* (NuTeV), Moriond 98, hep-ex/9806013.
11. S. Jadach *et al.*, Proc. ICHEP96 (Warsaw), ed. Z. Ajduk and A. K. Wroblewski (World Sci., 1997) pp 1072-1076, E. Milgliori, *ibid.*, pp 1077-1081.
12. L. Arnaudon *et al.*, CERN SL/94-7 (BI) and other sources can be found at http://www.cern.ch/LEPECAL/reports/reports.html.

13. LEP ElectroWeak Working Group, see http://www.cern.ch/LEPEWWG/. Internal note 98-01 was used for the lectures, and plots and numbers used here were current for Vancouver; these have been updated on the web for Vancouver proceedings.
14. P. C. Rowson, Proc. Radiative Corrections, Gatlinburg 1994, ed. B. F. L. Ward (World Sci., 1995) pp 121-137.
15. K. Baird, Talk at ICHEP98 (Vancouver), *loc. cit.*
16. Partical Data Group, C. Caso *et al.*, Eur. Phys. Jour. **C3** (1998) pp 90-102, see http://pdg.lbl.gov/pdg.html.
17. S. D. Drell and T. M. Yan, Phys. Rev. Lett. **25**, 316 (1970).
18. C. Balazs *et al.*, Phys. Lett. **B355**, 548 (1995).
19. F. Abe *et al.* (CDF), subm. Phys Rev. Lett., FERMILAB-PUB-98/256-E, hep-ex/9809001, and the references therein.
20. A. Gordon *et al.* (CDF), Proc. Moriond Electroweak (1997), ed. J. Tran Thanh Van, p109.
21. B. Abbott *et al.* (DØ), Phys. Rev. **D58**, 012002 (1998).
22. D. Glenzinski, Talk at Moriond 98 (QCD), hep-ex/9805020.
23. J. Z. Bai *et al.* (BES), Phys. Rev. **D53**, 20 (1996).
24. S. Catani *et al.*, Phys. Lett. **B263**, 491 (1991).
25. S. Bethke *et al.*, Nucl. Phys. **B370**, 310 (1992).
26. ALEPH preprint 98-020 CONF 98-011, source of the figures, is updated to H. Przysieniak, Talk at ICHEP98 (Vancouver), ALEPH 98-058 CONF 98-030, see http://alephwww.cern.ch/WWW/.
27. DELPHI preprint 98-85 CONF 153, see http://delphiwww.cern.ch:8002/vancouver/, paper 341(WW).
28. F. Abe *et al.* (CDF), Phys. Rev. Letters **80**, 2767 (1998).
29. B. Abbott *et al.* (DØ), Phys. Rev. **D58**, 052001 (1998).
30. R. Partridge, Talk at ICHEP98 (Vancouver), *loc. cit.*
31. D. Amidei *et al.*, Report of the TeV2000 Study Group, FERMILAB-PUB-96-082.
32. M. Acciarri *et al.* (L3), Phys. Lett. **B431**, 437 (1998).
33. R. Peccei, Talk at ICHEP98 (Vancouver), *loc. cit.*
34. F. Abe *et al.* (CDF), Phys. Rev. Lett. **75**, 1017 (1995).
35. B. Abbott *et al.* (DØ), Phys. Rev. **D58**, 031102 (1998).
36. W. Hollik, Talk at ICHEP98 (Vancouver), *loc. cit.*
37. See *eg.* M. Acciarri *et al.* (L3), subm. Phys. Lett. B, CERN-EP/98-072 L3-148.

EXPERIMENTAL CHALLENGES AT THE LHC

FELICITAS PAUSS AND MICHAEL DITTMAR
Institute for Particle Physics (IPP), ETH Zürich, CH-8093 Zürich, Switzerland

1. Introduction

The Standard Model (SM) of particle physics – the theory of electroweak and strong forces – provides a remarkably successful theoretical picture [1]. The SM has been tested rigorously at LEP, the Tevatron and the linear collider at SLAC [2]. The four LEP experiments have already given a definitive answer to the number of fundamental building blocks of matter: there exist three families of quarks and leptons with a light neutrino.

One of the key questions in particle physics today is the origin of the spontaneous symmetry breaking mechanism. The electroweak sector of the SM postulates that the Higgs mechanism is responsible for this symmetry breaking, and predicts a scalar Higgs boson. Introducing this Higgs boson in the SM allows the masses of all particles to be expressed in terms of their couplings to the Higgs. In order to complete the SM prediction we therefore have to establish experimentally the existence of the *last* missing element: the *Higgs* boson [3].

Even though the SM describes existing data very well, and even if it successfully passes further tests, we know that this Model is incomplete, as it supplies no answer to some fundamental questions. One problem of the SM is the instability of the mass of an elementary scalar, such as the Higgs boson, under radiative corrections in the presence of a high scale, like for example the Planck scale ($\approx 10^{19}$ GeV). These divergences disappear in *Supersymmetry* (SUSY) [4], because of cancellations between the virtual effects of SM particles and their supersymmetric partners, which are introduced to every known fermion and boson of the SM. Furthermore, SUSY must be a broken symmetry because known particles have no super partner of the same mass. These must be heavier, and are therefore not yet discovered.

T. Ferbel (ed.), Techniques and Concepts of High Energy Physics X, 193–238.

Another problem originates from extrapolating the coupling strength of the fundamental forces measured at mass scales of a few 100 GeV to energy scales relevant for cosmology, i.e. energies of about 10^{15} to 10^{19} GeV. Performing this extrapolation within the SM does not lead to unification of forces at very high scales. Introducing however SUSY unification of the electromagnetic, weak and strong forces at the GUT scale ($\approx 10^{15}$ GeV) is predicted which is consistent with a SUSY mass scale of $\mathcal{O}$(TeV).

It is possible that the Higgs boson is an elementary particle as predicted in the SM and its supersymmetric extension. Alternatives to a fundamental scalar Higgs involve *new strong forces.* In models without a scalar Higgs, the W and Z masses could then be due to a dynamical symmetry breaking [5]. In such a scenario the symmetry breaking could lead to a strong interaction between the longitudinal components of the intermediate vector bosons (W_L, Z_L). This strong interaction may be resonant or not. Resonances may occur in analogy with $\pi\pi$–scattering, which leads to spin=1 ρ–like states, or spin=0 very broad resonances, expected to be in the TeV mass range.

In spite of the impressive success of the SM there is a general consensus that the SM is not the ultimate description of nature, and that new phenomena should manifest themselves in the energy region of order 1 TeV. The Large Hadron Collider (LHC), operating at a centre–of–mass energy of 14 TeV with a design luminosity of $10^{34}\mathrm{cm}^{-2}\mathrm{s}^{-1}$, will be the first machine to probe parton–parton collisions directly at energies $\approx$ 1 TeV [6]. Such energies will be essential to address, for example, the questions of the origin of spontaneous symmetry breaking.

Currently no experimental evidence exists for any exotic new phenomena, therefore our discussion on search strategies at the LHC should be quite general. We follow however today's fashion and focus mainly on the detector requirements necessary for future successful searches for the Higgs and for Supersymmetry. The presented ideas and methods should nevertheless provide a good guidance for more "exotic" searches.

2. Present Experimental Status of the Standard Model and Beyond

In the following we summarise briefly the present status of physics topics relevant for LHC and speculate about what one might know from future experimental results before the start–up of the LHC, presently foreseen in the year 2005.

2.1. THE HIGGS SECTOR

It is well known that the value of the Higgs mass is not predictable within the SM. On the other hand, the Higgs cannot be too heavy, otherwise the perturbative regime breaks down, and this leads to an upper bound on the Higgs mass of about 1000 GeV.

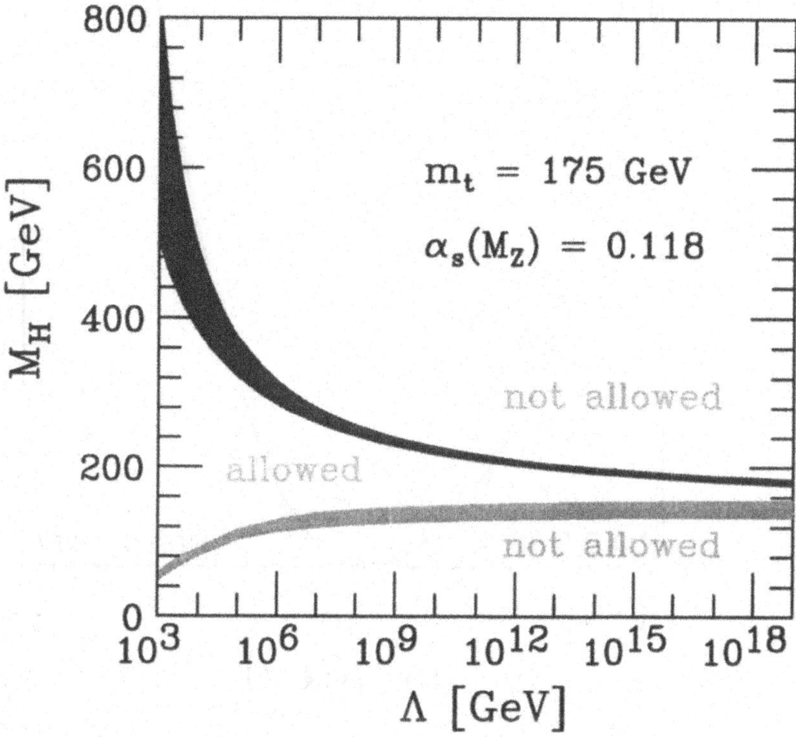

Figure 1. The area between the two curves shows the allowed Higgs mass range assuming the validity of the Standard Model up to a scale Λ [7].

The requirement of perturbative consistency of the theory up to a scale Λ sets an upper bound on the SM Higgs mass, while arguments of vacuum stability suggest a lower Higgs mass limit [7], depending also strongly on the top mass. Taking the measured value of the top mass (m_t= 174.1±5 GeV) and assuming that no new physics exists below the Planck scale, the Higgs mass should be around 160 ± 20 GeV, as shown in figure 1.

Particle physicists have been searching for many years for the Higgs boson, from zero mass up to the highest masses accessible at existing particle accelerators. At present, the four LEP experiments have ruled out the existence of a Higgs with a mass of less than 95 GeV [8] . From global fits to electroweak data one obtains an upper limit on the Higgs mass of 280 GeV (95% C.L.), as shown in figure 2 [9].

By the end of the year 2000 one expects to discover at LEP200 a Higgs boson up to $m_H \approx 106$ GeV, assuming 150 pb^{-1} per experiment at $\sqrt{s}$ =

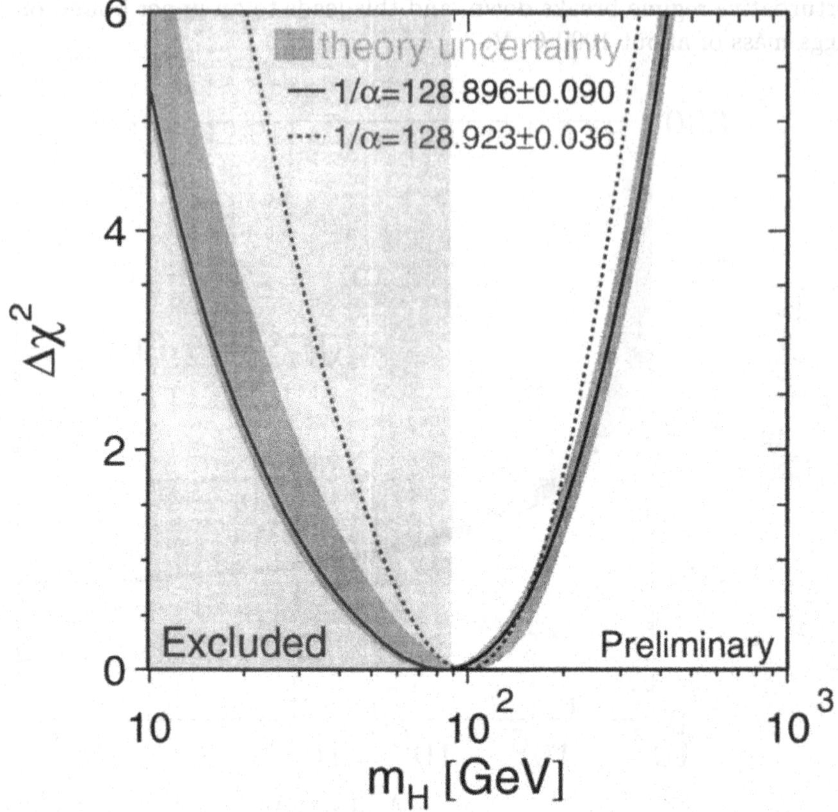

Figure 2. **$\Delta\chi^2$ result of a fit to all electroweak observables assuming to have the Higgs mass as the only remaining free parameter [9].**

200 GeV. In case no Higgs signal is found, a mass limit of $\approx$ 109 GeV (95% C.L.) will be placed [10].

In the supersymmetric extension of the SM a set of new particles should exist with a mass scale around 1 TeV. The minimal version of the supersymmetric SM (MSSM) contains three neutral and two charged Higgs bosons. One of the neutral ones is expected to have a mass around 100 GeV and is therefore of particular interest for searches at LEP200. So far the searches for this lightest Higgs boson resulted in a lower mass limit of about 80 GeV. The expectation for SUSY Higgs searches at LEP200 is summarised in figure 3 [11]. However, the discovery potential at LEP200 depend strongly on the available energy. Thus few GeV increase in $\sqrt{s}$ could change our understanding of the Higgs sector dramatically.

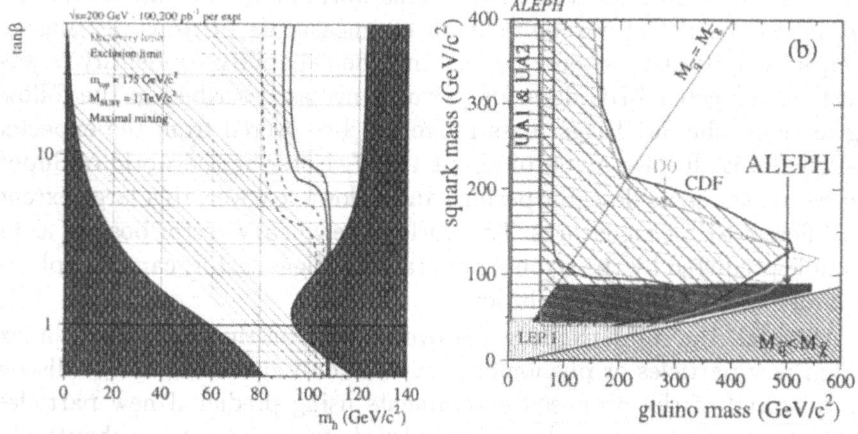

Figure 3. Expected sensitivity of SUSY Higgs searches (left figure) at LEP200 ($\sqrt{s}$= 200 GeV) and recent lower mass limits for squarks and gluinos (right figure) [11].

2.2. THE SPARTICLE SECTOR

Direct searches for sparticles at LEP200 have reached in most cases the kinematical limit, i.e. sparticle masses below $\approx$ 90 GeV are excluded [11]. Searches for sparticles at the Tevatron have excluded gluino and squark masses below about 250 GeV, as shown in figure 3. With the data collected during RunII ($\approx$ 1fb^{-1}/year) at the Tevatron, scheduled to start in 2000, one expects to reach gluino and squark masses of 300 to 400 GeV.

3. The World of Physics at the LHC

Discovering new phenomena in high energy physics experiments rely on the capability to separate *new* from *known* phenomena . The methods used exploit the different kinematics of signals and backgrounds in searching for new mass peaks, or comparing p_T spectra of leptons, photons and jets and their angular correlations with SM predictions. Other searches exploit the missing transverse energy signature which might originate from neutrinos or neutrino–like objects, or simply from detector imperfections. Depending on the particular physics process, different aspects of the detector performance parameters are important. The search for mass peaks requires in general excellent energy and momentum resolution for individual particles. Searches based on the missing transverse energy signature require detectors with hermetic calorimeter coverage up to $|\eta|$= 5.

Figure 4 shows the world of physics to be explored with multi–TeV proton-proton collisions at the LHC. This world is divided into sectors according to the detector requirements for measuring photons, leptons (e, μ, τ), missing transverse energy, jets and the capability to identify b–jets. With the expected LHC detector performance as described in the following sections, the SM Higgs mass range can be covered from the expected LEP200 limit all the way up to about 1 TeV. Entering the world of Supersymmetry, the same detector performance allows to cover, to a large extent, the different SUSY signatures. The world of new heavy vector bosons, as for example predicted by the strongly interacting Higgs sector, can be explored with an excellent lepton resolution.

Although the most exciting discoveries will be those of totally unexpected new particles or phenomena, one can only demonstrate the discovery potential of the proposed experiments using predicted new particles. However, the experiments designed under these considerations should also allow the discovery of whatever new phenomena might occur in multi–TeV pp collisions.

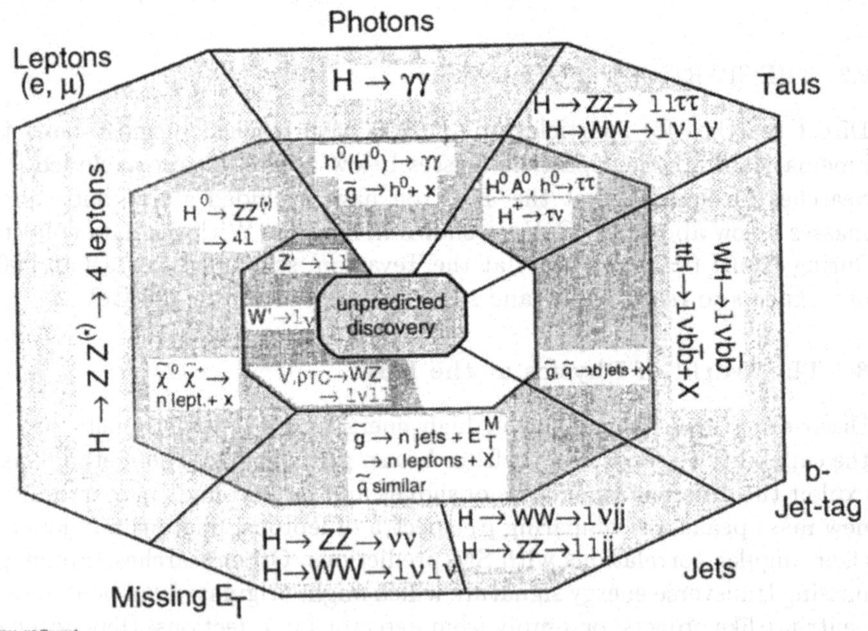

Figure 4. The world of physics at the LHC

3.1. THE EXPERIMENTAL CHALLENGE AT THE LHC

The total cross–section at hadron colliders is very large, i.e. about 100 mb at the LHC, resulting in an interaction rate of $\approx 10^9$ Hz at the design luminosity. Figure 5 shows the expected energy dependence of the total cross section and of some interesting physics processes which have much smaller cross sections. The detection of processes with signal to total cross–section ratios of about 10^{-12}, as for example for a 100 GeV Higgs decaying into two photons, will be a difficult experimental challenge.

Many of the above mentioned new particles decay into W and Z bosons, charged leptons or photons. Ws and Zs will have to be detected through their leptonic decays because hadronic decay modes will be overwhelmed by the QCD [12] background. These purely leptonic modes lead to very small branching fractions. In order to observe such signals, a machine with high constituent centre–of–mass energy and high luminosity is required.

The LHC fulfils these requirements, but the high luminosity leads to difficult experimental conditions: with an inter–bunch crossing time of 25 ns at design luminosity, on average 20 interactions ("minimum bias events") are expected per crossing, resulting in about 1000 charged tracks every 25 ns, in the pseudorapidity range of $|\eta| \leq 3$. Therefore, at peak luminosity, on average 2.2 charged particles are expected every 25 ns in a $2\times2\text{cm}^2$ cell at a distance of 7.5 cm from the interaction point at $\eta = 0$. This example shows that the inner tracking detectors have to operate in a hostile environment. Such high particle fluxes will make track reconstruction difficult. Simulation results make us believe that a very large number of electronic channels and good time resolution should nevertheless guarantee a high track–finding efficiency.

The expected 10^9 inelastic pp events per second at design luminosity will also generate a hostile radiation environment. This results in high radiation levels (high integrated dose) and in a large flux of low energy neutrons in the experimental area. As an example, figure 6 shows the radiation environment in CMS [13]. Radiation hard detectors and electronics are therefore required. Induced activity in the forward calorimeters has to be taken into account for long–term access and maintenance.

4. Design Objectives of ATLAS and CMS

An important aspect of the overall detector design is the magnetic field configuration. Large bending power is required to measure precisely high–momentum muons and other charged particles. The choice of the magnet structure strongly influences the remaining detector design. A *solenoid* provides bending in the transverse plane and thus facilitates the task of triggering on muons, which are pointing to the event vertex, so that one can

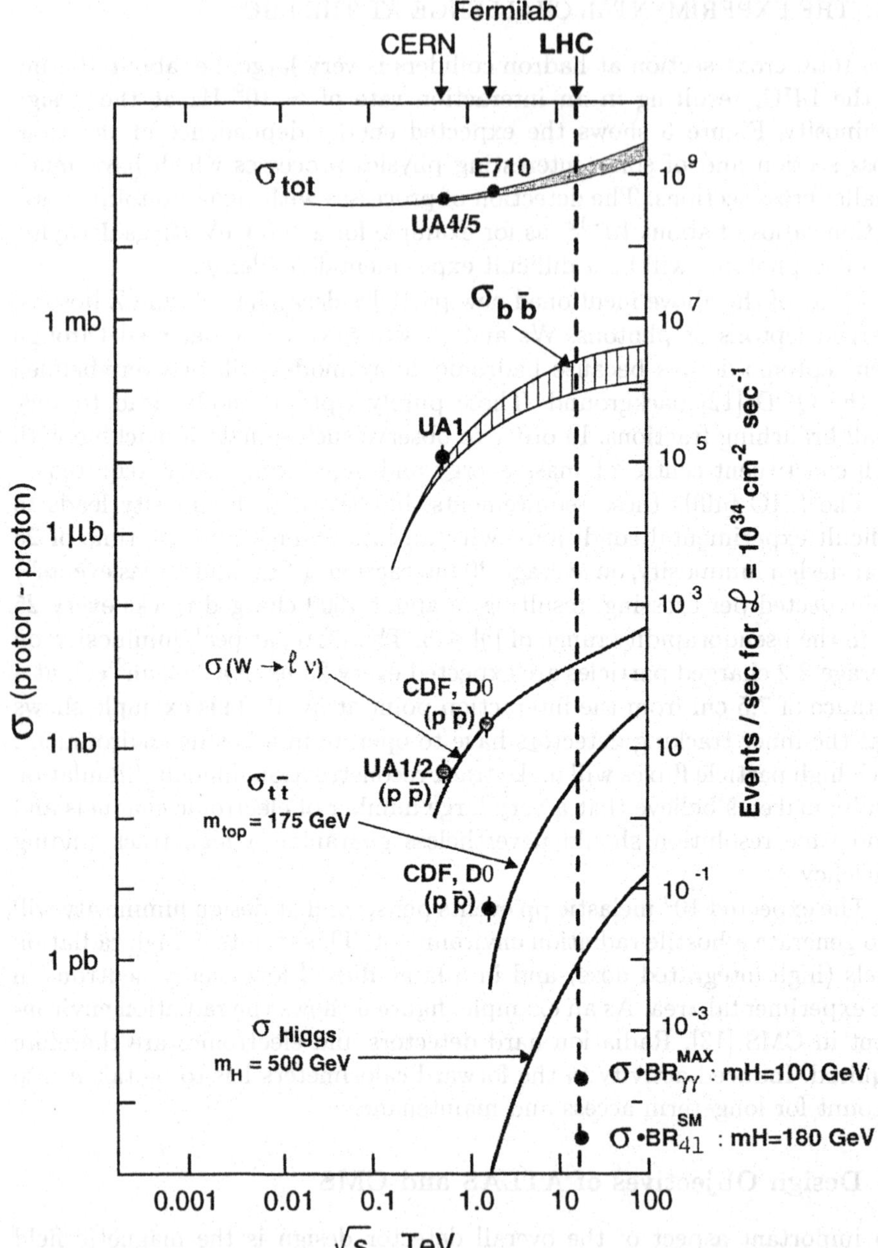

Figure 5. **Energy dependence of some characteristic cross–sections at hadron colliders.**

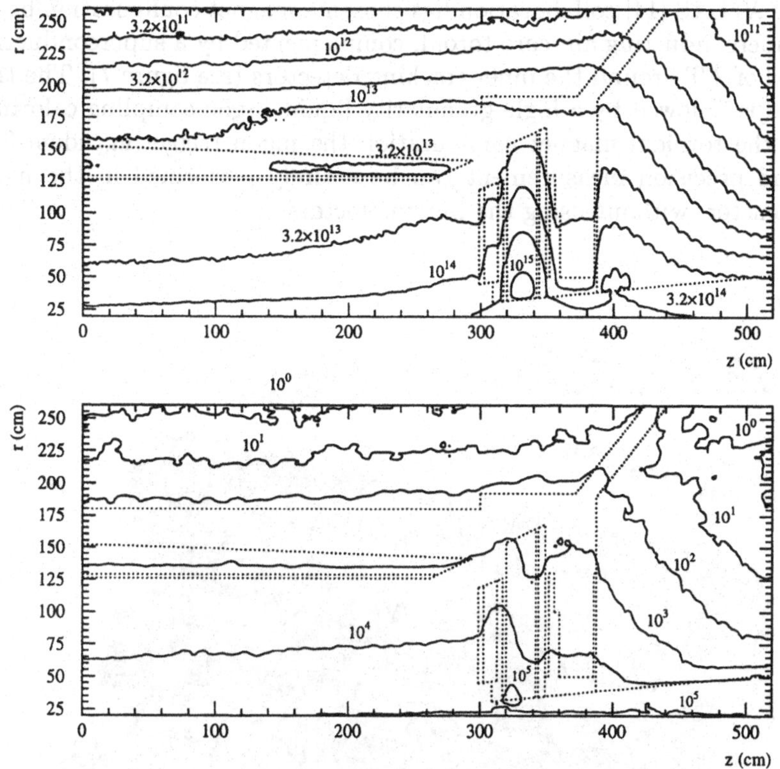

Figure 6. Fluence of neutrons and charged hadrons in cm^{-2} (upper plot) and radiation dose in Gy (lower plot) in the CMS calorimeter region. Values correspond to an integrated luminosity of 5×10^5 pb^{-1} [13].

take advantage of the small transverse dimensions of the beam (20μm). A drawback of a solenoid with limited length is the degradation of momentum resolution in the forward direction; therefore either a very long solenoid is required or an endcap toroid system has to be added. The main advantage of a *toroid* is a constant p_T resolution over a wide rapidity range. However, the closed configuration of a toroid does not provide magnetic field for the inner tracking, thus an additional solenoid is required to measure the momenta of charged tracks in the inner tracking detectors.

The identification and precise measurement of electrons, photons and muons over a large energy range, complemented by measurements of jets and missing transverse energy are the basic design goals of the ATLAS and CMS detectors. In addition, a good impact–parameter resolution and secondary vertex reconstruction will be important for b-tagging.

The ATLAS [14] collaboration has chosen a magnet configuration based on a superconducting air–core toroid, complemented by a superconducting solenoid of 2 T around the inner tracking detectors (see figure 7). The thin solenoid is followed by a high–granularity liquid–argon sampling calorimeter. In the toroidal magnet configuration the muon triggering, identification and precision measurement can be entirely performed in the muon spectrometer, without using the inner detectors.

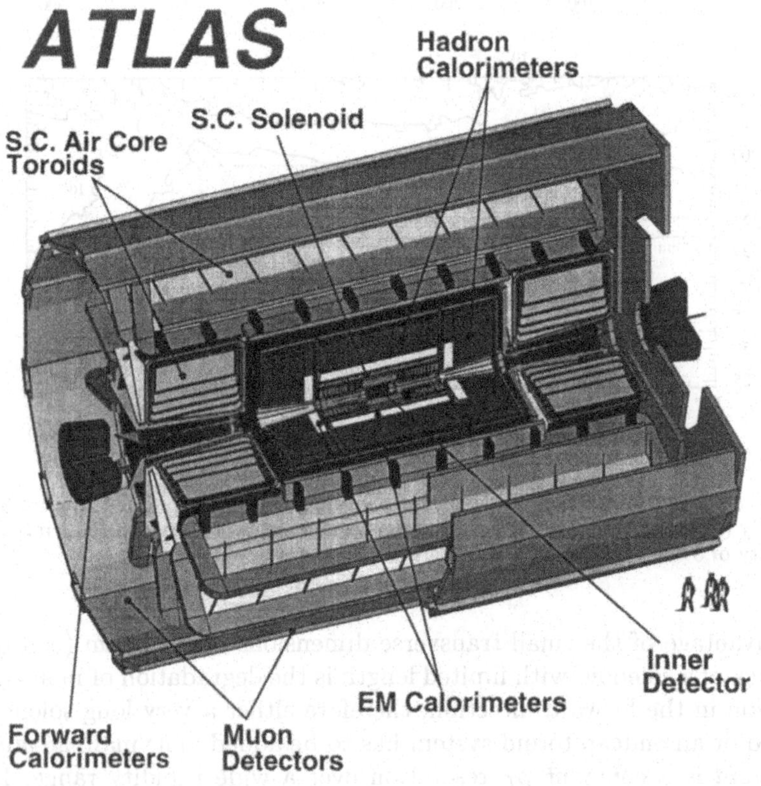

Figure 7. Schematic view of the ATLAS detector. The overall diameter is about 25 m and the overall length is about 40 m. The total weight amounts to about 8000 tons.

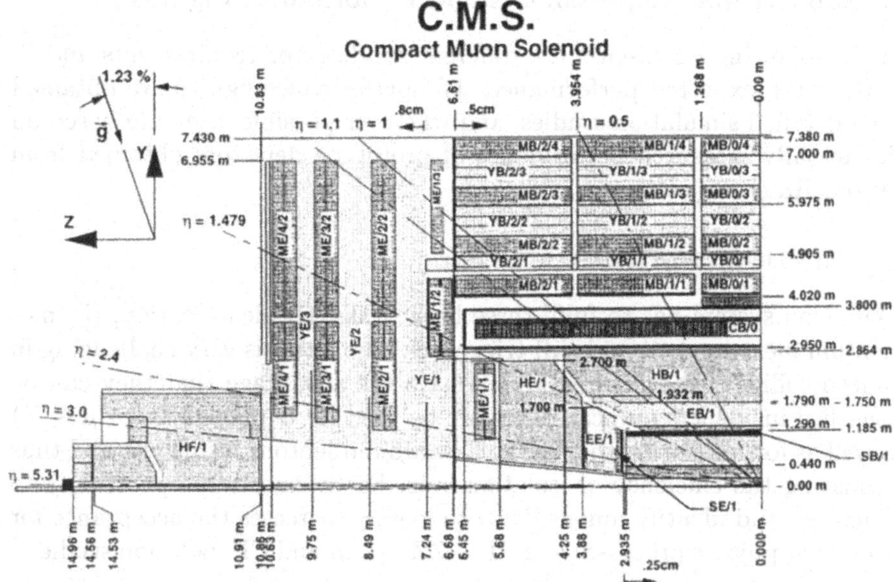

Figure 8. Schematic 1/4 view of the CMS detector. The total weight amounts to about 12500 tons.

The CMS [15] detector will use a high–field superconducting solenoid (4 T) allowing for a compact design of the muon spectrometer (see figure 8). The inner coil radius of about 3 m is large enough to accommodate the inner tracking system and the calorimeters. For the electromagnetic calorimeter $PbWO_4$ crystals have been chosen. The hadron calorimeter (also located before the coil) consists of copper absorber plates and scintillator tiles. Muons are triggered, identified and measured in four identical muon stations inserted in the return yoke. Their momenta are measured independently in the inner tracking chambers to improve the overall momentum measurement.

5. Subdetector Requirements and Performance Figures

In the following we discuss the different subdetector requirements and illustrate the expected performance. The performance figures are obtained from detailed simulation studies, and wherever possible, they are based on the actually measured performance of prototype detectors obtained from various R&D projects.

5.1. THE MUON SYSTEM

The muon system has to fulfil three basic tasks: (i) identification, (ii) momentum measurement and (iii) triggering. The latter is very challenging in hadron collider experiments. Muons have the advantage that they can be identified inside jets and can therefore be used for b–tagging ($b \rightarrow \mu + X$) as well as for measuring the energy flow around leptons inside jets and thus evaluating the efficiency of isolation cuts. Furthermore, the possibility to trigger on and identify muons down to low p_T increases the acceptance for important physics processes (e.g. $H \rightarrow ZZ^* \rightarrow 4\mu$ and CP violation studies).

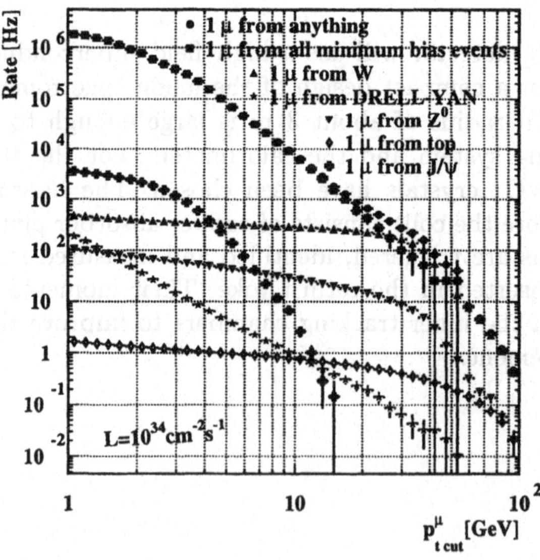

Figure 9. Predicted CMS muon trigger rates at the LHC design luminosity [16].

The performance of the μ–system is determined by (i) pattern recognition: hits from μ–tracks can be spoilt by correlated background (δ's, electromagnetic showers, punch-through) and uncorrelated background (neutrons and associated γs), (ii) momentum resolution: many factors influence the muon resolution, like for example, multiple scattering, fluctuations in the energy loss, accuracy of tracking devices in the muon spectrometer, alignment and the magnetic field map and (iii) 1st– level trigger: the μ–rate is dominated by π and K decays up to 4 GeV and by b– and c–quark decays from 4 to 25 GeV. Figure 9 shows the μ–trigger rate at the design luminosity. At p_T^μ = 10 GeV a 1st–level trigger rate of about 10^4 Hz is expected [16]. The trigger rate has to be adjustable by moving the threshold in a wide range of p_T^μ to account for different background conditions without however loosing efficiency for important physics channels.

The ATLAS superconducting air–core toroid system is optimised for stand–alone measurements [17]. The μ–spectrometer consists of precision chambers (monitored drift tubes and cathode strip chambers) which require a high–accuracy tracking (50μm/chamber), to be aligned to 30μm. Resistive plate chambers and thin gap chambers are used for triggering. The expected resolution ranges from about 2-3% at p_T^μ = 100 GeV to about 11% at p_T^μ = 1 TeV.

The CMS μ–system consists of 4 identical muon stations [18]. The

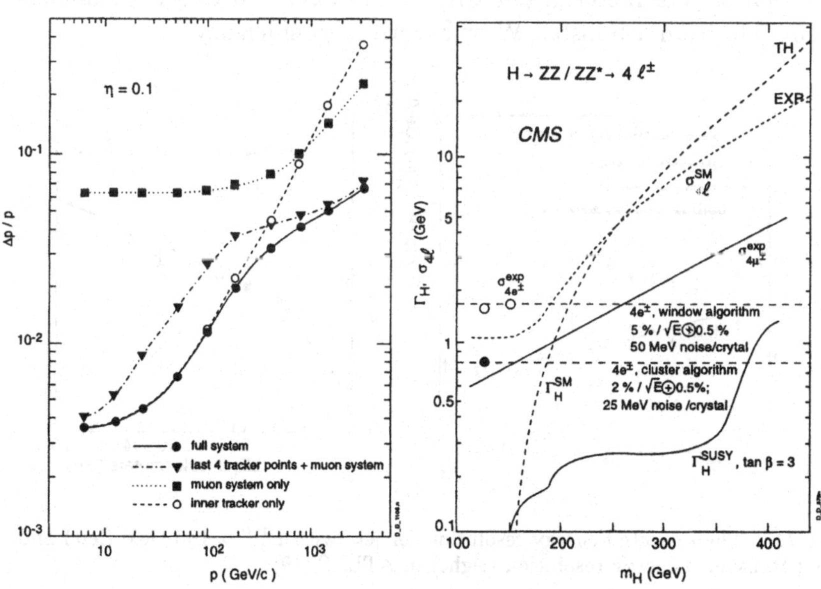

Figure 10. Expected muon momentum resolution in the CMS experiment (left) and the resulting experimental $H \rightarrow 4\mu$ mass resolution (right) compared with the expected width of the Higgs boson [18].

CMS μ–system consists of 4 identical muon stations [18]. In the barrel, 12 layers of drift tubes are used with σ = 250 μm per layer. Cathode strip chambers (6 layers) with σ = 75–150 μm are implemented in the endcaps. The chambers have to be aligned to 100–200 μm. The favourable aspect ratio (length/radius) of the superconducting coil allows good μ–momentum resolution up to pseudorapidity of 2.5 with a single magnet. The 4 T field provides a powerful combined measurement up to $|\eta| \approx 2.5$. Figure 10 shows the expected momentum resolution in the CMS experiment and illustrates the resulting experimental 4μ–resolution as a function of m_H. This resolution is compared to the width of the SM–Higgs and the expected 4e–resolution. The combined experimental 4l–mass resolution dominates the measured width up to $m_H \approx 250$ GeV.

5.2. HADRON CALORIMETER

The performance of the hadron calorimeter can be characterised by the jet–jet mass resolution and the missing transverse energy (E_T^{miss}) resolution. The mass resolution depends on the calorimeter resolution but also on the jet algorithm, fragmentation, energy pile–up and the cone–size for jet reconstruction. Studies of the effect on the mass resolution for W,Z $\rightarrow$ $jet + jet$, using the high–mass Higgs decay H $\rightarrow$ ZZ(WW) $\rightarrow$ $lljj$ have shown that a calorimeter granularity of $\Delta\eta \times \Delta\phi \approx 0.1 \times 0.1$ is sufficient to measure jets from a boosted W or Z with high efficiency.

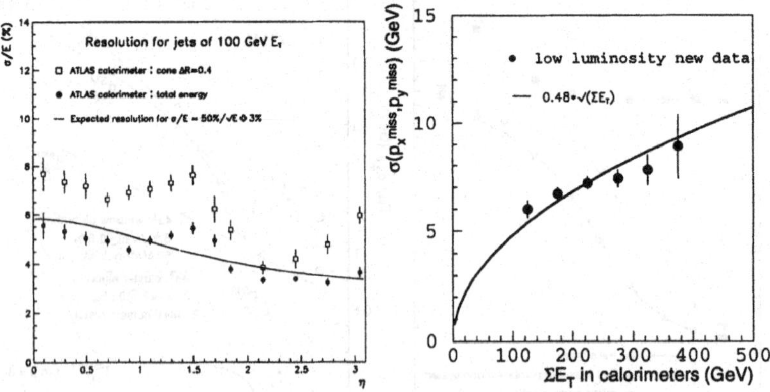

Figure 11. The expected energy resolution for jets with E_T^{jet} = 100 GeV (left) and the missing transverse energy resolution (right) in ATLAS [19].

A good E_T^{miss} resolution requires a hermetic calorimeter coverage up to $|\eta|$=5 (i.e. cracks and dead areas have to be minimized). This requirement has been taken into account in the design of the ATLAS [20], and CMS [21] calorimeter system. Figure 11 illustrates the jet resolution for $E_T^{jet} = 100$ GeV and the expected E_T^{miss} resolution in ATLAS at low luminosity.

The ATLAS barrel hadron calorimeter consists of iron with scintillating tiles. The endcaps use a parallel plate design for the copper liquid–argon calorimeter. The forward calorimeter is made of tungsten liquid–argon with very small gaps (0.25 to 0.5 m) to limit the ion build–up.

The barrel and endcaps hadron calorimeter in CMS is placed inside a 4 T field. CMS has chosen a copper/plastic–scintillator system. In the forward direction, due to the high radiation environment (absorbed dose $\sim$ MGy/year, neutron flux $\sim 10^9 cm^{-2}s^{-1}$), an iron/quartz fibre system was selected. More details about the proposed hadron calorimeters can be found in reference [22].

5.3. ELECTROMAGNETIC CALORIMETER

The performance of the electromagnetic calorimeter is best demonstrated using the $H \rightarrow \gamma\gamma$ reaction. For $m_H = 100$ GeV the Higgs width is only a few MeV, therefore the measured mass resolution is entirely dominated by the experimental resolution.

The CMS design goal requires a high resolution electromagnetic calorimeter, therefore a fully active (homogeneous) calorimeter consisting of $PbWO_4$ crystals has been chosen [13]. A GEANT simulation of a $H \rightarrow \gamma\gamma$ ($m_H =$ 100 GeV) in the CMS detector is shown in figure 12. The photon resolution expected in CMS is summarised in the following table:

Photon energy resolution in CMS		
	Barrel (η=0)	Endcap (η =2)
stochastic term	$2.7\%/\sqrt{E}$	$5.7\%/\sqrt{E}$
constant term	0.55%	0.55%
E_T noise ($L=10^{33}cm^{-2}s^{-1}$)	155 MeV	205 MeV
E_T noise ($L=10^{34}cm^{-2}s^{-1}$)	210 MeV	245 MeV

ATLAS has chosen a sampling calorimeter which provides longitudinal and transverse segmentation [20]. An accordion structure in liquid argon with lead absorbers is used. The expected γ–resolution as function of pseudorapidity for $E_T^{\gamma} = 50$ GeV is shown in figure 13. More details about the proposed electromagnetic calorimeters can be found in reference [22].

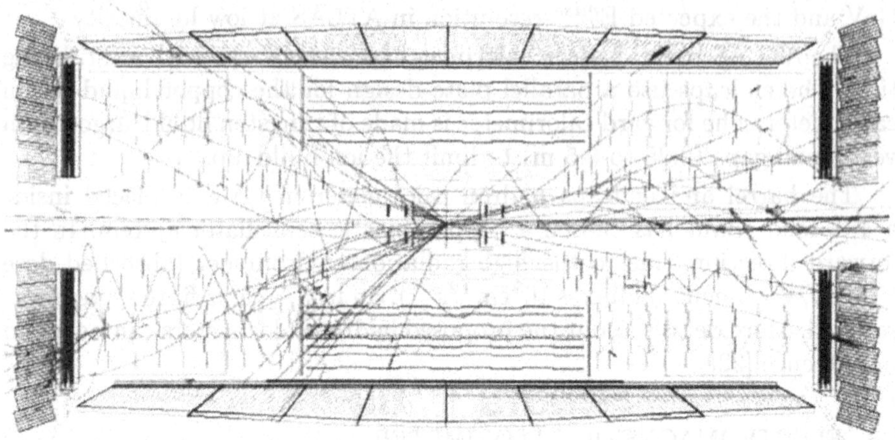

Figure 12. **GEANT simulation of a $H \to \gamma\gamma$ event in CMS (m_H = 100 GeV). To guide the eyes, the two electromagnetic clusters in the crystals are shown with associated dotted lines in the inner detector.**

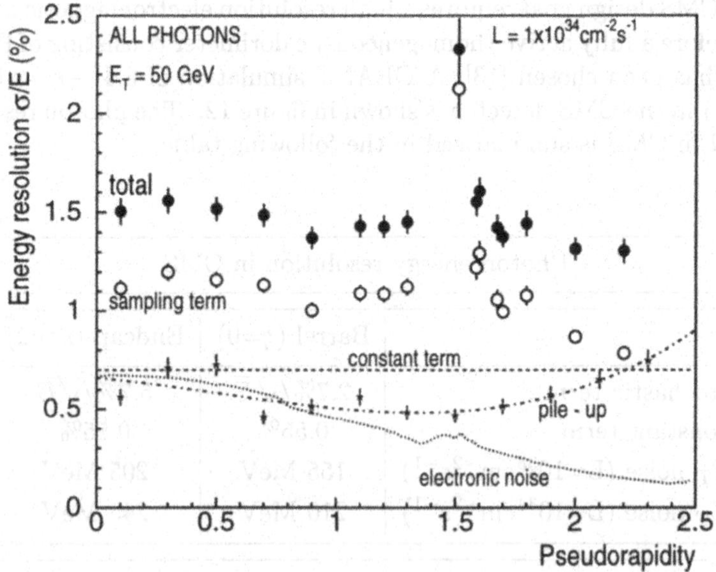

Figure 13. **γ–resolution as function of pseudorapidity for E_T^γ = 50 GeV in the ATLAS detector [19].**

5.4. INNER TRACKING DETECTORS

The inner tracking system provides precise momentum and impact parameter and secondary vertex measurements for charged particles. It is also essential for e and τ identification, and the calibration of the electromagnetic calorimeter with electrons, using the p/E matching.

Silicon detectors are already very successfully used in HEP experiments for vertex determination. However, a transition from detector sizes of $\mathcal{O}(\Updownarrow^{\in})$ with electronics at the periphery of the sensitive area to sizes of $\mathcal{O}(\triangledown/\Updownarrow^{\in})$ with electronics distributed over the sensitive area is needed. Furthermore, tracking at the LHC must cope with high instantaneous and integrated rates. The tracking system must operate at an integrated dose (500 fb^{-1}) of Mrad to 30 Mrad and a neutron fluence of 10^{14} to 10^{15} n/cm^2, therefore cooling is required. At the LHC silicon detectors are used for r $\leq$ 60 cm and gaseous detectors (microstrip gas chambers (MSGC) and transition radiation detectors (TRD)) are used for $\geq$ 60 cm. More details about silicon detectors can be found in reference [24].

ATLAS Barrel Inner Detector

H→b$\bar{\text{b}}$

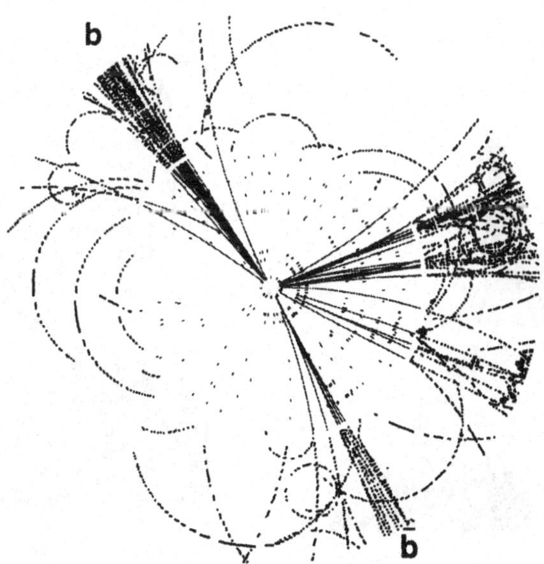

Figure 14. H → $b\bar{b}$ simulation in the ATLAS inner detector at low luminosity. Precision hits are shown for $0 < \eta < 0.7$; TRT hits are shown for z > 0 [25].

The design goals for the tracking system are:

- for isolated leptons in the CMS detector: $\Delta p_T/p_T \sim 0.1\ p_T$ (TeV). in ATLAS: $\Delta p_T/p_T \leq 30\%$ at 500 GeV, 1–2% at 20 GeV
- high–p_T track reconstruction efficiency: for isolated tracks: $\varepsilon > 95\%$, and within jets: $\varepsilon > 90\%$ (ghost tracks $< 1\%$ for isolated tracks).
- impact parameter resolution: at high p_T $\sim 20\mu$m (rϕ), $\sim 100\mu$m (z)

The ATLAS tracking system [25] is located inside a solenoidal field of 2 T and consists of 3 pixel layers ($\sigma_{r\varphi} \sim 12\mu$m), 4 silicon strip layers ($\sigma_{r\varphi} \sim 16\mu$m) and ~ 40 transition radiation tracking (TRT) layers ($\sigma_{r\varphi} \sim 170\mu$m). Figure 14 shows a GEANT simulation of the ATLAS barrel inner detector of a $H \to b\bar{b}$ decay ($m_H = 400$ GeV).

The CMS tracking [26] is inside a 4 T field and consists of 2 pixel layers ($\sigma_{r\varphi}$, $\sigma_z \sim 15\mu$m), 5 silicon strip layers ($\sigma_{r\varphi} \sim 15\mu$m) and 6 MSGC layers ($\sigma_{r\varphi} \sim 50\mu$m).

The pattern recognition and momentum resolution is effected by conversion and bremsstrahlung. A low material budget is desirable also in order to maintain the electromagnetic calorimeter performance. The ATLAS and CMS material budget is plotted in figure 15 and shows that the inclusion of support structures and cables increases the material budget beyond the desirable value. Figure 16a and b show the expected p_T resolution as a

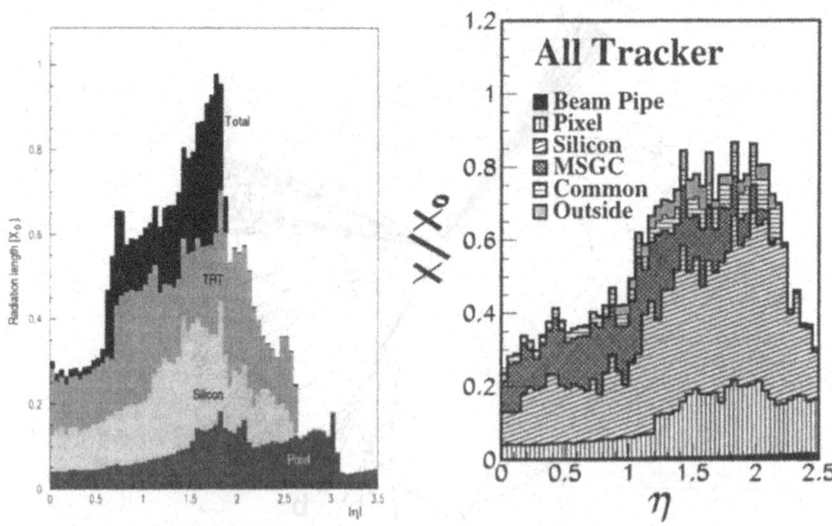

Figure 15. Material budget of the inner tracking system in ATLAS (left) [25] and in CMS (right) [26].

function of η in ATLAS and the impact parameter resolution in CMS as a function of p_T for different η–values.

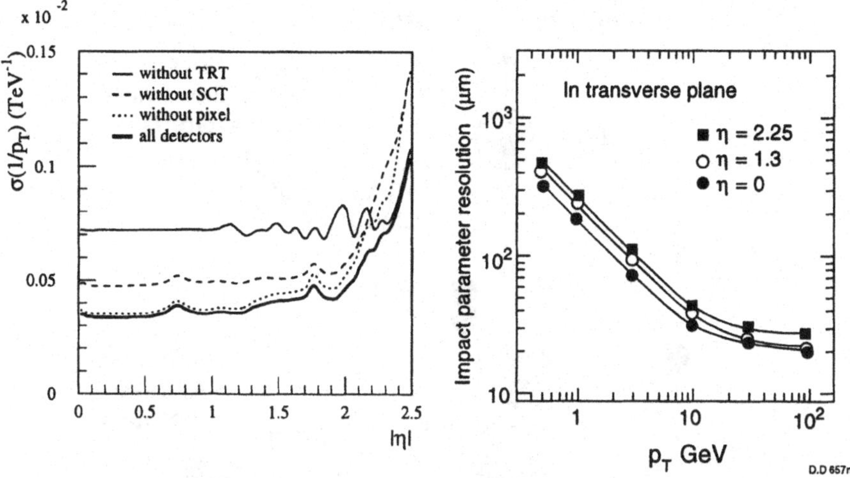

Figure 16. The expected p_T resolution for high–p_T tracks in the ATLAS tracking system (left). The effect of removing complete subdetectors is also shown [25].
The figure on the right shows the impact parameter resolution in CMS as a function of p_T for three different η–values [26].

5.5. DATA ACQUISITION AND TRIGGER

The challenge at the LHC will be the reduction of 10^9 Hz interaction rate to about 100 Hz output rate on tape for further off–line analysis. The on–line data reduction will proceed via different trigger levels. At the first level, local pattern recognition and energy evaluation on prompt macro–granular information will provide particle identification such as high–p_T electrons, muons and missing E_T. Level–1 will select events at 10^5 Hz. For level–2, finer granularity and more precise measurements will be used together with event kinematics and topology. By matching different subdetectors, clean particle signatures should be selected (e.g. W, Z, etc.), resulting in a level–2 rate of 10^3 Hz. Finally, event reconstruction and on–line analysis will result in physics process identification, leading to an output rate for further off–line analysis of about 100 Hz. Figure 17 shows the trigger/DAQ evolution in terms of level-1 rate and event size, from existing experiments to experiments at the LHC [27]. To illustrate the very demanding requirements for the trigger/DAQ system, we recall some interesting numbers:

- the CERN (Lab–wide) computing power available in 1980 was comparable to that of a modern desktop computer (1995),

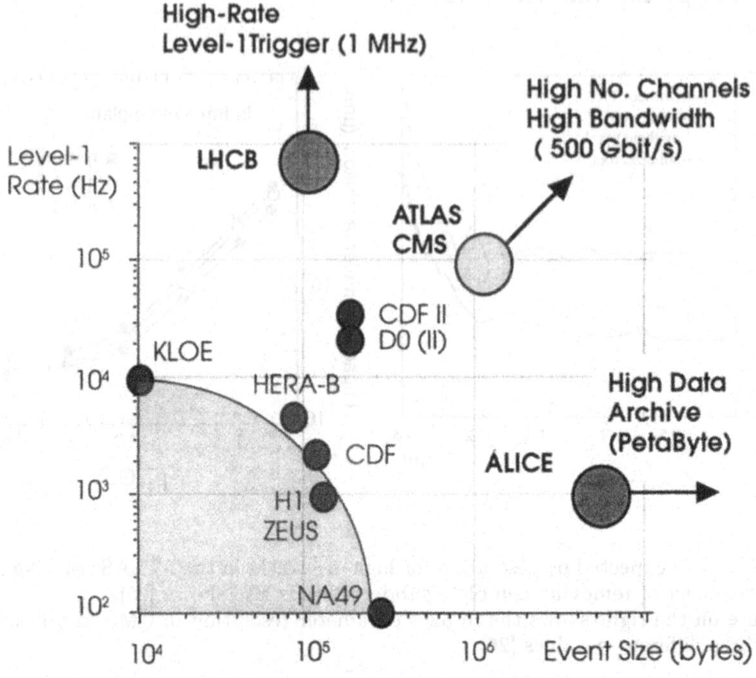

Figure 17. Expected evolution for trigger and data acquisition from existing experiments to future experiments at the LHC [27].

- the total number of processors in the LHC event filter equals the number of workstations and personal computers running at CERN in 1995 (4000),
- during one second of LHC running, the data volume transmitted through the readout network is equivalent to the amount of data moved in *one day* by CERN network system (FDDI, Ethernet, local nets) in 1995,
- the 'slow-control' data rate of an LHC experiment (temperature, voltage, status, etc.) is comparable to the current LEP experiment readout rate (100kByte/s),
- the data rate handled by the LHC event builders (500 Gbit/s) is equivalent to the amount of data exchanged by WORLD TELECOM (today).

6. TeV–scale Physics at LHC

Exploiting the LHC physics potential means that we can answer or shed considerable light on fundamental open questions such as the mass problem or unification of fundamental interactions. First we discuss some issues related to parton luminosities, i.e. the expected accuracy in SM cross–section measurements. This is important for establishing signals for new physics which require a comparison of the measured SM cross–sections with those from beyond the SM processes. The section on parton luminosities is followed by the discussion of the Higgs sector and selected topics in sparticle searches in order to demonstrate the discovery potential of the proposed pp detectors[1].

6.1. PARTON LUMINOSITIES

Accurate cross–section measurements for different SM processes are an important part of the LHC programme. Previous studies have concluded that accuracies of ±5% can be achieved. These estimates are based essentially on the possibility to measure the proton–proton luminosity and the subsequent uncertainties from the different parton distribution functions. However, a different proposed method might eventually lead to cross–section measurements which could approach ±1% accuracies [28]. The basic idea of the proposed new method is based on:

- Experiments at the LHC will study the interactions between fundamental constituents of the proton, the quarks and gluons, at energies where these partons can be considered as quasi–free. Thus, the important quantity is the parton–parton luminosity at different values of x_{parton} and not the traditionally considered proton–proton luminosity. Assuming collisions of essentially free partons, the production and decay of weak bosons, $u\bar{d} \to W^+ \to \ell^+\nu$, $d\bar{u} \to W^- \to \ell^-\bar{\nu}$ and $u\bar{u}(d\bar{d}) \to Z^0 \to \ell^+\ell^-$ are, in lowest order, known to at least a percent level. Cross–section uncertainties from higher order QCD corrections are certainly larger, but are included in the measured weak boson event rates. Similar higher order QCD corrections to other $q\bar{q}$ scattering processes at different Q^2, like $q\bar{q} \to W^+W^-$, can be expected. Thus, assuming that the Q^2–dependence can in principle be calcu-

[1]One usually assumes that "one" LHC year with a peak luminosity of L=10^{33}cm^{-2}s^{-1} and a running time of 10^7s produces an integrated luminosity of 10 fb^{-1}. A more realistic estimate would use an average run luminosity and includes losses due to machine and detector efficiencies. It would thus be more conservative to assume that a running time of 10^7s per year with the initial luminosity requires about 2–3 years to accumulate 10 fb^{-1}.

lated, very accurate theoretical predictions for cross–section ratios like $\sigma(pp \rightarrow W^+W^-)/\sigma(pp \rightarrow W^\pm)$ should be possible.

- It is a well known fact that the $W^\pm$ and Z^0 production rates at the LHC, including their leptonic branching ratios into electrons and muons, are huge and provide relatively clean and well measurable events with isolated leptons. For instance, assuming L=10^{33} cm^{-2} s^{-1}, one expects per day about 10^6 W$\rightarrow$ $l\nu$ events and about 700 WW$\rightarrow$ $l\nu l\nu$ events. Using the well known $W^\pm$ and Z^0 masses, possible x values of quarks and antiquarks are constrained by $m^2_{W^\pm,Z^0} = s x_q x_{\bar{q}}$ with $s = 4E^2_{beam}$. The product $x_q x_{\bar{q}}$ at the LHC is therefore fixed to $\approx 3 \times 10^{-5}$. Thus, the rapidity distributions of the weak bosons are directly related to the fractional momenta x of the quarks and antiquarks. Consequently, the observable η–distributions of the charged leptons from the decays of $W^\pm$ and Z^0 bosons are also related to the x–distributions of quarks and antiquarks. The shape and rate of the lepton η–distributions provide therefore the key to precisely constrain the quark and antiquark structure functions and their corresponding luminosities.

Using a PYTHIA simulation it could be shown that the rapidity distributions of W^+, W^- and Z^0 events, identified through their clean leptonic decays, determine directly and very accurately the x–distribution of quarks and antiquarks and their corresponding luminosity. The sensitivity of these lepton distributions to recent parton distribution functions is demonstrated in figure 18 for W and Z decays. Furthermore, it was shown that cross–sections of other $q\bar{q}$ related processes are strongly correlated with the single $W^\pm$ and Z^0 production. Ignoring remaining theoretical uncertainties from missing higher order calculations, one finds that ratios like $\sigma(q\bar{q} \rightarrow W^+W^-)/\sigma(q\bar{q} \rightarrow W^\pm)$ show stability within better than 1%. Thus, the shape and height of $W^\pm$ and Z^0 rapidity distributions provide a precise LHC luminosity monitor for $q, \bar{q}$ parton x–distributions at a $Q^2 = m^2_{W,Z}$. A similar analysis using qg$\rightarrow$ γ + jet and qg$\rightarrow$ Z + jet has been performed, and shows that the x–distribution of gluons can be extracted with similar precision [29].

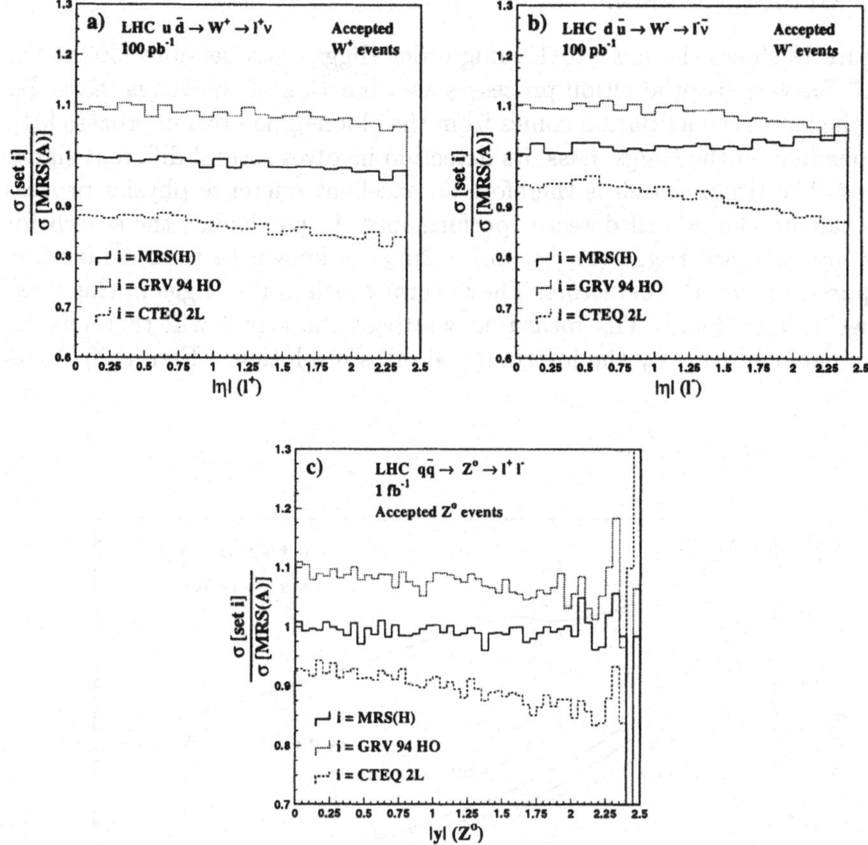

Figure 18. Rapidity dependence of the $\ell^{\pm}$ cross–section predictions for $W^{\pm} \to \ell^{\pm}\nu$ with different sets of structure functions relative to the one obtained from the MRS(A) parametrisation; a) for ℓ^{+}, b) for ℓ^{-}. The reconstructed rapidity distribution for $Z^0 \to \ell^{+}\ell^{-}$ is shown in c) [28].

6.2. SM HIGGS SEARCH

Figure 19 shows the next–to–leading order Higgs cross–sections [30] at the LHC for various production processes as a function of the Higgs mass. By far the largest contribution comes from the gluon–gluon fusion process [31]. Depending on the Higgs mass, its detection involves several different signatures. The Higgs search is therefore an excellent reference physics process to evaluate the overall detector performance. In particular, the search for the intermediate Higgs ($m_Z \leq m_H \leq 2m_Z$) is known to pose demanding requirements on the detectors. The natural width of the Higgs in this mass range is very small. The measured width of the signal will therefore be dominated entirely by the instrumental mass resolution. Figure 20 shows

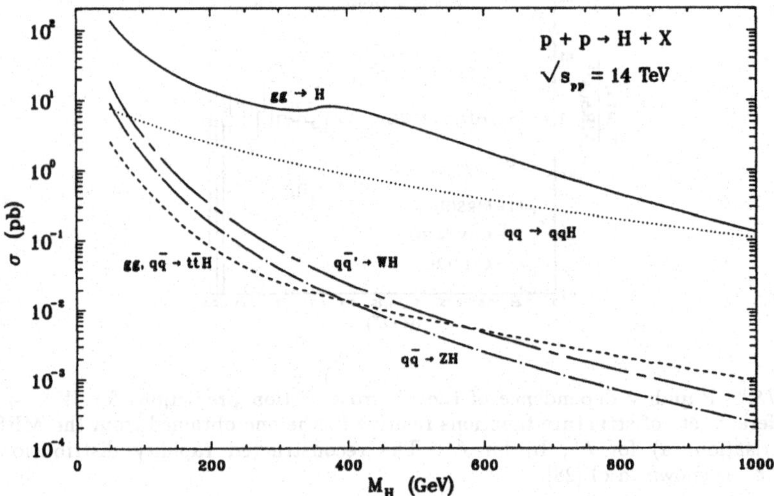

Figure 19. Next–to–leading order cross–section calculations for the SM Higgs [30].

the $\sigma \times BR$ [32] for the most promising Higgs search channels: $H \to \gamma\gamma$, $H \to ZZ^{(*)} \to 4\ell^{\pm}$, and $H \to WW^{(*)} \to \ell^+\nu\ell^-\bar{\nu}$.

Figure 21 summarises the expected observability of the SM Higgs in ATLAS[2] [33] and CMS [34], assuming 100 fb^{-1}.

The most promising signature for a SM Higgs with masses between the expected LEP200 limit and 130 GeV is the decay $H \to \gamma\gamma$ with a branching

[2]The expected Higgs signal significance from ATLAS does not yet include the channel $H \to WW^{(*)} \to \ell^+\nu\ell^-\bar{\nu}$.

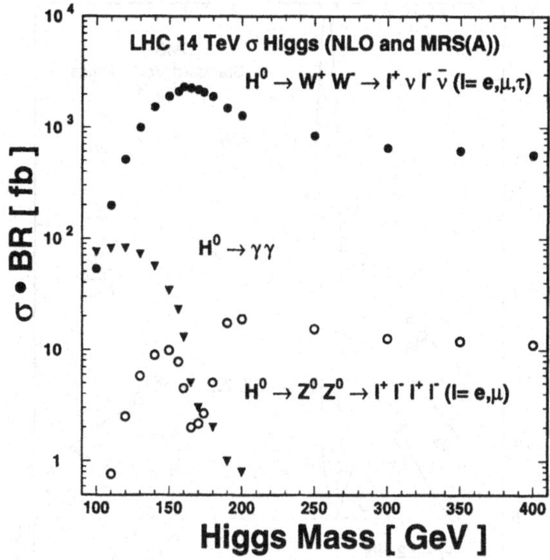

Figure 20. Expected $\sigma \times BR$ for different detectable SM Higgs decay modes [32].

ratio of only $\approx 2 \times 10^{-3}$. As can be seen from figure 22, this signal has to be detected above a large background from continuum $\gamma\gamma$ events. The detection of such a signal requires an excellent $\gamma\gamma$ mass resolution of $\leq 1\%$ (i.e. ≤ 1 GeV for $m_H = 100$ GeV and a very good π^0 rejection capability. More details about this channel can be found in section 6.2.1.

For Higgs masses between 130 GeV and 200 GeV the sensitivity of the $4\ell^{\pm}$ signature suffers from very low branching ratios as illustrated in figures 20 and much smaller signals, like the ones shown in figure 23, are expected. Consequently, a 5 standard deviation signal requires integrated luminosities of at least 30–100 fb^{-1}. A recent study has demonstrated that this Higgs mass region can also be covered by the $H \rightarrow WW(^*) \rightarrow \ell^+\nu\ell^-\bar{\nu}$ decay [35]. The performed analysis, described in section 6.2.2, shows that this channel should allow to discover a SM Higgs with 5 standard deviation for a Higgs mass between 140–200 GeV and integrated luminosities below 5 fb^{-1}.

For $2\times m_Z \leq m_H \leq 500$ GeV the decay $H \rightarrow ZZ \rightarrow 4\ell^{\pm}$ provides the experimentally easiest discovery signature as the events should contain four isolated high p_T leptons. A CMS simulation of a Higgs search using the 4 lepton invariant mass distribution is shown in figure 24, obvious Higgs mass peaks are visible. Furthermore, a Z mass constraint can be used for both lepton pairs to suppress other backgrounds. Estimates from ATLAS and

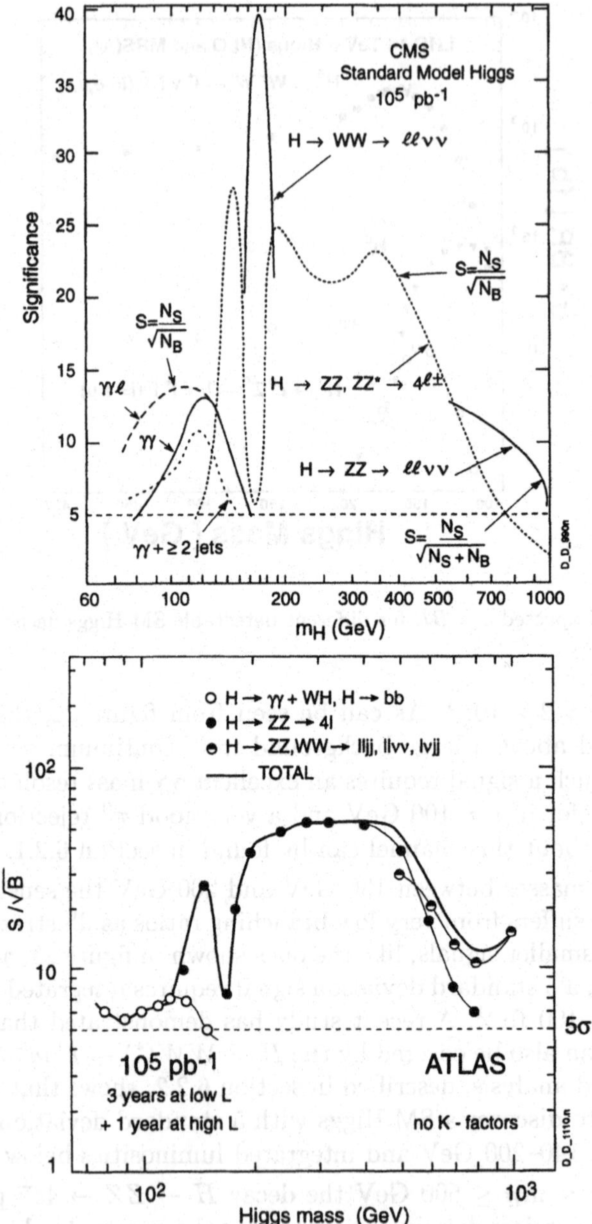

Figure 21. **Expected signal significances for the SM Higgs search in ATLAS [33] and CMS [34], assuming 100 fb^{-1}.**

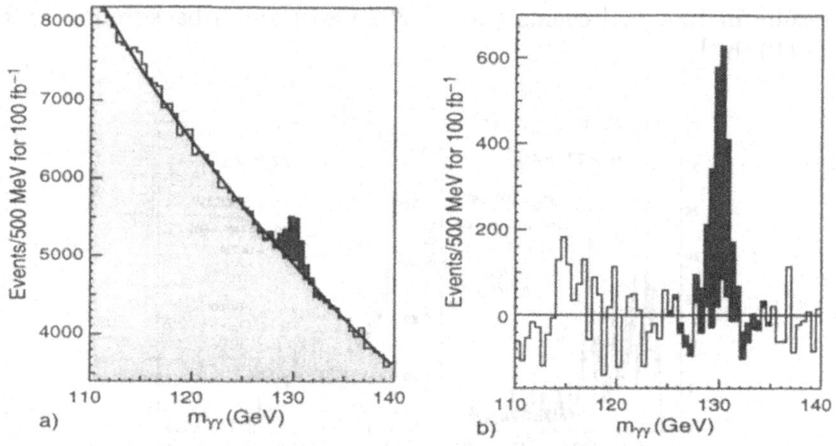

Figure 22. CMS simulation for $H \to \gamma\gamma$ ($m_H = 130$ GeV) before and after background subtraction [13].

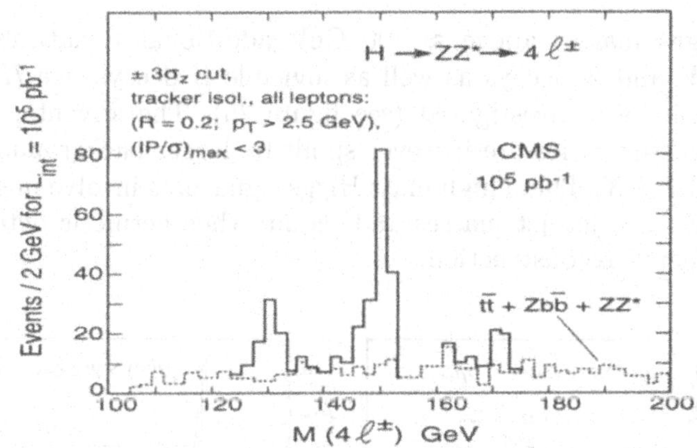

Figure 23. CMS simulation for $H \to ZZ^* \to \ell^+\ell^-\ell^+\ell^-$ and $m_H = 130, 150$ and 170 GeV.

CMS indicate that an integrated luminosity of about 10 fb^{-1} is required to discover a SM Higgs in this mass range with at least 5 standard deviation [36]. For example, an ATLAS study [37] shows that a Higgs ($m_H = 300$ GeV and $H \to ZZ \to 4\ell^{\pm}$) should be seen with 35 signal events above a continuum background of $\approx 13 \pm 4$ events, assuming 10 fb^{-1}. This study indicates also that the signal–to–background rate can be significantly improved by requiring that one reconstructed Z has a $p_T \geq m_H/2$. Using this cut results in 13 signal events (m_H = 300 GeV) and a background of 0.6 events (10 fb^{-1}).

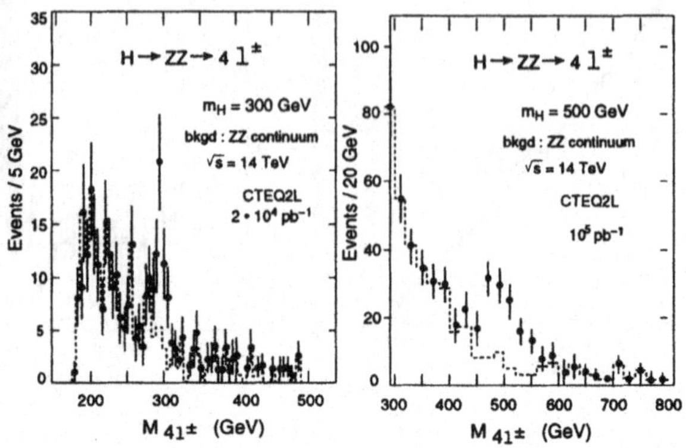

Figure 24. CMS simulation results for $H \to ZZ \to \ell^+\ell^-\ell^+\ell^-$ and $m_H = 300$ GeV and $m_H = 500$ GeV.

For Higgs masses above $\approx$ 400 GeV additional signatures involving hadronic W and Z decays as well as invisible Z decays like $H \to ZZ \to \ell^+\ell^-\nu\bar{\nu}$ have been investigated (see figure 25). The advantages of much larger branching ratios are however spoilt by larger backgrounds from $t\bar{T}$, $W+X$ and $Z+X$. These high mass Higgs signatures involve missing transverse energy and jet–jet masses and require thus hermetic detectors with good jet–energy reconstruction.

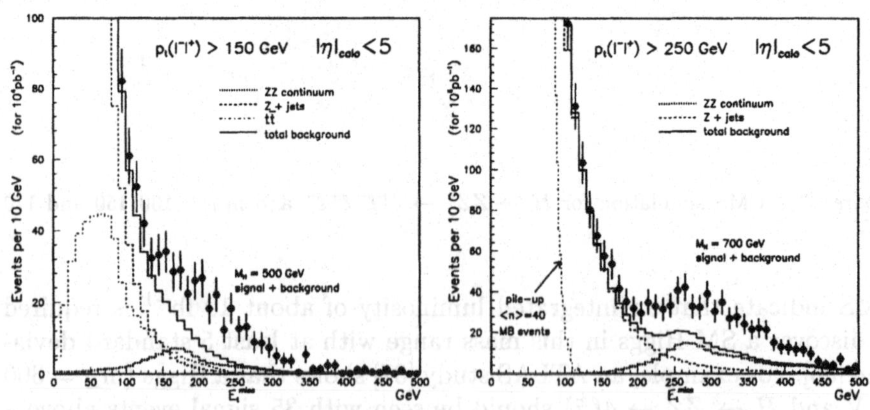

Figure 25. ATLAS simulation results for $H \to ZZ \to \ell^+\ell^-\nu\bar{\nu}$.

6.2.1. *The SM $H \rightarrow \gamma\gamma$ channel*

The $\gamma\gamma$ mass resolution depends upon the energy resolution and the resolution on the measured angle between the two photons. As regards the angle between the photons, the issue is the possible uncertainty on the knowledge of the position of the production vertex. Although very localised in the transverse plane, the interaction vertices have a r.m.s. spread of about 53 mm along the beam axis. If no other knowledge were available such a spread would contribute about 1.5 GeV to the mass resolution. Detailed studies suggest that the correct vertex can be located using charged tracks, even at the highest luminosities, where there are on average nearly 20 inelastic interactions per bunch–crossing. This method of using tracks for the vertex localisation is based on the expectation that the Higgs production events are harder than minimum–bias events and that they contain more high–p_T tracks. Using this fact it is possible to devise an algorithm to select the vertex of the Higgs event from the background of other primary vertices in the same bunch–crossing. This method is used by CMS [13].

ATLAS can use in addition the 1st and 2nd sampling of the electromagnetic calorimeter to measure the photon direction. In this case a contribution to the $\gamma\gamma$–mass resolution of about 530 MeV is expected [23].

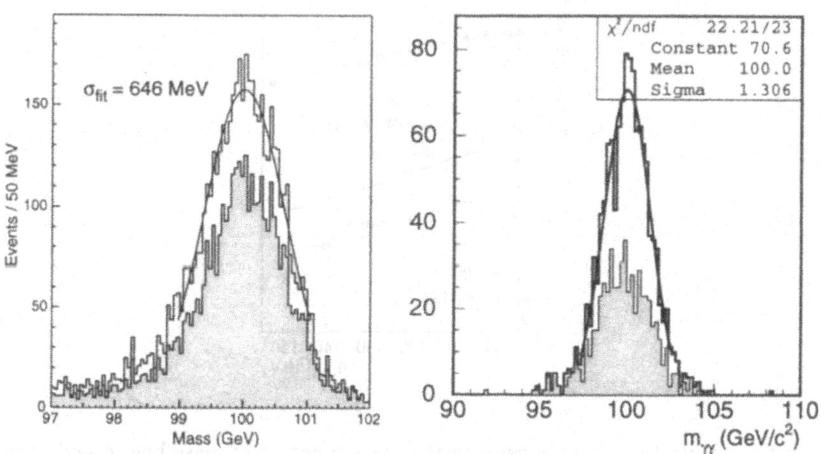

Figure 26. **Higgs mass resolution including converted photons: in the left plot (CMS) [13] the unconverted photons are shown as shaded area and in the right plot (ATLAS) [19] the shaded area corresponds to the converted photons**

Due to the material in front of the electromagnetic calorimeter (beam–pipe, inner tracking detector with support structures) photons will convert. In both experiments about 50% of the $H \rightarrow \gamma\gamma$ events have one or both of the photons converted. Detailed simulation studies have shown that a large fraction of these converted photons can be recovered with only a small degradation in resolution. Figure 26 shows the Higgs mass resolution taking the converted photons into account. This figure shows further that CMS expects about a factor 2 better mass resolution than ATLAS, which demonstrates the potential superior performance of a crystal calorimeter.

The dominant jet–background to the $H \to \gamma\gamma$ signal comes from jet–γ events, where the jet fragments to a leading π^0, carrying a large fraction of the jet transverse momentum. Isolation criteria using calorimeter and/or charged tracks are very powerful tools to reduce this potentially large background. In addition isolated π^0s can be rejected by detecting the presence of two close–by electromagnetic showers rather than one. This can be achieved using the lateral shower shape of the electromagnetic cluster. The resulting rejection factor depends strongly on the π^0 transverse momentum, e.g. rejection factors larger than 3 for $p_T < 40$ GeV can be achieved with a small γ–efficiency loss. Figure 27 illustrates that isolation cuts together with a π^0–rejection algorithm reduce the γ–jet background well below the intrinsic $\gamma\gamma$ background.

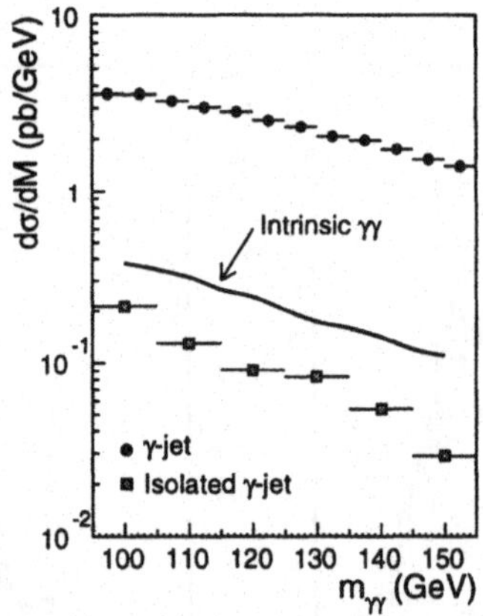

Figure 27. γ–jet background cross–section as a function of mass before and after isolation. The line shows the level of the irreducible di–photon background expected in CMS [13].

The signal significance ($N_S/\sqrt{N_B}$) for a SM Higgs decaying to two photons has been evaluated using events within a $\pm$ 1.4 σ mass window. Figure 28 shows the expected signal significance from CMS [13], as a function of the Higgs mass, for 30 fb^{-1} and 100 fb^{-1}. This figure demonstrates further that a luminosity of 30 fb^{-1} should enable CMS to detect the Higgs in the mass range between 100-150 GeV with more than five standard deviation in the decay $H \to \gamma\gamma$.

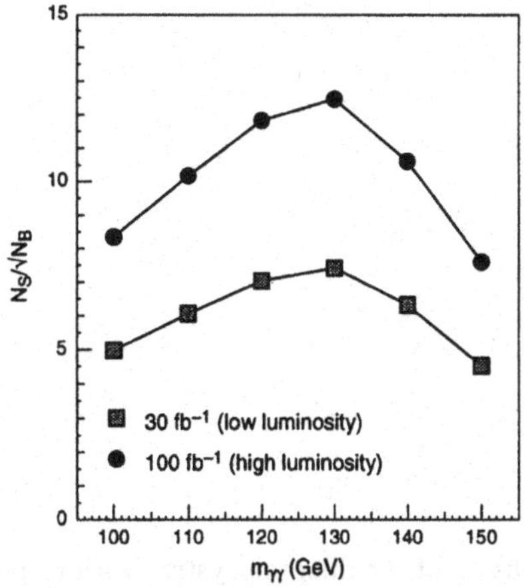

Figure 28. Signal significance as a function of m_H, for H$\to \gamma\gamma$ seen after 30 fb^{-1} and 100 fb^{-1} collected in CMS at low and high luminosity respectively [13].

6.2.2. *The SM $H \to W^+W^- \to \ell^+\nu\ell^-\bar{\nu}$ channel*

A recent simulation has demonstrated that the $H \to W^+W^- \to \ell^+\nu\ell^-\bar{\nu}$ channel can be used to observe a statistically significant signal in the Higgs mass range of 130–200 GeV. This analysis [35] exploits two important differences between a Higgs signal and the non–resonant background from $pp \to W^+W^-X$. As shown in figure 29, the signal events from gluon–gluon scattering are more central than the W^+W^- background from $q\bar{q}$ scattering. This difference is exploited by the requirement that the polar angle θ of the reconstructed dilepton momentum vector, with respect to the beam direction, satisfies $|\cos\theta| < 0.8$. As a result, both leptons are found essentially within the barrel region ($|\eta| < 1.5$) of the experiments. The $\cos\phi$ distribution in figure 29 shows the effect of W^+W^- spin correlations and the V–A structure of the W decays which results in a distinctive signature for W^+W^- pairs produced in Higgs decays. For a Higgs mass close to $2\times m_W$, the $W^\pm$ boost is small and the opening angle between the two charged leptons in the plane transverse to the beam direction is very small.

Finally, the lepton p_T spectra, which are sensitive to the Higgs mass as shown in figure 30, can further be used to improve the signal to background ratio and to determine the Higgs mass with an accuracy of $\delta m_H \approx \pm 5$ GeV, assuming 5 fb^{-1}.

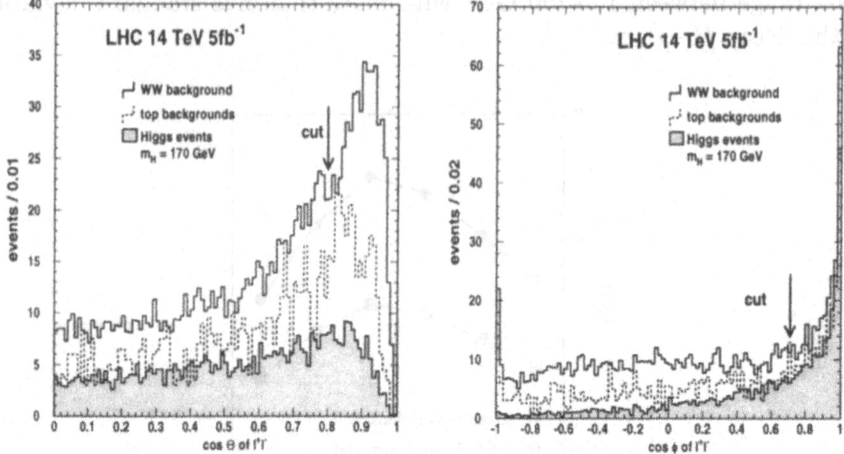

Figure 29. $|\cos\theta|$ distribution of the dilepton system with respect to the beam direction and $\cos\phi$ distribution of the dilepton system in the plane transverse to the beam direction for Higgs signal and background events [35].

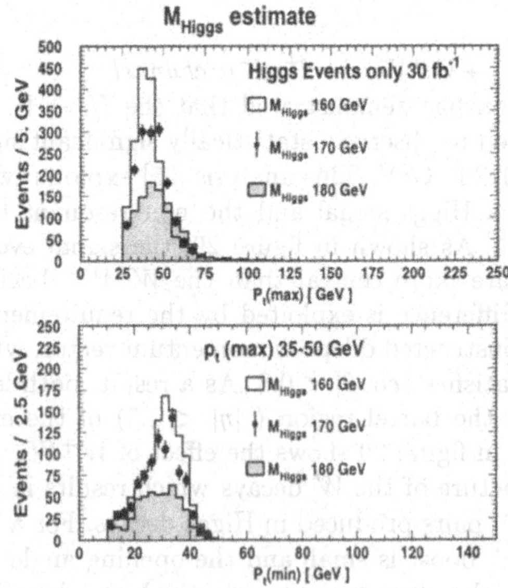

Figure 30. Expected lepton p_T spectra for $H \to W^+W^- \to \ell^+\nu\ell^-\bar{\nu}$ and three different Higgs masses [35].

6.3. SUSY SEARCHES

The attractive features of the MSSM are very well described in a Physics Report by H. P. Nilles [4] in 1984. We repeat here some of his arguments given in the introduction of the report:

"Since its discovery some ten years ago, Supersymmetry has fascinated many physicists. This has happened despite the absence of even the slightest phenomenological indication that it might be relevant for nature. .... Let us suppose that the Standard Model is valid up to a grand unification scale or even the Planck scale of 10^{19} *GeV. The weak interaction scale of 100 GeV is very tiny compared to these two scales. If these scales were input parameters of the theory the (mass)*^{2} *of the scalar particles in the Higgs sector have to be chosen with an accuracy of* 10^{-34} *compared to the Planck Mass. Theories where such adjustments of incredible accuracy have to be made are sometimes called unnatural.... Supersymmetry might render the Standard Model natural... To render the Standard Model supersymmetric a price has to be paid. For every boson (fermion) in the Standard Model, a supersymmetric partner fermion (boson) has to be introduced and to construct phenomenological acceptable models an additional Higgs supermultiplet is needed."*

SUSY signatures are excellent benchmark processes to evaluate the physics performance of LHC detectors and they thus have influenced the detector optimisation. In order to cover the largest possible parameter space in the Higgs sector the searches are more challenging compared to the SM Higgs because: (i) one low mass Higgs (h) *must* exist, (ii) there are 5 Higgs bosons: h, H^0, A^0 , $H^{\pm}$ and (iii) the expected $(\sigma \cdot \mathrm{BR})_{SUSY}$ for the $\gamma\gamma$- and $4l$–channel are smaller than for the SM Higgs. For the simulation results discussed below, all sparticle masses are assumed to be heavy enough such that Higgs bosons decay only into SM particles.

In the sparticle sector many different signatures have been studied [38] in the framework of the MSSM and mSUGRA. These studies include inclusive and exclusive signatures. Particular emphasis was given to the E_T^{miss} and b–jet signatures. In the following we briefly summarise the Higgs sector and discuss some selected topics in sparticle searches.

6.3.1. *The MSSM Higgs sector*

The MSSM Higgs sector requires the existence of two Higgs doublets, resulting in five physical Higgs bosons [39]. Within this model, at least one Higgs boson, the h, should have a mass smaller than $m_h \leq 125$ GeV [40]. The upper mass limit depends via radiative correction on the top quark mass, the a priori unknown value of tanβ and also via a mixing parameter on the mass of the stop quark. One expects that such a MSSM Higgs boson should be found soon at LEP200 if nature has chosen a $\tan\beta$ value of smaller than 4. The masses of the other four Higgs bosons A, H^0 and $H^\pm$ are less constrained but should essentially be degenerate once their mass is larger than $\approx$ 200 GeV.

Current LHC studies show that the sensitivity to the MSSM Higgs sector is somehow restricted. This is illustrated in figure 31, where the sensitivity of different signatures is shown in a rather complicated two–dimensional multi–line contour plot.

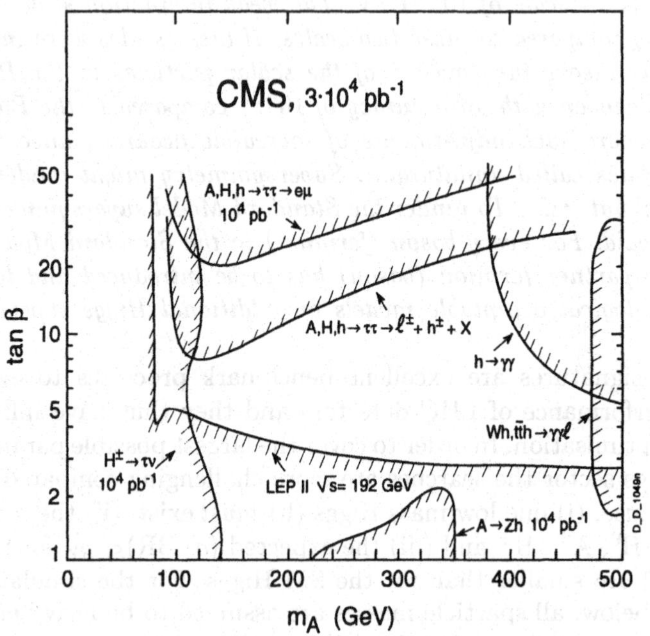

Figure 31. CMS 5 sigma significance contour plot for the MSSM Higgs sector in the m_A – $\tan\beta$ plane [34]. Each curve indicates the sensitivity for a specific Higgs search mode. No mixing in the stop sector is assumed.

The lightest neutral Higgs h. For the lightest Higgs the only established signature appears to be the decay $h \to \gamma\gamma$. For large masses of m_A ($m_A \geq$ 400 GeV) one finds essentially SM rates and sensitivity. For much smaller masses of m_A, the branching ratio $h \to \gamma\gamma$ is too small to observe a statistically significant signal. The combination of the $h \to \gamma\gamma$ search with other h decay modes, like $h \to b\bar{b}$, might help to enlarge the 5 sigma domain. A particular interesting aspect of an inclusive higgs search in SUSY events with the decay $h \to b\bar{b}$ decay will be discussed in the section on sparticle searches.

The heavy neutral Higgs Bosons H^0, A^0 *and small values of* $\tan\beta$. Assuming $\tan\beta$ values below 4, for some H^0 masses and decays, significant signals might be visible. For example, a H^0 with a mass close to 170 GeV appears to be detectable in the channel $H^0 \to WW^* \to \ell\nu\ell\nu$. Other studies indicate the possibility to observe the decays $H^0 \to hh \to \gamma\gamma b\bar{b}$ and $A \to Zh \to \ell^+\ell^- b\bar{b}$ for masses between 200 – 350 GeV. We will not go into further details here, as the relevance of such studies for low values of $\tan\beta$ depend very strongly on the forthcoming LEP200 results.

The heavy neutral Higgs bosons H^0, A^0 *and large values of* $\tan\beta$. For large values of $\tan\beta$, the Higgs production cross sections, especially the ones for $b\bar{b}H^0$ and $b\bar{b}A^0$, are much larger than for the SM Higgs of similar mass. The only relevant Higgs decays are $H^0, A^0 \to \tau\tau$ and $H^0, A^0 \to b\bar{b}$.

Assuming large $\tan\beta$ values, the rare decay $A, H \to \mu\mu$ might show up as a resonance peak above a large background[3]. Assuming excellent mass resolution in the $\mu\mu$ channel of about 0.01–0.02$\times$m$_{Higgs}$ [in GeV], a Higgs signal in this channel is detectable for an integrated luminosity of 30 fb^{-1} and $\tan\beta \geq 20$.

The charged Higgs $H^\pm$. Depending only slightly on $\tan\beta$, the relevant charged Higgs decay modes are $H^+ \to \tau^+\nu$ for masses below the $t\bar{b}$ threshold, and $H^+ \to t\bar{b}$ above. Inclusive $t\bar{t}$ events might thus provide a good experimental signature for $H^\pm$ with a mass below $m_{top} - 10$ GeV. One has to search for $t\bar{t}$ events with isolated τ candidates which originate from the decay chain $t\bar{t} \to bW^\pm bH^\pm$ and $H^\pm \to \tau\nu$.

Another interesting process might be the production of a heavy $H^\pm$ in association with a top quark, $gb \to tH \to ttb \to WWbbb$. A parton level analysis of this channel [41] indicates the possibility to obtain $H^\pm$ mass peaks with reasonable signal–to–background ratios.

Summary for the MSSM Higgs sector. Figure 32 illustrates the sensitivity of the ATLAS experiment to various Higgs decay channels as discussed above [42]. CMS reports very similar results. These two–dimensional multi–line 5 sigma (statistical) significance plots, especially in the logarithmic version, indicate sensitivity over almost the entire MSSM parameter space. However, it is worth to remind the reader that only statistical errors and

[3]The branching ratio is expected to be about a factor of 300 smaller than the one for the decay to $\tau\tau$, as it scales with $(m_\mu/m_\tau)^2$.

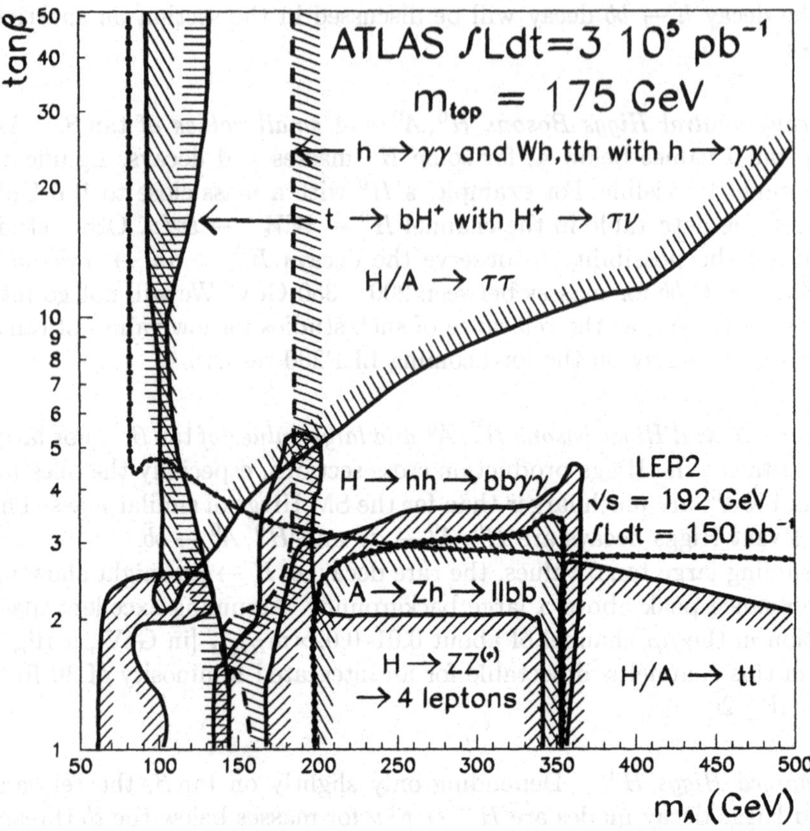

Figure 32. **Estimated ATLAS sensitivity (5 sigma) for MSSM Higgs searches, assuming 300 fb^{-1} [42]. The sensitivity of the different search signatures are shown in the m$_A$−tan β plane.**

only decays into SM particles are included in the evaluation of the discovery potential. As a consequence, the obtained curves, especially when extrapolated to larger integrated luminosities and combined for ATLAS and CMS are doubtful. This is especially the case for channels like $WH \to \ell\nu b\bar{b}$ and for $H^0, A^0 \to \tau\tau$ where the proposed signatures suffer certainly from the very poor signal–to–background ratio.

6.3.2. *Searches for SUSY Particles*

The MSSM contains 124 free parameters including those of the SM. That many free parameters do not offer a good guidance for experimentalists. Additional assumptions to constrain the parameter space are therefore desirable. The simplest approach is the so called mSUGRA (minimal Super–Gravity) model with only five additional new parameters ($m_0, m_{1/2}, \tan\beta, A^0$ and sign(μ)). This SUSY model is used for most of the simulation studies for which very advanced Monte Carlo programmes [43], [44], and [45] exist. This pragmatic choice of one approach to investigate the discovery potential appears to be sufficient, as essentially all required detector features can be tested.

6.3.3. *mSUGRA predictions*

Signatures related to the MSSM and in particular to mSUGRA searches are based on the consequences of R–parity conservation. R–parity is a multiplicative quantum number like ordinary parity. R–parity of the known SM particles is +1, while it is –1 for sparticles. As a consequence, sparticles have to be produced in pairs. Sparticles decay either directly or via some cascade processes to SM particles and the lightest supersymmetric particle (LSP), which is a neutral, massive and stable object with neutrino–like properties. This LSP should have been abundantly produced after the Big Bang and is an excellent candidate for cold dark matter. Usually one assumes the LSP to be the lightest neutralino $\tilde{\chi}_1^0$ which escapes detection. Large missing transverse energy is thus the prime SUSY signature in collider experiments.

Within the mSUGRA model, the masses of sparticles are strongly related to the universal fermion and scalar masses $m_{1/2}$ and m_0. The masses of the spin–1/2 sparticles are directly related to $m_{1/2}$. One expects approximately the following mass hierarchy: $\tilde{\chi}_1^0 \approx 0.5 \cdot m_{1/2}$, $\tilde{\chi}_2^0 \approx \tilde{\chi}_1^{\pm} \approx m_{1/2}$ and $\tilde{g} \approx 3 \cdot m_{1/2}$. The masses of the spin–0 sparticles are related to m_0 and $m_{1/2}$ and allow for some mass splitting between the "left" and "right" handed scalar partners of the degenerated left and right handed fermions. One finds the following simplified mass relations:

- $m(\tilde{q})(\tilde{u}, \tilde{d}, \tilde{s}, \tilde{c}$ and $\tilde{b}) \approx \sqrt{m_0^2 + 6 \cdot m_{1/2}^2}$,
- $m(\tilde{\nu})_{left} \approx m(\tilde{\ell^{\pm}})_{left} \approx \sqrt{m_0^2 + 0.52 \cdot m_{1/2}^2}$ and
- $m(\tilde{\ell^{\pm}})_{right} \approx \sqrt{m_0^2 + 0.15 \cdot m_{1/2}^2}$.

The masses of the left and right handed stop quarks ($\tilde{t}_{\ell,r}$) can have a large splitting. As a result, the right handed stop quark might be the lightest of all squarks.

Following the above mass relations and using the known SUSY couplings, possible SUSY decays and the related signatures can be calculated.

As an example we consider the $\tilde{\chi}_2^0$ decay $\tilde{\chi}_2^0 \to \tilde{\chi}_1^0 + X$ with X being: $\gamma^* Z^* \to \ell^+\ell^-$ or $h^0 \to b\bar{b}$ or $Z \to f\bar{f}$. Other possible $\tilde{\chi}_2^0$ decay chains are $\tilde{\chi}_2^0 \to \tilde{\chi}_1^{\pm(*)} + \ell^{\mp}\nu$ and $\tilde{\chi}_1^{\pm(*)} \to \tilde{\chi}_1^0\ell^{\pm}\nu$ or $\tilde{\chi}_2^0 \to \tilde{\ell}^{\pm}\ell^{\mp}$. This example clearly demonstrates the complexity of sparticle signatures.

For higher sparticle masses, even more decay channels might open up. It is thus not possible to define all search strategies a priori. Furthermore, possible unconstrained mixing angles between neutralinos lead to a model dependent search strategy for squarks and gluinos, as will be discussed below.

The discovery potential is usually given in the m_0–$m_{1/2}$ parameter space. Despite the model dependence, it allows to compare the sensitivity of different signatures. Having various proposed methods, the resulting figures are – as in the Higgs sector – rather complicated and require some time for appreciation. A typical example is shown in figure 33 [46], where the different curves indicate the LHC sensitivity for different signatures and different sparticles. It is usually assumed that the maximum information about SUSY can be extracted in regions which are covered by many different signatures. The meaning of the various curves should become clear in the following section.

6.3.4. *Squark and Gluino searches*

The cross–section for strongly interacting sparticles are large at the LHC, as can be seen in figure 34 [46]. For example the pair production cross–section of squarks and gluinos with a mass of ≈ 1 TeV has been estimated to be as large as 1 pb resulting in 10^4 produced SUSY events for 10 fb^{-1}. This high rate, combined with the possibility to observe many different decay modes, could turn LHC into a “SUSY–factory”.

Depending on the SUSY model parameters, a large variety of massive squark and gluino decay channels and signatures might exist. A search for squarks and gluons at the LHC should consider the various signatures resulting from the following decay channels:

- $\tilde{g} \to \tilde{q}q$ including $\tilde{g} \to \tilde{t}t$
- $\tilde{q} \to \tilde{\chi}_1^0 q$ or $\tilde{q} \to \tilde{\chi}_2^0 q$ or $\tilde{q} \to \tilde{\chi}_1^{\pm} q'$
- $\tilde{\chi}_2^0 \to \ell^+\ell^-\tilde{\chi}_1^0$ or $\tilde{\chi}_2^0 \to Z^0\tilde{\chi}_1^0$ or $\tilde{\chi}_2^0 \to h^0\tilde{\chi}_1^0$
- $\tilde{\chi}_1^{\pm} \to \tilde{\chi}_1^0\ell^{\pm}\nu$ or $\tilde{\chi}_1^{\pm} \to W\tilde{\chi}_1^0$.

The various decay channels can be separated into at least three distinct event signatures.

- Multi–jets + missing transverse energy; these events should be spherical in the plane transverse to the beam.
- Multi–jets + missing transverse energy + n(=1,2,3,4) isolated high p_T leptons; these leptons originate from cascade decays of charginos and neutralinos.

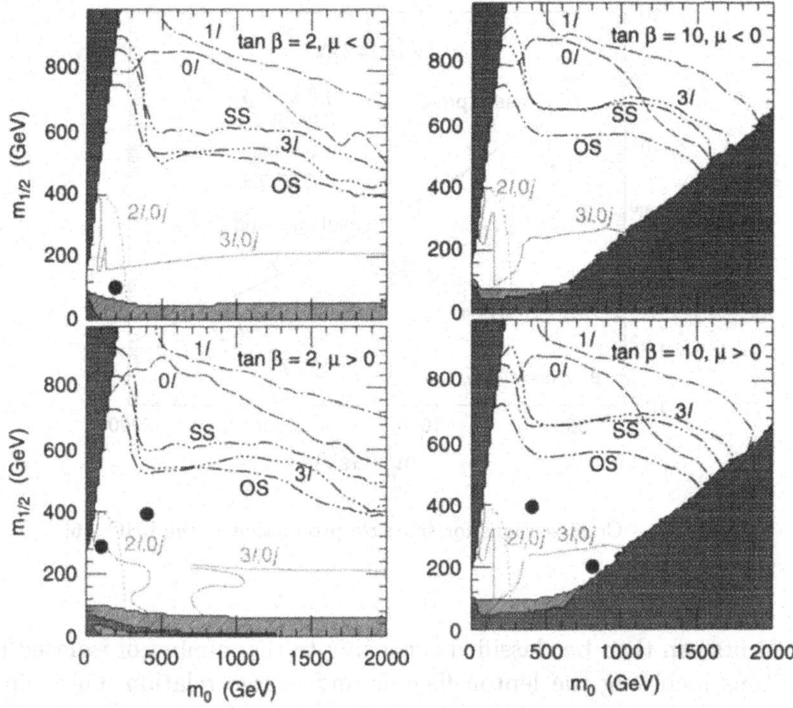

Figure 33. **Expected mSUGRA sensitivity of various signatures in the $m_0 - m_{1/2}$ plane at the LHC, assuming an integrated luminosity of 10 fb^{-1} [46]. The different curves indicate the expected sensitivity for SUSY events with n leptons (ℓ) and for events with lepton pairs with same charge (SS) and opposite charge (OS).**

– Multi-jets + missing transverse energy + lepton pairs of the same charge; such events can originate from $\tilde{g}\tilde{g} \to \bar{\tilde{u}}\bar{\tilde{u}}\tilde{d}\tilde{d}$ with subsequent decays of the squarks to $\tilde{u} \to \tilde{\chi}^+ d$ and $\bar{\tilde{d}} \to \tilde{\chi}^+ u$ followed by leptonic chargino decays $\tilde{\chi}^+ \to \tilde{\chi}_1^0 \ell^+ \nu$.

The observation and detailed analysis of different types of SUSY events might allow the discovery of many sparticles and should help to measure some of the SUSY parameters.

A search strategy for squarks and gluinos at the LHC would select jet events with large visible transverse mass and missing transverse energy.

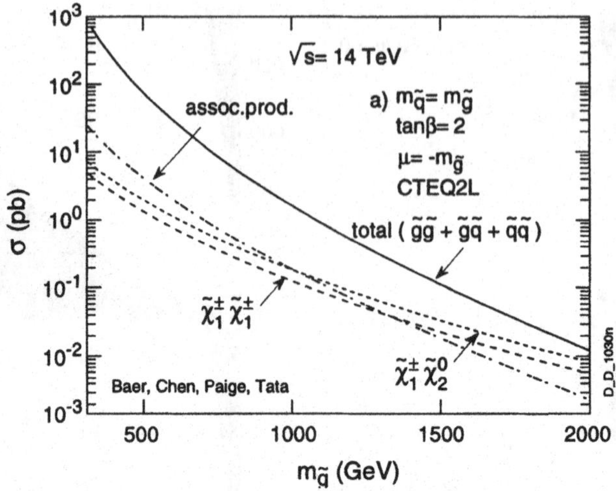

Figure 34. Cross-sections for sparticle production at the LHC [46].

Such events can then be classified according to the number of isolated high p_T leptons including the lepton flavour and charge relation. Once an excess above SM backgrounds is observed, one would try to interpret the events and measured cross-section(s) in terms of $\tilde{g}, \tilde{q}$ masses and decay modes for various SUSY models. Concerning the SM background processes, the largest backgrounds originate mainly from W+jet(s), Z+jet(s) and $t\bar{t}$ events. Using this approach, very encouraging signal-to-background ratios, combined with sizable signal cross-sections, are obtainable for a large range of squark and gluino masses. The simulation results of such studies indicate, as shown in figure 35, that LHC experiments are sensitive to squark and gluinos masses up to about 2 TeV, assuming 100 fb^{-1} [47]. Figure 35 also illustrates that detailed studies of branching ratios using the different decay chains are possible up to squark or gluino masses of about 1.5 TeV, where significant signals can be observed for many different channels. Another consequence of the expected large signal cross-sections is the possibility that at the LHC start-up, for an integrated luminosity of only $\approx$ 100 pb^{-1}, squarks and gluinos up to masses of about 600 to 700 GeV can be discovered, which is well beyond the most optimistic Tevatron Run III accessible mass range.

Given this exciting squark and gluino discovery potential for many different channels, one has to keep in mind that all potential signals depend strongly on the understanding of the various background processes and thus on the detector systematics. A thorough study of backgrounds including the

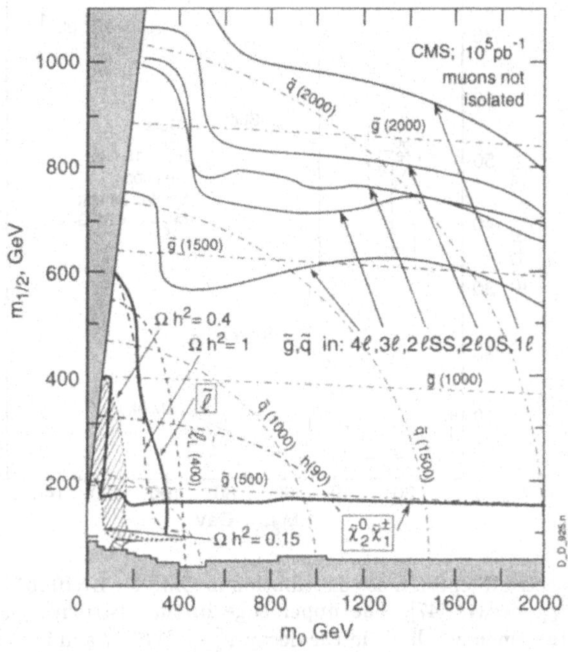

Figure 35. Expected CMS sensitivity for squarks and gluinos, sleptons and for $\tilde{\chi}_2^0\tilde{\chi}_1^\pm$ in the $m_0 - m_{1/2}$ plane, assuming 100 fb^{-1} [47]. The different solid lines show the expected 5 sigma signal, estimated from $S/\sqrt{S+B_{SM}}$, coverage domain for the various signatures using isolated high p_T leptons. The dashed–dotted lines indicate the corresponding squark and gluino masses.

shapes of background distributions is therefore mandatory. In particular, the requirements of very efficient lepton identification and good missing transverse energy measurement demand for a well understood detector response. This will require certainly some time, given the complexity of the LHC detectors.

Our discussion of the SUSY discovery potential has illustrated the sensitivity of the proposed ATLAS and CMS experiments. The next step after a discovery is the determination of SUSY parameters, thus obtaining a deeper insight of the underlying theory.

The production and decays of $\tilde{\chi}_2^0\tilde{\chi}_1^\pm$ provide enough rates for masses below 200 GeV and should allow, as shown in figure 36, to measure accurately the dilepton mass distribution and their relative p_T spectra. The mass distribution and especially the edge in the mass distribution has been shown to be sensitive to the mass difference between the two neutralinos

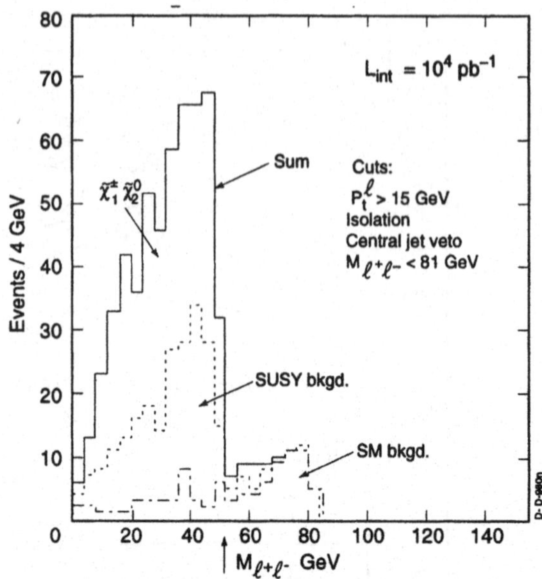

Figure 36. **Expected dilepton mass distribution in CMS for L=10 fb^{-1} and for trilepton events from $\tilde{\chi}_2^0\tilde{\chi}_1^{\pm}$ decays [47]. The upper edge in the distribution at about 50 GeV corresponds to the kinematic limit in the decay $\tilde{\chi}_2^0 \rightarrow \ell^+\ell^-\tilde{\chi}_1^0$ and is thus sensitive to the mass difference of $\tilde{\chi}_2^0 - \tilde{\chi}_1^0$.**

involved in the decay.

In contrast to the rate limitations of weakly produced sparticles at the LHC, detailed studies of clean squark and gluino events are expected to reveal more information. One finds that the large rate for many distinct event signatures allows to measure masses and mass ratios for several SUSY particles, produced in cascade decays of squarks and gluons. Many of these ideas have been discussed at the 1996 CERN SUSY–Workshop [48]. An especially interesting proposal is the detection of the Higgs boson, h, via the decay chain $\tilde{\chi}_2^0 \rightarrow \tilde{\chi}_1^0 h \rightarrow \tilde{\chi}_1^0 b\bar{b}$. The simulated mass distribution for $b\bar{b}$–jets in events with large missing transverse energy is shown in figure 37 [47]. Higgs mass peaks above background are found for various choices of $\tan\beta$ and $m_0, m_{1/2}$.

Another interesting approach to determine the SUSY mass scale has been suggested in [49]. The idea is to define an effective mass, using the scalar p_T sum of the four jets with the largest transverse energy, plus the missing transverse energy of the event. This effective mass shows an approximately linear relation with the underlying SUSY mass–scale, defined as the minimum of the squark or gluino mass.

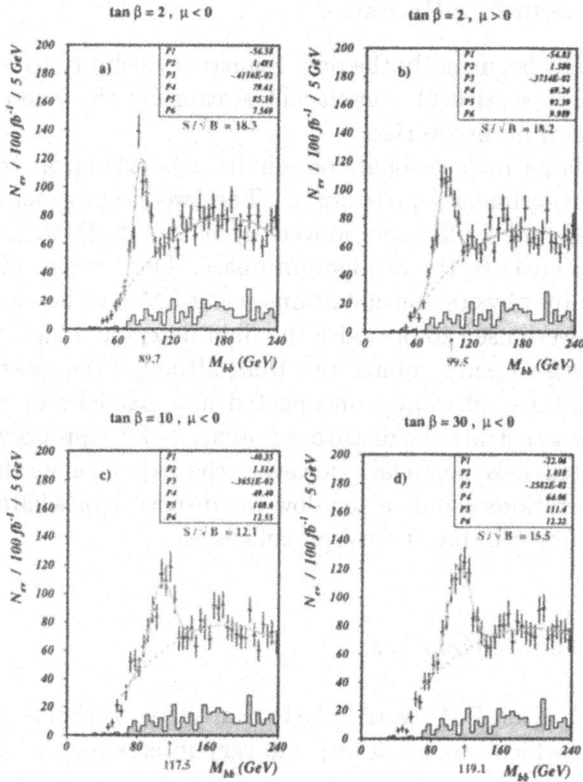

Figure 37. Higgs signals in squark and gluino events, reconstructed from the invariant mass of $h \to b\bar{b}$ and assuming 100 fb^{-1} in the CMS experiment [47].

The possibility to extract model parameters within the mSUGRA framework was illustrated using the following method: for each point in the parameter space a set of experimental constraints on sparticle masses is derived and a fit to parameters of mSUGRA is performed. For example, assuming $m_0 = m_{1/2} = 400$ GeV, $A_0 = 0$, $\tan\beta = 2$ and sign$(\mu) = +$, one obtains for 30 fb^{-1} the following sensitivity: $m_0 = 400 \pm 100$ GeV, $m_{1/2} = 400 \pm 10$ GeV and $\tan\beta = 2 \pm 0.8$.

7. Concluding Remarks

The LHC is currently the only realistic possibility to reach the TeV energy range in constituent–constituent scattering; this energy range is expected to be rich in discoveries.

A large international community is working on the realization of the LHC experimental programme. The two large general–purpose detectors ATLAS and CMS have moved from the R&D to preproduction and in certain areas to the production phase. The concept of both detectors was driven by physics considerations using SM and beyond the SM processes, which were used to optimise the final detector design.

It was already pointed out that, although the most exciting discoveries will be those of totally unexpected new particles or phenomena, one can only demonstrate the discovery potential of the proposed experiments using predicted new particles. However, the experiments designed under these considerations should also allow the discovery of whatever new phenomena might occur in multi–TeV pp collisions.

Acknowledgements

One of us (F.P.) would like to thank Tom Ferbel for the invitation to this school, which was in all respects very interesting and superbly organised.

References

1. For a theoretical review of the SM and further references see: M.J. Herrero, these proceedings.
2. For a review of the experimental tests of the SM and further references see: L. Nodulman, these proceedings.
3. For a review and further references see: The Higgs Hunter Guide, UCD-89-4,SCIPP-89/13,BNL-41644.
4. H.P. Nilles, *Phys. Reports* **110**, 1 (1984).
5. A. Dobado *et al.* , Proc. LHC Workshop, Aachen, 1990, CERN 90-10(1990), Vol.II; R.Casalbuoni *et al.* , ibid.
   E. Eichten *et al.* , *Phys. Lett.* B **405**, 305 (1997).
6. The LHC Study Group, The Large Hadron Collider: Conceptual Design, CERN/AC/95-05(LHC), 1995.
7. T. Hambye and K. Riesselmann, *Phys. Rev.* D **55**, 7255 (1997) and hep-ph/9708416.
8. Open presentation of the four LEP Collaborations at the LEPC meeting, CERN, 12 November 1998.
9. See for example Günter Quast, LEPEWWG Status Report at the LEPC meeting, 15 September 1998.
10. See for example E. Gross, A.L. Read and D. Lellouch, CERN-EP/98-094 (1998).
11. Recent search results are summarised in: D. Treille, Plenary Talk at ICHEP 98, Vancouver, British Columbia, Canada, July 23-30, 1998.
12. For a discussion about jet rates and QCD at the LHC see: J. Stirling, these proceedings.
13. CMS Collaboration, Electromagnetic Calorimeter Project, Technical Design Report, CERN/LHCC 97-33, 15 December 1997.
14. ATLAS Collaboration, Technical Proposal, CERN/LHCC 94-43, LHCC/P2, 15 December 1994.
15. CMS Collaboration, Technical Proposal, CERN/LHCC 94-38, LHCC/P1, 15 December 1994.
16. G. Wrochna; CMS NOTE/1997 - 096.
17. ATLAS Collaboration, The Muon Spectrometer Technical Design Report, CERN/LHCC 97-22, 5 June 1997.
18. CMS Collaboration, The Muon Project Technical Design Report, CERN/LHCC 97-32, 15 December 1997.
19. ATLAS Collaboration, Calorimeter Performance Technical Design Report, CERN/LHCC 96-40, 13 January 1997.
20. ATLAS Collaboration, Tile Calorimeter Technical Design Report, CERN/LHCC 96-42, 15 December 1996.
21. CMS Collaboration, The Hadron Calorimeter Project Technical Design Report, CERN/LHCC 97-31, 20 June 1997.
22. For a discussion about electromagnetic and hadron calorimeters in high energy experiments see: T. Virdee, these proceedings.
23. ATLAS Collaboration, Liquid Argon Calorimeter Technical Design Report, CERN/LHCC 96-41, 15 December 1996.
24. For a discussion about silicon tracking detectors see: I. Abt, these proceedings.
25. ATLAS Collaboration, Inner Detector Technical Design Report, CERN/LHCC 97-16 and 97-17, 30 April 1997.
26. CMS Collaboration, The Tracker Project Technical Design Report, CERN/LHCC 98-6, 15 April 1998.
27. P. Sphicas and S. Cittolin, private communication.
28. M.Dittmar, F. Pauss and D. Zürcher; *Phys. Rev.* D **56**, 7284 (1997) and hep-ex/9705004.
29. F. Behner, M. Dittmar, F. Pauss and D. Zürcher; contributed paper 418, 1997 EPS conference in Jerusalem, Israel.

30. Z. Kunszt, S. Moretti and W. J. Stirling; *Z. Phys.* C **74**, 479 (1997) and hep-ph/9611397.
31. The large Higgs production rate from gluon–gluon fusion has originally been pointed out by H. Georgi, S. Glashow, M. Machacek and D. Nanopoulos; *Phys. Rev. Lett.* **40**, 692 (1978).
32. The assumed Higgs cross section (NLO) are taken from [30] while the branching ratios are taken from the program HDECAY written by A. Djouadi, J. Kalinowski and M. Spira; e-Print Archive: hep-ph/9704448.
33. ATLAS Collaboration, in [19] page 25.
34. The CMS Higgs sensitivity studies are summarised in R. Kinnunen and D. Denegri; CMS Note 1997/057 (1997).
35. M. Dittmar and H. Dreiner; *Phys. Rev.* D **55**, 167 (1997) and hep-ph/9608317; and contributed paper 325, 1997 EPS conference in Jerusalem, Israel and CMS NOTE-1997/083.
36. Justifications about the expected performance figures can be found at http://atlasinfo.cern.ch/Atlas/Welcome.html for ATLAS and at http://cmsdoc.cern.ch/cms.html for CMS.
37. E. Richter-Was *et al.* ATLAS Internal Note Phys–No–048, 17/7/1995 (unpublished) and private communication E. Richter-Was.
38. For a summary of early studies of SUSY signatures at the LHC see: F. Pauss, Proc. LHC Workshop, Aachen, 1990, CERN 90-10(1990), Vol.I.
39. For a summary on supersymmetric Higgs searches at the LHC and a discussion of the notation see: Z. Kunszt and F. Zwirner, Proc. LHC Workshop, Aachen, 1990, CERN 90-10(1990), Vol.II.
40. For a recent calculation of the upper limit for the lightest MSSM Higgs mass and further references therein see: M. Quiros and J. R. Espinosa CERN-TH-98-292, Sep 1998 and hep-ph/9809269.
41. V. Barger, R.J.N. Phillips and D.P. Roy; *Phys. Lett.* B **324**, 236 (1994) and hep-ph/9311372.
42. E. Richter-Was *et al.* ; Int. J. Mod.Phys. **A13**, 1371 (1998) and CERN-TH-96-111.
43. ISAJET 7.40, F. E. Paige, S. D. Protopescu, H. Baer and X. Tata; BNL-HET-98-39, and hep-ph/9810440
44. SPYTHIA, S. Mrenna; *Comput. Phys. Commun.* **101**, 232 (1997) and e-Print Archive: hep-ph/9609360
45. T. Sjöstrand, *Comput. Phys. Commun.* **82**, 74 (1994).
46. H. Baer, C. Chen, F. Paige and X. Tata; *Phys. Rev.* D **53**, 6241 (1996) and hep-ph/9512383.
47. S. Abdullin *et al.* ; CMS Note-1998/006.
48. For an overview of LHC SUSY studies see the transparencies of the 1996 LHCC SUSY Workshop, October 29-30; CMS document 1996-149 Meeting.
49. I. Hinchliffe, F. Paige, G. Polesello and E. Richter-Was; ATLAS-Phys-No-107.

PARTICLE PHYSICS IN THE EARLY UNIVERSE

EDWARD W. KOLB
Fermi National Accelerator Laboratory
Batavia, Illinois 60510 USA

1. Introduction

Perhaps the most striking illustration of the true unity of science is the development of the interdisciplinary field of "particle cosmology." Particle physics examines nature on the smallest scales, while cosmology studies the universe on the largest scales. Although the two fields are separated by the scales of the objects they study, they are unified because it is impossible to understand the origin and evolution of large-scale structures in the universe without understanding the "initial conditions" that led to the structures. The initial data was set in the very early universe when the fundamental particles and forces acted to produce the perturbations in the cosmic density field. A complete understanding of the present structure of the universe will also be impossible without accounting for the dark component in the density field. The most likely possibility is that this ubiquitous dark component is an elementary particle relic from the early universe.

The study of the structure of the present universe may reveal insights into events which occurred in the early universe, and hence, into the nature of the fundamental forces and particles at an energy scale far beyond the reach of terrestrial accelerators. Perhaps the early universe was the ultimate particle accelerator, and will provide the first glimpse of physics at the scale of Grand Unified Theories (GUTs), or even the Planck scale.

As a cosmologist I am interested in events that happened a long time ago. But in studying the past, I believe it is best to take the approach of a historian rather than an antiquarian. Now an antiquarian and a historian are both interested in things from the past. But an antiquarian is interested in old things just because they are old. To an antiquarian, there is no difference between a laundry list from June 1215 and the Magna Carta: they are both equally old. A historian, on the other hand, is interested in

T. Ferbel (ed.), Techniques and Concepts of High Energy Physics X, 239–261.

the past because it shapes the present. The job of a historian is to sort through events of the past and see which are important and which are not. I am not interested in the early universe just because it happened a long time ago, or it was really hot, or it was a bang (a really, really big one). The real reason I study the early universe is that events which occurred in the early universe left an imprint upon the present universe.

In these lectures I will concentrate on two events which occurred in the early universe. The first is the generation of perturbation in the density field during an early period of rapid expansion known as cosmic inflation. The second is the genesis of dark matter. The record of these events is written in the arrangement of galaxies, galaxy clusters, and imperfection in the isotropy of the cosmic microwave background radiation. If we really understood particle physics, we could predict the nature of those patterns. If we really knew how to read the story in the structures, we would learn something about particle physics. The story is there on the sky, patiently waiting for our wits to become sharp enough to read it.

In these lectures I will discuss the early universe. So the first thing we must do is to follow the procedure outlined by William Shakespeare [1]:

Now entertain conjecture of a time
When creeping murmur and the poring dark
Fills the wide vessel of the universe.

2. The Density Field of the Universe

The universe is not exactly homogeneous and isotropic, but it is a sufficiently accurate description of the universe on large scales that it is useful to consider homogeneity and isotropy as a first approximation, and discuss departures from this idealized smooth universe.

Let us begin by considering the density field, $\rho(\vec{x})$. If the average density of the universe is denoted as $\langle\rho\rangle$, then we can define a dimensionless density contrast $\delta(\vec{x})$ as

$$\delta(\vec{x}) = \frac{\rho(\vec{x}) - \langle\rho\rangle}{\langle\rho\rangle} . \tag{1}$$

Of course we cannot predict $\delta(\vec{x})$, but we can hope to predict the statistical properties of $\delta(\vec{x})$. The correct arena to discuss the statistical properties of the density field is in Fourier space, where one decomposes the density contrast into its various Fourier modes $\delta_{\vec{k}}$:

$$\delta(\vec{x}) = V \int d^3k\, \delta_{\vec{k}}\, e^{-i\vec{k}\cdot\vec{x}}, \tag{2}$$

where V is some irrelevant normalization volume. After a little Fourier manipulation and some mild assumptions about the density field, it is easy to show that the two-point autocorrelation function of the density field can be expressed solely in terms of $\left|\delta_{\vec{k}}\right|^2$:

$$\langle\delta(\vec{x})\delta(\vec{x})\rangle = A\int_0^\infty \frac{dk}{k}\,k^3\left|\delta_{\vec{k}}\right|^2 , \tag{3}$$

where A is yet another irrelevant constant.

So long as the fluctuations are Gaussian, all statistical information is contained in a quantity known as the *power spectrum*, which can be defined as either

$$\begin{aligned} \Delta^2(k) &= k^3\left|\delta_{\vec{k}}\right|^2 , \qquad \text{or} \\ P(k) &= \left|\delta_{\vec{k}}\right|^2 . \end{aligned} \tag{4}$$

The first choice is much more physical, as it represents the power per logarithmic decade in the fluctuations. Although the first choice makes much more sense, the second choice is what is usually used. It turns out that graphs of $P(k)$ have a nicer form (but less physical content) than corresponding graphs of $\Delta^2(k)$. Since the widespread availability of color graphics, presentation seems to be everything, and information content of secondary concern.

2.1. THE POWER SPECTRUM FROM LARGE-SCALE STRUCTURE

The power spectrum is related to the *rms* fluctuations in the density on scale $R = 2\pi/k$. The exact relationship depends upon sampling procedure, window functions, *etc.* But for a simple intuitive feel, imagine we have mass points spread throughout some sample volume. Now place a sphere of radius R in the volume and count the number of points within the sphere. Then repeat as often as you have the time or patience to do so. There will be an average number $\langle N\rangle$, and an *rms* fluctuation $\langle(\delta N/\langle N\rangle)^2\rangle^{1/2}$. The power spectrum is related to that *rms* fluctuation: $\Delta(k = 2\pi R^{-1}) \propto \langle(\delta N/\langle N\rangle)^2\rangle^{1/2}$. Repeating the procedure for many values of R will give Δ as a function of R—the power spectrum.

Now how does one go about observing the mass within a sphere of radius R? Well, it is difficult to measure the mass. It is easier to count the number of galaxies. So one assumes that the galaxy distribution traces the mass distribution. Although it seems reasonable that regions of high density of galaxies correspond to regions of high mass density, since most of the mass is dark, the proportionality might not be exact. Thus, we have to allow for a possible *bias* in the power spectrum. Other problems also arise. The

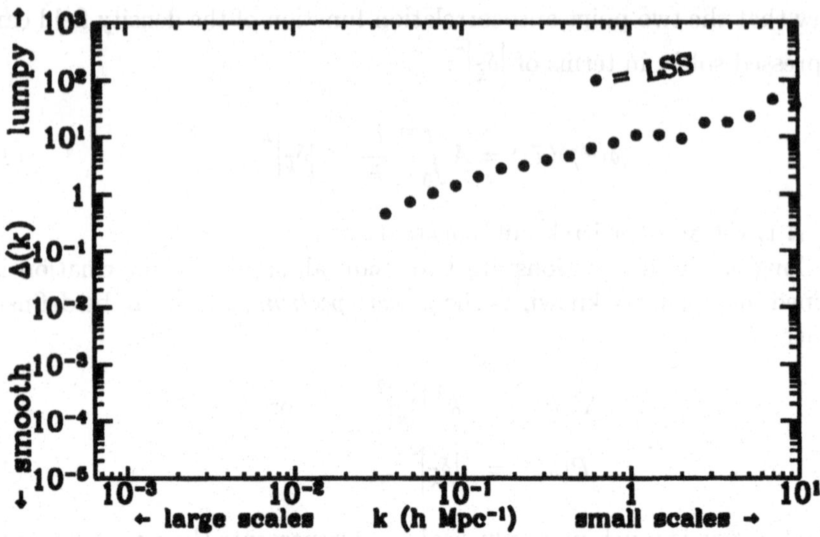

Figure 1. **An example of a power spectrum deduced from a large-scale structure (LSS) survey. A megaparsec (Mpc) is 3.26×10^{24}cm, and h is the dimensionless Hubble constant, $H_0 = 100h$ km $\mathrm{s}^{-1}\mathrm{Mpc}^{-1}$. A wavenumber k is roughly related to a length scale of $2\pi/k$.**

distance to an object is not measured directly; what is measured is its redshift. The redshift is determined by the distance to the object, as well as its peculiar motion. In regions of large overdensity the peculiar motions may be large, resulting in what is known as redshift distortions. Another problem is nonlinear evolution, which distorts the power spectrum in regions of large overdensity. Thus, if one wants to compare the observed power spectrum with the linear power spectrum generated by early-universe physics, it is necessary to make corrections for bias, redshift-space distortions, and nonlinear evolution.

Deducing the power spectrum from galaxy surveys is a tricky business. Rather than go into the details, uncertainties, and all that, I will just present a representative power spectrum in Fig. (1). Note that $\Delta(k)$ decreases with increasing length scale (decreasing wavenumber). The universe is lumpy on small scales, but becomes progressively smoother when examined on larger scales.

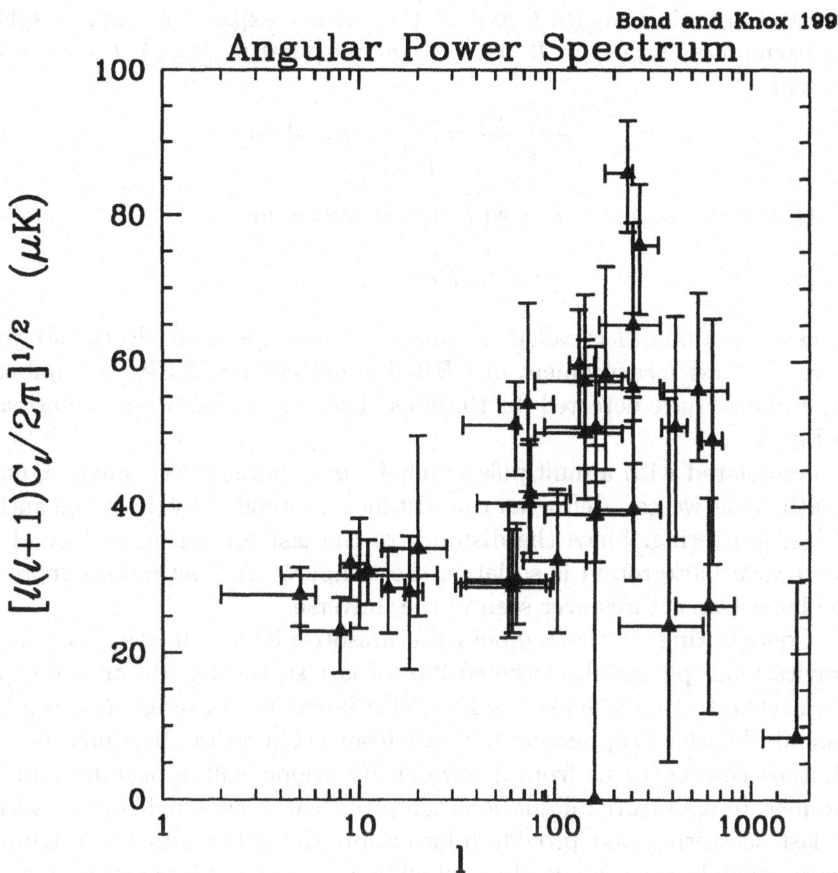

Figure 2. The angular power spectrum of CBR fluctuations (courtesy of Dick Bond and Llyod Knox).

2.2. THE POWER SPECTRUM DEDUCED FROM THE COSMIC BACKGROUND RADIATION

The microwave background is isotropic to about one part in 10^3. If one removes the anisotropy caused by our motion with respect to the cosmic background radiation (CBR) rest frame, then it is isotropic to about 30 parts-per-million. But as first discovered by the Cosmic Background Explorer (COBE), there are intrinsic fluctuation in the temperature of the CBR.

Just as perturbations in the density field were expanded in terms of Fourier components, a similar expansion is useful for temperature fluctuations. Because the surface of observation about us can be described in terms of spherical angles θ and ϕ, the correct expansion basis is spherical harmonics, $Y_{lm}(\theta,\phi)$. If the average temperature is $\langle T\rangle$, then one can expand

$$\frac{\Delta T(\theta,\phi)}{\langle T\rangle} = \sum_{l,m} a_{lm} Y_{lm}(\theta,\phi) \ . \tag{5}$$

Of course $\langle a_{lm}\rangle = 0$, but with proper averaging,

$$\langle |a_{lm}|^2\rangle \equiv C_l \neq 0 \ . \tag{6}$$

C_l as a function of l is called the angular power spectrum. In the six years since the first measurement of CBR fluctuations by COBE, a number of experiments have detected fluctuations. The present situation is illustrated in Fig. 2.

Associated with a multipole number l is a characteristic angle θ, and a length scale we can define as the distance subtended by θ on the surface of last scattering. Since the distance to the last scattering surface of the microwave background is so large, the temperature fluctuations represent the largest structures ever seen in the universe.

Contributing to the temperature anisotropies are fluctuations in the gravitational potential on the surface of last scattering. Photons escaping from regions of high density will suffer a larger than average gravitational redshift, hence will appear to originate from a cold region. In similar fashion, photons coming to us from a low-density region will appear hot. In this manner, temperature fluctuations can probe the density field on the surface of last scattering and provide information about the power spectrum on scales much larger than can be probed by conventional large-scale structure observations.

The region of wavenumber and amplitude of the power spectrum probed by COBE is illustrated in Fig. 3. There are now measurements of CBR fluctuations on smaller angular scale, corresponding to larger k.

Finally, Fig. 4 combines information from both large-scale structure surveys and CBR temperature fluctuations. The trend is obvious: on small distance scales the power spectrum is "large," which implies a lot of structure. Matter is clustered on small scales. But on "large" scales the power spectrum decreases. As one examines the universe on larger scales, homogeneity and isotropy becomes a better and better approximation.

The data shown is only illustrative of many data sets. Although combining different data sets is uncertain and risky (problems with normalization, *etc.*) the qualitative features are the same. Figure 4 is best regarded as an impressionist representation of the situation.

Another thing to keep in mind is that the power spectrum may not be the entire story. The power spectrum contains all statistical information

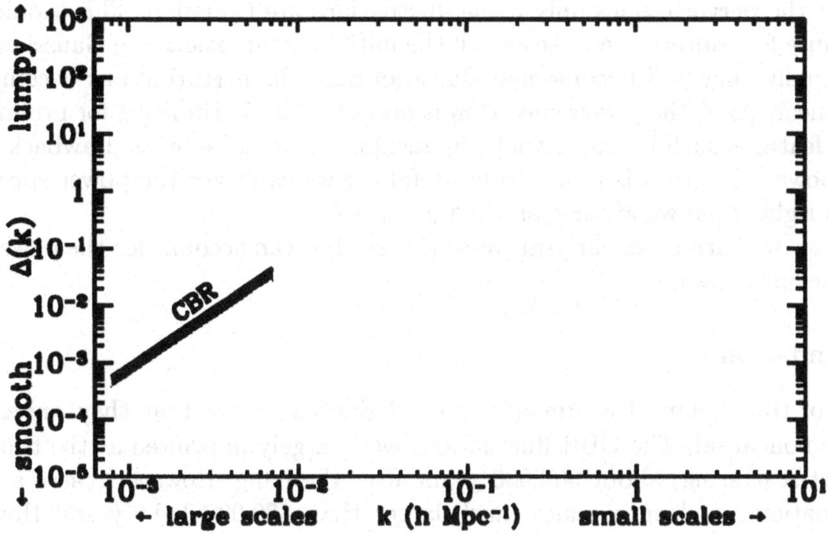

Figure 3. The power spectrum deduced by measurements of large angular scale CBR temperature fluctuations.

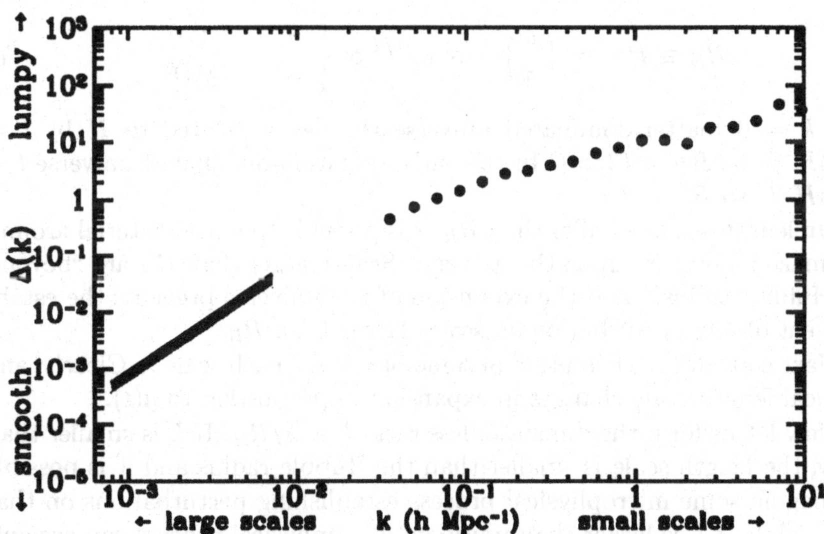

Figure 4. The "grand unified" power spectrum, including determinations from large-scale structure surveys (the points), and deduced from CBR temperature fluctuations (the box).

about the perturbations only if the fluctuations are Gaussian. This should be cause for concern, because even if the initial perturbations are Gaussian, eventually they will become non-Gaussian once the perturbations become nonlinear. Also, the power spectrum is not a useful discriminant for prominent features such as walls, voids, filaments, *etc.* In spite of its drawbacks, the power spectrum is remarkably useful—if we can't get the power spectrum right, then we are not on the right track.

Now we turn to an early-universe theory that can account for the power spectrum: inflation

3. Inflation

One of the striking features of the CBR fluctuations is that they *appear* to be noncausal. The CBR fluctuations were largely imprinted at the time of last-scattering, about 300,000 years after the bang. However, there are fluctuations on length scales much larger than 300,000 light years! How could a causal process imprint fluctuations on scales larger than the light-travel distance since the time of the bang? The answer is inflation, but to see how that works, let's define the problem more exactly.

First consider the evolution of the Hubble radius with the scale factor $a(t)$:[1]

$$R_H \equiv H^{-1} = \left(\frac{\dot{a}}{a}\right)^{-1} \propto \rho^{-1/2} \propto \begin{cases} a^2 & \text{(RD)} \\ a^{3/2} & \text{(MD)}. \end{cases} \tag{7}$$

In a $k = 0$ matter-dominated universe the age is related to H by $t = (2/3)H^{-1}$, so $R_H = (3/2)t$. In the early radiation-dominated universe $t = (1/2)H^{-1}$, so $R_H = 2t$.

On length scales smaller than R_H it is possible to move material around and make an imprint upon the universe. Scales larger than R_H are "beyond the Hubble radius," and the expansion of the universe prevents the establishment of any perturbation on scales larger than R_H.

Next consider the evolution of some physical length scale λ. Clearly, any physical length scale changes in expansion in proportion to $a(t)$.

Now let us form the dimensionless ratio $L \equiv \lambda/R_H$. If L is smaller than unity, the length scale is smaller than the Hubble radius and it is possible to imagine some microphysical process establishing perturbations on that scale, while if L is larger than unity, no microphysical process can account for perturbations on that scale.

Since $R_H = a/\dot{a}$, and $\lambda \propto a$, the ratio L is proportional to $\dot{a}$, and $\dot{L}$ scales as $\ddot{a}$, which in turn is proportional to $-(\rho + 3p)$. There are two

[1]Here and throughout the paper "RD" is short for radiation dominated, and "MD" implies matter dominated.

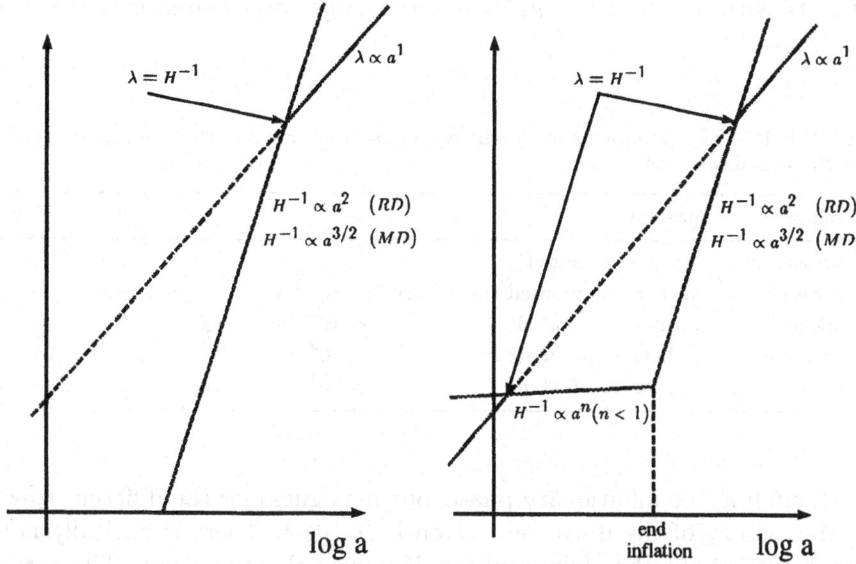

Figure 5. **Physical sizes increase as $a(t)$ in the expanding universe. The Hubble radius evolves as $R_H = H^{-1} = (8\pi G\rho(a)/3)^{1/2}$. In a radiation-dominated or matter-dominated universe (left) any physical length scale λ starts larger than R_H, then crosses the Hubble radius ($\lambda = H^{-1}$) only once. However, if there was a period of early inflation (right) when R_H increased more slowly than a, it is possible for a physical length scale to start smaller than R_H, become larger than R_H, and after inflation ends become once again smaller than R_H. Periods during which the scale is larger than the Hubble radius are indicated by the dotted line.**

possible scenarios for $\dot{L}$ depending upon the sign of $\rho + 3p$:

$$\dot{L}\begin{cases} < 0 \rightarrow & R_H \text{ grows faster than } \lambda, & \text{happens for } \rho + 3p > 0 \\ < 0 \rightarrow & R_H \text{ grows more slowly than } \lambda, & \text{happens for } \rho + 3p < 0. \end{cases} \tag{8}$$

In the standard scenario, $\rho + 3p > 0$, R_H grows faster than λ. This is illustrated by the left-hand side of Fig. 5.

For illustration, let us take λ to be the present length $\lambda_8 = 8h^{-1}$Mpc, the scale beyond which perturbations today are in the linear regime. The physical length scale, which today is λ_8, was smaller in the early universe by a factor of $a(t)/a_0$, where a_0 is the scale factor today. Today, the Hubble radius is $H_0^{-1} \sim 3000h^{-1}$Mpc. Of course, today λ_8 is well within the current Hubble radius. But in the standard picture, R_H grows faster than λ, and there must therefore have been a time when the comoving length scale that corresponds to λ_8 was larger than R_H

Sometime during the early evolution of the universe the expansion was such that $\ddot{a} > 0$, which as we have just seen, requires an unusual equation of state with $\rho + 3p < 0$. This is referred to as "accelerated expansion" or "inflation."

TABLE 1. Different epochs in the history of the universe and the associated tempos of the expansion rate.

tempo	passage	age	ρ	p	$\rho + 3p$
prestissimo	string dominated	$< 10^{-43}$s	?	?	?
presto	vacuum dominated (inflation)	$\sim 10^{-38}$s	ρ_V	$-\rho_V$	$-$
allegro	matter dominated	$\sim 10^{-36}$s	ρ_ϕ	0	+
andante	radiation dominated	$< 10^4$yr	T^4	$T^4/3$	+
largo	matter dominated	$> 10^4$yr	$\rho_{\rm matter}$	0	+

Including the inflationary phase, our best guess for the different epochs in the history of the universe is given in Table 1. There is basically nothing known about the stringy phase, if indeed there was one. The earliest phase we have information about is the inflationary phase. As we shall see, the information we have is from the quantum fluctuations during inflation, which were imprinted upon the metric, and can be observed as CBR fluctuations and the departures from homogeneity and isotropy in the matter distribution, *e.g.*, the power spectrum. A lot of effort has gone into studying the end of inflation. It was likely that there was a brief period of matter domination before the universe became radiation dominated. Very little is known about this period after inflation. Noteworthy events that might have occurred during this phase include baryogenesis, phase transitions, and generation of dark matter. We do know that the universe was radiation dominated for almost all of the first 10,000 years. The best evidence of the radiation-dominated era is primordial nucleosynthesis, which is a relic of the radiation-dominated universe in the period 1 second to 3 minutes. The earliest picture of the matter-dominated era is the CBR.

Here, I am interested in events during the inflationary era. The first issue is how to imagine a universe dominated by vacuum energy making a transition to a matter-dominated or radiation-dominated universe. A simple way to picture this is by the action of a scalar field ϕ with potential $V(\phi)$. Let's imagine the scalar field is displaced from the minimum of its potential as illustrated in Fig. 6. If the energy density of the universe is dominated by the potential energy of the scalar field ϕ, known as the *inflaton*, then $\rho+3p$ will be negative. The vacuum energy disappears when the scalar field evolves to its minimum.

If the inflaton is completely decoupled, then it will oscillate about the minimum of the potential, with the cycle-average of the energy density

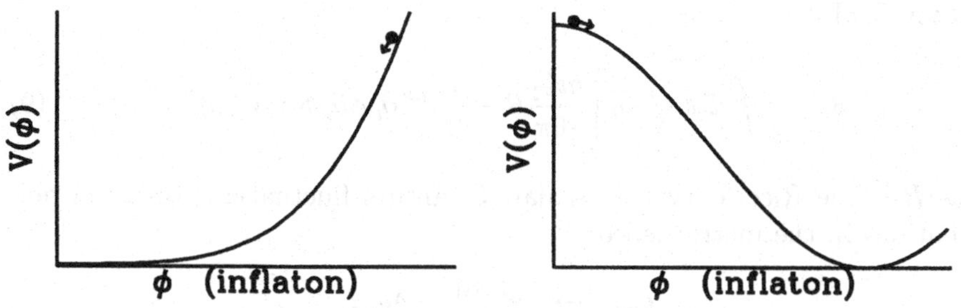

Figure 6. Schematic illustrations of the inflaton potential energy. The potential on the left is a "large-field" model, where the inflaton field starts large and evolves toward its minimum. The right figure illustrates a "small-field" model. A more accurate description of large-field and small-field potential is the sign of the second derivative of the potential: large-field models have $V'' > 0$ while small-field models have $V'' < 0$.

decreasing as a^{-3}, *i.e.*, a matter-dominated universe. But at the end of inflation the universe is cold and frozen in a low-entropy state: the only degree of freedom is the zero-momentum mode of the inflaton field. It is necessary to "defrost" the universe and turn it into a "hot" high-entropy universe with many degrees of freedom in the radiation. Exactly how this is accomplished is still unclear. It probably requires the inflaton field to be coupled to other degrees of freedom, and as it oscillates, its energy is converted to radiation either through incoherent decay, or through a coherent process involving very complicated dynamics of coupled oscillators with time-varying masses. In either case, it is necessary to extract the energy from the inflaton and convert it into radiation.

3.1. QUANTUM FLUCTUATIONS

During inflation there are quantum fluctuations in the inflaton field. Since the total energy density of the universe is dominated by the inflaton potential energy density, fluctuations in the inflaton field lead to fluctuations in the energy density. Because of the rapid expansion of the universe during inflation, these fluctuations in the energy density are frozen into super-Hubble-radius-size perturbations. Later, in the radiation or matter-dominated era they will come within the Hubble radius as if they were *noncausal* perturbations.

The spectrum and amplitude of perturbations depend upon the nature of the inflaton potential. Mukhanov [3] has developed a very nice formalism for the calculation of density perturbations. One starts with the action for gravity (the Einstein–Hilbert action) plus a minimally-coupled scalar inflaton field ϕ:

$$S = -\int d^4x \sqrt{-g}\left[\frac{m_{Pl}^2}{16\pi}R - \frac{1}{2}g^{\mu\nu}\partial_\mu\phi\partial_\nu\phi + V(\phi)\right] . \tag{9}$$

Here R is the Ricci curvature scalar. Quantum fluctuations result in perturbations in the metric tensor

$$\begin{aligned} g_{\mu\nu} &\rightarrow g_{\mu\nu}^{FRW} + \delta g_{\mu\nu} \\ \phi &\rightarrow \phi_0 + \delta\phi , \end{aligned} \tag{10}$$

where $g_{\mu\nu}^{FRW}$ is the Friedmann–Robertson–Walker metric, and $\phi_0(t)$ is the classical solution for the homogeneous, isotropic evolution of the inflaton. The action describing the dynamics of the small perturbations can be written as

$$\delta_2 S = \frac{1}{2}\int d^4x \left[\partial_\mu u \partial^\mu u + z^{-1}\frac{d^2z}{d\tau^2}\, u^2\right] ; \qquad z = a\dot{\phi}/H , \tag{11}$$

i.e., the action in conformal time τ ($d\tau^2 = a^2(t)dt^2$) for a scalar field in Minkowski space, with mass-squared $m_u^2 = -z^{-1}d^2z/d\tau^2$. Here, the scalar field u is a combination of metric fluctuations $\delta g_{\mu\nu}$ and scalar field fluctuations $\delta\phi$. This scalar field is related to the amplitude of the density perturbation.

The simple matter of calculating the perturbation spectrum for a non-interacting scalar field in Minkowski space will give the amplitude and spectrum of the density perturbations. The problem is that the solution to the field equations depends upon the background field evolution through the dependence of the mass of the field upon z. Different choices for the inflaton potential $V(\phi)$ results in different background field evolutions, and hence, different spectra and amplitudes for the density perturbations.

Before proceeding, now is a useful time to remark that in addition to scalar density perturbations, there are also fluctuations in the transverse, traceless component of the spatial part of the metric. These fluctuations (known as tensor fluctuations) can be thought of as a background of gravitons.

Although the scalar and tensor spectra depend upon $V(\phi)$, for most potentials they can be characterized by Q_{RMS}^{PS} (the amplitude of the scalar and tensor spectra on large length scales added in quadrature), n (the scalar spectral index describing the best power-law fit of the primordial scalar spectrum), r (the ratio of the tensor-to-scalar contribution to C_2 in the angular power spectrum), and n_T (the tensor spectral index describing the best power-law fit of the primordial tensor spectrum). For single-field, slow-roll inflation models, there is a relationship between n_T and r, so in fact there are only three independent variables. Furthermore, the amplitude of the fluctuations often depends upon a free parameter in the potential, and the spectra are normalized by Q_{RMS}^{PS}. This leads to a characterization of a wide-range of inflaton potentials in terms of two numbers, n and r.

In addition to the primordial spectrum characterized by n and r, in order to compare to data it is necessary to specify cosmological parameters (H_0, the present expansion rate; Ω_0, the ratio of the present mass-energy density to the critical density—a spatially flat universe has $\Omega_0 = 1$; Ω_B, the ratio of the present baryon density to the critical density; Ω_{DM} the ratio of the present dark-matter density to the critical density; and Λ, the value of the cosmological constant), as well as the nature of the dark matter.

The specification of the dark matter is by how "hot" the dark matter was when the universe first became matter dominated. If the dark matter was really slow at that time, then it is referred to as cold dark matter. If the dark matter was reasonably hot when the universe became matter dominated, then it is called hot dark matter. Finally, the intermediate case is called warm dark matter. Neutrinos with a mass in the range 1 eV to a few dozen eV would be hot dark matter. Light gravitinos, as appear in gauge-mediated supersymmetry breaking schemes, is an example of warm dark matter. By far the most popular dark matter candidate is cold dark matter. Examples of cold dark matter are neutralinos and axions.

4. The Flavor of the Month

There are exactly 31 different combinations of n, r, cosmological parameters, and dark matter mixes.[2] For this reason, different cosmological models are like flavors of ice cream at Baskin Robbins. There is always a flavor of the month that everyone seems to like. Flavors come in and out of taste/fashion, with some adherents always choosing the same, while others like to sample a wide variety. A menu of the six most popular flavors are given in Table 2.

Obviously, other combinations are possible. A comparison of the power spectrum in these models to our impressionist version of the observationally determined power spectrum is shown in Fig. 7. Obviously CDM has too much power on small scales. Hot dark matter is a disaster because it has *no* power on small scales. Tilted dark matter does better than CDM. Mixed dark matter does somewhat better, as does Λ dark matter (not shown). Rather than χ-by-eye, I quote the results of one statistical analysis, including many data sets, in Table 3 [2].

[2]This statement clearly is not true.

TABLE 2. Different flavors of cosmological models.

flavor	n	r	H_0	Ω_0	Ω_B	Ω_{COLD}	Ω_{HOT}	Ω_Λ
CDM	1	0	50 km s^{-1}Mpc^{-1}	1	0.05	0.95	0	0
HDM	1	0	50 km s^{-1}Mpc^{-1}	1	0.05	0	0.95	0
MDM	1	0	50 km s^{-1}Mpc^{-1}	1	0.10	0.70	0.20	0
TCDM	0.8	0	50 km s^{-1}Mpc^{-1}	1	0.05	0.95	0	0
OCDM	1	0	50 km s^{-1}Mpc^{-1}	0.5	0.05	0.45	0	0
ΛCDM	1	0	50 km s^{-1}Mpc^{-1}	1	0.05	0.45	0	0.50

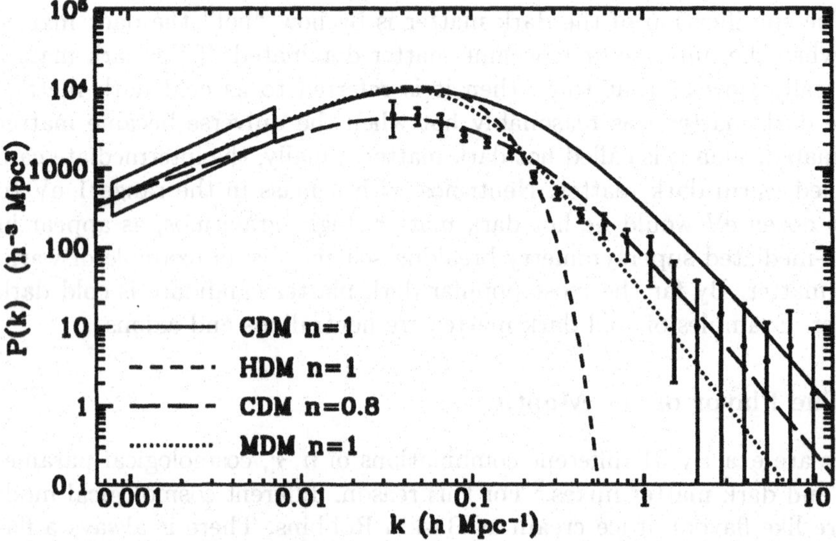

Figure 7. The empirically determined power spectrum of density perturbations and the (linear-theory) predictions of several models. The models shown are cold dark matter; hot dark matter; tilted cold dark matter; and mixed dark matter.

My reading of the comparison between data and experiment is that the results of Table 3 should be regarded as a *relative* measure of the agreement between models and present data. For instance, it is fair to say that MDM is a much better fit than CDM. But one should be very careful before rejecting a model based upon these numbers.

Although one might get the best χ^2 with a model having 10% baryons, 30% cold dark matter, 30% hot dark matter with 15% each in two species of neutrinos, 20% cosmological constant, 10% warm dark mater, and seasoned

TABLE 3. One analysis of the comparison of data and models.

model	$\chi^2/d.o.f.$
CDM	3.8
TCDM	2.1
ΛCDM	1.9
OCDM	1.8
MDM	1.2

with a little tilt, it doesn't mean that is the way the universe is constructed.

Clearly, what is needed are better observations: finer-scale observations of CBR fluctuations, as well as larger-scale determinations of the power spectrum from large-scale structure surveys. In the next few years such experiments will be done.

There is now an aggressive campaign to measure CBR anisotropies on fine angular scales. The culmination of this program will be the launch of two satellites—MAP by NASA and Planck by ESA.

Large-scale structure surveys are also progressing. The largest (three-dimensional) survey to date is the Las Campanas Redshift Survey, containing the three-dimensional location of over 30,000 galaxies. In the next two years another survey, called 2dF, will be completed. This survey will have about 100,000 galaxies in its catalog. Finally, the Sloan Digital Sky Survey will map π steradians of the north galactic cap and find the location of 1,000,000 galaxies, along with 150,000 quasars.

By the time these experiments/observations are complete we will be in the age of precision cosmology, and we should really be able to compare theory and observation.

The remaining missing piece of the puzzle may be the identity of the dark matter.

5. Dark Matter

5.1. WIMPY THERMAL RELICS

In this school, the matter of neutrino masses has been reviewed in great detail. A neutrino of mass m_ν contributes to $\Omega_\nu h^2$ an amount

$$\Omega_\nu h^2 = \left(\frac{m_\nu}{92\text{eV}}\right) . \tag{12}$$

If the mass of the neutrino is significantly less than 0.1 eV, then its contribution to Ω_0 is dynamically unimportant.

More promising than neutrino hot dark matter is cold dark matter. The most promising candidate for cold dark matter is the lightest supersymmetric particle, presumably a neutralino. Neutralino dark matter has been well studied and reviewed [4].

The next most popular dark-matter candidate is the axion. Although the axion is very light, since its origin is from a condensate, it is very cold. Axion dark matter has also been well studied and well reviewed [5].

There are presently several experiments searching for cosmic neutralinos and cosmic axions. Both types of searches seem sensitive enough to discover the relic dark matter, although it will take quite some time (and probably another generation of experiments) to completely cover the parameter space.

Neutralinos are an example of a thermal relic. A thermal relic is assumed to be in local thermodynamic equilibrium (LTE) at early times. The *equilibrium* abundance of a particle, say relative to the entropy density, depends upon the ratio of the mass of the particle to the temperature. If we define the variables $Y \equiv n_X/s$ and $x = M_X/T$, where n_X is the number density of WIMP (weakly interacting massive particle) X with mass M_X and $s \sim T^3$ is the entropy density, $Y \propto \exp(-x)$ for $x \gg 1$, while $Y \sim$ constant for $x \ll 1$.

A particle will track its equilibrium abundance so long as reactions which keep the particle in chemical equilibrium can proceed rapidly enough. Here, rapidly enough means on a timescale more rapid than the expansion rate of the universe H. When the reactions becomes slower than the expansion rate, then the particle can no longer track its equilibrium value and thereafter Y is constant. When this occurs, the particle is said to be "frozen out." A schematic illustration of this is given in Fig. 8.

The more strongly interacting the particle, the longer it stays in LTE, and the smaller its freeze-out abundance. Thus, the more weakly interacting the particle, the larger its present abundance. The freeze-out value of Y is related to the mass of the particle and its annihilation cross section (here characterized by σ_0) by

$$Y \propto \frac{1}{M_X m_{Pl} \sigma_0} \, . \tag{13}$$

Since the contribution to Ω is proportional to $M_X n_X$, which in turn is proportional to $M_X Y$, the present contribution to Ω from a thermal relic is (to first approximation) *independent* of the mass, and only depends on the mass indirectly through the dependence of the annihilation cross section on mass. The largest that the annihilation cross section can be is roughly M_X^{-2}. This implies that large-mass WIMPS would have such a small annihilation

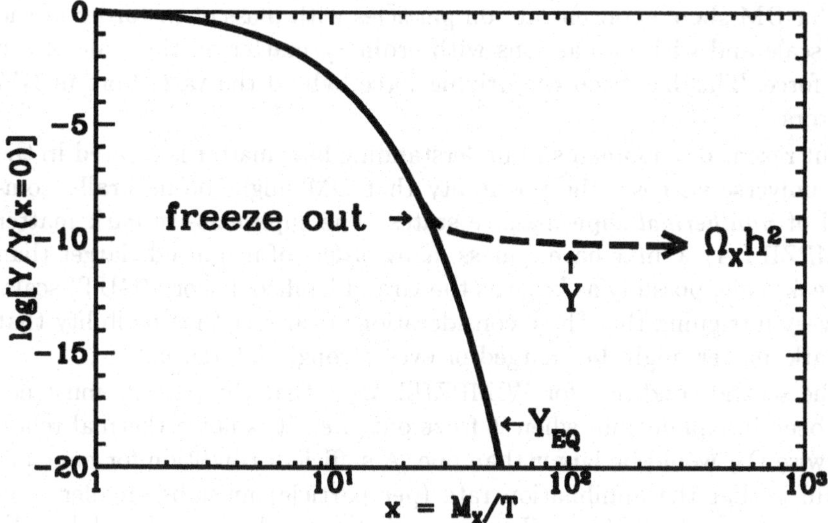

Figure 8. A thermal relic starts in LTE at $T \gg M_X$. When the rates keeping the relic in chemical equilibrium become smaller than the expansion rate, the density of the relic relative to the entropy density becomes constant. This is known as *freeze out.*

cross section that their present abundance would be too large. Thus, one expects a maximum mass for a thermal WIMP, which turns out to be a few hundred TeV.

The mass of WIMPS usually considered for dark matter run from a microvolt for axions to several dozen GeV for neutralinos. With the exception of massive magnetic monopoles, the possibility of dark matter particles of GUT-scale mass is not usually considered, because thermal relics of this mass would be expected to be over abundant by several orders of magnitude.

Recently, the idea that dark matter may be *supermassive* has received a lot of attention. Since wimpy little dark matter particles with mass less than a TeV are called WIMPS, dark matter particles of really hefty mass of 10^{12} to 10^{16} GeV seem to be more than WIMPS, so they are referred to as WIMPZILLAS.

5.2. WIMPZILLAS—SIZE DOES MATTER

The simple assumption that the dark matter (DM) is a thermal relic is surprisingly restrictive. The limit $\Omega_X \lesssim 1$ implies that the mass of a DM relic must be less than about 500 TeV [6]. The standard lore is that the hunt for DM should concentrate on particles with mass of the order of the weak scale and with interactions with ordinary matter on the scale of the weak force. This has been the driving force behind the vast effort in DM detectors.

But recent developments in understanding how matter is created in the early universe suggests the possibility that DM might be naturally composed of *nonthermal* supermassive states. The supermassive dark matter (WIMPZILLA) X may have a mass many orders of magnitude larger than the weak scale, possibly as large as the Grand Unified Theory (GUT) scale. It is very intriguing that these considerations resurrect the possibility that the dark matter might be charged or even strongly interacting!

The second condition for WIMPZILLAS is that the particle must not have been in equilibrium when it froze out (*i.e.*, it is not a thermal relic), otherwise Ω_X would be larger than one. A sufficient condition for nonequilibrium is that the annihilation rate (per particle) must be smaller than the expansion rate: $n\sigma|v| < H$, where n is the number density, $\sigma|v|$ is the annihilation rate times the Møller flux factor, and H is the expansion rate. Conversely, if the WIMPZILLA was created at some temperature T_* *and* $\Omega_X < 1$, then it is easy to show that it could not have attained equilibrium. To see this, assume Xs were created in a radiation-dominated universe at temperature T_*. Then Ω_X is given by $\Omega_X = \Omega_\gamma(T_*/T_0)M_X n_X(T_*)/\rho_\gamma(T_*)$, where T_0 is the present temperature (ignoring dimensionless factors of order unity.) Using the fact that $\rho_\gamma(T_*) = H(T_*)M_{Pl}T_*^2$, we find $n_X(T_*)/H(T_*) = (\Omega_X/\Omega_\gamma)T_0 M_{Pl} T_*/M_X$. We may safely take the limit $\sigma|v| < M_X^{-2}$, so $n_X(T_*)\sigma|v|/H(T_*)$ must be less than $(\Omega_X/\Omega_\gamma)T_0 M_{Pl} T_*/M_X^3$. Thus, the requirement for nonequilibrium is

$$\left(\frac{200\,\mathrm{TeV}}{M_X}\right)^2 \left(\frac{T_*}{M_X}\right) < 1\,. \tag{14}$$

This implies that if a nonrelativistic particle with $M_X \gtrsim 200$ TeV was created at $T_* < M_X$ with a density low enough to result in $\Omega_X \lesssim 1$, then its abundance must have been so small that it never attained equilibrium. Therefore, if there is some way to create WIMPZILLAS in the correct abundance to give $\Omega_X \sim 1$, nonequilibrium is guaranteed.

An attractive origin for WIMPZILLAS is during the defrosting phase after inflation. It is important to realize that it is not necessary to convert a significant fraction of the available energy into massive particles; in fact, it must be an infinitesimal amount. If a fraction ϵ of the available energy density is in the form of a massive, stable X particle, then $\Omega_X = \epsilon\Omega_\gamma(T_{RH}/T_0)$, where T_{RH} is the "reheat" temperature. For $\Omega_X = 1$, this leads to the limit $\epsilon \lesssim 10^{-17}(10^9\,\mathrm{GeV}/T_{RH})$.

In one extreme we might assume that the vacuum energy of inflation is immediately converted to radiation, resulting in a reheat temperature T_{RH}. In this case Ω_X can be calculated by integrating the Boltzmann equation with initial condition $N_X = 0$ at $T = T_{RH}$. One expects the X density to be suppressed by $\exp(-2M_X/T_{RH})$; indeed, one finds $\Omega_X \sim 1$ for $M_X/T_{RH} \sim 25 + 0.5\ln(M_X^2\langle\sigma|v|\rangle)$, in agreement with previous estimates [7] that for $T_{RH} \sim 10^9$GeV, the WIMPZILLA mass would be about 2.5×10^{10}GeV.

A second (and more plausible) scenario is that reheating is not instantaneous, but is the result of the decay of the inflaton field. In this approach the radiation is produced as the inflaton decays. The WIMPZILLA density is found by solving the coupled system of equations for the inflaton field energy, the radiation density, and the WIMPZILLA mass density. The calculation has been recently reported in Ref. [8], with result $\Omega_X \sim M_X^2\langle\sigma|v|\rangle(2000T_{RH}/M_X)^7$. For a reheat temperature as low as 10^9GeV, a particle of mass 10^{13}GeV can be produced in sufficient abundance to give $\Omega_X \sim 1$.

The large difference in WIMPZILLA masses in the two reheating scenarios arises because the peak temperature is much larger in the second scenario, even with identical T_{RH}. Because the temperature decreases as $a^{-3/8}$ (a is the scale factor) during most of the reheating period in the second scenario, it must have once been much greater than T_{RH}. The evolution of the temperature is given in Fig. 9. If we assume the radiation spectrum did not depart grossly from thermal, the effective temperature having once been larger than T_{RH} implies that the density of particles with enough energy to create WIMPZILLAS was larger. Denoting as T_2 the maximum effective temperature for the second scenario, we find $T_2/T_{RH} \sim (M_\phi/\Gamma_\phi)^{1/4} \gg 1$, where Γ_ϕ is the effective decay rate of the inflaton. See [8] for details.

Another way to produce WIMPZILLAS after inflation is in a preliminary stage of reheating called "preheating" [9], where nonlinear quantum effects may lead to an extremely effective dissipational dynamics and explosive particle production. Particles can be created in a broad parametric resonance with a fraction of the energy stored in the form of coherent inflaton oscillations at the end of inflation released after only a dozen oscillation periods. A crucial observation for our discussion is that particles with mass up to 10^{15} GeV may be created during preheating [10, 11, 12], and that their distribution is nonthermal. If these particles are stable, they may be good candidates for WIMPZILLAS.

To study how the creation of WIMPZILLAS takes place in preheating, let us take the simplest chaotic inflation potential: $V(\phi) = M_\phi^2\phi^2/2$ with $M_\phi \sim 10^{13}$ GeV. We assume that the interaction term between the WIMPZILLA and the inflaton field is of the type $g^2\phi^2|X|^2$. Quantum fluctuations of the X field with momentum $\vec{k}$ during preheating *approximately* obey the

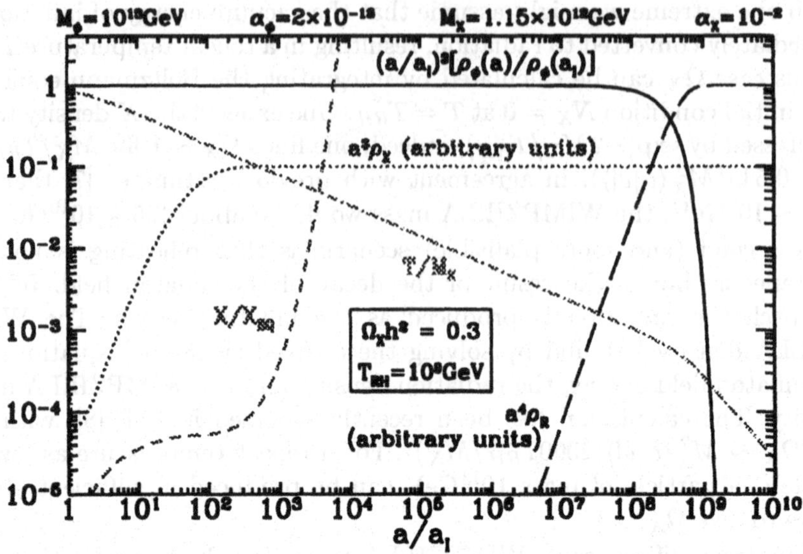

Figure 9. The evolution of energy densities and T/M_X as a function of the scale factor. Also shown is X/X_{EQ}.

Mathieu equation, $X_k'' + [A(k) - 2q\cos 2z]X_k = 0$, where $q = g^2\phi^2/4M_\phi^2$, $A(k) = (k^2 + M_X^2)/M_\phi^2 + 2q$ (primes denotes differentiation with respect to $z = M_\phi t$). Particle production occurs above the line $A = 2q$ in an instability strip of width scaling as $q^{1/2}$ for large q. The condition for broad resonance, $A - 2q \lesssim q^{1/2}$ [10], becomes $(k^2 + M_X^2)/M_\phi^2 \lesssim g\bar{\phi}/M_\phi$, which yields $E_X^2 = k^2 + M_X^2 \lesssim g\bar{\phi}M_\phi$ for the typical energy of particles produced in preheating. Here $\bar{\phi}$ is the amplitude of the oscillating inflaton field [9]. The resulting estimate for the typical energy of particles at the end of the broad resonance regime for $M_\phi \sim 10^{-6}M_{\rm Pl}$ is $E_X \sim 10^{-1}g^{1/2}\sqrt{M_\phi M_{\rm Pl}} \sim g^{1/2}10^{15}$ GeV. Supermassive X bosons can be produced by the broad parametric resonance for $E_X > M_X$, which leads to the estimate that X production will be possible if $M_X < g^{1/2}10^{15}$ GeV. For $g^2 \sim 1$ one would have copious production of X particles as heavy as 10^{15}GeV, *i.e.*, 100 times greater than the inflaton mass, which may be many orders of magnitude greater than the reheat temperature. Scatterings of X fluctuations off the zero mode of the inflaton field considerably limits the maximum magnitude of X fluctuations to be $\langle X^2\rangle_{\rm max} \approx M_\phi^2/g^2$ [13]. For example, $\langle X^2\rangle_{\rm max} \lesssim 10^{-10}M_{\rm Pl}^2$ in the case $M_X = 10\,M_\phi$. This restricts the corresponding number density of created X-particles.

For a reheating temperature of the order of 100 GeV, the present abundance of WIMPZILLAS with mass $M_X \sim 10^{14}$ GeV is given by $\Omega_X \sim 1$ if $\epsilon \sim 10^{-10}$. This small fraction corresponds to $\langle X^2 \rangle \sim 10^{-12} M_{\rm Pl}^2$ at the end of the preheating stage, a value naturally achieved for WIMPZILLA mass in the GUT range [13]. The creation of WIMPZILLAS through preheating and, therefore, the prediction of the present value of Ω_X, is very model dependent. The inflaton might preferably decay through parametric resonance into very light boson fields so that the end of the preheating stage and of the corresponding value of $\langle X^2 \rangle$ depends upon the coupling of the inflaton field not only to the WIMPZILLA, but also to other degrees of freedom. It is encouraging, however, that it is possible to produce supermassive particles during preheating that are as massive as $10^{12} T_{RH}$. Details of WIMPZILLA production in preheating can be found in [14].

Another possibility which has been recently investigated is the production of very massive particles by gravitational mechanisms [15, 16]. In particular, the desired abundance of WIMPZILLAS may be generated during the transition from the inflationary phase to a matter/radiation dominated phase as the result of the expansion of the background spacetime acting on vacuum quantum fluctuations of the dark matter field [15]. A crucial side-effect of the inflationary scenarios is the generation of density perturbations. A related effect, which does not seem to have attracted much attention, is the possibility of producing matter fields due to the rapid change in the evolution of the scale factor around the end of inflation. Contrary to the first effect, the second one contributes to the homogeneous background energy density that drives the cosmic expansion, and is essentially the familiar "particle production" effect of relativistic field theory in external fields.

Very massive particles may be created in a nonthermal state with sufficient abundance to achieve critical density today by the classical gravitational effect on the vacuum state at the end of inflation. Mechanically, the particle creation scenario is similar to the inflationary generation of gravitational perturbations that seed the formation of large-scale structures. However, the quantum generation of energy density fluctuations from inflation is associated with the inflaton field, which dominated the mass density of the universe, and not a generic sub-dominant scalar field.

If $0.04 \lesssim M_X/H_e \lesssim 2$ [15], where H_e is the Hubble constant at the end of inflation, DM produced gravitationally can have a density today of the order of the critical density. This result is quite robust with respect to the fine details of the transition between the inflationary phase and the matter-dominated phase. The only requirement is that

$$\left(\frac{H_e}{10^{-6} M_{Pl}}\right)^2 \left(\frac{T_{RH}}{10^9 \mathrm{GeV}}\right) \gtrsim 10^{-2} \,. \tag{15}$$

The observation of anisotropy in the cosmic background radiation does not fix H_e uniquely, but using $T_{RH} \lesssim \sqrt{M_{Pl}H_e}$, we find that the mechanism is effective only when $H_e \gtrsim 10^9$GeV (or, $M_X \gtrsim 10^8$GeV).

The distinguishing feature of this mechanism [15] is the capability of generating particles with mass of the order of the inflaton mass even when the WIMPZILLA interacts only extremely weakly (or not at all!) with other particles, including the inflaton. This feature makes the gravitational production mechanism quite model independent and, therefore, more appealing to us than the one occurring at preheating.

WIMPZILLAS can also be produced in theories where inflation is completed by a first-order phase transition [17]. In these scenarios, the universe decays from its false vacuum state by bubble nucleation [18]. When bubbles form, the energy of the false vacuum is entirely transformed into potential energy in the bubble walls, but as the bubbles expand, more and more of their energy becomes kinetic and the walls become highly relativistic. Eventually the bubble walls collide.

During collisions, the walls oscillate through each other [19] and the kinetic energy is dispersed into low-energy scalar waves [19, 20]. If these soft scalar quanta carry quantum numbers associated with some spontaneously broken symmetry, they may even lead to the phenomenon of nonthermal symmetry restoration [21]. We are, however, more interested in the fate of the potential energy of the walls, $M_P = 4\pi\eta R^2$, where η is the energy per unit area of the bubble with radius R. The bubble walls can be imagined as a coherent state of inflaton particles, so that the typical energy E of the products of their decays is simply the inverse thickness of the wall, $E \sim \Delta^{-1}$. If the bubble walls are highly relativistic when they collide, there is the possibility of quantum production of nonthermal particles with mass well above the mass of the inflaton field, up to energy $\Delta^{-1} = \gamma M_\phi$, γ being the relativistic Lorentz factor.

Suppose now that the WIMPZILLA is some fermionic degree of freedom X and that it couples to the inflaton field by the Yukawa coupling $g\phi\overline{X}X$. One can treat ϕ (the bubbles or walls) as a classical, external field and the WIMPZILLA as a quantum field in the presence of this source. This amounts to ignoring the backreaction of particle production on the evolution of the walls, but this is certainly a good approximation in our case. The number of WIMPZILLA particles created in the collisions from the wall's potential energy is $N_X \sim f_X M_P/M_X$, where f_X parametrizes the fraction of the primary decay products that are WIMPZILLAS. The fraction f_X will depend in general on the masses and the couplings of a particular theory in question. For the Yukawa coupling g, it is $f_X \simeq g^2 \ln\left(\gamma M_\phi/2M_X\right)$ [20, 22]. Supermassive particles in bubble collisions are produced out of equilibrium and they never attain chemical equilibrium. Assuming $T_{RH} \simeq 100$ GeV, the present abundance of WIMPZILLAS is $\Omega_X \sim 1$ if $g \sim 10^{-5}\alpha^{1/2}$. Here

$\alpha^{-1} \ll 1$ denotes the fraction of the bubble energy at nucleation which has remained in the form of potential energy at the time of collision. This simple analysis indicates that the correct magnitude for the abundance of X particles may be naturally obtained in the process of reheating in theories where inflation is terminated by bubble nucleation.

In conclusion, a large fraction of the DM in the universe may be made of WIMPZILLAS of mass greatly exceed the electroweak scale—perhaps as large as the GUT scale. This is made possible by the fact that the WIMPZILLAS were created in a nonthermal state and never reached chemical equilibrium with the primordial plasma.

Acknowledgements

This work was supported in part by the Department of Energy, as well as NASA under grant number NAG5-7092. The hospitality of Tom Ferbel and the inquisitiveness of the students were greatly appreciated.

References

1. W. Shakespeare, *Henry V.*
2. E. Gawiser and J. Silk, Science (1998).
3. For a review, see V. F. Mukhanov, H. A. Feldman, and R. H. Brandenberger. *Phys. Rep.* **215**, 203 (1992).
4. For a review, see G. Jungman, M. Kamionkowski and K. Greist, Phys. Rep. **267**, 196 (1995).
5. M. S. Turner, Phys. Rep. **197**, 67 (1990).
6. K. Griest and M. Kamionkowski, Phys. Rev. Lett. **64**, 615 (1990).
7. V. A. Kuzmin and V. A. Rubakov, Phys. Atom. Nucl. **61**, 1028 (1998).
8. D. J. Chung, E. W. Kolb, and A. Riotto, hep-ph/9809453.
9. L. A. Kofman, A. D. Linde and A. A. Starobinsky, Phys. Rev. Lett. **73**, 3195 (1994).
10. E.W. Kolb, A. D. Linde and A. Riotto, Phys. Rev. Lett. **77**, 4290 (1996).
11. B. R. Greene, T. Prokopec and T. G. Roos, Phys. Rev. D **56**, 6484 (1997).
12. E. W. Kolb, A. Riotto and I. I. Tkachev, Phys. Lett. **B 423**, 348 (1998).
13. S. Khlebnikov and I. I. Tkachev, Phys. Rev. Lett. **79**, 1607 (1997).
14. D. J. H. Chung, hep-ph/9809489.
15. D. J. Chung, E. W. Kolb and A. Riotto, hep-ph/9802238.
16. V. Kuzmin and I. I. Tkachev, hep-ph/9802304
17. D. La and P. J. Steinhardt, Phys. Rev. Lett. **62**, 376 (1989).
18. A. H. Guth, Phys. Rev. D **23**, 347 (1981).
19. S. W. Hawking, I. G. Moss and J. M. Stewart, Phys. Rev. D **26**, 2681 (1982).
20. R. Watkins and L. Widrow, Nucl. Phys. **B374**, 446 (1992).
21. E. W. Kolb and A. Riotto, Phys. Rev. D **55**, 3313 (1997); E. W. Kolb, A. Riotto and I. I. Tkachev, Phys. Rev. D **56**, 6133 (1997).
22. A. Masiero and A. Riotto, Phys. Lett. **B289**, 73 (1992).

LATEST NEWS ON NEUTRINOS

K. NAKAMURA
High Energy Accelerator Research Organization (KEK)
Oho, Tsukuba, Ibaraki 305-0801, Japan

1. Introduction

Neutrinos interact with matter through weak inteactions which is well prescribed by the Standard Model. For this reason, neutrinos provide a unique means to test the electroweak and strong interactions in, and to probe new physics beyond, the Standard Model.

Throughout the 1970s and 1980s, the main focus of the neutrino experiment was high-energy accelerator neutrino experiments to measure nucleon structure functions and to test electroweak interactions and perturbative QCD. In recent years, however, there is only one active experiment in this category, that is the NuTeV experiment at Fermilab. A comprehensive review article on these topics from 1980 until the present has been written by Conrad, Shaevitz and Bolton [1].

In spite of the fact that the neutrino has been extensively used as a probe to test the Standard Model, some of its basic properties, such as the mass and magnetic moment, are not well known. In the Standard Model, neutrinos are postulated to be massless. However, no basic principle is known to force the neutinos to be massless. Whether or not neutrinos have finite mass is the central issue of today's neutrino physics as well as one of the most important problems in particle physics.

Shortly before this NATO Advanced Study Institute at St. Croix, Neutrino '98 Conference was held at Takayama in Japan. The highlight of this conference was the "evidence for oscillation of atmospheric neutrinos" reported by the Super-Kamiokande (SK) Collaboration. Naturally this hot topic was central to this lecture entitled "Latest News on Neutrinos." SK also presented high-precision solar neutrino results and these considerably more constrain than before the parameter space allowed for MSW or vacuum-oscillation solutions for the solar-neutrino problem. Another news from Neutrino '98 was a candidate event for a tau neutrino interaction in

T. Ferbel (ed.), Techniques and Concepts of High Energy Physics X, 263–290.

emulsion found by the DONUT (Direct Observation of Nu Tau) experiment at Fermilab.

This lecture mainly focuses on these new results presented at Neutrino '98. Though the NuTeV Collaboration reported preliminary results on $\sin^2\theta_W$ and M_W from high-energy neutrino interactions at Neutrino '98, they are not discussed here. (However, Nodulman's lecture [2] includes this topic.) Double beta-decay experiments and observations of high-energy astrophysical neutrinos are not discussed either, though these are among major topics in neutrino physics.

2. Search for Tau Neutrino

Although the existence of three light species of neutrinos with mass less than half of Z-boson mass [3] was established, tau neutrinos have not been directly observed so far.

DONUT is a beam dump experiment with 800 GeV protons, using an iron-emulsion sandwich target (called an emulsion cloud chamber, ECC) designed to search for short-lived τ particles produced by charged-current (CC) ν_τ interactions. Only about 5% of the target mass is emulsion. Scintillating fiber trackers interleaved with emulsion target modules and downstream drift chamber spectrometer electronically locate interesting events to be scanned in emulsion.

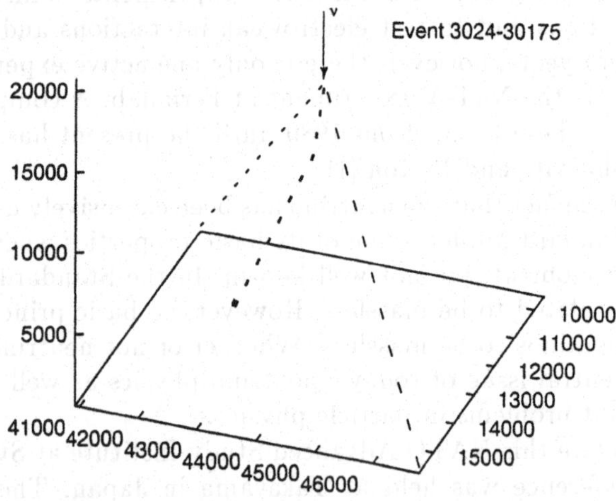

Figure 1. A candidate for a ν_τ event observed in the DONUT experiment. The central track has a kink after passing 4.5 mm from the interaction vertex. It is identified as an electron since an e^+e^- pair is seen along the track. The scale is shown in μm.

In the analysis, kinks of charged particle tracks due to tau decay have been searched for. At Neutrino '98, the first candidate for tau neutrino (see Fig. 1) was reported [4]. This event was found among 34 events of neutrino inteaction candidates. It is consistent with the expected fraction of ν_τ events (5%) in all neutrino interactions in the target. From the 1997 data run, 1,200 neutrino interactions of all flavors are estimated to have occurred in the target.

3. Direct Measurements of the Neutrino Masses

Direct measurements of the neutrino masses have been done using the kinematical relation in the decays of $^3\text{H} \to {}^3\text{He}\, e^- \bar{\nu}_e$ for ν_e, $\pi^+ \to \mu^+ \nu_\mu$ for ν_μ, and $\tau \to 3\pi^\pm \nu_\tau$ and $5\pi^\pm(\pi^0)\nu_\tau$ for ν_τ. Only upper limits of neutrino mass have been obtained so far. Table 1 lists the upper limits recommended in the 1998 edition of Review of Particle Physics [5]. Here, an upper limit for the electron neutrino mass is not quoted from tritium beta-decay experiments, but is taken as the "supernova limit" obtained from the spread of arrival times of neutrinos from SN1987A. This is because in many tritium beta-decay experiments the best-fit value of the square of the electron neutrino mass turned out to be significantly negative, and this poses a problem in setting a reliable limit below ~10 MeV [6].

At Neutrino '98, results were reported from Troitsk and Mainz tritium beta-decay experiments. Both showed anomalous accumulation of events near the endpoint of the electron energy spectrum, and this caused significantly negative $m(\nu_e)^2$. By assuming the existence of a bump at $5 \sim 15$ eV from the endpoint, $m(\nu_e) < 2.7$ eV was obtained at 95% confidence level (CL) from the combined 1994, 1996, and 1997 Troitsk data [7]. Also, with a similar assumption, 1998 Maintz data gave $m(\nu_e) < 3.4$ eV (95% CL) [8]. However, there is no justification for the existence of the assumed bump.

TABLE 1. Upper limits of the neutrino masses recommended in the 1998 edition of Review of Particle Physics.

	Upper limit	Confidence level	comment
ν_e	15 eV	—	TOF limit from SN1987A
ν_μ	170 keV	90%	PSI [9]
ν_τ	18.2 MeV	95%	ALEPH Collaboration [10]

4. Neutrino Oscillations — Phenomenology

In the direct neutrino-mass measurements, it is difficult to explore the mass region below ~1 eV. On the other hand, neutrino oscillations provide a means to explore the mass region much lower than 1 eV.

Neutrino oscillations occur if neutrinos have masses and a neutrino with definite flavor is not a mass eigenstate

$$\nu_\alpha = \sum_i U_{\alpha i}\nu_i, \tag{1}$$

where ν_αs are the flavor eigenstates ($\alpha = e, \mu, \tau$) and ν_is are mass eigenstates ($i = 1, 2, 3$). Often experimental results are analyzed in terms of two neutrino flavors. The mixing matrix in this case is given by

$$U = \begin{pmatrix} \cos\theta & \sin\theta \\ -\sin\theta & \cos\theta \end{pmatrix} \tag{2}$$

where θ is a mixing angle. The probability of $\nu_\alpha \to \nu_\beta$ for a neutrino born as ν_α after travelling a distance L in vacuum is given by

$$P(\nu_\alpha \to \nu_\beta) = \sin^2 2\theta \ \sin^2(1.27\Delta m^2 L/E) \tag{3}$$

where, $\Delta m^2 = |m_\alpha^2 - m_\beta^2|$ is in eV^2, neutrino energy E is in GeV and L is in km. The survival probability is given by $P(\nu_\alpha \to \nu_\alpha) = 1 - P(\nu_\alpha \to \nu_\beta)$.

If neutrinos travel in matter, resonant neutrino oscillations can occur. This effect is also called matter-enhanced neutrino oscillations or MSW (Mikheyev, Smirnov, and Wolfenstein) effect [11, 12]. This is a mechanism which gives large transition probabilities even with a small vacuum mixing. It is particularly important for the analyses of solar neutrino observations.

In the ordinary matter, neutrinos feel an effective potential due to coherent scattering with the particles in matter. This causes the effective mass and mixing angle in matter which are different from the mass and mixing angle in vacuum. In the matter, all neutrino species feel the common neutral-current (NC) potential. This potential causes a common overall phase which is not observable. However, ν_e and $\bar{\nu}_e$ feel the CC potential due to the CC interactions with electrons. Other neutrinos do not feel the CC potential. Consequently, the MSW effect occurs, for example, in $\nu_e \leftrightarrow \nu_\mu$ oscillations, whereas there is no difference between $\nu_\mu \leftrightarrow \nu_\tau$ oscillations in vacuum and in matter.

For $\nu_e \leftrightarrow \nu_\mu$ (and also $\nu_e \leftrightarrow \nu_\tau$) oscillations in matter, the resonance condition is given by

$$2EV_C = \Delta m^2 \cos 2\theta \tag{4}$$

$$V_C = \sqrt{2} G_F N_e, \tag{5}$$

where V_C is the CC potential, G_F is the Fermi constant and N_e is the electron number density in matter. Therefore, the electron number density where resonant neutrino oscillations occur is dependent on the neutrino energy, Δm^2, and the mixing angle in vacuum. For solar neutrinos, which are produced in the solar core, the MSW effect occurs if N_e determined from the resonance condition is less than N_e in the core region.

For more about neutrino-oscillation phenomenology, readers are referred to textbooks [13] or review articles [14].

5. Experimental Results on Neutrino Oscillations

5.1. ACCELERATOR AND REACTOR EXPERIMENTS

5.1.1. *CHORUS and NOMAD*

CHORUS and NOMAD are experiments at CERN designed to search for $\nu_\mu \to \nu_\tau$ oscillations in the Δm^2 range of > 1 eV^2 with a sensitivity to the mixing angle, $\sin^2 2\theta < 2 \times 10^{-4}$. This mass range is cosmologically interesting because relic neutrinos in this range comprize part of cosmological dark matter. Candidates for tau-neutrinos are identified through CC interactions $\nu_\tau N \to \tau^- X$, followed by the τ decay.

The CHORUS experiment exploits an emulsion hybrid spectrometer with 800 kg of emulsion target in the first run in 1994 – 1995 and additional 800 kg in the second run in 1996 – 1997. This technique allows a three-dimensional visual reconstruction of the trajectories of τ and its decay products (see Fig. 2).

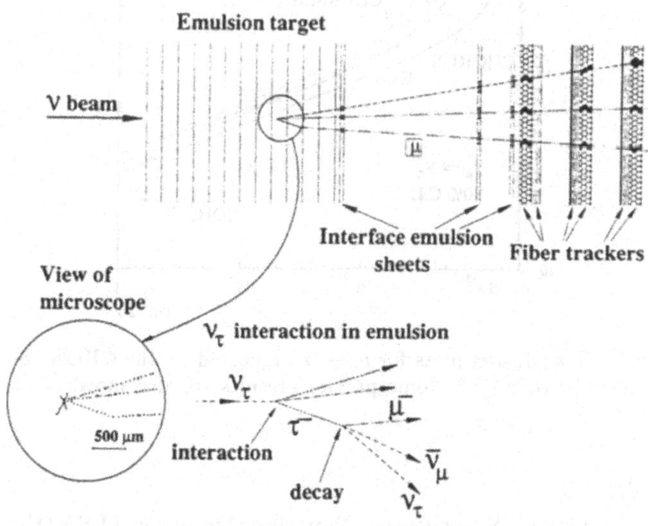

Figure 2. Schematic of the front part of the CHORUS detector. A typical signature of a ν_τ charged-current event is also shown.

Another $\nu_\mu \to \nu_\tau$ oscillation experiment NOMAD is conducted with a purely electronic detector. Its neutrino target is a set of drift chambers made of low-density and low-Z materials, placed in a uniform magnetic field of 4 kG provided by a recycled UA1 dipole. It may be called an "electronic bubble chamber." The fiducial mass is 3 tons over the volume of $2.6 \times 2.6\times \sim 4$ m^3. The NOMAD detector is equipped with a transition radiation detector, an electromagnetic calorimeter, a hadron calorimeter, and muon chambers. Thus, it has a good μ/e identification capability. The τ signature is searched for on the statistical basis using kinematical criteria to isolate the τ decay products from the remainder of the event and to characterize missing transverse momentum carried away by the neutrino(s) from τ decay.

Recently, both the CHORUS [15] and NOMAD [16] Collaborations published the results, but they were updated at Neutrino '98. No ν_τ candidates were found, and the 90% CL limit for $\sin^2 2\theta$ is 1.3×10^{-3} (CHORUS [17]) and 2.2×10^{-3} (NOMAD [18]) at large Δm^2. Figure 3 shows the regions excluded by CHORUS and by NOMAD together with those from previous experiments. Eventually, a limit of $\sin^2 2\theta < 2\times 10^{-4}$ will be reached, if no candidate signature is found in the further analyses.

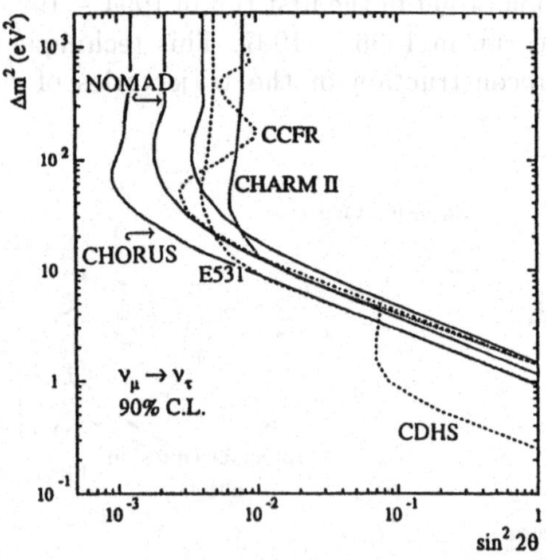

Figure 3. 90% CL exclusion plots for $\nu_\mu \to \nu_\tau$ reported by the CHORUS and NOMAD Collaborations at Neutrino '98. Some previous results are also shown.

5.1.2. *LSND*

Recently, the Liquid Scintillator Neutrino Detector (LSND) at LAMPF reported [19] evidence for $\nu_\mu \to \nu_e$ oscillations in addition to the previous positive results for $\bar{\nu}_\mu \to \bar{\nu}_e$ oscillations [20]. The allowed (Δm^2, $\sin^2 2\theta$)

regions are consistent for the two CP-conjugate oscillation channels (see Fig. 4).

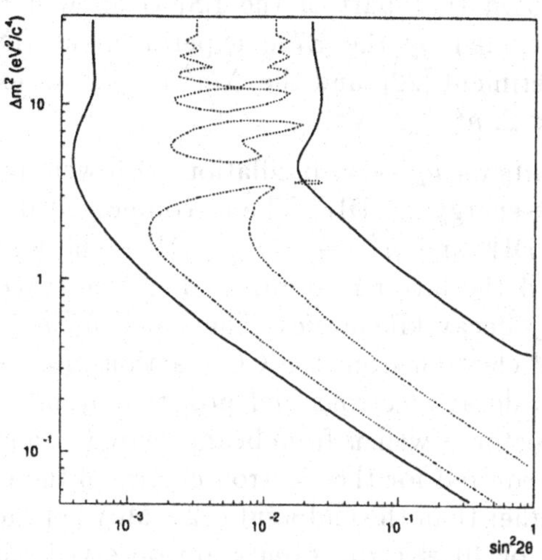

Figure 4. LSND results. The solid curves show the 95% CL region favored by the DIF (decay in flight) $\nu_\mu \to \nu_e$ measurement. Also, the region favored by the DAR (decay at rest) $\bar{\nu}_\mu \to \bar{\nu}_e$ measurement is shown by the dotted contours. The confidence level of this region is not defined exactly.

The LSND detector contains 167 tons of dilute liquid scintillator and is located about 300 m from the neutrino source. Neutrinos come from decay of π^+ produced by protons from the LAMPF 800 MeV linac. By an appropriate kinematical setting, one can suppress the $\bar{\nu}_e$ flux or ν_e flux in the beam.

In the $\bar{\nu}_\mu \to \bar{\nu}_e$ experiment [20] was utilized a neutrino flux produced by π^+ decay at rest (DAR) followed by $\mu^+ \to e^+\nu_e\bar{\nu}_\mu$ with a maximum $\bar{\nu}_\mu$ energy of 52.8 MeV. Only 3% of π^+ are allowed to decay in flight (DIF) due to short open space around the target. Thus, the main components of the neutrino flux with energy less than 52.8 MeV are ν_e, ν_μ, and $\bar{\nu}_\mu$. A $\bar{\nu}_e$ component in the beam comes from the π^- decay chain. It is suppressed to the flux ratio of $\bar{\nu}_e/\bar{\nu}_\mu \sim 7.8 \times 10^{-4}$ because of (i) smaller π^- production, (ii) large probability of absorption of π^- at rest in the beam stop, (iii) and also a large probability of capture of μ^- (which is produced by π^- DIF) in the beam stop.

In the LSND, $\bar{\nu}_e$s were detected through the reaction $\bar{\nu}_e p \to e^+ n$ followed by $np \to d\gamma$. The signal is defined by an electron signal followed by a 2.2 MeV γ correlated both in position and time. For a tight event analysis, an electron energy is required in the range $36 < E_e < 60$ MeV in order to suppress electrons from $\nu_e\,^{12}\mathrm{C} \to e^-\,^{12}\mathrm{N}$ which have energy $E_e < 36$ MeV. The experiment observed an excess of 17.4 ± 4.7 $\bar{\nu}_e$ events, which was interpreted as evidence for $\bar{\nu}_\mu \to \bar{\nu}_e$ oscillations with an oscillation probability of $(3.1 \pm 1.2 \pm 0.5) \times 10^{-3}$.

It should be noted that part of the LSND allowed region for $\bar{\nu}_\mu \to \bar{\nu}_e$ was already excluded by the E776 experiment at BNL [21] and the Bugey reactor experiment [22], and the Δm^2 region not excluded by these experiment is $0.3 < \Delta m^2 < 2$ eV^2.

The recent results on $\nu_\mu \to \nu_e$ oscillations [19] were obtained by using a ν_μ flux from high-energy π^+ DIF. The ν_e component in the beam comes from $\pi^+ \to e^+\nu_e$ DIF and $\mu^+ \to e^+\nu_e\bar{\nu}_\mu$ DIF. The former has a small branching ratio and the latter is suppressed by the longer muon lifetime and the three-body decay kinematics. The signal for $\nu_\mu \to \nu_e$ candidates is a single, isolated electron from the CC reaction $\nu_e\,\mathrm{C} \to e^-\,\mathrm{X}$. To avoid background due to decay electrons and positrons from cosmic-ray muons stopping in the detector as well as from beam-related ν_μ and $\bar{\nu}_\mu$ events, $60 < E_e < 200$ MeV is required for the electron energy of the candidate events. The lower cut is higher than the endpoint (52.8 MeV) of the Michel electron spectrum. A total of 40 electron events are observed with an estimated background of 21.9 ± 2.1 events from the ν_e component in the beam. The excess of events is consistent with $\nu_\mu \to \nu_e$ with an oscillation probability of $(2.6 \pm 1.0 \pm 0.5) \times 10^{-3}$. (Throughout this lecture, the first error represents the statistical error, and the second the systematic.)

5.1.3. *KARMEN*

KARMEN is an experiment at the pulsed spallation neutron facility, ISIS, of Rutherford Appleton Laboratory, with 56-ton segmented liquid-scintillator calorimeter. The neutrino beam is produced by 800 MeV protons from a rapid-cycle synchrotron stopped in the beam dump. During the period 1990 – 1995, this experiment searched for $\bar{\nu}_\mu \to \bar{\nu}_e$ oscillations with no evidence. However, this result was not sensitive enough to exclude entirely the $(\Delta m^2, \sin^2 2\theta)$ region favored by LSND. On the other hand, the KARMEN data showed an anomalous excess in the time distribution of neutrino-induced events at ~3.6 μs after beam-on-target [23]. This anomaly remained a puzzle.

Subsequently, the KARMEN detector was upgraded by strengthening a veto counter system. The KARMEN2 experiment is scheduled during 1997 – 1999. After three years, KARMEN2 sensitivity will cover most of the $(\Delta m^2, \sin^2 2\theta)$ region favored by LSND. At Neutrino '98, preliminary results were reported [24]: from the Feb. '97 – Feb. '98 run, no $\bar{\nu}_e$ events were found in contrast to the expected background of 2.88 ± 0.13 events. In the time distribution of neutrino-induced events, however, an anomalous excess showed up again.

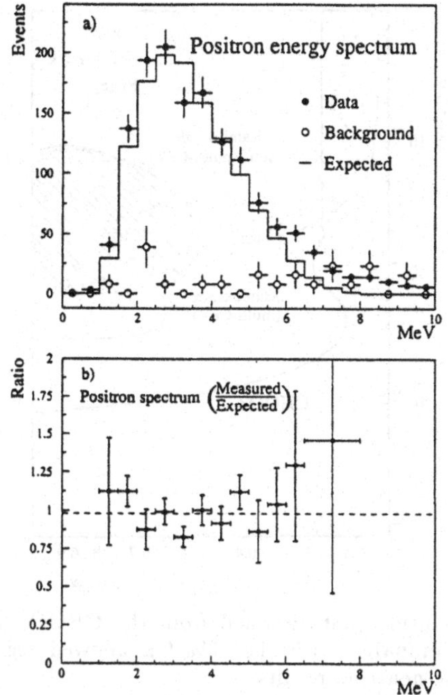

Figure 5. (a) Positron energy spectrum and the reactor-off background for the same live time. The expected positron spectrum is also shown. (b) The ratio of measured to expected positron spectrum. The background was subtracted.

5.1.4. *CHOOZ*

CHOOZ is a reactor neutrino-oscillation experiment [25] conducted with a liquid-scintillator calorimeter located in a 300 meters water equivalent (mwe) underground laboratory at a distance of about 1 km from the Chooz power station in the Ardennes region of France. With an average $E/L \sim$ 3MeV/1km $\sim 1/300$, this experiment is sensitive to Δm^2 down to 10^{-3} eV^2 for $\bar{\nu}_e$ disappearance, $\bar{\nu}_e \to \bar{\nu}_x$. This is an order of magnitude lower than the previous reactor experiments. The reactor-produced neutrinos are ~100% pure $\bar{\nu}_e$s and the intensity is known to better than 2%. Therefore, it is not a mandatory to measure the $\bar{\nu}_e$ flux with a detector closer to the reactors.

The detector consists of three concentric regions: a 5-ton central target region of Gd-loaded scintillator contained in a plexiglass vessel, a 17-ton intermediate "containment" region instrumented with photomultiplier tubes (PMTs), and an optically separated 90-ton outer region which vetoes cosmic-ray muons and provides additional shielding against external γ-rays and neutrons. The intermediate region protects the target from PMT radioactivity. Also, the γ-rays from neutron capture in the target are required to be contained in this region. The $\bar{\nu}_e$ signal is a delayed coincidence between the prompt e^+ + two annihilation γ-rays and 8-MeV γ-ray signal

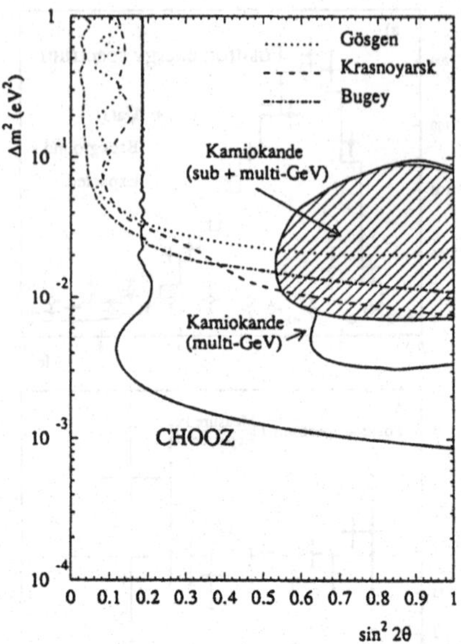

Figure 6. 90% CL exclusion plot obtained from the CHOOZ and some previous reactor experiments are compared with the 90% CL allowed region obtained from the Kamiokande atmospheric-neutrino results.

from neutron capture.

The initial results was presented [25] from 2718 live hours of data taken during the period of March – October, 1997. Figure 5a shows the positron energy spectrum obatined. The ratio of the measured to the expected (in the case of no oscillations) positron spectrum is shown in Fig. 5b. Here, the background is subtracted. The energy-independent ratio is 0.98±0.04±0.04. Thus, the CHOOZ found no evidence for $\bar{\nu}_e \to \bar{\nu}_x$ and the 90% CL excluded region is shown in Fig. 6. Specifically, this result excludes the possibility that the atmospheric-neutrino anomaly (see the next section) observed by Kamiokande is due to $\nu_\mu \to \nu_e$ oscillations.

5.2. ATMOSPHERIC NEUTRINO OBSERVATIONS

5.2.1. *Introduction*

Atmospheric neutrinos are produced by the decay of π and K mesons produced in the nuclear interactions of the primary component of cosmic rays in the atmosphere. Since pions are dominant and they decay according to $\pi^\pm \to \mu^\pm + \nu_\mu(\bar{\nu}_\mu)$, $\mu^\pm \to e^\pm + \nu_e(\bar{\nu}_e) + \bar{\nu}_\mu(\nu_\mu)$, roughly speaking $\nu_\mu : \nu_e \approx 2 : 1$ at low momenta where all unstable particles decay before reaching the Earth (hereafter ν represents $\nu + \bar{\nu}$).

Atmospheric neutrinos are observed by massive underground detectors.

Two different techniques have been used so far. One is a tracking calorimeter with mass provided by iron plates interleaved by a fine tracking device. NUSEX, Fréjus, and Soudan 2 detectors belong to this class. The other technique is a water-Cherenkov method. IMB, Kamiokande, and SK detectors utilize this method.

Neutrino events in these detectors are classified into fully contained (FC) events and partially contained (PC) events. Both types of events are required to have their vertex position inside the detector fiducial volume. The FC events have no visible energy deposited in the anti-coincidence region surrounding the fiducial volume. The PC events have some of the produced particles which deposit visible energy in the anti-coincidence region.

These detectors have a capability of μ/e identification. The electrons are identified by the characteristic shower. In water-Cherenkov detectors, the electrons show more diffused Cherenkov rings than muons due to shower and multiple scattering, and are discriminated from muons. The μ/e identification provides a means to tag the parent neutrino flavor. Those events in which a muon is identified are called "μ-like" events. On the other hand, "e-like" events have an identified electron. A Monte Carlo simulation of the SK detector shows that more than 85% (95%) of the FC single-prong e-like (μ-like) events are ν_e (ν_μ) CC interactions. For the FC multi-prong events, the contamination of NC interactions is significant, particularly at energies lower than 1 GeV (~30%). Also, the production cross sections of the multi-prong events have larger ambiguities than those of the single-prong events, and lepton identification in a multi-prong event is more difficult than in a single-prong event. For these reasons, the single-prong events are more relevant for the study of the atmospheric-neutrino fluxes than the multi-prong events, particularly for water-Cherenkov detectors. The Monte Carlo simulation of SK also shows that more than 98% of the PC events are produced by ν_μ. Therefore, PC events are safely classified as μ-like.

Since the calculated atmospheric-neutrino fluxes [26, 27] involve 20 – 30% uncertainty in the absolute normalization, it is more relevant to represent the result of observation using the double ratio defined by $R \equiv \frac{(\mu/e)_{\rm DATA}}{(\mu/e)_{\rm MC}}$, where $(\mu/e)_{\rm DATA} \equiv (\mu/e)_{\rm observed}$ is the observed ratio of the number of μ-like single-ring events to that of e-like single-ring events and $(\mu/e)_{\rm MC} \equiv (\mu/e)_{\rm predicted}$ is the corresponding Monte Carlo-predicted ratio.

5.2.2. *Old Kamiokande Results*

The Kamiokande Collaboration found that R is significantly less than unity for both sub-GeV events [28, 29] and multi-GeV events [30]. According to the Kamiokande's definition, the sub-GeV events are those events with visible energy cut of $E_{\rm vis} < 1.33$ GeV and a lower momentum cut of 100 (200)

MeV/c for e-like (μ-like) events. The multi-GeV events are FC events with $E_{\text{vis}} > 1.33$ GeV and all the PC events. The IMB Collaboration [31] and the Soudan 2 Collaboration [32] also observed R less than unity. However, the Fréjus Collaboration [34] and the NUSEX Collaboration [33] reported the value of R consistent with unity. These previous results on R are shown in Fig. 7.

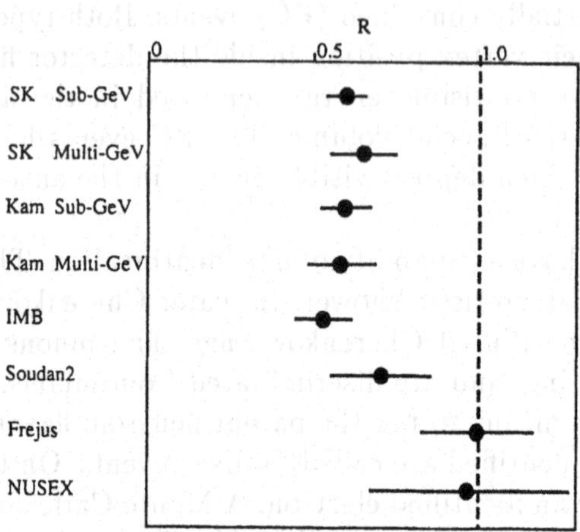

Figure 7. Results on the double ratio R. Each error bar represents the quadratic sum of the statistical and systematic errors.

More interesting results presented by the Kamiokande Collaboration is zenith-angle (θ_z) dependence of R for the multi-GeV events [30]. Since the angular correlation between the direction of the parent neutrino and that of the produced charged lepton is $15 \sim 20°$, the zenith-angle distribution of the leptons reflects the direction of neutrinos. The data suggested that while R of the down-going neutrinos with a typical travel distance of $\sim$15 km is unity, it decreases monotonically with $\cos\theta_z$ to $\sim$0.3 for upward-going neutrinos travelling all the way through the Earth.

It was suggested [29, 30] that a possible explanation for this "atmospheric-neutrino anomaly" in the value of R and its zenith-angle dependence for the multi-GeV events may be sought in neutrino oscillations. The favored value of Δm^2 is $\sim 10^{-2}$ eV^2 with a large mixing. Among the possibilities of $\nu_\mu \to \nu_e, \nu_\tau, \nu_s$ oscillations (ν_s represents a sterile neutrino. It is an SU(3)$\otimes$U(1) singlet, and, therefore, it has no weak interactions. A right-handed neutrino is an example.), $\nu_\mu \to \nu_e$ has been ruled out by the initial CHOOZ result (see Fig. 6).

5.2.3. *Super-Kamiokande Results: Evidence for Neutrino Oscillations*

Super-Kamiokande (SK) is a 50,000-ton water-Cherenkov detector located 1,000 m underground in the Kamioka mine in Japan. Its 22,000-ton fiducial mass is viewed by about 11,200 photomultiplier tubes with 50-cm diameter. Figure 8 shows a schematic of SK. The detector was completed and the data-taking was started in 1996.

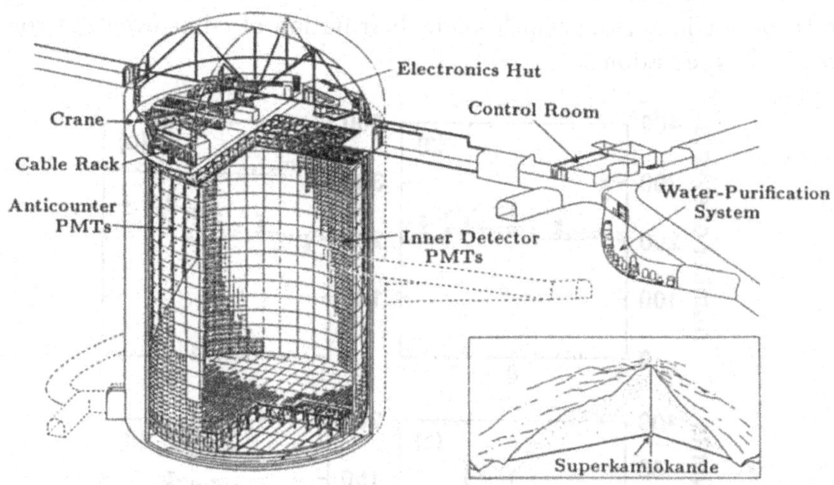

Figure 8. An artist's impression of the Super-Kamiokande detector.

At Neutrino '98, the SK group reported atmospheric-neutrino results based on an exposure of 33 kton·yr [35]. The values of R obtained are shown in Table 2, and compared with the Kamiokande results. The results are quite consistent in both sub-GeV and multi-GeV data.

TABLE 2. The values of R observed by Super-Kamiokande (SK) are compared to those observed by Kamiokande (Kam). In the Kam analysis of multi-GeV fully contained events, multi-prong events are used if charged lepton flavor is identified. However, in the SK analysis of multi-GeV envents, only single-prong events are used to deduce R.

	Experiment	Exposure (kt·yr)	R
sub-GeV	SK	33.0	$0.63 \pm 0.03 \pm 0.05$
	Kam	7.7	$0.60^{+0.06}_{-0.05} \pm 0.05$
multi-GeV	SK	33.0	$0.65 \pm 0.05 \pm 0.08$
	Kam	8.2(FC), 6.0(PC)	$0.57^{+0.08}_{-0.07} \pm 0.07$

Figure 9 shows the zenith-angle dependence of (a) sub-GeV e-like, (b) sub-GeV μ-like, (c) multi-GeV e-like, and (d) multi-GeV (fully contained and partially contained) μ-like events. The shaded boxes show the Monte

Carlo predictions with statistical error. These data indicate that μ-like events show a strong deviation from expectation in the shape of the zenith-angle distribution. In particular, multi-GeV μ-like events show strong up/down asymmetry in contrast to the calculated up/down ratio of near unity. On the other hand, the zenith-angle distribution of e-like events is consistent with expectations.

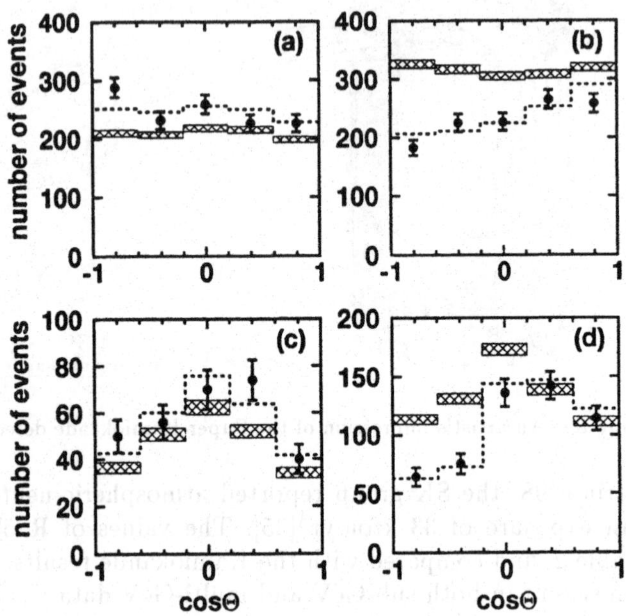

Figure 9. Multi-GeV zenith-angle distributions for (a) sub-GeV e-like, (b) sub-GeV μ-like, (c) multi-GeV e-like, and (d) multi-GeV (FC+PC) μ-like events observed in Super-Kamiokande. The Monte Carlo prediction and its statistical error is shown by the shaded boxes. The horizontal axis shows cosine of the zenith angle; 1.0 corresponds to the downward direction and -1.0 the upward direction. The dotted lines are calculated on the assumption of $\nu_\mu \rightarrow \nu_\tau$ oscillations with $\sin^2 2\theta = 1$ and $\Delta m^2 = 2.2 \times 10^{-3}$ eV^2.

Given this result, it is tempting to examine the hypothesis of two-flavor ($\nu_\mu \rightarrow \nu_\tau$) oscillations. An analysis of the combined sub-GeV and multi-GeV data from SK was made by fitting 70 data points (5 angular $\times$ 7 momentum bins for both e-like and μ-like events) with Δm^2 and $\sin^2 2\theta$ as well as an overall normalization taken as free parameters [35, 36]. For the flux model, that of Ref. [26] was used. The best fit was obtained at $\sin^2 2\theta = 1$ and $\Delta m^2 = 2.2 \times 10^{-3}$ eV^2 with χ^2_{min}/DOF (degree of freedom) = 65.2/67. The resulting zenith-angle distributions are shown by the dotted lines in Figure 9, which reproduce the observed behavior of the data well. The 68%, 90%, and 99% CL allowed regions in the $(\sin^2 2\theta, \Delta m^2)$ plane are shown in Fig. 10. The allowed region of Δm^2 is shifted toward the lower values compared to the Kamiokande's allowed region, but there is an overlap region, $\sim 5 \times 10^{-3}$ eV^2, at 90% CL level.

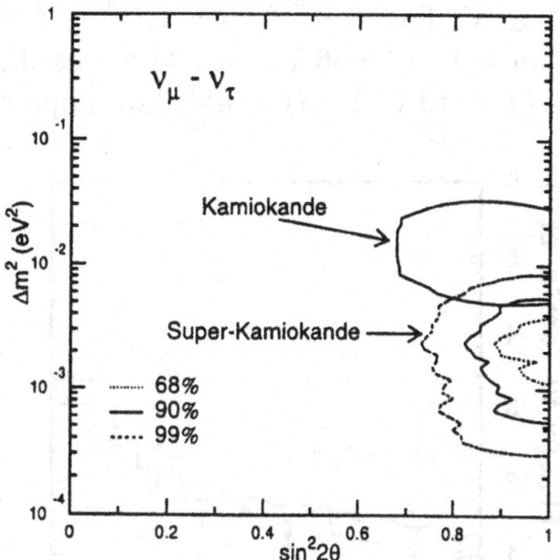

Figure 10. The 68%, 90%, and 99% CL allowed regions for the $\nu_\mu \rightarrow \nu_\tau$ hypothesis are shown for the Super-Kamiokande atmospheric-neutrino results. Also shown is the 90% CL allowed region for the Kamiokande results.

Another hypothesis, $\nu_\mu \rightarrow \nu_e$ was also tested taking the matter effect into account. In this case, the zenith-angle distribution of e-like events will show a deviation from the expectation with null oscillation. The result was χ^2_{min}/DOF $= 87.8/67$ at $\sin^2 2\theta = 0.93$ and $\Delta m^2 = 3.2 \times 10^{-3}$ eV^2. Therefore, this hypothesis is less favored. For null hypothesis ($\sin^2 2\theta = 0$ and $\Delta m^2 = 0$), χ^2/DOF $= 135/69$.

SK Collaboration concludes [35, 36] that the atmospheric-neutrino data give evidence for neutrino oscillations. The data are consistent with $\nu_\mu \rightarrow \nu_\tau$ with $\sin^2 2\theta > 0.82$ and $5 \times 10^{-4} < \Delta m^2 < 6 \times 10^{-3}$ eV^2 at 90% CL. Though the SK Collaboration did not show an analysis to test the $\nu_\mu \rightarrow \nu_s$ hypothesis, an equally good fit to the data is expected. Therefore, other measures are needed to discriminate between the $\nu_\mu \rightarrow \nu_\tau$ and $\nu_\mu \rightarrow \nu_s$ possibilities. This will be discussed later in section 6.

5.2.4. *Upward-Going Muons: Supporting Evidence for Oscillations*

The fluxes of high-energy atmospheric ν_μs can be measured with underground detectors through the detection of muons produced by the CC interactions of ν_μs in the rock surrounding the detector. However, the neutrino-induced downward-going muons are difficult to differentiate from cosmic-ray muons. On the contrary, upward-going muons are considered to be neutrino-induced, since upward-going cosmic-ray muons range out in the Earth.

Having observed the zenith-angle dependence of μ-like events indicating neutrino oscillations, that of higher-energy ν_μ-induced events is interesting to see. The mean energy of neutrinos which produce upward through-going muons is about 100 GeV. For $\Delta m^2 \sim 5 \times 10^{-3}$ eV^2 and large mixing, almost horizontal muons ($L \sim 1,000$ km) would not oscillate, but vertically upward-going muons ($L \sim 10,000$ km) would show appreciable oscillations.

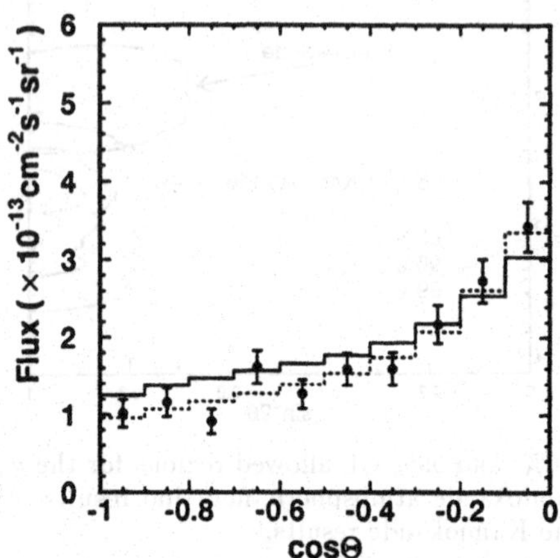

Figure 11. Super-Kamiokande results of the zenith-angle distribution of upward through-going muons. The solid line shows the expected distribution with no neutrino oscillation effects, while the dashed line shows that for $\sin^2 2\theta = 1.0$ and $\Delta m^2 = 2.5 \times 10^{-3}$ eV^2.

At Neutrino '98 were presented not only the SK [35], but also the Kamiokande's new results [35, 37] on the zenith-angle distributions of upward through-going muons. Both results show the zenith-angle distribution expected from neutrino oscillations. Figure 11 shows the SK data. A total of 617 events were observed during 537 live days. The 90% CL allowed parameter regions are shown in Figure 12. These regions are wider than the region allowed by the SK sub-GeV and multi-GeV data. However, they nicely overlap. Here, the flux model of Ref. [26] is used with an overall flux normalization taken to be a free parameter in the fits.

The SK Collaboration also presented the data of those upward-going muons that stop in the detector. The upward stopping muons have a mean energy of ~10 GeV, substantially lower than that of the upward through-going muons. For $\Delta m^2 \sim 5 \times 10^{-3}$ eV^2 and large mixing, the stopping/through-going flux ratio is expected to be smaller than that calculated with null hypothesis. In fact, the observed ratio was $0.22 \pm 0.023 \pm 0.014$, compared to the predicted ratio of 0.39 ± 0.05 for no oscillations. The 90% CL allowed parameter region was calculated based on the zenith-angle distribution of this ratio, and is shown in Figure 12. This region is also wider than, but consistent with, the region allowed by the SK sub-GeV and multi-GeV data.

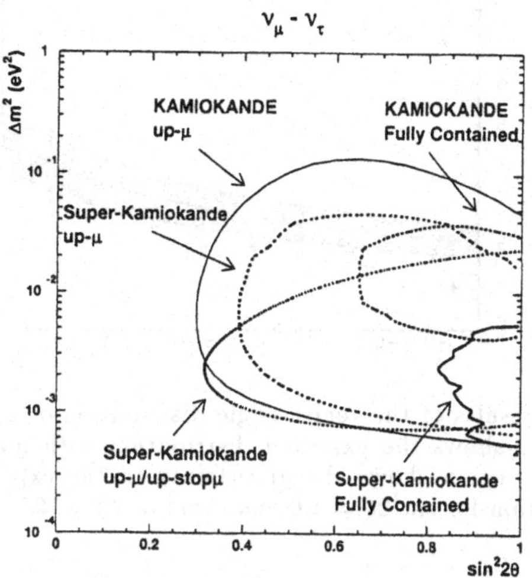

Figure 12. The neutrino-oscillation parameter regions allowed for $\nu_\mu \rightarrow \nu_\tau$ at 90% CL by the Kamiokande and Super-Kamiokande atmospheric-neutrino data.

Yet another supporting evidence for neutrino oscillations came from the MACRO underground experiment [38, 39] at Gran Sasso. Figure 13 shows the MACRO results of zenith-angle distribution of upward through-going muons. A total of 479 events were observed including 28 estimated background events. The ratio of the observed to the expected number of events based on the flux model of Ref. [27] is $0.74 \pm 0.036 \pm 0.046 \pm 0.13$. The last error is a theoretical uncertainty. The $\nu_\mu \rightarrow \nu_\tau$ neutrino-oscillation hypothesis was tested based both on the number of events and on the angular distribution. Both methods gave $\sin^2 2\theta = 1.0$ and Δm^2 of a few times 10^{-3} eV^2, supporting the SK and Kamiokande results on neutrino oscillations.

5.3. SOLAR NEUTRINO EXPERIMENTS

5.3.1. *Introduction*

The Sun is a main-sequence star at a stage of stable hydrogen burning. It produces an intense flux of electron neutrinos as a consequence of nuclear fusion reactions which generate solar energy, and whose combined effect is

$$4p + 2e^- \rightarrow {}^4\text{He} + 2\nu_e + 26.73 \text{ MeV} - E_\nu \tag{6}$$

where E_ν represents the energy taken away by neutrinos, with an average value being $< E_\nu > \sim 0.6$ MeV. Each neutrino-producing reaction and the

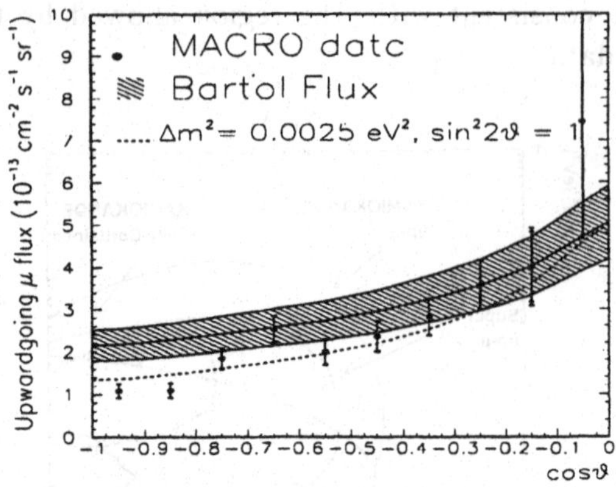

Figure 13. MACRO results of the zenith-angle distribution of upward through-going muons. The solid curve shows the expected distribution with no neutrino oscillation effects and the shaded region shows the uncertainty in the expectation. The dashed curve shows the prediction for $\sin^2 2\theta = 1.0$ and $\Delta m^2 = 2.5 \times 10^{-3}$ eV^2.

resulting flux predicted by the standard solar model (SSM) calculation by Bahcall and Pinsonneault (BP95) [40] —the best solar model, as they claim, with helium and heavy-element diffusion— are listed in Table 3. Bahcall *et al.* (BP98) [41] recently updated this SSM calculation. Table 3 also lists the BP98 fluxes. The ^{8}B solar-neutrino flux has been changed appreciably. This is principally due to the adoption of the lower ^{7}Be $(p,\gamma)\,^8$B cross section [41]. Figure 14 shows the energy spectra of solar neutrinos from these reactions.

TABLE 3. Neutrino-producing reactions in the Sun (the first column) and their abbreviations (second column). The neutrino fluxes predicted by Bahcall and Pinsonneault (BP95) [40] and Bahcall *et al.* (BP98) [41] are listed in the third and fourth columns, respectively.

Reaction	Abbr.	BP95	BP98
$pp \to de^+\nu$	pp	$5.91(1.00^{+0.01}_{-0.01}) \times 10^{10}$	$5.94(1.00^{+0.01}_{-0.01}) \times 10^{10}$
$pe^-p \to d\nu$	pep	$1.40(1.00^{+0.01}_{-0.02}) \times 10^{8}$	$1.39(1.00^{+0.01}_{-0.01}) \times 10^{8}$
^{3}He $p \to {}^4$He $e^+\nu$	hep	1.21×10^3	2.10×10^3
^{7}Be $e^- \to {}^7$Li $\nu(\gamma)$	^{7}Be	$5.15(1.00^{+0.06}_{-0.07}) \times 10^{9}$	$4.80(1.00^{+0.09}_{-0.09}) \times 10^{9}$
^{8}B $\to {}^8$Be$^* e^+\nu$	^{8}B	$6.62(1.00^{+0.14}_{-0.17}) \times 10^{6}$	$5.15(1.00^{+0.19}_{-0.14}) \times 10^{6}$
^{13}N $\to {}^{13}$C $e^+\nu$	^{13}N	$6.18(1.00^{+0.17}_{-0.20}) \times 10^{8}$	$6.05(1.00^{+0.19}_{-0.13}) \times 10^{8}$
^{15}O $\to {}^{15}$N $e^+\nu$	^{15}O	$5.45(1.00^{+0.19}_{-0.22}) \times 10^{8}$	$5.32(1.00^{+0.22}_{-0.15}) \times 10^{8}$
^{17}F $\to {}^{17}$O $e^+\nu$	^{17}F	$6.48(1.00^{+0.15}_{-0.19}) \times 10^{6}$	$6.33(1.00^{+0.12}_{-0.11}) \times 10^{6}$

Observations of solar neutrinos directly addresses the SSM and, more generally, the theory of stellar structure and evolution which is the basis of

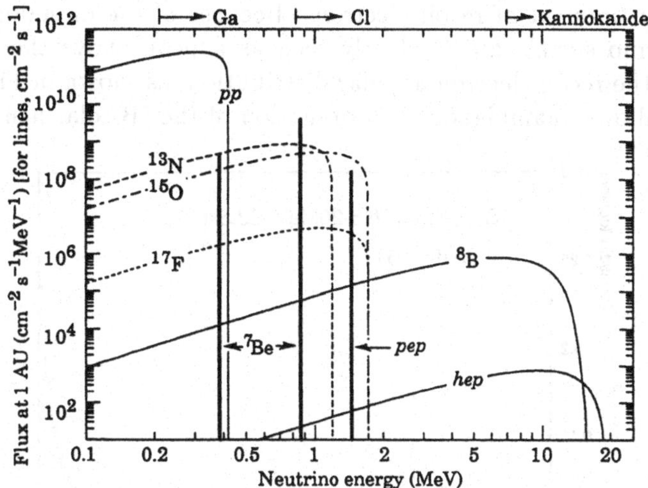

Figure 14. The solid curves show the energy spectra of solar neutrinos from the pp chain and the dashed curves show those from the CNO cycle. The detectable energy regions in various solar neutrino experiments are also shown. The fluxes are given at 1 AU (astronomical unit, the average Sun-Earth distance).

the SSM. The Sun as a well defined neutrino source also provides extremely important opportunities to investigate neutrino oscillations because of the wide range of matter density and the very long distance from the Sun to the Earth.

5.3.2. *Average Flux*

Solar neutrinos have been observed by five experiments, Homestake chlorine experiment ($\nu_e\,^{37}\mathrm{Cl} \to e^-\,^{37}\mathrm{Ar}$), SAGE and GALLEX gallium experiments ($\nu_e\,^{71}\mathrm{Ga} \to e^-\,^{71}\mathrm{Ge}$), and Kamiokande and SK νe scattering experiments.

In the chlorine and gallium experiments, the number of solar neutrinos $N_{\rm obs}$ detected in unit time is given by

$$N_{\rm obs} = N_{\rm atom} \sum_i \int_{E_{\rm th}} \phi_i(E)\sigma(E)dE \tag{7}$$

where $N_{\rm atom}$ is the number of target atoms contained in the detector, $E_{\rm th}$ the reaction threshold, $\phi_i(E)$ the solar-neutrino flux from a fusion reaction (specified by a suffix i) in the Sun, and $\sigma(E)$ the neutrino-capture cross section. The capture rate is given by $N_{\rm obs}/N_{\rm atom}$, and is represented by a conventional unit, SNU (Solar Neutrino Unit, 1 SNU $= 10^{-36}$ capture/atom/s).

Kamiokande and SK solar-neutrino observations by means of νe scattering are characterized by (i) realtime observation, (ii) strong directional correlation between the incoming neutrino and recoil electron, and (iii) energy measurement of recoil electrons. Because of the directionality, the solar-neutrino signal can be clearly seen as a peak above the flat background in the recoil-electron angular distribution, as shown in Fig. 15 [42]. This fact allows unambiguous determination of the ^{8}B solar-neutrino flux.

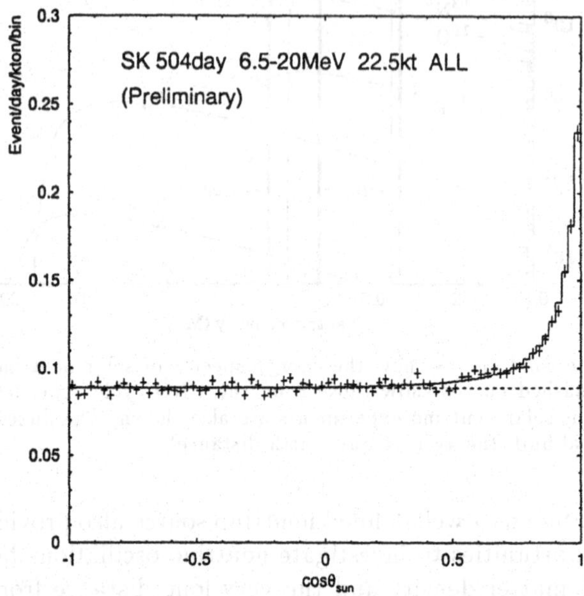

Figure 15. The directional distribution to the Sun of the recoil electrons with energies above 6.5 MeV from 504 live days of the Super-Kamiokande observation. The angle θ_{sun} is that between the electron direction and the radius vector from the Sun. One sees a clear peak of the solar-neutrino signal. The solid line shows the best fit to the data, and the dashed line shows the background.

Table 4 shows the latest average solar-neutrino flux (or capture rate for the chlorine and gallium experiments) reported from the five experiments. All these experiments have observed less solar neutrinos than expected from the calculations based on the SSM.

Two problems have been identified so far. As can be seen from Fig. 14, Kamiokande and SK observes purely ^{8}B solar neutrinos. This leads to the famous solar neutrino problem, *i. e.*, the deficit of the ^{8}B solar-neutrino flux. Analyzing both chlorine and Kamiokande results, it follows that there is strong suppression of the contribution from ^{7}Be solar neutrinos to the chlorine capture rate [47]-[52]. If the suppression factor R is defined as the ratio of the observed flux to the SSM prediction, $R(\phi_\nu(^7\mathrm{Be})) \ll R(\phi_\nu(^8\mathrm{B}))$. This conclusion is also supported by the results of the low counting rates observed by the gallium experiments. The strong suppression of the ^{7}Be solar neutrinos is the other solar neutrino problem, and it makes any astrophysical arguments to solve the solar-neutrino problems untenable simply

TABLE 4. Results of the solar-neutrino flux measurements compared with the BP95 [40] and BP98 [41] SSM calculations. The Homestake, GALLEX, and SAGE results and the corresponding calculations are given in terms of neutrino capture rates in units of SNU. The Kamiokande and Super-Kamiokande results and the corresponding calculations are given in terms of the ^{8}B solar-neutrino flux in units of $10^6 \mathrm{cm}^{-2}\mathrm{s}^{-1}$.

Experiment	Ref.	Result	BP95	BP98
Homestake	[43]	$2.56 \pm 0.16 \pm 0.16$	$9.3^{+1.2}_{-1.4}$	$7.7^{+1.2}_{-1.0}$
GALLEX	[44]	$77.5 \pm 6.2^{+4.3}_{-4.7}$	137^{+8}_{-7}	129^{+8}_{-6}
SAGE	[45]	$66.6^{+6.8}_{-7.1}{}^{+3.8}_{-4.0}$	137^{+8}_{-7}	129^{+8}_{-6}
Kamiokande	[46]	$2.80 \pm 0.19 \pm 0.33$	$6.62(1.00^{+0.14}_{-0.17})$	$5.15(1.00^{+0.19}_{-0.14})$
Super-Kamiokande	[42]	$2.44 \pm 0.05^{+0.09}_{-0.07}$	$6.62(1.00^{+0.14}_{-0.17})$	$5.15(1.00^{+0.19}_{-0.14})$

because ^{8}B is produced from ^{7}Be in the pp chain of the solar-energy generation.

Resonant neutrino oscillations in matter between two flavors, on the other hand, provide a natural and elegant solution to both of the solar neutrino problems without any *a priori* assumptions or fine tuning. Oscillations into both active and sterile neutrinos are considered. Several authors made extensive MSW anlyses using all the existing data and ended up with similar results. Two solutions called the small and large mixing-angle solutions are known. In a recent analysis, Hata and Langacker [53] used the solar-neutrino data as of 1996. They incorporated the Kamiokande's recoil-electron spectrum and day-night flux data in addition to the average fluxes. Preliminary SK data on the average flux was also taken into account. They obtained viable solutions for the BP95 SSM [40]: the small-mixing solution ($\Delta m^2 \sim 1.6 \times 10^{-5}$ eV^2 and $\sin^2 2\theta \sim 8 \times 10^{-3}$) and the large mixing-solution ($\Delta m^2 \sim 1.6 \times 10^{-5}$ eV^2 and $\sin^2 2\theta \sim 0.6$).

Yet another solution to the solar-neutrino problems, called "just-so" [54] solution, is provided by two-flavor neutrino oscillations in vacuum [53]. With BP95 SSM fluxes [40], disconnected regions of the acceptable solution exist at $\Delta m^2 \sim (5-8) \times 10^{-11}$ eV^2 and $\sin^2 2\theta = 0.65 - 1$.

5.3.3. *Further Results from Super-Kamiokande*

The Kamiokande and SK solar-neutrino observations produced not only the average flux but also other interesting data. Among them, the day-night flux difference and the recoil electron energy spectrum have important bearing on the discrimination of various possibilities to solve the solar-neutrino problems. For instance, the MSW small-mixing solution would be positively identified by the distortion of the recoil-electron energy spectrum.

The day-night flux difference, if any, would be caused by the regeneration of ν_es by the Earth. Should the MSW large-mixing solution be correct, an observable day-night flux difference would be expected. However, the SK result on the day-night flux difference from the 504 live days of observation is $(D-N)/(D+N) = -0.023 \pm 0.020 \pm 0.014$ [42]. No statistically significant day-night flux difference was observed and, therefore, the MSW large-mixing solution is not favored though it is not yet completely excluded [42].

The SK recoil-electron energy spectrum normalized to the prediction of BP95 SSM calculation [40] is shown in Fig. 16. Here, the inner error bars show statistical errors and the outer error bars show a quadratic sum of the statistical and systematic errors. Although errors are large, the data points above 13 MeV are appreciably higher than the data below 13 MeV. This is suggestive of the spectrum distortion, and, therefore, neutrino oscillations. The "just-so" vacuum solution is favored at ~5% CL, but the MSW small-mixing solution is still acceptable at ~1% CL [42]. To conclude anything, the SK needs more statistics and significance.

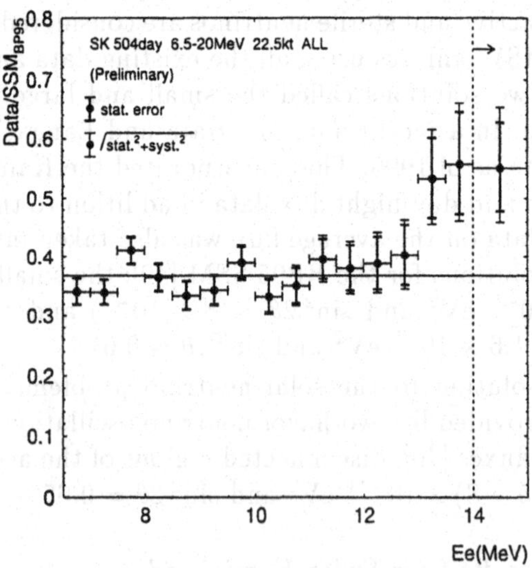

Figure 16. Recoil-electron energy spectrum observed by Super-Kamiokande with 504 live days. The inner error bars are the statistical errors and the outer error bars are the quadratic sum of the statistical and systematic errors.

6. Implications

The results of the recent neutrino-oscillation experiments are summarized in Table 5. As has been discussed, there are three evidences for neutrino oscillations; $\nu_\mu \to \nu_e$ with $\Delta m^2 \sim 1$ eV^2 (LSND), $\nu_\mu \to \nu_\tau$ or ν_s with $\Delta m^2 \sim$ a few $\times 10^{-3}$ eV^2 (SK contained events and upward-going muons observed by Kamiokande, SK, and MACRO), and $\nu_e \to \nu_x$ with $\Delta m^2 \sim 10^{-5}$ eV^2 or 10^{-10} eV^2 (solar-neutrino experiments). However, with the three neutrino species there are only two independent Δm^2 values. Consequently, if all these evidences are real, an additional neutrino is required. This neutrino should be sterile so as not to couple to the Z-boson. There are at least two possibilities: (i) atmospheric, $\nu_\mu \to \nu_\tau$, and solar, $\nu_e \to \nu_s$ and (ii) atmospheric, $\nu_\mu \to \nu_s$, and solar, $\nu_e \to \nu_\tau$.

TABLE 5. Summary of the recent neutrino-oscillation exeriments. For experiments with "No evidence," the limit for $\sin^2 2\theta$ is given at large Δm^2 and that for Δm^2 is given at $\sin^2 2\theta = 1$.

Experiment			$\sin^2 2\theta$	Δm^2 (eV^2)
CHORUS	$\nu_\mu \to \nu_\tau$	No evidence	$< 1.3 \times 10^{-3}$	< 1
NOMAD	$\nu_\mu \to \nu_\tau$	No evidence	$< 2.2 \times 10^{-3}$	< 1
LSND	$\bar\nu_\mu \to \bar\nu_e$	Evidence	$2 \times 10^{-3} \sim 3 \times 10^{-2}$	$0.3 \sim 2$
	$\nu_\mu \to \nu_e$	Evidence	$5 \times 10^{-4} \sim 3 \times 10^{-2}$	$0.3 <$
KARMEN2	$\bar\nu_\mu \to \bar\nu_e$	No evidence	$< 3 \times 10^{-3}$	$< 7 \times 10^{-2}$
CHOOZ	$\bar\nu_e \to \bar\nu_x$	No evidence	< 0.2	$< 9 \times 10^{-4}$
SK	$\nu_\mu \to \nu_\tau$	Evidence	$0.82 <$	$5 \times 10^{-4} \sim 6 \times 10^{-3}$
	$\nu_\mu \to \nu_s$			
up-going μ	$\nu_\mu \to \nu_\tau$	Evidence	~ 1	a few $\times 10^{-3}$
solar ν	$\nu_e \to \nu_x$	Evidence	$\sim 8 \times 10^{-3}$	$\sim 10^{-5}$, or
			$0.65 \sim 1$	$\sim 10^{-10}$

Recently, there have appeared a number of theoretical attempts to explain all the observed features of neutrino-oscillation experiments by introducing one or more sterile neutrino species. However, it is out of the scope of this lecture to go into these theoretical attempts. Here, let me mention how the possibilities of $\nu_\mu \to \nu_\tau$ and $\nu_\mu \to \nu_s$ can be distinguished in the atmospheric-neutrino observations.

These two cases are distinguishable based on the fact that ν_s does not interact with matter. Consequently, while there are no observable matter effects for $\nu_\mu \to \nu_\tau$, the $\nu_\mu \to \nu_s$ oscillations induce matter effects that may be observable. The matter effects modify the energy and zenith-angle dependence of the oscillation probability from that of vacuum oscillations.

High statistics studies of multi-GeV atmospheric-neutrino events [55] or upward-going muons [56] will allow to distinguish the two cases.

Another method is the study of isolated π^0 events [35, 57] which are mainly produced in the NC interactions. For $\nu_\mu \rightarrow \nu_\tau$, the ratio $(\pi^0)_{\mathrm{DATA}}/(\pi^0)_{\mathrm{MC}} = 1$, whereas for $\nu_\mu \rightarrow \nu_s$, $(\pi^0)_{\mathrm{DATA}}/(\pi^0)_{\mathrm{MC}} < 1$. With an exposure of 33 kton·yr, SK observed 210 isolated π^0 events compared to 192.5 events predicted by Monte Carlo [35]. At present, a limitation of this method comes from the uncertainties in the neutrino-induced single π^0 production cross section. In the K2K experiment (see the next section), this cross section will be measured with a water-Cherenkov detector similar to SK. This leads to a significant reduction of the present systematic uncertainty. Therefore, with the future increase of the observed isolated π^0 events in SK, this method will be promising.

7. Future of Neutrino-Oscillation Experiments

7.1. ACCELERATOR EXPERIMENTS

Neutrino beams produced by high-energy accelerators have energies of 1 $\sim$ a few times 10 GeV. To test the evidence of neutrino oscillations obtained from the atmospheric-neutrino observations with accelerator-produced neutrino beams, a baseline distance of 100 $\sim$ 10,000 km is required. The τ production threshold of the neutrino beam energy is 3.5 GeV. Below this energy, the $\nu_\mu \rightarrow \nu_\tau$ experiment is actually a disappearance experiment, $\nu_\mu \rightarrow \nu_x$. There are three long baseline neutrino-oscillation experiments either under construction or under planning.

K2K (KEK-to-Kamioka) is a $\nu_\mu \rightarrow \nu_e$ appearance and $\nu_\mu \rightarrow \nu_x$ disappearance experiment with a distance of 250 km and an average neutrino energy of 1.4 GeV. This is a two-detector experiment. The far detector is SK and the near detector consists of a 1,000-ton water-Cherenkov counter and additional fine-grained components. At 90% CL, K2K is sensitive to Δm^2 down to 2×10^{-3} eV^2 at $\sin^2 2\theta = 1$. The near detector also measures NC single π^0 production cross section. This experiment is under preparation, and data-taking is scheduled to start in 1999.

MINOS is a long baseline experiment from Fermilab to the Soudan mine in Minnesota. The baseline distance is 730 km. A 120-GeV proton beam from Main Injector will be used for the neutrino production. It is planned that MINOS should start data-taking in 2002. As of mid-1998, the best beam energy and detector configuration are still under discussion.

In Europe, a long baseline experiment between CERN and Gran Sasso with a distance of again 730 km is under discussion. A 400-GeV proton beam from SPS will produce a neutrino beam with an average energy of $\sim$25 GeV. For the far detector, there are several proposals. Among them,

ICARUS is a multi-kton liquid-argon TPC and OPERA is an emulsion-based detector, both capable of positive identification of τ and, therefore, $\nu_\mu \to \nu_\tau$ appearance experiment. It is hoped that the experiment will start in 2003.

To test the LSND evidence for $\bar{\nu}_\mu \to \bar{\nu}_e$ and $\nu_\mu \to \nu_e$, the BooNE experiment is proposed to use a $1 \sim 2$ GeV neutrino beam from the Fermilab 8-GeV booster. The detector is dilute liquid scintillator, similar to LSND. Phase I of BooNE (MiniBooNE) was approved by Fermilab, and it is hoped to start data-taking in 2001. MiniBooNE expects ~1,000 events if the LSND evidence is real.

7.2. REACTOR EXPERIMENTS

CHOOZ will further accumulate statistics and reduce systematic uncertainties. The Palo Verde experiment in Arizona is another reactor neutrino-oscillation experiment to search for $\bar{\nu}_e \to \bar{\nu}_x$ in the parameter region similar to CHOOZ. The location of this experiment is much shallower (32 mwe) than CHOOZ. It uses a segmented liquid-scintillator detector with a fiducial mass of 12 tons. The Palo Verde detector started running at the end of May, 1998.

The construction of a new 1,000-ton, ultra-pure liquid-scintillation detector, Kamland has started in 1998 at Kamioka. There are power reactors distributed at $150 \sim 200$ km from Kamioka. The sensitivity of Kamland to Δm^2 therefore approaches to the solar-neutrino region, 10^{-5} eV^2. The completion of the detector construction is scheduled at the end of 2000.

7.3. SOLAR-NEUTRINO EXPERIMENTS

Assuming that the solution to the solar-neutrino problem is really given by neutrino oscillations, how can one discriminate different scenarios? The measurements of the solar-neutrino energy spectrum and the day-night flux difference, and the measurement of solar-neutrino flux by utilizing NC reactions are the key issues. The MSW small-mixing solution causes the energy-spectrum distortion. If this is the true solution, the high-statistics of SK would allow to identify it in a few years.

SNO uses 1,000 tons of heavy water (D_2O) to measure solar neutrinos through both inverse beta decay ($\nu_e d \to e^- pp$) and the NC interaction ($\nu_x d \to \nu_x pn$). In addition, νe scattering events will also be measured. SNO can directly measure the energy spectrum of the ^{8}B solar neutrinos through CC interctions. Furthermore, SNO can measure the ^{8}B solar-neutrino flux with both CC and NC interactions. If the flux observed by the CC interactions is smaller than the flux observed by the NC interactions, undoubtedly it signifies neutrino oscillations either in matter or in vacuum. If the CC flux is identical with the NC flux, it means either the standard neutrinos (in this case, other mechanisms must be sought for the cause of the solar-neutrino problems) or neutrino oscillations into sterile neutrinos. The SNO detector has been completed in April and water fill started in May, 1998.

Borexino is an experiment at Gran Sasso with 300 tons of ultra-pure liquid scintillator. The primary purpose of this experiment is the measurement of the ^{7}Be solar-neutrino flux, whose possible deficit is now a key question, by lowering the detection threshold for the recoil electrons to 250 keV. Also, vacuum neutrino oscillations cause seasonal variation of the ^{7}Be solar-neutrino flux. The Borexino detector is under construction, and the experiment is expected to start in the year of 2000.

It is hoped that these second-generation experiments, SK, SNO, and Borexino, will finally provide the key to solving the different solar-neutrino problems raised by the first-generation experiments.

8. Conclusion

After about 70 years since the Pauli's hypothesis on its existence and more than 40 years since its discovery, we are finally approaching the elusive neutrino to unveil its secret of mass. There are now three evidences for neutrino oscillations. The most striking evidence is the zenith-angle distribution of the atmospheric-neutrino-induced μ-like events observed in Super-Kamiokande. However, three independent values of Δm^2 cannot be accommodated by the known three species of neutrino. Sterile neutrino which mixes with some species of ordinary neutrino has been invoked for rescue.

One may feel something unnatural with the mixing with sterile neutrino. Since the evidences are all challenged by other new experiments, we look forward to learning outcomes of these new experiments as well as further results from the ongoing experiments, which will eventually settle the problem and may bring new surprizes. Once established, the neutrino mass will lead us not only to a new era of neutrino physics, but also to a wealth of new physics beyond the Standard Model.

Acknowledgments

I would like to thank Tom Ferbel for inviting me to this stimulating and enjoyable School and for his kind hospitality. I also appreciate numerous discussions with, and useful information from, the members of the Super-Kamiokande and K2K Collaborations.

References

1. J.M. Conrad, M.H. Shaevitz, and T. Bolton, Rev. Mod. Phys. 70 (1998) 1341.

2. L. Nodulman, in these Proceedings.
3. D. Karlen, Eur. Phys. J. C 3 (1998) 319.
4. M. Nakamura, talk presented at Neutrino '98, Takayama, June, 1998.
5. Particle Data Group, Eur. Phys. J. C 3 (1998) 1.
6. D.E. Groom, Eur. Phys. J. C 3 (1998) 312.
7. V.M. Lobashev, talk presented at Neutrino '98, Takayama, June, 1998.
8. C. Weinheimer, talk presented at Neutrino '98, Takayama, June, 1998.
9. K. Assamagan *et al.*, Phys. Rev. D 53 (1996) 6065.
10. ALEPH Collaboration, R. Barate *et al.*, Eur. Phys. J. C 2 (1998) 395.
11. S.P. Mikheyev and A. Yu. Smirnov, Yad. Fiz. 42 (1985) 1441 [Sov. J. Nucl. Phys. 42 (1985) 913]; Nuovo Cimento 9C (1986) 17.
12. L. Wolfenstein, Phys. Rev. D 17 (1978) 2369; Phys. Rev. D 20 (1979) 2634.
13. See, for example, C.W. Kim and A. Pevsner, *Neutrinos in Physics and Astrophysics* (Harwood Academic Publishers, 1993).
14. See, for example, T.K. Kuo, Rev. Mod. Phys. 61 (1989) 937.
15. E. Eskut *et al.*, Phys. Lett. B 424 (1998) 202.
16. J. Altegoer *et al.*, Phys. Lett. B 431 (1998) 219.
17. O. Sato, talk presented at Neutrino '98, Takayama, June, 1998.
18. J.J. Gómez-Cadenas, talk presented at Neutrino '98, Takayama, June, 1998.
19. C. Athanassopoulos *et al.*, Phys. Rev. Lett. 81 (1998) 1774.
20. C. Athanassopoulos *et al.*, Phys. Rev. Lett. 77 (1996) 3082.
21. L.Borodovsky *et al.*, Phys. Rev. Lett. 68 (1992) 274.
22. B. Achkar *et al.*, Nucl. Phys. B 434 (1995) 503.
23. B. Armbruster *et al.*, Phys. Lett. B 348 (1995) 19.
24. B. Zeitnitz, talk presented at Neutrino '98, Takayama, June, 1998.
25. M. Apollonio *et al.*, Phys. Lett. B 420 (1988) 416.
26. M. Honda *et al.*, Phys. Rev. D 52 (1995) 4985.
27. G. Barr *et al.*, Phys. Rev. D 39 (1989) 3532; V. Agrawal *et al.*, Phys. Rev. D 53 (1996) 1314.
28. K.S. Hirata *et al.*, Phys. Lett. B 205 (1988) 416.
29. K.S. Hirata *et al.*, Phys. Lett. B 280 (1992) 146.
30. Y. Fukuda *et al.*, Phys. Lett. B 335 (1994) 237.
31. D. Casper *et al.*, Phys. Rev. Lett. 66 (1991) 2561; R. Becker-Szendy *et al.*, Phys. Rev. D 46 (1992) 3720.
32. W.W.M. Allison *et al.*, Phys. Lett. B 391 (1997) 491.
33. M. Aglietta *et al.*, Europhys. Lett. 8 (1989) 611.
34. Ch. Berger *et al.*, Phys. Lett. B 245 (1990) 305.
35. T. Kajita, talk presented at Neutrino '98, Takayama, June, 1998.
36. Y. Fukuda *et al.*, Phys. Rev. Lett. 81 (1998) 1562.
37. S. Hatakeyama *et al.*, Phys. Rev. Lett. 81 (1998) 2016.
38. F. Ronga, talk presented at Neutrino '98, Takayama, June, 1998.
39. M. Ambrosio *et al.*, Phys. Lett. B 434 (1998) 451.
40. J.N. Bahcall and M.H. Pinsonneault, Rev. Mod. Phys. 67 (1995) 781.
41. J.N. Bahcall, S. Basu, and M.H. Pinsonneault, Phys. Lett. B 433 (1998) 1.
42. Y. Suzuki, talk presented at Neutrino '98, Takayama, June, 1998.
43. B.T. Cleveland *et al.*, Astrophys. J. 496 (1998) 505.
44. T. Kirsten, talk presented at Neutrino '98, Takayama, June, 1998.
45. V.N. Gavrin, talk presented at Neutrino '98, Takayama, June, 1998.
46. Y. Fukuda *et al.*, Phys. Rev. Lett. 77 (1996) 1683.
47. N. Hata, S. Bludman, and P. Langacker, Phys. Rev. D 49 (1994) 3622.
48. V. Castellani *et al.*, Phys. Lett. B 324 (1994) 425.
49. X. Shi, D.N. Schramm, and D.S.P. Dearborn, Phys. Rev. D 50 (1994) 2414.
50. W. Kwong and S.P. Rosen, Phys. Rev. Lett. 73 (1994) 369.
51. J.N. Bahcall, Phys. Lett. B 338 (1994) 276.
52. S. Parke, Phys. Rev. Lett. 74 (1995) 839.

53. N. Hata and P. Langacker, Phys. Rev. D 56 (1997) 6107.
54. S.L. Glashow and L.M. Krauss, Phys. Lett. B 190 (1987) 199.
55. R. Foot, R.R. Volkas, and O. Yasuda, Phys. Rev. D 58 (1998) 013006.
56. P. Lipari and M. Lusignoli, Phys. Rev. D 58 (1998) 073005.
57. F. Vissani and A. Yu. Smirnov, Phys. Lett. B 432 (1998) 376.

WHEN CAN YOU HAVE PARTICLE OSCILLATIONS?

AMANDA J. WEINSTEIN
Stanford Linear Accelerator Center
Stanford University, Stanford, California 94309

At this year's Summer Institute, someone asked why we consider the phenomenon of neutrino oscillation, yet never consider the possibility of oscillations between the electron, muon, and tau leptons themselves. To answer this question, we must reexamine under what conditions the concept of particle oscillation is meaningful. It is helpful to begin by considering the analogous problem of the double well potential.

In the case of the double well potential, the system has two eigenstates: an even parity eigenstate, and an odd parity eigenstate. If we take a coherent superposition of eigenstates $\psi_L = |E\rangle + |0\rangle$ we can see that this superposition mainly populates the lefthand well and leaves the righthand well empty. Likewise the coherent superposition of eigenstates $\psi_R = |E\rangle - |O\rangle$ mainly populates the righthand well and leaves the lefthand well empty. If we start, at time $t = 0$, with the ψ_L eigenstate, we find that it time evolves as

$$|E\rangle e^{\frac{-i(E_E - E_O)t}{2}} + |O\rangle e^{\frac{i(E_E - E_O)t}{2}} \tag{1}$$

up to an overall phase. Since the two eigenstates evolve with separate phases, we find that at time $t = \frac{\pi}{(E_E - E_O)}$ this state will have evolved into the coherent superposition ψ_R.

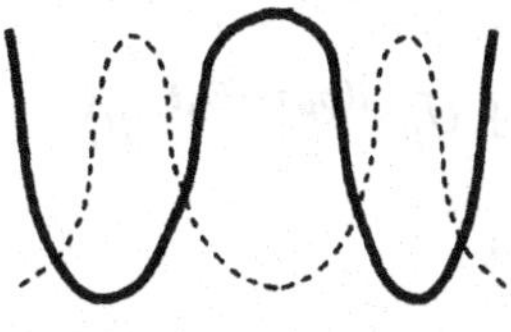

Even Parity Solution

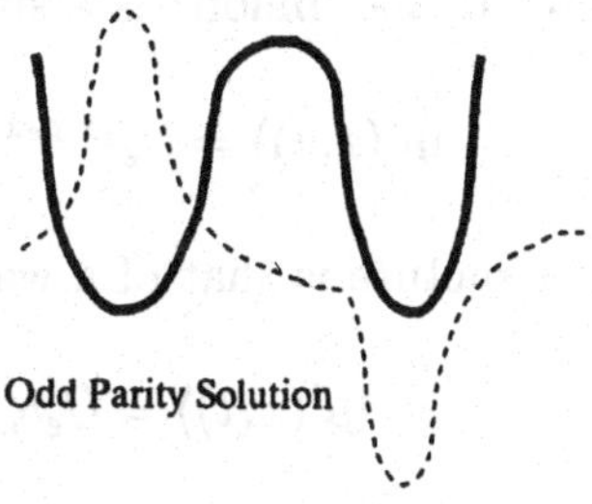

Odd Parity Solution

T. Ferbel (ed.), Techniques and Concepts of High Energy Physics X, 291–293.

So if we have a detector that detects only particles in the lefthand well (i.e. is sensitive only to the ψ_L eigenstate) at $t = 0$ that particle will initially appear to be in the lefthand well. If we look at some later time t, the state of the system will have evolved to ψ_R, which means the particle will be in the righthand well. To a person able only to detect particles in the lefthand well, the particle will appear to have vanished, only to reappear at a later time. We say that the particle represented by the state ψ_L oscillates.

Let us now turn to the familar case of neutrino oscillations. In the Standard Model, we label the electron, muon, and tau neutrinos by their association in SU(2) doublets with the electron, muon, and tau mass eigenstates. Those assignments are made on the assumption that the neutrino is massless, and therefore if the neutrinos have mass, the mass eigenstates and flavor eigenstates would not coincide. However, Standard Model production processes for neutrinos (such as decay) produce flavor eigenstates, giving a natural way, within the Standard Model, to produce a coherent superposition of neutrino mass eigenstates. Likewise, the interaction processes involved in detection also differ for different flavor eigenstates, so our detection process is also biased in favor of a particular coherent superposition of eigenstates; i.e. we are looking in one side of the well.

On the other hand, the idea of muon-electron oscillation within the Standard Model is obviously unreasonable. First the electron, muon, and tau are normally defined to be the mass eigenstates. in which case they do not oscillate. We could of course do our bookkeeping in some different way, and define the electron to be some coherent superposition of lepton mass eigenstates, but there is still no Standard Model process which would produce this coherent superposition. So such bookkeeping is, in the context of the Standard Model, merely perverse.

Now let us presume that there is some way to extend the Standard Model so our bookkeeping is somewhat less perverse. For instance, let us define the 'electron' to be some superposition of electron and muon mass eigenstates, and let us define the 'muon' to be an orthogonal superposition, and let us presume that there is some decay process which preferentially produces that coherent superposition of mass eigenstates that we call the 'muon.' Can we now observe oscillation of the 'muon' into an 'electron?'

Here is where we must be careful. While it is common to represent the state vector of the 'muon' as a superposition of plane wave states:

$$|\mu'(x,t)\rangle = U_e e^{i(p_e x - E_e t)}|e\rangle + U_\mu e^{i(p_\mu x - E_\mu t)}|\mu\rangle \tag{2}$$

the correct picture is that of a wave packet

$$|\mu'(x,t)\rangle = U_e b_e(x,t)|e\rangle + U_\mu b_\mu(x,t)|\mu\rangle \tag{3}$$

where for $i = e, \mu$

$$b_i(x,t) = \oint_{-\infty}^{\infty} dp a_i(p) e^{i(p_i x - E_i t)} \tag{4}$$

and $a_i(p)$ is the amplitude to create an electron or muon mass eigenstate with momentum p. That amplitude is then defined in such a way as to give us a wave localized in space with width $\Delta\bar{p}$.

There is no time to go into detail here, but we find, if we follow the prescription in [1] that in order for oscillation to occur,

$$\sqrt{(\bar{p}+\Delta\bar{p})^2 + {m_e}^2} > \sqrt{(\bar{p}-\Delta\bar{p})^2 + {m_\mu}^2} \tag{5}$$

and vice versa. It is clear that if the masses of the two mass eigenstates were comparable (as they are in the case of the neutrinos) this condition is easily satisfied for a reasonable packet width, but in the case of the electron and muon, which are widely split in mass, the condition cannot be so satisfied.

Put in more physical terms, within the 'muon' wave packet, the states of different mass move with different momenta, and the wave packet rapidly separates into more than one packet. Because that decoherence is so rapid with respect to what would be the oscillation length of our 'muon,' it is not meaningful to talk about observing 'muon' oscillation. On the other hand, the mass splitting between the neutrinos is extremely small, so the wave packet for a particle flavor of neutrino remains coherent for a long time on the order of its oscillation length. Thus even if we could conceive of a model which allowed for preferential creation and detection of a particular superposition of lepton mass eigenstates, the mass splitting between those eigenstates is too large; the particle's wave packet rapidly decoheres and oscillation cannot be observed.

References

1. Yoshihiro Takeuchi, Yuichi Tazaki, S.Y. Tsai and Takashi Yamazaki, NUP-A-98-15,1998

SILICON DETECTORS

Technology and Applications

I. ABT
Max-Planck-Institut für Physik
Föhringer Ring 6, D-80805 München

1. Introduction

Silicon detectors have had an enormous impact on the field of high energy physics over the last 15 years. They are usually used to provide high precision tracking information. A relatively recent addition to the standard equipment of high energy physics experiments, they are now crucial for many measurements. This lecture series tries to explain what silicon detectors are, what they can do and what their future might be. No attempt of completeness is made. There are certainly many developments and applications that could or perhaps should be mentioned. However, a selection has to be made and so the author apologizes only half-heartedly. One goal of these lectures is to clarify the terms that are frequently used in connection with silicon detectors. Another goal is to explain the complexity of constructing a real device using silicon detectors and to show that many decisions have to be taken. Some guidelines on how to make the relevant decisions are also given. The intricacies of the design of a real silicon detector and its production are not discussed in full technical detail. Some selected applications are presented instead. At the end, the limitations of silicon and some commonly mentioned alternatives are discussed.

2. What are Silicon Detectors?

In principle, a silicon detector is a solid state ionization chamber. Thus it is a member of the large family of detectors based on ionization. While most of the family members work with ionization in gases, a silicon detector takes advantage of the special electronic structure of a semi-conductor.

T. Ferbel (ed.), Techniques and Concepts of High Energy Physics X, 295–333.

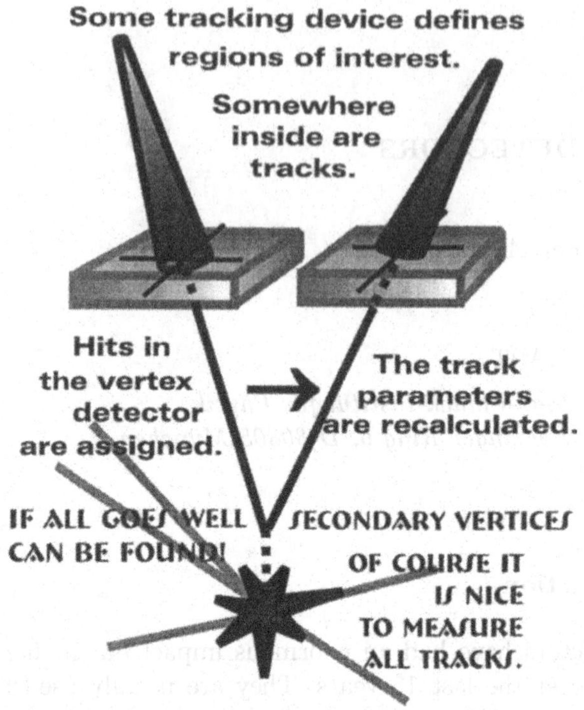

Figure 1. Principle of a vertex detector.

2.1. USAGE AS VERTEX DETECTORS

The most common application of silicon detectors in high energy physics is as active elements of vertex detection systems. Figure 1 illustrates the concept of such systems. Vertex detectors are the detector component positioned closest to the primary interaction point, also called primary vertex. Some tracking device finds tracks and these are extrapolated towards the vertex region. The extrapolations are translated into regions of interest where hits are searched for. These hits are assigned to the tracks and the track parameters are recalculated. By this, the precision is improved such that secondary vertices become distinguishable, hence the name vertex detector. Such secondary vertices are associated with the decay of particles and secondary interactions. Of course there are also tertiary and higher order vertices, all of which should and can, in principle, be identified.

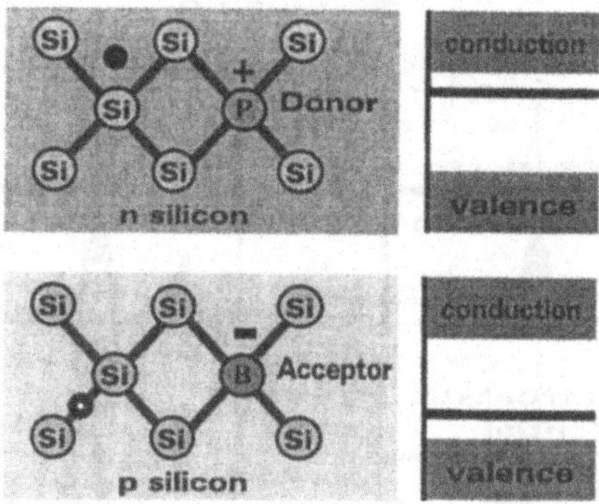

Figure 2. Semiconductor properties of defects in silicon.

2.2. SILICON, THE MATERIAL

For a good introduction into the solid state physics of semiconductors, please have a look at Ref. [1] or Ref. [2]. At room temperatures, the properties of silicon are determined by impurities(see Fig.2). Totally pure silicon would be an interesting material, but is basically unobtainable. It is easier to use the impurities and control them by doping. Silicon has four valence electrons and forms a hexagonal crystal. Defect atoms with 5 valence electrons, like phosphor, act as so called donors, as they donate an electron to the crystal. In the band structure, this electron sits within the band-gap, but close to the conduction band. Silicon with excess donors is called n-type silicon. Defect atoms with three valence atoms, like boron, act as so called acceptors. Here an electron is caught by the boron and it then also sits in the band-gap, but close to the valence band. In this case, somewhere else a hole is created due to the missing electron. Silicon with excess acceptors is called p-type silicon.

2.3. CONSTRUCTING A DIODE

A junction between p- and n-type silicon creates a diode. Figure 3 shows the electron and hole densities as well as the electrostatic potentials in a diode close to the junction. The application of an external potential as shown in Fig. 3 is called reverse biasing. The currents in an ideal diode are also described.

In an unbiased diode there are small and equal generation and recombination currents. Some holes and electrons diffuse through the potential barrier, and as a result there is a certain electron density in the p- and hole

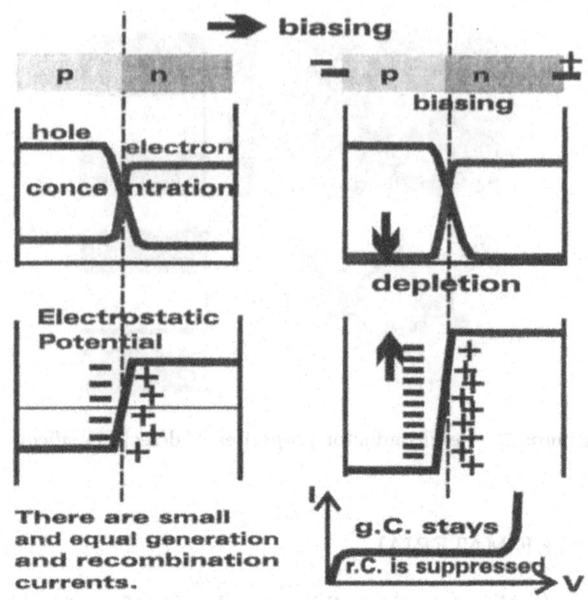

Figure 3. Electron and hole densities[top] and electrostatic potential[center] of unbiased[left] and biased[right] diode close to the junction. The behavior of generation and recombination current is indicated at the bottom.

density in the n-region. When an external potential is applied, the potential barrier becomes higher and the diffusion and thus the recombination current is suppressed. A zone depleted of all carriers forms and starts to grow. The external bias voltage at which the whole diode is depleted is called the full depletion voltage. The generation current in a perfect diode stays constant up to a voltage called break-through voltage, at which the field becomes too high for the internal structure of the diode. This current is referred to as leakage current.

2.4. ELECTRONS AND HOLES

When a charged particle traverses silicon, it produces ionizing and non-ionizing energy loss. The non-ionizing energy loss creates radiation damage(Sec. 2.5.8) and the ionization loss causes the creation of electron-hole pairs which produce the signal (see Fig. 4). The number of pairs created depends on the amount of ionization, and thereby on the absolute value of the charge and momentum of the particle, and on the thickness of the crystal. Silicon has a band gap of 1.12eV at 300°K, and a minimum ionizing particle creates on average 8000 electron-hole pairs in 100μm of silicon crystal.

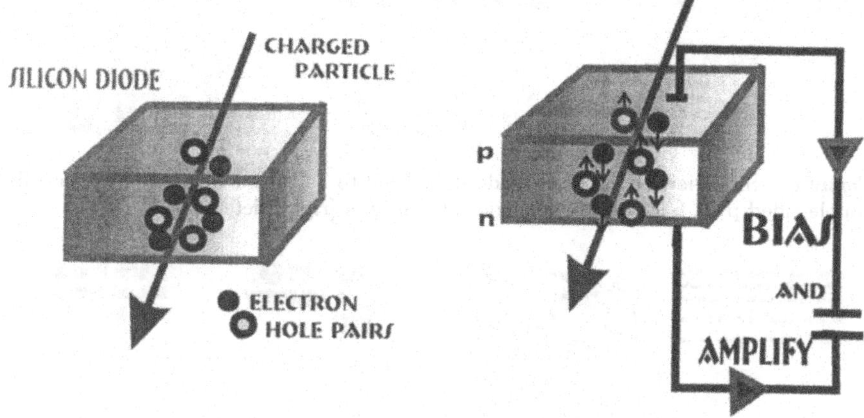

Figure 4. Creation of electron hole pairs and charge collection after biasing.

Figure 5. Side-view of a p on n diode. A typical detector is around 300μm thick. The p-implantation is around 1μm deep.

Once an external bias voltage exceeds the full depletion voltage all the created charge can be collected. The holes drift to the p-side of the diode, the electrons to the n-side.

2.5. DETECTORS

Basically, all silicon detectors are constructed as so called p on n diodes. Nothing else will be discussed In these lectures. Such a diode consists typically of an around 300μm thick n-type bulk, where on one side a layer of p^+-doped material of about 1μm thickness is implanted(see Fig. 5). p^+ denotes that the defects of the n bulk are over-compensated. Depending on the doping of the n material, a certain electron density is intrinsic before biasing. When the diode is biased, the highest fields occur on the p-side of the diode and the depletion zone grows from the p- towards the n-side.

Detectors are typically made from 4 inch wafers. The necessity to handle the wafers, and the increase of wafer imperfections near the edge, limit the

Figure 6. Side-view of a p on n diode structured to become a single-sided p on n [left], double-sided p on n [center] or a single-sided n on n [right] detector.

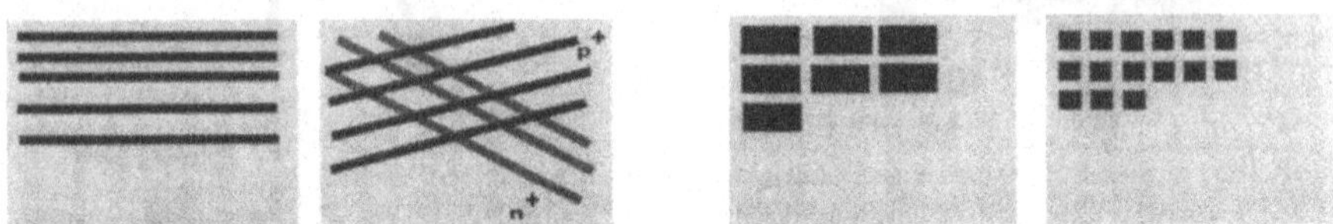

Figure 7. Top view of [from left to right] strips on a single-sided, strips on a double-sided, pads on a single-sided, pixels on a single-sided detector.

possible sizes of the resulting detectors. A maximum size of $6\times6\text{cm}^2$ or $5\times7\text{cm}^2$ is possible, depending whether a square or rectangular shape is needed.

2.5.1. *Single- and Double-sided p on n Detectors*

In order to obtain spatial resolution, the p implantation of a simple p on n diode can be structured(see left of Fig. 6). The result is a so called single-sided p on n detector. It is also possible to add a structured n^+ implantation on the n-side of the diode. The resulting double-sided detector can measure two independent projections(see Fig. 6 and 7). It is also possible to only structure the n-side of the detector, resulting in a n on n single-sided detector. However, the construction of n on n single-sided detectors requires work on the p-side of the diode. Thus, the technology of a double-sided detector is needed and has to be paid for. Figure 7 gives an overview of the most common structures:

Strips are the most common structures used on silicon detectors. They can be equally spaced or not, parallel to the edges or not, and have typical pitches between 25μm and 200μm. In principle, strips could have any form and can wind arbitrarily all over the detector. For some applications, this makes sense. However, straight strips with equal spacing are definitely the most common.

Pads and Pixels are featured on single-sided detectors. A pixel is a small pad. Typical pad sizes are $200\times200\mu\text{m}^2$ to $2\times2\text{mm}^2$. Pixels are typically $50\times\ 50\mu\text{m}^2$ to $200\times200\mu\text{m}^2$. Again, it would be easy to make any shape

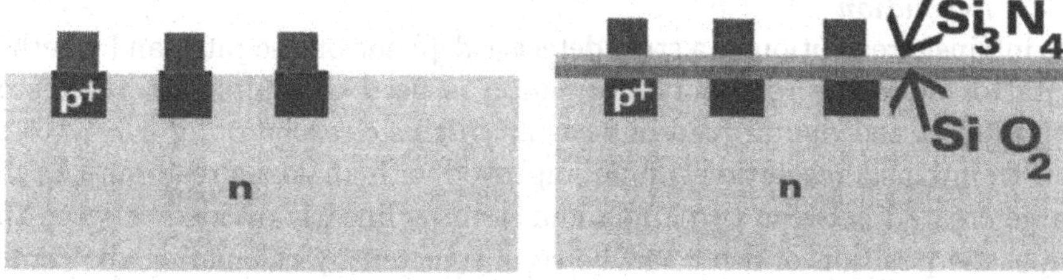

Figure 8. Side view of strips contacted directly [left] or through a SiO_2 and a Si_3N_4 layer [right].

of pad or pixel, but in practice people mostly choose rectangular pads and pixels. A special case of pixel detectors are charged-coupled devices, CCDs (see Ref. [3]).

2.5.2. *Signal Retrieval*

In order to collect a signal from an implanted structure that structure, has to be connected to the outside world(see Fig. 8). The easiest is to just put down a metal strip on top of the implantation. The disadvantage is that any current generated in the diode flows through that contact, and, if no external capacitor is used, right into the amplifier. Many amplifiers don't really like that! Some amplifiers do not mind, but generally this current still creates unwanted noise. Therefore, in most cases, capacitive coupling is chosen. In modern applications, the capacitors are integrated into the detector. This is desirable, as external capacitors are often difficult to fit, and they double the number of contacts. In the following, I will always talk about detectors with internal capacitors. Internal capacitors are built by having an oxide layer separating the implantation and the aluminum strip. Such a layer is about 50nm to 200nm thick. A failure in the oxide is called a pin-hole. Such a failure is very undesirable, as the resulting currents affect more than one strip. An extra layer of silicon nitrite (around 50nm to 100nm) can provide extra security. The resulting capacitors are good for voltage differences of up to 100 Volts(see Sec. 2.5.9). In many applications, the voltage drop across the capacitors is controlled and kept to a few volts during normal operation. For a double-sided detector, that means that the electronics has to be floating on at least one side. Only if a double-sided detector is operated at bias voltages small compared to the break-through voltage of the internal capacitors, can the electronics on both sides work with the same ground.

2.5.3. *Resolution*

The intrinsic resolution of a strip detector depends on the pitch and whether digital or analogue read-out information is used. A simple strip detector with pitch a and digital read-out has an intrinsic resolution of $\sigma = a/\sqrt{12}$.

The intrinsic resolution can be improved with analogue read-out. As the charge created between two implanted strips is linearly divided between the strips, the position of a hit can be reconstructed by calculating the center of gravity of the observed charges. However, if the implantation itself is hit, basically all the charge remains within this strip. Only a small portion is capacitively coupled to the two neighboring strips. In order to optimize the resolution for a given pitch, the implantation width should thus be small. However, that creates large gaps between implantations, where the potential is influenced by the back-side. That can cause the loss of some of the charge. In addition, the field at the point indicated by an arrow in the left picture of Fig. 9 increases. At a certain gap width, the internal p-n-junction breaks before the full depletion voltage is reached and the detector becomes inoperable. Thus the implantation width cannot be very small compared to the pitch. As a result, one has to decrease the read-out pitch in order to achieve better intrinsic resolution, or introduce intermediate strips (see Fig. 9). The pitch cannot be made arbitrarily small, because that increases the input capacitance that the read-out electronics sees, and thereby the noise(see Sec. 2.5.11).

The pitch of the silicon strip detectors for a particular application is usually chosen such, that the intrinsic resolution of the detectors is irrelevantly small. The overall resolution of the system is limited by multiple scattering. Very small pitches are mainly used to separate tracks close to each other. For pattern recognition, it is undesirable to have more than one track hit the same strip.

2.5.4. *Intermediate Strips*

Intermediate strips allow the construction of detectors with improved resolution at fixed read-out pitch.

In general, it is desirable to keep the number of read-out channels as small as possible, because read-out channels are expensive and work intensive. It is also technically difficult to achieve read-out pitches below 50μm, as there is basically no lateral space left for contacts to the outside world. In addition, it is hard to get front-end pre-amplifier chips that have input pitches of less than 50μm.

Detectors with one, two or even three intermediate strips have been used. The signal is capacitively coupled to the 2 nearest read-out strips. That distributes the signal ofr a particle traversing a region between two read-out

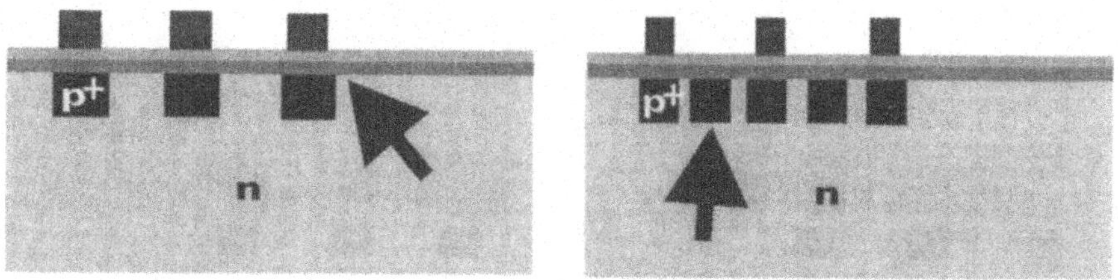

Figure 9. Side view of a detector without [left] and with [right] intermediate strips.

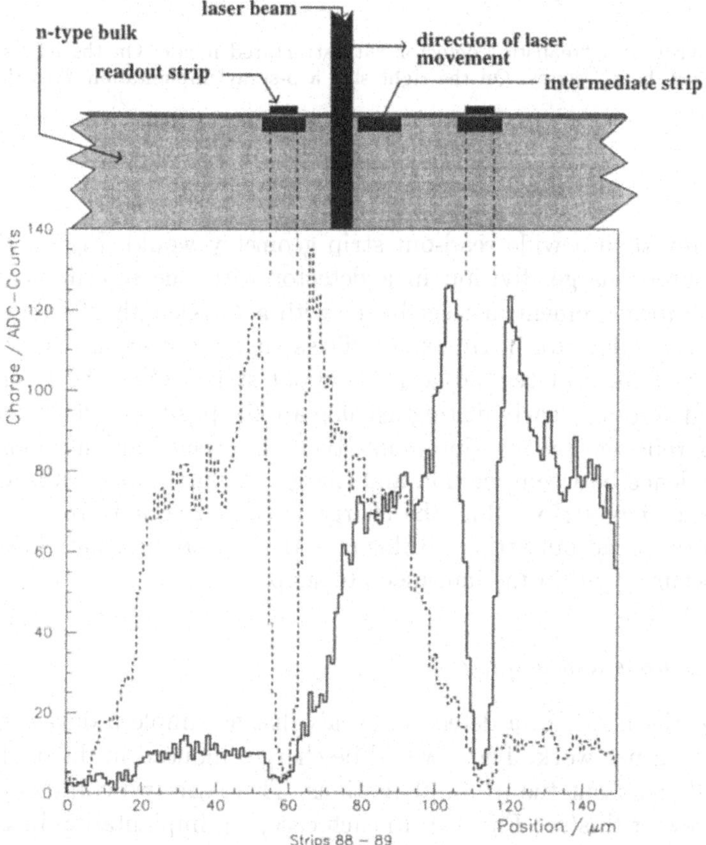

Figure 10. Measured charged division in a detector with one intermediate strip. The schematic of the detector is shown on top. A laser beam is moved across the detector. The dashed line is the charge seen by the strip on the left [#88]. The full line is the charge seen by the strip on the right [#89]. The x axis is the position of the laser beam. The deep minima right on the read-out strips result from the reflection of the laser light from the aluminum read-out strip.

Figure 11. Side view of a "realistic" detector with structured n-side. On the left, so called p-stops isolate the n^+ strips. On the right side a p-spray implantation does the same job.

strips more evenly than a wide read-out strip geometry would. Figure 10 shows the measured charge division in a detector with one intermediate strip [4]. For the measurement, a laser-beam with a wavelength of 963nm was moved across the detector in 1μm steps. The strip pitch is 55μm. Shown are the collected charges of two adjacent read-out strips while the laser-beam was moved across. Clearly distinguishable are the positions where the laser beam gets reflected by the aluminum strips. Between implantations, the linear dependence between position and charge sharing is well realized. When the intermediate strip is hit, the charge is shared equally between the two neighboring read-out strips. In this case there is no information for where the hit occured within the intermediate strip.

2.5.5. *n-sides: p-stops and p-spray*

A closer look at the n-side of a detector reveals that as simple a device as described above cannot work. There would be charges induced in the oxide layer that would short out the n^+ implantations. There are two commonly used ways to prevent this(see Fig. 11). In each case, a p implantation insulates the n^+-strips. In the p-stop version, separate implantations are used. That basically doubles the number of structures to be made. Especially with intermediate strips, that can get very crowded. In the p-spray option the whole area is implanted. This option has the principle advantage of not introducing more structures, and it has the lower internal fields. However, the doping in p-stops is easier to control during manufacturing, and at this point in time p-stops are more common than p-sprays.

2.5.6. *Biasing*

A detector has to be fully depleted in order to deliver the full signal. The full depletion voltage depends on the resistivity of the material used. It is typically between 30 Volts and 120 Volts. It is advisable to run at least 10 to 20 Volts above full depletion voltage, because a higher field speeds up the signal, which is very important for applications with fast electronics [peaking times of less than 100ns].

The individual strips are taken to the desired potential through bias resistors. This resembles any bus system used to distribute voltages. The voltage is brought from the external world to a bias structure, usually a ring, from which it is internally distributed. There are three different choices on the market: polysilicon resistors, punch-through structures and implanted resistors. Table 1 lists advantages and disadvantages of the three options.

Biasing Choice	Advantages	Disadvantages
Polysilicon	radiation hard easy to operate	takes space expensive
Punch-Through	cheap	not radiation hard difficult to operate
Implantation	cheap radiation hard	works only on p-side ⇒ single-sided detectors only

TABLE 1. Advantages and disadvantages of different bias methods.

The choice of biasing method really depends on the application and on the budget available. One of the main inputs to the decision is the radiation dose that the detector is most likely to see.

2.5.7. *I-V Characteristics*

Ever since the beginning of Sec. 2.5, I claimed that a silicon detector is basically a structured diode. Therefore, the leakage current it draws versus the bias voltage should show the typical diode behavior(see Sec. 2.3). In practice, most detectors are not perfect. And this is reflected in their IV-curves. Figure 12 shows IV-curves for three double-sided detectors. Each one has a break-through voltage much higher than the full depletion voltage. A "perfect" detector shows a small rise in current at the beginning and then a long plateau up to the break-through voltage. The small rise at the beginning is connected to the oxide charges. Some imperfections can cause the current to rise with voltage. Quite often, these rises are linear, i.e., resistive behavior is observed. Imperfections on the p-side are immediately visible, while n-side imperfections only take effect after the detector is fully depleted. Some imperfections only take effect at voltages much higher than

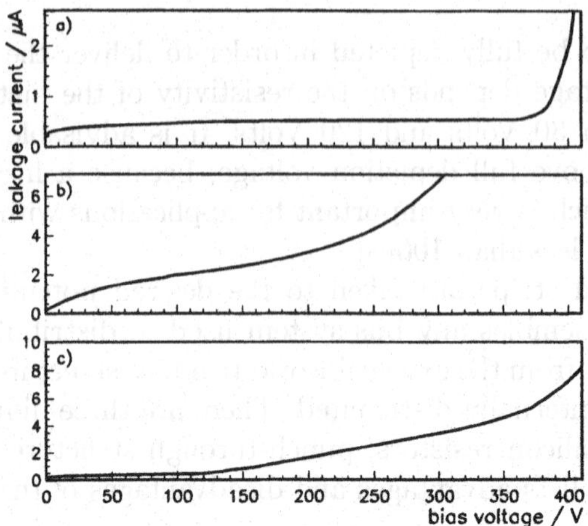

Figure 12. Typical IV-curves for silicon detectors: The curves shown are those of $5\times7\text{cm}^2$ double-sided detectors designed for the HERA-B experiment(Sec. 3.2.2). All three detectors are fully depleted at around 110V. Detector a) is as close to a diode as they come. Detector b) has an imperfection on the p-side, reflected in a steady rise of current even before it is fully depleted. Detector c) has an imperfection on the n-side, which becomes visible as soon as the depletion zone reaches it.

full depletion voltage. Such imperfections are irrelevant for detectors that are not subjected to irradiation.

2.5.8. *Radiation Damage and Full Depletion Voltage*

Silicon detectors are damaged by charged and neutral particles. The damage is caused by non-ionizing energy loss. Charged particles mainly damage the bulk material. Neutral particles, especially soft ones, also damage the surface structures.

For strip detectors, the bulk damage is the more important effect. The crystal itself is damaged such that donors are removed and acceptors are created. This happens through the dislocation of lattice atoms. A damaged crystal has some self-healing ability called annealing. However, on a longer time-scale it also gets even sicker. This is called anti-annealing. Annealing can be quite well understood and is connected to diffusion. There are several models for anti-annealing. However, none of them is quite complete or convincing.

The operational consequence of radiation damage is that at any fixed voltage the leakage current goes up as $I=I_0+I_d\cdot\Phi$, where soon the original current I_0 becomes totally negligible. The size of I_d depends on many factors, but, even after a moderate radiation dose Φ, the resulting leakage current I

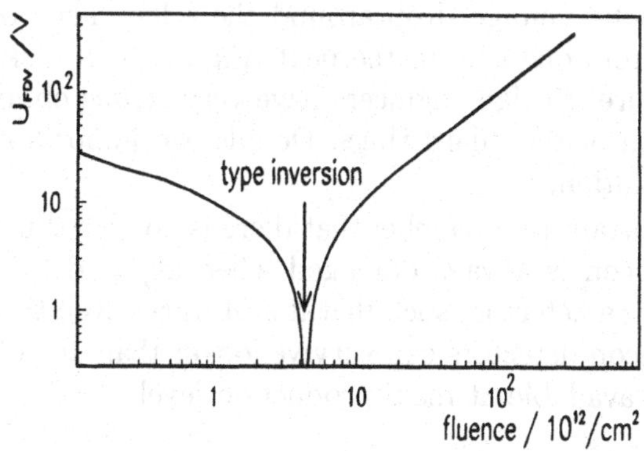

Figure 13. **Development of the full depletion voltage with integrated radiation dose, here given as the number of minimum ionizing particles per cm^2.**

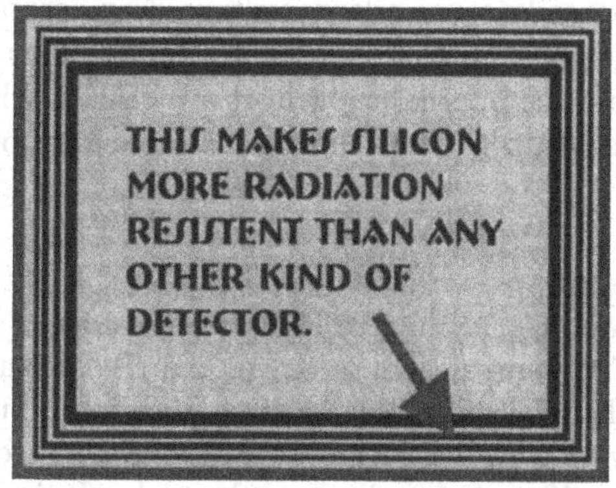

Figure 14. **Multiple guard-rings gradually take down the voltage between the active area and the edge.**

can easily increase by a factor of 100. In addition, the full depletion voltage changes. Figure 13 shows that the full depletion voltage first decreases and then, after a point called type inversion, increases. In order to still fully deplete the device, higher and higher voltages have to be applied. As real detectors are usually not perfect diodes(see Sec. 2.5.7), the leakage current increases accordingly. By how much, depends on the quality of the device. The full depletion voltage after a severe radiation dose, lets say 5×10^{14} minimum ionizing particles, can be 500 Volts or more. Detectors are made to be able to work at that kind of voltage by guard-rings that shield the active area from the voltage drop around the edge. The voltage is taken down gradually from one ring to the next. Figure 14 is a sketch of such a guard-ring structure. Design engineers have very strong opinions about the number and width of the guard-rings. Details are important, but there is more than one solution.

It is very important to remember that there is no such thing as "radiation hard" silicon. Silicon is always damaged when exposed to radiation. The trick is to design the detectors such that it still works even though damaged. Well designed silicon detectors can survive longer than any other ionization devices currently available at mass production levels.

2.5.9. *Radiation Bursts and Pin-Holes*

In some applications, silicon detectors are periodically exposed to bursts of radiation. In accelerators, this is often related to beam losses. Such bursts can create an enormous amount of charge inside the detector. This charge can dissipate only through the resistive bias structure. The time constants involved can temporarily cause a large voltage drop across the silicon-oxide and nitrite layers, which might then locally break. This creates so called pin-holes. Therefore, the internal capacitors are built to withstand relatively high voltage drops of up to 100V. External capacitors break down at far smaller voltages.

2.5.10. *Read-Out*

Some remarks about retrieving the signal were already made in Sections 2.5.2 and 2.5.4. The general goal is to get the resolution needed with the fewest read-out channels possible. Most applications, but by far not all, are best served with a capacitively coupled silicon detector and a charge integrating pre-amplifier. Depending on the technological choices and the needed resolution, a detector may have intermediate strips or not. However, quite often a read-out pitch of around 50μm or 100μm turns out to be optimal. The pre-amplifiers are usually packaged in custom made chips. The peaking time has to match the repetition rate of the experiment. The faster the pre-amplifiers have to be, the more difficult they are to get. Modern chips have typically 64 or 128 channels. Some chips have integrated pipelines that store the data for a while. This is needed for high frequency experiments where the silicon is read out only after a first level trigger has decided that the event under consideration is worth it.

The silicon detectors that are right now being conceived, or already in use, usually involve quite large systems. A double-sided detector can have 2000 or more channels, and a complete system has often more than 100 [LHC more than 1000] detectors. This creates hundreds of thousands of channels, all of which have to be connected to pre-amplifier chips, which then have to be connected to the outside world. The connections between detector and chip is made by wire bonding, often through a fan-out. A fan-out doubles the number of wire bonds, but it allows for pitch adjustment, which is sometimes necessary. Even though chips are custom made for high energy physics, and very often for a specific experiment, they are used under slightly different geometrical conditions. In most applications, there is very little space available for the pre-amplification electronics, which adds to the fun of designing a complete system.

The connections to the outside world also have to be considered carefully, especially in 4π detectors. Here the cables have to be threaded through the outer shells of the detector. The cables create holes in the acceptance and can add material in front of the outer shells. It is therefore desirable to multiplex the signals before routing them to the outside world. An intrinsic multiplexing is done for the pixels in a CCD. But CCDs are quite slow. In all applications with tight timing conditions, the speed of the link, has to increase with the number of multiplexed channels. The higher the speed of a link the more difficult it becomes. A reasonable compromise has to be made depending on the constraints of the application.

A fundamental decision to be made for any detector system is whether to read out the analogue information, and, if so, to what accuracy. In principle, the digital information "hit" or "no hit" would be enough for most applications. However, in practice a phenomenon called "common mode" makes digital read-out often unusable. When real silicon detectors are connected to real amplifier chips, the baseline of the chips can jump for some events. These jumps can be higher than the signal from a minimum-ionizing particle. This makes it impossible to adjust a single threshold for digital read-out. The reason for common mode is not well understood. The ground of the detector couples in one way or another to the ground of the chips, and the whole assembly acts as an antenna or signal generator. It is believed that the problem is best controlled by "perfect" grounds. There is no firm belief, let alone knowledge, on how to achieve these "perfect" grounds in an experimental hall. If a sufficient number of bits is read out, usually 8 is fine, it is possible to monitor the common mode and subtract it online.

2.5.11. *Signal to Noise*

The distribution of the size of the signals in a 300μm silicon detector corresponds to a Landau distribution with a mean of 24000 electrons.

The electron-equivalent noise as seen by the pre-amplifier has several components: white series, white parallel and 1/f:

$$ENC^2 = a_1 \frac{C_{inp}^2}{T_p} + a_2 I_L T_p + a_3 C_{inp}^2$$

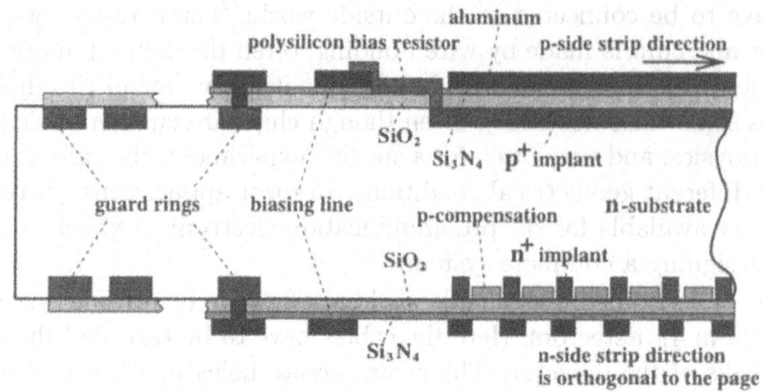

Figure 15. Cross section of a double-sided detector with polysilicon bias resistors

C_{inp} is the input capacitance, and determined by the layout of the detector and fan-out. T_p is the peaking time of the pre-amplifier; it has to be chosen to match the repetition rate of the experiment. The constants a_i are dependent on the technology of the amplifier chip. The only parameter that can be tuned during operation is I_L, the leakage current. At room temperature the leakage current through the bulk is reduced by roughly a factor 2 by cooling down by 7°C. This becomes important after irradiation.

Typically, the systems are configured such that at room temperature the signal to noise ratio for an undamaged detector comes out to 15 to 20 for fast [$T_p \approx$ 50ns], and 50 to 100 for slow, electronics.

2.5.12. *Detector Production and Prices*

The production of silicon detectors is not your typical "do it yourself job". For a small scale production of a medium-complicated device, you need a $\mathcal{O}$($15M) silicon laboratory and a lot of expertise. Even though the principles of a silicon detector are relatively straightforward, the details are very involved. Figure 15 is a drawing showing a little more technical detail. In reality, it is even more complicated, and small changes in layer thicknesses, distances between structures, doses or process temperatures, result in catastrophic failures.

There are quite reliable commercial suppliers like Hamamatsu, CSEM or Sintef [this list is incomplete by definition]. There are also unreliable suppliers which will not be listed, because my legal insurance leaves a lot to be desired. The price of a detector varies widely, depending on size and requirements. For maximum area detectors from 4 inch wafers, the following numbers are a very loose guideline. single-sided p on n detectors are between $500 and $2000. Double-sided p on n detectors are between $1000 and

\$4000. single-sided n on n detectors are almost as expensive as double-sided detectors, as both sides of the wafer have to be worked on. The set-up costs for a new line of detectors is significant. A double-sided detector requires 10 or even more layers, and for each layer a so-called mask is required. Each mask, depending on the needed accuracy, can cost more than $\mathcal{O}(\$2000)$. Total set-up charges of more than \$30,000 are not unheard of. It is clear that a large silicon system is not inexpensive. Single detectors, however, are per cm^2 even more expensive due to the set-up charge.

2.5.13. *System Costs*

In the last section it became clear that silicon costs money. But pieces of silicon don't make a detector system. In addition, a mechanical support structure and a control and read-out system is required. That also costs money.

Mechanical support structures are made out of low-Z materials, preferably with thermal expansion coefficients close to silicon and a lot of strength. Carbon fiber and graphite constructions are common. Beryllium supports are also often used. All these materials have in common the fact that they are expensive. Detector cooling is often difficult and involved. The resulting systems also have a tendency to cost a lot of money.

The read-out systems consist of pre-amplifiers, perhaps with pipelines, pitch adaptors, optical links or twisted pair cables, A/D converters, plenty of control electronics, and probably fast processors to deal with the raw data. The detectors need power-supplies, and so does all the read-out electronics. All this has to be controlled and monitored.

It can easily happen that the silicon itself does not dominate the system cost. In the case of the HERA-B silicon vertex detector system, for example(see Sec. 3.2.2), the total cost is around \$3M, while the actual silicon detectors only cost \$300k.

2.5.14. *Choices*

In the previous sections words like "typically" and "usually" were not uncommon. When building a silicon detector system, there are plenty of choices. In many cases, there is no clear "best choice". But there are usually some very bad choices. In large collaborations it can take longer to make the choices than to actually build the device. Figure 16 can be used as a guideline on how to choose the "correct" piece of silicon. It should certainly enable the reader to hold his/her own in any collaboration or other such meeting.

The standard solution is to use silicon strip detectors with a read-out pitch adjusted to the resolution requirement of the application. If the track density or the radiation level are too high, the prescribed cure is to go for

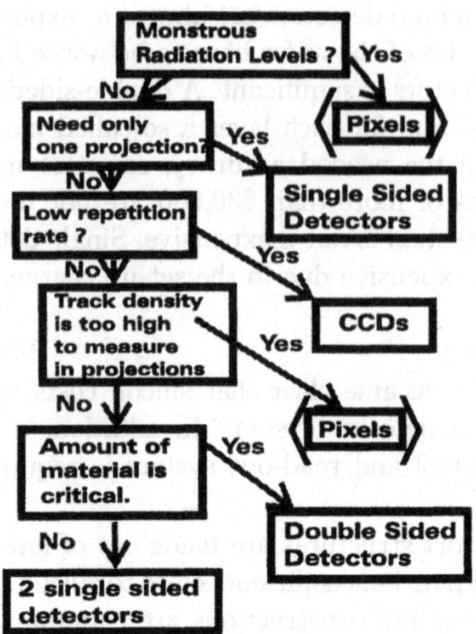

Figure 16. Decision Flow Chart. This is an extreme simplification.

pixels. It has to be noted, however, that, at the moment, no totally functional read-out scheme for pixel detectors other than CCDs exits. CCDs can only be used if the read-out rate is sufficiently low. In such a case of low read-out rate, it is always worth to explore the CCD option. The measurement of space-points is very attractive, and it makes track reconstruction easier and more efficient. Even with perfect hits in two projections, false assignments occur that can be prevented by measuring space-points(see Fig. 17).

The evaluation of different solutions involving strip detectors very often focuses on double sided versus single-sided. Many people are afraid of double-sided detectors because their operation demands a little more care and thought. Some suppliers also have problems making double-sided detectors while they are more successful with single-sided devices. Nevertheless, if multiple scattering limits the overall performance of the envisioned system, double-sided detectors should clearly be the choice.

3. Applications

Looking outside our field, the range of applications of silicon detectors is quite large. CCDs, in particular, are used in many scientific and commercial detection devices or cameras. This ranges from video cameras to X-ray detectors in satellites [5]. A review of charge-coupled devices as particle

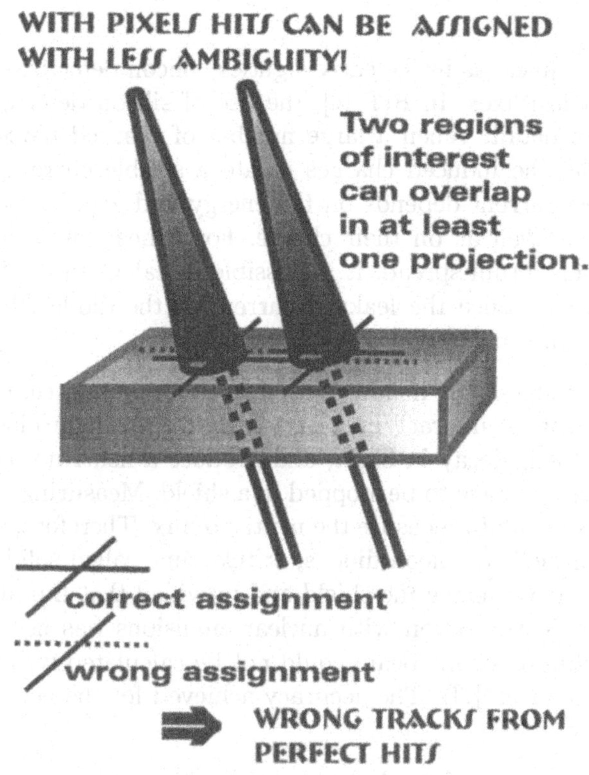
WITH PIXELS HITS CAN BE ASSIGNED
WITH LESS AMBIGUITY!
Two regions
of interest
can overlap
in at least
one projection.
correct assignment
wrong assignment
WRONG TRACKS FROM
PERFECT HITS

Figure 17. The advantage of measuring space-points. A hit, even perfectly measured in projections, has a higher probability of being assigned to the wrong track than a hit measured as a space-point.

tracking detectors is given in [3]. Strip detectors are not widely used in industry, but they are much more common in high energy physics. Basically, every modern high energy physics experiment has a silicon based component, an exception being neutrino detectors. Silicon is used mostly to satisfy the requirements of high precision tracking close to the interaction point, and as the active material in high density calorimeters.

3.1. SOME HISTORY

Silicon detectors are a relatively recent invention. Their development went in parallel to the development of integrated circuits. Without the revolutionary progress made in the last 15 years in the packaging of pre-amplifiers, the wide spread usage of silicon detectors would not have been possible.

3.1.1. *Fluxes*

In "historical" times, as in the early eighties, silicon detectors were used to measure particle fluxes. In Ref. [6], the use of silicon detectors at CERN is described in detail. When a large number of charged particles traverse a silicon diode, the induced charges create a sizable current that can be measured. The current depends on the energy and angular distribution of the particles, as well as on their charge. For a beam with a well known energy and momentum spread, it is possible to calculate the flux from the generated current, once the leakage current of the diode (that has to be subtracted) is known.

This principle was, for instance, used to calibrate neutrino fluxes. Here, the incident protons interact in a beryllium target to produce pions and kaons. The mesons decay in flight, and produce a neutrino beam. The accompanying muons have to be stopped in a shield. Measuring the muon flux in the shield is a way to measure the neutrino flux. Therefore, at the CERN neutrino beam facility, silicon diodes, at that time called solid state detectors, were positioned inside the shield and provided that flux measurement. However, a cross-calibration with nuclear emulsions was necessary, as the angular distribution of the beam could not be calculated well enough from Monte Carlo(see Ref. [7]). The accuracy achieved for the neutrino flux was about 3%.

Silicon diodes are still used as warning devices. Many experiments place diodes close to the beam-pipe and monitor their currents. Either the long-term dose is deduced from the increased leakage current, or sudden increases in current are used as an early warning against special problems such as beam-losses.

3.1.2. *Dawn of the Age of Silicon: Charm Lifetimes*

The first true silicon vertex detector was constructed in 1983 for ACCMOR [NA11], an experiment at CERN designed to measure charm lifetimes [8]. The experiment used 8 single-sided detectors with 20μm strip pitch. At that time, the read-out was a major problem. It was basically not possible to work with such a pitch in the external world. Therefore, only every third strip was read out actively. These detectors, with their analogue read-out actually achieved a spatial resolution of 5μm. Thus, from the very beginning, the point resolution of the devices was not the limiting factor for the overall performance. Figure 18 gives a schematic view of the NA11 setup, while Fig. 19 shows an event and the reconstructed charm decay. It should be noted that another part of the same collaboration pioneered the usage of CCDs as tracking detectors [9].

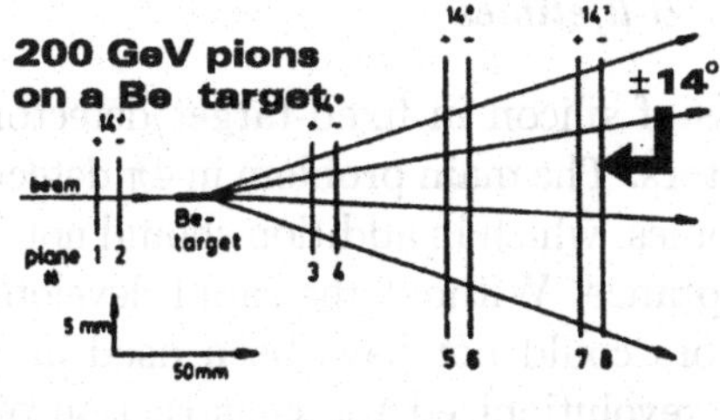

1983

8 layers of single sided Si detectors
strip pitch: 20 μm
readout pitch: 60 μm
spatial resolution: 5 μm

FIRST SILICON VERTEX DETECTOR

CONSTRUCTED TO MEASURE CHARM LIFETIMES

NA 11 → NA 32 → many more

Figure 18. Schematic of the vertex region of the NA11 detector.

RECONSTRUCTING CHARM

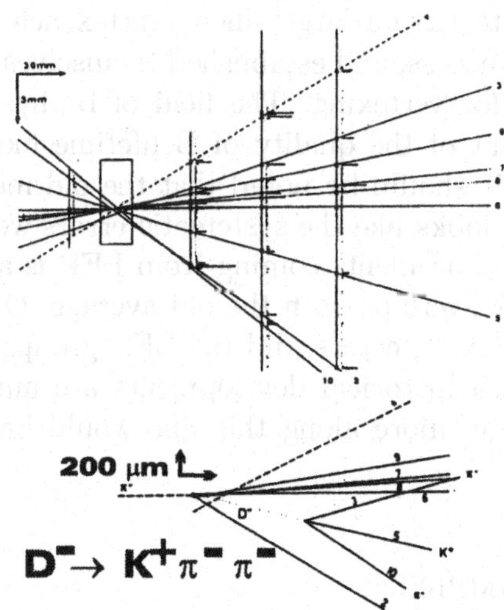

Figure 19. Display of an NA11 event and its reconstructed charm decay.

3.1.3. *Silicon goes 4π: B-lifetimes*

The immediate success of silicon in fixed-target detectors was followed by their usage in 4π detectors. The main problem in 4π detectors is to find space for the read-out electronics, which in addition should not introduce too much material into the acceptance. Without the rapid development in integrated circuits, silicon detectors could not have been used in a 4π geometry. Integrated circuits have revolutionized the construction of all detectors, not just silicon detectors. It should never be forgotten that the electronics is as important as the detector itself. For simplicity, at the beginning most experiments used single-sided detectors. The first 4π experiment with a vertex detector constructed from double-sided silicon was ALEPH [10] at LEP. The barrel shaped detector was constructed out of 27 faces deployed in two layers(see Fig. 20). Each face carried 4 detectors, each with an area of $5\times5\text{cm}^2$. The detectors had a strip pitch of 25μm on both sides. The read-out pitch was 50μm in the $r-\phi$ and 100μm in the z-direction. The intrinsic resolution was 12μm and 17μm, respectively. Again, the overall performance of the detector was limited by multiple scattering. For muons, residuals of 20μm in $r-\phi$ and 40μm in z were achieved. A special chip, the CAMEX64 [11] was developed for the read-out.

All of the LEP detectors, as well as all 4π detectors everywhere, were eventually upgraded to have a large silicon vertex detector system. These systems were a huge success, and established themselves very quickly as the standard technology for vertexing. The field of b-physics was revolutionized. The improvement of the quality of B lifetime measurements can be seen in Fig. 21 [12]. It should be noted that the old measurements are all systematically low. It looks like the systematic errors were underestimated. The first group of measurements coming from LEP is a very tight cluster, where every error bar overlaps with the old average. Only after really understanding their vertex detectors, did the LEP groups dare to measure a longer B lifetime. Such historical developments are unfortunately not uncommon. However, any more along this line would belong in an entirely different lecture.

3.2. SELECTED EXAMPLES

The selected examples are in no way representative. They are rather extreme cases where one particular choice of technology is pursued almost to the limit. There are many other interesting and challenging systems in operation, production, or in design. Just pick up any proposal or detector paper for a LEP, B-factory, LHC, or a Tevatron detector!

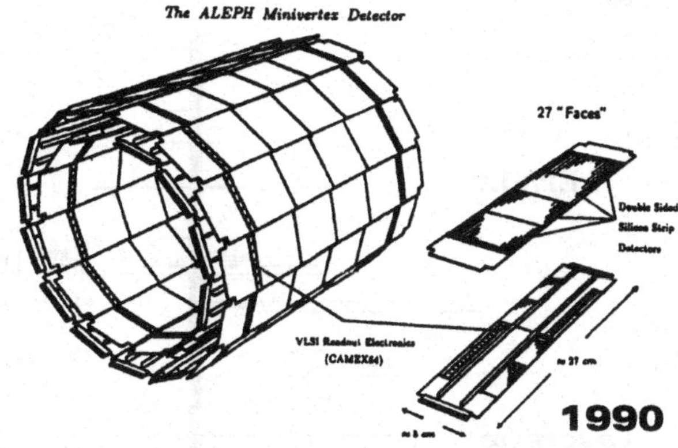

FIRST TIME DOUBLE-SIDED DETECTORS WERE USED IN A 4π GEOMETRY.

2 layers , total of 27 faces
strip pitch 25 μm
readout pitch 50/100 μm
intrinsic resolution 12/17 μm
alignment with tracks
achieved residuals (muons) 20/40 μm
limited by multiple scattering

NOW ALL 4π DETECTORS HAVE SILICON.

Figure 20. Schematic view of the first ALEPH vertex detector.

3.2.1. *SLD*

The SLD [Stanford Linear Detector] operates at the SLC [Stanford Linear Collider] at the Stanford Linear Accelerator Center [SLAC]. SLD is designed to operate at the Z^0 resonance, which determines its size, and at SLC, which determines its overall timing. The SLC has a low, repetition rate of 120Hz. As, in addition, the occupancy per beam crossing is very low and thus hits from 26 beam crossings can be sorted out later, a CCD system that takes more than 200ms to read out can be used. A full description of the SLD vertex detector system can be found in Ref. [13]. The first complete system called VXD2 started to take data in early 1992, and was a 120 Mpixel device. Since 1996, the upgraded version, VXD3, a 307 Mpixel device, is in use. Its point resolution is of the order of 4μm. The SLD vertex detector upgrade(see Fig. 22) is similar to many other upgrades at 4π detectors, for example, at LEP. The 4 detectors operating at LEP at CERN were

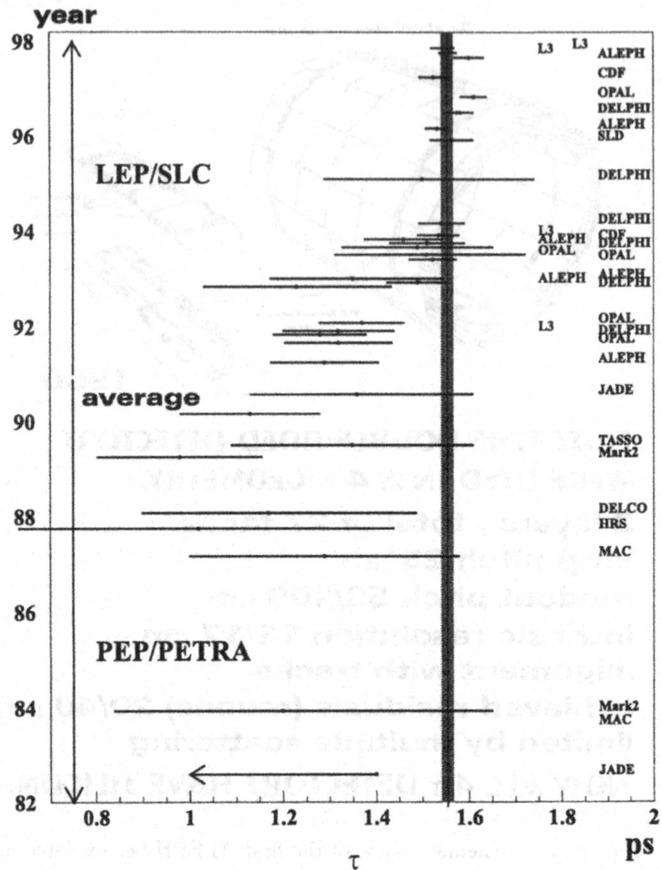

Figure 21. Development of B lifetime measurements: Measurements without silicon detectors dated before 1990 are averaged separately. Silicon detectors significantly reduced the error bars.

also designed for Z^0 physics. All these detectors look very similar in design. Some technical choices are different, but the principle layers of the onion are equivalent. All LEP detectors also have vertex detectors, tracking devices, particle identification devices, calorimeters and muon chambers. And the first vertex detector built for all of them turned out to be too short. As many interesting physics phenomena occur predominantly in forward-backward direction, a large angular coverage, i.e., a long barrel, is desirable. The parameters achieved in the second version of a detector become often only possible through the experience gathered while building the first detector.

The power of the SLD vertex detector is demonstrated in Fig. 23. Tracks are shown as reconstructed with the central drift chamber, as well as with

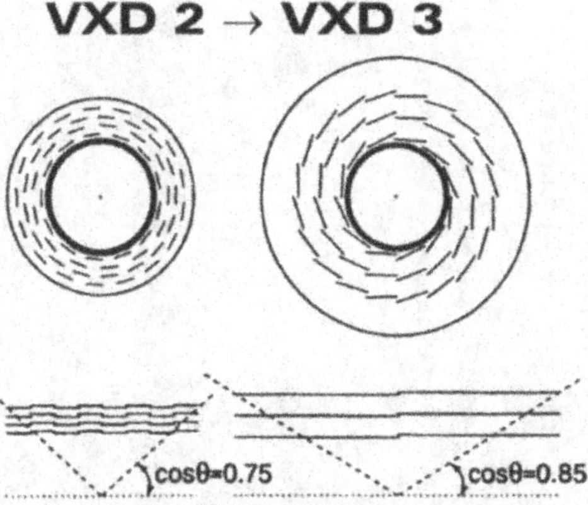

Figure 22. Layout of the VXD2 and VXD3 SLD Vertex Detectors. VXD3 is longer than VXD2, thus increasing angular coverage. In addition, the placement of layers is improved.

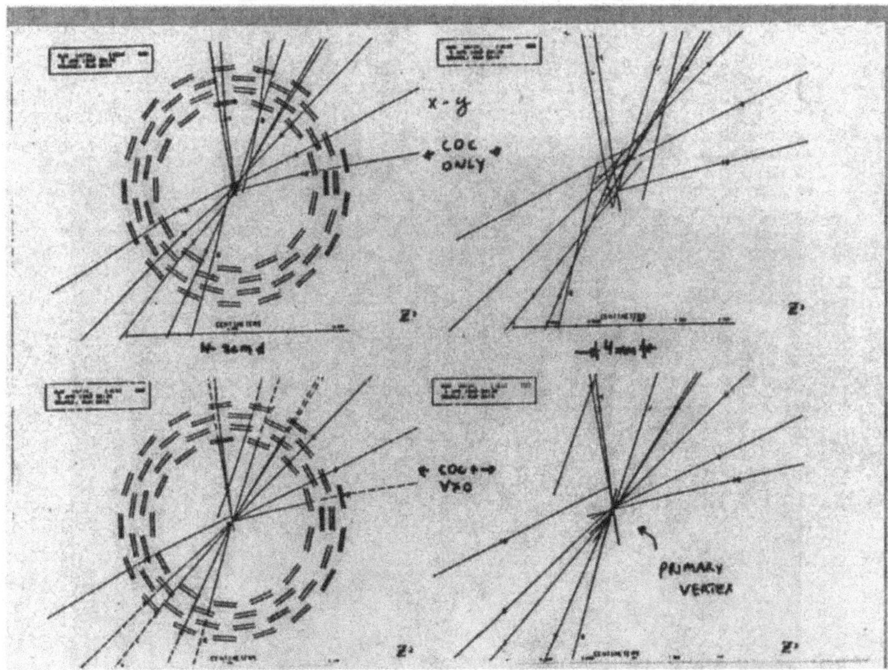

Figure 23. Demonstration of the power of the SLD VXD2 system: On the left are the tracks as they penetrate the layers. On the right, a close up of the interaction region is seen. The x-y projection is displayed. The top pictures show the tracks as reconstructed with the central drift chamber, while the bottom ones depict them as reconstructed with the help of the vertex detector.

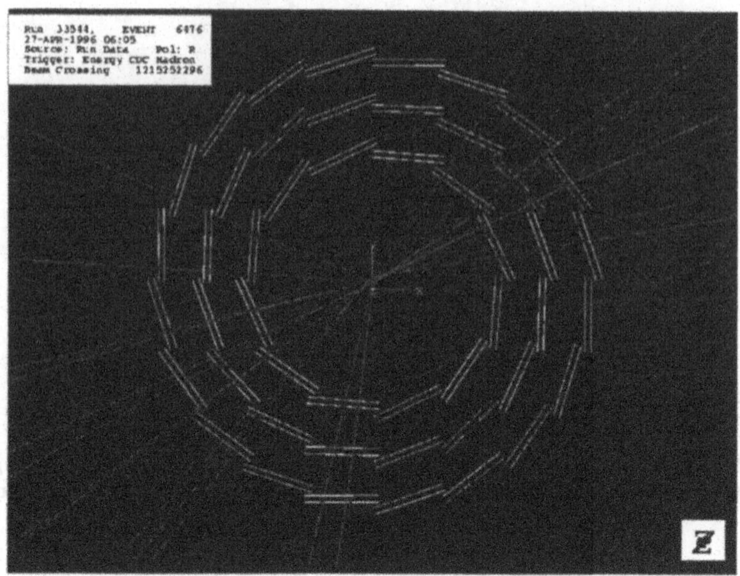

Figure 24. Event as seen by the VXD 3 detector.

Figure 25. **Revealing secondary vertices: In both event hemispheres, the secondary vertices can be seen clearly.**

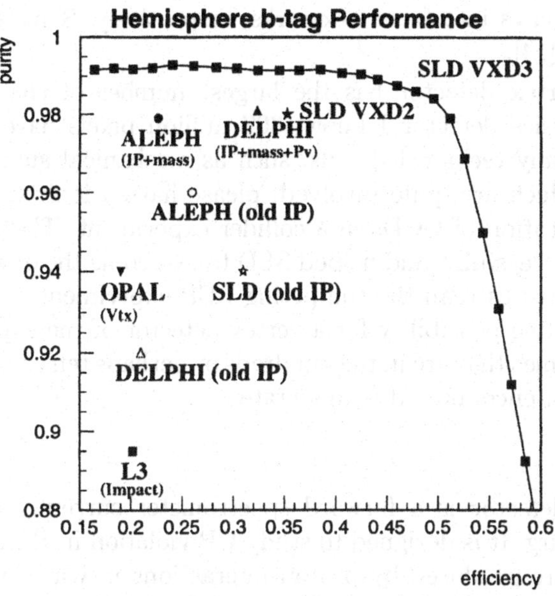

Figure 26. Comparison of b-tag performance: The purity versus efficiency for the SLD VXD3 detector is compared with results from VXD2 and the 4 LEP detectors.

the improved track reconstruction after inclusion of the information from the vertex detector [VXD2]. Figure 24 and Fig. 25 show the x-y projection of an event as seen by VXD3. The close-up in Fig. 25 clearly reveals secondary vertices.

The ability to identify secondary vertices opens up a wide field of b-quark physics. The events containing b-quarks are identified[tagged] using the visible decay length associated with B-mesons and b-baryons. All 4 LEP detectors, as well as SLD, have widely explored that possibility. In the tagging of b-quarks, it is also where SLD's pixels pay off. Figure 26 gives a comparison between different LEP results and SLD [14]. In all tagging efforts there is a trade-off between purity and efficiency. As the c-quark also has a significant lifetime, any b-quark sample is threatened by c-quark contamination. With its VXD3 detector, SLD achieves excellent purity for up to almost 50% efficiency. However, it should also be noted that SLD has the additional advantage of small initial beam spots and a small distance(28mm), between the beam axis and the first layer of CCDs. Good b-tags translate into good results on measurements such as R_b and R_c, the fractions of Z decays into b- or c- quarks, respectively. Some comparisons can be found in [14].

The SLD vertex detector has the largest number of channels of any high energy physics detector I know. 307 million pixels have to be dealt with. For the many technical details, such as mechanical support, cooling and read-out, which are quite involved, please have a look at Ref. [13]. It is the only application of CCDs in a collider experiment. The detector was operated very successfully, and helped SLD to overcome the disadvantage of having lower statistics than the competing LEP experiments. CCDs are an extremely attractive possibility for a vertex detector, if the experiment can allow read-out times that are in the hundreds or perhaps tens of milliseconds, and the radiation encountered is moderate.

3.2.2. *HERA-B*

The HERA-B detector is a forward spectrometer currently operating at DESY in Hamburg. It is designed to study CP violation in B-meson decays. The B-mesons are produced by proton interactions in wires placed within the beam-pipe of the HERA proton ring. As at LEP, a silicon vertex detector is used to identify the B-mesons through their visible decay lengths, which at HERA-B are about 1cm.

All particles are boosted into the extreme forward direction. Therefore, the HERA-B vertex system takes the so-called "Roman pot" design to an extreme. Figure 27 illustrates the idea. When a particle leaves the beam-pipe at a small angle, it traverses a lot of material. The amount of matter traversed is reduced by inserting pots into the beam-pipe. That also provides a way to get closer to the beam axis. The HERA-B silicon detectors are operated as close as 1cm to the beam axis. Unfortunately, the HERA proton beam moves around during injection and acceleration, and therefore the Roman pots have to be movable. The resulting mechanical system is quite involved. A 2.5m long vessel carries the Roman pots and manipulators for 32 silicon detector modules(see Fig. 28 for a schematic and Fig. 29 for a picture). In order to further reduce the material in front of the silicon detectors, the aluminum walls of the Roman pots are reduced to 150μm. As 150μm thick aluminum cannot withstand atmospheric pressure, this requires a secondary vacuum inside the pots. The engineering of the modules themselves is also not at all trivial. They have to fit into the very limited space inside the pots and, due to the vacuum, they have to be cooled through their support structures. As the electronics produces much more heat than the silicon detectors do, different cooling paths for electronics and detectors have to be provided. Everything has to be made out of carbon fiber and kapton because the system is still limited by multiple scattering. The resulting construction is depicted in Fig. 30. A picture of a mounted module, ready to be inserted

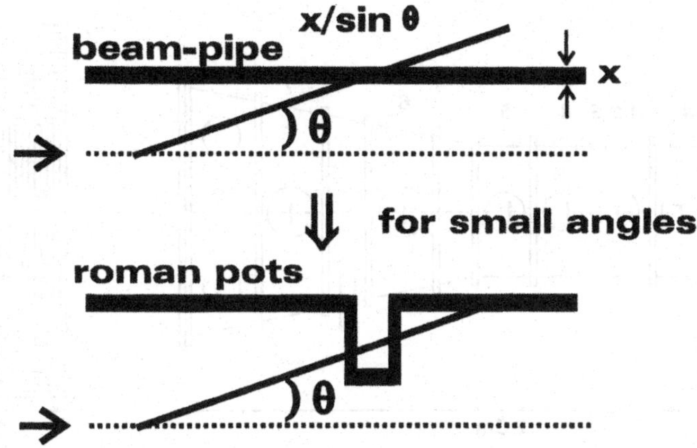

Figure 27. Schematics of a Roman pot: Particles produced at small angles θ relative to the beam axis have to traverse too much material to be tracked by normal detectors placed outside of a beam-pipe. Roman pots are inserted into the aperture to circumvent this problem.

into its pot, is shown in Fig. 31. Each module has two silicon detectors and has more than 10,000 wire bonds. It costs about $12,000. The actual silicon detectors, when purchased commercially make up about 30% of that cost. The total system cost is such that the silicon contributes only about 10% to the total. And it should be recognized thatthese silicon detectors are very expensive and very complicated devices for mass production runs.

The silicon detectors for HERA-B are pushed to the limit in radiation resistance. The expected dose at 1cm distance to the beam axis is 10 Mrad per year, mainly caused by a flux of $\approx 3\times10^{14}/\text{cm}^2$ minimum ionizing particles. This is about the limit of what is feasible with current technology. The signal to noise will be reduced to almost the limit of usability, depending on the ability to reduce the leakage current through cooling(see Sec. 4.3). It is not feasible to cool the silicon to optimum values. It is therefore foreseen to replace the silicon every year. Even if it were possible to cool the silicon to its optimum operational temperature of about -10°C, it would not survive more than two years, as the full depletion voltage would increase to unmanageable values. For more information check Ref. [15].

3.2.3. *Silicon Drift Detector in CERES*

As mentioned before, all kinds of structures are possible on a silicon wafer. It is also possible to construct a real drift detector, where the time of arrival of the charge is used to measure one of the coordinates. Figure 32 shows a circular detector [16] constructed for the CERES[NA45] experiment at

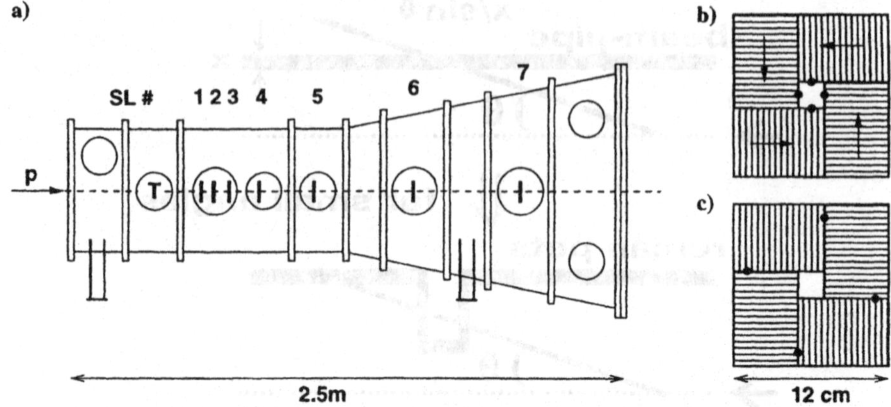

Figure 28. a) Schematic of the VDS vessel with the positions of the wire targets (T) and the VDS super-layers (SL 1-7). b,c) Arrangement of the detectors around the beam axis. The detectors are switched between positions b and c, so that the point of highest irradiation [black dots] on the detector is changed regularly.

Figure 29. Picture of the HERA-B vertex vessel.

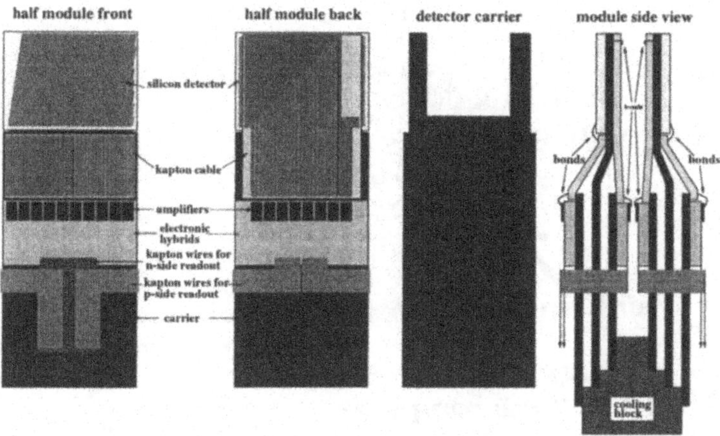

Figure 30. Schematic of the HERA-B silicon vertex detector module: Two half-modules are mounted together in one Roman pot.

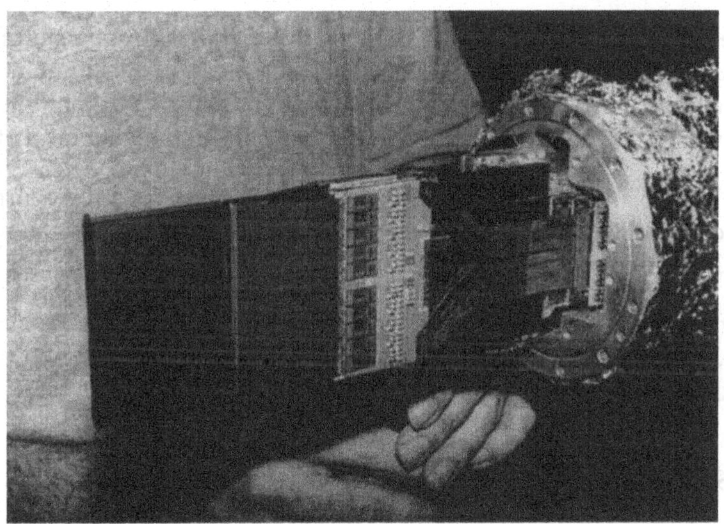

Figure 31. Picture of a mounted module to be inserted into its Roman pot.

CERN. In this experiment, the beam passes through the hole in the middle of the silicon detector. The radius and the angle Φ is measured for scattered particles hitting the detector. The charge is pulled to the edge of the detector by the field induced by 240 circular field electrodes. The resolution in Φ is given through 360 signal anodes. The resolution in R is given by the drift

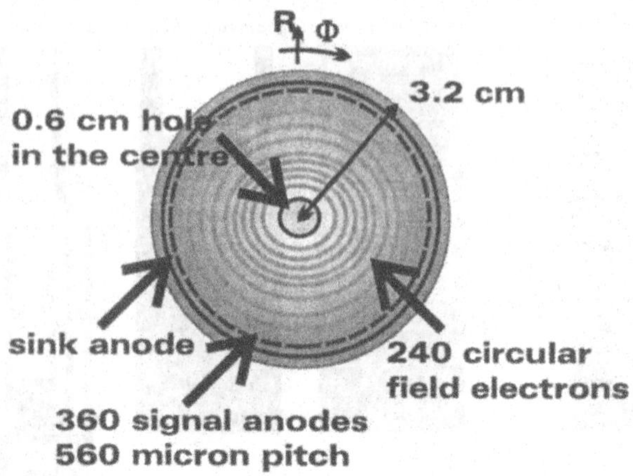

Figure 32. Schematic of the silicon drift detector used in the CERES experiment.

time measurement. The nominal drift field is 500 Volts/cm resulting in a maximum drift time of $4\mu s$. The resolution is less than $20\mu m$ in R and Φ. The special trick in the design are the sink anodes. Without them, the current would flow into the signal anodes and cause too high a level of noise.

The big advantage of silicon drift detectors is that a relatively large area can be covered with very few read-out channels. The big disadvantage is that such a device is necessarily rather slow.

Very specialized designs, such as the one presented here, can be very efficient. However, they don't come cheap, and can usually never be used for anything else.

3.2.4. *Silicon-Tungsten Calorimeters*

Related to the historical flux measurements are the modern calorimetric applications of silicon. Wherever there is little space and/or a lot of radiation, calorimeters made of dense materials, with silicon as the active element, can be considered. The dense material is usually tungsten. The silicon detectors usually feature pads. Such calorimeters are often used to measure the luminosity of an electron ring by looking for electrons scattered at small angles. Consequently, they can be found for instance at LEP,SLC and HERA. The standard geometry depicted in Fig. 33 is only one possibility. Wedge shaped objects forming rings, and other more exotic constructions, can be used.

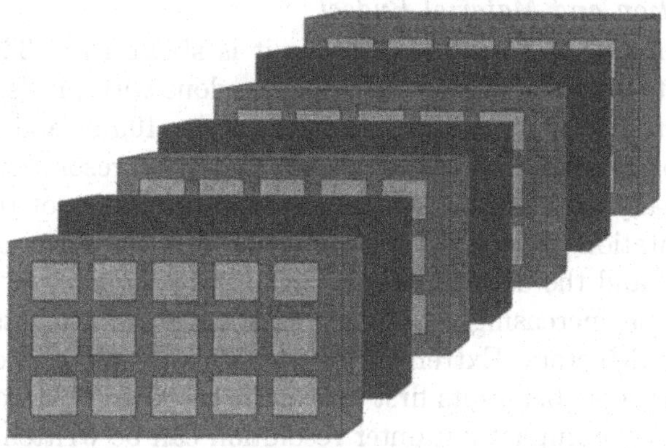

Figure 33. Schematic of a standard silicon-tungsten calorimeter: Layers of tungsten and silicon pad detectors form a sampling calorimeter.

4. Limitations of Silicon Detectors

4.1. BASIC PARAMETERS

4.1.1. *Speed*

The speed of silicon detectors will start to become an issue if event rates will continue to rise. The speed depends on the drift field and thus on the bias voltage, but, at normal operational parameters, electrons take about 3ns to traverse $100\mu s$, while holes need about 8ns for the same distance. Thus, 25ns is the minimum time needed when the p-side is read out, and the full signal is required for a detector is that is $300\mu m$ thick. In cases of very low occupancy, several events can be read out together. Hits from different events are then separated through additiona; information. This option has rarely been used, but it should be looked into more often.

4.1.2. *Size*

Many applications call for very large areas of silicon. Square-meters of silicon are planned for LHC, and this trend will continue. Very often the segmentation into small individual wafers causes problems. Basically, all detectors today are manufactured from 4 inch [10cm] wafers. However, 6 inch wafers have been used to produce detectors, and there is no physical law preventing 10 inch wafers. However, the over-all properties of a detector can be ruined by a single defect. The probability for a defect is at least proportional to the area of the device. It will be very difficult to have a good yield for very large detectors, and that will most likely result in forbidding costs per cm^2.

4.1.3. *Resolution and Material Budget*

As far as resolution is concerned, the limit is about 1μm. That has been achieved for strip detectors [17] and could be done with pixels. The corresponding structures on the silicon are of the order 10μm, and pose no real problem to good manufacturers. However, the actual resolution of a silicon system is usually not limited by the intrinsic resolution of the detectors. The main limitation of vertex detectors come from the material needed for the beam pipe and the detector itself. This is why Roman pot systems(see Fig. 27) become increasingly popular, and some experiments try to use thinner silicon detectors. Extremely important is the amount of material a track has to traverse before its first hit can be recorded. The corresponding contribution to the impact paramter resolution can be written as:

$$\sigma_{ms} = 13.6 \frac{MeV}{c} \frac{1}{p} D \sqrt{\frac{X}{X_0}}$$

where p is the particle momentum, D the distance from the interaction point, X/X_0 the fraction of a radiation length traversed, and ms is a reminder that multiple scattering is responsible. For somewhat normal values like D =20cm and p =30×10^{3}MeV/c, and a detector with 10μm intrinsic resolution, multiple scattering starts to dominate at X/X_0 of 0.012. That translates into about 1mm of aluminum.

Material pile-up is quite a problem. A typical system has more than one layer, and following layers are affected by the first layers. Therefore, all mechanical support structures, the read-out electronics close to the detectors, and cooling devices, have to have as little material as possible. Many designs start out being based on beryllium [100μm ≈ 0.03% X_0] and beryllium oxide [100μm ≈ 0.09% X_0]. However, both materials are difficult to handle and are very expensive. So most people use carbon fiber or graphite support structures. Typical values for those are ≈ 0.3% X_0 for a thickness of ≈ 700μm. For a 300μm detector, the silicon itself adds 0.3% X_0 In principle, thinner detectors can be made, however, they are too fragile for mass production, and thus cannot be used for large systems. In addition, the size of the signal is proportional to the thickness of the detector. 300μm is usually a good choice. Generally, it can be argued that anything less than 1% of a radiation length per layer is very good.

4.2. RADIATION DAMAGE

4.2.1. *Integral Dose*

The amount of integral radiation a silicon detector can digest and still function is its serious limitation. As explained in Sec. 2.5.8, a detector can in principle function as long as it can hold the voltage necessary to fully deplete it. The well designed guard ring structures can certainly be made to hold 1kV or more. However, a single defect on the n-side can cause a single strip to cause a break-through. Thus, perfect n-sides are needed in addition to good guard structures. Unfortunately, it is basically impossible to conclusively test the n-sides before type inversion. There are indications for n-side defects in the IV-curves(see Sec. 2.5.7). However, it is impossible to predict whether and at what voltage the device will fail. However, we should not forget that silicon detectors are by far the most radiation resistant detectors we have in large-scale production right now. The current generation of experiments expect to be able to use their silicon after a dose of up to $3 \times 10^{14}/\text{cm}^2$ minimum ionizing particles or 10Mrad.

4.2.2. *Radiation Bursts*

As explained in Sec. 2.5.9, radiation bursts can create pin-holes. The internal capacitors cannot be made to withstand significantly more than 100V without creating other problems. Thus, a detector will get destroyed if it gets exposed to strong bursts creating voltage drops larger than 100V. The system usually tolerates a couple of pin-holes, but at a certain point the detector becomes unusable. It is necessary to control the environment such that bursts do not become a habit.

4.3. COOLING

As mentioned before(see Sec. 2.5.11), the current generated in the bulk of a silicon detector at room temperature can be reduced by a factor of 2 through cooling by 7°C. This is important after a detector is damaged by irradiation and the increased leakage current reduces the signal to noise ratio. Unfortunately, the geometrical and mechanical realities of a detector system, as well as the heat produced by the read-out electronics, can limit the ability to cool the silicon. The cooling capacity thus can limit the results that can be achieved for the signal to noise ratio.

Cooling also suppresses anti-annealing(see Sec. 2.5.8). That is beneficial. However, it also suppresses annealing, which is bad. Fortunately, the two effects occur on different time scales, days for annealing, months for anti-annealing. Thus, it is useful to slightly warm up detectors from time to time to let them anneal, and cool them down again before they can anti-anneal. It is also useful to adjust the operation temperature such that annealing is not totally suppressed. The optimum temperature turns out to be around -10°C. This is in many cases below the temperature achievable with a reasonable and affordable technical effort.

5. The Future of Silicon Detectors

5.1. SHORT-TERM FUTURE

Almost all of the next generation of experiments have a silicon detector component. Some silicon systems are pure vertex detectors, where other systems define tracks, and hits in the silicon are attached to these tracks. Others are trackers in their own right. They have many layers, and they are used for stand-alone tracking. Many designs feature the classical strips, some call for pads, and some for pixels. Some of the trackers will use several square meters of silicon, and some vertex detectors will use the most refined pieces of silicon ever made. Collider experiments generally want huge silicon trackers. These many-layer designs are generally built because a drift chamber could not operate in the environment at hand. Their resolution is totally dominated by multiple scattering, and the silicon technology itself can be rather crude.

A true vertex detector is used only when some other component already finds the tracks, and the information from the silicon is used only to refine the track parameters. Such a detector is designed with minimal material and optimized silicon detectors. The most delicate silicon detector designs can be found in fixed-target applications, where a single detector can be placed at a very special location.

Unfortunately, the design of many of the devices currently under construction is not very well motivated. The systems are hybrids between trackers and vertex detectors. Quite often they are built before anybody had the time to clearly specify what is needed or wanted. In some collaborations, especially the very large ones, decisions may be more influenced by political than by technical and physics considerations. This is not only true for silicon detectors, but it is a clear trend in detector contruction that should be reversed as soon as possible!

5.2. LONG-TERM FUTURE

DISCLAIMER:
Any prediction the author made in the past turned out to be wrong!

Silicon is actually not cheap, requires some expertise, and is not easy to handle. Therefore, lots of people would like to replace it with something else. However, there is no well developed something-technology at hand. On top of this, all the technologies that are at the moment considered as alternatives(see below) are also expensive and difficult. Therefore, I predict that silicon detectors are going to stay, no matter what. Even in 50 years they will have a wide range of applications in high energy physics, if there will be high energy physics in 50 years.

The question remains whether a mode of operation can be found for silicon detectors that allows their usage after radiation doses equivalent to more than $10^{15}/\text{cm}^2$ minimum ionizing particles. Irradiated silicon at nitrogen temperatures could be the way. At low temperatures, silicon itself becomes an insulator with a small band-gap. The original defects in the material are compensated by radiation defects. The resulting material is something new. The research is ongoing [18] and we will have to see what comes of it.

5.3. ALTERNATIVES

The first two of the following "alternatives" are listed only because the discussion about them resurfaces every time a silicon detector seems to be too expensive or too difficult. The other two technologies are not ready to be used for the construction of a large device. However, they show some promise.

5.3.1. *GaAs*

IThis was advertised as the technology of the future. As far as detectors are concerned, it is now a technology of the past. It was supposed to be radiation hard. However, that is only true for neutral irradiation. It is worse than silicon under charged irradiation. Another draw-back of GaAs is that the signal is small to start with. The average number of electron-hole pairs is 3000 for each 0.1% of a radiation length instead of 8000. Everything considered, GaAs cannot any longer be counted as an alternative to silicon.

5.3.2. *Scintillating Fibers*

Scintillating fibers are used in tracking devices, for instance, in the D0 upgrade [19]. However, they cannot achieve the resolution wanted for vertex detectors. They are also hard to read out, and they are not radiation hard. So while they can be useful in a particular tracking device I do not consider them an alternative to silicon.

5.3.3. *Scintillating Capillaries*

There is an impressive number of technological problems yet to be solved. Most importantly, there is no clear scheme how to read them out efficiently. However, it might work one day.

5.3.4. *Diamonds*

The small size of the signal remains a problem, because the total thickness of the substrate cannot contribute to it. However, the thickness of the layer contributing is being constantly improved. There are still a number of technical problems to be solved before a large system can be designed. However, diamond seems the most promising alternative at the moment [20].

6. Conclusions

Silicon detectors are an extremely powerful, tool widely used in high energy physics. Over the last ten years, they have become a standard piece of equipment and they will continue to be so over the next ten years. I risk the prediction that, as long as particles are tracked, there always will be a place for silicon. At the moment, silicon detectors are the most radiation resistant detectors that are available for large scale projects. That may remain so for quite some time to come. Anybody designing a detector should study diligently what requirements the vertex detector has to fulfill. In order to get the best possible detector, all choices have to be made carefully. That is only possible when the requirements are known and clearly stated.

7. Acknowledgments

I would like to thank Tom for inviting me to his school. It was a wonderful experience. And I would like to thank Mrs. Petra Strube for her help in preparing this write-up.

References

1. Kittel C. (1953) *Introduction to Solid State Physics*, John Wiley & Sons, New York.
2. Tyagi, M.S. (1991) *Semiconductor Materials and Devices*. John Wiley & Sons, New York.
3. Damerell, C.J.S. (1998), Charged-Coupled Devices as Particle Tracking Detectors, *Review of Scientific Instruments*, **Vol 69** pp. 1549-1573
4. Abt, I. et al (1998) Characterization of Silicon Microstrip Detectors Using an Infrared Laser System, **MPI-PhE/98-13**.
5. Holl, P. et al (1997), A 36 cm^2 Monolithic pn-CCD for X-ray Detection on the XMM and ABRIXAS Satellites,*97 IEEE Nuclear Science Symposium*, submitted to Transactions on Nuclear Science.
Vol. A 235 pp. 85-90
6. Heijne, E.H.M. (1983), *CERN Yellow Report*, **83-6**
7. Abt, I. et al (1985), An Absolute Calibration of the Solid State Detectors in the Narrow Band Neutrino Beam at CERN, *Nuclear Instruments & Methods in Physics Research*,
8. Bailey, R. et al (1984), A Silicon Strip Detector Telescope for the Measurement of Production and Decay of Charmed Particles, *Nuclear Instruments & Methods in Physics Research*, **Vol. 226** pp. 56-58
9. Damerell, C.J.S. et al (1981), *Nuclear Instruments & Methods in Physics Research*, **Vol. 185** pp. 33-42
10. Schwarz, A. (1990), Construction, Operation and First Results for the ALEPH Double-sided Silicon Strip Vertex Detector, *Proceedings of the 25th International Conference on High Energy Physics*, **Vol. 2** pp. 1345-1347
11. Butler, W. et al (1988), Low Noise - Low Power Monolithic Readout Electronics for Silicon Strip Detectors, *Nuclear Instruments & Methods in Physics Research*, **Vol.**

A 273 pp. 778-783
12. Moser, H.-G. (1997), History of B-lifetime Measurements, *private communication*
13. Abe, K. et al (1997), Design and Performance of the SLD Vertex Detector: a 307 Mpixel tracking system, *Nuclear Instruments & Methods in Physics Research*, **Vol. A 400** pp. 287-343
14. Dong, S. (1998), Comparison of b-tag Performance, *http://www.slac.stanford.edu/~sudong/*
15. Riechmann, K. for the HERA-B collaboration (1998), Overview of the HERA-B Vertex Detector System and First Results from Prototype Runs, *Nuclear Instruments & Methods in Physics Research*, **Vol. A 408** pp. 221-328
16. Chen, W. et al (1993), Performance of the Multianode Cylindrical Silicon Drift Detector in the CERES NA45 Experiment: First Results, *Nuclear Instruments & Methods in Physics Research*, **Vol. A 326** pp. 273-278
17. Straver, J. et al (1994), One Micron Spatial Resolution with Silicon Strip Detectors, *Nuclear Instruments & Methods in Physics Research*, **Vol. A 348** pp. 485-490
18. Palmieri,V.G. et al (1998), Evidence for Charge Collection Efficiency Recovery in Heavily Irradiated Silicon Detectors Operated at Cryogenics Temperatures, *Nuclear Instruments & Methods in Physics Research*, **Vol. A 413** pp. 475-478
19. Wayne, M. (1993), A Scintillating Fiber Detector for the D0 Upgrade, *FERMILAB-CONF-93-043-E*,
20. Smith, K.M. (1996) Progress in Diamond and GaAs Detectors, *Nuovo Cim.*, **Vol. 109A**, pp. 1239-1252

CALORIMETRY

Tejinder S. Virdee
EP Division, CERN and
Imperial College of Science Technology and Medicine,
London, UK

1. INTRODUCTION

The aim of Particle Physics is to answer the two following questions: what are the fundamental constituents of matter? and what are the fundamental forces that control their behaviour at the most basic level? Experimentally this involves the study of hard particle interactions, determining the identity of the resulting particles and measuring their momenta with as high a precision as possible. Some thirty years ago a single detection device, the bubble chamber, was sufficient to reconstruct the full event information. At the current high centre of mass energies no single detector can accomplish this even though the number of particles whose identity and momenta need to be determined is limited [electrons, muons, photons, single charged hadrons, jets of hadrons, b-jets, taus and missing transverse energy $E_t(\nu)$]. This leads to a familiar onion-like structure of present day high energy physics experiments.

Starting from the interaction vertex the momenta (and sometimes the identity) of charged particles is determined in the inner tracker which is usually immersed in a solenoidal magnetic field. Identification of b-jets can be accomplished by placing high spatial resolution detectors such as Si pixel or microstrip detectors close to the interaction point. Following the tracking detectors are calorimeters which measure the energies, and identity, of electrons, photons, single hadrons or jets of hadrons. With the absorption of these particles only muons and neutrinos penetrate through the calorimeters. The muons are identified and measured in the outermost sub-detector, the muon system, which is usually immersed in a magnetic field. The presence of neutrinos is deduced from the apparent imbalance of transverse momentum or energy.

These lectures deal with calorimeters and rely heavily on previous reviews of calorimetry [1-6]. The emphasis is placed on their use at the future Large Hadron Collider and examples from ATLAS [7] and CMS [8] are extensively used.

T. Ferbel (ed.), Techniques and Concepts of High Energy Physics X, 335–386.

2. CALORIMETRY

Neutral and charged particles incident on a block of material deposit their energy through creation and destruction processes. An example of such phenomena is illustrated in Figure 1 in which a 50 GeV electron is incident on the BEBC Ne/H_2 (70%/30%) bubble chamber in a 3 T magnetic field. The deposited energy is rendered measurable by ionisation or excitation of the atoms of matter in the active medium. The active medium can be the block itself (*totally active or homogeneous calorimeter*) or a sandwich of dense absorber and light active planes (*sampling calorimeter*). The measurable signal is usually linearly proportional to the incident energy.

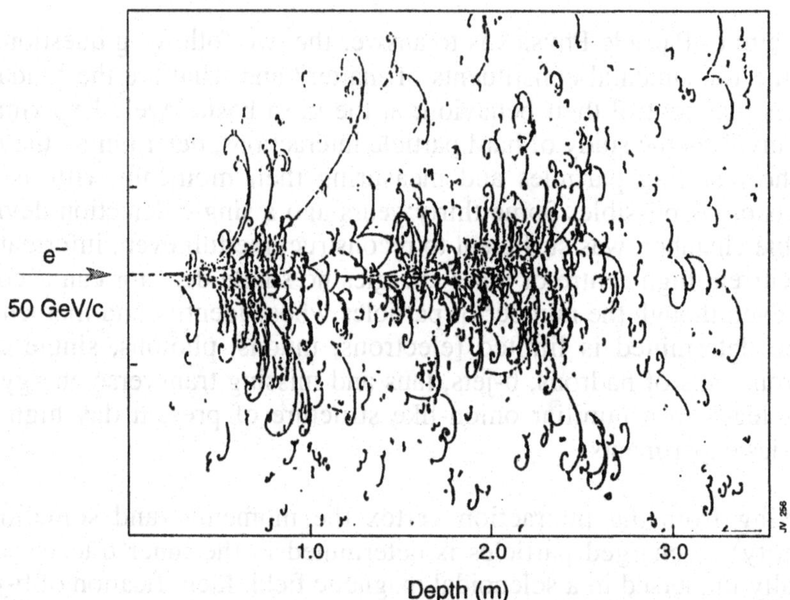

Figure 1: An example of a 50 GeV electron shower in a Ne/H_2 (70%/30) filled BEBC bubble chamber. The radiation length is ≈ 34 cm.

Calorimeters are key detectors in present-day experiments for the following reasons:

1 Calorimeters can measure energies of both neutral and charged particles.

2 The absorption of energy of incident particles is via a cascade process that leads to a number of secondary particles, *n*, where *n* is proportional to the incident energy. The cascade development is statistical in nature and the uncertainty on the measurement of energy (σ) is governed by the statistical fluctuation on *n*. Hence the relative energy resolution improves with energy as $\sigma/E \propto 1/\sqrt{n} = 1/\sqrt{E}$. This contrasts with momentum measurement of charged particles with tracking devices (in a magnetic field) where the relative momentum resolution dp/p worsens

with increasing p.

3. The longitudinal depth required to contain the cascades only increases logarithmically whereas for magnetic spectrometers the size scales as √p for a constant dp/p.
4. Calorimeters are essentially the only devices that can measure the energies of jets. More emphasis is now placed on the measurement of global characteristics such as jets and missing transverse energy. These arise from processes occurring at the constituent level.
5. Full geometric coverage enables the determination of missing transverse energy, which, if significant, signals the presence of weakly interacting particles such as neutrinos.
6. The cascade develops differently, longitudinally and laterally, for electrons/photons, hadrons and muons. This difference can be exploited to determine the identity of particles.
7. Calorimeters are devices with potentially fast response.
8. The pattern of energy deposit in a calorimeter with good lateral and longitudinal segmentation allows fast, efficient and very selective triggering on e/γ, jets and missing transverse energy.

3. INTERACTION WITH MATTER

3.1 Energy Loss by Charged Particles

Moderately relativistic charged particles, other than electrons, lose energy in matter through the Coulomb interaction with the atomic electrons. The energy transferred to the electrons causes them either to be ejected from the parent atom (*ionisation*) or to be excited to a higher level (*excitation*). The energy loss is given by the Bethe-Bloch equation :

$$-\frac{dE}{dx}\Big|_{ion} = N_A \, \frac{Z}{A} \, \frac{4\pi\alpha^2(\hbar c)^2}{m_e c^2} \, \frac{Z_i^2}{\beta^2}\left[\ln\frac{2m_e c^2\gamma^2\beta^2}{I} - \beta^2 - \frac{\delta}{2}\right] \quad (1)$$

where E is the kinetic energy of the incident particle with velocity β and charge Z_i, I (≈ 10×Z eV) is the mean ionization potential in a medium with atomic number Z. A very useful quantity is *areal density* measured in units of $g.cm^{-2}$. The energy loss of relativistic particles of unit electric charge per unit areal density is found to be roughly the same in all materials with

$$\frac{1}{\rho}\frac{dE}{dx} \approx 1.5 - 2 \; \frac{MeV}{g.cm^{-2}}$$

where ρ is the density of the medium. The energy loss rate in liquid hydrogen, gaseous helium, carbon, aluminium, tin and lead is shown in Figure 2 [9]. It can be seen that the above approximation is valid for all solids.

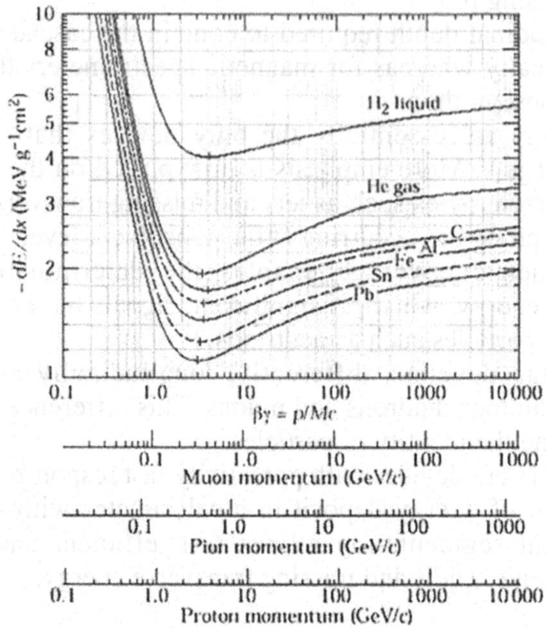

Figure 2: Energy loss rate in liquid hydrogen, gaseous helium, carbon, aluminium, tin and lead.

3.2 Energy Loss by Electrons

Above ≈1 GeV radiative processes dominate energy loss by electrons and photons. In the intense electric field of nuclei relativistic electrons radiate photons (*bremstrahlung*) and photons are converted into electron-positron pairs (*pair creation*).

In dealing with electrons and photons at high energies striking blocks of material (e.g. calorimeters) it is convenient to measure the depth and radial extent of the resulting cascades in terms of *radiation length* (X_0) and *Moliére radius* (R_M).

Consider the process of bremstrahlung. A free electron cannot radiate a photon. However a charged particle emits radiation when it is subjected to acceleration or deceleration. The acceleration/deceleration is greater the lighter the particle. The Feynman diagram for the bremstrahlung process is shown in Figure 3. The cross section for the process comprises the coupling constant at the three vertices and the propagator term (α $1/m^2$)

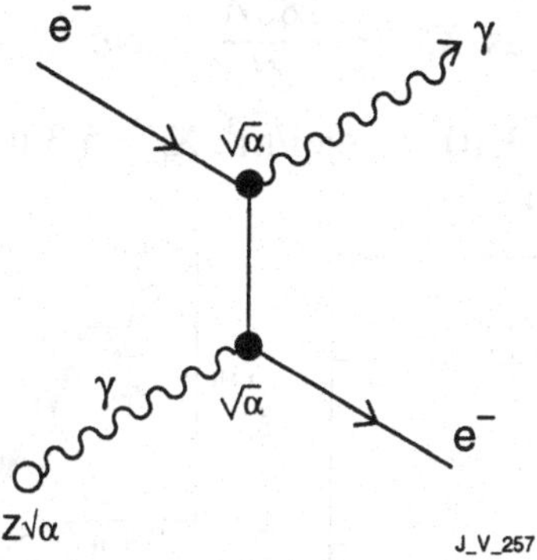

Figure 3: The Feynman diagram for bremstrahlung

$$\sigma \propto \frac{Z^2\alpha^3}{m_e^2 c^4}$$

We are interested in dσ/dν where ν is the energy of the emitted photon. We can make a guess for the expression using dimensional arguments:

$$\frac{d\sigma}{d\nu} \;\alpha\; \frac{Z^2\alpha^3}{m_e^2 c^4} \frac{(\hbar c)^2}{\nu}$$

Turning this to energy loss per unit distance traversed by the electrons gives

$$-\frac{d\sigma}{dx}\Big|_{rad} = n \int_{\nu_{min}}^{\nu_{max}} \nu \frac{d\sigma}{d\nu}\, d\nu = n \frac{Z^2\alpha^3(\hbar c)^2}{m_e^2 c^4} \left(\nu_{max} - \nu_{min}\right)$$

where ν_{max} = kinetic energy of electron, $\nu_{min} \approx 0$ and n is no. of nuclei/unit volume. A numerical factor [4 ln(183/$Z^{1/3}$)] has to be added describing the effect of the possible range of impact parameters of the electron. At large impact parameters the protons are shielded by atomic electrons. Hence

$$-\frac{dE}{dx}\Big|_{rad} = \left[4n \frac{Z^2\alpha^3(\hbar c)^2}{m_e^2 c^4} \ln\frac{183}{Z^{1/3}}\right] E \qquad (2)$$

Since $-\frac{dE}{dx} \propto E \Rightarrow \frac{dE}{E} = -\,B\,dx \Rightarrow E = E_0\, e^{-Bx}$

where B is a constant.

The *radiation length* is defined to be the distance over which the electron loses, on average, all but 1/e of its energy i.e. $X_0 = 1/B$ i.e

$$X_0 = \left[4n \frac{Z^2\alpha^3(\hbar c)^2}{m_e^2 c^4} \ln\frac{183}{Z^{1/3}}\right]^{-1}$$

and can be approximated as $X_0 \approx \dfrac{180A}{Z^2}\ g.cm^{-2}$

e.g. for Pb, Z = 82, n = $3.3.10^{28}$ nuclei/m^3, $X_0 \approx 5.3$ mm which is close to the PDG [9] value of 5.6 mm.

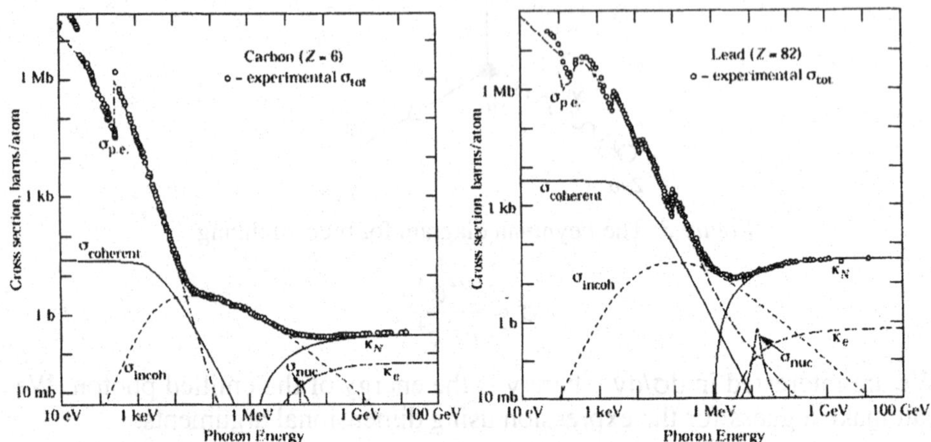

Figure 4: The photon total cross-sections as a function of energy in carbon and lead.

3.3 Energy Loss by Photons

Photons lose energy through photoelectric effect and Compton scattering at low energies and by pair production at relativistic energies. The cross-section for *photoelectric effect* is given by

$$\sigma_{pe} \approx Z^5\alpha^4\left(\frac{m_e c^2}{E_\gamma}\right)^n \quad n = \frac{7}{2}\ at\ E_\gamma << m_e c^2\ and\ n \to 1\ at\ E_\gamma >> m_e c^2$$

with a strong dependence on Z. The cross-section for *Compton scattering* has been calculated by Klein and Nishina :

$$\sigma_C \approx \frac{\ln E_\gamma}{E_\gamma}\ per\ electron\ and\quad \sigma_C^{atom} = Z\,\sigma_C\ per\ atom$$

If the energy of the photon is >> $m_e c^2$ then the dominant energy loss mechanism is *pair production* and its probability can be deduced, as done in Equation (1) for bremstrahlung. It is given by:

$$\sigma_{pair} \approx \frac{7}{9}\frac{A}{N_A}\frac{1}{X_0}$$

The probability of a pair conversion in 1 X_0 is $e^{-7/9}$. Since the photon disappears on producing a pair a mean free path length can be defined as

$$L_{pair} = \frac{9}{7} X_0 \text{ independent of energy.}$$

The photon total cross-sections as a function of energy in carbon and lead are shown in Figure 4 [9] which shows the above mentioned dependences.

3.4 Critical Energy and Moliére Radius

The *critical energy*, ε, is defined to be the energy at which the energy loss due to ionisation (at its minimum i.e. at $\beta \approx 0.96$) and radiation are equal (over many trials) i.e.

$$\frac{(dE/dx)_{rad}}{(dE/dx)_{ion}} = \frac{Z\alpha}{\pi\ m_e c^2}\ E\ \beta^2\ \frac{\ln 183/Z^{1/3}}{\ln\left[\frac{(2m_e c^2\beta^2)}{I\ (1-\beta^2)}\right] - \beta^2} = 1$$

which simplifies to

$$\Rightarrow\ \varepsilon \approx \frac{560}{Z} \quad (E\ in\ MeV)$$

The Moliére radius gives the average lateral deflection of critical energy electrons after traversal of 1 X_0 and is parameterized as:

$$R_M = \frac{21_{MeV} X_0}{\varepsilon} \approx \frac{7A}{Z}\ g.cm^{-2}$$

3.5 Hadronic Interactions

A high energy hadron striking an absorber interacts with nuclei resulting in multi-particle production consisting of secondary hadrons (e.g. $\pi^{\pm}$, π^0, K etc.). A simple model treats the nucleus, mass number A, as a black disc with radius R. Then

$$\sigma_{\text{int}} = \pi\ R^2 \propto A^{2/3} \quad where\ R \approx 1.2 \times A^{1/3}\ fm$$

$$\text{infact } \sigma_{\text{inel}} = \sigma_0\ A^{0.7} \quad \text{where } \sigma_0 = 35\ mb$$

The nuclear interaction length can be defined as $\lambda_{\text{int}} = \dfrac{A}{N_A \sigma_{\text{int}}} \propto A^{1/3}$

In dealing with hadrons it is convenient to measure the depth and radial extent of the resulting cascades in terms of *interaction length* (λ_{int}).

The values of the above mentioned parameters for various materials are listed in Table. 1.

Table 1: Physical properties of some materials used in calorimeters.

	Z	ρ g.cm^{-3}	I/Z eV	(1/ρ)dT/dx MeV/g.cm^{-3}	ε MeV	X_0 cm	λ_{int} cm
C	6	2.2	12.3	1.85	103	≈ 19	38.1
Al	13	2.7	12.3	1.63	47	8.9	39.4
Fe	26	7.87	10.7	1.49	24	1.76	16.8
Cu	29	8.96		1.40	≈ 20	1.43	15.1
W	74	19.3		1.14	≈ 8.1	0.35	9.6
Pb	82	11.35	10.0	1.14	6.9	0.56	17.1
U	92	18.7	9.56	1.10	6.2	0.32	10.5

4. THE ELECTROMAGNETIC CASCADE

4.1 Longitudinal Development

A high energy electron or photon incident on a thick absorber initiates a cascade of secondary electrons and photons via bremstrahlung and pair production as illustrated in Figure 5. With increasing depth the number of secondary particles increases while their mean energy decreases. The multiplication continues until the energies fall below the critical energy, ε. Ionization and excitation rather than generation of more shower particles dominate further dissipation of energy.

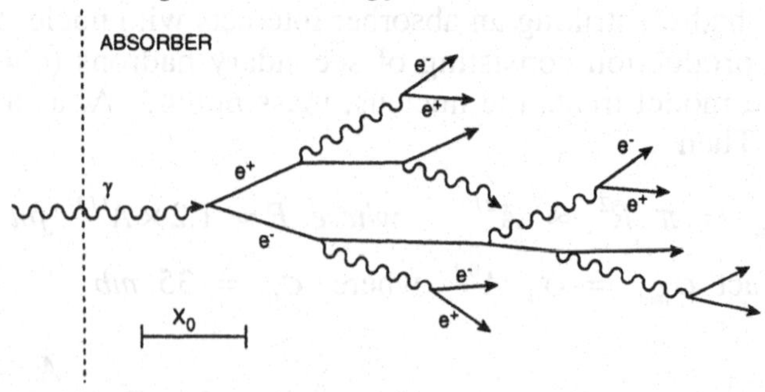

Figure 5: Schematic development of an electromagnetic shower.

Consider a simplified model of development of an electromagnetic shower initiated by an electron or a photon of an energy E. A universal description, independent of material, can be obtained if the development is described in terms of scaled variables:

$$t = \frac{x}{X_0} \quad and \quad y = \frac{E}{\varepsilon}$$

Since in 1 X_0 an electron loses about 2/3rd of its energy and a high energy photon has a probability of 7/9 of pair conversion, we can naively take 1 X_0 as a generation length. In each generation the number of particles increases by a factor of 2. After t generations the energy and number of particles is

$$e(t) = \frac{E}{2^t} \quad and \quad n(t) = 2^t$$

respectively.

At shower maximum where $e \approx \varepsilon$, the no. of particles is

$$n(t_{max}) = \frac{E}{\varepsilon} = y \text{ and } t_{max} = \ln\frac{E}{\varepsilon} = \ln y$$

Critical energy electrons do not travel far ($\leq 1X_0$). After the shower maximum the remaining energy of the cascade is carried forward by photons giving the typical exponential falloff of energy deposition caused by the attenuation of photons. Longitudinal development of 10 GeV showers in Al, Fe and Pb is shown in Figure 6 [3]. It can be noted that the shower maximum is deeper for higher Z materials because multiplication continues down to lower energies. The slower decay beyond the maximum is due to the lower energies at which electrons can still radiate. Both of the above effects are due to lower ε for higher Z materials.

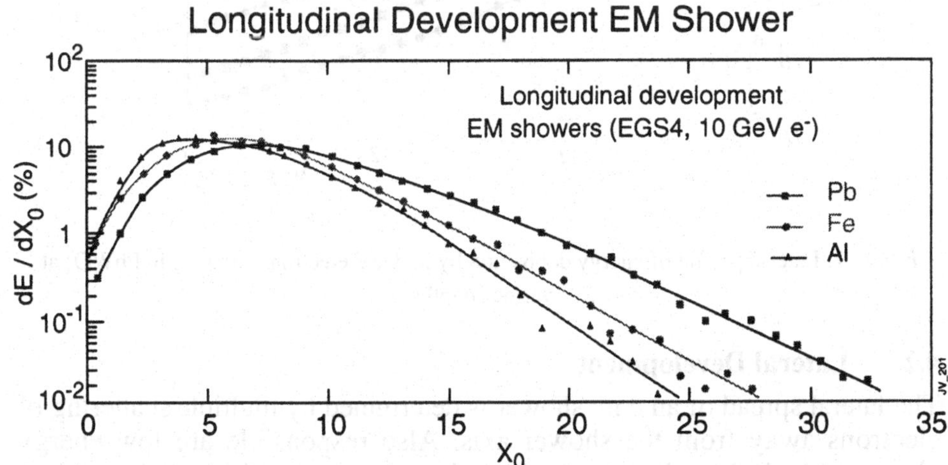

Figure 6: Simulation of longitudinal development of 10 GeV electron showers in Al, Fe and Pb.

The mean longitudinal profile of energy deposition is given by:

$$\frac{dE}{dt} = Eb\frac{(bt)^{a-1}e^{-bt}}{\Gamma(a)}$$

The maximum of the shower occurs at $t_{max} = (a-1)/b$. Fits to t_{max} give

$t_{max} = \ln y - 0.5$ for electron-induced cascades and

$t_{max} = \ln y + 0.5$ for photon-induced cascades.

The coefficient a can be found using t_{max} and assuming $b \approx 0.5$. The photon induced showers are longer since the energy deposition only starts after the first pair conversion has taken place. The m.f.p. length for pair conversion of a high energy photon is $X_\gamma = (9/7)\ X_0$.

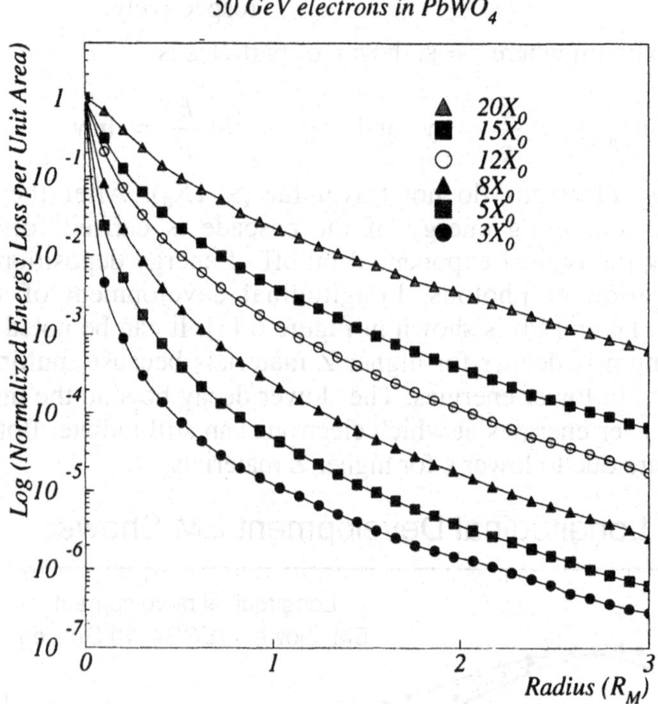

Figure 7: Lateral profile of energy deposition by 50 GeV electrons showers in $PbWO_4$ at various depths.

4.2 Lateral Development

The lateral spread of an e.m. shower is determined by multiple scattering of electrons away from the shower axis. Also responsible are low energy photons which deposit their energy a long way away from their point of emission, especially when emitted from electrons that already travel at large angles w.r.t. the shower axis. The e.m. shower begins, and persists, with a narrow core of high energy cascade particles, surrounded by a halo of soft particles which scatter increasingly as the shower depth increases. This is shown in Figure 7 for 50 GeV electrons incident on lead tungstate [10]. In different materials the lateral extent of e.m. showers scales fairly accurately with the Moliére radius. An infinite cylinder with a radius of ≈ 1 R_M contains ≈ 90% of the shower energy. For lead tungstate, and a depth of 26 X_0, the amount of energy contained in a cylinder of a given radius is shown in Figure 8. The fact that e.m. showers are very narrow at the start can be used to distinguish single photons from pizeros (see Section 7.2).

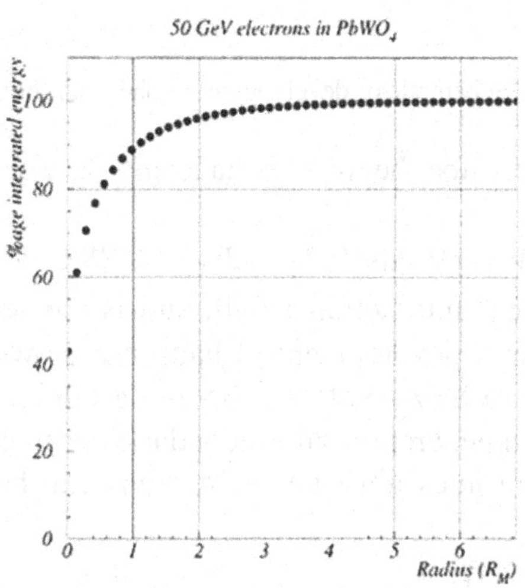

Figure 8: The percentage of energy contained in a cylinder of lead tungstate of different radii.

5 THE HADRONIC CASCADE

5.1 Longitudinal Development

A situation analagous to that for e.m. showers exists for hadronic showers. The interaction responsible for shower development is the strong interaction rather than electromagnetic. The interaction of the incoming hadron with absorber nuclei leads to multiparticle production. The secondary hadrons in turn interact with further nuclei leading to a growth in the number of particles in the cascade. Nuclei may breakup leading to spallation products. The cascade contains two distinct components namely the electromagnetic one (π^0s etc.) and the hadronic one ($\pi^{\pm}$, n, etc) one. This is illustrated in Figure 8.

The multiplication continues until pion production threshold is reached. The average number, *n*, of secondary hadrons produced in nuclear interactions is given by $n \propto \ln E$ and grows logarithmically. The secondaries are produced with a limited transverse momentum of the order of 300 MeV.

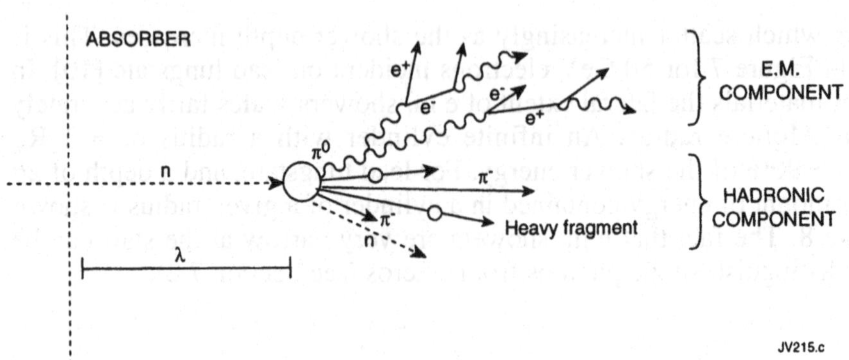

Fig. 9: Schematic of development of hadronic showers.

It is convenient to describe the average hadronic shower development using scaled variables

$$\nu = x/\lambda \quad and \quad E_{th} \approx 2m_{\pi} = 0.28\ GeV$$

where λ is the nuclear interaction length and is the scale appropriate for longitudinal and lateral development of hadronic showers. The generation length can be taken to be λ. Note $\lambda \approx 35\ A^{1/3}\ g.cm^{-2}$. Furthermore, if it is assumed that for each generation <n> secondaries/primary are produced and that the cascade continues until no more pions can be produced then in generation ν

$$e(\nu) = \frac{E}{\langle n\rangle^{\nu}}$$

$$e(\nu_{max}) = E_{th} \quad \therefore\ E_{th} = \frac{E}{\langle n\rangle^{\nu_{max}}}$$

$$n^{\nu_{max}} = \frac{E}{E_{th}} \Rightarrow \nu_{max} = \ln(E/E_{th})/\ln\langle n\rangle$$

The number of independent particles in the hadronic cascades compared to electromagnetic ones is smaller by E_{th}/ε and hence the intrinsic energy resolution will be worse at least by a factor $\sqrt{(E_{th}/\varepsilon)} \approx 6$. The average longitudinal energy deposition profiles are characterised by a sharp peak near the first interaction point (from π^0s) followed by a exponential fall-off with scale λ. This is illustrated in Fig. 9. The maximum occurs at $\nu_{max} \approx 0.2\ lnE + 0.7$ (E in GeV).

It can be seen that over 9λ are required to contain the energy of high energy hadrons. A parameterisation for the depth required for almost full containment (95%) is given by $L_{0.95}(\lambda) \approx t_{max} + 2\lambda_{att}$ where $\lambda_{att} \approx \lambda\ E^{0.13}$.

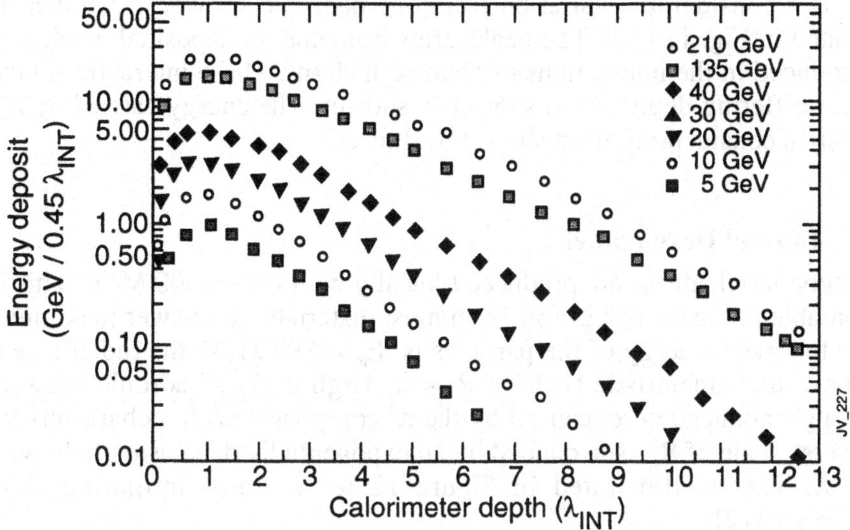

Fig. 10: longitudinal profile of energy deposition for pion showers of different energies.

270 GeV Incident Pions in Copper

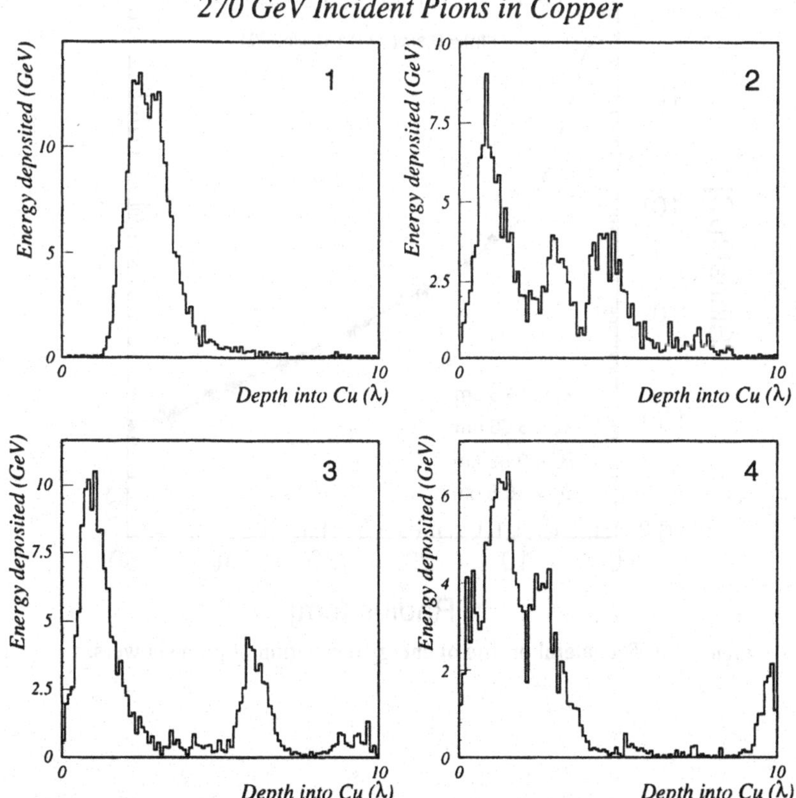

Figure 11: A simulation of the development of four representative pion showers in a block of copper.

There is a considerable variation from one hadronic shower to another as illustrated in Fig. 11 [10]. The peaks arise from energy deposited locally by π^0s produced in the interactions of charged hadrons. These interactions take place at differing depths from shower to shower. The energy carried by π^0s also varies considerably from shower to shower.

5.2 Lateral Development

The secondary hadrons are produced typically with $<p_t> \approx 300$ MeV. This is comparable to the energy lost in 1λ in most materials. At shower maximum, where the mean energy of the particles is $E_{th} \approx 280$ MeV, the radial extent will have a characteristic scale of $R_\pi \approx \lambda$. High energy hadronic showers show a pronounced core, caused by the π^0 component with a characteristic transverse scale of R_M, surrounded by an exponentially decreasing halo with scale λ. This is illustrated in Figure 12 for a lead/scintillating fibre calorimeter [12].

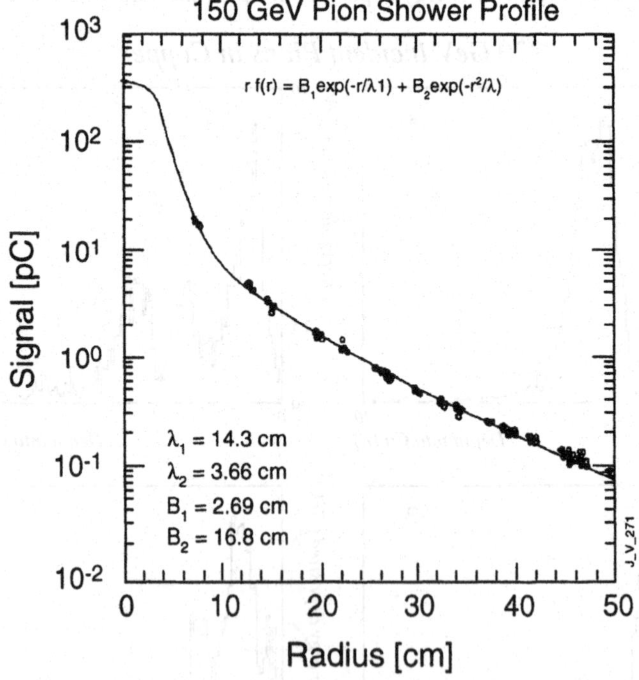

Figure 12: The lateral profile of energy deposition of pion showers.

6. THE ENERGY RESOLUTION

The energy resolution of calorimeters is usually parameterised as :

$$\frac{\sigma}{E} = \frac{a}{\sqrt{E}} \otimes \frac{b}{E} \otimes c$$

where the r.h.s. is assumed to be the square root of the quadratic sum of the three terms.

The first term, with coefficient a, is the *stochastic or sampling* term and accounts for the statistical fluctuation in the number of primary and independent signal generating processes, or any further process that limits this number. An example of the latter is the conversion of light into photo-electrons by a photo-device.

The second term, with coefficient b, is the *noise* term and includes:
- the energy equivalent of the electronics noise and
- the fluctuation in energy carried by particles, other than the one(s) of interest, entering the measurement area. This is usually labeled pileup.

The last term, with coefficient c, is the *constant* term and accounts for:
- imperfect quality of construction of the calorimeter
- non-uniformity of signal generation and/or collection
- cell-to-cell inter-calibration error
- the fluctuation in the amount of energy leakage from the front, the rear and the sides of the volume used for the measurement of energy,
- the fluctuation in the amount of energy deposited in dead areas in front or inside the calorimeter,
- the contribution from the fluctuation in the e.m. component in hadronic showers.

The tolerable size of the three terms depends on the energy range involved in the experiment. The above parametrisation allows the identification of the causes of resolution degradation. The quadratic summation implies that the three types of contributions are independent which may not always be the case.

6.1 Intrinsic Electromagnetic Energy Resolution

It is instructive to look at homogeneous calorimeters in which all the energy is deposited in the active medium. If the shower is fully contained then the intrinsic energy resolution is determined by the fluctuation in the number, n, of ions or photons produced. If W is the mean energy required to produce an electron-ion pair (or a photon) then $n = E/W$, and

$$\frac{\sigma}{E} = \frac{\sqrt{n}}{n} = \sqrt{\frac{W}{E}}$$

However the fluctuation is smaller as the total energy deposited (= incident energy) does not fluctuate. The improvement in resolution is characterised by the Fano factor, F, as

$$\frac{\sigma}{E} = \sqrt{F} \times \sqrt{\frac{W}{E}} = \sqrt{\frac{FW}{E}}$$

F is dependent on the nature of processes that lead to energy transfer in the detector including ones that do not lead to ionisation e.g. phonon excitations.

Consider calorimeters used for the spectroscopy of low energy (≈MeV). gamma rays. The two commonly used detectors are inorganic scintillators (e.g. NaI) and semiconductor detectors (e.g. Ge). The energy resolution of the Ge detector is superior and is measured to be $\sigma \approx 180$ eV for photons carrying 100 keV. The above formula gives $\sigma = \sqrt{(FEW)} \approx 195$ eV where $F_{Ge}=0.13$ and W=2.96 eV. It should be noted that without the Fano factor $\sigma \approx 540$ eV!

Another illustration employs noble liquids for the energy measurement. In principle a precision similar to that for Ge should be possible. However, the ^{207}Bi electron conversion line at 976 keV in liquid argon yields $\sigma \approx 11$ keV whereas the above formula would give

$$\sigma = \sqrt{(FEW)} = \sqrt{(0.11 x 23.7 x 976 x 10^3)} \approx 1.6\ keV.$$

An additional source of fluctuation is in the amount of energy going into mechanisms other than one being used for measurement e.g. scintillation when ionisation charge is collected. Not all the created electron-ion pairs contribute to the collected charge. In the absence of electric field about half of the pairs recombine and give scintillation light through molecular de-excitation. If $n = n_{ion} + n_{scint}$ and only charge is collected then

$$\sigma_{ion} = \sqrt{n \frac{n_{ion}}{n} \frac{n_{scint}}{n}} = \sqrt{\frac{n_{ion}(n - n_{ion})}{n}}$$

Measuring both light and charge can improve the resolution e.g. if $n_{ion}/n = 0.9$ then the resolution improves by a factor 3 w.r.t. the Poisson expectation ($\sqrt{n_{ion}}$). The improvement is illustrated in Figure 13 [13]

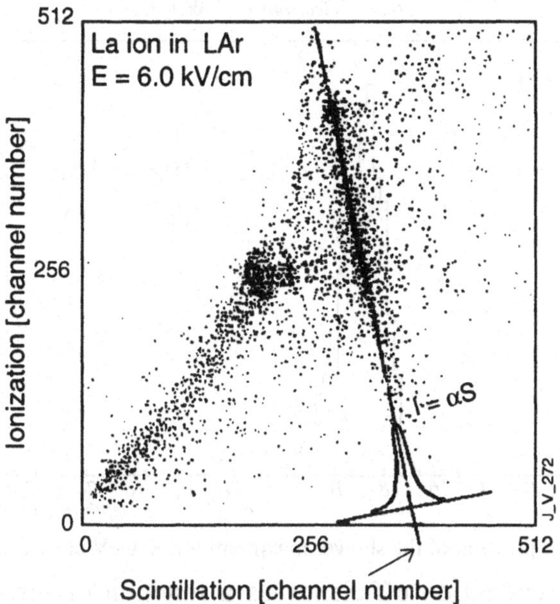

Figure 13: The anti-correlation between the ionization signal and the scintillation light in liquid argon.

Other phenomena may limit the number of signal generating events. Lead glass shower detectors are based on the detection of Cerenkov light, produced by the electrons and positrons with kinetic energies greater than ~ 0.7 MeV. This means that at most 1000 / 0.7 ~ 1400 independent particles, per GeV of deposited energy, produce Cerenkov light. The resolution is then dominated by the fluctuation in this number and thus cannot be better than $\sigma/E = \sigma_n \geq 3\% / \sqrt{E}$. This is further limited by photo-electron statistics as only about 1000 photo-electrons are generated when using photomultipliers to detect the scintillation light. This leads to an additional loss of resolution given by $\sigma_{pe} \approx 3\% / \sqrt{E}$.

6.2 Electromagnetic Energy Resolution – Constant Term

6.2.1 Longitudinal Non-Uniformity

Longitudinal non-uniformity of signal generation and/or collection either intrinsically or through radiation damage, when folded with the shower-to-shower fluctuation in the longitudinal profile (at a fixed energy) leads to a loss of energy resolution. Since the fluctuation is essentially independent of energy a contribution to the constant term arises. The fluctuation of the shower maximum is plotted in Figure 14 [14] for 50 GeV electrons giving $\sigma \approx 1\ X_0$.

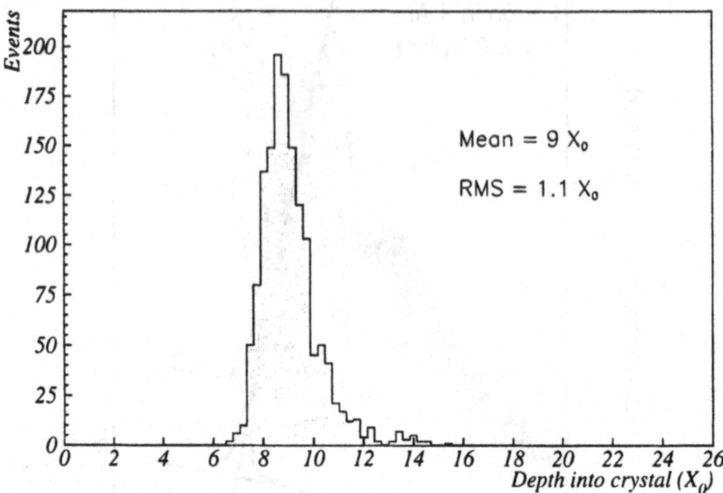

Figure 14: The position of the shower maximum for 50 GeV electrons in $PbWO_4$.

The ideal light collection efficiency as a function of distance from the photo-device end for $PbWO_4$ crystals for CMS is shown in Figure 15a [14]. The collection efficiency in the region of the shower maximum (5-10 X_0) should be constant with a slight increase at the rear of the crystals (photo-sensor end) to compensate for the energy leakage in showers developing late. The measured correlation between the slope of the light collection efficiency function in the region of shower maximum and the induced constant term is shown in Figure 15b.

6.2.2 *Cell-to-cell Intercalibration Error*

Electromagnetic showers are narrow and usually the central cell, or at most, the central 4 cells contain most of the energy of the shower. Since the lateral shower shape is nearly independent of energy, any effect due to imperfect cell-to-cell inter-calibration will end up in the constant term. If the reconstructed energy is $E = \Sigma\, g_i.E_i$ and if the r.m.s. error on g_i is δ then the constant term will range from $\delta /\sqrt{N}$ where N is the number of cells with significant energy i.e. from $\delta /2$ and $\delta /4$. This implies that the cell-to-cell intercailbration error should be substantially better than the desired constant term.

At the LHC the rate of isolated electrons from the electronic decays of W and Z bosons is large: 45 Hz at $L=10^{34}$ $cm^{-2}s^{-1}$ with $p_T > 20$ GeV in the region $|\eta| < 1.5$. These electrons will be used to establish the calibration. In a study by CMS [8, ECAL TDR] stringent cuts on isolation parameters are made to select electrons that have lost only a small amount of energy due to bremstrahlung. The isolation condition requires that the energy contained in a matrix of 3x3 crystals, centred on the electron impact, is more than 92%

of the energy contained in a larger matrix of 7x11 crystals. The energy, E, is then compared with the electron momentum, p, measured in the inner tracker. As an example the resolution on the parameter E/p is found to be $\sigma_{E/p}\approx 1.5\%$ at $\eta=0.9$ with a reconstruction efficiency of about 45%. The statistical precision on the calibration coefficient improves as $\sigma_{E/p}/\sqrt{N_e}$ where N_e is the number of usable electrons. Hence about 25 good electrons per crystal will be sufficient to achieve the design goal of a calibration (and inter-calibration) uncertainty of 0.3%.

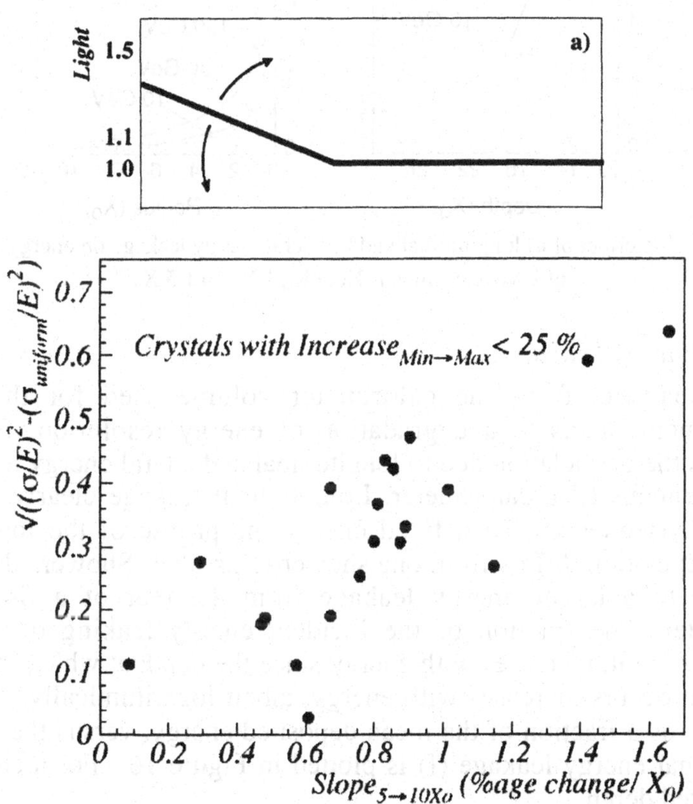

Figure 15: a) The ideal light collection efficiency as a function of distance from photosensor end for $25X_0$ deep crystals, b) The measured correlation between the slope in the region of shower maximum and the induced constant term.

The calibration of the hadronic calorimeter can be established using selected events containing high p_T photons or Z bosons balanced by a single jet. It may also be possible to use single pions from $\tau^{\pm} \rightarrow \pi^{\pm}\nu_{\tau}$ decays. The momentum of the charged pion, measured in the tracker, can be comparewd with the energy measured in the calorimeter.

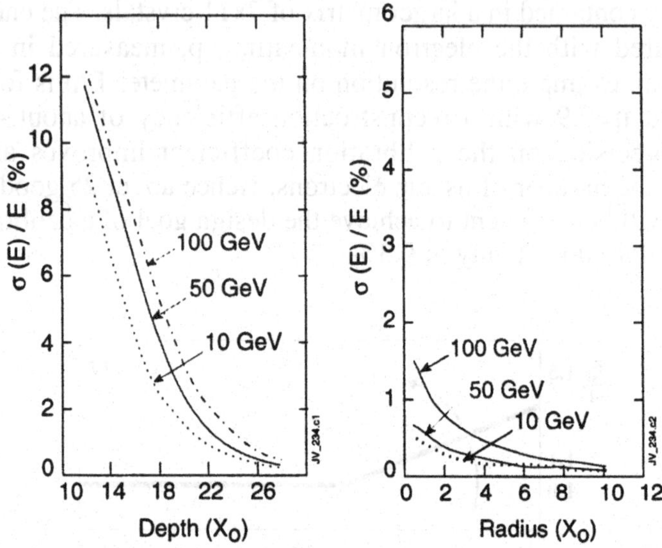

Figure 16: The effect of a) longitudinal and b) lateral energy leakage on energy resolution of LXe calorimeter. Note R_M(LXe) ≈ 1.5 X_0.

6.2.3 Energy Leakage

Energy leakage, from the calorimeter volume used for the energy measurement, leads to a degradation of energy resolution. Figure 16 illustrates the degradation due to longitudinal and lateral energy leakage for a homogeneous LXe calorimeter. Longitudinal leakage clearly has more serious consequences. At a fixed energy the profile of the longitudinal energy deposition differs from one shower to another. Showers developing late lead to a larger energy leakage from the rear of a finite depth calorimeter. The fraction of the incident energy leaking out, and the fluctuation on it, increases with energy since the depth at which the shower maximum occurs increases with energy, albeit logarithmically. The r.m.s. deviation, as a fraction of the mean deposited energy, versus the fractional longitudinal energy leakage (f) is plotted in Figure 16. For a calorimeter with a fixed depth

$$\frac{\sigma_{rms}}{\langle E_{dep} \rangle} = \frac{f}{2} \quad for \quad f < 20\%$$

The loss of energy resolution due to the lateral energy leakage is smaller since the lateral profile of energy deposition differs much less from one shower to another. The energy dependence of the fluctuation is also weak as the lateral shower shape is almost independent of energy especially at high energies.

6.3 Energy Resolution of Sampling Electromagnetic Calorimeters

When the very best energy resolution is not required, sampling calorimeters are employed. The shower energy is measured in active layers, often of low Z, sandwiched in between passive absorber layers of high Z materials. Only a fraction of the shower energy is dissipated in the active medium and the energy resolution is dominated by the fluctuation in this fraction. If the energy loss in an active layer is much smaller than that in the absorber layer then the number of independent charged particles crossing an active layer can be approximated by $n=E/\Delta E_{abs}$ where ΔE_{abs} is the energy lost by a minimum ionizing particle (m.i.p.) in the absorber layer.

Now $\Delta E_{abs} = t_{abs} \times (dE/dx)$ where t_{ab} is measured in units of X_0. Hence

$$\frac{\sigma}{E} = \frac{\sqrt{n}}{n} \propto \frac{\sqrt{t_{abs}}}{\sqrt{E}}$$

For a fixed thickness of an active layer the energy resolution improves with decreasing absorber thickness. The above formula is not valid if the crossings between consecutive active layers are correlated, i.e. when t_{abs} is small. A generally valid formula is:

$$\frac{\sigma_s}{E} = \frac{5\%}{\sqrt{E}}\left(1-f_{samp}\right)\,\Delta E_{cell}^{0.5\left(1-f_{samp}\right)}$$

where ΔE_{cell} is the energy deposited in a unit sampling cell i.e. 1 active and 1 absorber layer. f_{samp} is labeled the *sampling fraction* and is the fraction of the total energy that is deposited in the active medium.
As $f_{samp} \to 1$, σ_s = 'a' $\to 0$ (usually $a \neq 0$ due to imperfections in calorimeter systems) and as $f_{samp} \to 0$, $\sigma_s \;\alpha\; \sqrt{\Delta E_{cell}} \;\alpha\; \sqrt{\Delta E_{abs}}$
The sampling fraction can be calculated as follows. If d is the thickness of active layer then

$$f_{samp} = 0.6 f_{mip} = 0.6\,\frac{d\left(\frac{dE}{dx}\right)_{act}}{\left[d\left(\frac{dE}{dx}\right)_{act} + t_{abs}\left(\frac{dE}{dx}\right)_{abs}\right]}$$

For a sampling calorimeter with 1 cm Pb and 1 cm scintillator plates $f_{mip} \approx 2/(12.75+2) \approx 13.5\%$. The fractional energy resolution as a function of $\sqrt{(d/f_{samp})}$ is shown in Figure 17 [15]. Clearly the energy resolution of gas calorimeters will be poor as the sampling fraction tends to be very low.

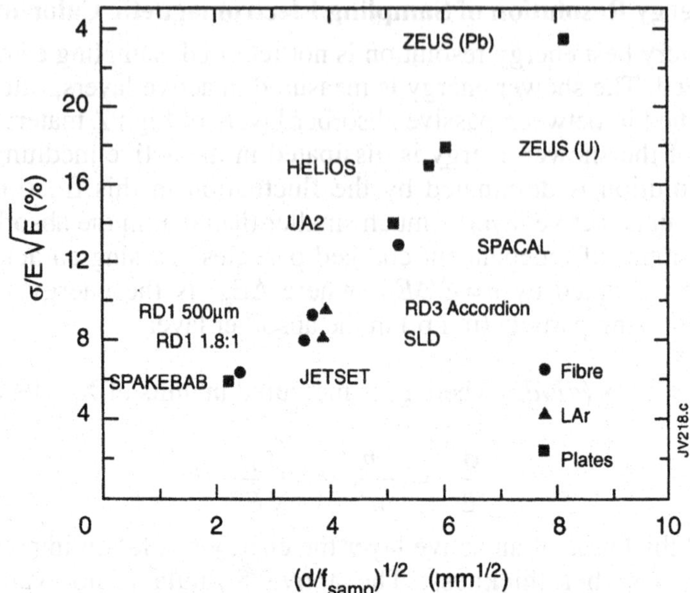

Figure 17: The fractional energy resolution of some calorimeters as a function of $\sqrt{(d/f_{samp})}$

6.4 Energy Resolution of Hadronic Calorimeters

Hadronic calorimeters, because of the large depth required ($\approx 10\lambda$), are by necessity sampling calorimeters. The response of a sampling electromagnetic calorimeter can be expressed as

$$E_{vis} = e\ E$$

where E, E_{vis} are incident and visible energies respectively and $e = f_{samp}$, the lectromagnetic sampling fraction. Similarly the response of a hadronic sampling calorimeter is

$$E_{vis} = e\ E_{em} + \pi\ E_{ch} + n\ E_n + N\ E_{nucl}$$

where E_{em}, E_{ch}, E_n, E_{nucl} are respectively the energy deposited by electromagnetic component, charged hadrons, low energy neutrons and energy lost in breaking up nuclei. Each component has its own sampling fraction. N is normally very small but E_{nucl} can be large e.g. it is $\approx$ 40% in Pb calorimeters. Hence the ratio of the response to electromagnetic and hadronic showers i.e. e/h is usually > 1 and the hadronic calorimeter is said to be *non-compensating*.

In hadronic calorimeters the fluctuation in the visible energy has two sources :

- sampling fluctuations as in the e.m. case which can be reduced by finer sampling and

- intrinsic fluctuation in the shower components (δE_{em}, δE_{ch} etc.) from shower to shower as seen in Figure 11.

Therefore the stochastic term is given by

$$a_h = \sigma_{samp} \oplus \sigma_{intr}$$

$$\sigma_{samp} = \frac{a}{\sqrt{E}} \quad where \quad a \approx 10\% \sqrt{\Delta E_{cell}}$$

$$\sigma_{intr} = \frac{a_{intr}}{\sqrt{E}} + c$$

where c is the constant term which depends on e/h and vanishes for a compensating calorimeter.

6.4.1 *The Neutral Component in Hadronic Cascade and the Role of e/h*

The hadronic showers have an electromagnetic component (F_0) which is determined essentially by the first interaction. There is a considerable event-to-event fluctuation in F_0. On average the neutral e.m. energy per interaction is f_0=1/3 as roughly one-third of the mesons produced in hadronic interations will be neutral [e.g. $(\pi^0/(\pi^+ + \pi^- + \pi^0)) \approx 1/3$]. If sufficiently energetic the 2nd generation $\pi^\pm$ may also produce π^0s. The value of F_0 after n generations can be estimated as follows. For

n=1 $F_0 \rightarrow f_0$
n=2 $F_0 \rightarrow f_0 + f_0 (1 - f_0)$
n=3 $F_0 \rightarrow f_0 + f_0 (1 - f_0) + f_0 (1 - f_0)^2$
....
$F_0 = [1 - (1-f_0)^\nu]$

This leads to $F_0 \rightarrow f_0$ at low energies and $F_0 \rightarrow 1$ at very high energies as n is large. This can be seen in Figure 18 [16] which shows a simulation of pions of 20 GeV and 200 GeV incident on lead. A large event-to-event fluctuation in the neutral fraction is evident. The increase in F_0 with energy is due to the fact that neutral pions, developing as e.m. showers, do not produce any hadronic interactions.

It usually turns out that the response to electrons and photons i.e. the e.m. component (labeled *e*) differs from that due to charged hadrons i.e. the non-e.m. component (labeled *h*). If E is the incident energy the response to electrons (E_e) and charged pions (E_π) can be written as :

$E_e = e\,E$, $\quad E_\pi = [\,e\,F_0 + h\,(1 - F_0)\,]\,E$ leading to

$$\frac{e}{\pi} = \frac{(e/h)}{[(e/h)F_0 + (1 - F_0)]}$$

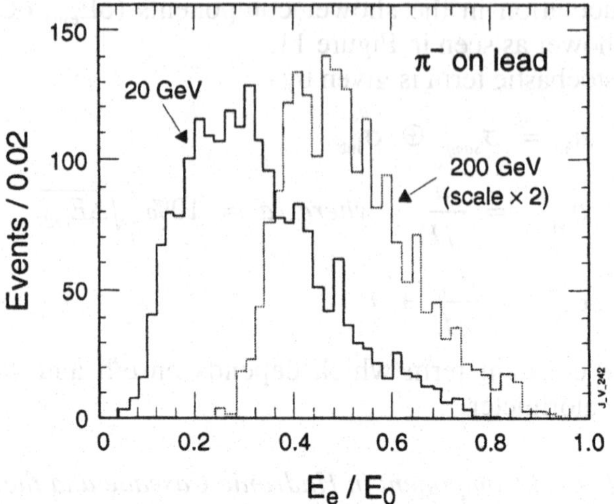

Figure 18: Distribution of e.m. energy fraction for charged pions incident on lead.

If e/h = 1 the calorimeter is said to be compensating.
Consider $dE_\pi = [\,(e\ -h\,)\,dF_0\,]\,E$. Then

$$\frac{dE}{E} = \frac{dF_0\,|(e/h) - 1|}{[(e/h)F_0 + (1 - F_0)]}$$

Hence the fractional error depends on e/h, F_0 and dF_0. If e/h=1 then there is no contribution due to the fluctuation dF_0. For example:

$$\frac{dF_0}{F_0} = \frac{df_0}{f_0} \sim \frac{1}{\sqrt{f_0\langle n\rangle}}$$

i.e. for a 200 GeV hadron, <n> ≈ 9, $dF_0 \approx 0.6 \Rightarrow (dE/E)_{comp} \approx 3.5\%$.

$$\left.\frac{dE}{E}\right|_{comp} \sim \frac{1}{\sqrt{\ln E}}\ \ and\ \ \to 0\ \ as\ \ E \to \infty\ \ \text{since}\ \langle n\rangle \propto \ln E$$

This aspect is illustrated by calorimeters using quartz fibres as active media. Charged particles traversing the fibres generate Cerenkov light which is guided to photomultipliers by the fibres themselves. Such a technique is employed by CMS for calorimetry in the very forward region ($3 < |\eta| < 5$) [8]. The aim is to measure the energies of, and tag, high energy jets from the WW fusion process. The signal in the calorimeter arises predominantly from the electromagnetic component as charged hadrons have a very high Cerenkov threshold when compared to that of electrons. Hence e/h is very large and the energy resolution at high energies will be dominated by the fluctuation in F_0. The resolution should improve as 1/lnE rather than as $1/\sqrt{E}$. Figure 19 shows the measured energy resolution of the CMS copper/quartz fibre calorimeter. Also shown is the resolution after subtraction of the contribution from photostatistics. It should be noted that the photostaistics contribution is sizeable as only about 1 photoelectron per GeV is generated.

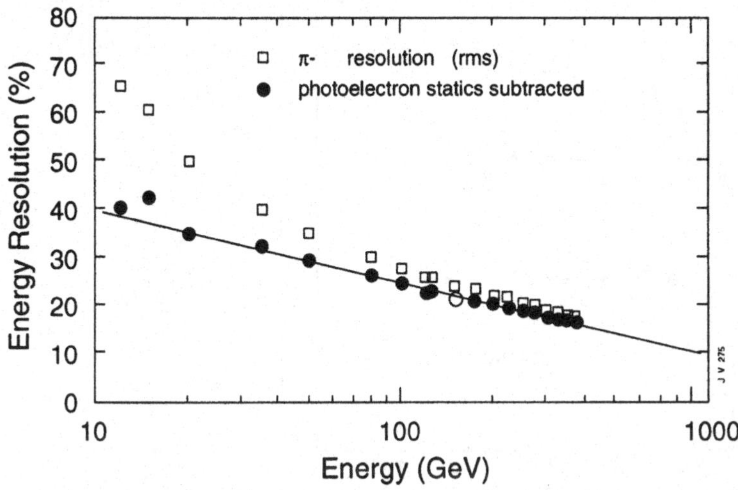

Figure 19: The measured pion energy resolution of a copper/quartz fibres calorimeter.

If |e/h| ≥ 10% the performance of the calorimeter is compromised because of the fluctuation in the π^0 content of the cascades. This leads to:
- a non-Gaussian measured energy distribution for mono-energetic hadrons,
- an e/π ratio that is different from unity and that varies with energy,
- a non-linear response in energy to hadrons,
- an additional contribution to the relative energy resolution (σ/E),
- a σ/e that does not improve as $1/\sqrt{E}$.

These effects have been observed and are illustrated in Figure 20 [3].

6.4.2 *Compensation*

The degree of (non-) compensation is expressed by the energy independent ratio *e/h*.. Th *e/h* ratio cannot be measured directly but can be inferred from the energy dependent e/π signal ratios. Two relations between the signal ratio e/π(E) and e/h by Groom [16] and Wigmans [17] are :

$$\frac{e}{\pi} = \frac{e/h}{1 + (e/h - 1)F_0}$$

$$F_0 = 1 - (E/0.76)^{-0.13} \qquad \textit{D. Groom}$$

$$\textit{or} \quad F_0 = 0.11 \ln E \qquad \textit{R. Wigmans}$$

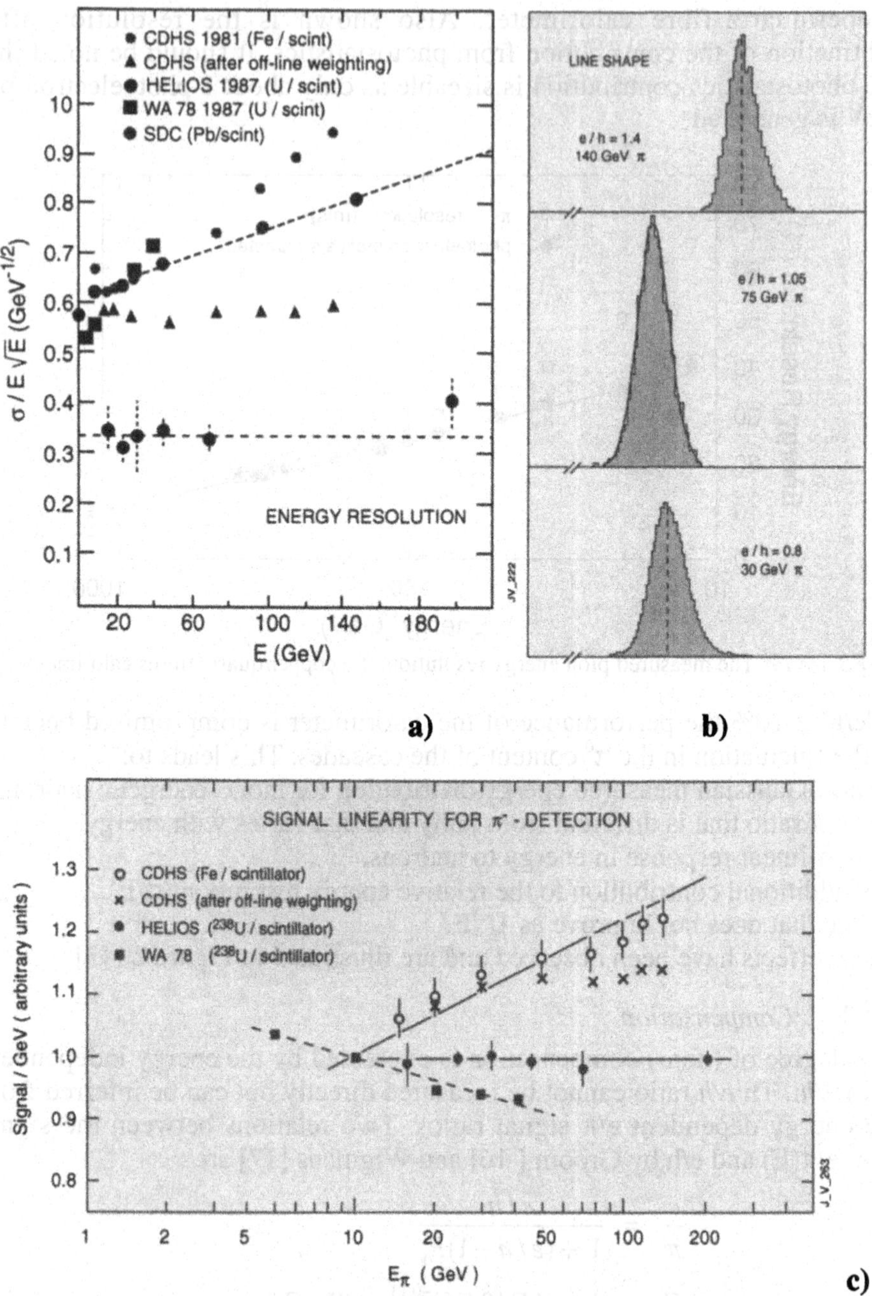

Figure 20: Experimental observation of the consequences of e/h≠1 [3]: a) the energy resolution: (σ/E).√E is plotted to show deviations from scaling for non-compensating devices, b) line-shape for monoenergetic pions is only Gaussian for the compensating calorimeter, c) the signal/GeV plotted as a function of pion energy, showing signal non-linearity for non-compensating calorimeters.

It is instructive to see how the energy is dissipated by a hadron in a Pb absorber. The breakdown of the dissipated energy is as follows:

42% in breaking up nuclei and not rendered measurable (invisible)
43% by charged particles
12% by neutrons with kinetic energy ~ 1 MeV
3% by photons with an energy ~ 1 MeV.

The sizeable amount of invisible energy loss means that hadronic calorimeters tend to be under-compensating ($e/h > 1$).

Compensation can be achieved in three ways;

boost the non-e.m. response using depleted uranium,
suppress e.m. response
boost the detectable response to low energy neutrons.

The ZEUS Collaboration[18] have found that achieving compensation for U/scintillator and Pb/scintillator calorimeters requires absorber/scintillator plate thickness ratios given by 1:1 and 4:1 respectively. They also used the technique of interleaved calorimeters to determine the intrinsic energy resolution of U and Pb calorimeters. This is accomplished by reading out odd and even scintillator layers separately. The results are as follows:

hadrons	Pb	$\sigma_{samp} = 41.2 \pm 0.9\%/\sqrt{E}$	$\sigma_{intr} = 13.4 \pm 4.7\%/\sqrt{E}$
	U	$\sigma_{samp} = 31.1 \pm 0.9\%/\sqrt{E}$	$\sigma_{intr} = 20.4 \pm 2.4\%/\sqrt{E}$
electrons	Pb	$\sigma_{samp} = 23.5 \pm 0.5\%/\sqrt{E}$	$\sigma_{intr} = 0.3 \pm 5.1\%/\sqrt{E}$
	U	$\sigma_{samp} = 16.5 \pm 0.5\%/\sqrt{E}$	$\sigma_{intr} = 2.2 \pm 4.8\%/\sqrt{E}$

The intrinsic fluctuations in a compensating Pb calorimeter are smaller than those for a U one. However the sampling has to be much coarser for Pb calorimeter leading to a much poorer e.m. energy resolution. ZEUS therefore chose U as the absorber material. It can also be seen that for compensating Pb and U calorimeters the energy resolution is dominated by sampling fluctuations and is given by

$$\sigma_{samp} = \frac{11.5\% \sqrt{\Delta E_{cell}(MeV)}}{\sqrt{E(GeV)}}$$

The sampling fluctuations for hadrons are larger than those for e.m. showers by a factor of 2. From the above it is evident that very good e.m. energy resolution is incompatible with $e/h = 1$.

6.5 Jet Energy Resolution

Hadronic calorimeters are primarily used to measure the energies of jets and hence the quantities that characterize their are:

• jet energy resolution and energy linearity,

• missing transverse energy resolution.

The jet energy resolution is limited by effects from
• algorithms used to define jets (energy is dependent on cone radius, lateral segmentation of cells etc.),
• the fluctuation in the particle content of jets due to differing fragmentation from one jet to another,
• the fluctuation in the underlying event,
• the fluctuation in energy pileup in high luminosity hadron colliders
• magnetic field.

In experiments on e+e- machines the jet energy resolution can be improved as the centre of mass energy can be used to constrain the energies of jets if the jet directions are measured relatively precisely.

6.5.1 Jet Energy Resolution

Jet energy resolution can be deduced using single particle resolution in the limit that either stochastic or constant terms dominate. Consider two cases: one in which a single particle with energy E and the second in which a jet of particles, each carrying energy $k_i = z_i E$ where $\Sigma z_i = 1$, is incident on the calorimeter. Assume first that the stochastic terms dominate. Since

$$\frac{\sigma(E)}{E} = \frac{a}{\sqrt{E}} \oplus c$$

$$\sum_i z_i = 1, \ \sum_i k_i = E \qquad \frac{dk_i}{k_i} = \frac{a}{\sqrt{k_i}} \oplus c$$

$$dk_i = a\sqrt{k_i}$$

$$dE_J = \sqrt{\sum_i (dk_i)^2} = \sqrt{\sum_i a^2 k_i} = a\sqrt{E}$$

$$\therefore \ \frac{dE}{E} = \frac{a}{\sqrt{E}}$$

Therefore an ensemble of particles act, with respect to errors, as a single particle. In the high energy regime where the constant term dominates

$$dE_J \approx \sqrt{(cz_i E)^2} = cE\sqrt{z_i^2}$$

Assuming that there is a leading particle, *l*, with energy fraction *zl*, then

$$\frac{dE_J}{E} \approx c\ E\ z_l$$

For fragmentation function $zD(z) = (1-z)^2$ $\langle z_l \rangle \approx 0.23$ implying that the constant term is reduced. For a calorimeter with a = 0.3 and c = 0.05, in which a 1 TeV jet fragments into 4 hadrons of equal energy, the error on the energy decreases form 50 GeV to 25 GeV.

6.5.2 Di-Jet Mass Resolution v/s Cone Size

One figure of merit of a hadron calorimeter is di-jet mass resolution. For the purposes of measuring the jet energy resolution low p_t di-jets (50 $<p_t<$60 GeV), high p_t di-jets (500 $<p_t<$600 GeV) and high mass di-jets (3 $< m_{Z'} <$4 TeV) at the LHC can be used [19]. The mass resolution for the three categories v/s cone size, ΔR, where $\Delta R = \sqrt{(\Delta\eta^2+\Delta\phi^2)}$ in pseudorapidity (η) and ϕ space, is shown in Figure 21a for a perfect calorimeter with no underlying event. It can be seen that the mass resolution improves with increasing cone size. However when running at high luminosity there are ≈<30> minimum bias events which accompany the event of interest. The fractional mass resolution as a function of cone size is plotted in Figure 21b.

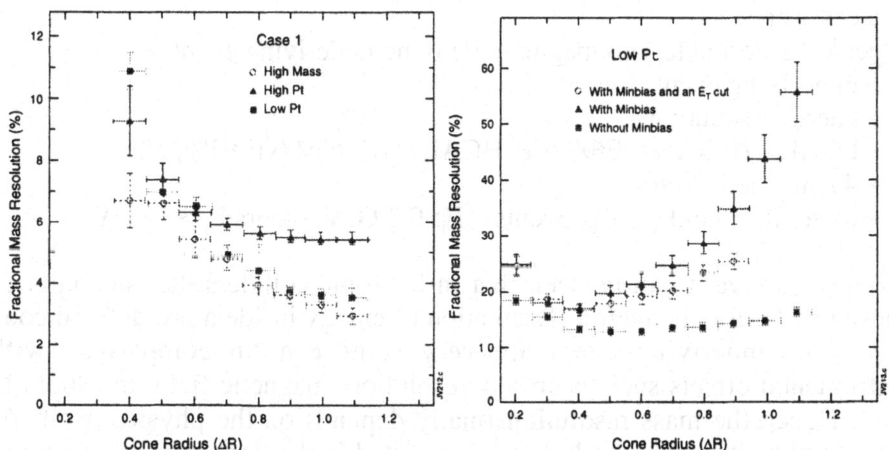

Figure 21: The fractional jet-jet mass resolution as function of cone radius a) perfect calorimeter, b) with 30 minimum bias events overlapped with the event of interest.

The mass resolution for low and high p_t di-jets is tabulated in Table 2 for different conditions.

When running at high luminosity at the LHC there are ≈ <25> minimum bias events that accompany the event of interest. The fractional mass resolution as afunction of the cone size is plotted in Figure 21.

Table 2: Mass resolution (in%) for low (top) and high p_t di-jets for different conditions (see text)

ΔR	Case 1	Case 2	Case 3	Case 4	Case 5
0.4	10.9	10.3	12.2	13.2	12.9
0.5	7.0	10.1	11.9	12.8	12.6
0.6	5.5	10.9	13.1	12.8	13.1
0.7	4.9	11.2	13.7	13.6	13.3
0.8	4.4	12.0	13.7	13.6	13.3
0.9	3.7	13.0	14.4	14.3	13.8
1.0	3.6	14.3	16.0	14.8	-

ΔR	Case 1	Case 2	Case 3	Case 4	Case 5
0.4	6.7	-	7.3	-	6.4
0.5	6.6	6.6	7.1	6.4	5.9
0.6	5.4	5.8	6.4	5.6	5.6
0.7	4.8	5.2	6.0	5.2	5.6
0.8	3.9	5.1	6.0	4.9	5.5
0.9	3.6	4.9	5.7	4.9	4.7
1.0	3.3	4.9	5.7	4.9	4.8

The cases are :
1: perfect colorimeter, no magnetic field, no underlying event,
2: + underlying event,
3: + energy resolution,
ECAL – $\sigma/E=3\%/\sqrt{E}\oplus0.5\%$, HCAL: $\sigma/E=60\%/\sqrt{E}\oplus3\%$, e/h=1
4: + 4T magnetic field
5: + tower threshold (low p_t events – E_t>0.3 GeV, others E_t > 1 GeV)

From the above it can be seen that in hadronic colliders the uncertainties caused by jet fragmentation (fluctuation of energy inside a pre-defined cone size) and underlying event are very significant in comparison with instrumental effects such as energy resolution, magnetic field, threshold E_t etc.). Hence the mass resolution finally depends on the physics itself. At high luminosities the resolution is degraded if the cone-size is too small (some signal energy is excluded) or if the cone size is too large (significant pileup energy is included). In order to obtain the best mass resolution the cone size has to be optimised for each process and instantaneous luminosity.

6.5.3 *Di-jet Mass Resolution v/s Calorimeter Lateral Segmentation*

The mass resolution due to the angular error, dθ, in defining the jet axis is given by:

$$\frac{dM}{M} = \frac{p_T}{M} d\theta$$

Only highly boosted and low mass di-jets (e.g. boosted Zs from H→ZZ) will have a significant contribution from the angular error. This is illustrated in Figure 22 [19].

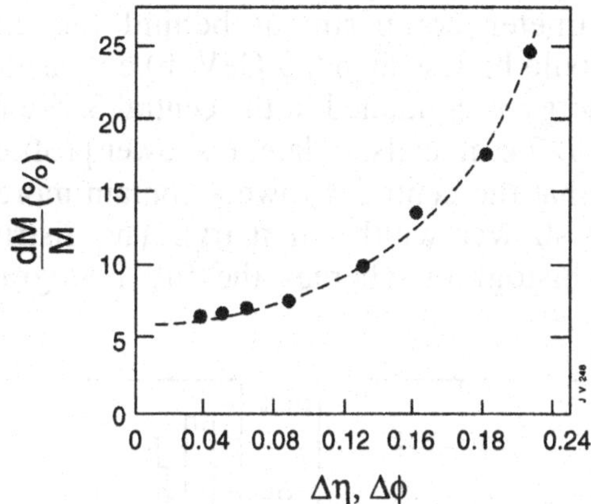

Figure 22: The fractional jet-jet mass resolution as a function of the tower size.

7. PARTICLE IDENTIFICATION USING CALORIMETERS

Several channels from potential new physics at the LHC may appear through final states containing leptons or photons e.g. some decay modes of the Higgs boson such as $H \to \gamma\gamma$ or $H \to ZZ^* \to 4l$. Such modes have very small cross-sections in the range of 10-100 fb. However the backgrounds from QCD processes can be large. Hence a large rejection factor against the background is required while maintaining a high efficiency for the signal. Below we consider some ways in which calorimeters can be used to identify isolated electrons and photons from hadrons and jets.

7.1 Isolated electromagnetic shower-jet separation

The largest source of electromagnetic showers is from the fragments of jets, especially π^0s. A leading π^0 taking most of the jet energy can fake an isolated photon. There are large uncertainties in jet production and fragmentation. Furthermore the ratio of production of di-jets to irreducible di-photon background is ≈ 2.10^6 and γ-jet/irreducible γγ is ≈ 800. Hence a rejection of ≈ 5000 against jets is needed.

Jets can be distinguished from single electromagnetic showers by

• demanding an energy smaller than some threshold in the hadronic compartment behind the electromagnetic one

• using isolation cuts

• demanding a lateral profile of energy deposition in the ECAL consistent with that from an electromagnetic shower.

Using these criteria ATLAS [7] estimates that the rejection factor against jets can be ≈ 1500 for a photon efficiency of 90%. This is illustrated in Figure 23 where the effect of various cuts is shown: a) the energy (E_T^{had}) in the hadron calorimeter compartment behind the e.m. one of size $\Delta\eta\times\Delta\phi$=0.2x0.2 should be less than 0.5 GeV, b) e.m. isolation (R_{isol})– more than 90% of the energy is contained in the central 3×5 e.m. cells compared with that in central 7×7 e.m. cells, c) lateral shower profile ($R_{lateral}$)– look for an e.m. core such that the central 4 towers contain more than 65% of the shower energy, d) shower width in η (σ_η). The distribution for jets is shown as dashed histogram whereas the full histograms depict single photons.

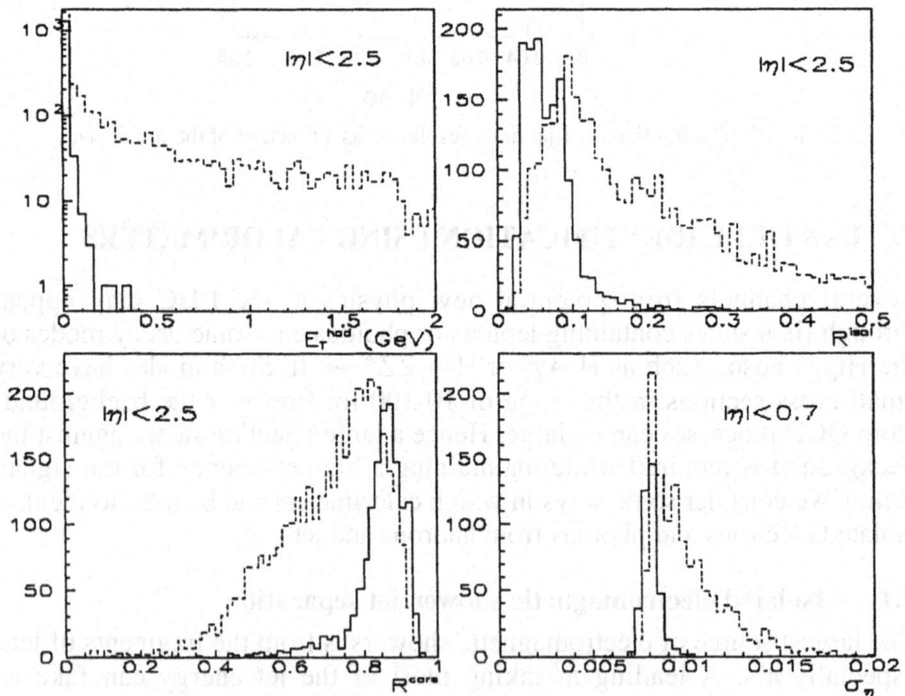

Figure 23: The distributions used to cut against jets. Solid histogram is for photons and the dashed one for jets. See text for explanation.

7.2 Photon – π^0 separation

After the application of the above criteria only jets resulting in leading π^0s can fake genuine single photons. Further rejection can only be achieved by the recognition of two e.m. showers close to each other. CMS [8] uses the fine lateral granularity (≈2.2cm×2.2cm) of their crystals and a neural network algorithm that compares the energy deposited in each of the 9 crystals in a 3×3 crystal array with that expected from a single photon. Variables are constructed from the 9 energies, x and y position of impact and a pair measuring the shower width. The fraction of π^0s rejected is shown in Figure 24.

The narrowness of the e.m. shower in the early part can be used to reject events consisting of two close-by e.m. showers. Planes of fine pitch orthogonal strips after a pre-shower, placed at a depth of $\approx 2.5\ X_0$, can also be used to distinguish π^0s from single photons. Results using 2mm pitch strips are shown in Figure 24.

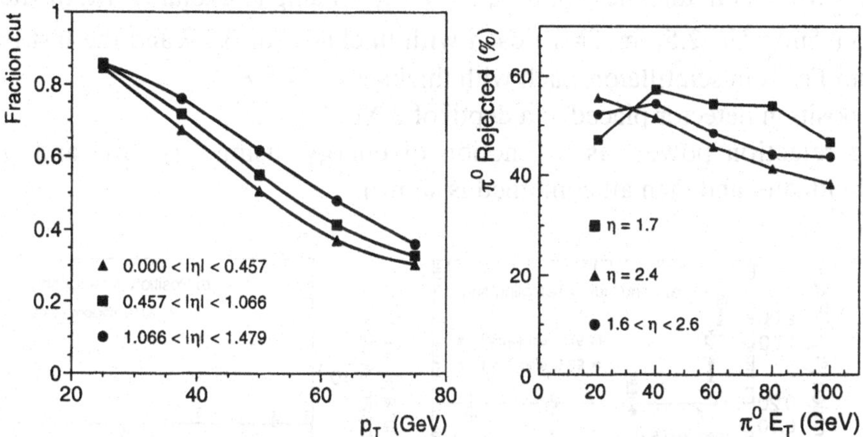

Figure 24: lhs) Fraction of pizeros rejected using lateral shape of energy deposit as a function of p_T rhs) Variation of pizero rejection as a function of η using two planes of orthogonally oriented 2mm pitch Si strips after 2 and 3 X_0.

Electron-hadron Separation

A high energy pion faking an electron leads to the contamination of signals using prompt electrons. At LHC in order to bring down the rate of fake electrons from this source to a factor $\approx$ 10 below that from the genuine sources (e.g. b –> e X, W –> eν etc.) an e-π separation of ≥ 1000 is required for $p_T \geq 10$ GeV/c.

The electron–hadron separation is usually based on the difference in the longitudinal and lateral development of showers intiated by electrons and charged hadrons. One or more of the following can be used to achieve the desired pion rejection power when detecting electrons :

- a preshower detector between $\approx 1.5 - 4\ X_0$
- lateral segmentation
- longitudinal segmentation including a hadron calorimeter
- energy - momentum matching

The ultimate rejection power is limited by the charge exchange process or the first hadronic interaction, which results in one or several π^0's taking most of the energy of the incoming hadron. The shower from such hadrons then looks like an e.m. shower. Therefore sampling of showers early in their longitudinal development is important.

The separation power for single particle, using (i - iii) is shown in Figure 25 [20]. The structure of the calorimeter consisted of :

• towers of a lateral size of ~11 x 11cm (effective $X_0 \approx$ 8mm),
• 8-fold longitudinal segmentation, the first four samplings (2mm U / 2.5 mm TMP) with thickness of 3, 6, 10, 7 X_0 leading to a total of 1λ, the next two (5mm U/ 2.5mm TMP) each with thickness of 0.7 λ and the last two (5cm Fe/ 1cm scintillator) each with thickness of 2.5 λ.
• a position detector placed at a depth of 3 Xo.

The rejection power, as a function of energy, using (ii), (iii) and (iv) individually and then all combined is shown.

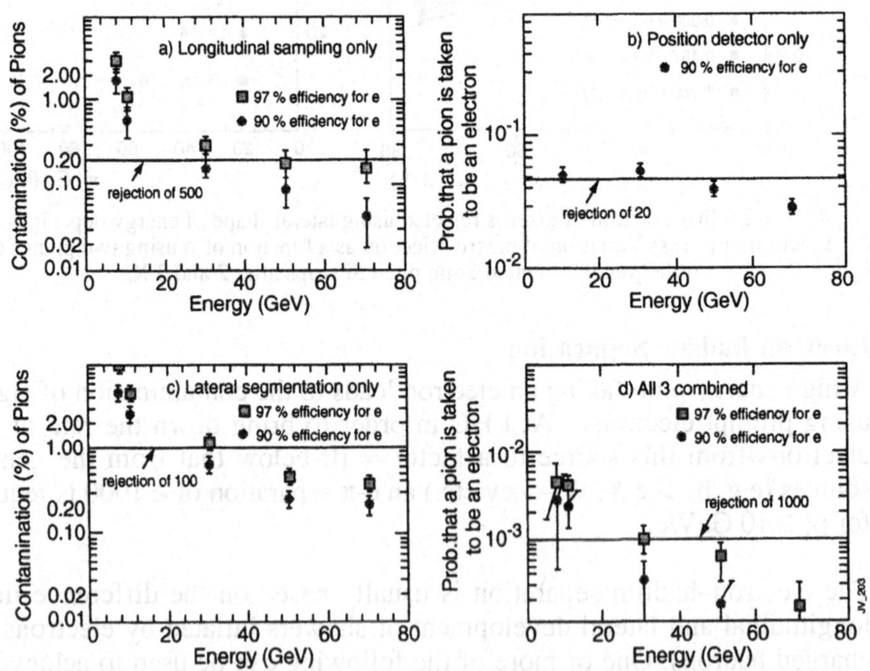

Figure 25: The probability that a single pion is taken to be an electron using a) longitudinal profile only, b) preshower detector only, c)lateral profile only and d) all three combined.

8. ELECTRONICS NOISE

Noise is any unwanted signal that obscures the desired signal. Therefore noise degrades the accuracy of the measurement. There are two types of noise: *intrinsic* and *extrinsic* noise. The intrinsic noise is generated in the detector or electronics and cannot be eliminated though possible reduced. The extrinsic noise is due to pickup from external sources or unwanted feedback (e.g. ground loops, power supply fluctuations etc.) and is usually eliminated by proper design.

Intrinsic noise has two principal components namely :

• thermal noise (Johnson or Nyquist noise) – *series noise*

Any resistor, R, will develop a voltage across its ends whose average value is zero but r.m.s. is

$$\langle v^2 \rangle = 4ktR.\Delta f$$

• shot noise – *parallel noise*

This source arises from fluctuation in the charge carriers and is given by

$$ENC^2 = \frac{4ktR_s(C_d + C_{in})^2}{\tau} I_s + I_n \tau I_p$$

where C_d is detector capacitance, C_{in} is input capacitance of the amplifier, I_n is leakage current, τ is the shaping time and I_s, I_p are series and parallel noise integrals ($\approx$ 1 for $(RC)^2$ shaping). For example, for τ = 50 ns, and a leakage current of 1 μA, ENC $\approx$ 800 electrons. Further examples are considered in Sections 9.3 and 10.2.

9. INORGANIC SCINTILLATORS

The desirable properties of a scintillator are:

• a high efficiency of conversion of deposited energy into scintillation light,
• a conversion to light that is proportional to the energy deposited,
• a high light output,
• a medium that is transparent to its emitted light,
• a short luminescence decay time,
• a refractive index n $\approx$ 1.5 for efficient coupling to photosensors
• radiation hardness for LHC operation.

No material simultaneously meets all these criteria. Inorganic scintillators (e.g. sodium iodide) have the best light output and linearity whilst organic scintillators (e.g. plastic scintillator) have faster light output but smaller light yield and display saturation of output for radiation with high linear energy transfer. Two types of light emission are possible: *fluourescence* resulting in prompt emission of light in the visible wavelength range and *phosphorescence* resulting in slower emission of light at longer wavelengths.

The most demanding physics channel for an electromagnetic calorimeter at the LHC is the two-photon decay of an intermediate-mass Higgs boson. The background is large and the signal width is determined by the calorimeter performance. The best possible performance in terms of energy resolution only possible using fully active calorimeters such as inorganic scintillating crystals.

Inorganic scintillators have crystalline structure. The valence band contains electrons that are bound at the lattice sites whereas electrons in the conduction band are free to move throughout the crystal. Usually in a pure crystals the efficiency of scintillation is not sufficiently large. A small amount of impurity, called an activator, is added to increase the probability of emission of visible light. Energy states within the forbidden gap are created through which an electron, excited to the conduction band, can de-excite. Passage of a charged particle through the scintillator creates a large number of electron-hole pairs. The electrons are elevated to the conduction band whereas the +ve holes quickly drift to an activator and ionize it. The electrons migrate freely in the crystal until they encounter ionised activators. The electrons drop into the impurity sites creating activator excited energy levels which de-excite typically with $T_{1/2} \approx 100$ ns. In a wide category of materials the energy required to create an electron-hole pair is $W \approx 3E_g$ e.g. in NaI, $W \approx 20$ eV, NaI(Tl) $N_\gamma \approx 40000\gamma$/MeV of ≈ 3 eV.

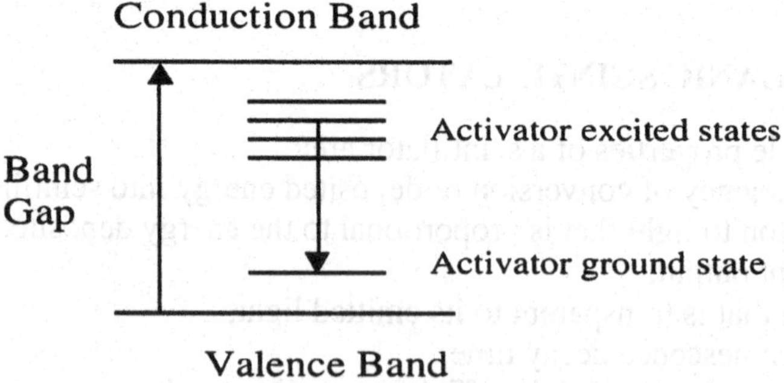

Figure 26: The energy level diagram for a scintillating crystal containing an activator

The consequence of luminescence through activator sites is that the crystal is transparent to its own scintillation light. In this case the emission and absorption bands do not overlap and self-absorption is small. The shift towards longer wavelengths is known as *Stokes' shift*.

The scintillation mechanism in crystals without activators is more complex. For example, in lead tungstate the intrinsic emission in the blue is through excitons localized on the Pb site whereas the green emission is due to defects in the crystalline structure linked to oxygen vacancies [21].

The properties of various crystals used in high energy experiments are given in Table 3. The parameters of some of the recently designed crystal calorimeters are given in Table 4 [22].

Table 3: Properties of various scintillating crystals.

Crystal		NaI(Tl)	CsI(Tl)	CsI	BaF_2	BGO	CeF_3	$PbWO_4$
Density	$g.cm^{-2}$	3.67	4.51	4.51	4.89	7.13	6.16	8.28
Rad. length	cm	2.59	1.85	1.85	2.06	1.12	1.68	0.89
Moliére radius	cm	4.5	3.8	3.8	3.4	2.4	2.6	2.2
Int. length	cm	41.4	36.5	36.5	29.9	22.0	25.9	22.4
Decay Time	ns	250	1000	35 6	630 0.9	300	10-30	<20>
Peak emission	nm	410	565	420 310	300 220	480	310- 340	425
Rel. Light Yield	%	100	45	5.6 2.3	21 2.7	9	10	0.7
d(LY)/dT	%/°C	≈ 0	0.3	- 0.6	- 2 ≈ 0	- 1.6	0.15	-1.9
Refractive Index		1.85	1.80	1.80	1.56	2.20	1.68	2.16

Table 4: Parameters of various experiments using scintillating crystals.

Experiment		KTeV	BaBar	BELLE	CMS
Laboratory		FNAL	SLAC	KEK	CERN
Crystal Type		CsI	CsI(Tl)	CsI(Tl)	$PbWO_4$
B-Field	T	-	1.5	1.0	4.0
Inner Radius	m	-	1.0	1.25	1.3
No. of crystals		3,300	6,580	8,800	76,150
Crystal Depth	X_0	27	16-17.5	16.2	26
Crystal Volume	m^3	2	5.9	9.5	11
Light Output	p.e./MeV	40	5,000	5,000	2
Photosensor		PMT	Si PD	Si PD	APD*
Gain of photosensor		4,000	1	1	50
Noise / channel	MeV	Small	0.15	0.2	30
Dynamic Range		10^4	10^4	10^4	10^5

* APD: Si avalanche photodiode

9.1 Radiation Damage in Crystals

All crystals suffer from radiation damage at some level. It is rare that irradiation affects the scintillation mechanism itself. However formation of colour centres takes place leading to absorption bands. A colour centre is a crystal defect that absorbs visible light. A high concentration of blue light colour centres makes crystals yellowish. The simplest colour centre is an F-centre where an electron is captured in an anion vacancy. The consequence of colour centre production is a decrease in the light attenuation length leading to a decrease in the amount of light incident on the photosensor. This is illustrated in Figure 27 for various samples of $PbWO_4$ crystals grown under differing conditions. The crystals were irradiated using γs, incident at the front of the crystal, from a ^{60}Co source.

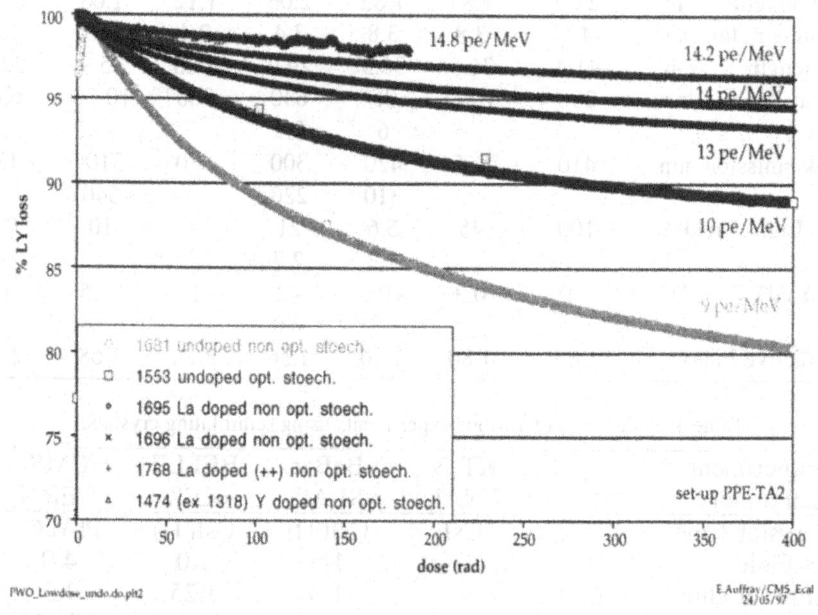

Figure 27: The loss in the collected light as a function of dose (delivered at ≈ 0.15 Gy/hr) for crystals grown under various conditions.

Extensive R&D has been carried out over the last 5 years in order to improve the radiation hardness of $PbWO_4$ crystals [21]. Generally the strategy has been to decrease the concentration of defects that lead to colour centre production by optimizing the stoechiometry (the concentration of PbO and WO_3 in the melt) and annealing after the growth of the crystal. The remaining defects are compensated by specific doping, e.g. by pentavalent elements on the W site and trivalent on the Pb site, and by improving the purity of the raw materials. The levels of improvement can be seen from Figure 27. The most recent crystals of lead tungstate have shown very good resistance to irradiation. This is illustrated in Figure 28. The loss of collected light, for crystals doped with both niobium and yttrium, show a decrease in the collected light of less than 2% at saturation. The effect of irradiation can also be dose-rate dependent.

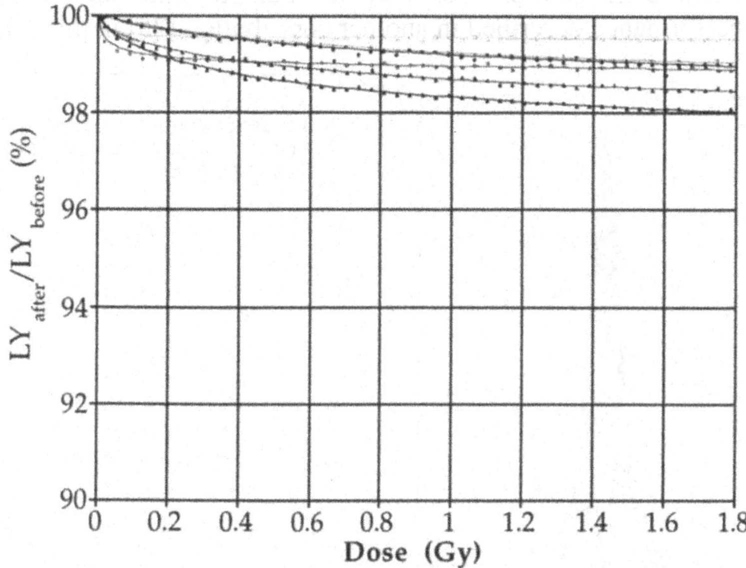

Figure 28: The loss in the collected light as a function of dose (delivered at ≈ 0.15 Gy/hr) for recent $PbWO_4$ crystals doped with Niobium and Yttrium.

The radiation dose expected at the shower maximum for the lead tungstate calorimeter of CMS, over the first ten years of LHC operation, is below 4,000 Gy in the barrel region ($|\eta| < 1.5$), ≈ 70,000 Gy at $|\eta| \approx 2.5$ rising to 200,000 Gy at $|\eta| \approx 3.0$. Furthermore the expected dose rate at design luminosity, and shower maximum, is below 0.3 Gy/h in the barrel region, ≈ 6 Gy/h at $|\eta| \approx 2.5$ rising to 15 Gy/h at $|\eta| \approx 3.0$.

9.2 Performance of CMS Lead Tungstate Crystals

Several matrices of improving quality have been tested in electron beams over the last few years [23,24]. Radiation damage leads to a decrease in the attenuation length and hence in the collected light. As the efficiency of the scintillation mechanism is not affected by irradiation the energy resolution will not be affected as long as the attenuation length does not fall below ≈ 2-3 time the length of the crystal. The small loss of light can be corrected by regularly measuring the response to a known amount of light injected into crystals. This has been demonstrated in beam tests [24].

Results from a recently tested prototype are shown in Figure 29. The distribution of the sum of energy in 9 crystals for electron of an energy of 280 GeV is shown. An excellent energy resolution is measured without significant tails. The measured energy resolution is also shown. The stochastic term is expected to be < 3% in the final calorimeter since the surface area of the photosensor will be doubled. This should result in physics performance described in another contribution [25] in this school.

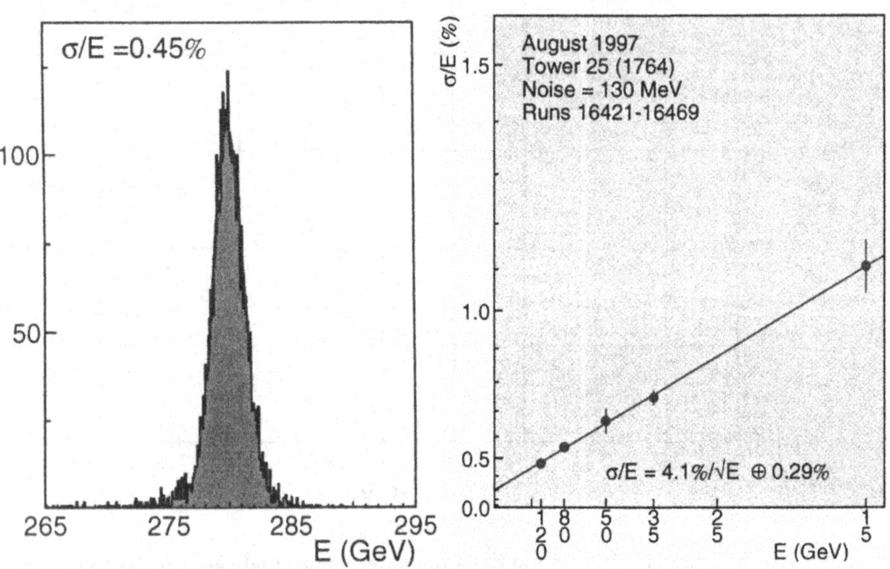

Figure 29: a) The distribution of the sum of energy in 9 crystals for an electron of an energy of 280 GeV, b) the measured energy resolution

9.3 Photosensors

9.3.1 Photomultipliers

The contribution to the energy resolution from the process of conversion of light to photoelectrons can be significant. For example, in a lead glass calorimeter about 10,000 Cerenkov photons/GeV impinge on the photomultiplier. The conversion leads to about 1000 photoelectrons/GeV and hence the contribution to the stochastic term will be

$$\sigma_{pe} = \frac{\sigma}{E} = \frac{\sqrt{1000}}{1000} \approx 3.2\%$$

The maximum number of independent $e^{\pm}$ particles, given that the Cerenkov threshold is 0.7 MeV, is 1000/0.7 per GeV i.e. n = 1400 $e^{\pm}$. This leads to an additional contribution to the energy resolution i.e. $\sigma_n = (\sqrt{1400})/1400 \approx 2.7\%$. The observed resolution then becomes

$$\frac{\sigma}{E} = \sqrt{\sigma_n^2 + \sigma_{pe}^2} \approx 4.5\%$$

An energy resolution of σ/E ~ 5% / √E for e.m. showers has been attained in a large lead glass array [26].

9.3.2 Silicon Avalanche Photodiodes

The light output from $PbWO_4$ crystals is low. These crystals are deployed by CMS in a 4T transverse magnetic field and the use of photomultipliers is excluded. Unity gain Si photodiodes cannot be used since even the small rear shower leakage from 25 X_0 deep crystals considerably degrades the energy resolution [23]. This is due to the fact that the response to ionising radiation is significant compared to the signal due to scintillation light. Hence CMS use Si avalanche photodiodes (APDs) with a gain of about 50. The particularity of these devices, over and above photomultipliers, is the noisy amplification process. The working principle of these devices is shown in Figure 30.

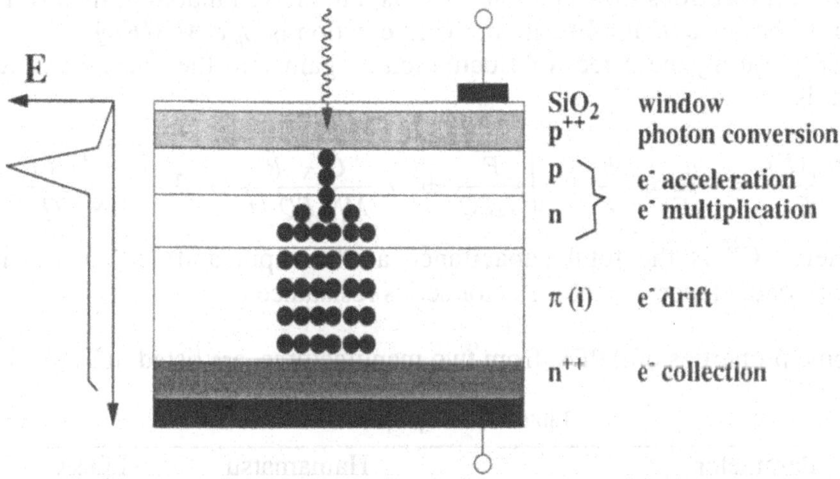

Figure 30: The working principle of a Si avalanche photodiode.

Consider a crystal with a light yield of N_γ photons/MeV. N_γ.E photons hit the APD for an energy deposit E. Assuming a quantum efficiency Q (which can easily be ≈ 85% for APDs),

No. of photoelectrons is $N_{pe} = N_\gamma E.Q$

Then the photostatistics fluctuation is $\pm\sqrt{N_{pe}}$

If there is no fluctuation in the gain process then the no. electrons transferred to the amplifier is (M=gain) $M.N_{pe} \pm M\sqrt{N_{pe}}$

BUT if the multiplication process is noisy and the gain itself has a fluctuation, σ_M, then the no. of electrons is $M.N_{pe} \pm \sqrt{(M^2 + \sigma_M^2)}\sqrt{N_{pe}}$

Hence the photostatistics contribution to the energy resolution becomes

$$\frac{\sigma_{pe}(E)}{E} = \frac{1}{\sqrt{N_\gamma EQ}}\sqrt{\frac{M^2+\sigma_M^2}{M^2}} = \frac{1}{\sqrt{N_\gamma EQ}}\sqrt{F}$$

where F is called the 'excess noise factor' and quantifies the induced degradation in the energy resolution due to fluctuations in the amplification process. Typically for APDs F≈ 2 and for photomultipliers F ≈ 1.2.

Another source of energy resolution degradation arises when APDs are damaged by irradiation. APDs behave as conventional Si devices under irradiation and the leakage current increases with the same damage constant ($\alpha = 2.10^{-17}$ A/cm). The leakage current can have two sources: surface and bulk current. The surface leakage (I_s) current does not undergo multiplication whereas the bulk current generated in the amplification region (I_b) does.

For I_s, if electrons flow at a rate of 1 e/s, but arrive randomly, then $\sigma_s \,\alpha\, \sqrt{I_s}$.
For I_b, because of amplification, the fluctuation is $\sigma_b \,\alpha\, M\, \sqrt{(FI_b)}$
Incorporating the effect of fluctuation in gain into the energy resolution yields

$$\frac{\sigma_{pe}(E)}{E} = \frac{a}{\sqrt{E}} \oplus \frac{b}{E} \oplus \sqrt{\frac{F}{N_\gamma EQ}} \oplus \alpha\, \frac{"C"\sqrt{"R"}}{MN_\gamma EQ\sqrt{\tau}} \oplus \beta\, \frac{\sqrt{I_s + M^2 FI_b}}{MN_\gamma EQ}\sqrt{\tau}$$

where "C" is the total capacitance at the input and "R" is the input transconductance + photodetector series resistance.

Some properties of APDs, from two manufactures, are listed in Table 5.

Table 5: Some properties of APDs

Parameter	Hamamatsu	EG&G
Active Area	25 mm^2	25 mm^2
Quantum Efficiency at 450nm	80%	75%
Capacitance	100 pF	25 pF
Excess Noise Factor, F	2.0	2.3
Operating Bias Voltage	400-420 V	350-450 V
dM/dV x 1/M at M=50	5%	0.6%
dM/dT x 1/M at M=50	-2.3%	-2.7%
Passivation Layer	Si_3N_4	Si_3N_4

9.4 System Aspects

A real calorimeter is a system comprising active media, electronics chain, mechanical structure, all enclosed in an environment that must be kept stable. Hence many other factors have to be considered in order to maintain the resolution achieved in beam tests. For example, in the case of the CMS ECAL, the temperature of the crystals has to be maintained to within 0.1% since both the crystal and the photosensor have a temperature dependence of the output signal of d(Signal)/dT ≈ -2%/°C. This requires a powerful cooling system and a hermetic environmental shield. To maintain uniformity of response across crystals the mechanical structure has to be thin and preferably made of low-Z material. Furthermore, no load from one crystal should be transferred to its neighbours. A 300 μm glass fibre alveolar structure has been chosen by CMS. The electronics system has to provide a stable response, deliver high resolution of digitization (12-bits) and a large dynamic range (≈16-bits) whilst preserving a low electronics noise per channel (<40 MeV/channel). Furthermore, the on-detector electronics must be radiation hard and have as low a power consumption as possible.

More information on the systems aspects of calorimeters can be found in the ATLAS [7] and CMS [8] Technical Design Reports.

10 CALORIMETRY USING NOBLE LIQUIDS

Calorimeters using liquid filled ionization chambers as detection elements have several important advantages. The absence of internal amplification of charge results in a stable calibration over long periods of time provided that the purity of the liquid is sufficient. The number of ion pairs created is large and hence the energy resolution is not limited by primary signal generating processes. The considerable flexibility in the size and the shape of the charge collecting electrodes allows high granularity both longitudinally and laterally.

The desirable properties of liquids used in ionization chambers are:

- a high free electron or ion yield leading to a large collected charge,
- a high drift velocity and hence a rapid charge collection,
- a high degree of purity. The presence of electron scavenging impurities leads to the reduction of electron lifetime and consequently a reduction in the collected charge.

The properties of noble liquids are given in Table 6.

10.1 Charge Collection in Ionisation Chambers

Ionisation chambers are essentially a pair of parallel conducting plates separated by a few mm and with a potential difference in an insulating liquid (e.g. liquid argon).

Consider what happens when a single ion-pair is created at a distance (d-x) from the +ve electrode (Figure 31a). The electron drifts towards the +ve electrode and induces a charge

$$Q = -e\,\frac{(d-x)}{d}$$

where d is the width of the gap. Assuming that the electron drifts with a velocity v, and the time to cross the full gap is v_d, then the induced current is

$$i(t) = \frac{dQ}{dt} = -e\,\frac{v}{d} = -\frac{e}{t_d}$$

The contribution from the drifting ions can be neglected as their drift velocity is about three orders of magnitude smaller than that for electrons.

Table 6: Properties of noble liquids.

		LAr	LKr	LXe
Density	g/cm^3	1.39	2.45	3.06
Radiation Length	cm	14.3	4.76	2.77
Moliere Radius	cm	7.3	4.7	4.1
Fano Factor		0.11	0.06	0.05
Scintillation Properties				
Photons/MeV		-	1.9 10^4	2.6.10^4
Decay Const. Fast	ns	6.5	2	2
Slow	ns	1100	85	22
% light in fast component		8	1	77
λ peak nm		130	150	175
Refractive Index @ 170nm		1.29	1.41	1.60
Ionization Properties				
W value	eV	23.3	20.5	15.6
Drift vel (10kV/cm)	cm/μs	0.5	0.5	0.3
Dielectric Constant		1.51	1.66	1.95
Temperature at triple point	K	84	116	161

Now consider the case where a charged particle traverses the gap (Figure 31b). Suppose N ion-pairs are produced and are uniformly distributed across the gap. The fraction of electrons still moving at a time t after traversal is $(t_d-t)/t_d$ for $t_d < t$. Therefore

$$i(t) = -Q_0\,\frac{v}{d}\left(1-\frac{t}{t_d}\right)$$

where Q_0 = Ne and the current is at its maximum at time t = 0 and disappears once all the charges have crossed the drift gap. This time is about 400 ns for a 2 mm LAr gap.Hence

$$q(t) = \int_0^t i(t)\, dt = -Q_0 \left(\frac{t}{t_d} - \frac{t^2}{2t_d^2} \right) \quad for \; t < t_d$$

The total collected charge (for t >t_d) is

$$Q_c = \frac{Q_0}{2}$$

The factor two is due to uniform distribution of ionisation. During drift the electrons can be trapped by impurities. Then the induced current will be reduced. In fact if the electron lifetime is τ

$$i(t) = \frac{Q_0}{t_d} \left(1 - \frac{t}{t_d} \right) e^{-t/\tau} \quad for \; t < t_d$$

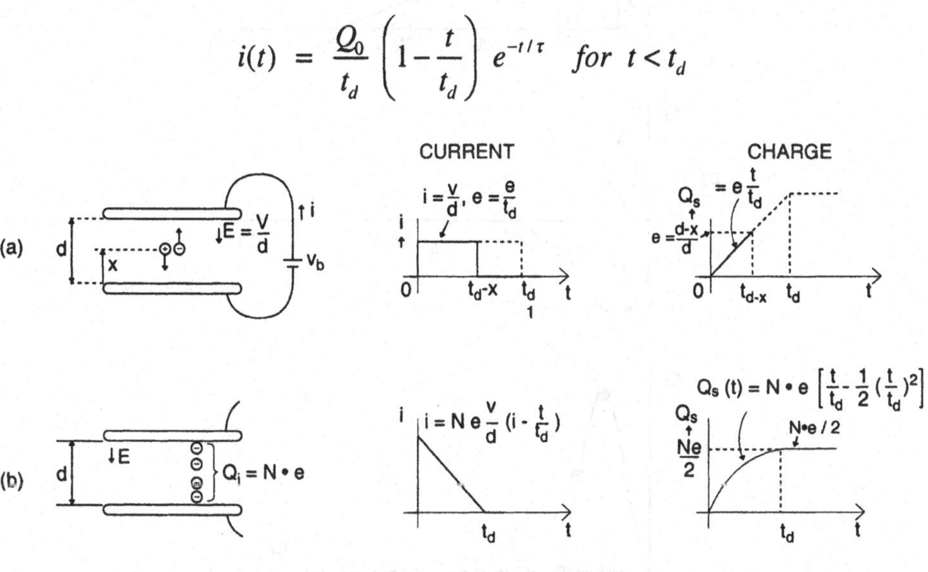

Figure 31: The distributions for the current and charge for a) a single e-ion pair, b) uniformly distributed e-ion pairs.

10.2 Signal Shapes

As discussed above, for a long electron 'lifetime' the induced current has a triangular shape with a duration equal to the electron drift time t_d (Figure 32a [15]). The total collected charge is also shown. It is clear that a device based on full charge collection will be slow, and hence not suitable for use at the LHC. However the energy information is contained in the initial current i_0. The information can be extracted and high rate operation established by clipping the signal with a fast bipolar shaping (Figure 32b). If the system impulse response has zero integrated area then pileup does not produce a baseline shift. For a peaking time, t_p that is much faster than drift time i.e. $t_p << t_d$, the output response becomes the first derivative of the current pulse (Figure 32c). The height of the output pulse is proportional to the initial current. However, with respect to full charge collection, the energy equivalent of the electronics noise will increase as this scales with $1/\sqrt{\tau}$, where τ (=RC) is the shaper time constant. At high luminosities, pileup also influences the choice of the value of τ. Pileup scales as $\sqrt{\tau}$. As an example, the optimized value for τ gives $t_p \approx 40$ ns for the ATLAS "accordion" e.m. calorimeter..

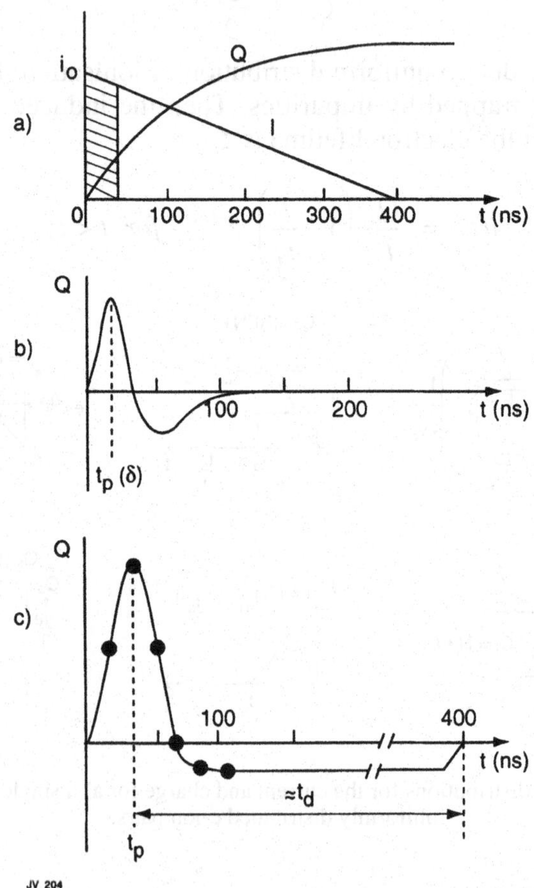

Figure 32: a) Induced current and integrated charge, b) bipolar shaping function and c) the shape of the output pulse, all as a function of time.

10.3 Examples of Noble Liquid Calorimeters

Conventionally ionization chambers are oriented perpendicularly to the incident particles. However in such a geometry it is difficult to

- realize fine lateral segmentation with small size towers, which in addition need to be projective in collider experiments,

- implement longitudinal sampling,

without introducing insensitive regions, a large number of penetrating interconnections, and long cables which necessarily introduce electronics noise and lead to significant charge transfer time. To overcome these shortcomings a novel absorber-electrode configuration, known as the 'accordion' (Figure 33, [27]), has been introduced, in which the particles traverse the chambers at angles around 45^0.

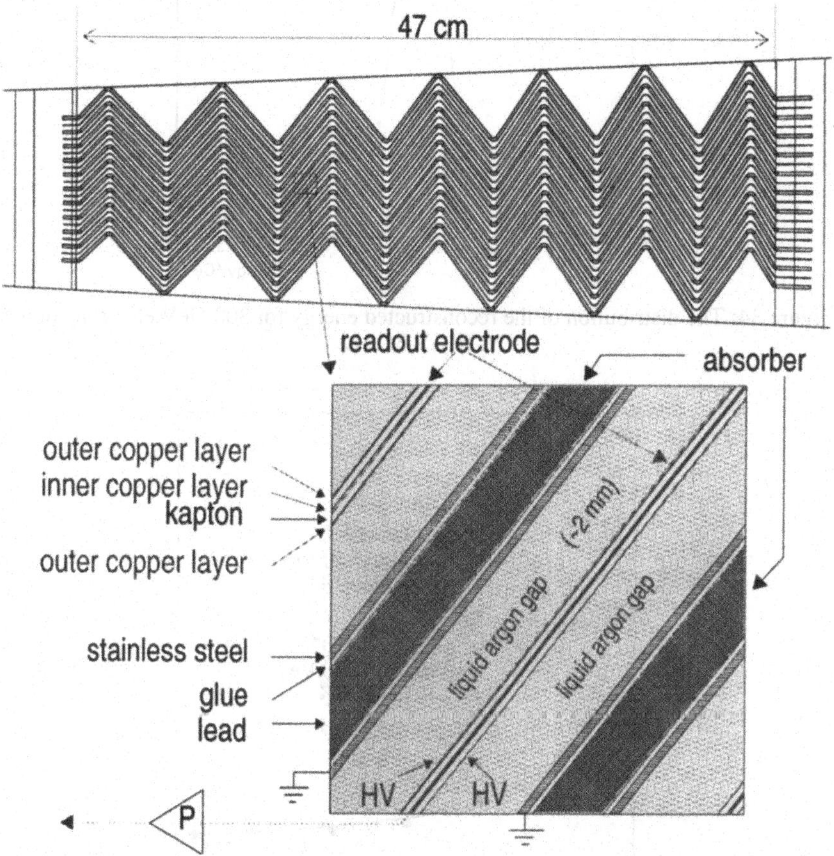

Figure 33: Top) the "accordion" structure of absorber plates of the ATLAS ECAL, below) details of the electrode structure.

In a variant, the NA48 [28] experiment has chosen an arrangement of electrodes that is almost parallel to the incident particles. With such structures the electrodes can easily be grouped into towers at the front or at the rear of the calorimeters. In ATLAS the absorber is made of lead plates, clad with thin stainless steel sheets for structural stiffness and corrugated to the shape shown in Figure 33. Details of the sampling structure are also shown. The read-out electrodes are made out of copper clad kapton flexible foil and kept apart from the lead plates by a honeycomb structure.

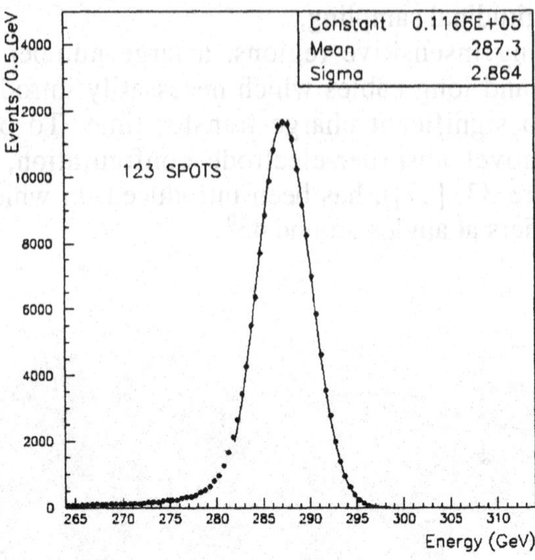

Figure 34: The distribution of the reconstructed energy for 300 GeV electrons in the ATLAS ECAL.

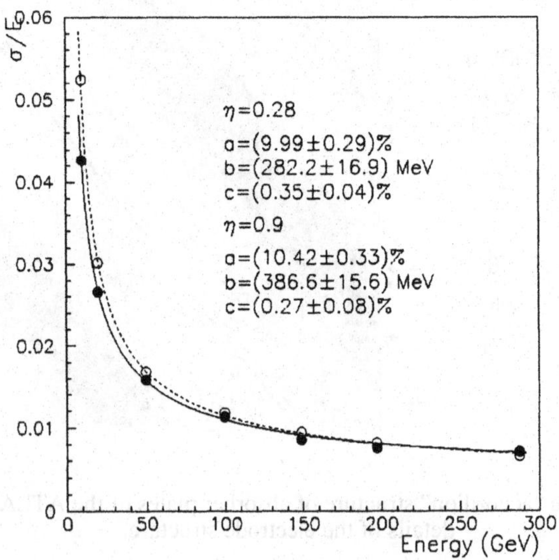

Figure 35: The fractional energy resolution for the ATLAS barrel prototype ECAL.

The results from a beam test of a large ATLAS prototype are shown in Figures 34 and 35. The electron shower is reconstructed using a region of 3x3 cells each of a size of ≈ 3.7 cm x 3.7 cm. The distribution of reconstructed energy for 300 GeV electrons, over a large area, is shown in Figure 34. The fractional energy resolution is shown in Figure 35 and can be parameterised as

$$\frac{\Delta E}{E} \approx \frac{10\%}{\sqrt{E}} \oplus \frac{0.28_{GeV}}{E} \oplus 0.35\%$$

where E is in GeV. The response of more than 150 cells over a large area has also been measured. The cell-to-cell non-uniformity is measured to be ≈0.58%. The major contributions come from mechanics (residual ϕ-modulation, gap non-uniformity, variation of absorber thickness) and calibration (amplitude accuracy). The large flux of isolated electrons from W or Z decays will be used to establish cell-to-cell intercalibration.

COMBINED E.M. AND HADRONIC CALORIMETRY

The LHC pp-experiments have put more emphasis on high precision e.m. calorimetry. This is not compatible with perfect compensation. For example the electromagnetic energy resolution of the ZEUS U-calorimeter is modest. The e.m. (σ_E) and hadronic (σ_h) resolutions are given by

$$\frac{\sigma_E}{E} = \frac{17\%}{E} \quad and \quad \frac{\sigma_h}{E} = \frac{35\%}{E}$$

Nevertheless it is very important to ensure:

- a Gaussian hadronic energy response function (a moderate energy resolution is acceptable),
- hermiticity
- linearity of response, especially for jets.

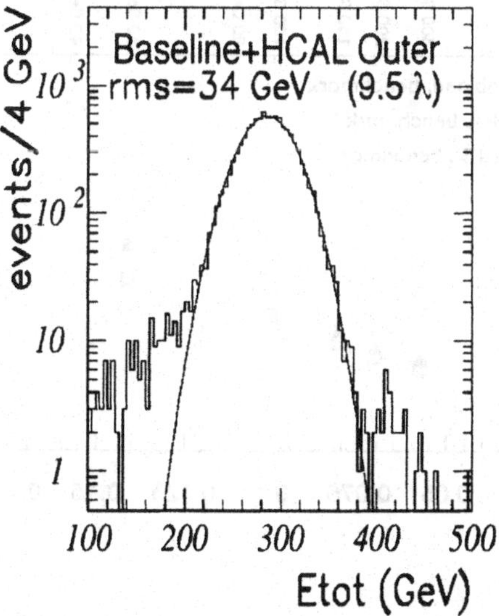

Figure 36: Distribution of reconstructed energy of 300 GeV pions [8, HCALTDR].

As an example in CMS this is done by introducing multiple longitudinal samplings. Reading out separately the first scintillator plate, placed behind the e.m. calorimeter, allows to distinguish between the cases where an e.m. shower has developed in the crystals (little signal from the first scintillator) and the ones where a hadronic shower has started (signal from the first scintillator). The energy observed in the first scintillator serves to make a correction. Infact the correction can be somewhat 'hard-wired' by choosing an appropriate thickness for the scintillator. The longitudinal leakage can also be up-weighted by increasing the thickness of the last scintillator. The measured energy distribution for 300 GeV pions in the CMS baseline is shown in Fig. 36. The tails are kept below a few percent.

The test beam results of the combined calorimetry of ATLAS (LAr ECAL and Fe/Scintillator HCAL) are shown in Figure 37. The data are compared with results from two simulation codes namely Fluka and GCALOR. Use is made of three energy-independent corrections for the:

- intercalibration between the e.m. and hadronic calorimeter
- energy lost in the cryostat wall separating the two calorimeters
- non-compensating behaviour of the e.m. calorimeter. A quadratic correction is made.

The above procedure minimizes the fractional energy resolution resulting in a systematic underestimation of the reconstructed energy: by 20% at 30 GeV and decreasing to ≈10% at 300 GeV. Other weighting methods, which have the effect of simultaneously minimizing the non-linearity and the energy resolution can also be employed.

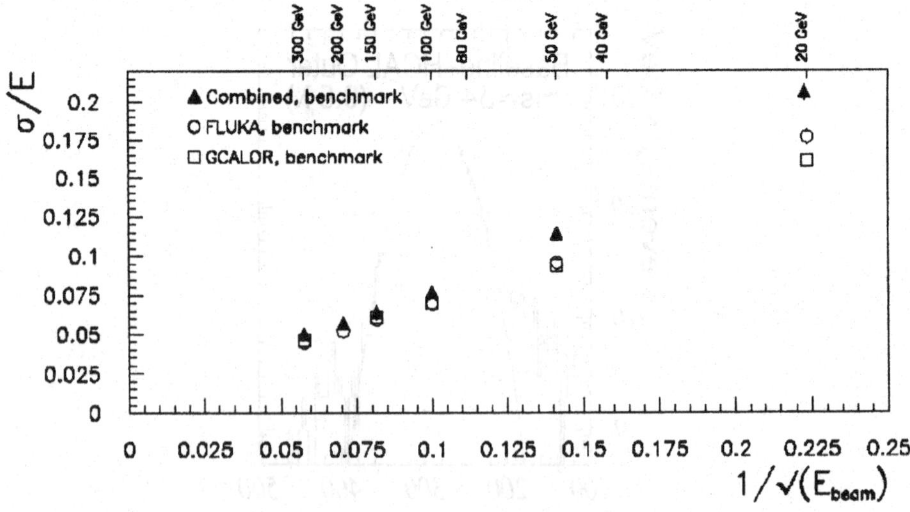

Figure 37: The energy resolution for pions compared with FLUKA and GCALOR simulation codes.

The hadronic cascade simulation codes such as FLUKA, GHEISHA and GCALOR have improved substantially and can now be used with some confidence in the design of hadron calorimeters.

CONCLUSIONS

Calorimeters are key detectors in present day high energy physics experiments. Large precision electromagnetic calorimeters are operating well (CLEO, KTeV, NA48), will soon be commissioned (BaBar, BELLE) or are under construction (ATLAS, CMS). High performance combined e.m. and hadronic calorimeters are also operating at DESY (H1 and ZEUS). Calorimeter systems have played a major role in many of the recent discoveries in particle physics e.g. W,Z in UA1/UA2, top quark in CDF/D0 and undoubtedly will play a similarly crucial role in the next generation of experiments at e^+e^- and p-pbar/pp machines.

Acknowledgements

We would like thank Tom Ferbel for the invitation to lecture at this school. Tom and his team organized a wonderful school with an enthusiastic participation of the students and in great surroundings.

References

1 U. Amaldi, *Exptl. Tech. In High Energy Physics*, p. 257,ed. T. Ferbel, Addison Wesley, 1987. Also in *Phys. Scripta* **23**(1981)409.

2 C. W. Fabjan, *Exptl. Tech. In High Energy Physics*, p. 325,ed. T. Ferbel, Addison Wesley, 1987.

3 R. Wigmans, *Ann. Rev. Nucl. And Part. Sci.* **41**(1991)133.

4 C. W. Fabjan and R. Wigmans, CERN-EP/89-64 (1989).

5 T. S. Virdee, Proc. 2nd Intl. Conf. On Calorimetry in High Energy Physics, p3, Capri, 1991, ed. A Ereditato, World Scientific.

6 R. Wigmans, Proc. 2nd Intl. Conf. On Calorimetry in High Energy Physics, p24, Capri, 1991, ed. A Ereditato, World Scientific.

7 ATLAS Technical Design Reports: Liquid Argon Calorimeter, CERN/LHCC/96-?? (1996), Tile Calorimeter, CERN/LHCC/96-42 (1996), http://atlasinfo.cern.ch/Atlas/GROUPS/notes.html

8 CMS Technical Design Reports, Electromagnetic Calorimeter, CERN/LHCC 97-33 (1977), Hadron Calorimeter, CERN/LHCC 97-31 (1997), http://cmsdoc.cern.ch/LHCC.html

9 Review of Particle Physics, C. Caso et al., *Euro. Phys. Journal* **C3**(1998)1, http://pdg.lbl.gov/

10 D. Barney, private communication.

11 M. De Vincenzi et al., WA78, *Nucl. Instr. and Meth.*, **A243**(1986)348.

12 D. Acosta et al., SPACAL, *Nucl. Instr. and Meth.*, **A294**(1990)193.

13 T. Doke et al., *Nucl. Instr. Meth.*, **A237**(1985)475.

14 D. J. Graham and C. Seez, CMS Note 1996/002 (1996).

15 D. Fournier and L. Serin, p. 291, 1995 European School of High Energy Physics, CERN 96-04, 1996, eds. N. Ellis and M. Neubert.

16 D. Groom, To appear in Proc. of Intl. Conf. On Calorimetry in High Energy Physics, Tucson, 1998.

17 R. Wigmans, *Nucl. Instr. and Meth.*, **A259**(1987)389.

18 G. Drews et al., *Nucl. Instr. and Meth.*, **A290**(1990)335 and H. Tiecke (ZEUS Calorimeter Group) *Nucl. Instr. and Meth.*, **A277**(1989)42.

19 A. Beretvas et al., CMS TN/94-326 (1994).

20 R. Apsimon et al., *Nucl. Instr. and Meth.*, **A305**(1991)331.

21 A. Annenkov et al., CMS NOTE 1998/041 and references therein.

22 G. Gratta et al., *Ann. Rev. Nucl. Part. Sci.* **44**(1994)453.

23 J. P. Peigneux et al., *Nucl. Instr. and Meth.*, **A378**(1996)410.

24 E. Auffray et al., *Nucl. Instr. and Meth.*, **A412**(1998)223.

25 F. Pauss, these proceedings.

26 M. A. Akrawy et al., *Nucl. Instr. and Meth.*, **A290**(1990)76.

27 D. Fournier, *Nucl. Instr. and Meth.*, **A367**(1995)5.

28 D. Schinzel, Proc. of Intl. Wire Chamber Conference, Vienna, 1998.

INTRODUCTION TO MICROWAVE LINACS

D.H. WHITTUM
Stanford Linear Accelerator Center, Stanford University
Stanford, California

Abstract. The elements of microwave linear accelerators are introduced starting with the principles of acceleration and accelerating structures. Considerations for microwave structure modelling and design are developed from an elementary point of view. Basic elements of microwave electronics are described for application to the accelerator circuit and instrumentation. Concepts of beam physics are explored together with examples of common beamline instruments. Charged particle optics and lattice diagnostics are introduced. Considerations for fixed-target and colliding-beam experimentation are summarized.

1. Introduction

The microwave linear accelerator or *linac* is a hybrid of microwave electronics, mechanical engineering, metallurgy, and craftsmanship. The primary component is a *structure* consisting of cups cut from copper, stacked, brazed, baked, mounted and fed with a copper tube, a *waveguide*. Every eight milliseconds the structure is pulsed with microwave power, megawatts for microseconds. If the copper cups have been shaped and tuned to one part in ten-thousand, they resonate in concert, driven by the incident wave. Sparks or *beams* fly through at about the speed of light, surfing the electromagnetic wave. Beams are wayward things and *magnets* are used to guide them. At the end of the linac, high-energy physicists pan through the debris, wondering when the statistics will improve. In these notes we contemplate why a linac looks as it does, and how it works.

In Sec. 2 we start with the problem of acceleration and develop the logic behind the multi-cell structure, and the electromagnetic fields it can support. Detailed calculations aren't really necessary here, but may be found in [1]. Appreciating the character of the fields we may provide for acceler-

T. Ferbel (ed.), Techniques and Concepts of High Energy Physics X, 387–486.

ation, we determine what manner of beam may benefit from them. In Sec. 3 we develop the subject of beam-physics, consisting mostly of a view of a beam as a collection of single particles, communicating, if at all, through their collective fields. We include discussion of beamline instrumentation, magnetic optics, and an upstream view of fixed-target and collider experiments. The topography of the whole subject is sketched in Fig. 1. We will concentrate only on the "main linac", the region of the machine where the beam is highly relativistic. With electrons in mind, this is most of the machine. For definiteness, examples are drawn for the most part from the Two-Mile Accelerator [2, 3].

2. Principles of Acceleration

2.1. ELECTRODYNAMICS

Classical electrodynamics is the foundation for the principles of acceleration, and is worth a short review. We consider a particle of charge q, and associate with it a position $\vec{R}$ and a velocity $\vec{V} = d\vec{R}/dt$. Mechanical momentum is given by $\vec{P} = m\gamma\vec{V}$, with m the particle mass,

$$\gamma = \left(1 - \frac{\vec{V}^2}{c^2}\right)^{-1/2}, \tag{1}$$

the Lorentz factor, and c the speed of light. Kinetic energy is

$$\varepsilon = mc^2(\gamma - 1) = \left(m^2c^4 + c^2\vec{P}^2\right)^{1/2} - mc^2. \tag{2}$$

Particle motion is governed by

$$\frac{d\vec{P}}{dt} = \vec{F}, \tag{3}$$

where the Lorentz force

$$\vec{F} = q\left(\vec{E} + \vec{V} \times \vec{B}\right), \tag{4}$$

is determined from $\vec{E}$, the electric field in units of V/m, and $\vec{B}$, the magnetic induction or flux density in units of Wb/m^2 or T. This relation defines the fields, abstracts them from their sources, describes the motion of particles, and the response of media. It is half of electrodynamics.

Where the particle is relativistic, V is close to c, and the effect of the magnetic induction can be appreciable. It is easier to produce 1 T, than it is to produce the equivalent 3×10^8 V/m. For this reason magnets are used

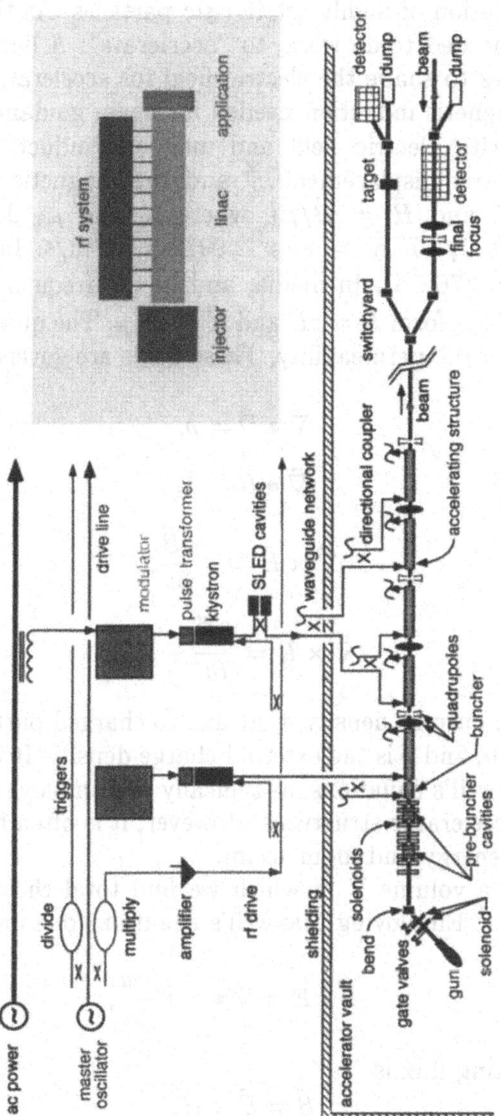

Figure 1. A microwave accelerator consists of a number of sub-systems working together. These include the microwave system—colloquially, the "rf" system—and the injector. These incorporate accelerating structures, requiring cooling water, mechanical support and alignment, and a vacuum system. The injector includes a gun, and may incorporate a laser system or an rf bunching system. Beam transport requires magnets, and their associated power supplies and cooling. Reliable operation requires rf and beam-monitoring by means of instrumentation circuits. In these notes we introduce the essential features of the linac, the beam, and the instrumentation.

to control the motion of highly relativistic particles. At the same time, an electric field is needed to do work, to "accelerate". A linac includes accelerating structures to shape the electric field for acceleration, and magnets to shape the magnetic induction needed for beam guidance.

Associated with electric field and magnetic induction, are two constructs, the electric displacement $\vec{D}$ and the magnetic field $\vec{H}$. In vacuum, $\vec{D} = \varepsilon_0 \vec{E}$ and $\vec{H} = \vec{B}/\mu_0$, where ε_0 and μ_0 define a choice of units, subject to $1/\sqrt{\mu_0 \varepsilon_0} = c \approx 2.9979 \times 10^8$m/s. In practical units, $\sqrt{\mu_0/\varepsilon_0} = Z_0 \approx 376.7\ \Omega$. In media, and in the frequency domain, these expressions take the form $\vec{D} = \varepsilon \vec{E}$ and $\vec{H} = \vec{B}/\mu$. The quantity ε is the permittivity and μ is the permeability. These fields are governed by Maxwell's equations,

$$\vec{\nabla} \bullet \vec{D} = \rho, \tag{5}$$

$$\vec{\nabla} \bullet \vec{B} = 0, \tag{6}$$

$$\vec{\nabla} \times \vec{E} = -\frac{\partial \vec{B}}{\partial t}, \tag{7}$$

$$\vec{\nabla} \times \vec{H} = \frac{\partial \vec{D}}{\partial t} + \vec{J}. \tag{8}$$

$\vec{J}$ is the external current density, that due to charged particles not already incorporated in μ, and ρ is the external charge density. It is rare in practice to drag out Maxwell's equations, as typically one employs a derived circuit-equivalent for accelerator structures. However, it is often helpful to refer to conservation of energy and momentum.

We consider a volume V in which we find total charge density ρ and current density $\vec{J}$. Employing Maxwell's equations one may show that

$$-\vec{J} \bullet \vec{E} = \vec{\nabla} \bullet \vec{S} + \frac{\partial u}{\partial t}, \tag{9}$$

where the Poynting flux is

$$\vec{S} = \vec{E} \times \vec{H}, \tag{10}$$

and energy density is

$$u = \frac{1}{2}\left(\varepsilon_0 \vec{E}^2 + \mu_0 \vec{H}^2\right). \tag{11}$$

In practice energy conservation is employed as a check of one's understanding of a circuit. In steady-state, power incident on an accelerator circuit should be balanced by what is reflected, transmitted, dissipated, and taken up by the beam.

Momentum conservation also figures prominently. The Lorentz force law may be expressed as

$$\frac{d\vec{P}_{mech}}{dt} = \int\limits_V dV \left(\rho\vec{E} + \vec{J} \times \vec{B}\right),$$

where $\vec{P}_{mech}$ is the mechanical momentum of the constituent particles within the volume V. Applying Maxwell's equations one may cast this in the form

$$\left(\frac{d\vec{P}_{mech}}{dt} + \frac{d\vec{P}_{em}}{dt}\right)^a = \int\limits_{\partial V} T^{ab} dS_b, \tag{12}$$

where summation is implied over repeated indices and $a, b = 1, 2, 3$ index Cartesian spatial coordinates. The momentum associated with the fields in the volume is

$$\vec{P}_{em} = \frac{1}{c^2} \int\limits_V dV\, \vec{S},$$

and $\vec{S}$ is the Poynting vector of Eq. (10). The flux of momentum through ∂V, the bounding surface, is given by the electromagnetic stress tensor,

$$T^{ab} = \varepsilon_0 E^a E^b + \mu_0 H^a H^b - u\delta_{ab}, \tag{13}$$

with δ_{ab} the Kronecker delta. This result will be helpful for structure design.

2.2. MATERIALS NEEDED

As to the matter of acceleration, Eq. (4) implies that a particle's energy ε varies according to

$$\frac{d\varepsilon}{dt} = q\vec{V} \bullet \vec{E}, \tag{14}$$

and with this we may classify accelerators according to the method of producing $\vec{E}$: electrostatic, inductive or fully electromagnetic. The first two are prominent in the history of accelerators, and to the present-day. However, for high-energy particles there is no substitute for the last, the microwave accelerator. It is not obvious at first how to employ an electromagnetic wave for acceleration. In the 1930's this was considered a research project [4]. In free-space, we are familiar with the mechanism of radiation pressure. Particles jitter transversely in a passing wave, and their jitter velocity causes them to be deflected in the forward direction by the transverse magnetic field. In this way particles may be pushed along by the wave, accelerated. This is just Thomson scattering. One intuits however that high-energy particles, with large relativistic inertia, jitter little. Thus radiation pressure or "second order acceleration" doesn't work well at high-energy, at least not directly. One may in principle circumvent the problem by introducing another external field to enforce the requisite transverse motion. This could be another wave, or a static magnetic field, as is seen in the *inverse free-electron laser* and the *inverse cyclotron maser*. The problem is solved in *laser-plasma accelerators*, by arranging that the radiation pressure act on low-energy plasma electrons, radial currents from which then induce an electric field, parallel to the high-energy beam and adequate for acceleration.

Historically, however, and perhaps logically, there is a more direct approach, "first-order acceleration". To see what this entails, we consider first a wave in free-space, and first-order particle motion in the wave. The wave may be decomposed into plane-waves, and at first-order in the applied fields, the impulse received by the particle is just the sum of the impulse from each wave. Thus it is enough to analyze motion in a plane wave of a particular angular frequency ω. We denote the component of wavenumber parallel to the particle motion, $k_{\|} = \hat{V} \bullet \vec{k}$, with $\vec{k}$ the wavevector, and $\vec{V}$ the particle velocity. Net energy gain of the particle is given by the integral over the particle displacement s

$$\Delta\varepsilon = \int ds\ E_{\|} \cos\left(\omega t - k_{\|} s\right),$$

with $E_{\|}$ the parallel component of the electric field. Energy gain depends on the phase $\psi = \omega t - k_{\|} s$ witnessed by the particle. Notice however that

$$\frac{d\psi}{ds} = \frac{\omega}{V} - k_{\|} > 0$$

since in free-space $\omega/|\vec{k}| = c > V$. Wave-crests continually flow past the particle alternately accelerating or decelerating it. On average $\Delta\varepsilon = 0$. Secular first-order acceleration in infinite free-space is not possible.

There are two solutions to this problem. One is to terminate the interaction of the particle with the wave, the other is to modify the dispersion characteristics of the wave to obtain synchronism with the particle. In the former case, the wave is trapped in the accelerator, forming a *standing-wave*; in the latter case, one has a *travelling-wave*. Both require the introduction of media. While dielectrics, plasmas, and other media are conceivable, and quite interesting as research projects, Hansen's original concept for a geometry employed *conducting* boundaries [4], and in fifty years has proved hard to beat. To see how microwaves within a conducting geometry may provide acceleration, it is enough at first to consider only a single cavity.

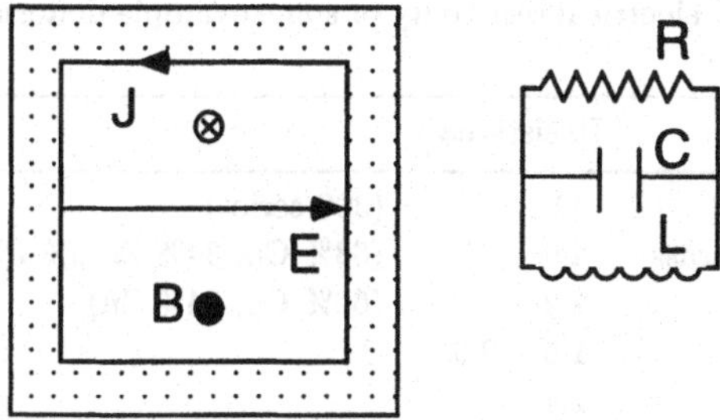

Figure 2. Maxwell's equations applied to a conductor-vacuum geometry reduce to a prescription for field lines, and wall currents. In the first approximation, the system behaves as an LC circuit. Equivalent resistance is low.

2.3. THE ACCELERATOR CAVITY

We consider how to describe and design a cavity useful for acceleration. First let us reduce Maxwell's equations to a simpler, pictorial form for a perfect conductor-vacuum geometry, as in Fig. 2. As we will see, structures are made from material of high-conductivity so as to minimize power loss. A high-conductivity material will quickly "short" any tangential electric field at its surface. Thus due to the ease with which charge redistributes itself over a conducting surface, electric field lines in a cavity terminate normal to a boundary, or not at all, closing on themselves. Similarly, currents are induced and flow to cancel the magnetic field within the conductor. There is no magnetic charge, however and so magnetic field lines never terminate and they can only close on themselves. This much we obtain from Gauss's Law, Eq. (5), and the solenoidal condition, Eq. (6). Faraday's Law, Eq. (7), and Ampere's Law, Eq. (8), combined describe the mutual excitation of electric and magnetic field, and dictate oscillation in a hollow conductor. In circuit terms, a region of space permeable to magnetic field presents an inductance, and a region permitting electric field presents a capacitance, and where both are present one expects to see behavior as in an LC circuit, oscillation at angular frequency $\omega = 1/\sqrt{LC}$. These considerations, summarized in Fig. 2, suggest that a cavity may be useful not only for field shaping, and thus first-order acceleration, but that the LC character of the system might be exploited for resonant excitation.

Yet conductors are not perfect. Wall currents flow through a lossy medium, resistance is present. Our LC circuit is really an RLC circuit. To quantify resistance, let us examine the matter of rf-dissipation for a normal conductor. Ohm's law takes the microscopic form $\vec{J} = \sigma \vec{E}$, where σ is the conductivity. Preferring higher conductivity, we might consult a listing of materials as in Table 1, and find that, considering cost, there is nothing better than the item labelled "OFE copper", the stuff of linacs. This is

TABLE 1. DC electrical resistivity of some example materials in units of $1.7 \times 10^{-8}\Omega$-m.

Material	Resistivity	
stainless steel	42	(300-series)
free-cutting brass	4.0	(63% Cu, 34% Zn 3% Pb 0.15% Fe)
brass	2.2	(66% Cu, 34% Zn)
aluminium	1.5 – 2.9	
gold	1.4	
copper	1.0	(OFE)
silver	0.93	

99.99% pure, "oxygen-free electronic-grade" copper, "C10100" in the Unified Numbering System for Metals. Conductivity is $\sigma \approx 5.8 \times 10^7$ mho/m.

In the Drude model the figure σ arises from electron motion in the applied field, subject to drag due to collisions with the ions. With collisions, electrons undergo a random-walk, and thus it is not surprising to find that the magnetic field, in the conductor satisfies a diffusion equation. In a random-walk process, diffusion depth is proportional to square-root of time, and in fact, the penetration depth, or *skin-depth*, for an rf-excitation takes the form $\delta = \sqrt{2/\mu\sigma\omega}$, with ω the angular frequency of the excitation. At 3 GHz ("S-Band") in copper, $\delta \approx 1\ \mu$m.

Knowing the skin-depth, one may determine power dissipation in a conducting boundary. The wall current density J should be sufficient to cancel the local magnetic field H within the conductor, on the order of $J \approx H/\delta$. This results in a power dissipation per unit volume J^2/σ within the conductor, so that net dissipated Poynting flux is $\delta\, J^2/\sigma \approx R_s H^2$. The surface resistance is, $R_s = 1/\sigma\delta$, 14 mΩ for copper at S-Band, and varying as $\omega^{1/2}$. Net dissipated power may be expressed then in terms of an integral over the conducting boundary, and quantified in terms of a *wall quality factor*, Q_w,

$$P_w = \frac{1}{2} R_s \int_{wall} \vec{H}^2 dS \equiv \frac{\omega U}{Q_w}. \tag{15}$$

A factor of 1/2 arises from time-averaging, and $\vec{H}$ is the peak magnetic field. The energy stored in the cavity, U, may be expressed in terms of $\vec{H}$, from Eq. (11), with a volume integral, permitting us to solve for

$$\frac{1}{Q_w} = \frac{\delta}{2} \frac{\int_{wall} \vec{H}^2 dS}{\int_{volume} \vec{H}^2 dV}. \tag{16}$$

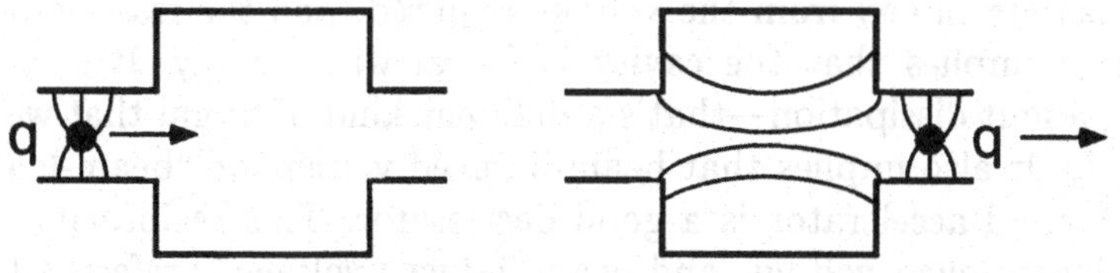

Figure 3. We consider the interaction of a charged particle with a cavity.

Roughly speaking, in each oscillation in a volume V, with a conducting surface area A, a fraction of the energy is dissipated, in proportion to the lossy volume $A\delta$. Quality factor is $Q_w \approx V/2A\delta$ in order of magnitude. Since cavity dimensions are of order the free-space wavelength λ one can see that $Q_w \approx \lambda/4\delta$ will be a large number. The corresponding decay-time for fields is

$$T_0 = \frac{2Q_w}{\omega}, \tag{17}$$

as one can check by solving $dU/dt = -P_w = -\omega\, U/Q_w$. For an S-Band accelerator $\lambda \approx 0.1\,\mathrm{m}$ and $Q_w \approx 10^4$; the decay-time may be quite long, $1\,\mu\mathrm{s}$, even while the oscillation period is short, 0.3 ns.

This discussion provides a description of a cavity as an *RLC* circuit. Next let us consider just what kind of accelerator our cavity can be. We picture an initially unexcited cavity, through which a beam-tube has been cut, as in Fig. 3. We suppose the beam tube is small enough that the fields cannot propagate into it, but remain confined. We consider the transit of a relativistic charge q through the cavity.

After the particle has left the cavity we suppose that energy U has been deposited in the the accelerating mode of the cavity. From the law of energy conservation, we expect a quadratic relation $U = k_l q^2$ for some choice of constant k_l characteristic of the cavity mode. This *loss-factor* has units of V/C. After the particle has left the cavity, the accelerating mode oscillates with some amplitude, V. To relieve the term "amplitude" of any ambiguity, we define V in units of voltage such as would be witnessed by a trailing particle. Notice that the energy loss by the charge q itself may be seen as arising through the action of this voltage. In transit through the cavity the charge witnesses an electric field that starts at 0 and rises to its maximum value. Thus the effective self-induced voltage is $V/2$, so that $U = qV/2$, or $V = 2k_l q$. Single-bunch beam-induced voltage is determined by the bunch charge and the loss-factor. A bit of algebra shows then that $U = V^2/4k_l$. This last relation, between energy and voltage must be independent of the means by which the voltage was produced. Thus if we wish to employ this cavity to accelerate, we must provide some energy, and the amount of energy is determined from the voltage required, and the loss-factor. Large loss-factor implies that the cavity is frugal with energy. It says though nothing about dissipation—that's a different kind of frugal that we'll come to shortly. It also implies that beam-induced voltage or "beam-loading" is large. A good accelerator is a good decelerator. This reciprocity between externally-supplied voltage, and beam-induced voltage is referred to as the fundmantal theorem of beam-loading.

If large loss-factor is good, how large can loss-factor be? We consider the energetics of the accelerator cavity in more detail. Suppose the length of the cavity is L and the cross-section is A. In terms of peak on-axis electric field E stored energy is roughly

$$U \approx \frac{1}{2}\varepsilon_0 E^2 \times \frac{1}{4} \times AL,$$

with a factor of $1/4$ to account for the radial variation in E^2. Next, let us relate E to "gap-voltage" V. For an electron crossing a gap of length L, with oscillating field of amplitude E, voltage gain is *not* $E \times L$, rather it is

$$V = \int dsE\left(t = t_0 + \frac{s}{V}\right)$$

for an electron entering at time t_0. For an electric field varying sinusoidally in time, uniformly across the gap, the result is that maximum voltage gain is $V \approx E \times L \times T$, where $T = \sin(\theta/2)/(\theta/2)$ is the *transit-time factor*, and $\theta = \omega L/V$ is the *transit-angle*. For a fixed field amplitude, maximum voltage gain corresponds to $\theta \approx \pi$. In this case, the electron enters the cavity when the field $E = 0$ and leaves just as $E = 0$ a half-cycle later.

With this we may compute

$$k_l = \frac{V^2}{4U} \approx \frac{L}{\varepsilon_0 A} T^2,$$

roughly the inverse of the capacitance of two parallel plates. At first one might think that large loss-factor would be easy to obtain by lengthening the cavity, or decreasing the plate area. However, loss-factor has a maximum at $L \approx 0.371\lambda$; greater lengths lead to lower k_l due to the transit-time factor. At the same time, cross-section cannot be arbitrarily small, since the electric field must vanish at the outer boundaries. For simple geometries, such as circular or rectangular pipe, this implies minimum $A \approx \lambda^2/2$. This corresponds to loss factor $k_l \approx 1/\varepsilon_0\lambda$, or 1 V/pC for an S-Band cavity. Shorter wavelength accelerators can make do with lower stored energy, but they exhibit heavier beam-loading, and thus must be operated with lower

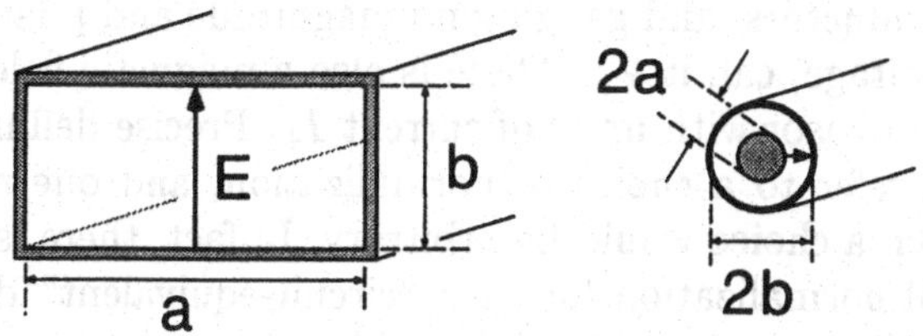

Figure 4. The two most common choices for waveguide are rectangular guide and coaxial line.

bunch charge. Associated with loss-factor is a second, more commonly employed quantity, $[R/Q] = V^2/\omega U = 4k_l/\omega$, "R-over-Q". This is an awkward notation, but it is conventional. Our estimate corresponds to

$$\left[\frac{R}{Q}\right] \approx \frac{8}{\pi^2} Z_0 \frac{\sin^2(\theta/2)}{(\theta/2)}$$

with a broad maximum in the vicinity of 220 Ω for cavity length in the range $L \approx \lambda/4 - \lambda/2$. $[R/Q]$ is a single thing, not two as the notation unfortunately seems to imply, and it is a function only of the *shape* of the cavity, not the scale.

Notice that $[R/Q]$ and k_l have no connection with wall conductivity or dissipation. They characterize energy-loss of a short bunch, one much shorter than a period, and very much shorter than a damping time. These quantities know nothing of dissipation. They determine the energy that must be stored to establish a prescribed gap voltage.

Understanding the geometric scalings for the energetics of the conductor-vacuum geometry forming an accelerator cavity, we can turn next to the practical matter of putting energy into the cavity. For this we need a short excursion into the subject of waveguide.

2.4. WAVEGUIDE FOR ALL OCCASIONS

To move a signal from point A to point B, one needs "waveguide", two-conductor cable (coaxial, stripline) or hollow waveguide (rectangular, circular, elliptical). The two most common gometries are illustrated in Fig. 4, and we'll visit with these at length in Sec. 2.14. However, just now we are concerned with the *function* of the guide; independent of the particular guide geometry employed, this may be pictured quite simply. There are two directions in the guide, forward and reverse, or "+" and "−". We consider a single-frequency, steady-state excitation, and inspect first the case of a wave propagating in one direction, call it the "+" direction. We may describe the amplitude of the wave by considering the electric field between the conductors, and gauging its magnitude and phase by a phasor with units of voltage, call it V_+. There is also a magnetic field and we may describe it by a phasor with units of current I_+. Precise definitions of these phasors would refer to a choice of normalization, and one might be concerned that such a choice would be arbitrary. In fact, there is arbitrariness in the choice of normalization for such "circuit-equivalent" descriptions of wave systems. We can take a bit of the arbitrariness out by asking that the quantity $P_+ = \Re V_+ I_+^* /2$ should correspond to the power flow through the guide, i.e., the Poynting flux integrated across the waveguide cross-section. This still leaves one with the freedom to choose a normalization corresponding to something with units of impedance $Z_w = V_+/I_+$, referred to as the waveguide impedance or waveguide mode impedance.

Next we consider the more general situation consisting of both forward and reverse waves together on the line. The net voltage and current phasors at a point z along the guide are given by

$$\tilde{V}(z,\omega) = V_+(\omega)\, e^{-j\beta z} + V_-(\omega)\, e^{j\beta z}, \tag{18}$$

$$Z_w \tilde{I}(z,\omega) = V_+(\omega)\, e^{-j\beta z} - V_-(\omega)\, e^{j\beta z}, \tag{19}$$

where β is the wavenumber in the guide geometry. Let us abbreviate $R = V_-(\omega)/V_+(\omega)$, thinking of this as a reflection coefficient due to a device placed downstream on the cable. The actual impedance at the plane z is then

$$Z = \frac{\tilde{V}(z,\omega)}{\tilde{I}(z,\omega)} = Z_w \frac{1+e^{2j\beta z} R}{1-e^{2j\beta z} R}. \tag{20}$$

This result indicates that for a short length of cable, with $\beta z << 1$, spatial phase-shift through the cable is small, and the guide is a lumped element. Longer cable constitutes a transmission line for waves. Generically we may call such a circuit, one comparable to a wavelength in size, a *microwave circuit*. Above 1 GHz most circuits are microwave circuits. A linac is a microwave circuit.

Returning to our forward and reverse waves, and the resulting standing-wave on the line, consider that if one probes the voltage associated with this standing-wave, as a function of z, one finds maximum voltage $|V_+| + |V_-|$, and minimum $|V_+| - |V_-|$. The ratio of maximum to minimum is

$$VSWR = \frac{1+|R|}{1-|R|}. \tag{21}$$

This *voltage standing-wave ratio* may be viewed as a property of the device to which the waveguide is attached, and it is a function of frequency. Typically microwave devices come with a specification for maximum VSWR over

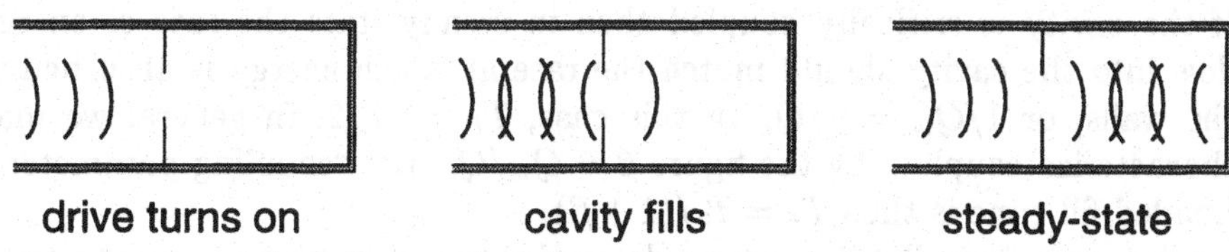

Figure 5. **In the transient filling of a cavity by a forward wave in the connecting guide, the reverse wave is a superposition due to reflection and diffraction.**

a range of frequencies. Let us consider then the microwave device foremost in our considerations just now, the accelerator cavity.

2.5. VIRTUES OF A RESONATOR

To power our cavity we consider a picture like that of Fig. 5. We make a small hole in the cavity wall and attach a waveguide to it. The waveguide transmits a wave through the hole or "coupling iris". This is just an electromagnetic version of the "ripple tank". The wave incident on the iris is initially mostly reflected. However, over time, the cavity gradually fills and as it does it begins to radiate; waves diffract through the coupling hole back into the external guide. Eventually, due to losses in the cavity walls, the system reaches steady-state, and the diffracted wave may partially or totally cancel the reflected wave. If the coupling iris geometry is such that the cancellation is total, the cavity is said to be "critically coupled". For a critically coupled cavity in steady-state, the incident power flows down the guide and into the cavity walls.

The effect of wall-losses, and external coupling may be quantified in terms of the wall quality factor, Q_w, and a "diffractive" or "external" quality factor, Q_e. If the energy stored in the cavity volume is U, then the power flowing into the walls is $P_w = \omega U/Q_w$. If the incident power is turned off, then the power flowing out of the cavity and down the waveguide is $P_e = \omega U/Q_e$. The net rate at which energy is leaving the cavity volume—with rf drive off— is the sum of these rates, and may be characterized by the "loaded Q",

$$\frac{1}{Q_L} = \frac{1}{Q_e} + \frac{1}{Q_w}. \tag{22}$$

Just as Q_e limits the rate at which energy may leave the cavity, it also limits the rate at which energy may be put into the cavity by means of external rf-drive. In general, the response-time of the cavity to external drive alone is the loaded fill-time

$$T_f = \frac{2Q_L}{\omega}. \tag{23}$$

If the cavity is critically coupled then in steady-state the rate of energy flow into the cavity should match the rate at which energy is absorbed in the walls, or $1/Q_w = 1/Q_e$ In this case, $T_f = T_0/2$. In general we may characterize coupling by the figure $\beta = Q_w/Q_e$, the "coupling parameter". Loaded fill time is then $T_f = T_0/(1+\beta)$.

Appreciating all this we can determine the power requirements for a critically-coupled cavity in steady-state. The forward power P_F in the connecting guide is just the power dissipated in the walls

$$P_w = \frac{\omega}{Q_w} U = \frac{\omega}{Q_w} \frac{V^2}{\omega\,[R/Q]} = \frac{V^2}{R_{shunt}}, \tag{24}$$

or $V = \sqrt{R_{shunt} P_w}$, where $R_{shunt} = Q_w\,[R/Q]$ is the *shunt-impedance*.

According to our calculations, a single normal-conducting S-Band cavity may have a shunt impedance of at most about 3 MΩ. Thus to achieve 1 MV of acceleration, we need a 0.3 MW source with a pulse length of about 1 μs. This steady-state power requirement is set by dissipation in copper. Thus if one had a less lossy material—a superconductor, for example— peak power requirements would be lower. To emphasize the point, consider that if one had a perfectly lossless cavity, and were running the machine at 120 Hz pulse repetition frequency, then with 8.3 ms between pulses, rf drive at the level of 30 W would be adequate to store the 1/4 J needed to establish 1 MV. What this means is that the shape, size, cost, and almost every other attribute of normal conducting microwave linacs is determined by losses in copper.

TABLE 2. Scaling of cavity parameters with frequency. No beam-tubes yet.

band	frequency	Q_w	T_0	$[R/Q]$	R_{shunt}
UHF	714 MHz	3.2×10^4	14 μs	$2.2\times10^2\Omega$	7 MΩ
S	2.856 GHz	1.6×10^4	1.8 μs	$2.2\times10^2\Omega$	3.5 MΩ
X	11.424 GHz	8.0×10^3	220 ns	$2.2\times10^2\Omega$	1.8 MΩ
W	91.392 GHz	2.8×10^3	9.8 ns	$2.2\times10^2\Omega$	0.6 MΩ

The resulting scalings for cavity parameters versus wavelength are summarized in Table 2; values listed there are for an idealized cylindrical cavity with negligibly small beam-tubes. To summarize our progress thus far, we have pursued the logic illustrated in Fig. 6, starting with a terminated interaction, and forming a cavity amenable to resonant excitation. We have arrived at the concept of shunt-impedance, something much bigger than

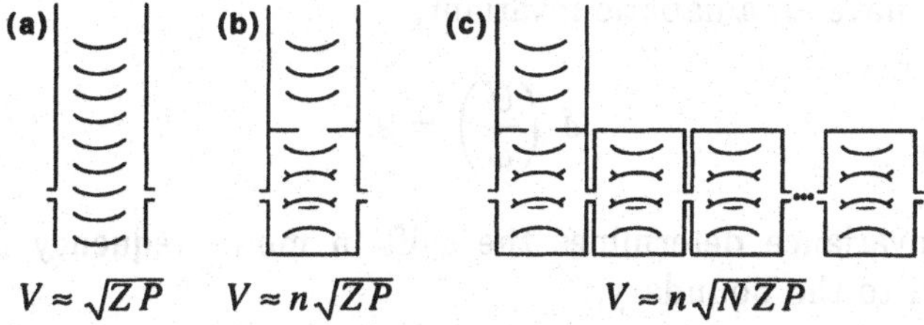

Figure 6. Summarizing the logic leading to the multi-cell linac. Accelerating voltage V is determined from the steady-state incident power P. The ratio V^2/P ("shunt-impedance") will be of order $Z \approx Z_0$ in free-space, as in (a). This can be improved greatly by use of resonant excitation, as in (b), as quantified by transformer ratio n, and by use of N cells as in (c). The beam travels to the right in this view.

the impedance of free-space, and limited only by losses, since it is derived from resonant energy storage. One sees in Fig. 6(c) the hint that we can do better still, than a single-cavity, and we will come to this in Sec. 2.13. Before that, a few particulars of the single cavity bear elaboration.

2.6. SLATER'S THEOREM

Performance of a resonant circuit depends on tune and quality factor. The matter of tune relates directly to tolerances on cell-manufacture and assembly. Let us consider a closed lossless pillbox that has been excited in a particular mode. Let the mode frequency be ω, and the stored energy U. Next we slowly make a localized inward displacement of the conducting boundary by an amount $\delta\xi$—a *dimple*. Work done on the mode in the course of dimpling is given by

$$\delta U = -\int T\, dS\, \delta\xi,$$

where

$$T = \frac{1}{2}\varepsilon_0 \vec{E}^2 - \frac{1}{2}\mu_0 \vec{H}^2,$$

is the local electromagnetic stress on the surface, just Eq. (13), evaluated with the help of conducting boundary conditions. The signs and magnitudes in this expression are easily checked, for example, for the cases of a capacitor and an inductor. A charged capacitor feels compressive stress, and "wants" to collapse. An energized inductor wants to explode. One can also see this in the field line picture of the cavity mode, where electric field lines are strung from surface to surface pulling them inward, and magnetic field lines are coiled up inside, pushing out.

As we apply this displacement $\delta\xi$ we are perturbing a harmonic oscillator. This is like varying the length of a pendulum while it is swinging. In this case we have an adiabatic invariant,

$$\delta\left(\frac{U}{\omega}\right) = 0.$$

Adiabatic invariance determines the shift in mode frequency due to the perturbation to the boundary,

$$\frac{\delta\omega}{\omega} = \frac{\delta U}{U} = \frac{1}{2U}\int\left(\mu_0\vec{H}^2 - \varepsilon_0\vec{E}^2\right)\delta V. \tag{25}$$

The integral extends over the volume excluded from the cavity by the perturbation, and the perturbed fields are employed in the integral. This is Slater's theorem [5]. Pushing inward on a cavity wall where E is large does negative work, lowers the stored energy, and therefore the frequency. Pushing inward on a wall where H is large requires that work be done, raises the stored energy, and therefore the frequency.

This result provides the quantitative basis for some practical accelerator bench-work, cavity tuning, and cavity field mapping. An S-Band cavity tunes at about 1 MHz/mil, where 1 mil $= 0.001'' = 25.4\ \mu$m. With $Q \approx 10^4$, allowable mistuning is a fraction of a MHz, i.e., a fraction of a mil. This is a constraint on machining precision and assembly, mostly assembly. The tolerance is sufficiently severe that it implies the need for post-assembly tuning in conjunction with bench measurements. Such microwave measurements also make use of Slater's theorem in various ways. For example one may "map" the electric-field profile in a structure by insertion of a small bead on the beam-axis. The bead perturbs the accelerating mode frequency by an amount proportional to E^2. Thus pulling the bead through the cavity and recording cavity frequency versus bead-position permits one to map the electric field. The external coupling of a cavity to a waveguide may be gauged by placing a shorting plunger on the input line, and plotting resonance frequency versus short position. Tune of a cavity with low Q_e will depend strongly on short position. An additional consequence of Slater's Theorem is the need for temperature regulation. The coefficient of thermal expansion for copper is $\alpha \approx 1.7 \times 10^{-5}$/K, and thus temperature control at the level of 1 K is required.

On this subject we could also mention the matter of *joints*. As seen in Fig. 2, the function of the copper boundary is to carry currents. If, in assembly, two copper boundaries should be joined so as to conduct current, then they should be *bonded* (brazed or diffusion bonded) so as to form a good current-carrying joint. Meanwhile, two flat-looking surfaces merely placed and held together may contact at as few as three points. Thus clamped structures, often tried, often don't work, exhibiting poor tune, and low Q_w.

This is easy to understand from the picture of magnetic field lines coiled inside the copper, waiting for the chance to bulge out. A small crack looks like home to any nearby magnetic field, and it will move right in. Slater's theorem tells us that this lowers the mode frequency. Dissipation in the crack tends to lower Q_w. It is possible to make a clamped structure, however the cuts must not interrupt the flow of wall current.

2.7. EQUIVALENT CIRCUIT

To understand bench-measurements involving amplitude and phase, we need a bit more than the energetics we have considered thus far. Let us set down a mathematical description of the single-cavity system coupled to a beam and a waveguide. We concentrate on a single-mode.

The waveguide excitation may be described by the forward-going voltage in the connecting guide V_F and the reverse voltage V_R. These are coefficients of transverse electric field in the guide, i.e., wave amplitudes graced with units of voltage. Continuity of the field implies a condition $V_c = V_F + V_R$, with V_c the cavity voltage, where we choose units again such that this is the voltage witnessed by a relativistic particle traversing the cavity. Maxwell's equations may be expressed in the form

$$\left(\frac{d^2}{dt^2} + \omega_0^2\right) V_c = [lossy\ walls] + [waveguide] + [beam],$$

where terms on the right are currents. The current corresponding to the excitation from the waveguide is just the current that is "missing" due to the coupling iris aperture. This in turn is proportional to the magnetic field at the reference plane. On the other hand, the magnetic field polarity is opposite for the forward and the reverse signals in the waveguide, since the direction of power flow is opposite for the two. Thus the virtual current associated with the waveguide excitation is proportional to $V_F - V_R$. The minus sign here is the same one appearing in Eq. (19). On general grounds then, our cavity-waveguide equation may be expressed as,

$$\left(\frac{d^2}{dt^2} + \omega_0^2\right) V_c = -\frac{\omega_0}{Q_w}\frac{dV_c}{dt} + \frac{\omega_0}{Q_e}\frac{d}{dt}(V_F - V_R) - 2k_l \frac{dI_b}{dt}$$

It is not obvious that the re-introduction of Q_w and Q_e in this way is consistent with our previous discussion, specifically Eqs. (15) and (22). However, this may be checked by considering steady-state energetics. Similarly one may check the beam-current term by considering a single bunch with reference to the principle of superposition.

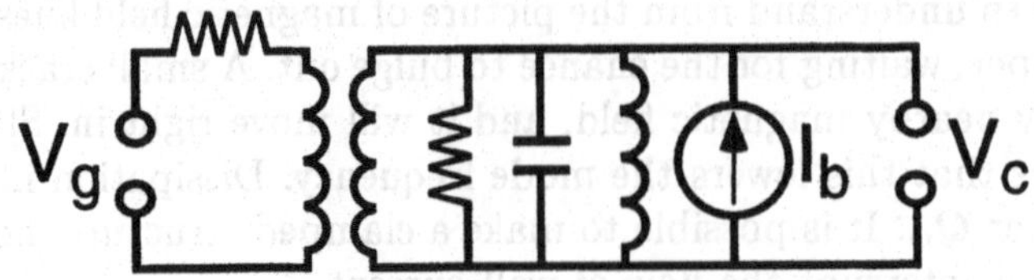

Figure 7. An equivalent circuit for a cavity coupled to a waveguide and a beam.

Employing the continuity condition to eliminate reverse voltage, one may describe the cavity excitation as a driven, damped harmonic oscillator.

$$\left(\frac{d^2}{dt^2}+\frac{\omega_0}{Q_L}\frac{d}{dt}+\omega_0^2\right)V_c=2\frac{\omega_0}{Q_e}\frac{dV_F}{dt}-2k_l\frac{dI_b}{dt}. \tag{26}$$

This result permits one to compute transient and steady-state excitation by external drive through the waveguide, or by the beam. If one prefers, one may express this result in terms of an equivalent circuit, for example, as in Fig. 7. In steady-state, with beam off, and external drive at angular frequency ω, $V_F=\Re\tilde{V}_Fe^{j\omega t}$, one may express the cavity voltage as $V_c=\Re\tilde{V}_ce^{j\omega t}$, with

$$\frac{\tilde{V}_c}{\tilde{V}_F}=\frac{2\beta}{1+\beta}\cos\psi\, e^{j\psi}.$$

The *tuning-angle* ψ is given by

$$\tan\psi=Q_L\left(\frac{\omega_0}{\omega}-\frac{\omega}{\omega_0}\right). \tag{27}$$

Analysis of energy conservation in the steady-state provides one with the relation of the equivalent circuit quantities V_F and V_R to power. Forward power in the connecting guide is

$$P_F=\beta\frac{\left|\tilde{V}_F\right|^2}{R_{shunt}}. \tag{28}$$

This shows that the forward and reverse voltage variables we are using should be referred to an impedance $R_{shunt}/2\beta$. Since this will differ from the waveguide impedance Z_w, we put a transformer in the equivalent circuit model of Fig. 7. The equivalent turns ratio n is given by $n^2=R_{shunt}/2\beta Z_w$. One can go on to express cavity voltage in terms of forward power,

$$\left|\tilde{V}_c\right|=\frac{2\beta^{1/2}}{(1+\beta)}\left|\cos\psi\right|\left(R_{shunt}P_F\right)^{1/2}. \tag{29}$$

This relation is to be contrasted with the expression in terms of dissipated power P_w, developed in Sec. 2.5, $V = \sqrt{R_{shunt}P_w}$. The two results are general, but we emphasize that $P_F \neq P_w$ in general. Optimum cavity voltage, at fixed power, with negligible beam-loading, occurs for critical coupling, $\beta = 1$ and perfect tune $\psi = 0$. To appreciate the effect of errors, consider that at $\psi = 26°$ one has lost 10% of the gap voltage; this corresponds to fractional detuning $\delta = (\omega_0 - \omega)/\omega_0 \approx 1/4Q_L$. For a critically-coupled S-Band cavity this is a detuning of 100 kHz, equivalent to a dimension error at the 3-μm level.

For comparison with measurement, one may express the steady-state complex reflection coefficient as

$$S_{11} = \frac{\tilde{V}_R}{\tilde{V}_F} = \frac{2\beta}{1+\beta} \cos\psi e^{j\psi} - 1. \tag{30}$$

This quantity was referred to as R in the discussion of waveguide and both notations are common. The notation here refers to the *scattering-matrix* or "S-matrix". As we will see in Sec. 2.14, this is in general an $N \times N$ matrix for an N-port device. Here $N = 1$. This quantity is directly measurable as a function of frequency and in modern times special instruments, vector-network analyzers, are routinely employed for this purpose. A measurement of S_{11} permits one to infer ω_0, β, and Q_L, and to assess systematics associated with temperature regulation and vacuum. Thanks to Slater's theorem, repeated measurement of S_{11}, as a small dielectric bead is pulled along the beam-axis, permits one to map the electric field. Equation (30) represents the contribution to S_{11} from a single mode. In practice one may expect that measurement of S_{11} for a microwave cavity will exhibit both narrow and broad resonances. The latter are often resonances on the connecting guide, the former are the resonant modes residing within the cavity proper, and they are very narrow, since Q_L is very large.

Appreciating what a cavity mode and a waveguide can do to each other, let us consider the effect they may have on the beam.

2.8. ELECTROMAGNETIC MULTIPOLES

We consider a particle passing through a beam-tube with a structure of some kind enclosing the tube, as seen in Fig. 8. We approximate the particle trajectory as a straight line centered on transverse coordinate $\vec{r}_\perp$ and passing some reference plane at time t_0. The voltage experienced by the beam in transit through the geometry may be expressed in terms of axial electric field E_s as

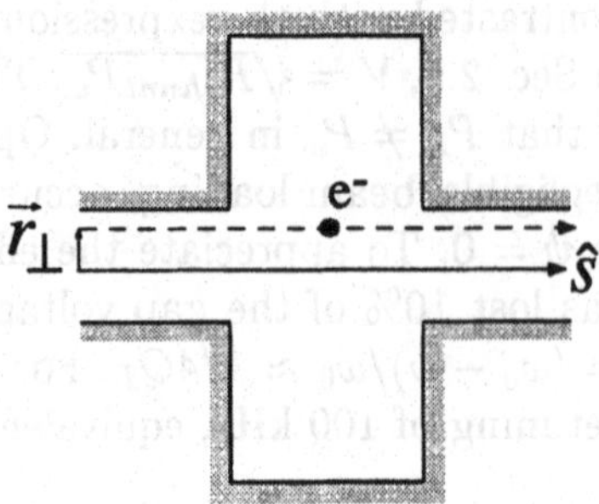

Figure 8. **The interaction of a beam with a cavity depends on its transverse coordinate $\vec{r}_\perp$.**

$$V_c(\vec{r}_\perp, t_0) = \int_{-\infty}^{+\infty} ds\, E_s\left(\vec{r}_\perp, s, t_0 + \frac{s}{V}\right),$$

where V is the particle speed. We may express the electric field in terms of Fourier components,

$$E_s(\vec{r}_\perp, s, t) = \int_{-\infty}^{+\infty} \frac{d\omega}{\sqrt{2\pi}} e^{j\omega t} \tilde{E}_s(\vec{r}_\perp, s, \omega),$$

with a corresponding result for voltage,

$$V_c(\vec{r}_\perp, t_0) = \int_{-\infty}^{+\infty} \frac{d\omega}{\sqrt{2\pi}} e^{j\omega t_0} \tilde{V}_c(\vec{r}_\perp, \omega),$$

where

$$\tilde{V}_c(\vec{r}_\perp, \omega) = \int_{-\infty}^{+\infty} ds\, \tilde{E}_s(\vec{r}_\perp, s, \omega)\, e^{j\omega s/V}.$$

Applying Maxwell's equations in the vicinity of the beam-orbit,

$$\left(\nabla_\perp^2 + \frac{\partial^2}{\partial s^2} + \frac{\omega^2}{c^2}\right) \tilde{E}_s(\vec{r}_\perp, s, \omega) = 0,$$

we can, after an integration by parts, establish that

$$\left(\nabla_\perp^2 - \Gamma^2\right) \tilde{V}_c(\vec{r}_\perp, \omega) = 0, \tag{31}$$

where

$$\Gamma = \left(\frac{\omega^2}{V^2} - \frac{\omega^2}{c^2}\right)^{1/2} \approx \frac{\omega}{\gamma V}. \tag{32}$$

The voltage witnessed by a particle depends in general on its displacement from the cavity axis. For a relativistic beam $\Gamma \to 0$. In this limit, the voltage imparted by a cavity mode is a harmonic function of the transverse coordinate.

A harmonic function regular in the vicinity of the center-axis may be expressed in the form of a multipole expansion.

$$\tilde{V}_c(\vec{r}_\perp, \omega) = \sum_{m=0}^{\infty} r^m \{b_m \cos(m\phi) - a_m \sin(m\phi)\}, \tag{33}$$

introducing polar coordinates, $\vec{r}_\perp = (x, y) = r(\cos\phi, \sin\phi)$. Where the geometry respects circular symmetry, and for the case of a single excited mode, the sum reduces to single terms, pure multipoles. In particular, the $m = 0$ term is independent of position. One prefers such a "monopole" mode for acceleration, since then gap voltage is independent of the beam orbit. On the other hand, real structures require a power feed, and this implies a deviation from cylindrical symmetry. Thus one is interested to quantify azimuthal mode purity in structures.

Generally one may classify modes based on their leading-order azimuthal harmonic. To appreciate the effects of such harmonics, it helps to consider not merely the longitudinal kicks associated with a structure excitation, but the transverse kicks as well. Transverse impulse per unit charge may be expressed as

$$\vec{P}_\perp(\vec{r}_\perp, t) = \int_{-\infty}^{+\infty} ds \left\{ \frac{1}{V}\vec{E}_\perp + \hat{s} \times \vec{B}_\perp \right\}_{(\vec{r}_\perp, s, t+\frac{s}{V})}. \tag{34}$$

Employing Faraday's Law and the ballistic-trajectory assumption, one may show that

$$\frac{\partial \vec{P}_\perp}{\partial t} = -\vec{\nabla}_\perp V_c, \tag{35}$$

the Panofsky-Wenzel theorem [6].

To make these considerations more concrete, consider an accelerator structure respecting inversion symmetry in y, but not in x. The monopole-mode voltage will take the form, at leading order,

$$V_c(x, y, t_0) = \Re V_0 e^{j\psi_0} \left\{ 1 + B_1 x + B_2 \left(x^2 - y^2 \right) + \ldots \right\},$$

where V_0, ψ_0 are the amplitude and phase seen on-axis. An off-axis trajectory corresponds to a different amplitude and phase. Keeping only the dipole ($m = 1$) correction, and expressing $B_1 = |B_1|\, e^{j\psi_1}$, one can see that

$$V_c(x, y, t_0) \approx \Re V_0 e^{j\psi_0} e^{j\delta\psi} \left(1 + 2|B_1|\, x \cos\psi_1 + |B_1|^2 x^2 \right)^{1/2}$$

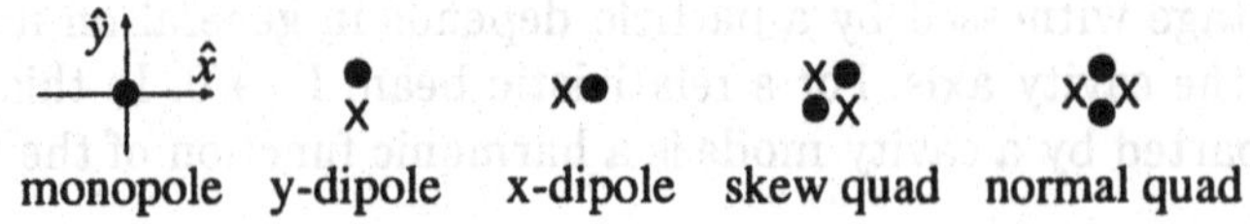

Figure 9. **Electromagnetic multipoles depicted according to the symmetry of the characteristic voltage witnessed by a relativistic beam, travelling into the page.**

where

$$\tan \delta\psi = \frac{|B_1|\, x \sin \psi_1}{1 + |B_1|\, x \cos \psi_1}$$

Thus the absence of a reflection symmetry *permits* a phase and an amplitude asymmetry in the accelerating voltage. An additional consequence is the presence of transverse deflections associated with this mode,

$$\vec{P}_{\perp} = \vec{\nabla}_{\perp} \Re V_0 e^{j\psi_0} \frac{j}{\omega} \left\{ 1 + B_1 x + B_2 \left(x^2 - y^2 \right) + \ldots \right\} \tag{36}$$

or, at lowest order $P_x \approx -\left(V_0 \left|B_1\right| / \omega_0\right) \sin\left(\psi_0 + \psi_1\right)$. Thus particles phased "on-crest" ($\sin \psi_0 = 0$) experience a transverse deflection determined by the imaginary part of B_1, the ***phase-asymmetry*** of the mode. Such an asymmetric geometry permits acceleration at an angle. This is undesirable in an accelerator cavity. This problem appears in practice wherever structures are asymmetrically fed, as with a single waveguide feed, for example.

While the $m = 0$ accelerating mode represents the most important example of a multipole, there are others, as seen in Fig. 9. The dipole modes are essential in understanding the transverse dynamics of a high-current beam, and arise naturally in the context of *wakefields*.

2.9. WAKEFIELDS

We have seen in Sec. 2.3 that the accelerating mode is excited by the beam, with voltage determined by the loss-factor k_l. In fact, each mode of the structure is excited by the beam, to the extent that the mode has some parallel electric field component, specifically, non-zero loss factor evaluated on the beam orbit. This is not cause for immediate alarm, because it takes some effort actually to design a mode to interact well with the beam. Nevertheless, when beam intensity is high, one may be concerned. Beam-excitation of parasitic modes is described by an elegant formulation in terms of "wakefields" [7].

To illustrate, let us return to the single-cavity problem of Sec. 2.7, and solve explicitly for the beam-induced voltage in a particular mode, using Eq. (26), with no external drive $V_F = 0$. One can check by differentiation that

$$V_c(t) = -\int_{-\infty}^{t} dt'\, G(t-t')\, \frac{dI_b}{dt}(t') .$$

where the Green's function is

$$G(\tau) = 2k_l \frac{\sin(\Omega\tau)}{\Omega} \exp\left(-\tfrac{1}{2}\nu\tau\right) H(\tau) ,$$

with H the step-function, $\Omega = (\omega_0^2 - \frac{1}{4}\nu^2)^{1/2}$, and $\nu = \omega_0/Q_L$. If we suppose cavities are placed along the beamline at separation L, we may translate this result into loss of voltage per unit length, or gradient,

$$\frac{d}{ds} mc^2\gamma = e \int_{-\infty}^{\tau} d\tau' W_{\parallel}(\tau - \tau')\, I_b(\tau') , \tag{37}$$

for an electron located at $\tau = t - z/c$. One can check, with an integration by parts, that $W_{\parallel} = G'/L$. More generally the "longitudinal wakefield", $W_{\parallel}$ includes contributions from other modes. In the frequency domain $\tilde{V}_c(\omega)/L = -Z_{\parallel}(\omega)\, \tilde{I}_b(\omega)$, where $Z_{\parallel}$ is the contribution to the "longitudinal impedance" per unit length from this mode, a Lorentzian line-shape with width set by Q_L,

$$Z_{\parallel}(\omega) = \frac{2j\, k_l' \omega\, \omega_0}{\omega_0^2 - \omega^2 + j\, \omega\, \omega_0/Q_L} , \tag{38}$$

where $k_l' = k_l/L$. In general, impedance receives contributions from this and all other high-Q modes with longitudinal electric field non-zero on the beam-axis. In addition to such narrow-band terms in the impedance, there are broad-band terms at frequencies above cut-off in the beam-tube. At very high-frequencies quasi-optical phenomena appear, as nearby boundaries cause the space-charge fields of the beam to diffract. This effect is particularly severe when the bunch-length is short, for then high-frequencies are well-represented in the beam-distribution. In accounting for impedance one looks for any deviation from uniformity in the beam-tube. The most important constriction in an accelerator is the beam-passing aperture between accelerator cells. Other common pipe-irregularities include beam monitoring instrumentation, collimators, bellows sleeves, vacuum port shields, and transitions in beam-pipe dimension. Even for a smooth pipe, wall resistivity results in energy loss, and thus one is concerned wherever the beam-passing aperture is small and lossy. These effects are acute in circular machines, where the beam orbits through an astronomical length of pipe. However, even in a linac, longitudinal wakefields may account for a reduction in beam energy of several percent, when single-bunch charge is high, and bunch length is short. To characterize such effects in simple terms, one may refer to the total loss factor per unit length,

$$k'_{tot} = \frac{1}{Q_b^2} \int_{-\infty}^{+\infty} d\omega \, \left| \tilde{I}_b(\omega) \right|^2 \Re Z_{\|}(\omega), \tag{39}$$

in terms of a form-factor quantifiying coherence at angular frequency ω, and the impedance describing the coupling to the beam-environment. So for example, in an S-Band structure fundamental-mode loss factor would be 1 V/pC/cell, while total loss factor might be 4–8× larger, depending on the bunch length and beam-port radius [7].

A more baneful effect in a linac is the dipole wakefield excited by an off-axis beam. To illustrate, we return to the single cavity mode, and apply the Panofsky-Wenzel theorem, to the case of a dipole-mode excitation,

$$P_x = \Re\left(\frac{j}{\omega_0} \frac{\partial \tilde{V}_c}{\partial x} e^{j\omega_0 t} \right). \tag{40}$$

We may characterize the efficacy of this mode as a beam-deflector in much the same way that we characterized the monopole mode as a beam-accelerator. We define

$$\left[\frac{R_a}{Q}\right] = \frac{\left|\tilde{V}_c(x=a)\right|^2}{\omega_0 U},$$

just the $[R/Q]$ for this mode evaluated on an off-axis trajectory at $x = a$. In the limit $a \to 0$, we obtain the figure of merit

$$\left[\frac{R_\perp}{Q}\right] = \lim_{a\to 0} \left[\frac{R_a}{Q}\right] \frac{1}{a^2} = \frac{\left|\partial \tilde{V}_c / \partial x\right|^2}{\omega_0 U}. \tag{41}$$

For example, for the lowest dipole mode of a circular cavity of radius R one may show that maximum $[R_\perp/Q] \approx 3Z_0/R^2$.

In terms of $k_\perp = \omega_0 [R_\perp/Q]/4$, one may express the energy deposited in this mode as $U = k_\perp Q_b^2 x_b^2$ for a beam offset x_b. One wouldn't mind a small amount of energy loss due to such a mode, particularly if the beam orbit were well-centered — except that with this mode comes a deflecting magnetic field. Consequently the head of a short bunch may deflect the tail. A rocking motion develops, and this motion is unstable, since the source of deflections is the beam offset itself. This effect was first observed in operation of linacs in multi-bunch mode, and appeared as a truncated current waveform on a current monitor. It was referred to at first as "pulse-shortening", and later, as "beam break-up" (BBU) when it was realized that transverse oscillations were the cause of beam loss [8]. Because of this wakefield, and the potential for bunch-to-bunch excitation, modern structure design entails elaborate measures to cause parasitic modes to damp and decohere [9]. Structure and lattice alignment have become subjects of the first importance. Ground motion, girder mechanical resonances, and thermal-bowing of girders have become subjects for research [10].

We can extract from our analysis a prescription for the transverse motion through the linac. For the case of the single cavity, we may describe dipole-mode excitation according to

$$\left(\frac{\partial^2}{\partial t^2} + \frac{\omega_0}{Q_L}\frac{\partial}{\partial t} + \omega_0^2\right)\frac{\partial V_c}{\partial x} = -2k_\perp \frac{\partial I_b}{\partial t} x_b,$$

and the circuit parameters here refer to the dipole mode, not the accelerating mode. Applying the Panofsky-Wenzel theorem, Eq. (35), we may express this more directly in terms of the impulse,

$$\left(\frac{\partial^2}{\partial t^2} + \frac{\omega_0}{Q_L}\frac{\partial}{\partial t} + \omega_0^2\right) P_x = -2k_\perp I_b x_b.$$

The solution is

$$P_x(t) = -\int_{-\infty}^{t} dt'\, G(t-t')\, I_b(t')\, x_b(t'),$$

where the Green's function

$$G(\tau) = 2k_\perp \frac{\sin(\Omega\tau)}{\Omega} \exp\left(-\tfrac{1}{2}\nu\tau\right) H(\tau).$$

Notational conventions differ on the subject of transverse wakefields. We will refer to the wakefield per unit length W_x, with the identification $G/L \rightarrow W_x$. The rate of change of momentum per unit length may then be expressed as

$$\frac{dp_x}{ds} = e\int_{-\infty}^{\tau} d\tau'\, W_x(\tau-\tau')\, I_b(\tau')\, x_b(\tau'), \tag{42}$$

in the absence of other transverse forces.

As with the longitudinal wakefield, the transverse wakefield picks up contributions due to any deviation from uniformity on the beamline, as well as wall resistivity. The narrow-band components of the corresponding transverse impedance account for the original beam break-up problems observed in multi-bunch mode on the linacs of old. These narrow-band terms,

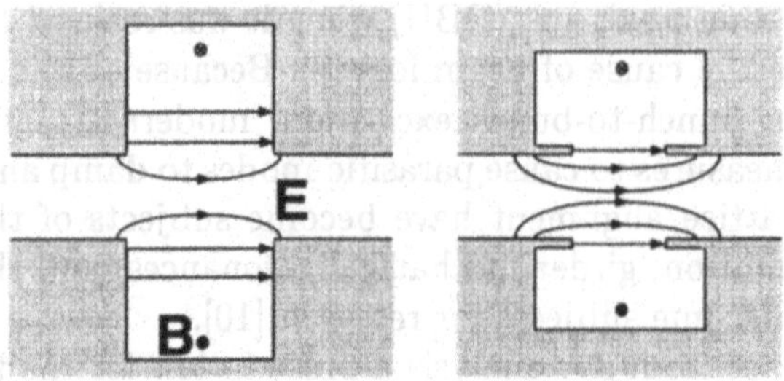

Figure 10. **Cutting of the beam port in the cylindrical pillbox on the left reduces $[R/Q]$. This can be fixed with nose-cones as seen on the right.**

corresponding to the "long-range" wakefield, are diminished by detuning of successive cells, so as to arrange destructive interference between the dipole modes of the structure as a whole. This was achieved, accidentally, but happily, in the Two-Mile accelerator by virtue of constant-gradient design, followed by additional detuning *in situ* [11]. The broad-band terms, corresponding to the "short-range" wakefield, account for single-bunch beam break-up. This is controlled only by lowering bunch charge, or strengthening the magnetic lattice. At the design-stage one may reduce the short-range wakefield by enlarging the beam port. However, a larger beam-port aperture easily lowers the $[R/Q]$. The conjunction of these two problems leads us directly to the matter of structure design.

2.10. CAVITY DESIGN

Thus far we have seen more or less what is desired of a cavity geometry, in terms of coupled-cavity circuit performance, and we can appreciate that there are consequences for beam dynamics. However, the parameters that determine the dynamics of the cavity and the beam are determined themselves by cuts in copper. Let us consider then how to accomplish the desired circuit parameters, and field-shape, by design, for an illustrative problem as illustrated in Fig. 10.

As we have seen, an accelerator cavity requires a beam-port. Cutting the port causes electric field lines to extend into the beam-tube, and develop transverse components. This reduces the integrated longitudinal field, and therefore $[R/Q]$. Reduction of the cavity radius and introduction of "nose-cones" attracts electric field lines to the gap, and improves the field-line shape. These considerations are illustrated with a simple estimate.

When the beam tube radius and gap are small, such a *re-entrant* cavity is approximately a coaxial line with a gap in the center conductor. Let us denote the beam tube radius a, inner conductor outer radius R_1, outer conductor inner radius R_2, gap length L_1, and cavity length L_2 as seen in Fig.

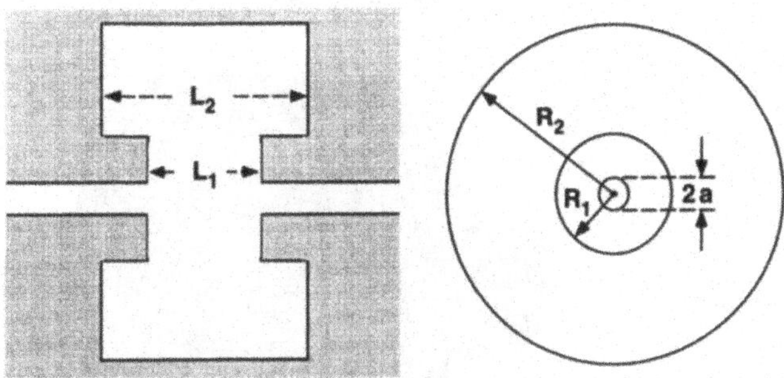

Figure 11. We analyze a reentrant cavity.

11. The resonant frequency of the circuit may be determined from the capacitance and inductance. Capacitance is determined by the energy stored in the electric field. If L_2 is short compared to a wavelength, most of this energy resides in the gap, and capacitance is $C \approx \varepsilon_0 \pi R_1^2/L_1$. Inductance is determined by magnetic energy stored in the coaxial line, arising from displacement current $I = \varepsilon_0 \omega \pi R_1^2 E$, and this is $L \approx (\mu_0/2\pi)L_2 \ln(R_2/R_1)$. Resonant frequency is then $\omega = 1/\sqrt{LC}$. Using $[R/Q] = V^2/\omega U \approx 2/\omega C$, we find

$$\left[\frac{R}{Q}\right] \approx \frac{2}{\pi} Z_0 T^2 \sqrt{\frac{L_1 L_2}{2R_1^2} \ln\left(\frac{R_2}{R_1}\right)},$$

with T the transit angle factor for the gap. Inspection of these results for frequency and $[R/Q]$ suggests that one can in fact recover the high impedance of the closed pillbox by judicious design. This was accomplished, in the early days of linacs, by approximate calculations along the lines of our example [5]. In practice, the analytic approach was supplemented with "cut and try" on the bench, to improve on the guidance of the design equations. In modern times it is common to employ software to solve Maxwell's equations and thereby optimize cavity dimensions.

To illustrate, Fig. 12 shows an example of a re-entrant cavity, the phase-monitor cavity, designed by Altenmueller and Brunet for use on the Two-Mile Accelerator circa 1965 [12]. These still see use today, for fixed-target experiments [13]. The numerical geometry has been set-up with the code *GdfidL* [14]. In the notation of Fig. 11, dimensions are $L_1 = 1.0235''$, $L_2 = 1.7005''$, $2a = 0.8000''$, $2R_1 = 1.0000''$, $2R_2 = 2.6525''$. These dimensions were arrived at after iteration on the bench, and chosen to provide a frequency 2 MHz low prior to brazing, so that frequency could be adjusted by dimpling. Altenmueller and Brunet infer from their measurements $Q_w \approx 9600$, $Q_L \approx 1200$, coupling parameter $\beta \approx 7.0$, $[R/Q]/T^2 \approx 370\ \Omega$ (adjusting a factor of two conventional difference), $T \approx 0.819$, $[R/Q] \approx 248$

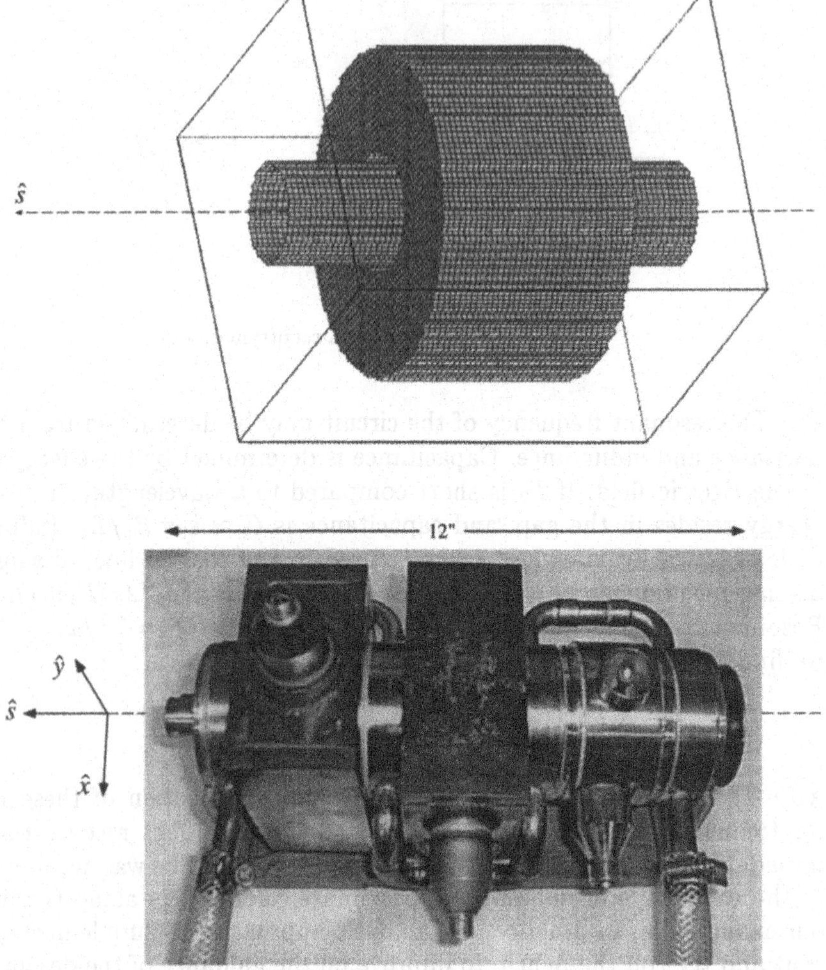

Figure 12. An example re-entrant cavity, the phase-monitor cavity of the original cavity beam-position monitor system on the Two-Mile Linac. Shown is the the numerical geometry employed for calculation and the actual cavity, part of a triplet.

Ω. The corresponding numerical result for S_{11} is seen in Fig. 13. Numerical calculation with GdfidL indicates $[R/Q] \approx 236\ \Omega$, and $Q_w \approx 1.51 \times 10^4$. Observed temperature detuning coefficient is 25 kHz/°F, 1° RF for 0.5 °F.

Nose-cones are but one feature of interest in cavity design. Symmetric input and output couplers help to eliminate dipole kicks. To control long-range wakefields, slots may be added to cells to externally couple higher modes, effectively damping them. Wakefields can be caused to decohere

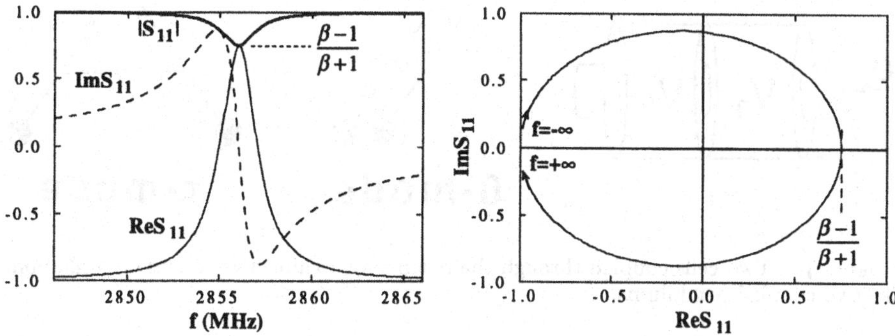

Figure 13. S_{11} for the phase cavity ($\beta \approx 7.0$) plotted versus frequency and in *Smith-Chart* form.

by judicious detuning of cells. Such design work takes place under the constraint of good shunt-impedance for the accelerating mode. Interesting design problems also are found beyond the structures themselves, and include windows, couplers, mode-converters and other components of the linac rf system [15]. For high-power handling it is important to maintain low insertion loss, good match, and low peak field, everywhere but in the accelerating cavities themselves. Numerous codes are available and the designer has quite a few tools to choose from [16].

2.11. TWO COUPLED CAVITIES

If we were to take the work of Secs. 2.3-2.5, on a single cavity with a single waveguide-feed, and extrapolate it to a 20 GeV machine, we would find that we need more than 10^4 power feeds, a lot of copper-waveguide. A more economical scheme turns out to be feasible and consists of a series of cavities placed along a common beam-tube. Power is fed by waveguide to a single-cavity, and finds its way to the other cavities, through the beam-tube. Such a multi-cell accelerator behaves as a chain of coupled pendula. Coupling constants, periods, and Q's are determined from the geometry sculpted in copper. The concept may be designed as a standing-wave accelerator, with a single feed. Or it may be implemented as a travelling-wave accelerator, with an input waveguide at one end, and output waveguide and load at the other end.

We consider first a two-cell structure as seen in Fig. 14. Recognizing the similarity with the problem of two coupled pendula we expect to find two modes of oscillation, a mode with $V_1 = V_2$ ("0-mode"), and a mode with $V_1 = -V_2$, ("π-mode"). For the 0-mode each field line from cell 1 connects to a field line from cell 2, and no field lines terminate on the coupling iris. Displacement current from cell 1 flows directly into cell 2 and no current

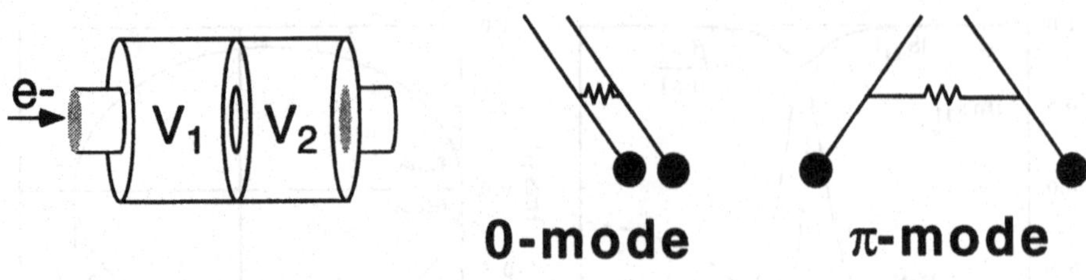

Figure 14. Two cells coupled through the beam-tube exhibit two modes of oscillation, like two coupled pendulums.

flows through the shared common wall. Since no current flows, one could remove the common wall without disturbing the modal field pattern. In this case one has simply smooth pipe terminated at each end. This is just the original geometry of the uncoupled cells, but with twice the cavity length. However for the accelerating mode, as seen in Fig. 2, frequency depends only on radius. Thus the 0-mode frequency Ω_0 is just that of the uncoupled cells. In the π-mode field-lines terminate on the coupling iris, and we expect the tune to be sensitive to the iris geometry. From Slater's theorem, we expect that removal of copper in the vicinity of large electric field, and small magnetic field, will reduce the capacitance of the circuit, and therefore raise the frequency. Thus we expect the π-mode frequency $\Omega_\pi > \Omega_0$.

To treat the problem more quantitatively, let us describe the modes with amplitudes ξ_0 and ξ_π. That is to say, when the structure is excited in the zero mode, the electric field exhibits a certain spatial pattern that varies in time in proportion to $\xi_0(t)$. In general, fields may be a superposition of the two modes. According to our reasoning these mode amplitudes evolve according to

$$\left(\frac{d^2}{dt^2} + \Omega_0^2\right) \xi_0 = 0, \quad \left(\frac{d^2}{dt^2} + \Omega_\pi^2\right) \xi_\pi = 0.$$

We may express this in terms of the original cell voltages by identifying $\xi_0 = V_1 + V_2$, $\xi_\pi = V_1 - V_2$. These equalities assume merely a choice of units for ξ_0 and ξ_π. Substituting these expressions, and re-arranging, one finds

$$\left(\frac{d^2}{dt^2} + \Omega_0^2\right) V_1 = \frac{1}{2}\left(\Omega_\pi^2 - \Omega_0^2\right)(V_2 - V_1),$$

$$\left(\frac{d^2}{dt^2} + \Omega_0^2\right) V_2 = \frac{1}{2}\left(\Omega_\pi^2 - \Omega_0^2\right)(V_1 - V_2).$$

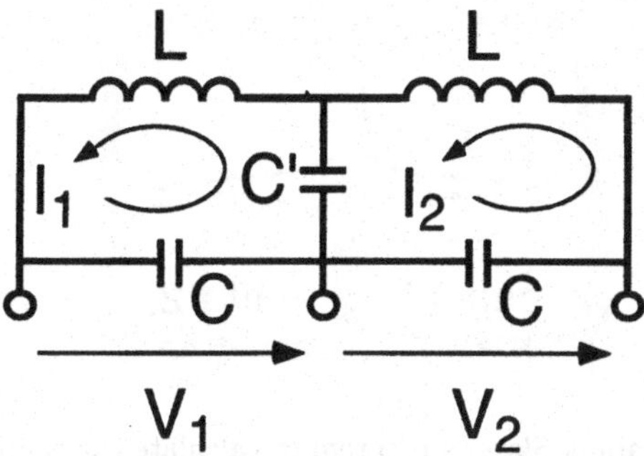

Figure 15. Equivalent circuit for two coupled cells.

This result describes the time-evolution of two-coupled cavities in terms of the frequency shift between 0 and π modes. The strength of the coupling is quantified by the dimensionless figure

$$\kappa = 2\frac{\Omega_\pi^2 - \Omega_0^2}{\Omega_\pi^2 + \Omega_0^2}.$$

In terms of this cell-to-cell coupling constant, and

$$\omega_0^2 = \frac{1}{2}\left(\Omega_\pi^2 + \Omega_0^2\right),$$

we may express our result as

$$\left(\frac{d^2}{dt^2} + \omega_0^2\right) V_1 = \frac{1}{2}\kappa\omega_0^2 V_2, \tag{43}$$

$$\left(\frac{d^2}{dt^2} + \omega_0^2\right) V_2 = \frac{1}{2}\kappa\omega_0^2 V_1. \tag{44}$$

With some algebra one can show that

$$\Omega_0^2 = (1 - \kappa/2)\omega_0^2, \quad \Omega_\pi^2 = (1 + \kappa/2)\omega_0^2.$$

If it is helpful, one may think of the equivalent circuit of Fig. 15. One can check that the circuit parameters satisfy $\Omega_0^2 = 1/LC$ and $C/C' \approx \kappa/2$.

At this point, having examined the kinematics of modes, we have determined that there must be a figure κ, but we haven't determined what this circuit parameter is—how it depends on the cell geometry. To calculate cell-to-cell coupling from the geometry we apply Slater's Theorem. We suppose the cells are excited in the π-mode and sketch a cylindrical volume

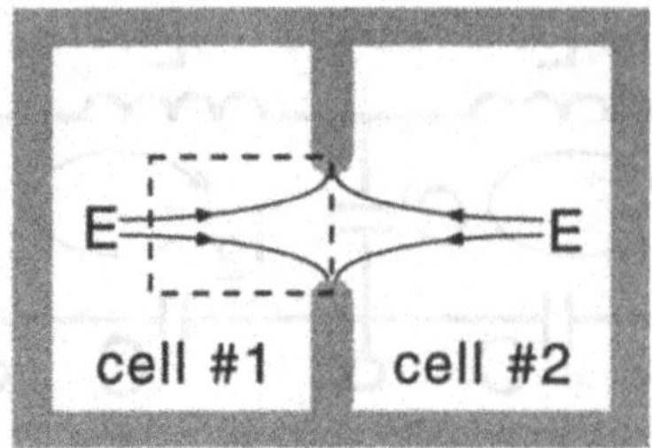

Figure 16. We apply Slater's theorem to calculate the coupling of two-cells.

with radius equal to that of the coupling iris a, as indicated in Fig. 16. In this accelerating mode, electric field is large in the vicinity of the iris, and magnetic field is small. As a result, the iris experiences mechanical tension due to the pull of the electric field lines. This tension is transmitted by the field lines through the endcaps of our imaginary cylinder to the wall separating the cells. We may express the time-averaged balance of forces as

$$\int_{endcap} dS\, T^{zz} = -\int_{sidewall} dS\, T^{zr} = \int_{sidewall} dS\, T^{rr}.$$

The first equality expresses conservation of axial (z) momentum, and the second conservation of radial (r) momentum. Integrals over the cylinder endcap centered in the iris vanish since the axial field vanishes there. Thus the axial force may be estimated with an integral of $T^{zz} \approx \varepsilon_0 E^2/2$ over the left endcap, in terms of the peak electric field E,

$$\int_{sidewall} dS\, T^{rr} = \int_{endcap} dS\, T^{zz} \approx \pi a^2 \frac{1}{2} \varepsilon_0 E^2 \times \frac{1}{2}$$

where the last factor of 1/2 results from time-averaging. If we permit the iris to relax adiabatically, deforming in radius by an amount $\delta a < 0$, the mode will do work, and the stored energy U, will be reduced

$$\delta U = \delta a \int_{sidewall} dS\, T^{rr} \times 2 \approx \pi a^2 \delta a \frac{1}{2} \varepsilon_0 E^2$$

where the factor of two in the first equality results from counting both the left and right halves of the iris. Closing the iris, the mode loses energy

$$\Delta U = -\frac{\pi}{6} a^3 \varepsilon_0 E^2.$$

Total stored energy is meanwhile $U \approx \varepsilon_0 E^2 V/8$, with V the total volume of the two cells. Adiabatic invariance of U/ω then implies that with the closing of the iris, the π-mode frequency shifts by an amount

$$\frac{\Delta\omega}{\omega} = -\frac{4\pi}{3}\frac{a^3}{V}.$$

Once the iris is closed we know that the frequency must be restored to the unperturbed, closed-cavity value, and this is just Ω_0. Thus we have computed the frequency separation of the 0 and π modes, $\Delta\omega \approx \Omega_0 - \Omega_\pi$. The coupling constant is then

$$\kappa \approx \frac{8\pi}{3}\frac{a^3}{V}. \tag{45}$$

Subsituting $V = 2\pi LR^2$ with L the length of a single cell, and R the cell radius one obtains $\kappa \approx 1.3a^3/LR^2$. A more precise calculation provides the coefficient 1.57 [1].

Appreciating the behavior of the modes let us characterize their properties for acceleration. The 0-mode is just the mode of a single cavity, and thus its $[R/Q]$ can be no better than the single-cell case. For the π-mode on the other hand we see that a particle accelerated in cell 1, and leaving this cell at about the time the field is reversing, arrives in cell 2 just as the field in that cell is ready to accelerate. Thus one expects the net voltage to be the sum of each cell voltage. Stored energy meanwhile is roughly the sum of the energies stored in each cell. With twice the voltage of a single cell, for twice the energy, we see that $[R/Q] = V^2/\omega U$ is twice the $[R/Q]$ for a single cell. As for wall Q, the magnetic field is small near the iris, thus for a given single-cell voltage, the dissipated Poynting flux in the walls is unchanged by the coupling. The wall area has increased by a factor of 2, so the dissipated power is twice as large. However, the stored energy is also larger by 2, so Q_w is unchanged. This implies that shunt impedance $R_s = Q_w[R/Q]$ is twice the single-cell case. Thus if our microwave power source was capable of providing a 2 MV "gap-voltage" with one cell, we should be able to arrange 3 MV with a two-cell structure.

2.12. MULTI-CELL STRUCTURES

This brush with the 2-cell structure encourages us to continue right on to the case of N-cells, expecting shunt-impedance larger by a factor of about N. Actually though it is simpler to proceed directly to the infinite structure. According to our analysis of the effect of a single-coupling iris, as summarized in Eqs. (43) and (44), we expect that cell voltages in the

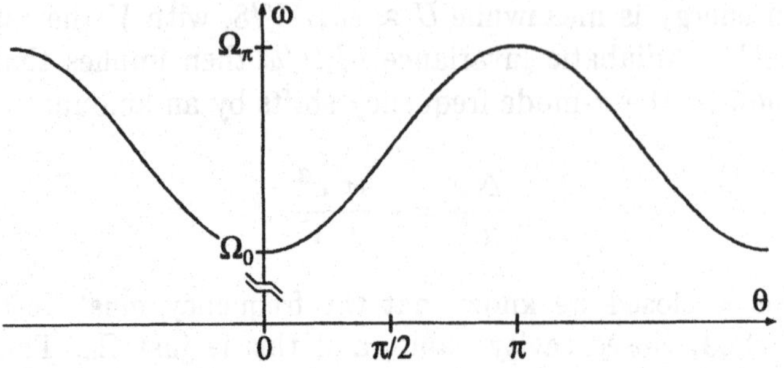

Figure 17. Brillouin curve for a periodic structure.

infinite structure will satisfy

$$\left(\frac{d^2}{dt^2}+\omega_0^2\right)V_n=\frac{1}{2}\kappa\omega_0^2\left(V_{n-1}+V_{n+1}\right), \tag{46}$$

where n indexes the cells, and we continue to neglect Ohmic losses. Since we have made two iris cuts in each cell, our circuit parameter ω_0 differs from the 2-cell case, as it is shifted by twice as much relative to the unperturbed cell frequency Ω_0, $\omega_0^2=\Omega_0^2/\left(1-\kappa\right)$.

To appreciate the behavior of this coupled system consider the case of a single-frequency excitation, imposed somehow on one particular cell. In steady-state, a general solution for cell voltages may be expressed as $V_n=\Re\tilde{V}_n e^{j\omega t}$, for some choice of cell phasors $\tilde{V}_n$. The phasors describe, among other things, the flow of energy along the waveguide. One could compute this from the Poynting flux integrated over the coupling hole, however, it's enough for us to recognize that conservation of energy requires $|\tilde{V}_{n+1}|=|\tilde{V}_n|$. Thus adjacent cells are related by a phase-factor, and from periodicity, this factor must be the same for any two successive cells. Let us denote this ***cell-to-cell phase-advance*** $\theta\left(\omega\right)$. To the right of the cell being excited we observe a rightward travelling wave,

$$\tilde{V}_{n+1}=e^{-j\theta}\tilde{V}_n. \tag{47}$$

Substituting this in Eq. (46) we find

$$\omega^2=\omega_0^2\left(1-\kappa\cos\theta\right), \tag{48}$$

the ***dispersion relation*** for our periodic line. This is illustrated in Fig. 17. Given that the result is periodic in θ, and symmetric it is enough to plot the result on the interval $(0,\pi)$. The vertical scale in this sketch has a break to indicate that the relative separation of 0 and π modes may be quite small.

This separation is significant as it is the full-width of the *passband*, the range in frequency, associated with the original cavity mode, over which the structure will permit microwave propagation.

To construct a finite periodic structure, it is enough to know that interior cells in the structure support forward and backward waves with the dispersion characteristics of the infinite structure. A general solution for cell voltages may consist of both left and right travelling waves, and then must take the form

$$\tilde{V}_n = \tilde{V}^+ e^{-jn\theta} + \tilde{V}^- e^{jn\theta}.$$

End cells determine the boundary conditions on forward and reverse amplitudes $\tilde{V}^{\pm}$.

For a standing-wave structure, one has left and right-going waves of equal amplitude, with relative phase determined by the end cells. As we saw in the case of the two-cell structure, maximum shunt-impedance for a standing-wave structure occurs for π phase-advance per cell. Structure length is limited by mode density near π-mode, as mode separation should be greater than the natural $1/Q_w$ mode-width.

A travelling-wave structure employs a power feed at one end, and an output-waveguide at the other. End cells are matched, designed to insure that in steady-state there is no reflection on the input line, and no standing-wave in the structure. The rate at which energy flows into the coupling cell matches the rate at which energy flows through the structure and into the walls. Propagation through the structure is governed by the group velocity, as determined from the slope of the Brillouin curve, and is given by

$$\beta_g = \frac{V_g}{c} = \frac{L}{c}\frac{d\omega}{d\theta} = \frac{1}{2}\left(\frac{\omega_0 L}{c}\right)\frac{\kappa \sin\theta}{\sqrt{1-\kappa\cos\theta}}, \tag{49}$$

with L the cell period. Group velocity is maximum and dispersion is minimum in mid-band, and group velocity approaches zero at 0 and π phase-advance. In steady-state, interior cell-voltages are

$$V_n(t) = \Re \tilde{V}^+ e^{j\omega t - jn\theta},$$

and the net voltage experienced by a particle entering the structure at time t_0 is

$$V_{NL}(t_0) = \sum_n V_n\left(t_0 + \frac{nL}{c}\right) = \Re \tilde{V}^+ e^{j\omega t_0} \sum_n \exp jn\left(\frac{\omega L}{c} - \theta(\omega)\right).$$

Thus secular energy-gain requires synchronism between the particle, and the forward-wave in the structure. The transit angle must equal the geometric phase-advance per cell,

$$\frac{\omega L}{c} = \theta(\omega). \tag{50}$$

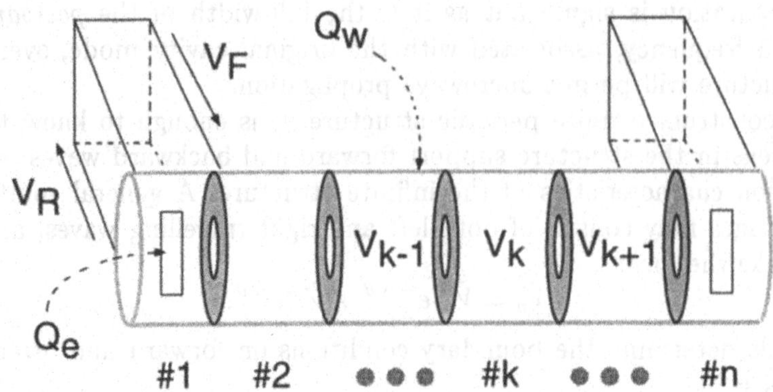

Figure 18. An N-cell travelling-wave accelerator structure may be modelled as N coupled resonators, or, more simply, as a transmission line.

To appreciate the scalings for multi-cell structures we should include losses. For the standing-wave structure this is straightforward, amounting to an effective wall-Q for the operating mode. For the travelling-wave structure we could proceed by fashioning a multi-cell circuit model, as depicted in Fig. 18. However a simpler approach is possible. We view the structure as a transmission line, with input and output matched at the operating frequency. We suppose a steady-state power P_{in} is provided to the input, and compute the net no-load voltage.

The steady flow of power P through the structure at location z, at group velocity V_g, implies stored energy per unit length $u = P/V_g$. Due to losses P may diminish through the line. Picking a particular cell one may say that if P_- is the power flowing into the cell, and P_+ the power flowing out, the difference must be due to power flowing into the walls,

$$P_- - P_+ \approx -L\frac{dP}{dz} = \frac{\omega}{Q_w}(uL),$$

where L is the cell-length, uL is the energy stored in the cell, and z is the displacement from the input cell. This may be expressed as

$$\frac{d}{dz}(V_g u) = -\frac{\omega u}{Q_w}. \tag{51}$$

This result determines the profile for stored energy in the structure, in terms of a group-velocity profile determined by the variation in beam-port radius, with initial condition set by $u(0) = P_{in}/V_g(0)$. Voltage may be determined from the shunt-impedance per unit length $r \equiv R/L$, with R the shunt-impedance associated with a single-cell. Identifying the local gradient

$G \equiv V/L$, with V the local single-cell voltage, one finds

$$\tau = \frac{Q_w}{\omega u} G^2 = \frac{G^2}{-dP/dz}$$

To illustrate, consider a *constant-impedance* structure, one with constant iris-radius, and therefore constant V_g. One can show that

$$G(z) = \left(\frac{\omega u}{Q_w}\tau\right)^{1/2} = G(0)\exp(-\alpha z),$$

where the attenuation constant is

$$\alpha = \frac{\omega}{2 Q_w V_g}.$$

The net voltage is then obtained by integration

$$V = \int_0^{L_s} G(z) dz = G(0)\frac{1 - e^{-\alpha L_s}}{\alpha},$$

with $L_s = NL$ the structure length, and N the number of cells. In terms of the *attenuation parameter*,

$$\tau = \alpha L_s = \frac{\omega L_s}{2 Q_w V_g}, \tag{52}$$

one may express this result as

$$V = (P_{in} R_{shunt})^{1/2} (1 - e^{-\tau}) \left(\frac{2}{\tau}\right)^{1/2}, \tag{53}$$

where $R_{shunt} = rL$, is larger by N than the shunt-impedance for a single-cell. One can show that maximum no-load voltage occurs for $\tau \approx 1.26$ and is given by $V \approx 0.9\,(P_{in}\,R_s)^{1/2}$. The attenuation parameter also determines the power to the load $P_{out} = P_{in} e^{-2\tau}$, and the fill-time, $T_f = L_s/V_g$, is just

$$T_f = \frac{2 Q_w}{\omega}\tau. \tag{54}$$

For example, the Mark III accelerator employed constant impedance structures, operated at 2856 MHz, and $\pi/2$ phase-advance per cell [17]. Group velocity was $0.01c$ and shunt-impedance per unit length was $r \approx 47.3$ MΩ/m. Structure length was 3.05 m, and fill-time was 1μs. With $Q_w \approx 1.0 \times 10^4$, attenuation parameter was $\tau \approx 0.90$. For their fixed-target experiments, the Mark III group developed a 20 MW klystron, and designed and built a linac powered by 21 such tubes. With one tube powering one structure, no-load voltage of 47.5 MV/tube could be achieved, and maximum beam energy was 1 GeV. Subsequent adventures, with a 20-GeV beam and a thousand or so *constant-gradient* structures can be found in [2].

2.13. SPACE-HARMONICS

It is not always adequate to have only the simple description of the multi-cell structure in terms of cell voltages as sampled by a relativistic beam. Where the sampling is done by a non-relativistic beam (in an injector), or by a bench-measurement device (a perturbing bead), one would appreciate a prescription for understanding the observations, and this means a more detailed picture of the fields in a multi-cell structure. This picture can be approached quite readily based on the analysis of the last section. We consider a right-travelling wave, in steady-state, at angular frequency ω. Based on the analysis of cell voltages leading to Eq. (47), we may infer that the steady-state axial electric field phasor satisfies the phase-advance boundary condition

$$\tilde{E}_z(\vec{r}_\perp, z+L) = e^{-j\theta}\tilde{E}_z(\vec{r}_\perp, z). \tag{55}$$

Defining wavenumber $\beta_0 = \theta/L$, and

$$\tilde{e}_z(\vec{r}_\perp, z) = e^{-j\beta_0 z}\tilde{E}_z(\vec{r}_\perp, z),$$

we observe that $\tilde{e}_z(\vec{r}_\perp, z+L) = \tilde{e}_z(\vec{r}_\perp, z)$, $\tilde{e}_z$ is a periodic function of z. This implies that we may expand it in a Fourier series,

$$\tilde{e}_z(\vec{r}_\perp, z) = \sum_{n=-\infty}^{+\infty} e_n(\vec{r}_\perp)\, e^{j2\pi n z/L}.$$

This in turn provides a decomposition for the electric field

$$\tilde{E}_z(\vec{r}_\perp, z) = \sum_{n=-\infty}^{+\infty} e_n(\vec{r}_\perp)\, e^{j\beta_n z}, \tag{56}$$

where the wavenumber

$$\beta_n = \beta_0 + \frac{2\pi n}{L}. \tag{57}$$

This result is Floquet's theorem. Applying Maxwell's equations to the electric field, we find that the harmonic amplitudes satisfy

$$\left(\nabla_\perp^2 + \frac{\omega^2}{c^2} - \beta_n^2\right) e_n = 0.$$

For a structure with circular symmetry, for example

$$e_n\left(\vec{r}_\perp\right) = E_n I_0\left(r\sqrt{\beta_n^2 - \frac{\omega^2}{c^2}}\right), \tag{58}$$

with I_0 the modified Bessel function. In general higher multipole content may be present, when not excluded by symmetry. With this "space-harmonic" decomposition, the problem of describing the fields is reduced to a determination of the space-harmonic amplitudes E_n.

Meanwhile, a relativistic beam on a ballistic orbit in effect performs a spatial Fourier transform on the fields in the structure, selecting out the synchronous component at each frequency ω, corresponding to wavenumber ω/c. For the powered structure, the voltage gain of a speed-of-light particle is determined by the $n = 0$ space-harmonic, as this satisfies $\beta_0 = \omega/c$ by design. This is why the model of the multi-cell structure in terms of single-cell voltages is adequate, from the descriptive point of view. The space-harmonic decomposition is helpful for design. In the early days, multi-cell structure design proceeded by analytic calculation [18] and bench-measurements on test stacks of a few cells [2]. Such calculations and measurements invoke the space-harmonic decomposition. In modern times one may in addition employ numerical design with prescribed phase-advance boundary conditions, as in Eq. (55).

To summarize, in pursuing the theory of the microwave accelerator we have followed the logic illustrated in Fig. 6. starting with a terminated interaction, and forming a cavity amenable to resonant excitation. Having arrived at the concept of shunt-impedance, we went on to find that shunt-impedance can be improved by use of multiple cells. One can achieve a high accelerating voltage, with a low peak power, limited only by losses, and the structure tune. These are the basic principles of the microwave linac as formulated by Ginzton, Hansen and Kennedy [19]. They leave a great deal of freedom to the structure designer as one can see in the literature [20, 21].

2.14. MICROWAVE CIRCUITS

To appreciate the practical aspects of a linac rf system, and associated instrumentation, it is helpful to take a look at a few common rf components, as depicted schematically in Fig. 19. The most conspicuous items in the figure are the lines, indicating waveguide and the circles, indicating connectors. We'll consider these first, and then look at common circuit elements.

Considerations in the choice of waveguide include: power-handling, multi-moding, attenuation, phase-stability, coefficient of thermal expansion,

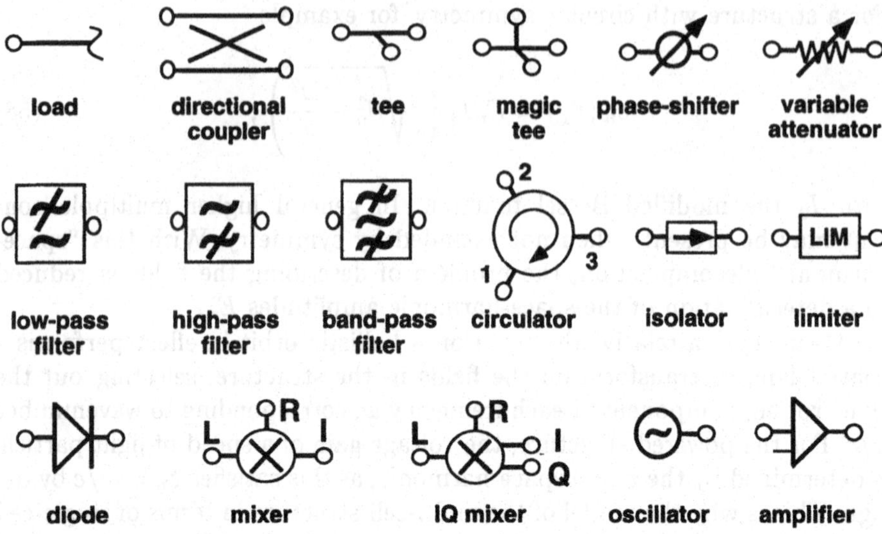

Figure 19. Components commonly found in microwave circuits.

convenience and cost. Waveguide appears in two applications in an accelerator system. One is for transport of high-power, and the other is for monitoring of the beam and the rf system, "instrumentation". For instrumentation work, one typically employs coaxial cable or "coax", as this is cheapest to obtain and install. The accelerator itself is a high-power application, demanding low-attenuation, and resilience against "breakdown". For this, hollow rectangular waveguide is a common choice.

Attenuation includes skin-effect losses, dominant at low-frequency, and, in coax, bulk dielectric losses, dominant at high-frequency. Attenuation is accompanied by dispersion thanks to the Kramers-Kronig relations [22]. Cumulative phase-shift through a length of line depends on the guide length, and thus the temperature. Where phase between two outputs is the concern, one may require temperature-stabilized water-flow, a thermal blanket, and/or special low-expansion coefficient cable. For instrumentation work, if one can arrange that phase difference between two cable outputs is the relevant quantity, it may be enough that the cables are of roughly equal length, and traverse the same run. Phase is in general sensitive to mechanical flexure, and guide should not be unintentionally crushed, dented, bent, made to bear a load or stepped on. Intentional dimpling may be employed to adjust phase-length of rectangular guide, which for that matter is used to bear a load, commonly its own weight. Where precise phase relationships are required within an instrumentation circuit, semi-rigid cable may be employed, and the circuit enclosed in a box to preserve mechanical integrity, as well as to provide shielding against electromagnetic interference and pulsed noise—"technical noise".

2.14.1. *Rectangular Guide*

A selection of common rectangular waveguide dimensions is indicated in Table 3. Due to the proliferation of "standards" for band nomenclature, there really isn't a single standard band designation as such, and the nomenclature of Table 3 may be considered vernacular. More detailed tables are available from waveguide manufacturers, including tolerances on guide dimension, and an assortment of standard flanges [23]. Accelerators have been built and operated in all of the bands listed in Table 3.

TABLE 3. Common rectangular waveguide ("WR") interior dimensions, recommended operating band, and cut-off frequency. Also indicated is the larger band designation into which the waveguide may be classed.

Guide	Size (inch)	Rec. (GHz)	f_c (GHz)	Band	(GHz)
WR650	6.500×3.250	1.12 - 1.70	0.91	*L*	(1.0 - 2.0)
WR284	2.840×1.340	2.60 - 3.95	2.08	*S*	(2.0 - 4.0)
WR187	1.872×0.872	3.95 - 5.85	3.15	*C*	(4.0 - 8.0)
WR90	0.900×0.400	8.20 - 12.40	6.56	*X*	(8.0 - 12.0)
WR62	0.622×0.311	12.40 - 18.00	9.49	*Ku*	(12.0 - 18.0)
WR42	0.420×0.170	18.00 - 26.50	14.05	*K*	(18.0 - 27.0)
WR28	0.280×0.140	26.50 - 40.00	21.08	*Ka*	(27.0 - 40.0)

The mode of rectangular guide with the lowest cut-off frequency is the TE_{10} mode. This mode has electric field oriented parallel to the b-dimension, as in Fig. 4, peaking in the guide center, and dropping monotonically to zero at the edges of the guide, so as to satisfy conducting boundary conditions. The TE_{10} mode dispersion relation is, neglecting losses,

$$\left(\frac{\omega}{c}\right)^2 = \left(\frac{\pi}{a}\right)^2 + \beta^2, \tag{59}$$

relating angular frequency ω to wavenumber β and guide long dimension a, as in Fig. 4. This is to say that the spatial variation of a forward wave in the guide is characterized by guide wavelength $2\pi/\beta$ different from the free-space wavelength $2\pi c/\omega$. The dispersion relation, Eq. (59) looks different from that of free-space, and the reason for this is that the mode is a superposition of *two* free-space waves, each propagating at an angle to the guide-axis, and phased in such a way as to satisfy conducting boundary conditions. Cut-off occurs at frequency $f_c = c/2a$ and corresponds to a

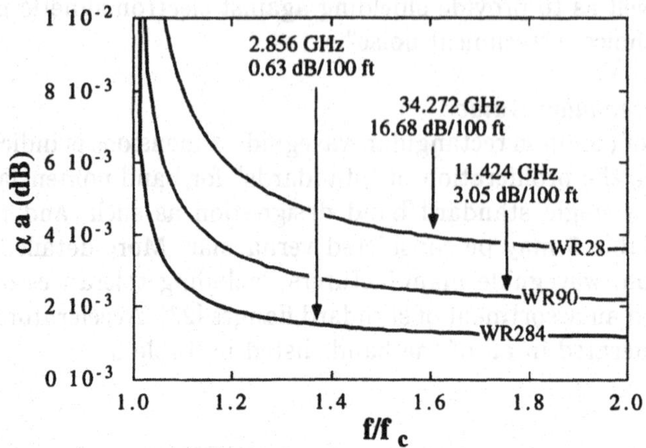

Figure 20. Attenuation α normalized to waveguide width a, versus frequency for three sizes of waveguide.

propagation angle of 90 °, i.e., no propagation. The lower frequency listed in column 3 of Table 3, $1.25 \times f_c$, is recommended to insure that the group velocity, $d\omega/d\beta$, is not too low. The upper frequency in column 3, $1.90 \times f_c$ is just below the cut-off for the next mode(s) in the waveguide.

Waveguide is not lossless, of course, and finite surface resistance damps wave amplitude, amounting to attenuation per unit length given by [24]

$$\alpha = \frac{R_s}{Z_0}\left[\frac{\omega^2}{c^2}\frac{1}{b} + \frac{2\pi^2}{a^3}\right]\frac{c}{\omega}\frac{1}{\beta}. \tag{60}$$

At the lower end of the operating band, the wave has lower group velocity and spends more time in a given length of guide. Thus attenuation becomes large near cut-off. At higher frequencies, surface resistance is increasing, and so attenuation ultimately must increase. Minimum attenuation occurs at $2.3 - 2.4 \times f_c$, outside the single-mode operating band. In Fig. 20 one can see Eq. (60) illustrated for three example waveguide sizes. As a point of reference, OFE copper WR284 has attenuation of 0.629 dB/100 ft at 2856 MHz.

As a practical matter, waveguide insertion loss must be accounted for in the layout of the accelerator microwave network. As an example, the "thick-wall" WR284 network for two successive tube stations in the Two-Mile Accelerator is sketched in Fig. 21. Waveguide arm phase-lengths range from 118 to 139 λ_g, where the guide wavelength $\lambda_g = 2\pi/\beta \approx 15.3$ cm. Over these 60-ft lengths, one expects attenuation in the range of 0.37-0.43 dB. The measured results lie in the range 0.41-0.68 dB, including coupler and flange insertion loss [2]. This means that 10–15% of the power is

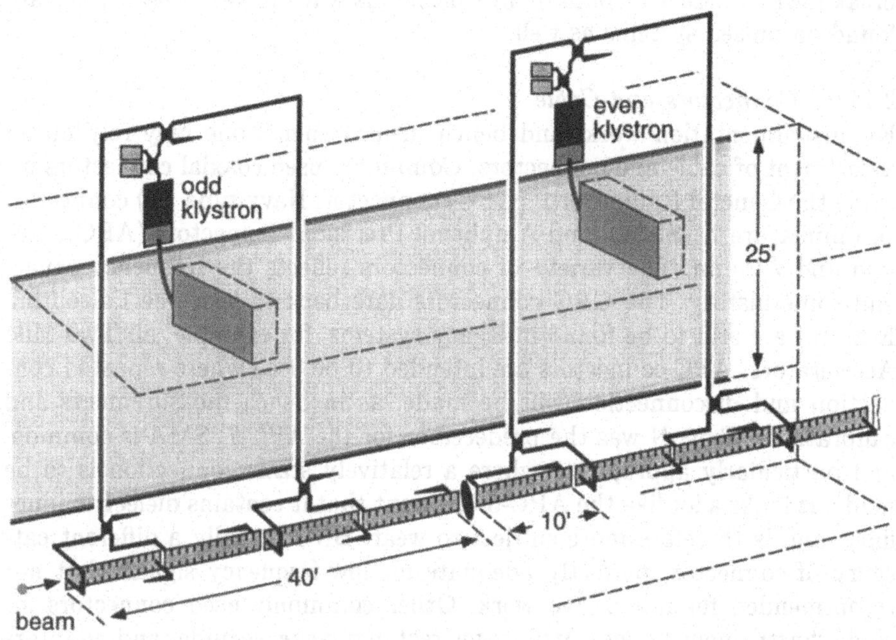

Figure 21. Waveguide layout for a standard pair of klystrons on the Two-Mile Accelerator. This configuration "babaabab" repeats along the linac, as discussed in connection with coupler asymmetry.

lost before it reaches the accelerator. Theoretically one could reduce the attenuation loss by a factor of about 2.3, by employing nearly-overmoded custom-made guide. However, the cost-savings for this accelerator wouldn't have been worth it. Even in modern-times, with a utility savings accruing at 3 cents/kW-hr it would take years to "pay" for the cost of the mode-converters from the custom-guide to standard guide. For future machines, however, where rf power consumption may approach several hundred megawatts, waveguide loss is taken quite seriously. Much interest attaches to the design of overmoded or "quasi-optical", low-loss, high-power waveguide components [25].

Fig. 21 also indicates the waveguide coupler layout employed. This layout can be understood in part from the discussion of Sec. 2.8, and beam deflection due to phase asymmetry. This effect was reduced by offset of the coupling cells, and waveguide feed layout designed to provide partial cancellations of rf kicks. Kick cancellation is not local, however, but is completed only over the 80' length spanned by eight structures, and then only approximately.

One may also find, for high-power systems, pressurized waveguide; these are intended to function with electronegative gas, such as SF_6, to inhibit breakdown. These are common in continuous-wave (CW) systems, but are found on pulsed systems as well.

2.14.2. *Connectors and Cable*

For instrumentation work, and bench measurements one may rely on an assortment of cable and connectors. Commonly used coaxial connectors include the General Radio Corp. (GRC) connector, Navy-type (N) connector, sub-miniature A (SMA), and Amphenol Precision Connectors (APC), 3.5-mm and 7.0-mm. The variety of connectors reflects the frequency range, and some history. The GRC connectors date back to near the Paleolithic Era, but are still to be found in legacy systems, for example, all Two-Mile Accelerators. APC connectors are intended to be used where repeated connection and disconnection will be made, as in bench measurements and calibrations. Type N was the predecessor for the APC-7. SMA is common, and particularly appropriate where a relatively static connection is to be made. It looks a lot like the APC-3.5, except that it contains dielectric inner liner, and is therefore more subject to wear. BNC is really a different category of connector, perfectly adequate for low-frequency signals, but not recommended for microwave work. Other commonly used connectors include "instrument flange" fittings for rectangular waveguide, and adaptors from waveguide instrument flange to standard coax, and *precision* adaptors. For accelerator work one has in addition vacuum-waveguide flanges that mate to knife-edge gaskets to form a crush-seal. These and associated adaptors to instrument flange are not commercial standards.

While folks often refer to cable in terms of the connector they are accustomed to seeing on it, the cable itself may have a name, here are a few.

- *RG-214* is often referred to as "N-type cable" and sees routine use for microwave signals in an S-Band linac system. It has a black polyvinylchloride jacket with 0.425" outer diameter (OD), and a polyethylene dielectric. Attenuation constant is 18 dB/100 ft at 3 GHz and 27 dB/100 ft at 5 GHz.
- *Heliax*^{TM} is a brand name of Andrews Corporation, for coax with a dielectric that is mostly air, with outer conductor supported on a helical foam ribbon. As a result of the diminished volume of lossy-dielectric, attenuation per unit length is low. It can be delivered, already cut to length and connectorized, in a variety of sizes. It is excellent for high-frequency beamline instrumentation, particularly where low signal level and noise are concerns.
- *RG-223/U* is often referred to as "BPM cable" as it is commonly used to transmit signals from beam position monitor (BPM) striplines. It shouldn't be used for microwave work, unless attenuation and disper-

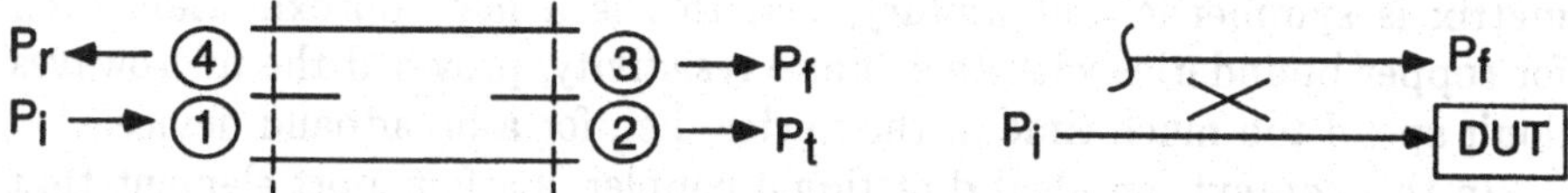

Figure 22. A directional coupler permits one to monitor a circuit. The circuit sketch on the right illustrates a typical use, to monitor forward power to a device-under-test (DUT) attached to port 2, with a matched load on the reverse arm, port 4.

sion are desired. Attenuation constant is 3.2 dB/100 ft at 100 MHz, 16.5 dB/100 ft at 1 GHz and 46.0 dB/100 ft at 5 GHz.

- *RG-58C* is often referred to as "BNC cable" and isn't to be used for microwave frequencies. RG-58C has an attenuation constant of 4.9 dB/100 ft at 100MHz.

RG-214, RG-58C and RG-223/U have a group velocity of 0.66c. In addition to these, there is a wealth of specialty cable available commercially, designed to provide either low loss, low thermal expansion, insensitivity to flexure, or mechanical ruggedness.

Having employed low attenuation cable to extract a marvelous beam-induced or other signal, one will probably need an *attenuator*. These can be fixed, or variable, for example, mechanically variable, by knob or screw adjustment. Beams are pretty variable, so variable attenuators are good to have. Like most microwave components, attenuators come with a rating for VSWR and insertion loss over a specified bandwidth.

To understand a circuit it often helps to disconnect a new device and replace it with a *matched load* . Matched loads come in all shapes and sizes, with a specification for VSWR and bandwidth. A cavity with $\beta = 1$ is an example of a matched load. Generally however one prefers a match over a broad band, and this implies use of high-loss material, equivalent to very low Q_w in the parlance of cavities. S-Band structure output loads may employ a Kanthal-loaded tapered geometry, for example.

2.14.3. *Directional Coupler*

To monitor the waves in a circuit it is helpful to tap off a small bit of the forward and reverse waveforms, while not interfering appreciably with the circuit. One needs a *directional coupler* as depicted in Fig. 22.

The function of many basic components can be understood from their description in terms of an S-matrix, and the directional coupler is a good example of this. The S-matrix for an N-port device is an $N \times N$ matrix defined with respect to reference planes on the connecting waveguide, and a common normalization for waveguide impedance. With phasor $\tilde{V}_k^+$ incident on port k, and all other ports terminated in a matched load, the outgoing phasor on port l is $S_{lk}\tilde{V}_k^+$. For elements that are free of sources the S-

matrix is symmetric and unitary, and this is a fair approximation even for copper boundaries with their finite resistivity, provided the microwaves don't spend too much time in the device, i.e., for a broadband device.

In this context, an ideal directional coupler is a four-port element that is symmetric upon reflection, is lossless and is perfectly matched. These conditions, together with unitarity restrict the S-matrix to the form

$$S = \begin{pmatrix} 0 & \alpha & j\sqrt{1-\alpha^2} & 0 \\ \alpha & 0 & 0 & j\sqrt{1-\alpha^2} \\ j\sqrt{1-\alpha^2} & 0 & 0 & \alpha \\ 0 & j\sqrt{1-\alpha^2} & \alpha & 0 \end{pmatrix}, \quad (61)$$

for some parameter $\alpha \leq 1$, and some choice of reference planes. A real directional coupler is described by coupling C and directivity D, defined with respect to the quantities indicated in the sketch: power incident on port 1, P_i, transmitted power to port 2, P_t, forward power on port 3, P_f, and reverse power, P_r, on port 4. The coupling and directivity are, in units of decibels (dB),

$$C = 10\log_{10}\left(\frac{P_i}{P_f}\right), \quad D = 10\log_{10}\left(\frac{P_f}{P_r}\right). \quad (62)$$

The isolation is defined according to $I = 10\log_{10}(P_i/P_r)$, so that $C+D = I$. The 3-dB coupler is rather special in that, with a matched load on one arm, it can distribute power equally to the two output arms. Couplers come in high-power, vacuum waveguide and low-power, instrument-connectorized versions.

2.14.4. *Tee*

The simplest-looking element is the *tee* as illustrated in Fig. 23. Normalizing all ports to a common impedance, one may apply unitarity to determine the form the S-Matrix must take. After a choice of reference planes (L and D in the figure), one can show that the S-Matrix may be parameterized in terms of two real variables ϕ and θ, such that [25]

$$S = \begin{pmatrix} \frac{1}{2}\left(e^{j\phi} - \cos\theta\right) & \frac{1}{2}\left(-e^{j\phi} - \cos\theta\right) & \frac{\sin\theta}{\sqrt{2}} \\ \frac{1}{2}\left(-e^{j\phi} - \cos\theta\right) & \frac{1}{2}\left(e^{j\phi} - \cos\theta\right) & \frac{\sin\theta}{\sqrt{2}} \\ \frac{\sin\theta}{\sqrt{2}} & \frac{\sin\theta}{\sqrt{2}} & \cos\theta \end{pmatrix}. \quad (63)$$

The ideal lossless tee cannot be matched ($|S_{11}| \neq 0$), except in the degenerate case, $\sin\theta = 0$.

The tee forms the basis for many useful devices. With a shorting plane on arm 3, and a signal incident on port 1, the signal transmitted through

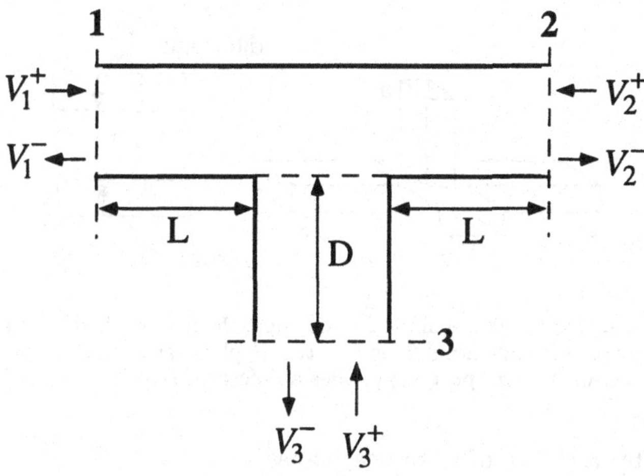

Figure 23. A tee is a three-port element with symmetry between two arms.

from port 1 to port 2 is a superposition of a wave transmitted directly to port 2, and one that travels down arm 3, and, after multiple reflections in arm 3, returns to be split between arms 1 and 2. The impedance seen looking into arm 1 then depends on the position of the shorting plane on arm 3. A "stub-tuner" consists of one or more such tees, with moveable shorting planes, and can be quite helpful in providing a match when arm 2 is connected to a DUT. A tee with moveable short can also function as a reflective switch. Considering the frequency dependence of the constructive or destructive intereference, one recognizes that this device may also function as a tunable band-pass or a band-reject ("notch") filter.

Speaking of which, we have it seems already touched on the basic elements needed to fashion filters. Hollow waveguide is a high-pass filter, due to the phenomenon of cut-off. A tee can function as a band-pass filter. A resonant cavity can serve as a narrow-band filter. More generally an assortment of tee's and cavities ("poles") can be employed to fashion a two-port element with a custom-made S_{21}. One can find a wide assortment of commercial tubular, cavity and waveguide microwave filters, fixed filters, mechanically-tuned, or voltage-controlled.

2.14.5. *Magic Tee*

A hybrid tee is a four-port device as illustrated in Fig. 24. In the ideal case, it is lossless and the symmetry seen in the figure precludes coupling of ports 1 and 4. A *magic tee* is a hybrid tee that is matched on all ports; in practice, matching relies on obstacles situated in the junction. Constraints of symmetry and unitarity reduce the possible forms of the S-matrix to

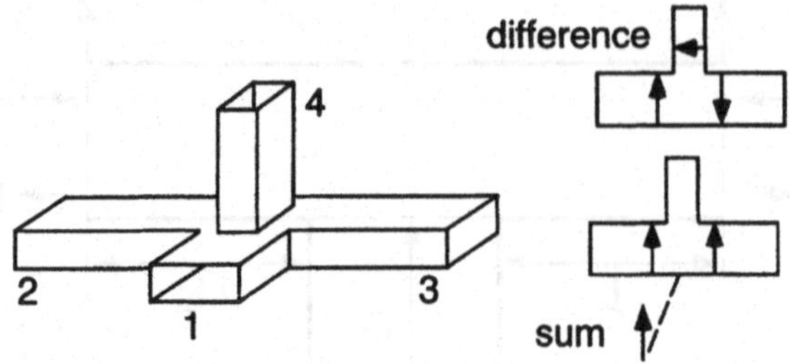

Figure 24. A magic tee is a symmetric four-port device matched on all ports. With signals incident on through arms 2 and 3, the H-plane arm, port 1, provides a sum signal, and the E-plane arm, port 4, provides a difference signal.

just one, after a choice of reference planes,

$$S = \frac{1}{\sqrt{2}} \begin{pmatrix} 0 & 1 & 1 & 0 \\ 1 & 0 & 0 & -1 \\ 1 & 0 & 0 & -1 \\ 0 & 1 & -1 & 0 \end{pmatrix}. \tag{64}$$

From this result, or just from the symmetry evident in Fig. 24, one can show that for two incident voltages, on ports 2 and 3, port 1 provides a an output proportional to the sum, and port 4 an output proportional to the difference. In conjunction with a phase-shifter and a variable attenuator, the magic tee can be employed to balance two large signals, permitting easy inspection of a small residual.

2.14.6. *Circulator, Isolator*

An ideal *circulator* is a three-port element described by

$$S = \begin{pmatrix} 0 & 0 & 1 \\ 1 & 0 & 0 \\ 0 & 1 & 0 \end{pmatrix}.$$

Inspecting this S-matrix one can see that the device is matched on all ports, and a signal incident on one arm is circulated to only one adjacent arm. Based on our analysis of the tee, one might think that such a device would be impossible to construct; however, our analysis of the tee assumed a reciprocal device, one with symmetric permeability and permittivity tensors. A circulator meanwhile incorporates μ-material and a static magnetic field, and is non-reciprocal, by design. A real circulator has noticeable insertion loss. With one arm terminated in a matched load, a circulator becomes a two-port device, then called an *isolator*. These devices are almost essential for isolating a source from a load, to avoid pulling of the source frequency or amplitude, and in isolating one part of a circuit from another, to avoid an inadvertent standing-wave and resulting modification of the circuit operating point. Mention of operating point hints that some circuits are non-linear. Let us consider the basic non-linear elements.

2.14.7. *Crystal Detector*

To monitor the time-variation of a low-frequency voltage one may use an oscilloscope. This method doesn't work well with microwaves, however, since typical scope bandwidths are lower than 1 GHz, and typical microwave frequencies are above 1 GHz. To monitor amplitude one needs a microwave diode, a ***crystal detector***. A diode is a non-linear element, with ideal characteristic $i = i_s(e^{eV/k_BT} - 1)$, where $k_B \approx 1\mathrm{eV}/(38.7 \times 300\,\mathrm{K})$ is Boltzmann's constant, T is temperature, and i_s is the saturation current. A microwave diode is a diode encapsulated in a matched waveguide mount (coaxial or other). The problem of designing such a mount was once a major research problem [26]; nowadays there is a variety of commercial detectors to choose from, depending on one's needs for dynamic range, bandwidth, sensitivity, and output polarity. Microwave diodes are ubiquitous elements, appearing in all microwave detection circuits. In addition to power-detectors, there are also special-purpose diodes, including step-recovery diodes for generating harmonics, as in a frequency-multiplier, and PIN diodes for fast switching or phase-shifting.

The first characteristic to know about a diode is that it can be destroyed by too high an input level. If routine operation will involve extreme signal ranges, one needs also attenuation and a *limiter*. A limiter is a two-port element that changes state ("breaks down") above a certain power level, absorbing or reflecting the incident signal. A limiter is essential where expensive test equipment is employed with beam-induced signals. At high-power limiters may generate un-wanted harmonics, and this is not usually desirable; they are intended to function like a circuit-breaker.

It is instructive to consider a simple model of the ideal microwave diode. Let us apply a single-frequency voltage waveform to the ideal diode characteristic. We express $eV/k_BT = A\cos\phi$ with $\phi = \omega t + \phi_0$, t is time, ω is the angular frequency of the signal, and ϕ_0 is the phase of the signal. Using the identity,

$$e^{A\cos\phi} = \sum_{k=-\infty}^{+\infty} I_k(A)\, e^{jk\phi}, \tag{65}$$

with I_k the modified Bessel function, we may express the current through the diode as

$$i = \sum_{k=-\infty}^{+\infty} i_k(A)\, e^{jk\phi},$$

for some coefficients i_k. The "DC" component is

$$i_0 = i_s \left(I_0 (A) - 1 \right) \approx \frac{1}{4} i_s A^2,$$

and in the last equality we employ the small-argument expansion for I_0. Combined with a low-pass filter characteristic (integral to the crystal mount or in the output circuit) crystals evidently provide a "square-law" response, for small signal amplitude, with current output proportional to incident microwave power. Where the crystal is being employed as the primary power-monitor, it could be labelled, calibrated, and re-calibrated occasionally. For best results, the calibration waveform should mimic as closely as possible the waveform to be monitored, because a crystal mount will not transmit all frequencies equally. Crystals may be employed beyond the range of square-law response; this requires a polynomial fit based on the calibration data covering the full range of power-levels to be employed.

As an example we could consider monitoring an S-Band klystron-amplifier output. A typical setup would include a 50 or so dB directional coupler, a low-pass filter to block out harmonics generated by the klystron, attenuators, and a crystal mount, with a replaceable 1N21B cartridge. Where high-precision is required, an occasional check of calibrations might employ calorimetry, inferring average power based on temperature variation of cooling water to a load. In addition, there are several instruments that can greatly simplify circuit tests, as well as instrumentation problems themselves. These include the power-meter, frequency-counter, spectrum analyzer, and the scalar or vector (phasor) network analyzer. Items such as these come with documentation roughly proportional to cost, and thus the manufacturers' literature provides a wealth of additional information.

2.14.8. *Mixer*

Where phase-information is of interest it is common to employ a *mixer*, a three-port device, accepting two high-frequency inputs, and providing output at the difference frequency. The simplest concept for a mixer consists of a tee to sum two signals, labelled R and L, and a diode to detect the output. The voltage across the diode takes the form $eV/k_BT = A_R \cos \phi_R + A_L \cos \phi_L$, where $\phi_R = \omega_R t + \phi_{0R}$, and $\phi_L = \omega_L t + \phi_{0L}$. Using the identity, Eq. (65), twice, we may express the diode current as

$$i = i_s \sum_{m,n=-\infty}^{+\infty} I_m (A_R) I_n (A_L) e^{j(m\phi_R + n\phi_L)} - i_s.$$

Thus one may expect the output to include all frequencies of the form $m\omega_R + n\omega_L$. This isn't usually satisfactory, since one is interested in the component at the intermediate frequency ("IF") $\omega_{IF} = \omega_R - \omega_L << \omega_R, \omega_L$. To remove the "DC" offset

$$i_0 \approx i_s \left(I_0 \left(A_R\right) I_0 \left(A_L\right) - 1\right)$$

the mixer may incorporate two diodes, exposed to opposite polarity versions of the R signal, with outputs subtracted. This can be accomplished for example with a magic tee, the L-signal incident on the H-plane arm, the R-signal on the E-plane arm, and diode mounts on the thru-arm ports, with balanced outputs. The lowest frequency component remaining is then

$$i_{IF} = 2 i_s I_1 \left(A_R\right) I_1 \left(A_L\right) \cos\left(\phi_R - \phi_L\right)$$

In phasor-form we may write this as $i_{IF} = \Re \tilde{i}_{IF} e^{j\omega_{IF} t}$, where

$$\tilde{i}_{IF} = 2 i_s I_1 \left(A_R\right) I_1 \left(A_L\right) e^{j\left(\phi_{Ro} - \phi_{Lo}\right)}.$$

This last result is quite helpful, for conceptual purposes. Ideal operation of a mixer invokes a CW signal of known, constant amplitude and phase for the L port. If one thinks of the R input in simple phasor terms, as $V_R = \Re \tilde{V}_R e^{j\omega_R t}$, with phase referred to the L signal, then the IF voltage produced by the circuit $V_{IF} = \Re \tilde{V}_{IF} e^{j\omega_{IF} t}$ is simply a phasor $\tilde{V}_{IF} \approx k \tilde{V}_R$ providing an approximately linear analog of the input signal, referred in phase to the L port, with IF carrier frequency. The last equality is written with the approximation $A_R << 1$, in mind, and the constant k depends on the mixer specification, and the L-amplitude. This conversion, as well as features of the unwanted higher-frequency terms will be described in plots provided by the manufacturer for the particular mixer.

Mixers come in a variety of forms more sophisticated than a simple balanced mixer. The "IQ" or "dual-output" mixer is a four-port device, providing two IF outputs representing the in-phase and quadrature components of the R-signal referred to the L-signal. One output provides $V_{IF-I} = \Re \tilde{V}_{IF} e^{j\omega_{IF} t}$, and the other $V_{IF-Q} = \Im \tilde{V}_{IF} e^{j\omega_{IF} t}$. In practical terms, mixers permit one to take a high-frequency signal down below 1 GHz, where filters, amplifiers and other components are cheap, at the expense of some additional caution in designing the circuit. Related to the mixer concept is the "modulator", where a pulsed bias voltage is applied to shift the phase or amplitude of a high-frequency signal, to produce a modulated high-frequency signal.

2.14.9. *Oscillators and Amplifiers*

Just a few words about oscillators and amplifiers, as there are libraries on the subject. When using old, cast-off, or "don't-know-where-it's-been" sources or amps, one should keep in mind that front-panel displays are for monitoring not measuring. A frequency-counter and power-meter are helpful to check what an instrument is actually putting out. Since diodes have a temperature-dependent characteristic, most solid state devices accept temperature as a "control knob". So it is good practice to exert some control over temperature. Silicon is a good particle detector, so it is preferable not to situate solid-state components in an accelerator vault.

We could also say a word about "noise". For a system in equilibrium at temperature T, thermodynamics tells us to expect, in a bandwidth Δf noise-power $P_n = k_B T \Delta f$. Most systems aren't yet aware of this equilibrium. Accelerator vaults and associated areas are flooded with all manner of signals, ranging from ionizing radiation, to rf frequencies leaking out of unterminated couplers, to pulsed noise from the modulator circuits, and ac noise at the power-line frequency and harmonics. Diurnal variations round out the low-frequency end of the "noise" spectrum.

Amplifiers also don't do as well as the black-body, and so usually come with a specification for "noise figure" or NF. Given power gain G for the amplifier, one may determine the noise power output by the device P_N, above that expected at the operating temperature T, according to

$$NF = 1 + \frac{P_N}{G k_B T \Delta f}.$$

Operationally, in terms of loss in the ratio of signal-to-noise,

$$NF = \frac{(S/N)_{input}}{(S/N)_{output}}.$$

By convention NF is specified at 290 K (17°C), usually in decibels, as $10 \log_{10} NF$, or in terms of equivalent noise temperature $T_{eq} = (NF - 1)T$.

With microwave circuit components in-hand, we can turn to instrumentation; but first let us examine what there is to instrument.

3. Physics of Beams

Thus far we have discussed the microwave accelerator in detail, and left the beam pictured as a little point charge. In this section, we take a closer look at the beam, first some formal-looking details and then the more colloquial aspects of beam physics. The theoretical development of the subject of beams is extensive, but what is actually known about beams is limited by the instrumentation available.

3.1. FORMAL-WEAR FOR BEAMS

For purposes here a beam is a collection of electrons of comparable momenta, moving at about the speed of light in the lab-frame. One may be concerned that this definition excludes discussion of low-voltage beams, but even an 80 keV electron is moving at $0.5c$. And typical accelerating fields are on the order of several MeV/m, so we omit at most about an inch of linac, and actually, not even that.

To describe our beam, we might at first be inclined to pull out quantum electrodynamics and set to work. Then we realize as we observe collisions, that we need to pull out the whole Standard Model. At this point we realize that we really can't describe beams completely until we have figured out the Theory of Everything. Fortunately such a complete description of beams isn't needed. Accelerator phenomena take place at low center of mass energy where electrodynamics rules. If we trace the evolution of a beam from emission through collisions we find that it's behavior is well-described by classical mechanics and electrodynamics, with only a few exceptions. For the present work, we will limit ourselves to the linac, a classical venue.

3.1.1. *Kinetic Theory*

Adopting a classical description then, we may associate with each electron a position $\vec{R}$ and a velocity $\vec{V} = d\vec{R}/dt$, and momentum $\vec{P}$. Let us suppose our beam contains N electrons. The beam is completely specified given the positions and momenta $\left\{\vec{R}_i(t), \vec{P}_i(t)\right\}$, with $i = 1, ..., N$ at some instant of time t.

In principle, to describe the evolution of this system, one need merely determine the fields at this time, including fields due to the beam, and then solve $6N$ ordinary differential equations coupled to Maxwell's equations. This is really too much work though, and all to produce too much information. Such a solution would include the Coulomb field of each and every electron, and detailed information on electron-electron collisions mediated by these fields. And electrons don't even collide too often, so one is storing quite a lot of information in RAM, in order to accurately model something that is really a higher-order effect. There is a simpler, approximate method of describing such a beam.

We start from the *Klimontovich* distribution,

$$F(\vec{r}, \vec{p}, t) = \sum_{k=1}^{N} \delta^3 \left[\vec{r} - \vec{R}_k(t)\right] \delta^3 \left[\vec{p} - \vec{P}_k(t)\right]$$

the density in the six-dimensional space with coordinates $(\vec{r}, \vec{p})$. We observe by differentiation, and use of the Lorentz force law, that

$$\frac{\partial F}{\partial t} + \vec{v} \bullet \vec{\nabla}_{\vec{r}} F - \frac{e}{m}\left(\vec{E} + \vec{v} \times \vec{B}\right) \bullet \vec{\nabla}_{\vec{p}} F = 0,$$

where $\vec{v}$ is velocity. We emphasize that this is a rather odd equation, not because of its form, a continuity equation, but because F is a spiky thing, and $\vec{E}, \vec{B}$ also behave badly in the vicinity of each electron. A clearer picture of the system may be formed by selecting a length-scale that distinguishes between the microscopic and macroscopic features. We average on this microscopic length-scale,

$$f = \langle F \rangle, \quad \vec{e} = \left\langle \vec{E} \right\rangle, \quad \vec{b} = \left\langle \vec{B} \right\rangle,$$

permitting us to express,

$$F = f + \delta f, \quad \vec{E} = \vec{e} + \delta \vec{e}, \quad \vec{B} = \vec{b} + \delta \vec{b}.$$

The smoothed distribution satisfies,

$$\frac{\partial f}{\partial t} + \vec{v} \bullet \vec{\nabla}_{\vec{r}} f - \frac{e}{m}\left(\vec{e} + \vec{v} \times \vec{b}\right) \bullet \vec{\nabla}_{\vec{p}} f = \frac{e}{m}\left\langle \left(\delta\vec{e} + \vec{v} \times \delta\vec{b}\right) \bullet \vec{\nabla}_{\vec{p}} \delta f \right\rangle.$$

In the absence of fluctuations in the actual density, the right-side vanishes. Such fluctuations disappear however only in the formal limit, $e \to 0, m \to 0$, e/m constant. In this limit, all cross-sections vanish, there are no single-particle emissions or collisions as such. With finite e, fluctuations represent a correction, but a small correction, and not a dominant effect. Thus we do well to describe the distribution in iterative fashion. At lowest-order, neglecting two-particle correlations, we have the *Vlasov* equation,

$$\frac{\partial f}{\partial t} + \vec{v} \bullet \vec{\nabla}_{\vec{r}} f - \frac{e}{m}\left(\vec{e} + \vec{v} \times \vec{b}\right) \bullet \vec{\nabla}_{\vec{p}} f = 0. \tag{66}$$

At this order too, we have Maxwell's Equations (for the smoothed fields) determined by moments of f,

$$\rho(\vec{r}, t) = -e \int d^3\vec{p} f(\vec{r}, \vec{p}, t),$$

$$\vec{J}(\vec{r}, t) = -e \int d^3\vec{p}\,\vec{v} f(\vec{r}, \vec{p}, t).$$

One can check that

$$\frac{\partial \rho}{\partial t} + \vec{\nabla} \bullet \vec{J} = 0.$$

To simplify our notation we will denote the smoothed fields by $\vec{E}$, $\vec{B}$.

The Vlasov equation is a continuity equation in phase-space. It states that the convective derivative of f along a physical orbit vanishes. If one moves with an electron in phase-space, one finds that the local density of electrons, f, is constant. Said differently, trajectories in phase-space don't cross.

Let us emphasize a few points to put aside some common misconceptions. Beam-induced fields are not omitted in this treatment, merely microscopic fluctuations in these fields. Thus fields of electrostatic character ("space-charge" fields) fit perfectly well in this description. But let us not call them "Coulomb fields", since we have averaged out the fluctuations that this phrase suggests, and neglected the collisions that these fields mediate. Collisions of course would exert a force specific to the scattering particles and cause phase-space trajectories to cross. In addition to intra-beam scattering, we have also omitted other collisions, for example, between an electron and an ion resident in the beamline. Radiation is not omitted in this treatment, merely fluctuations in the radiation fields, radiation seen by one electron, but not its neighbor. But let us not say that "incoherent radiation" is omitted, since an externally-produced wave with poor coherence fits easily in this description. Finally we emphasize, as a caveat, that important exceptions to this picture appear in beam-physics problems other than the electron-linac. These are fluctuation-dissipation problems, associated with scattering, radiation, or cooling. In these problems, corrections to the right-side of the Vlasov equation may often be approximated as in the Fokker-Planck equation.

3.1.2. *Vlasov Equation*

To fully enjoy the Maxwell-Vlasov equations we should consider a practical computational interpretation. Suppose that we start at $t = 0$, and solve these equations for the distribution and the fields, taking note of the fields at every step in time. Then let us return to $t = 0$, and generate a finite collection of "test-particles", i.e., coordinates $(\vec{r}, \vec{p})$. We'll suppose this initialization or "loading" in phase-space represents a sampling of the original initial Vlasov distribution, when smoothed on some appropriately small length scale ("binned", "allocated"). We then "track" these individual test particles by integrating the equations of motion in the fields already determined. As far as the particles are concerned, these are prescribed external fields. Along the way, we compute charge and current densities associated with the test particles, averaged on the microscopic length-scale. When we are finished tracking, we should find that we have computed the evolved Vlasov distribution, with precision limited by the statistics of our original sampling. Actually we realize that we could return to $t = 0$ again, and track the particles while computing the fields from the "macroparticle" distribution itself. We should arrive at the same result, if we have adequately modelled the charge and current densities with our sampling.

This is the basis for "particle-in-cell"simulations of beams. It is also the basis for a Hamiltonian description of beams, in which we picture individual particles with motion described by a single Hamiltonian, H. Particles do not interact with each other directly, they simply move in a Hamiltonian prescribed by the electromagnetic fields. The Vlasov equation may then be expressed simply in terms of the Poisson bracket in canonical coordinates $(\vec{q},\vec{p})$,

$$[f,H]=\frac{\partial H}{\partial\vec{p}}\bullet\frac{\partial f}{\partial\vec{q}}-\frac{\partial H}{\partial\vec{q}}\bullet\frac{\partial f}{\partial\vec{p}}, \tag{67}$$

as

$$\frac{df}{dt}=[f,H]+\frac{\partial f}{\partial t}=0. \tag{68}$$

Viewing each particle as an independent Hamiltonian system, and the beam as a statistical ensemble of such non-interacting systems, the Vlasov equation amounts to a statement of *Liouville's theorem*.

3.2. BEAMS AT WORK

3.2.1. *Time-Structure of Beams*

In describing the beam and the accelerator it is helpful to distinguish between the *main-linac*, the region of linac where $V \approx c$, and the *injector*, where $V < c$. Our work on the accelerator structure dictates some basic features we must require of our beam distribution in the main-linac. In this section we describe these requirements.

At high-gradient it is not economically (nor thermally) feasible to operate the linac continuously, thus it is *pulsed*. Machine pulses repeat at some frequency f_{rep} with 10–360 Hz a typical range. The length of the rf pulse should be enough to fill the structure, at least about 1 μs at S-Band. The maximum length is constrained by the klystron modulator circuit, with 1–10 μs typical. Within an rf pulse, beam electrons should reside near the accelerating peak of the voltage. In this way one arrives at the picture of Fig. 25, depicting the time-structure of the beam, as dictated by the accelerating mechanism.

Some freedom still remains within this picture. Depending on the capabilities of the injector, one may operate with every bucket filled, where then bunches are spaced at the rf period $1/f_{rf}$, one may skip buckets, or one may fill just one bucket. The beam pulse duration T_p may be the full length of the rf pulse, or shorter, 50 ns–3 μs being typical. Intrapulse current is set by the requirements for experimentation, and limited by collective effects, longitudinal and transverse wakefields, beam-loading and beam break-up. For definiteness we often picture a single bunch as a Gaussian current profile,

$$I_b(t)=\frac{Q_b}{\sqrt{2\pi}\sigma_t}\exp\left(-\frac{t^2}{2\sigma_t^2}\right), \tag{69}$$

as a function of time t, corresponding to bunch charge Q_b and bunch length $c\sigma_t$. Real bunch profiles exhibit a variety of shapes, and may contain fine

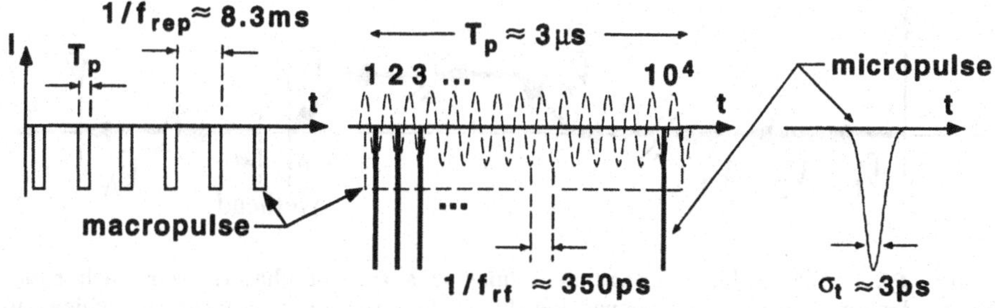

Figure 25. Typical time-structure of a beam in a normal conducting rf linac, viewed on three different time-scales. For illustration here pulse repetition frequency $f_{rep} = 120$ Hz, macropulse length $T_p \approx 3$ μs, with every bucket filled, so that micropulses or "bunches" are spaced at the rf period $1/f_{rf} \approx 350$ ps.

structure. In general one should be alert to the fact that there are typically three currents that may be quoted in connection with a beam. The peak current within a bunch is quite high, approaching 10^3 A. For a bunch-train, the average current, within the pulse, is just $Q_b f_{rf}$, when every bucket is filled, and this may be on the order of 10^0 A. Finally the average current is just the total charge per pulse multiplied by the machine repetition frequency, and this may be on the order of 10^{-5} A.

In the limit of low-current, where collective phenomena may be neglected, the function of the linac rf system is quite simple. Let us consider an electron entering the main-linac at time t. We suppose there are N klystrons powering the linac, and each klystron may be powering one or several structures. Consider the net voltage witnessed by the relativistic electron, absent beam-loading ("no-load" voltage),

$$V = \sum_{n=1}^{N} V_n \cos(\omega t + \varphi_n),$$

where each klystron is responsible for a voltage-phasor $\tilde{V}_n = V_n e^{j\varphi_n}$ established in the structure(s) it powers. This may be expressed more simply as $V = \Re \tilde{V} e^{j\omega t}$, where

$$\tilde{V} = \sum_{n=1}^{N} \tilde{V}_n.$$

Thus we may express $V = |\tilde{V}| \cos(\omega t + \varphi)$, with $\tilde{V} = |\tilde{V}| e^{j\varphi}$. Evidently, in the absence of beam-induced voltage, and considering only the longitudinal phase-space, the linac is simply a sinusoidal kick. It is quantified by a single-phasor, the no-load voltage of the linac. Considering then an electron bunch, one can see that if one tube is improperly phased, then the center energy

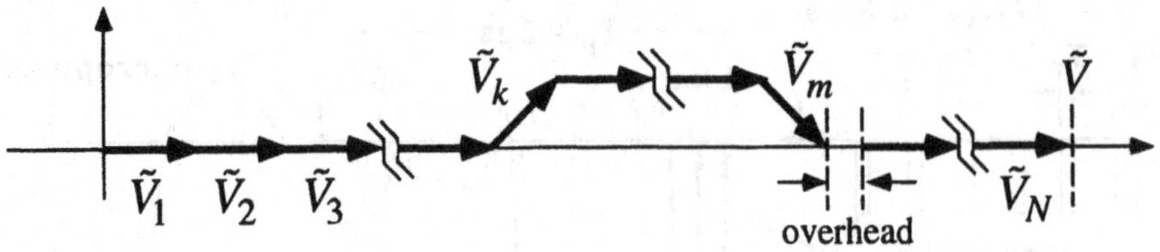

Figure 26. Net no-load voltage in a linac is a sum of phasors from each tube. Phase-closure need not be achieved locally. It does require "overhead" in achievable voltage.

and the energy spread deviate from the design after propagating through the structure(s) powered by this tube. However, since the result at the end of the linac is determined simply by a sum of phasors, a later tube may be re-phased to correct the error and achieve "phase-closure", as illustrated in Fig. 26.

Notice that the effect of the phase-error is to reduce the net length of the no-load voltage phasor. This implies that phase-closure requires "overhead" in maximum voltage available to the linac from the tube complement. This may be achieved for example with spare tubes on "stand-by", operating at the machine repetition rate, triggered after the beam has passed, but available to be put on the beam on short notice.

In the meantime, phase-closure is no guarantee that the beam hasn't noticed the mis-phased tube, for the transverse motion depends on energy, and the consequence of mis-phasing is a change to the energy profile along the linac. In fact, to understand the beam dynamics in detail, it is important to appreciate the *local* variation in beam energy as the simple picture of "linac as sine-wave" does not provide sufficient detail for tracking particles through the machine. An electron's local energy variation (absent wakefields) may be described by

$$\frac{d}{ds}mc^2\gamma = eG\cos(\psi + \phi), \tag{70}$$

where the voltage produced by the local structure has been quantified in terms of local gradient G, and phase ϕ. The electron phase relative to a properly-phased ($\phi = 0$) accelerating wave is $\psi = \beta s - \omega t$, and varies according to

$$\frac{d\psi}{ds} = \beta - \frac{\omega}{V}. \tag{71}$$

With $V \approx c$ and phase velocity of the accelerating mode $V_\phi = \omega/\beta \approx c$, the phase ψ is constant. The electron is "frozen" longitudinally, unless it is bumped by a mis-phased tube.

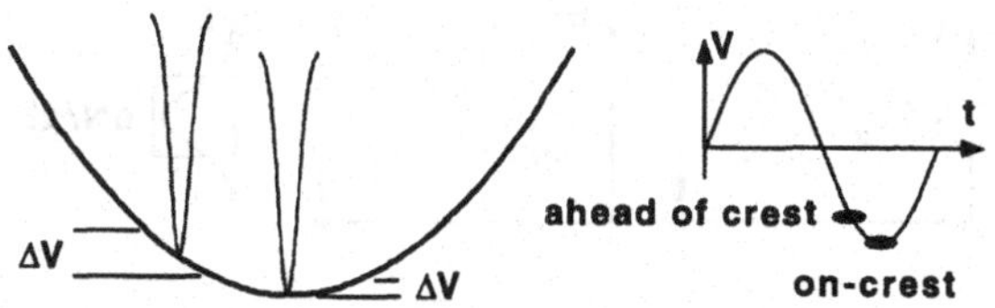

Figure 27. Energy spread of the beam is determined in the first approximation by bunch length and phasing.

This model is helpful in understanding the structure required of a micropulse, the collection of charge residing within one bucket, a single bunch. The longitudinal structure of a single bunch is constrained in the first approximation simply by the sinusoidal shape of the accelerating waveform. The problem is sketched in Fig. 27. Let us suppose for illustration that the initial distribution for the bunch takes the form of Eq. (69), and let us express this in terms of phase,

$$\frac{dQ_b}{d\psi} = \frac{Q_b}{\sqrt{2\pi}\sigma_\psi} \exp\left\{-\frac{(\psi - \bar{\psi})^2}{2\sigma_\psi^2}\right\}.$$

Using this distribution and Eq. (70), and performing integrals one can check the simple rule of thumb relating root-mean-square (rms) phase-width and asymptotic rms energy spread

$$\sigma_\psi\,(^\circ RF) \approx 6.8\sqrt{\sigma_\gamma\,(\%)}. \tag{72}$$

This is just the first approximation to the beam energy distribution, valid at low charge. The actual voltage witnessed by beam particles is a superposition of the applied (or "no-load") voltage, and an effective voltage due to collective radiation, the longitudinal wakefield. The longitudinal structure determines the frequency components present in the beam spectrum, and thus the character of coherent radiation in the linac. Where single-bunch charge is high and bunch length is short, the wakefield results in a significant loss in mean gradient, of order a few percent. At the same time, due to wakefields, phasing ahead of crest can actually *reduce* energy spread, below that of Eq. (72). Long and short-range wakefields are important in determining the energy distribution of the beam, be it a single bunch, or a train of bunches. Energy spread is not all bad, by the way, provided it is erased before the high-energy experiment can see it. Temporary use of energy-spread, along some region of the linac, may help to inhibit beam break-up ("BNS damping").

In connection with energy-spread, transverse oscillations and the like, it is important to distinguish between the distribution in 6-dimensional

Figure 28. Illustrating projection of the phase-space distribution

phase-space, to which Liouville's theorem applies, and *projected* distributions. This is illustrated in Fig. 28, showing a sampling of two distributions, in energy and time. Such a sketch invites one to distinguish between *projected* distributions and the full distribution. As one can see, a projection of the distribution onto the energy-axis (histogram by energy) shows merely energy-spread. In the full distribution, energy-spread may in fact be correlated or uncorrelated with another phase-space dimension. The two samples seen in Fig. 28 have the same "microscopic" density in 2-dimensional phase-space, but this fact is lost in projection. We could add that a distribution in E, t is itself a projection, so the actual phase-space density is not made clear in this figure, either.

This matter of microscopic density in the full-phase space is sometimes referred to in terms of "correlated" and "uncorrelated" beam attributes, energy-spread being just one example. Liouville's theorem tells us that we cannot increase the density in 6-dimensional phase-space. Uncorrelated energy-spread cannot be removed then, in the absence of fluctuation-dissipation effects. However, where a correlation exists, it may be exploited. The phase-closure problem illustrates this practical difference. If uncorrelated energy spread were present on injection, it could not be reduced by re-phasing a tube.

In this section we have glimpsed the requirements on the time-structure of the beam. Let us consider next requirements on the distribution projected in the transverse variables.

3.2.2. *Transverse Phase-Space*

High rate for an experiment imposes requirements on the transverse phase-space permitted to the beam. However, before considering such things, there is a logical prerequisite: What is required to get the beam through the linac? Let us illustrate the first and most basic ingredient, *emittance*. We pose the problem illustrated in Fig. 29 of transmitting a beam through a pipe.

We expect at first to pull out the Lorentz force law. However, let us suppose there are no externally applied forces. Let us neglect also forces generated by the beam itself. For a relativistic beam, electrostatic and magnetostatic forces cancel to within a factor of $1/\gamma^2$. Radiative forces, wakefields, are present but we may neglect them if charge is sufficiently

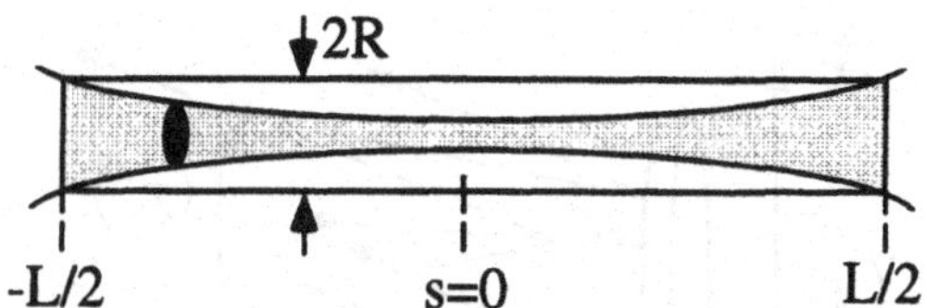

Figure 29. We consider the problem of propagating a beam through a pipe.

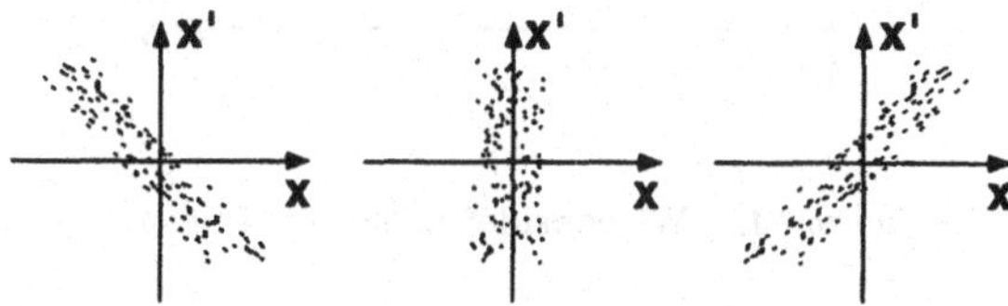

Figure 30. Beam motion in a drift corresponds to shear in trace-space.

low. We consider then simply a relativistic, low-intensity, ballistic beam. What determines the behavior of such a beam in a drift?

Since motion is ballistic, particle trajectories are straight lines determined by initial position x_0 and initial angle x_0',

$$x = x_0 + x_0's, \quad x' = x_0',$$

where s is the displacement down the pipe, and the angle of the trajectory is $x' = dx/ds = p_x/p_s$. One can predict the outcome of the experiment pictured in Fig. 29 with some simple sketches as seen in Fig. 30. There we see a sampling of the beam distribution, projected onto the $x - x'$ plane. This is sometimes referred to as a "trace-space" depiction of the beam, since with it, one may trace out the evolution of the beam. A particle's angle is constant, and the transverse position coordinate varies linearly with displacement s. Asymptotically one can see that the size of the beam will diverge with s, and the beam will scrape the pipe if the drift length L is large enough. Analyzing the geometry of this bundle of rays, we may quantify this divergence.

Let us define the *beam position* in the horizontal plane as $x_b = \langle x \rangle$. The brackets $\langle ... \rangle$ indicate an average over some subset of the beam distribution. Oftentimes this subset will refer to a single bunch. However, what subset we choose depends on the application, and given the variety in the time structure of the beam, we have many choices. In practice one may consult what is measurable, but for now we will refer to this subset simply as a "beam-slice" or just "the beam" for short, with a picture as in Fig. 31. The

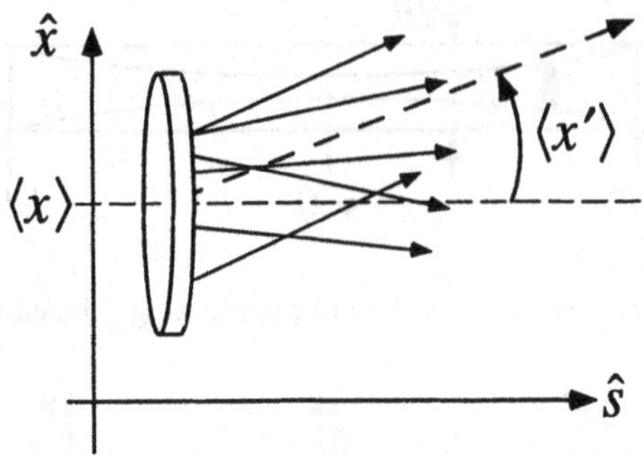

Figure 31. We consider a "slice" of beam.

beam position behaves as a single-particle travelling at angle $x'_b = \langle x' \rangle$. Steered properly one could imagine a single particle making it through a great length of pipe. However, unlike a single particle, the beam has a finite size, one that eventually grows with s. Let's see this. Second moments of the beam are

$$\sigma_x^2 = \left\langle x^2 \right\rangle - \langle x \rangle^2, \quad \sigma_{x'}^2 = \left\langle x'^2 \right\rangle - \langle x' \rangle^2, \quad \sigma_{xx'} = \langle xx' \rangle - \langle x' \rangle \langle x \rangle .$$

In our drift, the second moments vary according to

$$\left\langle x^2 \right\rangle = \left\langle x_0^2 \right\rangle + 2 \langle x_0 x_0' \rangle s + \left\langle x_0'^2 \right\rangle s^2, \quad \langle x\, x' \rangle = \langle x_0\, x_0' \rangle + \left\langle x_0'^2 \right\rangle s$$

Completing the square in the expression for σ_x^2 one can show that

$$\sigma_x^2 (s) = \frac{\varepsilon^2}{\sigma_{x'}^2} + \sigma_{x'}^2 (s - s_*)^2$$

where

$$\varepsilon^2 = \sigma_x^2 \sigma_{x'}^2 - \sigma_{xx'}^2. \tag{73}$$

The quantity ε is referred to as the root-mean-square *emittance*. One can show that it is a constant in such a drift. The beam forms a waist at $s_* = -\sigma_{xx'0}/\sigma_{x'}^2$, and the spot-size there is $\sigma_{x*} = \varepsilon/\sigma_{x'}$. This expression shows that to make a small spot, one must arrange a high angle of convergence $\sigma_{x'}$. Equivalently one may write

$$\sigma_x^2 (s) = \sigma_{x*}^2 + \frac{\varepsilon^2}{\sigma_{x*}^2} (s - s_*)^2 = \sigma_{x*}^2 \left\{ 1 + \frac{(s - s_*)^2}{\beta_*^2} \right\}, \tag{74}$$

where $\beta_* = \sigma_{x*}^2/\varepsilon$. This form emphasizes that a small-spot implies rapid divergence, on a length-scale varying quadratically with the minimum spot-size.

This divergent behavior ("hourglass") in a drift is similar to the diffraction of light, also a bundle of rays. Diffraction of coherent light causes the intensity-weighted beam-size to double after a waist in a Rayleigh length $L_R = 4\pi\sigma_x^2/\lambda$, where the rms width may be expressed in terms of the laser waist w as $\sigma_x = w/2$, and w determines the field amplitude variation $\propto exp\,(-r^2/w^2)$. Roughly speaking then, a beam of electrons behaves like a beam of photons, with wavelength $\lambda \approx 4\pi\,\varepsilon$. This might seem odd given that diffraction in the case of photons is sometimes described as a characteristic feature of their wave-nature. In fact, we have seen, one doesn't need waves to see diffraction. The wave nature of photons lies instead in the statement that the photon beam has emittance no less than $\lambda/4\pi$. Similarly, one may infer that the emittance of an electron beam can be no less than $\lambda_D/4\pi$, where λ_D is the de Broglie wavelength. Beams today are orders of magnitude away from this quantum-limited emittance, leaving us plenty of room for advances in beam-physics.

We may translate this analysis of the beam drifting through the pipe, into a statement about the beamline consisting of the pipe and the drift, and what emittance it can "accept". Given that a beam with too large an emittance will be scraped by the pipe, let us determine how to maximize transmission. We suppose that by adjustment of the incoming beam moments, we may place the waist at the pipe center. We adjust the waist σ_{x*} so as to minimize the beam size at the pipe exit, and thereby minimize scraping. One can show then that maximum beam-size is $\sigma^2 = \varepsilon L$, with L the pipe length. So, for example, if our beam is round, of uniform density with radius $r_b(s)$, then $\sigma = r_b/2$, and scraping occurs if $r_b > R$. Stated differently, scraping occurs if $\varepsilon > A$, where $A = R^2/4L$, with R the pipe radius and L the pipe length L. This is the *acceptance* of a circular pipe for a uniform beam, with waist centered in the pipe.

This notion of acceptance is a useful figure of merit, and may be generalized to more elaborate beamlines. Evidently specification of the beamline for this purpose means a description of the physical aperture as a function of s, and the beam "optics". While in principle one may configure an aperture and optics to "select" an emittance, this is not common on microwave linacs. Intentional scraping appears only in special apertures, "collimators". These may be employed, for example, to define energy or to remove beam "halo". Halo refers to a small population of electrons separated by several σ from the beam "core", and may be produced by scattering upstream. Other than the aperture of the collimator jaws, the smallest aperture on a microwave linac is typically the cell-to-cell coupling iris of the accelerating structure, about $1''$ in diameter at S-Band. Good transmission through this narrow constriction requires magnetic focusing.

Appreciating the significance of emittance, one might like to generalize the notion, to devise a quantity that is invariant and descriptive of limitations on focusing the beam. However, while our beam may be described by some Hamiltonian, there are quite a few we could cook up, and they may generate different invariants. Nevertheless, in the vicinity of the target, or the beam-collision point, the beam still passes through a drift, and its optical behavior still refers to the emittance already defined. The *normalized emittance*, ε_n,

$$\varepsilon_n^2 = (\sigma_x^2 \sigma_{p_x}^2 - \sigma_{x-p_x}^2)/m^2c^2. \tag{75}$$

is a useful figure of merit, as one can show that this quantity is a constant, in the presence of acceleration that is uniform throughout the beam-slice. This one can understand since x and p_x are unchanged by a Lorentz boost. Normalized emittance is also constant for a linear restoring force in x. For a high-energy beam then, and given ε_n, one has $\varepsilon \approx \varepsilon_n/\gamma$ and this is useful for simple calculations. This is to say that if one knows the normalized emittance at the 1 GeV point in the linac, one may hope that the same figure applies at the 45 GeV point, so that the emittance is smaller by ×45. In practice, one will find that normalized emittance grows through the linac. To appreciate how this occurs, one must contemplate a number of features of the beam dynamics: the choice of beam-slice, and the non-linear and energy-dependent ("chromatic") character of the forces at work. While Liouville may assure us that microscopic density in phase-space is constant, high-energy experiments represent projections of phase-space. As we saw in Fig. 28, Liouville's theorem does not prevent a projected distribution from becoming diluted.

Before we get too far out in phase-space describing beams, let's take a look at what is observable in practice.

3.3. INSTRUMENTATION

As we have seen a beam may be described in theory by a distribution in six-variables, 3 position variables, and 3 momentum variables — a distribution in phase-space. The task of the machine physicist is to produce beams with the phase-space desired by the high-energy experimentalist. This implies a requirement to monitor the relevant attributes of the phase-space distribution.

In fact, many attributes are not monitored, due to technical difficulties, and the problem of beamline instrumentation is a very active area of research. It is quite feasible to monitor intensity and first moments on a pulse-to-pulse basis, transverse position, phase, and energy. Monitoring of second-moments from shot to shot is more of a challenge. Quantities typically available over several machine pulses include transverse size (rms), transverse asymmetry (skew), and energy spread.

To diagnose attributes of a beam-distribution, we must cause the beam to couple to an external circuit. This coupling may be electrostatic, inductive or electromagnetic, and we may classify instruments accordingly. Inductive pickups may be coreless (fast) or employ a high-μ core. The workhorses of beamline instrumentation are typically low-frequency electrostatic or inductive pickups. "Low-frequency" refers to the frequencies actually employed in the output circuit; all pickups exhibit parasitic high-frequency coupling impedance (wakefields) to some degree. Within the category of radiative diagnostics there are coherent and incoherent radiators, including microwave cavities, gaps, and other structures, as well as sources of synchrotron, Cherenkov, transition or other radiation. At frequencies much lower than $1/\sigma_t$ one might employ a microwave detection circuit, and at much higher frequencies a camera, or streak-camera. This classification is not all-inclusive, as there are diverse ways to employ foils, secondary emission, kickers and screens.

Let us illustrate some common instruments, and for definiteness we may think of the beam in terms of the picture of Fig. 25. At 120 Hz, one beam pulse arrives every 8.3 ms. A beam pulse may consist simply of one bunch of charge with a length of 1 mm, and transverse size of 100 μm - 1 mm, or a pulse may consist of a train of such bunches, a train lasting from 100 ns to 3 μs.

3.3.1. *Ion-Chamber*

The most basic instrument for an accelerator is one that registers *beam loss*. Without such a monitor, symptoms of a mis-steered beam may still be visible, appearing on other instrumentation, that we will discuss, toroids and beam-position monitors. Without instrumentation of some kind, deposition of MW/cm^2 in a beamline component might appear first on a temperature monitor, and perhaps not long after that, a vacuum pressure readout (e.g., an ion pump current). Protection of the machine favors a dedicated diagnostic of beam-loss, and an ion-chamber [27] is the most common instrument for this purpose. Slightly more elaborate is a length of gas-filled coaxial line, as proposed by Panofsky, with a high-voltage suitable for operation as a proportional counter. Stretched the length of the linac such a "Panofsky long ion-chamber" (PLIC), can serve as a monitor of *localized* loss, with resolution on the order of a meter. In addition to machine protection (shut-off or rate-limit), a PLIC can assist in beam-steering. At the SLC there are eight PLIC systems, implemented with 250 V applied to 1/2" Heliax^{TM} (Andrews Corp.), loaded with 95% Ar gas, with Freon and CO_2 [28]. The primary concern in the design of such a system is to avoid shadowing of the cable by beamline components, and this may dictate pairs of cables, or threading of cable through magnet bores where feasible.

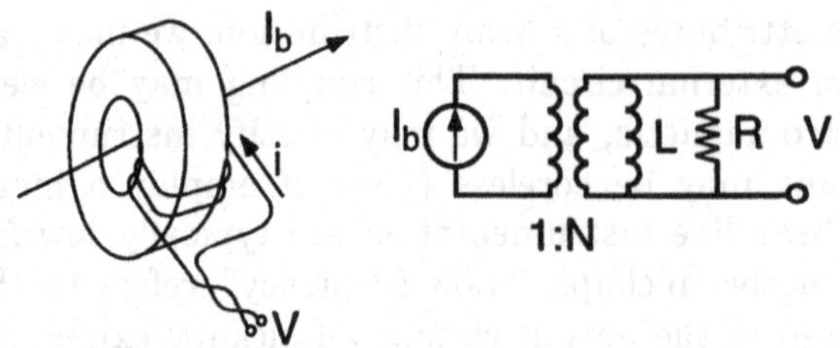

Figure 32. A toroid current monitor and equivalent circuit.

3.3.2. *Toroid*

One step up from a beam-loss monitor is a current monitor, most commonly for microwave linacs, a toroid. As seen in Fig. 32, a simple current monitor can be fashioned from a toroidal ring wrapped with a coil of N turns, coupled out through cable of characteristic impedance R. The azimuthal magnetic field produced by the beam current I_b and the toroid circuit current i is, from Ampere's Law,

$$H = \frac{(I_b - iN)}{2\pi r}$$

with r the mean radius of the toroid. The magnetic flux through the toroid circuit is then

$$\phi = \mu AN \frac{(I_b - iN)}{2\pi r},$$

with A the cross-section of the toroid. Faraday's law then determines the voltage developed across the terminals $V = d\phi/dt$. With output terminated in impedance R, the circuit current satisfies

$$\frac{di}{dt} + \frac{R}{L}i = \frac{1}{N}\frac{dI_b}{dt},$$

where the inductance is $L = \mu AN^2/2\pi r$. If L/R is large compared to the beam-pulse length then $V \approx I_b R/N$. In this limit the circuit is equivalent to a current-source providing $i \approx I/N$ to the cable. Example implementations may be found in the literature [29, 30].

3.3.3. *Screens and Radiative Diagnostics*

Screens or "paddles" are among the simplest and most common instruments employed on a linac, usually in combination with a camera. A screen is generically a piece of metal or ceramic with a phosphorescent coating, and a graticule to register scale in the tranverse dimensions x, y. The light from the screen is imaged and employed to infer the transverse beam distribution.

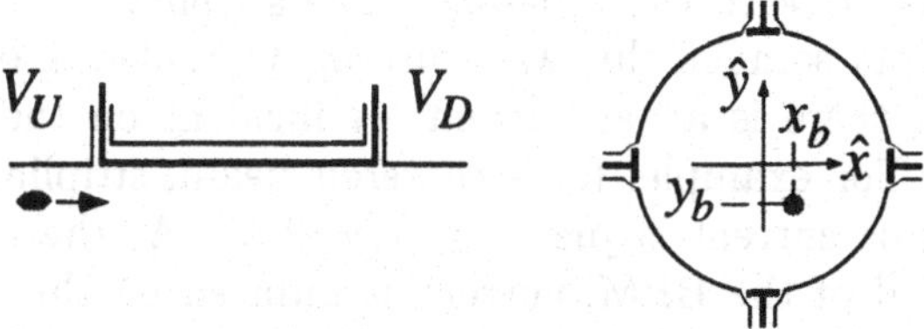

Figure 33. A stripline BPM, side-view of one stripline, and assembly end-view.

As a monitor of position the screen aids in steering, and as a check of beam energy, when employed after a bend magnet. Placing a screen at an image point for a target permits one to tune up the beam on the screen first, before delivery to the target. The drawbacks to screens are that they are destructive to the beam and are eventually destroyed themselves by an intense beam. Operational issues include burn-spots on the screen and saturation of the camera. In addition, beam-jitter on the screen affects the inference of beam-moments. Examples of different screen, graticule and camera configurations can be seen in [31, 32, 33]. Analysis of beam profile data from a screen may be found in [34].

Other radiative diagnostics include transition radiation (e.g., from the front or rear of a screen), synchrotron radiation from a bending magnet or wiggler magnet, and Cherenkov radiation. Depending on the application, diagnosed spectra may extend from the infrared through the ultra-violet and X-ray. Time-resolution, if needed, can be aided by filtering the light. Detection may be through a scintillator and photo-multiplier tube in the case of ionizing radation, or a photo-diode for wavelengths up to 20 μm. A more elaborate system may employ a gated CCD [35] or streak-camera [36, 37]. The latter permits resolution of features on the scale of a single 3 ps bunch. Time resolution is not always important though. An early application of synchrotron and Cherenkov radiation for monitoring of a beam spot in the mm-cm range can be seen in [38]. At the other extreme one finds modern work employing Compton-scattering of laser-light to measure a 100-nm beam spot [39].

3.3.4. *Beam-Position Monitor*

There are several different kinds of BPM in common use, button electrode, stripline, and resonant cavity. Let us consider the stripline as it is fairly common; the geometry is seen in Fig. 33.

A transient current (a beam) propagating through a conducting pipe is accompanied by equal and opposite current flowing through the pipe-wall. The distribution of current around the pipe depends however on the beam position relative to the pipe-axis. The stripline BPM cuts the pipe wall and intercepts some of this wall current. Considering one stripline, as seen in Fig. 33, there is a net current $-i$ incident on the upstream end of the stripline. For example, for a centered beam, stripline width w and pipe radius r, this current is just $i = I_b(w/2\pi r)$. As the beam arrives at the upstream end of the BPM, current is induced on the coaxial output, and the interior stripline. If the characteristic impedance of these lines is Z, then the voltage induced on each is $V = Zi/2$, since they appear in parallel. In this way two waves each with current $i/2$ are launched, one up the coax, the other down the interior stripline. As the beam reaches the downstream end of the BPM it induces an opposite polarity pulse on the output coaxial line corresponding to current $-i/2$. If the phase-velocity of the wave on the stripline matches the beam velocity, this signal cancels the forward TEM wave on the downstream coaxial output. A current $-i/2$ is then launched backward on the interior stripline. Thus the voltage appearing on the upstream coaxial output is bipolar in character,

$$V_U(t) = \frac{1}{2} Z \left[i(t) - i\left(t - \frac{2l}{V}\right) \right],$$

with l the stripline length. A matched stripline of this type is sometimes referred to as a directional coupler, since $V_D \approx 0$. This implies that such a device may be employed to monitor two counter-propagating beams in a common pipe.

To determine the position dependence of the wall current $-i$, we may employ the method of images [22], and expand the result as an infinite series

$$i = \frac{w}{2\pi r} I_b \left[1 + 2 \sum_{n=1}^{\infty} \left(\frac{\rho}{r}\right)^n \cos n(\phi - \theta) \right].$$

where ϕ specifies the position of the stripline, r is the pipe-radius, and the beam-position is given by $x = \rho\cos\theta$, $y = \rho\sin\theta$. The BPM illustrated in Fig. 33 shows four striplines at angles $\phi = 0$, $\pi/2$, π, and $3\pi/2$. We can illustrate the problem of position monitoring more simply however with two striplines at angles $\phi = 0$ and $\phi = \pi$. In this case we have wall currents

$$i_{\phi=0} \approx \frac{w}{2\pi r} I_b \left[1 + 2\frac{x}{r} + 2\frac{x^2 - y^2}{r^2} + 2\frac{x^3 - 3xy^2}{r^3} + \ldots \right],$$

$$i_{\phi=\pi} \approx \frac{w}{2\pi r} I_b \left[1 - 2\frac{x}{r} + 2\frac{x^2 - y^2}{r^2} - 2\frac{x^3 - 3xy^2}{r^3} + \ldots \right].$$

In processing the signals from the coaxial outputs, one may employ tees to form sum and difference signals

$$\Sigma i = i_{\phi=0} + i_{\phi=\pi}, \Delta i = i_{\phi=0} - i_{\phi=\pi} \tag{76}$$

The sum provides a normalization signal with which to extract position x from the difference. Evidently BPM's exhibit intrinsic non-linearity.

The electronics processing the stripline readouts will perform a low-pass ($\approx$50 MHz) filter on each channel, and add and subtract by means of tees. Following this will be a combination of limiters, amplifiers and digitization. The processor circuit will also exhibit non-linearity, as well as scale and offset errors. One may find too that BPM processor outputs are multiplexed ("MUX'd") for readout from the control system, and this can on occasion cause confusion, since orbits inferred from MUX'd BPM readouts may correspond to different machine pulses, and not a single physical orbit. Further discussion of BPM electronics may be found in [40, 41].

3.3.5. *Beam-Phase Monitor*

In a microwave linac one typically has a number of phases to control and this implies, to monitor. The injector system defines a reference phase and one wishes to control all tube phases relative to the injector, in principle. Maintenance of accurate phase-relationships in a multi-tube linac requires a single clock, a "master oscillator" from which all rf signals are derived. In addition, one would like this drive signal to be distributed to all tubes in a phase-stable fashion. Meanwhile, to set the phase of a tube, one requires a diagnostic of the tube phase relative to the beam. So for example, a beam traversing an unpowered structure will induce a microwave signal on a forward coupler placed on the structure output to the load. If the tube is phased to put the beam on the accelerating crest, then the beam-induced voltage as seen on the output will be π out of phase with the component due to the tube. This is the basis for the "induction" phasing technique employed for the original Two-Mile linac rf system [2]. Beam-phase relative to a tube can also be assessed with the help of a BPM placed downstream, after a bend magnet. In this case adjustment of an in-line phase-shifter at the tube input will produce a sinusoidal variation of beam-position. On-crest acceleration corresponds to the phase setting where the BPM reading is at an extremum. Thus there are a number of ways to get at the phase-information one requires.

To appreciate the systematic issues associated with a phase-monitor, consider the system seen in Fig. 34, comparing local beam-phase to klystron output phase. This could provide a pulse-to-pulse beam-phase reference, not an absolute figure, but one whose phase offset could be calibrated out by comparison with one of the other techniques. Cable lengths may vary due to temperature drifts associated with the diurnal cycle, or thermal equilibration of the accelerator vault. In this case the phase-monitor output will drift. In the SLC, the full range of daily temperature variation is 20°C upstairs in the klystron gallery and 1°C downstairs in the tunnel. To monitor

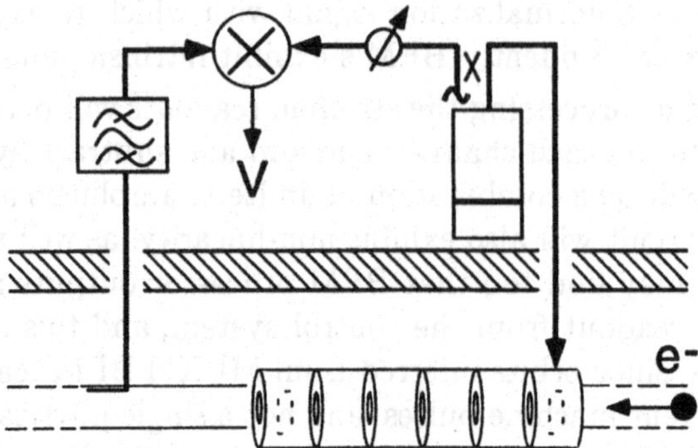

Figure 34. **A simple phase-monitor comparing the filtered signal taken from a stripline within the accelerator vault to a signal sampled from the klystron output waveform.**

for phase-drift one may employ resistive temperature detectors, and to control it one may employ phase-stabilized cable, self-regulating heater tape, and still more elaborate measures [42]. A more conventional solution is to provide for temperature control of electronics buildings. However, even in this case it is possible to "detect" the air-conditioner cycle with the help of the electronics in one's processor. If one actually employs a single stripline signal then the output is weakly dependent on position and this will appear in the phase signal. This can be corrected by combining the stripline outputs. The same principles can be employed to monitor the pulse-to-pulse timing jitter of two bunches [43].

3.3.6. *Bunch-Length Monitor*

The simplest monitor of bunch length employs a gap in the beam-tube to permit the beam to radiate coherently, as seen in Fig. 35, with gap length L and pipe radius R. The power spectrum radiated will depend on the geometry, and the beam spectrum. For example, for a Gaussian beam, the Fourier transform is

$$\tilde{I}_b(\omega) = \int_{-\infty}^{\infty} \frac{d\omega}{\sqrt{2\pi}} e^{-j\omega t} I_b(t) = \frac{Q_b}{\sqrt{2\pi}} \exp\left(-\tfrac{1}{2}\omega^2\sigma_t^2\right), \qquad (77)$$

Roughly half of the incident spectrum lies at wavelengths below $\lambda \approx 4\pi\sigma_z$, and this determines the frequency range of interest for the detection circuit. For a 1 mm bunch length, one is interested in frequencies in K-band. A crystal detector, looking into a filter blocking out low-frequencies, provides a monitor of power radiated. The crystal should be positioned outside the vault, well away from any pulsed-noise or ionizing radiation sources. One can then extract a signal to monitor bunch length, after normalization by a

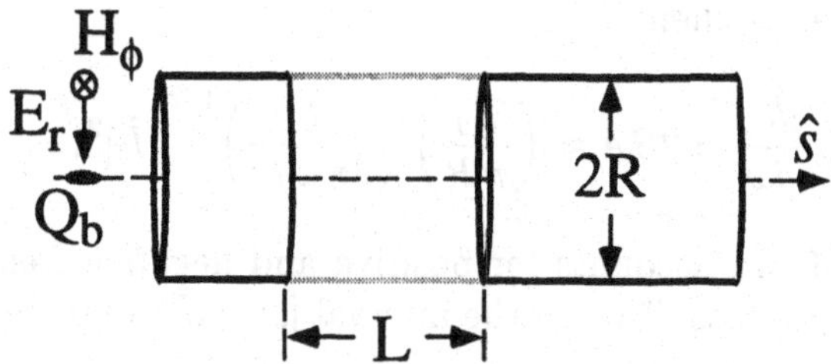

Figure 35. A gap-based monitor makes use of the radiation of a beam through an interruption in the return current-path in the beam-pipe, bridged by a dielectric gap.

beam-intensity signal derived from the low-frequency part of the spectrum or a toroid.

In addition to frequency, we require some appreciation of the signal level. In the first approximation, this system is just a gap with capacitance of order $C \approx \varepsilon_0 \pi R^2/L$ and loss-factor $k_l \approx 1/C$. However, for this application one is interested to grasp the frequency dependence of the coupling. In practice one will filter out the low frequency signal to permit clean detection of the high-frequency components. For typical parameters $R, L >> \sigma_z$, and we are interested in frequencies for which the geometry is no longer a lumped circuit, but an obstacle that serves to diffract the beam self-fields out of the pipe. We may estimate the power radiated by considering the beam self-fields in the frequency domain $E_r = Z_0 I/2\pi r$, $H_\phi = I/2\pi r$, and the Poynting flux near the pipe wall, $S = E_r H_\phi = Z_0(I/2\pi R)^2$. We need only compute the fraction of this flux that leaves the pipe, to gauge the signal level to be expected. We treat the dielectric liner as transparent, for simplicity. If we were to compute in detail the fields at the far side of the gap, we would find a superposition of waves with transverse wavenumber $k_\perp$ and longitudinal wavenumber $k_{||}$, satisfying $k_{||}^2 + k_\perp^2 = (\omega/c)^2$. To represent fields diffracting from a radius $r = R - \delta r$, we expect to require transverse wavenumbers $k_\perp \approx \pi/\delta r$. In the high-frequency limit ($\omega R/c >> 1$), all component plane-waves propagate at small angles and the deviation in $k_{||}$ from the unscattered value ω/c is small $\delta k_{||} << k_{||}$. Meanwhile, waves add constructively at the far side of the gap, when they are in phase $\delta\varphi \approx k_\perp \delta r + \delta k_{||} L \approx 0$. Combining these estimates one arrives at a figure for the annulus that leaves the pipe,

$$\delta r \leq \left(\frac{L}{4\pi\, k_{||}} \right)^{1/2}$$

This determines the area intercepted due to diffraction, $A \approx 2\pi R \delta r$. The radiated spectrum is then

$$\frac{dU}{d\omega} \approx 2SA \approx \left(\frac{Z_0}{\pi R}\right)\left(\frac{L}{4\pi\,\omega/c}\right)^{1/2} |\tilde{I}_b|^2,$$

and the factor of two accounts for positive and negative frequencies. Considering a flat high-pass filter on the interval $[\omega_c, \infty]$, we integrate to obtain

$$k_{eff}(\omega_c) = \frac{1}{Q_b^2}\int_{\omega_c}^{\omega} d\omega \frac{dU}{d\omega} = \frac{\Gamma(1/4)}{8\,\pi^{5/2}}\frac{Z_0 c}{R}\left(\frac{L}{c\sigma_t}\right)^{1/2} f(\omega_c\sigma_t). \tag{78}$$

The function

$$f(u) = \frac{1}{\Gamma(1/4)}\int_{u^2}^{\infty} dx\, x^{-3/4}e^{-x},$$

and $f(0) = 1$, with $\Gamma(1/4) \approx 3.6$. For results more precise than this one may employ a numerical field-solver. As an example, for $L \approx 1$ cm, $R \approx 1.27$ cm, and $c\sigma_t \approx 1$ mm, we have $k_{eff}(0) \approx 0.7$ V/pC. Thus a 1 nC bunch deposits about 0.7 μJ, in about 33 ps, or 20 kW peak. Even after dispersion in the connecting guide the peak power will be quite high, and this must be accounted for in the filtering. One may guess that from the point of view of nearby beamline instrumentation in the vault, a gap is also an excellent "noise" generator, unless it is enclosed in a conducting box. Further adventures with bunch length monitoring, and gap-pickups can be found in [43].

While we have computed the enengy radiated out of the pipe, in fact, half of the energy is radiated *into* the pipe. Associated with the gap there is a high-frequency wakefield that may interact with particles that follow [7]. This confirms that *any* discontinuity in the beam-pipe will interact with the beam, as claimed in Sec. 2.9. In fact, this *diffraction-model* impedance, evaluated for the accelerating structure iris, is the dominant contributor to the broadband impedance of a linac.

3.3.7. *Wire-Scanner*

A direct approach to extracting beam-profile information is to move a wire target through the beam path, over the course of several beam pulses. A proportional signal can be derived either from secondary electron emission or bremsstrahlung [44]. In the latter case, at each wire position, the beam produces bremsstrahlung radiation that may be detected with the help of a scintillator and photo-multiplier tube (PMT), and acquired with a gated analog to digital converter (GADC). The scheme is depicted in Fig. 36. The principle of operation is that a histogram of counts versus wire posi-

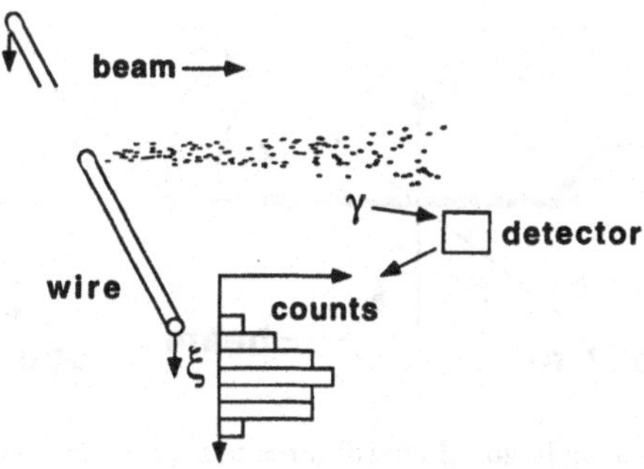

Figure 36. Use of a wire target to infer beam-size, for example, by detection of bremsstrahlung photons with a scintillator, PMT and GADC.

tion provides a root-mean-square (rms) width dependent on the beam size. Adopting coordinates u, v in the transverse plane, as seen in Fig. 37, let us denote by $N_e(v)$ the beam distribution in the transverse plane, integrated over a time-scale corresponding to the gate-width for the GADC, and projected onto the v-axis. The number of γ's generated by an infinitely thin wire is proportional to $N_e(v)$. Thus by moving such a wire in ξ, with small steps $\delta\xi$, inferring the step in v, $\delta v = \sin(\alpha)\delta\xi$, and determining $N_\gamma(v)$, one could in principle compute $N_e(v)$ up to an overall scale factor depending on the geometry traversed by the γ's and the detection system. This distribution $N_\gamma(v)$ will have the rms width σ_v of the actual beam distribution in v. The second moment of the inferred distribution referred to a coordinate origin v_w is then $\langle v^2\rangle = \langle (v_b - v_w)^2\rangle + \sigma_v^2$, where the brackets indicate an average over all wire positions weighted by N_γ. A wire of finite dimension may be thought of as a collection of infinitesimal wires each with a different position v_w with respect to the wire-centroid. The inferred distribution $N_\gamma(v)$ in this case is just the sum of the individual distributions from each wire and the rms width Σ_v may be determined from

$$\Sigma_v^2 = \left\langle v^2 \right\rangle - \langle v\rangle^2 = \left\langle (v_b - v_w)^2 \right\rangle + \sigma_v^2 = \sigma_v^2 + \sigma_w^2,$$

where the brackets again refer to an average over all (infinitesimal) wire positions weighted by N_γ. For example, $\sigma_w^2 = D^2/16$ for a wire with circular cross-section of diameter D. Thus a 50-μm diameter wire contributes 12.5-μm in quadrature.

These figures refer to the coordinate v normal to the wire-orientation. Typically several wires are employed and it is convenient to refer moments to a common coordinate system x, y. If the u, v axes are obtained from the x, y axes with a rotation by θ_w, so that $v = -x\sin(\theta_w) + y\cos(\theta_w)$ then

$$\sigma_v^2 = \sigma_x^2 \sin^2\theta_w - \sigma_{xy}\sin 2\theta_w + \sigma_y^2\cos^2\theta_w.$$

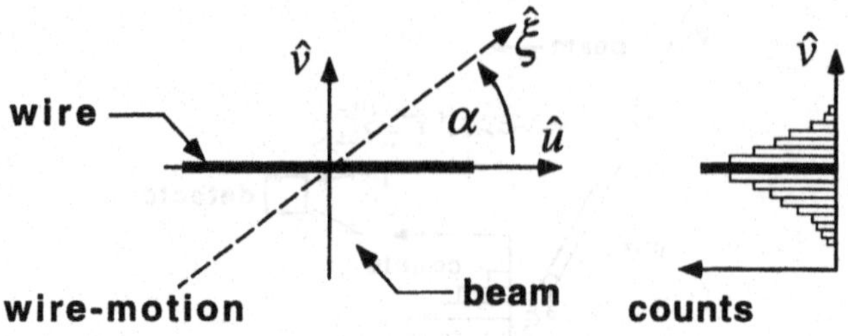

Figure 37. We consider orthogonal coordinates u, v in the transverse plane, with a wire oriented in the u-direction, and travelling along an axis ξ oriented at an angle α to the wire. The beam travels into the page.

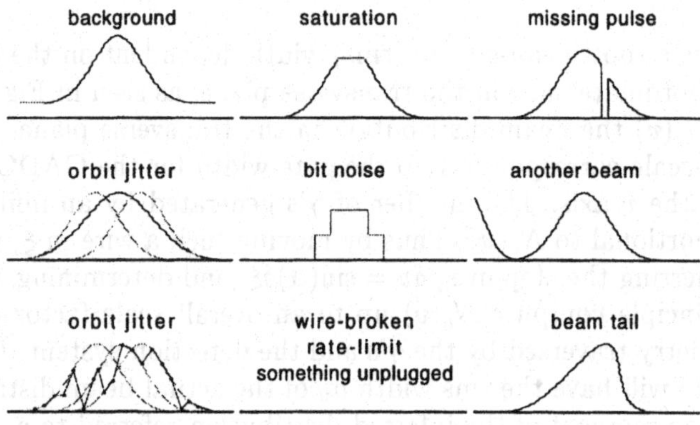

Figure 38. Some of the features one may see with a wire-scan.

Evidently one may infer three second-moments with three wires, provided the wire orientations θ_w are not degenerate, for example, $0°, 45°, 90°$. Fitting of inferred distributions can be quite elaborate in that asymmetric profiles reveal beam dynamics significant to the process of machine tune-up.

Systematic issues are illustrated in Fig. 38 and include background from upstream sources of ionizing radiation flux (e.g. beam-scraping and linac "dark current"), wire-breakage, wire-vibration, and beam orbit motion from pulse to pulse. The latter can be corrected with the help of the BPM system and the control software. The last sketch actually may be showing something related to beam dynamics, a beam profile asymmetry or "tail".

Due to problems with breakage, wire-scanners are an ongoing area for research. In the earliest wire-scanners, tungsten wire (gold-plated to permit soldering) was chosen for its high $Z = 74$. More recently, carbon wire has been employed, with detection of electrons at 90° to the beam-axis, rather

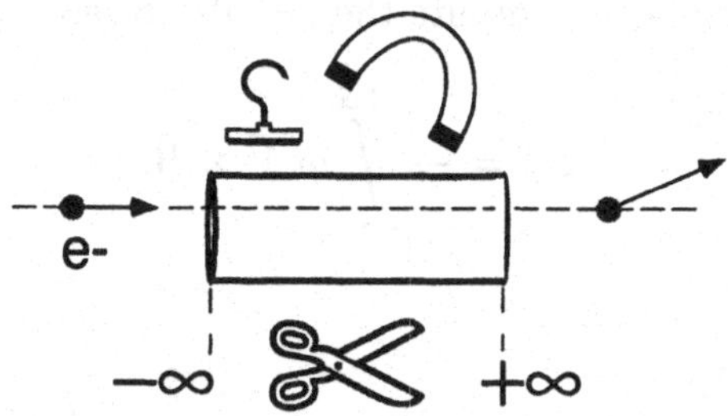

Figure 39. Deflection by *any* "thin magnet" may be decomposed into multipoles.

than small-angle bremsstrahlung photons. The difficulty of maintaining the small diameter wire needed to resolve micron-beam dimensions has motivated development of a "laser-wire" system based on Compton scattering rather than bremsstrahlung.

3.4. MAGNETIC MULTIPOLES

Appreciating what features of the beam are observable, let us consider next the matter of getting a relativistic beam through the main linac. We found in Sec. 3.2.2 that even a relativistic beam will diverge due to emittance and for this reason some beam-confining forces are necessary. We have already figured out that magnetic forces are preferable to the electric variety, in the case of a relativistic beam. What manner of magnetic forces may we apply?

Following the first rule of beamline design, we put the magnets on the beamline, not *in* the beam. In this case, the applied magnetic induction satisfies $\vec{\nabla} \times \vec{B} = 0$ and $\vec{\nabla} \bullet \vec{B} = 0$, within the beam-path. Thus we have a choice, we may express $\vec{B}$ as a curl or a gradient; we choose gradient,

$$\vec{B} = \vec{\nabla}\varphi. \tag{79}$$

The solenoidal condition then implies that the magnetic potential φ satisfies Laplace's equation, $\nabla^2\varphi = 0$. Boundary conditions then determine the character of the magnetic potential, and they are shaped by the magnetic materials and coil currents outside the beam-pipe, or at least outside the beam-path.

Laplace's equation is a strong constraint on the form the magnetic potential can take. To see this, consider a magnet of some kind as depicted in Fig. 39; let us compute the net kick to an incident electron, with the approximation that the electron orbit is ballistic through the magnet (a "thin" magnet). We integrate to compute the impulse received by the electron,

$$\Delta\vec{p} = -e\int_{-\infty}^{+\infty} dt\, \vec{V}\times\vec{B},$$

or

$$\Delta p_x = -e\int_{-\infty}^{+\infty} dt\, (V_y B_s - V_s B_y) = -e\int_{-\infty}^{+\infty} ds\, \left(\frac{V_y}{V_s}\frac{\partial\varphi}{\partial s} - \frac{\partial\varphi}{\partial y}\right).$$

Since we are calculating the impulse to first order in the fields, V_y may be taken to be constant and this term integrates to 0. Proceeding in a similar fashion with Δp_y, we find that

$$\Delta\vec{p} = -e\hat{s}\times\vec{\nabla}_\perp\psi, \tag{80}$$

where,

$$\vec{\nabla}_\perp = \hat{x}\frac{\partial}{\partial x} + \hat{y}\frac{\partial}{\partial y} = \hat{r}\frac{\partial}{\partial r} + \hat{\phi}\frac{1}{r}\frac{\partial}{\partial\phi},$$

$$\psi(\vec{r}_\perp) = \int_{-\infty}^{+\infty} ds\, \varphi(s, \vec{r}_\perp). \tag{81}$$

Thus the kinds of kicks we may impart to a paraxial beam at first order in the applied magnetic field are described by the function ψ. Notice that,

$$\nabla_\perp^2\psi(\vec{r}_\perp) = \int_{-\infty}^{+\infty} ds\, \nabla_\perp^2\varphi(s, \vec{r}_\perp) = \int_{-\infty}^{+\infty} ds\, \left(\nabla^2 - \frac{\partial^2}{\partial s^2}\right)\varphi(s, \vec{r}_\perp)$$

$$= \int_{-\infty}^{+\infty} ds\, \left(-\frac{\partial^2}{\partial s^2}\right)\varphi(s, \vec{r}_\perp) = -B_s(s, \vec{r}_\perp)\Big|_{-\infty}^{+\infty} = 0.$$

In the region of the beam-pipe then we may express ψ as

$$\psi(\vec{r}_\perp) = \sum_{m=0}^{\infty} r^m \{a_m\cos(m\phi) + b_m\sin(m\phi)\}, \tag{82}$$

or, in Cartesian coordinates,

$$\psi = a_0 + a_1 x + b_1 y + a_2(x^2 - y^2) + 2b_2 xy + a_3(x^3 - 3xy^2) + b_3(3x^2y - y^3) + \ldots.$$

These terms come with nomenclature as indicated in Table 4. Together

TABLE 4. Magnetic multipoles associated with a thin magnet, and the corresponding first-order kicks, per unit charge.

Term	Δp_x	Δp_y	Nomenclature
a_1	0	ea_1	vertical bend
b_1	$-b_1$	0	horizontal bend
a_2	$2a_2y$	$2a_2x$	skew quadrupole
b_2	$-2b_2x$	$2b_2y$	normal quadrupole
a_3	$6a_3xy$	$3a_3(x^2-y^2)$	skew sextupole
b_3	$-3b_3(x^2-y^2)$	$6b_3xy$	normal sextupole

with this mathematical description in terms of multipoles, one may construct a picture for each multipole, based on the corresponding magnetic equipotentials.

There is more variety to magnets than is illustrated by the simple kicks given here. A sufficiently thick or strong magnet can appreciably perturb the beam orbit within the magnet itself. In the case of linacs however one is interested simply to transport the beam from the injector to the application (target, collision point) controlling the beam-orbit and the beam-size. Typically one prefers that x and y be uncoupled. For simplicity one prefers that the orbit should lie in a plane. With these assumptions, one is interested primarily in the horizontal bend and the normal quadrupole as in Fig. 40. These two kinds of magnets are the starting point for the subject of beam-optics.

Applying Ampere's Law to the geometries shown one may relate the kick seen by the beam to the coil currents. For the horizontal bend magnet we have $\vec{B} = B_p\hat{y}$, and the pole-tip field $B_p = \mu_0(2NI/g)$, in terms of the gap g, and the product of coil current I, and the number of turns N.

A normal quadrupole field corresponds to $\vec{B} = \vec{\nabla}\{kxy\}$ where $k = \partial B_x/\partial y = \partial B_y/\partial x$ is the quadrupole gradient. This field is defined by an equipotential surface $xy = R^2/2$, a hyperbola, with R the radial distance from the center-axis to the pole-tip. The field at the pole-tip is then $B_p = kR$. We suppose that the magnetic flux is driven by coils carrying current NI placed with the symmetry indicated in Fig. 40. We may compute k in terms of the current and the geometry, using Ampere's Law,

$$-2NI = \oint \vec{H} \bullet d\vec{l} = \int_{pole} \frac{\vec{B}}{\mu} \bullet d\vec{l} + \int_{gap} \frac{\vec{B}}{\mu_0} \bullet d\vec{l}.$$

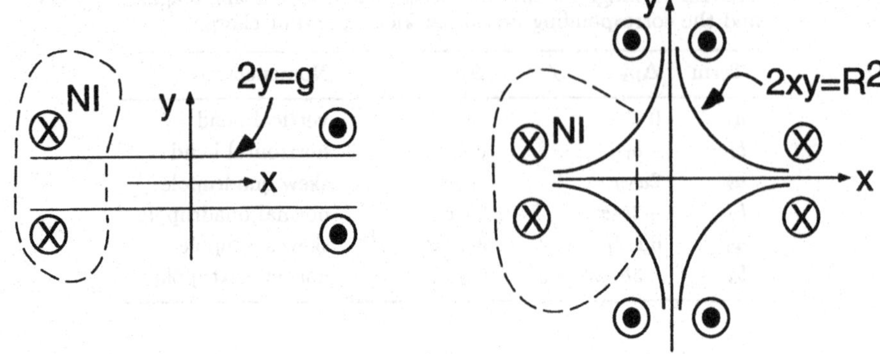

Figure 40. The horizontal bend, and the normal quad are the basic building blocks for optical systems.

Since $\mu >> \mu_0$, we need only the integral in the gap and this is

$$\int_{gap} \frac{\vec{B}}{\mu_0} \bullet d\vec{l} = \frac{1}{\mu_0} \int_{-R/\sqrt{2}}^{R/\sqrt{2}} B_y dy = \frac{kR^2}{\mu_0}.$$

Thus we find $k = \mu_0(2NI/R^2)$, expressing quadrupole gradient in terms of the coil current. The validity of this analysis requires that the pole tip field be small enough that the permeability is not driven to saturation. Beyond saturation, the differential response of the material to any further increase in coil current, is that of vacuum; additional magnetic flux ceases to be confined by the magnet material, and the on-axis gradient increases less rapidly with coil current. Moreover the $m = 2$ symmetry is no longer enforced by the pole-tip geometry and aberrations appear. For iron this implies a maximum field of 2 T, and for quadrupoles, a maximum pole-tip field of 1 T. These numbers are important constraints on our ability to confine beams. In practice the "strength" of the lattice is the ultimate limit on tolerable single-bunch charge, due to the effect of wakefields. Further reading on "iron-dominated" magnets may be found in [45].

3.5. MOTION IN A PLANE

To formulate linac optics it would be simplest, and nearly correct to consider only quadrupoles. However, an offset quad introduces a dipole component. In addition, dipoles are present in the form of *correctors*, used to compensate offset quads and to steer the beam. Dipoles are also useful in that they can be employed for energy analysis and longitudinal bunch compression. For these reasons it is useful to formulate a description of the motion incorporating dipoles, and not just quadrupoles. Higher multipoles, particularly sextupoles, are also handy; however, they may be treated as a higher-order correction to the motion governed by dipoles and quads. We assume bend magnets are separated from accelerating structures, so that we may take the electromagnetic fields to be zero in the following analysis.

We define a reference orbit in the machine corresponding to the motion of a particular particle, with position $\vec{r}_0$, velocity $\vec{v}_0$ and momentum $\vec{p}_0$, varying in time. We suppose this motion lies in a plane, and designate by $\hat{y}$ the unit vector normal to the plane. We denote the local tangent to the orbit $\hat{s} = \vec{v}_0/|\vec{v}_0|$, and we define $\hat{x} = \hat{y} \times \hat{s}$. These unit vectors form the basis for a local right-handed coordinate system. We may parameterize the orbit with variable s such that $ds/dt = |\vec{v}_0|$; s is arc-length traversed by the reference particle. Geometry dictates that there is a quantity κ, such that

$$\frac{d\hat{s}}{ds} = -\kappa\hat{x}, \quad \frac{d\hat{x}}{ds} = \kappa\hat{s}, \quad \frac{d\hat{y}}{ds} = 0.$$

κ is the local curvature of the orbit. We may view a length ds of the reference trajectory as a section of circular arc with radius $\rho = 1/\kappa$. Consulting the Lorentz force law,

$$\frac{d\vec{p}_0}{dt} = |\vec{v}_0| \frac{d\vec{p}_0}{ds} = |\vec{v}_0| \frac{d}{ds} |\vec{p}_0| \hat{v}_0 = |\vec{v}_0| |\vec{p}_0| (-\kappa\hat{x}) = -e\vec{v}_0 \times \vec{B}(\vec{r}_0),$$

we see that the assumption of motion in a plane requires $B_x = 0$ on the reference orbit. The curvature is determined by B_y,

$$\kappa = \frac{1}{\rho} = \frac{-eB_y(\vec{r}_0)}{|\vec{p}_0|}. \tag{83}$$

In practical units, $3.34 p_0(\mathrm{GeV/c}) = B(\mathrm{T})\rho(\mathrm{m})$.

Having determined the reference orbit, we may locate orbits of nearby particles relative to it, parameterizing their motion by s as well. We consider one such particle, finding it at local coordinates $\vec{r}_\perp = x\hat{x} + y\hat{y}$, as depicted in Fig. 41. We describe this particle's motion to linear order in deviations from the reference orbit. Taking into account the rotation of the horizontal axis due to curvature, we express the rate of change of position with reference arc-length s as

$$\frac{d\vec{r}}{ds} = x'\hat{x} + y'\hat{y} + (1 + \kappa x)\,\hat{s}.$$

Thus the rate at which length is traversed by the particle is

$$\frac{dS}{ds} = \sqrt{(1 + \kappa x)^2 + x'^2 + y'^2} \approx (1 + \kappa x).$$

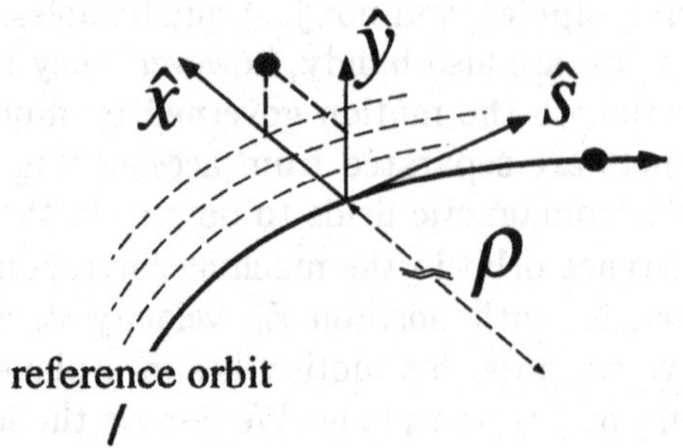

Figure 41. We describe particle motion relative to a reference orbit with local curvature κ.

This states that at linear order the particle orbit is tangent to the reference trajectory, and has the appearance of a concentric circular arc, with bending radius $\rho_1 = \rho + x$.

The Lorentz force law then takes the form

$$\frac{d}{ds}p_1\frac{dx}{ds} \approx \frac{p_1}{\rho_1} + e\left(B_y - y'B_s\right),$$

$$\frac{d}{ds}p_1\frac{dy}{ds} \approx -e\left(B_x - x'B_s\right),$$

and we permit a momentum deviation $p_1 = p_0\left(1+\delta\right)$.

With the assumption of motion in a plane , we have set $a_1 = 0$, in the notation of Eq. (82). We will make the additional assumption that $a_m = 0$ for all m, considering such skew terms to be small corrections to paraxial motion about the design orbit. This amounts to the assumption of *mid-plane symmetry*, $\varphi\left(x,y,s\right) = -\varphi\left(x,-y,s\right)$. In this case, on the reference trajectory, $\partial B_y/\partial y = B_s = \partial B_x/\partial x = 0$. Since $\partial B_x/\partial y = \partial B_y/\partial x$, from Ampere's Law, the local magnetic field is characterized by just two quantities, the dipole component B_y and the quadrupole gradient $\partial B_y/\partial x$ evaluated on the reference orbit. Paraxial fields are then

$$B_y\left(\vec{r}_0 + \vec{r}_\perp\right) \approx B_y\left(\vec{r}_0\right) + \frac{\partial B_y}{\partial x}x, \quad B_x\left(\vec{r}_0 + \vec{r}_\perp\right) \approx \frac{\partial B_y}{\partial x}y$$

The equations of motion reduce to

$$x'' + \left(1-n\right)\kappa^2 x = \kappa\delta, \tag{84}$$

$$y'' + n\kappa^2 y = 0, \tag{85}$$

where the *field-index*

$$n = -\frac{1}{\kappa B_y}\frac{\partial B_y}{\partial x},$$

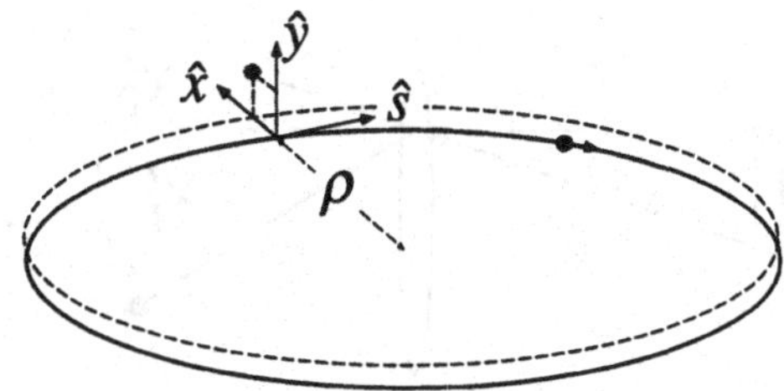

Figure 42. Two circles in a plane intersecting twice. One orbit is "focused" about the other.

is evaluated on the reference orbit.

The result for zero gradient, $n = 0$, may be understood with reference to Fig. 42, corresponding to constant curvature, i.e., motion in a uniform magnetic field. Where two circles are displaced from each other, one will appear, to be "focused" about the other. If the radii (momenta) are different, there is in addition a constant offset. Meanwhile, there is no way for particles to distinguish the reference plane from another displaced in y. Thus y perturbations are not restored to the reference trajectory.

A modest gradient, $0 < n < 1$, evidently provides focusing in both planes. Excursions in y are met with a horizontal field deflecting particles toward the axis. Excursions in x see a diminished vertical field, but are still deflected toward the reference orbit, provided the gradient is not too large.

Less evident is the possibility to confine a beam *without* a dipole field, employing only a quadrupole gradient alternating in sign. This concept of "alternating-gradient focusing" was first recognized (and patented) by Christofilos in 1950 [46]. Omission of the dipole field implies in practice a reduction in magnetic field energy, a smaller magnet, and a more compact, high-energy, "strong-focusing" machine. Historically the development of these concepts took place in the context of circular machines. For the linac lattice, we are interested exclusively in alternating gradient focusing.

A general solution for Eq. (84) may be expressed as a superposition of a particular solution of the inhomogeneous equation, and a solution of the associated homogeneous equation. Thus we may express

$$x = x_\beta + D\delta, \quad y = y_\beta. \tag{86}$$

The terms x_β and y_β are referred to as *betatron oscillations*, since they were first investigated for an accelerator of the same name. The term D is referred to as *dispersion* and quantifies the separation of off-momentum rays due to a bend. The betatron oscillations satisfy

$$\frac{d^2x_\beta}{ds^2} + \hat{K}_x x_\beta = 0, \tag{87}$$

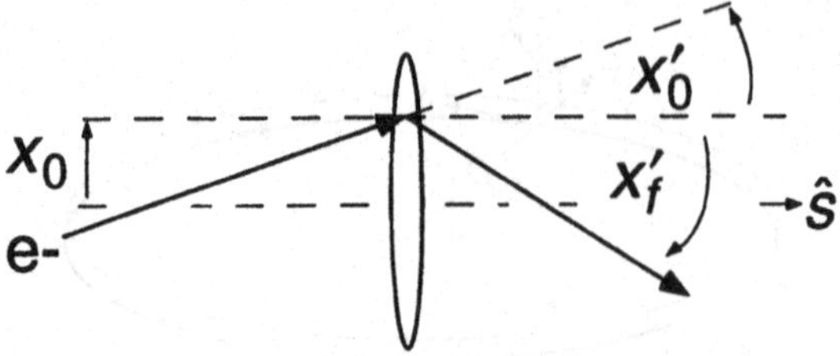

Figure 43. A quadrupole is a lens.

$$\frac{d^2 y_\beta}{ds^2} + \hat{K}_y y_\beta = 0, \tag{88}$$

where the focusing or "K-model" is specified by the reference momentum and the arrangement of dipoles and quadrupoles,

$$\hat{K}_x = -\frac{e}{p_0}\frac{\partial B_y}{\partial x}, \quad \hat{K}_y = \frac{1}{\rho^2} - \hat{K}_x. \tag{89}$$

with ρ the local radius of curvature. The dispersion satisfies

$$\frac{d^2 D}{ds^2} + \hat{K}_x D = \frac{1}{\rho}, \tag{90}$$

Dispersion may be computed directly by fashioning a Green's function from two independent solutions of Eq. (87), and convoluting it with the bending radius. That is to say, after a bend magnet, dispersion behaves as a betatron oscillation would, given the same initial conditions. Thus an appreciation of linear optics requires attention to the betatron motion. This motion is no more involved than what one finds in a normal quadrupole lattice, the problem of primary interest to us, in any case, for the linac.

3.6. LINEAR OPTICS

Let us examine beam motion with normal quads. Consider an electron travelling down the beam pipe and encountering a quadrupole field. The motion is depicted in Fig. 43. By convention we refer to a quad that deflects electrons toward the x-axis as a focusing or "F" quad. Notice then that an F quad is defocusing for electrons in the y-plane, and focusing for positrons in the y-plane. Just to the left of the quad the electron horizontal position is x_0, and its angle with respect to the beamline axis is x'_0. Just to the right of the quad its position is still x_0, since this is a thin quad, but its angle has changed to x'_f. If we know the normalized integrated field strength for this quad, at its present current setting, we may compute the kick $\Delta x' = -x_0/f$, where the length f is determined by the integrated field gradient, and the electron momentum,

$$\frac{1}{f} = -\frac{e}{p} \int_{-\infty}^{+\infty} ds \, \frac{\partial B_y}{\partial x}. \tag{91}$$

The motion as a whole is mapped through the element according to,

$$x'_f = x'_0 - \frac{x_0}{f}, \quad x_f = x_0$$

The quantity f is the focal length for this quad, at this momentum. Transport through the quad may be expressed in matrix form as

$$\begin{pmatrix} x \\ x' \end{pmatrix}_f = R \begin{pmatrix} x \\ x' \end{pmatrix}_0, \tag{92}$$

where

$$R = \begin{pmatrix} 1 & 0 \\ -1/f & 1 \end{pmatrix}$$

Similarly one may describe a drift of length L by the map,

$$R = \begin{pmatrix} 1 & L \\ 0 & 1 \end{pmatrix}. \tag{93}$$

Such "R-matrices" are handy in that one can concatenate them to determine the overall effect of a series of elements. Moreover one can show that any such matrix may be decomposed into an equivalent drift-quad-drift,

$$R = \begin{pmatrix} R_{11} & R_{12} \\ R_{21} & R_{22} \end{pmatrix} = \begin{pmatrix} 1 & q \\ 0 & 1 \end{pmatrix} \begin{pmatrix} 1 & 0 \\ -1/f_* & 1 \end{pmatrix} \begin{pmatrix} 1 & p \\ 0 & 1 \end{pmatrix}. \tag{94}$$

Thus any combination of quads and drifts may be characterized by the location of the 1st principal plane forward from the element entrance, p, the effective focal length f_*, and the location of the 2nd principal plane, backward from the element exit, q. For example, it is straightforward to show that the beamline depicted in Fig. 44, consisting of a quad with focal length f_1, followed by a drift of length L, followed by a quad of focal length $-f_2$ may provide *positive* f_*—focusing—in both planes. For this reason such a "doublet" is a handy building block for more elaborate optical systems. The simplest linac lattice is just a periodic arrangement of such doublets, a "FODO" lattice. The principal of such alternating-gradient focusing elements is that the beam envelope may be large in a focusing quad, and small in a defocusing quad, so that the average radial "pressure" provided by the magnetic fields is positive.

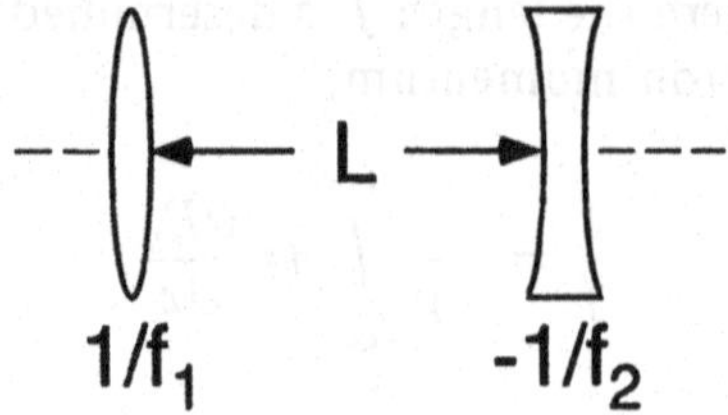

Figure 44. A quad doublet can provide focusing in both planes.

In general one is interested to know the beam-behavior throughout the lattice. This would seem to imply knowledge of the map between any two points. However, the most general description of transport is simpler than that. In an arbitrary lattice of quadrupoles and drifts, particle motion satisfies

$$\frac{d}{ds}\gamma\frac{dx}{ds} + K(s)\,x = 0, \tag{95}$$

where quad locations and settings determine $K \approx \gamma\hat{K}_x$, and we suppose $V \approx c$. For a prescribed energy profile $\gamma(s)$ there are two independent solutions to this homogeneous second order ordinary differential equation. Selecting an initial reference location, s_0 on the beamline, we may define a "cosine-like" solution $C(s, s_0)$ with initial conditions $C = 1$ and $C' = 0$, and a "sine-like" solution $S(s, s_0)$ with initial conditions $S = 0$ and $S' = 1$. In terms of these functions, an arbitrary "betatron" orbit may be represented as $x = x(s_0)C + x'(s_0)S$. The R-matrix for transport from s_0 to s is then

$$R(s_0 \to s) = \begin{pmatrix} C(s, s_0) & S(s, s_0) \\ C'(s, s_0) & S'(s, s_0) \end{pmatrix}. \tag{96}$$

Evidently, the C and S functions provide a complete description of the transport properties of the lattice. They also determine the Green's function with which one may incorporate perturbations to Eq. (95), due to correctors, quad-misalignments, rf kicks as in Eq. (36), and wakefields as in Eq. (42).

A complementary approach employs the "machine functions", or "Twiss parameters". We consider Eq. (95) in the absence of acceleration and look for a solution of the form

$$x = \Re\left(A(s)\,e^{j\phi(s)}\right),$$

with A and ϕ real functions, to be determined. Plugging this into the equation of motion we find

$$A'' + \hat{K}A - (\phi')^2 A = 0, \quad (\phi' A^2)' = 0,$$

where the prime denotes the derivative with respect to s. One can check that these equations are solved by

$$A(s) = c_1\sqrt{\beta(s)}, \quad \phi(s) = c_2 + \psi(s)$$

for any two constants c_1 and c_2, provided the functions β and ψ satisfy

$$\frac{d^2}{ds^2}\beta^{1/2} + \hat{K}\beta^{1/2} = \frac{1}{\beta^{3/2}}, \tag{97}$$

$$\psi(s) = \int_0^s ds' \frac{1}{\beta(s')}. \tag{98}$$

The most general solution of the equations of motion may then be expressed as

$$x = \Re\left\{\chi\sqrt{\beta(s)}e^{j\psi(s)}\right\},$$

where the phasor $\chi = c_1 \exp(jc_2)$ is a constant. This constant may be expressed, after a bit of algebra as

$$\chi e^{j\psi} = \frac{x}{\sqrt{\beta}}(1 - j\alpha) - jx'\sqrt{\beta},$$

where we introduce

$$\alpha = -\beta'/2. \tag{99}$$

One can go on to show that $\tan c_2 = -\alpha_0 - \beta_0 x'_0/x_0$, and $c_1^2 = J$, where

$$J = |\chi|^2 = \left(\frac{\alpha}{\sqrt{\beta}}x + \sqrt{\beta}x'\right)^2 + \frac{x^2}{\beta} = \gamma_T x^2 + 2\alpha x x' + \beta x'^2, \tag{100}$$

and we abbreviate

$$\gamma_T = \frac{1+\alpha^2}{\beta}. \tag{101}$$

The quantity J is an invariant of the motion, as noted by Courant and Snyder. This all implies that the constants c_1 and c_2 determine the orbit, and are determined by the initial conditions. The optical functions (or Twiss parameters) α and β describe on the other hand, the lattice, independent of a particular particle's initial conditions. The form of the Courant-Synder invariant indicates that a particle orbit in trace-space may be depicted as motion on an s-dependent ellipse as seen in Fig. 45. The optical functions specify the rotation and deformation of such ellipses as a function of s. Particle motion is specified by the action variable, labelling the particular ellipse the particle is on, and an angle variable labelling the location of

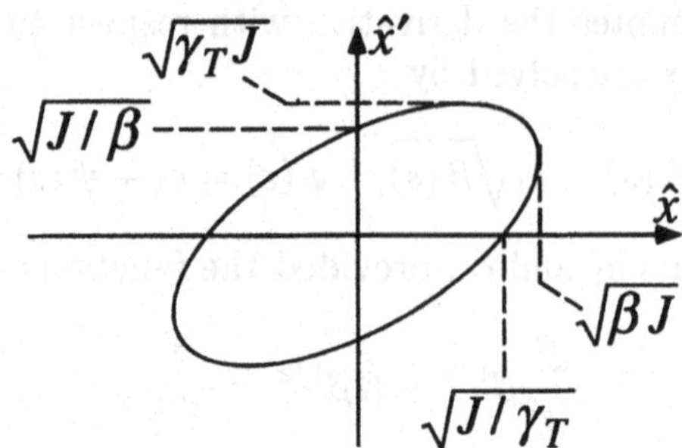

Figure 45. The Courant-Synder invariant defines an ellipse parameterized by β, α and γ_T.

the particle on the ellipse circumference. One then may picture the beam as a whole in terms of a collection of concentric ellipses, each ellipse with particles distributed around its circumference.

Since lattice transport is, we find, governed by α and β, we should be able to express the cosine-like and sine-like functions in terms of them. In fact,

$$C(s,0) = \sqrt{\frac{\beta(s)}{\beta(0)}}\left(\cos\psi + \alpha(0)\sin\psi\right), \tag{102}$$

$$S(s,0) = \sqrt{\beta(s)\,\beta(0)}\sin\psi, \tag{103}$$

as one may check by confirming that these expressions satisfy the equation of motion, and the appropriate initial conditions for the C and S functions. With this we may express the R-matrix as

$$R(s,0) = \begin{pmatrix} \sqrt{\frac{\beta(s)}{\beta(0)}}\left(\cos\psi + \alpha_0\sin\psi\right) & \sqrt{\beta(s)\,\beta(0)}\sin\psi \\ \frac{(\alpha_0-\alpha)\cos\psi-(1+\alpha_0\alpha)\sin\psi}{\sqrt{\beta(s)\beta(0)}} & \sqrt{\frac{\beta(0)}{\beta(s)}}\left(\cos\psi - \alpha\sin\psi\right) \end{pmatrix}. \tag{104}$$

One can show also that optical functions themselves may be "transported" according to

$$\begin{pmatrix} \beta \\ \alpha \\ \gamma_T \end{pmatrix}_s = \begin{pmatrix} C^2 & -2SC & S^2 \\ -CC & S'C + SC' & -SS' \\ C'^2 & -2S'C' & S'^2 \end{pmatrix} \begin{pmatrix} \beta \\ \alpha \\ \gamma_T \end{pmatrix}_0, \tag{105}$$

and for that matter, the beam second moments may be mapped using

$$\begin{pmatrix} \sigma_x^2 \\ \sigma_{xx'} \\ \sigma_{x'}^2 \end{pmatrix}_s = \begin{pmatrix} C^2 & 2SC & S^2 \\ CC & S'C + SC' & SS' \\ C'^2 & 2S'C' & S'^2 \end{pmatrix} \begin{pmatrix} \sigma_x^2 \\ \sigma_{xx'} \\ \sigma_{x'}^2 \end{pmatrix}_0. \tag{106}$$

While the C and S functions are unique, the β function is not. This might seem like a mere mathematical curiosity, in fact it is a physics problem, the problem of *matching*. For example, in a FODO lattice with quad separation L, and quad focal length f, alternating in sign, there is a periodic solution for the beta function with maxima and minima given by,

$$\beta_\pm = 2f\left(\frac{1 \pm \frac{L}{2f}}{1 \mp \frac{L}{2f}}\right)^{1/2}. \tag{107}$$

Phase advance μ per period (length $2L$) is given by $\sin(\mu/2) = L/2f$. One may check this using Eqs. (97) and (98). The existence of this solution, however, does not imply that the beam will be able to find it by itself. To see this, let us suppose that at $s = 0$ a beam is incident on our lattice. We follow the beam through the lattice and define at each point s

$$\hat{\beta} = \frac{\sigma_x^2}{\varepsilon}, \quad \hat{\gamma}_T = \frac{\sigma_{x'}^2}{\varepsilon}, \quad \hat{\alpha} = -\frac{\sigma_{xx'}}{\varepsilon}.$$

One can check that $\hat{\gamma}_T\hat{\beta} - \hat{\alpha}^2 = 1$. One can show that $\hat{\beta}$, $\hat{\alpha}$ are also solutions for the optical functions, by comparing Eqs. (105) and (106). Thus the beam will also define a choice of β-function, one that depends on the optics prior to injection. This suggests two questions. What initial conditions should be employed to insure that $\hat{\beta}(s)$ matches the intended $\beta(s)$, the design ellipse? And, is there a particular $\beta(s)$ that is better than others? We would like to design a linac lattice and match into it. The problem is not unlike that of an old television set, also a kind of accelerator. If the picture is fuzzy, one will need to turn some knobs ("matching"), and such tuning is predicated on the assumption that the machine has been designed to provide for a good picture within the range of knob settings ("lattice design"). First we describe matching.

3.7. MATCHING

The beam-optical functions $\hat{\beta}$, $\hat{\alpha}$ generate ellipses labelled by the invariant,

$$\hat{J} = \hat{\gamma}_T x^2 + 2\hat{\alpha}xx' + \hat{\beta}x'^2.$$

This is a single-particle quantity, one that depends, through $\hat{\beta}$ and $\hat{\alpha}$, on the initial conditions at injection into the linac, and the linac lattice. After averaging over the beam, we find that

$$\left\langle \hat{J} \right\rangle = \hat{\gamma}_T\sigma_x^2 + 2\hat{\alpha}\sigma_{xx'} + \hat{\beta}\sigma_{x'}^2 = 2\varepsilon^2.$$

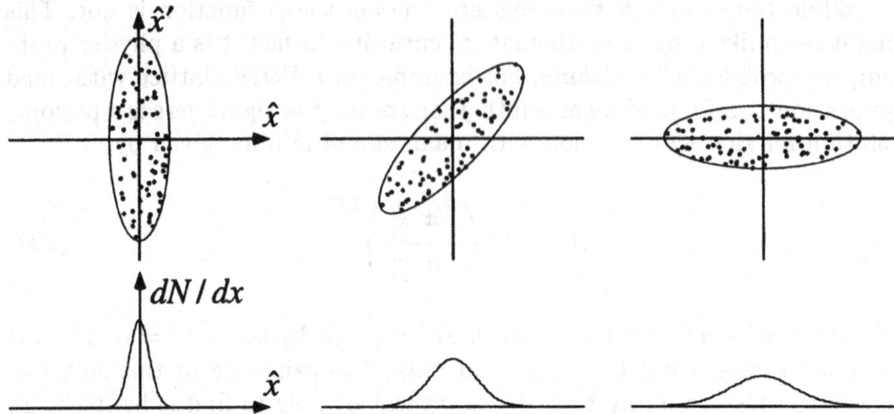

Figure 46. Rotation in phase-space and measurement of beam-size permit inference of emittance.

Thus the beam-based Courant-Snyder invariant is determined by the emittance. Let us compare this result to the lattice Courant-Synder invariant, averaged over the beam. With some algebra we see that

$$\begin{aligned}\langle J\rangle &= \gamma_T\sigma_x^2 + 2\alpha\sigma_{xx'} + \beta\sigma_{x'}^2 \\ &= \gamma_T\left(\varepsilon\hat{\beta}\right) + 2\alpha\left(-\varepsilon\alpha\right) + \beta\left(\varepsilon\gamma_T\right) = B_M\left\langle\hat{J}\right\rangle\end{aligned}$$

where

$$B_M = \frac{\langle J\rangle}{\left\langle\hat{J}\right\rangle} = \frac{\langle J\rangle}{2\varepsilon} = \frac{1}{2}\left(1+\alpha^2\right)\frac{\hat{\beta}}{\beta} + \frac{1}{2}\left(1+\hat{\alpha}^2\right)\frac{\beta}{\hat{\beta}} - \alpha\hat{\alpha}, \tag{108}$$

is the "beam magnification" factor. Minimum $B_M = 1$ corresponds to $\hat{\beta} = \beta$, $\hat{\alpha} = \alpha$, and a matched beam.

In practice the problem of matching, of arranging $B_M = 1$, is tied to the problem of emittance inference. Operationally it falls under the heading of "tune-up", preparation of the machine for experiment. There are several techniques employed for emittance inference, and the most common manage to rotate the beam in phase-space for display in coordinate space as seen in Fig. 46. This rotation may be achieved with a drift, by changing a quadrupole setting ("quad-scan"), or by some combination of these. Measurement of size at three non-degenerate points in a drift is enough to determine ε in principle, assuming no x-y correlations. One can see an example of this technique in [47]. Measurement of size at additional points, and in x and y, should be anticipated as it permits an assessment of errors.

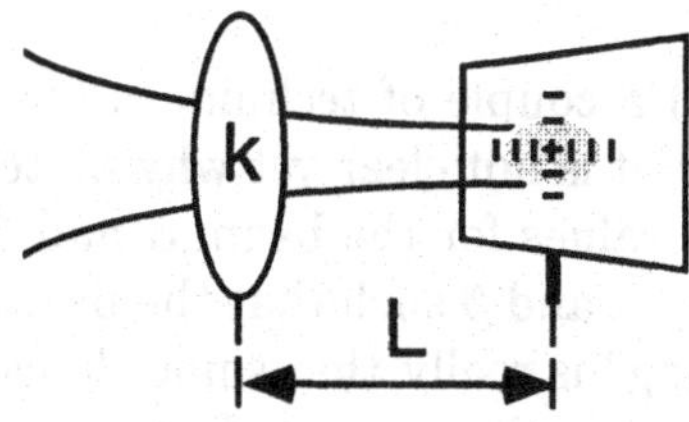

Figure 47. Emittance can be measured, and a beam matched, with the help of a quad--scan.

The quad-scan technique is illustrated in Fig. 47. The set-up consists of a thin focusing quad with focal length $1/k$, a drift of length L and a profile monitor (screen or wire-scanner). We are able to scan the quad strength, and measure beam-size at each quad setting, and we ask what this may tell us about emittance, and the beam optical functions. While the paraphernalia of linear optics may be brought to bear on this problem, let us pursue the more elementary approach employed to derive Eq. (74). In terms of beam-moments just after the quad (at $s = 0^+$), the beam-size at the screen may be expressed as

$$\sigma_x^2(k,L) = \sigma_x^2(0^+) + 2\sigma_{xx'}(0^+)L + \sigma_{x'}^2(0^+)L^2.$$

Beam moments just after the quad may be related to those just before the quad ($s = 0^-$) according to

$$\begin{aligned}\sigma_x^2(0^+) &= \sigma_x^2(0^-)\\ \sigma_{xx'}(0^+) &= \sigma_{xx'}(0^-) - k\sigma_x^2(0^-)\\ \sigma_{x'}^2(0^+) &= \sigma_{x'}^2(0^-) - 2k\sigma_{xx'}(0^-) + k^2\sigma_x^2(0^-)\end{aligned}$$

Putting these results together one can see then that beam-size at the screen is a quadratic function of quad-strength, with a minimum. Without loss of generality, we may refer quad-strength to the value for minimum spot, so that $k = 0$ refers to this minimum. This is equivalent to the assumption $\sigma_{xx'}(0^-) = -\sigma_x^2(0^-)/L$. After some algebra one can show that

$$\sigma_x^2(k,L) = L^2\left\{\frac{\varepsilon^2}{\sigma_x^2(0^-)} + k^2\sigma_x^2(0^-)\right\} = \sigma_x^2(0,L) + \frac{k^2L^4\varepsilon^2}{\sigma_x^2(0,L)}.$$

Thus the minimum size in the quad-scan, combined with the curvature of the parabola permit one to infer the emittance. With quad set for the minimum size ($k = 0$), one can determine

$$\hat{\beta} = \frac{\sigma_x^2(0,L)}{\varepsilon}, \quad \hat{\alpha} = -\frac{\hat{\beta}}{L},$$

at the screen location.

We have mentioned a couple of techniques for emittance measurement and matching; however, it is not clear yet why matching might be desirable. For example, for given values for the beam $\hat{\alpha}$ and $\hat{\beta}$ at injection, one could simply define the lattice α and β such that the beam is "matched". Thus the implication of "matching" is really that some choices for α and β are better than others. What are "good" choices of optical functions to be matched to? How to design a linac lattice?

3.8. CHROMATICITY

At the crudest level a lattice should insure a beam-size small enough to avoid current-loss. The more subtle issue in lattice design may be found in the circumstance that beams are not mono-energetic, they include a distribution in momenta, and particles of different momentum are deflected by different angles in a magnetic field. For example, a beam with zero emittance and zero momentum spread may be focused to a point in x by a lattice consisting of a quadrupole and a drift. With momentum spread the spot is blurred. In view of our discussion of the optics of a drift, in Sec. 3.2.2, we may infer that the beam has developed an rms emittance due to the lattice. The density in phase-space is unchanged, but in projection it is diluted. A "good" lattice should minimize projected emittance despite the momentum-dependence of the optics, "chromaticity".

Recalling $\delta = (p - p_0)/p_0$, the fractional deviation of a particle momentum p from the design momentum p_0, the chromatic properties of the lattice may be quantified in terms of $\beta_\delta = \partial\beta/\partial\delta$. Associated with this "beta-function chromaticity" one has also "phase-advance chromaticity" corresponding to $\psi_\delta = \partial\psi/\partial\delta$. Due to ψ_δ particles with different momenta advance through betatron oscillations at a different rate. Due to the beam collective behavior (wakefields), some amount of phase-advance chromaticity is quite helpful as it inhibits the beam from acting in a coherent manner in driving destructive microwave modes of the linac. The problem for lattice design is to maintain a narrowly bounded β-function for off-momentum particles, i.e., small β_δ.

Thus to control chromatic emittance growth one prefers that the β-function be well-behaved for momenta represented in the beam distribution. Taking the FODO lattice as an example, there is a distinguished choice for the β-function, namely the periodic solution. For the periodic lattice, there is a nearby solution for the β-function for off-momentum particles that is also periodic, and the net dependence of β on momentum at any point is not more than is accumulated over a single period. In the absence of a distinguished symmetry, lattice designers typically proceed by attempting to maintain quasi-periodic machine functions. Beyond the linac, to further diminish chromatic effects, one may employ *chromatic correction* derived from the interaction of dispersion with sextupole fields [49].

3.9. LATTICE DIAGNOSTICS

Having implemented an ingenious lattice design, and carefully matched to it, the next task in commissioning a new lattice is to figure out why the beam won't go through the pipe. One needs methods of diagnosing errors in the lattice. An excellent, broad review of the subject of lattice and other diagnostics can be found in a recent article by Zimmermann [48]. Uncertainties in this problem are of two types: model errors and BPM errors, and we discuss each in turn.

The description thus far, of the magnet system and its effect on the beam, has considered the ideal case, where magnets perform as designed. In fact a "lattice" in practice means a lattice with errors, and they can be sporadic, recurrent and time-consuming to identify as such, and locate. The most common errors are (a) electrical - backwards wiring of a magnet, (b) mechanical - positioning of a magnet (rotated, tilted), (c) sporadic operational failure - turn-to-turn short in a magnet coil. Item (a) is a particular concern during turn-on after work has been done on the beamline. Item (b) is always present to some degree. Item (c) can occur at any time. As a matter of routine, one may expect to account for hysteresis, the circumstance that a given coil current setting does not correspond uniquely to a particular magnetic flux. Rather the flux depends on the history of the magnet since it was last measured. In practice, one may adopt a procedure for "standardizing" magnets by cycling of the coil currents.

Typically magnets are carefully designed to reduce spurious multipole components, but what components are present, measured and tabulated we needn't consider an error, as they may be incorporated in our model, after generalizing the linear map R, to include higher-order terms, the "T" and "U" coefficients [49].

Mechanical errors are always present to some degree and it is up to the designer to determine, from their analysis of beam dynamics, to what precision mechanical surveying and alignment are required. In practice, beam-based orbit analysis is more accurate in locating quad and BPM centroids than mechanical alignment is capable of achieving or correcting [50]. As to sporadic operational failures, these can be time-consuming in a large accelerator complex. They occur because coil insulation is not impervious to radiation, because the earth moves, because electrical circuits do not function properly when wet, etc. To locate the problem one must examine the beam dynamics, compare observed orbits to those predicted from the model, and by means of sleuth-work narrow down the "bad" region in the lattice to one or a few elements. At that point one can stop the beam, enter the vault and employ a compass to look for wrong-polarity or visually inspect for misplaced magnetic materials. Failing this one may examine the voltage drop across the coils to locate the short, with due safety precautions. If all else fails, one may disassemble the questionable magnets one by one. Accurate lattice diagnosis from "upstairs" is appreciated.

To identify "good" and "bad" regions in a lattice, one analyzes orbit data, looking for discrepancies. To illustrate in the simplest fashion, suppose one observes a betatron oscillation on a collection of beam position monitors numbered $n = 1, 2, ..., N$. Accepting a model for the lattice, embodied in cosine-like and sine-like functions, one chooses launch variables (x_0, x_0') as fit parameters in such a way as to minimize the error in the fit

$$\Delta^2 = \frac{1}{N}\sum_{n=1}^{N}\left\{x_{BPM}(n) - x_0 C(s_n) - x_0' S(s_n)\right\}^2$$

One can show that this is accomplished with the choice,

$$\begin{pmatrix} x_0 \\ x_0' \end{pmatrix} = \frac{1}{\overline{C^2}\,\overline{S^2} - (\overline{SC})^2}\begin{pmatrix} \overline{S^2} & -\overline{SC} \\ -\overline{SC} & \overline{C^2} \end{pmatrix}\begin{pmatrix} \overline{Cx} \\ \overline{Sx} \end{pmatrix},$$

where the vertical bar represents an average over the BPM positions. In practice, pursuing such a procedure, one quickly develops an interest in BPM errors. These include scale, offset, linearity, rotation, cables mislabeled, etc. Thus for model-checking one must go to greater lengths. The weighting may be improved by incorporating known BPM errors and the expected β function. Averaging may be employed to beat down BPM errors. Errors in energy profile lead to errors in inferred quad focal length; thus δ is an important variable and one may want to include dispersion in the orbit-fit. For that matter, to allow for x-y coupling a 5x5 R-matrix may be in order, with fitting for $(x_0, x_0', y_0, y_0', \delta)$. Wakefields too affect the beam orbit, in a current-dependent fashion, with current varying from pulse to pulse. A single bunch passing through a misaligned structure will appear to kick itself. Multiple bunches talk to each other thanks to undamped dipole modes. Each structure provides a kick due to coupler asymmetry. The energy distribution is not merely a function of all klystron phases and amplitudes, but is coupled to bunch-length through the longitudinal wakefield. As the number of variables proliferates, and one realizes that most of them are hidden, the dynamical relations become obscure and one may call on singular-value decomposition to analyze large collections of variables at once. This effort is aided by new and better software and new beamline instruments.

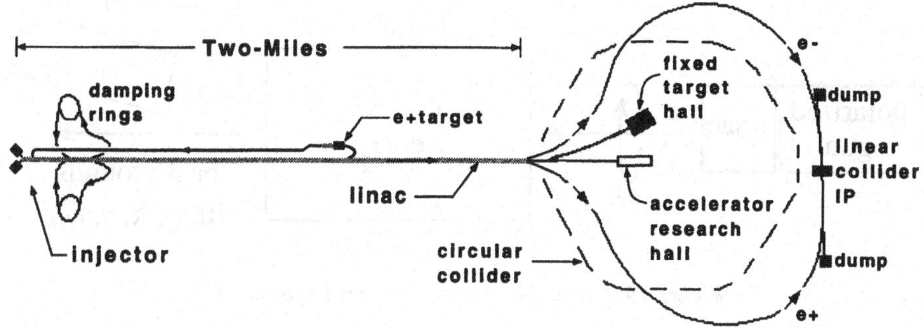

Figure 48. Layout of the Two-Mile Accelerator complex, indicating fixed-target and colliding beam configurations.

Having produced our high-current, low energy-spread, possibly low-emittance and rock-steady beam at high-energy let us consider how to arrange collisions. There are two kinds of schemes, fixed-target and collider. We consider each in turn, as they are implemented in the Two-Mile Accelerator complex seen in Fig. 48.

3.10. FIXED-TARGET

A generic fixed-target experiment is illustrated in Fig. 49, indicating a polarized electron gun, a diminutive linac, a significant-looking target, and a calorimeter. Not shown are magnets and collimation arranged to perform energy analysis and to remove backgrounds prior to the calorimeter. The energy available for collisions in the center-of-momentum (c.o.m) frame is $\sqrt{s}$, with

$$s = (E_1 + E_2)^2 - c^2 (\vec{p}_1 + \vec{p}_2)^2 , \tag{109}$$

where E_1 is the beam energy, E_2 is the energy of the target particle and $\vec{p}_i$ are the corresponding initial momenta. For a stationary target particle and a relativistic beam, $s \approx 2E_1E_2$, and c.o.m. energy varies as the square root of the incident beam energy.

To determine the event-rate we suppose the target is larger than the beam and uniform in the region of the beam, presenting a density n_t of scattering centers, distributed over a length l. If there are N_b beam electrons incident per pulse, there are $N_b n_t l \sigma$ events or counts per pulse for a process characterized by cross-section σ for scattering into the detector. With the experiment run at machine repetition frequency f_{rep} we may express the rate as $L\sigma$, where the *luminosity* is

$$L = (N_b f_{rep})\,(n_t l) , \tag{110}$$

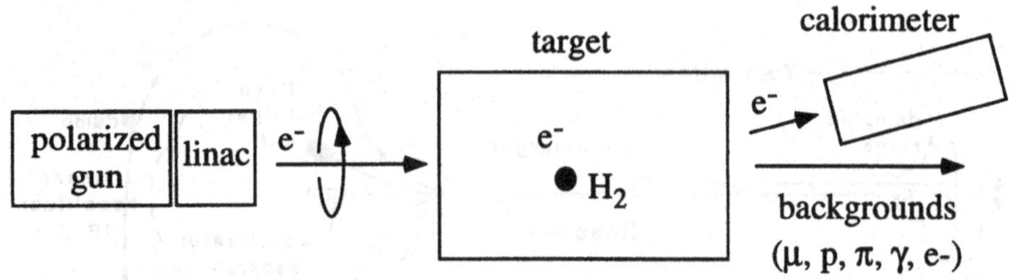

Figure 49. Scheme of a fixed-target experiment.

a product of two factors, one from the accelerator, and one from the target. The total number of events over the course of the run is then $N = TL\sigma$, where T is the time for which both the accelerator, the detector and all sub-systems are "up" and operating together. Luminosity, c.o.m. energy, and polarization circumscribe the capability of an accelerator for fixed-target experimentation. Beamline instrumentation may also be critical. We consider an example.

The E-158 experiment [13] proposes to scatter a polarized electron beam on the atomic electrons of an unpolarized liquid hydrogen target and to measure the asymmetry, $A_{LR} = (\sigma_R - \sigma_L)/(\sigma_R + \sigma_L)$, between the cross-sections for Moller scattering of right and left-handed electrons. The motivation for this is that the cross-section may be calculated precisely in the Standard Model, and thus a precision measurement may be employed to accurately discern *deviations* from the Standard Model. With the scheme depicted in Fig. 49, the left-right asymmetry may be inferred from the difference over many pulses between calorimeter readouts for a left-circularly polarized beam and a right-circularly polarized beam. One merely needs to count, precisely.

The setup consists of a Two-Mile linac, a transport line to the target (the "A-Line"), a liquid hydrogen target 1.5 m in length, and 3″ in diameter, magnets, and a calorimeter. With each pulse, a 5-kJ "blowtorch" of ejecta and disrupted beam will emerge from the target, and pass through a series of magnets that serve to bring scattered electrons with momenta from 12-24 GeV/c to the calorimeter, 60 m downstream. Most of the 500-kW average beam power will pass through the center hole of the annulus formed by the detector, and will dissipate itself in a beam dump. For this experiment $E_2 \approx$0.511 MeV, so that $\sqrt{s} \approx \sqrt{E_1}$, in units of MeV, and for $E_1 \approx 40 - 50$ GeV, $\sqrt{s} \approx$200 MeV. Luminosity is estimated from the target length $l \approx$ 1.5 m, and the number density of electrons in liquid hydrogen, $n_t \approx 4 \times 10^{22}$ cm^{-3}, so that $n_t l \approx 6 \times 10^{24}$ cm^{-2}. At 120 Hz, with $N_b \approx 1 \times 10^{11}$ electrons per pulse, $L \approx 8 \times 10^{37}$ cm^{-2}s^{-1}. To appreciate this number one must consult the statistical requirements for the experiment. How many Moller-scattered electrons are needed for an interesting experiment?

The theory of electroweak interactions predicts that $A_{LR} \approx 2 \times 10^{-7}$, and the goal of the experiment is to test this prediction at the 10%-level, requiring an uncertainty $\delta A_{LR} \approx 1 \times 10^{-8}$, or less. For a sample of size N, one expects statistical fluctuations with an rms $1/N^{1/2}$. Thus one should expect to make use of a sample at least as large as $N \approx 1/(\delta A_{LR})^2 \approx 1 \times 10^{16}$. Meanwhile, the Moller differential cross-section integrated over the angular acceptance of the detector gives a cross-section $\sigma \approx 14\,\mu$barn for a 50 GeV beam, where 1 barn$=10^{-24}$cm^2. The total number of Moller-scattered electrons expected in uptime $T \approx 1 \times 10^7\,s$ (4 months) is then $N = TL\sigma \approx 10^{16}$, with $N_b \approx 10^{11}$. A more detailed accounting of the statistics, and allowance for 50% down-time, puts the requirement on the pulse intensity at $N_b \approx 4 - 6 \times 10^{11}$.

Notice that none of these considerations consulted emittance. The target is big, and the beam may as well be big too. The high-charge requirement does strain the linac energy budget, and requires creative techniques for beam-loading compensation, to maintain a narrow energy spectrum. Perhaps most challenging though is the high-precision required. If we imagine repeating this experiment many times, our confidence in results inferred on any one experiment would depend on the scatter in A_{LR}. If one requires this scatter to be as 10% of A_{LR}, then with 10^9 machine pulses, one must bound single-pulse fluctuations at the level of 10^{-4} or so. So, to this level of precision, one must be able to bound or identify and tag any non-electroweak effects that could result in fluctuations in the combined probability of Moller electron production and transmission to the detector. Spurious effects could arise in principle, for example, if the laser on the gun jitters transversely in a manner correlated with laser polarization. In this case, the beam on the target might jitter, and so too would the scattered electrons. For this reason the BPM cavity seen in Fig. 12 is essential to the experiment.

From this example we see that for a precision fixed-target experiment, beamline instrumentation is crucial. Some practical limits are set by achievable no-load voltage with beam-loading compensation.

3.11. COLLIDING-BEAMS

To reach the highest energies it is desirable to collide a beam with another beam. For beams of equal energy E, the c.o.m. energy is then simply $\sqrt{s} = 2E$. The drawback in this concept is that luminosity is much lower than in a fixed target experiment. The physics goals for the Two-Mile Accelerator operated in this "SLC/SLD" mode are described in [51].

To determine luminosity, consider two beams with number densities n_1, n_2, interacting with cross-section σ to produce events at instantaneous rate per unit volume $dn/dt = 2\sigma c n_1 n_2$. We model each bunch as a tri-Gaussian,

$$n_k = \frac{N_k}{(2\pi)^{3/2}\,\sigma_{xk}\sigma_{yk}\sigma_{zk}} \exp\left\{-\frac{(x-x_k)^2}{2\sigma_{xk}^2} - \frac{(y-y_k)^2}{2\sigma_{yk}^2} - \frac{(s-V_k t)^2}{2\sigma_z^2}\right\},$$

with $k = 1, 2$ and $V_1 = c, V_2 = -c$. We suppose bunches collide at a rate f_{rep} and arrive at the average rate at which events are produced

$$\dot{N} = 2 f_{rep}\sigma c \int d^3\vec{r}\, dt\, n_1 n_2 = L\sigma.$$

After some integrations, luminosity may be expressed as

$$L = \frac{N_1 N_2 f_{rep}}{2\pi\Sigma_x\Sigma_y} \exp\left\{-\frac{(x_1-x_2)^2}{2\Sigma_x^2} - \frac{(y_1-y_2)^2}{2\Sigma_y^2}\right\}, \tag{111}$$

where

$$\Sigma_x^2 = \sigma_{x1}^2 + \sigma_{x2}^2, \quad \Sigma_y^2 = \sigma_{y1}^2 + \sigma_{y2}^2. \tag{112}$$

For illustration, values typical of the last SLC run are seen in Table 5 [52]. The rms beam sizes, omitting chromatic contributions, correspond to

TABLE 5. IP parameters SLC/SLD'97 [52].

parameter	e-x	e-y	e+x	e+y
ε_n (10^{-5} m-rad)	5.3	1.3	5.1	0.9
β_* (mm)	2.9	1.7	2.2	1.4
θ_* (μrad)	439	269	489	249
$\sqrt{\varepsilon\beta_*}$ (μm)	1.27	0.46	1.08	0.35

$\Sigma_x \approx 1.67\mu$m, $\Sigma_y \approx 0.58\mu$m, or $\Sigma_x \times \Sigma_y \approx 0.97\mu\text{m}^2$. For example, at 120 Hz, with $N \approx 3 \times 10^{10}$, this gives $L \approx 1.1 \times 10^{30}\text{cm}^{-2}\text{s}^{-1}$. On the Z-resonance, with a cross-section of 30 nb, this comes to 200 Z's per hour.

Equation (111) represents just the first approximation to the luminosity. In fact, chromatic effects are important and chromatic correction in the final focus is essential [49]. In addition, when such fine, intense beams collide it is important to account for the effect of beam-fields. Two oppositely-charged counter-propagating beams pinch each other, and this may result in luminosity enhancement, if they have enough time to constrict. Thus the bunch length should be long as long as is consistent with the depth of field, about 1 mm for the parameters of Table 5. This pinch effect accounted for a factor of two in luminosity during the last run. As particle trajectories are deflected by the oncoming beam, in the course of pinching, they will radiate synchrotron photons, or "beamstrahlung", and this is useful as a diagnostic [53]. Luminosity is also affected by slow drifts in machine optical parameters, and pulse-to-pulse orbit jitter, due for example to wakefield amplification of "noise" at the machine front-end. One may appreciate the tolerances on such effects based on the optics of a drift. Waist motion in s at the level of 1/3 of β_* provides a 10% reduction in luminosity. Typical orbit jitter is about 1/3 of the rms beam size, corresponding to a 20% reduction in luminosity. One can intuit that high luminosity requires a large constellation of systems to be functioning as one. Integrated luminosity requires exceptional attention to maintenance, and choreography by a team of expert operators, aided by beamline instruments, software and feedback [54].

4. Epilogue

We have covered a fair bit here, but there's more. A thorough discussion of the microwave aspects of colliders may be found in [55]. The subject of microwave linacs is also now available in textbooks [56, 57]. Beam optics is introduced in numerous texts [58, 59]. Features of collective beam dynamics are introduced in [7]. More specialized topics are easily located in the curricula of the US Particle Accelerator School, the CERN Schools, and the Joint Accelerator School. In between schools, there are many occasions to hear of the latest exploits, at the US Particle Accelerator Conference (PAC), the European PAC, the Asian PAC, and numerous other conferences. If one should tire of schools and conferences, there are many linacs in operation around the world where exciting work is taking place, for collider research, and for fixed-target experiments. And there is one linear collider.

Acknowledgements

These notes were prepared for the proceedings of the NATO Advanced Study Institute on Techniques and Concepts of High-Energy Physics, held in St. Croix, US Virgin Islands, June '98. This is a notoriously tough school, and it was only due to the invitation of Prof. Tom Ferbel that I was able to summon the courage to attend. These notes are the result and comments are welcome; please send them to *whittum@slac.stanford.edu*, unless I have moved to St. Croix.

Thanks to Perry Wilson and Mel Month for their encouragement. This work has benefited from collaboration and discussions with Marc Hill, Yury

Kolomensky, Eddie Lin, Dennis Palmer, David Pritzkau, Mike Seidel, Bill Spence and Frank Zimmermann. Work by Jennifer Burney and helpful discussions with Clive Field were instructive. I am indebted to David Pritzkau and Frank Zimmermann for helpful comments on the manuscript. Thanks to Angie Seymour for her support.

Participation in this school was supported in part by U.S. Department of Energy, Contract DE-AC03-76SF00515.

References

1. D.H. Whittum, "Introduction to Electrodynamics for Microwave Linear Accelerators", *Proceedings of the US-CERN-Japan School on RF Engineering for Particle Accelerators* , (KEK, Tsukuba, to be published), SLAC-PUB-7802.
2. R.B. Neal, *The Stanford Two-Mile Accelerator* (W. A. Benjamin, New York, 1968).
3. J.T. Seeman, "The Stanford Linear Collider", *Ann. Rev. Nucl. Part. Sci.* 41 (1991) pp. 389-428.
4. W.W. Hansen "A Type of Electrical Resonator", *J. Appl. Phys.* 9 (1938) pp. 654-663.
5. J.C. Slater, *Microwave Electronics* (D. Van Nostrand, Boston, 1950).
6. W.K.H. Panofsky and W.A. Wenzel, *Rev. Sci. Instrum.* 27 (1956) p. 967.
7. A.W. Chao, *Physics of Collective Beam Instabilities in High Energy Accelerators* (Wiley, New York, 1993).
8. J.C. Nygard and R.F. Post, "Recent advances in high power microwave electron accelerators for physics research", *Nucl. Instrum. Meth.* 11 (1961) pp. 126-135.
9. K.A. Thompson and R.D. Ruth, "Controlling transverse multibunch instabilities in linacs of high-energy colliders", *Phys. Rev. D* 41 (1990) pp. 964-977.
10. J.T. Seeman, *et al.*, "Alignment issues of the SLC linac accelerating structures", *Proceedings of the 1991 Particle Accelerator Conference* (IEEE, New York, 1991), pp. 2949-2951, SLAC-PUB-5439.
11. G.A. Loew, *et al.*, "Linac beam interactions and instabilities", *Proceedings of the VII International Conference on High Energy Accelerators*, (Academy of Sciences, Yerevan, 1970) pp. 229-252.
12. O. Altenmueller and P. Brunet, "Some RF Characteristics of the Beam Phase Reference Cavity" SLAC TN-64-51, Sept. 1964 (unpublished).
13. R. Carr, *et al.*,"A Precision Measurement of the Weak Mixing Angle in Moller Scattering", K.S. Kumar, coordinator, SLAC-Proposal-E-158, 1997 (unpublished).
14. W. Bruns, "GdfidL: A finite difference program for arbitrarily small perturbations in rectangular geometries", *IEEE Trans. Magn.* 32, No. 3 (1996).
15. W.R. Fowkes, *et al.*, "Reduced field TE_{01} X-band travelling-wave window", *Proceedings of the 1995 Particle Accelerator Conference* (IEEE, New York, 1995) pp. 1587-1589, SLAC-PUB-6777.
16. E. Nelson, "Cavity Design Programs", *Proceedings of the US-CERN-Japan School on RF Engineering for Particle Accelerators*, (KEK, Tsukuba, to be published).
17. M. Chodorow, E. L. Ginzton, W. W. Hansen, R. L. Kyhl, R. B. Neal, W. K. H. Panofsky and The Staff, "Stanford High-Energy Linear Electron Accelerator (Mark III)", *Rev. Sci. Instrum.* 26 (1955) pp. 134-204.
18. E.L. Chu, "The Theory of Linear Electron Accelerators", Microwave Laboratory Report No. 140, Stanford University, 1951 (unpublished).
19. E.L. Ginzton, W.W. Hansen, and W.R. Kennedy, "A Linear Electron Accelerator", *Rev. Sci. Instrum.* 19 (1948) pp. 89-108.
20. E.A. Knapp, *et al.*, "Accelerating Structure Research at Los Alamos", *Proceedings of the 1966 Linear Accelerator Conference* (LANL, Los Alamos, 1966) pp. 83-87,

LA-3609.
21. R.H. Miller, "Comparison of standing-wave and travelling-wave structures", *1986 Linear Accelerator Conference Proceedings* (SLAC, Stanford, 1986) SLAC-R-303, pp. 200-205.
22. J.D. Jackson, *Classical Electrodynamics* (John Wiley & Sons, New York, 1975).
23. Penn Engineering Components, 12750 Raymer St., North Hollywood, CA, 91605. Tel: 818-503-1511, Fax: 818-764-0195.
24. R.E. Collin, *Foundations for Microwave Engineering* (McGraw-Hill, Singapore, 1966).
25. S.G. Tantawi, *et al.*, "Active High Power RF Pulse Compression Using Optically Switched Resonant Delay Lines", *Advanced Accelerator Concepts*, AIP Proc. 398 (AIP, New York, 1997) pp. 813-821.
26. F. Seitz, "Research on silicon and germanium in World War II", *Physics Today*, Jan. 1995, pp. 22-34.
27. W.R. Leo, *Techniques for Nuclear and Particle Physics Experiments* (Springer-Verlag, Berlin, 1994).
28. J. Rolfe, *et al.*, "Long ion chamber systems for the SLC" *Proceedings of the 1989 Particle Accelerator Conference* (IEEE, New York, 1989) pp. 1531-1533.
29. T. Kobayashi, *et al.*, "Development of a beam current monitor by using an amorphous magnetic core", *Nucl. Instrum. Meth.* B79 (1993) pp. 785-787.
30. C. Nantista and C. Adolphsen, "Beam current monitors in the NLCTA", *Proceedings of the 1997 Particle Accelerator Conference* (IEEE, New York, 1997) pp. 2256-2258, SLAC-PUB-7524.
31. C. Nantista, *et al.*, "Beam profile monitors in the NLCTA", *Proceedings of the 1997 Particle Accelerator Conference* (IEEE, New York, 1998) pp. 2186-2188, SLAC-PUB-7523.
32. S. Yencho and D.R. Walz, "A high-resolution phosphor screen beam profile monitor", *Proceedings of the 1985 Particle Accelerator Conference*, IEEE Trans. Nucl. Sci. NS-32 (1985) pp. 2009-2011.
33. M.C. Ross, *et al.*, "High resolution beam profile monitors in the SLC", *Proceedings of the 1985 Particle Accelerator Conference*, *IEEE Trans. Nucl. Sci.* NS-32 (1985) pp. 2003-2005.
34. J.T. Seeman, F.-J. Decker, I. Hsu, and C. Young, "Characterization and monitoring of transverse beam tails", *Proceedings of the 1991 Particle Accelerator Conference* (IEEE, New York, 1991) pp. 1734-1736.
35. M. Minty, *et al.*, "Using a fast-gated camera for measurements of transverse beam distributions and damping times", *Accelerator Instrumentation: Fourth Annual Workshop*, AIP Conf. Proc. 281 (AIP, New York, 1992) pp. 158-167, SLAC-PUB-5993.
36. J.C. Sheppard, *et al.*, "Real time bunch length measurements in the SLC linac", *Proceedings of the 1985 Particle Accelerator Conference*, *IEEE Trans. Nucl. Sci.* NS-32 (1985) pp. 2006-2008.
37. R.L. Holtzapple, *Longitudinal Dynamics at the Stanford Linear Collider*, Ph.D. Thesis, Stanford University, June 1996.
38. R.W. Coombes and D. Neet, "Beam monitors based on light observation for the beam switchyard of the Stanford Two-Mile Linear Accelerator", *IEEE Trans. Nucl. Sci.*, NS-14 (1967) pp. 1111-1115.
39. T. Shintake, *et al.*, "Design of laser-Compton spot size monitor", *Proceedings of the XVth International Conference on High-Energy Accelerators*, *Int. J. Mod. Phys. A* (Proc. Suppl.) 2A (1993), pp. 215-218.
40. S.R. Smith, "Beam position monitor engineering", *Proceedings of the 7th Beam Instrumentation Workshop*, AIP Conf. Proc. 390 (AIP, New York, 1996), SLAC-PUB-7244.
41. H. Hayano, J.-L. Pellegrin, S. Smith and S. Williams "High resolution BPM for the FFTB", *Nucl. Instrum. Meth.* A320 (1992) pp. 47-52.

42. D. McCormick and M. Ross, "Control of coaxial cable propagation delay for a beam phase monitor", *Accelerator Instrumentation: Fourth Annual Workshop*, AIP Conf. Proc. 281 (AIP, New York, 1992) pp. 256-263, SLAC-PUB-6297.
43. F. Zimmermann, *et al.*, "Bunch-length and beam-timing monitors in the SLC final focus", *Proceedings of the Advanced Accelerator Concepts Workshop* (AIP, to be published).
44. C. Field, D. McCormick, P. Raimondi, and M. Ross "Wire Breakage in SLC Wire Profile Monitors", *Proceedings of the Eight Beam Instrumentation Workshop*, (SLAC, Stanford, to be published) SLAC-PUB-7832.
45. G.E. Fischer, "Iron dominated magnets", *Physics of Particle Accelerators* AIP Conf. Proc. 153 (1987) 1120.
46. A.C. Melissinos, "Nicholas C. Christofilos: His Contributions to Physics", *Proceedings of the CERN Accelerator School: Advanced Accelerator Physics Course*, S. Turner, ed., (CERN, Geneva, 1995) pp. 1067-1081, CERN-95-06.
47. K.D. Jacobs, J.B. Flanz, and T. Russ, "Emittance measurement at the Bates Linac", *Proceedings of the 1989 Particle Accelerator Conference*, (IEEE, New York, 1989) pp. 1526-1528.
48. F. Zimmermann, "Measurement and Correction of Accelerator Optics", *Proceedings of the Joint US-CERN-Japan-Russia School on Beam Measurement*, Montreux, Switzerland, May 11-20, 1998 (CERN, Geneva, to be published), SLAC-PUB-7844.
49. K.L. Brown, "A conceptual design of final focus systems for linear colliders", *Frontiers of Particle Beams*, M. Month and S. Turner, eds. (Springer-Verlag, Berlin, 1988) pp. 481-494.
50. T.L. Lavine, *et al.*, "Beam-based alignment technique for the SLC linac", *Proceedings of the 1989 Particle Accelerator Conference*, (IEEE, New York, 1989) pp. 977-979, SLAC-PUB-4902.
51. The SLD Collaboration, represented by M.J. Fero "First results from SLD with polarized electron beam at SLAC", *The Third Family and the Physics of Flavor*, (SLAC, Stanford, 1992) pp. 341-358, SLAC-PUB-6027.
52. R. Assmann, *et al.*, "Accelerator physics highlights in the 1997/98 SLC run", *Proceedings of the 1st Asian Particle Accelerator Conference* (KEK, to be published).
53. E. Gero, *et al.*, "Beamstrahlung as an optics tuning tool at the SLC IP", Proc. 1989 Particle Accelerator Conference (IEEE, New York, 1989) pp. 1542-1544.
54. T. Himel, "Feedback: Theory and Accelerator Applications", *Ann. Rev. Nucl. Part. Sci.* 47 (1997) pp. 157-192, SLAC-PUB-7398.
55. P.B. Wilson, "High Energy Electron Linacs: Application to Storage Ring RF Systems and Linear Colliders", *Physics of High Energy Particle Accelerators*, AIP Conf. Proc. 87 (AIP, New York, 1982) pp. 452-563, SLAC-PUB-2884.
56. T. Wangler, *RF Linear Accelerators* (John Wiley & Sons, New York, 1998)
57. H. Padamsee, J. Knobloch, and T. Hays, *RF Superconductivity for Accelerators* (Wiley, New York, 1998).
58. H. Wiedemann, *Particle Accelerator Physics*(Springer-Verlag, Berlin, 1993).
59. D.C. Carey, *The Optics of Charged Particle Beams*, (Harwood, Chur, 1987)

QUARK MIXING AND CP VIOLATION

R. ALEKSAN
DAPNIA/SPP, Centre d'Etudes Nucléaires, Saclay, F-91191 Gif-sur-Yvette Cedex, France

ABSTRACT

The violation of the CP symmetry is a phenomenon, the origin of which is not yet well established. It deserves a particular attention since it may be a fundamental property of nature with very important consequences for the evolution of the universe. In these lectures, we propose to give an overview of this phenomenon as we understand and explain it today. After introducing briefly the discrete space-time symmetries, we review the various observations related to the violation of the CP symmetry in the neutral kaon decays. We derive the general formalism describing a system composed of a neutral particle and its antiparticle and discuss the implications of CP non conservation. We then give a brief overview of the mass generation mechanism for the fermions in the framework of the Standard Model and show how it leads to the quark mixing phenomenon. Using this latter mechanism, we explain how CP violation is generated and study in some detail what are the relevant measurements which can help constraining the theory. We derive some of the predictions which can be made with particular emphasis on the expected effects in the decays of mesons involving the b quark. We review the various possibilities for observing these effects and calculate their possible magnitude. We also show how the consistency of the theory can be tested and discuss the theoretical difficulties for extracting the fundamental parameters from the measurements. Finally, we outline the experimental difficulties and prospects for studying CP non conservation at various facilities with emphasis on asymmetric B Factories. Many lectures and reviews have been made on the subjects discussed here. Very recently, the BaBar collaboration has produced a Physics Book in collaboration of many theorists [1]. The reader may find there extensive studies on CP violation as well as on B and D physics which will complement these lectures in a most comprehensive way.

T. Ferbel (ed.), Techniques and Concepts of High Energy Physics X, 487–596.

1. Introduction

The elaboration of a theory in Physics is greatly simplified by requiring the conservation of various symmetries. The ones that are most commonly known, are the conservation of energy, momentum and spin. They are particulary simple to understand because they involve invariance principles of the physical laws under continuous transformations of the space-time referential (by translation in time, space or by rotation respectively). There exists an other category of conservation laws tightly connected to the dynamic of the interaction. The most familiar examples are the conservation of the (electric, baryonic, leptonic) charges and the conservation of the strong isospin. These laws do involve continuous transformations as well, however they act on abstract spaces such as phases or isospin.
Finally, some laws which are also associated to the dynamic of the interaction, involve discrete transformations such as charge conjugation, parity and time reversal. In these lectures, we will focus on the latter symmetries and will study their consequences in the framework of the weak interactions.

1.1. C, P AND T TRANSFORMATIONS.

Let us first briefly recall what these transformations are. The reader will find more details in the references [2].

1.1.1. *Charge Conjugation, C.*

The action of charge conjugation on a particle described by the wave function $|f(\vec{p},\vec{s})\rangle$ where $\vec{p}$ and $\vec{s}$ are its momentum and spin respectively, is to change the particle into its antiparticle without modifying $\vec{p}$ and $\vec{s}$. Here, the antiparticle is defined by changing the sign of all the charges associated to the particle (electric, baryonic, leptonic...). Hence

$$C\,|f(\vec{p},\vec{s})\rangle = \eta_C\,|\bar{f}(\vec{p},\vec{s})\rangle \tag{1}$$

where η_C is a phase factor.

1.1.2. *The Parity, P.*

The transformation P changes the space vector $\vec{r}$ into $-\vec{r}$. Therefore, $\vec{p} = m\,\frac{d\vec{r}}{dt}$ becomes $-\vec{p}$ while the orbital momentum $\vec{L} = \vec{r} \times \vec{p}$ remains unchanged. By analogy, the cinetic momentum $\vec{s}$ is not modified. Thus

$$P\,|f(\vec{p},\vec{s})\rangle = \eta_P\,|f(-\vec{p},\vec{s})\rangle \tag{2}$$

where η_P is a phase factor often known as the intrinsic parity. Let us note that it is sometime convenient to use the helicity $\lambda = \frac{\vec{P}}{|\vec{P}|} \cdot \vec{s}$ instead of the spin and therefore λ becomes $-\lambda$.

1.1.3. *The Time Reversal, T.*

Under this transformation, t is replaced by $-t$ and therefore $\vec{p}$ and $\vec{s}$ are modified into $-\vec{p}$ and $-\vec{s}$ while the helicity is unaffected. Time reversal is defined as

$$T\,|f(\vec{p},\vec{s})\rangle = \eta_T^s\,|f(-\vec{p},-\vec{s})\rangle^* \tag{3}$$

where η_T^s is a phase factor depending on the spin. Not only $\vec{p}$ and $\vec{s}$ are modified under the transformation T but the wave function is also replaced by it complex conjugate. This is due to the fact that $\psi(\vec{r},t) = e^{\frac{i}{\hbar}(\vec{p}\cdot\vec{r}-Et)}$ satisfy the Schrödinger equation

$$i\hbar\,\frac{\partial\psi(\vec{r},t)}{dt} = -\frac{\hbar^2}{2m}\,\nabla^2\psi(\vec{r},t) \tag{4}$$

with $E = \frac{\vec{p}^2}{2m}$ while $\psi(\vec{r},-t)$ does not. The same solution is obtained by using $\psi(\vec{r},-t)^*$ which describes the same particle moving in the opposite direction. As a consequence, we can deduce that T is antiunitary since $\langle T\varphi|T\psi\rangle = \langle\psi|\varphi\rangle$.

1.2. CONSERVATION OF THE DISCRETE SYMMETRIES.

1.2.1. *The CPT Symmetry*

Using very general hypothesis based on locality and causality, it can be demonstrated [2] that field theories are conserved when the product C·P·T is applied. Important consequences are derived from this global symmetry; in particular the mass and the lifetime of a particle and its antiparticle are equal. So far no violation of this general symmetry has been observed from the measurements of the mass difference for particles and antiparticles. Some examples of experimental tests of CPT are

$$|m_{e^+} - m_{e^-}| \leq 0.02 \text{ eV}\ ,\ |m_P - m_{\bar{P}}| \leq 1 \text{ eV}\ ,\ |m_K - m_{\bar{K}}| \leq 3.6\ 10^{-10} \text{ eV} \tag{5}$$

1.2.2. *Conservation of Charge Conjugation.*

The conservation of C in the strong interactions is very difficult to test because one has to study systems in which the initial particles can be replaced by their antiparticles. Since this is technically not very easy to achieve, most of the tests are made by comparing the angular distributions of the charged pions in the final state (by example in the reaction $\overline{p} + p \rightarrow$

$\pi^+\pi^-\pi^o$...). From the observation $\frac{d\sigma}{d\Omega_{\pi^+}} = \frac{d\sigma}{d\Omega_{\pi^-}}$, one can deduce that the symmetry C is conserved at the percent level.
It is easier to verify the conservation of C in the electromagnetic interactions. A good test involves for example the decay of the meson η into $\mu^+\mu^-\pi^o$. The $\mu^+\mu^-$ pair is produced through electromagnetic interactions via the exchange of a virtual photon and is forbidden if C is conserved. No such final state is not observed and therefore an upper limit on the violation of C in the electromagnetic interaction is set at 5 10^{-6} level.

1.2.3. *Parity Conservation.*
A very precise test of Parity conservation is carried out by studing the reaction

$$\begin{array}{ccccc} p + F^{19} & \rightarrow & Ne^{20*} & \rightarrow & 0^{16} + \alpha \\ & & 1^+ & & 0^+ \quad 0^+ \end{array} \tag{6}$$

This reaction is forbidden if P is conserved. No such transition is observed leading to an upper limit of the order of 10^{-12} on the violation of Parity in the strong interactions.

1.2.4. *Conservation of Time Reversal.*
The simplest way to test this symmetry is to compare the process $a + b \rightarrow c + d$ to its time reversed one $c + d \rightarrow a + b$. As an example, the comparison of the reactions

$$p + Al^{27} \rightarrow Mg^{24} + \alpha \quad \text{and} \quad \alpha + Mg^{24} \rightarrow Al^{27} + p \tag{7}$$

allows one to obtain an upper limit of the order of 4 10^{-3} on the amplitude violating T is the strong interactions [3].
The reader may find a more complete list of tests concerning the symmetries C and P in [4].

1.3. VIOLATION OF C AND P SYMMETRIES IN THE WEAK INTERACTIONS: AN HISTORICAL REMAINDER

We have seen that Charge Conjugation and Parity are conserved in the strong and electromagnetic interactions within the experimental errors. However, one has to acknowledge that the world in which we live is made of matter, and therefore C had to be violated at some time scale. This fact is important enough to motivate the search of an interaction that does not conserve this symmetry. Weak interactions are playing a very particular role in this regard as we will see in the following.

1.3.1. *The Disintegration of the Meson $\pi^{\pm}$.*

Let us consider the decay of the π^+ and π^- mesons into a charged lepton and a neutrino. Since the lifetime of the pion is very long ($\sim 2.6\ 10^{-8}s$), one deduces that weak interaction is responsible for this decay. This is even more obvious considering the presence of a neutrino which is only produced through weak interactions. Since the π meson is a spin 0 boson and its decay products are spin 1/2 particles, we have, in principle, two possibilities as shown in the figure 1. The modes in the second line are

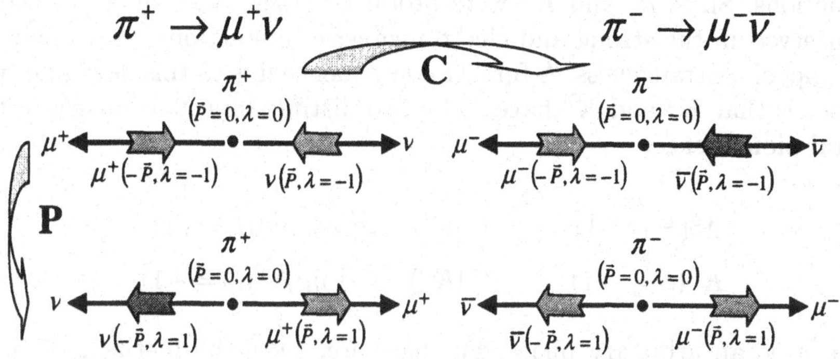

Figure 1. The two theoretical possibilities for the $\pi^{\pm}$ meson decay into a muon and a neutrino.

obtained by applying the Parity operation to the modes shown in the first line. Similarly, we can get the modes in the second column from the ones in the first column by using charge conjugation. Experiments show that only the first reaction in line 1 and the second one in line 2 exist in nature (the helicity of the muon is measured through the angular distribution of the electron produced in its decay). We therefore conclude that both Charge Conjugation and Parity are not conserved in the weak interactions. In fact, the C and P symmetries are both violated in a maximal way. However when C and P are applied consecutively, the CP transformation seems to be conserved. Let us note that the violation of Parity in the weak interactions has not been discovered using pion decays, but by studying the β disintegration of the Co^{60} nucleus [5].

1.3.2. *Strangeness.*

In the early fifties, "strange" phenomena were observed with the discovery of two new particles, the Λ^o baryon and the K^o meson. Both their lifetimes and the conditions in which they were produced, seemed somewhat peculiar.

- Both particles have very long lifetimes. For example the lifetime of Λ^o is of the order of 2.6 $10^{-10}s$ which is 10^{14} times larger than the one of the Δ baryon.
- Λ^o's were produced in association with the K^o particle.

$$\pi^- p \to \Lambda^o K^o + X \tag{8}$$

In order to explain these observations, Gell-man proposed to assign a new "charge" to these particles which he called Strangeness. He assumed that this new quantum number was conserved in the strong and electromagnetic interactions. Since Λ^o and K^o were produced simultaneously and since C is conserved in the strong and electromagnetic interactions, they ought to have opposite Strangeness. A further consequence due to this new quantum number is that K^o and $\overline{K^o}$ have to be two distinct objects contrary to the neutral pion. Hence,

$$\begin{aligned} \Lambda^o(S=-1) \quad &, \quad C|\Lambda^o\rangle = \eta_\Lambda \overline{\Lambda^o}(S=+1) \\ K^o(S=+1) \quad &, \quad C|K^o\rangle = \eta_K \overline{K^o}(S=-1) \end{aligned} \tag{9}$$

where η_i is an arbitrary phase. Furthermore, since the lifetimes of these particles are long, their disintegration is due to the weak interaction which therefore does not conserve Strangeness. The picture then becomes consistent.

But even more interesting, Gell-man and Païs [6] predicted that K^o and $\overline{K^o}$ have to be mixed (that is to say, K^o can oscillate into $\overline{K^o}$ and vice versa). The argumentation is the following. Since both K^o and $\overline{K^o}$ have a common final state $\pi^+\pi^-$, nothing prevents the process :

$$\underbrace{\underline{K^o \longrightarrow \pi^+\pi^- \longrightarrow \overline{K^o}}}_{K^o-\overline{K^o}\ \text{mixing}} \tag{10}$$

Gell-man and Païs made the conjecture that K^o and $\overline{K^o}$ are not the physical states (the mass eigenstates) but that these latter are superpositions of K^o and $\overline{K^o}$.

$$|K_1\rangle = \frac{1}{\sqrt{2}}\left[|K^o\rangle + |\overline{K^o}\rangle\right] \quad , \quad |K_2\rangle = \frac{1}{\sqrt{2}}\left[|K^o\rangle - |\overline{K^o}\rangle\right] \tag{11}$$

It is easy to verify that $|K_1\rangle$ and $|K_2\rangle$ are indeed CP eigenstates using the convention $CP|K^o\rangle = |\overline{K^o}\rangle$

$$CP|K_1\rangle = +|K_1\rangle \quad \text{and} \quad CP|K_2\rangle = -|K_2\rangle \tag{12}$$

We have previously said incorrectly that K^o and $\overline{K^o}$ decay into $\pi^+\pi^-$. Instead, one should have said that the physical states K_1 and K_2 decay into $\pi^+\pi^-$. However, since the $\pi^+\pi^-$ system is a CP eigenstate (CP $|\pi^+\pi^-\rangle = (-1)^{2l}\,|\pi^+\pi^-\rangle = +\,|\pi^+\pi^-\rangle$) with the CP parity +1, we conclude that only $K_1 \to \pi^+\pi^-$ is allowed if CP is conserved in the weak interactions. The decay $K_2 \to \pi^+\pi^-$ being forbidden, only more complicated final states (such as the 3 body decay $\pi^+\pi^-\pi^o$) can occur. As these final states are strongly suppressed due to the limited phase space ($M_K \approx 500$ MeV), the lifetime of K_2 should be much longer than the one of K_1. Indeed, this is what is observed and this is why K_1 and K_2 are commonly called K_S (Short) and K_L (Long).

$$\tau_{K_S} \approx 9.10^{-11} s \quad , \quad \tau_{K_L} \approx 5.10^{-8} s \tag{13}$$

1.4. CP NON CONSERVATION.

It is very fortunate that the lifetimes of K_S and K_L are so different. The main consequence is that even though one starts with a pure K^o beam, only the K_L component will remain if one "waits" long enough ($t \gg \frac{\hbar}{\Gamma_S}$). Indeed, since K_S and K_L are solutions of the Schrödinger equation, the time dependent wave functions are

$$|K_S^o(t)\rangle \;=\; e^{-(im_S+\Gamma_S/2)t}\,|K_S\rangle \quad , \quad |K_L^o(t)\rangle \;=\; e^{-(im_L+\Gamma_L/2)t}\,|K_L\rangle \tag{14}$$

where m_S, Γ_S and m_L, Γ_L are the mass and the width of K_S and K_L respectively. One can therefore write

$$|K^o(t)\rangle = \frac{e^{-(im_S+\Gamma_S/2)t}\,|K_S\rangle \;+\; e^{-(im_L+\Gamma_L/2)t}\,|K_L\rangle}{\sqrt{2}} \tag{15}$$

and the probability of the transition $K^o \to f$ is

$$\begin{aligned} |\langle f|T|K^o(t)\rangle|^2 &= \frac{1}{2}e^{-\Gamma_S t}\,|\langle f|T|K_S\rangle|^2 + \frac{1}{2}e^{-\Gamma_L t}\,|\langle f|T|K_L\rangle|^2 \\ &+ e^{-\left(\frac{\Gamma_S+\Gamma_L}{2}\right)t}\cos(\Delta m_K t - \Delta\varphi)\;|\langle f|T|K_S\rangle\,\langle f|T|K_L\rangle^*| \end{aligned} \tag{16}$$

where $\Delta m_K = m_L - m_S$ and $\Delta\varphi = \varphi_L - \varphi_S$. The phases of the strong interactions in the final states are denoted by φ_S and φ_L :

$$\langle f|T|K_S\rangle \;=\; e^{i\varphi_S}\,|\langle f|T|K_S\rangle| \quad , \quad \langle f|T|K_L\rangle \;=\; e^{i\varphi_L}\,|\langle f|T|K_L\rangle| \tag{17}$$

From Eq. 16, one concludes that for $t \gg \frac{\hbar}{\Gamma_S}$, only the transition $\langle f|T|K_L\rangle$ can be observed. For the intermediate times, one may also observe the

interference term if both the transitions $K_S \to f$ and $K_L \to f$ are allowed. It is easy to verify that for an initial $\overline{K^o}$, a similar formula is obtained with the exception of the interference term which will have the opposite sign. Since it is possible to obtain a pure K_L beam, it is "easy" to verify whether CP is conserved by searching for the forbidden $K_L \to \pi^+\pi^-$ transition.
In 1964, this transition was observed [7], establishing that CP is violated very slightly ($\sim 2\ 10^{-3}$ in amplitude) in the K^o decays. Thus, the equivalence between the CP eigenstates and the mass eigenstates for the K^o-$\overline{K^o}$ system has to be reconsidered. This deviation being very small, one writes

$$|K_S\rangle = \frac{|K_1\rangle + \varepsilon_K |K_2\rangle}{\sqrt{1+|\varepsilon_K|^2}} \quad , \quad |K_L\rangle = \frac{|K_2\rangle + \varepsilon_K |K_1\rangle}{\sqrt{1+|\varepsilon_K|^2}} \tag{18}$$

where ε_K is a complex number governing the strength of CP violation. During the following years, several experiments have measured this parameter more precisely and its present value [4] is $|\varepsilon_K| = (2.285 \pm 0.019) \times 10^{-3}$ (neglecting direct CP violation which is discussed later).
It will be useful for the following to write K_S and K_L in terms of K^o and $\overline{K^o}$. Using Eqs. 18 and 11, one gets

$$\left|K_{S(L)}\right\rangle = \frac{1}{\sqrt{2(1+|\varepsilon_K|^2)}} \left[(1+\varepsilon_K)\,|K^o\rangle \overset{(-)}{+} (1-\varepsilon_K)\left|\overline{K^o}\right\rangle \right] \tag{19}$$

1.5. THE CP VIOLATION MEASUREMENTS

The violation of the CP symmetry has now been observed in several decays. However all these decays involve the K^o particle. Therefore there is no evidence so far that this phenomenon is indeed a fundamental property of the weak interactions. We give in Table 1 a list of the different measurements together with the parameters which they allow one to extract.
All the above parameters can be measured either using a pure K_L state or a coherent superposition of K_S and K_L states.

1.5.1. *A Pure K_L Beam*

Two type of measurements can be made to observe CP violation with a K_L beam:

- either one compares the decay rate $K_L \to f$ to its CP conjugate one
- or one measures the K_L decay rate to a forbidden final state

K_L Decays to CP Conjugate Final States.
Let us compare the rate of positively charged leptons to the rate of negatively charged ones in the semileptonic decays of K_L and evaluate the

Studied quantity	Measured parameters
$\frac{\Gamma(K_L^o\to\ell^+\nu\pi^-)-\Gamma(K_L^o\to\ell^-\nu\pi^+)}{\Gamma(K_L^o\to\ell^+\nu\pi^-)+\Gamma(K_L^o\to\ell^-\nu\pi^+)} = \delta(\ell)$	$\delta(\ell) = (0.327 \pm 0.012)\%$
$\frac{A(K_L^o\to\pi^+\pi^-)}{A(K_S^o\to\pi^+\pi^-)} = \lvert\eta_{+-}\rvert e^{i\Phi_{+-}}$	$\lvert\eta_{+-}\rvert = (2.285 \pm 0.019) \times 10^{-3}$, $\Phi_{+-} = (43.7 \pm 0.6)^\circ$
$\frac{A(K_L^o\to\pi^o\pi^o)}{A(K_S^o\to\pi^o\pi^o)} = \lvert\eta_{oo}\rvert e^{i\Phi_{oo}}$	$\lvert\eta_{oo}\rvert = (2.275 \pm 0.019) \times 10^{-3}$, $\Phi_{+-} = (43.5 \pm 1.0)^\circ$
$\frac{A(K_L^o\to\pi^+\pi^-\gamma)}{A(K_S^o\to\pi^+\pi^-\gamma)} = \lvert\eta_{+-\gamma}\rvert e^{i\Phi_{+-\gamma}}$	$\lvert\eta_{+-\gamma}\rvert = (2.35 \pm 0.07) \times 10^{-3}$, $\Phi_{+-\gamma} = (44 \pm 4)^\circ$
$\frac{\Gamma(\overline{K^o}\to\pi^+\pi^-)-\Gamma(K^o\to\pi^+\pi^+)}{\Gamma(\overline{K^o}\to\pi^+\pi^-)+\Gamma(K_L^o\to\pi^+\pi^+)}$	$\lvert\eta_{+-}\rvert = (2.254 \pm 0.034) \times 10^{-3}$
$\alpha\ \lvert\eta_{+-}\rvert cos(\Delta m_K t + \Phi_{+-})$	$\Phi_{+-} = (43.6 \pm 0.5)^\circ$

TABLE 1. Reactions and measurements from which the violation of CP is inferred.

asymmetry

$$\delta(\ell) \; = \; \frac{\Gamma\left(K_L \to \ell^+\nu_\ell\pi^-\right) - \Gamma\left(K_L \to \ell^-\nu_\ell\pi^+\right)}{\Gamma\left(K_L \to \ell^+\nu_\ell\pi^-\right) + \Gamma\left(K_L \to \ell^-\nu_\ell\pi^+\right)} \tag{20}$$

From Eq. 19, one gets

$$\begin{aligned}\Gamma(K_L \to \ell^+\nu_\ell\pi^-) \;&\propto\; |(1+\varepsilon_K)\langle\ell^+\nu_\ell\pi^-|K^o\rangle - (1-\varepsilon_K)\langle\ell^+\nu_\ell\pi^-|\overline{K^o}\rangle|^2 \\ \Gamma(K_L \to \ell^-\nu_\ell\pi^+) \;&\propto\; |(1+\varepsilon_K)\langle\ell^-\nu_\ell\pi^+|K^o\rangle - (1-\varepsilon_K)\langle\ell^-\nu_\ell\pi^+|\overline{K^o}\rangle|^2\end{aligned} \tag{21}$$

Using the $\Delta S = \Delta Q$ rule which implies that $\langle\ell^-\nu_\ell\pi^+|K^o\rangle = \langle\ell^+\nu_\ell\pi^-|\overline{K^o}\rangle = 0$, one can easily verify that $\mathrm{Re}\varepsilon_K$ is obtained from the measurement of $\delta(\ell)$,

$$\delta(\ell) \; = \; \frac{2\mathrm{Re}\varepsilon_K}{1+|\varepsilon_K|^2} \tag{22}$$

From the present world average of this asymmetry ($\delta(\ell) = (3.27 \pm 0.12) \times 10^{-3}$ [4]) and the measurement of $|\varepsilon_K|$ discussed in the next subsection, one gets $\mathrm{Re}\varepsilon_\mathrm{K} = (1.63 \pm 0.06) \times 10^{-3}$. In addition, one deduces that the phase of ε_K is about $\pi/4$.

K_L Decays to Forbidden Final States.

Using Eq. 19, the ratio of the decay amplitudes for $K_{S(L)} \to \pi\pi$ is

$$\eta \; = \; \frac{A\left(K_L \to \pi\pi\right)}{A\left(K_S \to \pi\pi\right)} \; = \; \frac{(1+\varepsilon_K)\langle\pi\pi|K^o\rangle - (1-\varepsilon_K)\langle\pi\pi|\overline{K^o}\rangle}{(1+\varepsilon_K)\langle\pi\pi|K^o\rangle + (1-\varepsilon_K)\langle\pi\pi|\overline{K^o}\rangle} \tag{23}$$

Here, two possibilities have to be considered.

$\langle\pi\pi|K^o\rangle = \langle\pi\pi|\overline{K^o}\rangle$.

This is equivalent to say that *no* CP violation occurs in the direct decay $K^o \to \pi\pi$. It is then straightforward to deduce from Eq. 23

$$|\eta| \; = \; |\varepsilon_K| \tag{24}$$

Therefore both the measurements of $|\eta^{+-}| = (2.285 \pm 0.019) \times 10^{-3}$ and $|\eta^{oo}| = (2.275 \pm 0.019) \times 10^{-3}$ give directly the value of ε_K and CP violation is solely due to the $K^o - \overline{K^o}$ mixing, i.e

$$K^o \to \overline{K^o} \quad \not\equiv \quad \overline{K^o} \to K^o \tag{25}$$

The origin of CP violation in the K^o system might then come from a new interaction with $\Delta S = 2$. This conjecture is called the "Superweak" model [8].

$\langle\pi\pi|K^o\rangle \neq \langle\pi\pi|\overline{K^o}\rangle$.

This is possible if there are two amplidudes contributing to the direct decay $K^o \to \pi\pi$. This is indeed the case considering the different isospin states $A\,(K^o \to \pi\pi, I = 0)$ and $A\,(K^o \to \pi\pi, I = 2)$. In fact, if one assumes that the interaction responsible of the kaon decay (i.e. weak interaction) violates the CP symmetry and if (as we will see later) the K^o-$\overline{K^o}$ mixing can be generated due to a mechanism involving also the weak interaction, it would be natural to expect also a difference between the transition $K^o \to \overline{K^o}$ and $\overline{K^o} \to K^o$.

In order to see how the direct K^o decay might reveal CP violation, let us write the different isospin amplitudes :

$$\left.\begin{array}{l} A\,(K^o \to \pi\pi, I = 0) \\ \equiv A_0 = a_0 e^{i\Phi_0} e^{i\delta_0} \end{array}\right\} \xrightarrow{CP} \left\{\begin{array}{l} A\left(\overline{K^o} \to \pi\pi, I = 0\right) \\ \equiv \overline{A_0} = a_0 e^{i\overline{\Phi_0}} e^{i\overline{\delta_0}} \end{array}\right.$$

$$\left.\begin{array}{l} A\,(K^o \to \pi\pi, I = 2) \\ \equiv A_2 = a_2 e^{i\Phi_2} e^{i\delta_2} \end{array}\right\} \xrightarrow{CP} \left\{\begin{array}{l} A\left(\overline{K^o} \to \pi\pi, I = 2\right) \\ \equiv \overline{A_2} = a_2 e^{i\overline{\Phi_2}} e^{i\overline{\delta_2}} \end{array}\right. \tag{26}$$

where Φ and $\overline{\Phi}$ are the weak phases and δ and $\overline{\delta}$ are the final state strong phases. Should CP be violated in the weak interactions, one would get $\overline{\Phi} = -\Phi$. Similarly one would have $\overline{\delta} = -\delta$ if CP is violated in the strong interactions.

Let us consider the particular final state $\pi^+\pi^-$.

$$\begin{array}{rcl} A\,(K^o \to \pi^+\pi^-) & = & \sqrt{\frac{2}{3}} a_0 e^{i\Phi_0} e^{i\delta_0} + \sqrt{\frac{1}{3}} a_2 e^{i\Phi_2} e^{i\delta_2} \\ A\left(\overline{K^o} \to \pi^+\pi^-\right) & = & \sqrt{\frac{2}{3}} a_0 e^{i\overline{\Phi_0}} e^{i\overline{\delta_0}} + \sqrt{\frac{1}{3}} a_2 e^{i\overline{\Phi_2}} e^{i\overline{\delta_2}} \end{array} \tag{27}$$

where the numerical factors are Clebsch-Gordan coefficients. Assuming that CP is **only** violated in the weak interactions ($\overline{\Phi} = -\Phi$) and using Eq. 23, one gets

$$\eta^{+-} = \frac{i\left[\sqrt{\frac{2}{3}}a_0\sin\Phi_0+\sqrt{\frac{1}{3}}a_2e^{i(\delta_2-\delta_0)}\sin\Phi_2\right]+\varepsilon_K\left[\sqrt{\frac{2}{3}}a_0\cos\Phi_0+\sqrt{\frac{1}{3}}a_2e^{i(\delta_2-\delta_0)}\cos\Phi_2\right]}{\left[\sqrt{\frac{2}{3}}a_0\cos\Phi_0+\sqrt{\frac{1}{3}}a_2e^{i(\delta_2-\delta_0)}\cos\Phi_2\right]+i\varepsilon_K\left[\sqrt{\frac{2}{3}}a_0\sin\Phi_0+\sqrt{\frac{1}{3}}a_2e^{i(\delta_2-\delta_0)}\sin\Phi_2\right]} \tag{28}$$

This equation can be simplified since:

- Only $|\eta^{+-}|$ is measured. Therefore one can define an arbitrary global phase as it does not modify the measure parameter. We choose $\Phi_0 = 0$.
- The isospin 2 amplitude is suppressed ($\Delta I = 1/2$ rule). Thus $a_2 \ll a_0$ ($a_2 \simeq a_0/20$).

Using the simplification above, one obtains:

$$\eta^{+-} = \varepsilon_K + \underbrace{\frac{i}{\sqrt{2}}e^{i(\delta_2-\delta_0)}\frac{a_2}{a_0}\sin\Phi_2}_{\varepsilon'_K} \tag{29}$$

where $|\varepsilon'_K|$ is the observable that exhibits **direct CP violation**. A similar calculation for $K^o \to \pi^0\pi^0$ gives

$$\eta^{oo} = \varepsilon_K - 2\varepsilon'_K \tag{30}$$

and therefore by measuring the double ratio $|\eta^{oo}|^2/|\eta^{+-}|^2$, one extracts

$$Re\frac{\varepsilon'_K}{\varepsilon_K} = \frac{1 - \frac{|\eta^{oo}|^2}{|\eta^{+-}|^2}}{6} \tag{31}$$

Unfortunately, the present experimental measurements do not allow to establish direct CP violation:

$$Re\frac{\varepsilon'_K}{\varepsilon_K} = \begin{array}{ll} (2.3 \pm 0.7) \times 10^{-3} & [9] \\ (0.74 \pm 0.59) \times 10^{-3} & [10] \end{array} \tag{32}$$

and therefore as long as one does not observe a clear difference between the amplitude ratios

$$\eta^{\pm} = \frac{A(K_L \to \pi^+\pi^-)}{A(K_S \to \pi^+\pi^-)} \stackrel{?}{\neq} \frac{A(K_L \to \pi^o\pi^o)}{A(K_S \to \pi^o\pi^o)} = \eta^{oo}, \tag{33}$$

it is neither possible to attribute unambiguously the origin of CP violation in the kaon system to the "Superweak" model nor to the weak interactions.

We will see in the chapter 3 how to test these two models using a different system. The reader can get more details for the kaon system in the references [11].

1.5.2. *Coherent Superposition of K_S and K_L States*

Two complementary methods have been used to measure CP violation by studying the interference between the K_S and K_L component of a "K^o" beam.

- Starting from a K_L beam, it is possible to reintroduce a K_S component by regeneration. The idea is to let the beam traverse a thin foil on material. Because the elastic cross section is different for K^o and $\overline{K^o}$ $[\sigma(\overline{K^o}p) \rightarrow \overline{K^o}p) > \sigma(K^o p \rightarrow K^o p)]$, one obtains the superposition

$$|K_L\rangle \rightarrow |K_L\rangle + r|K_S\rangle \tag{34}$$

  where r is the amount of regenerated K_S.
- Alternatively, one could use directly a K^o and/or a $\overline{K^o}$ "beam" which are already superposition of K_S and K_L.

$$|K^o\rangle(|\overline{K^o}\rangle)\ \alpha\ (1 \mp \epsilon_K)[|K_S\rangle \pm |K_L\rangle] \tag{35}$$

  This is equivalent of setting $|r| = 1$. K^o "beam" are obtained for example by using the interactions $p\bar{p} \rightarrow \mathrm{K}^-(\mathrm{K}^+)\pi^+(\pi^-)\mathrm{K}^o(\overline{\mathrm{K}^o})$ where the flavor of the neutral kaon is deduced from the sign of the charged kaon.

Assuming CPT conservation and writing

$$\frac{A\,(K_L \rightarrow f_{CP})}{A\,(K_S \rightarrow f_{CP})} = |\eta_{f_{CP}}|e^{i\Delta\phi_{f_{CP}}} \tag{36}$$

one gets

$$\text{Rate}\,(``K^{o"}(t) \rightarrow f_{CP}) =$$
$$|r|^2 e^{-\Gamma_S t} + |\eta_{f_{CP}}|^2 e^{-\Gamma_L t} + 2|r\eta_{f_{CP}}|e^{-\left(\frac{\Gamma_S+\Gamma_L}{2}\right)t}\cos(\Delta m_K t - \Delta\phi_{f_{CP}} + \phi_r) \tag{37}$$

where ϕ_r is the regeneration phase.

Using the above formalism one can extract $|\eta_{f_{CP}}|$ and $\Delta\phi_{f_{CP}}$ by fitting the interference term from the data, ϕ_r being known from theoretical calculation or other experimental measurements. From the fits in figure 2, E773 [12] measures $\Delta\phi_{+-} = 43.53^\circ \pm 0.58^\circ \pm 0.49^\circ$ and $\Delta\phi_{oo} - \Delta\phi_{+-} = 0.62^\circ \pm 0.71^\circ \pm 0.75^\circ$ It should be noted that the fit results exhibit a dependence on Δm_K and τ_S, the K_S lifetime. The value of these parameter

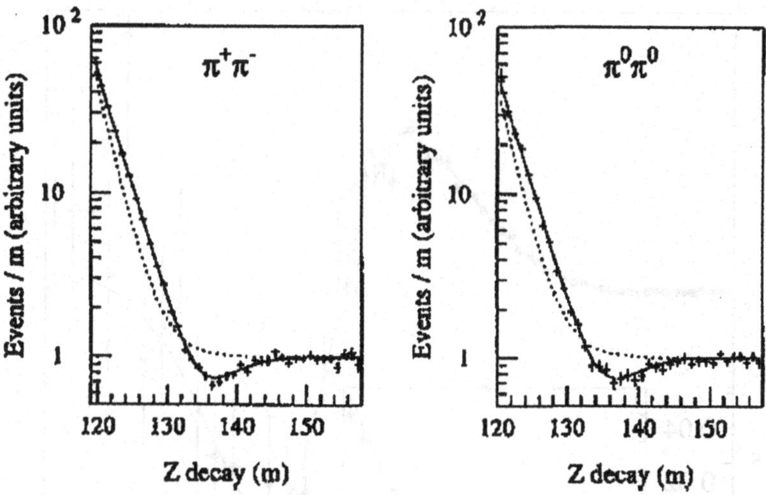

Figure 2. The decay distributions for "K^o"$\rightarrow \pi^+\pi^-$ and "K^o"$\rightarrow \pi^o\pi^o$ as function of the decay position from the E773 experiment. The solid (dotted) line is a fit with (without) the interference term.

could however be extracted when leaving them as free parameters in the fit.

1.5.3. *Time Dependent Asymmetry*

Finally, it is interesting to discuss a method involving the measurement of a time dependent asymmetry using K^o and $\overline{K^o}$ mesons. The time dependent asymmetry which has to be measured is

$$\mathcal{A}_{\pi\pi}(t) = \frac{\text{Rate}\left(\overline{K^o}(t) \rightarrow \pi^+\pi^-\right) - \text{Rate}\left(K^o(t) \rightarrow \pi^+\pi^-\right)}{\text{Rate}\left(\overline{K^o}(t) \rightarrow \pi^+\pi^-\right) + \text{Rate}\left(K^o(t) \rightarrow \pi^+\pi^-\right)} \tag{38}$$

The individual time dependent rates can be easily obtained from Eq. 16 by introducing CP violation.

$$\begin{aligned} &\text{Rate}\left(\overset{(-)}{K^o}(t) \rightarrow \pi^+\pi^-\right) = \\ &\frac{1}{2}\left(1 \overset{(+)}{-} 2\text{Re}\varepsilon_K\right)\left\{e^{-\Gamma_S t}\left|\langle\pi^+\pi^-|K_S\rangle\right|^2 + e^{-\Gamma_L t}\left|\langle\pi^+\pi^-|K_L\rangle\right|^2\right. \\ &\left.\overset{(-)}{+} 2e^{-\left(\frac{\Gamma_S+\Gamma_L}{2}\right)t}\cos\left(\Delta m_K t - \Delta\phi^{+-}\right)\left|\langle\pi^+\pi^-|K_S\rangle\langle\pi^+\pi^-|K_L\rangle^*\right|\right\} \end{aligned} \tag{39}$$

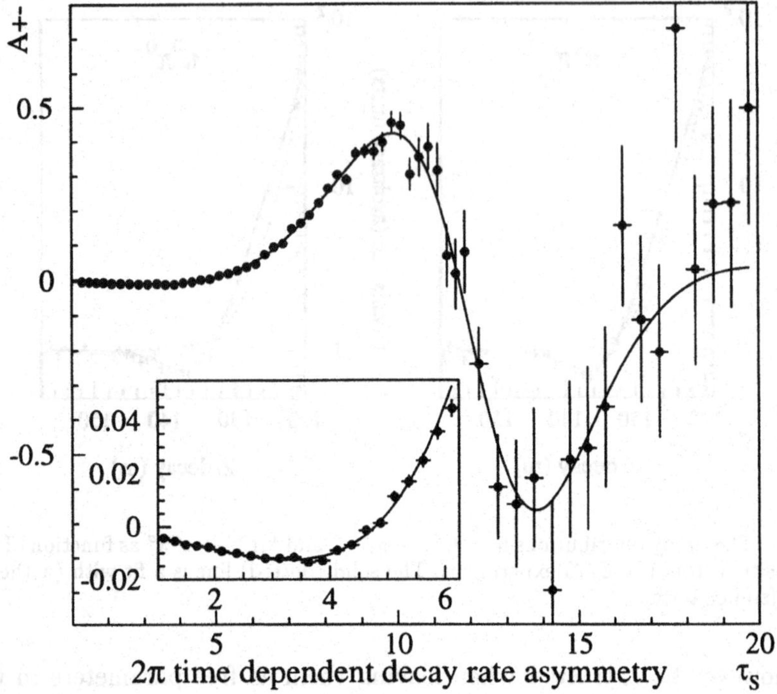

Figure 3. **The time dependent asymmetry $\mathcal{A}_{\pi^+\pi^-}(t)$ as observed by the CPLEAR experiment at CERN.**

The time dependent asymmetry then reads

$$\mathcal{A}_{\pi^+\pi^-}(t) \simeq 2\mathrm{Re}\varepsilon_K - \frac{2|\eta^{+-}|e^{\left(\frac{\Gamma_S-\Gamma_L}{2}\right)t}\cos\left(\Delta m_K t - \Delta\phi^{+-}\right)}{1 + |\eta^{+-}|^2 e^{(\Gamma_S-\Gamma_L)t}} \tag{40}$$

This asymmetry has been measured by the CPLEAR experiment [13] (see figure 3) and the value $|\eta^{+-}| = (2.254 \pm 0.024 \pm 0.024) \times 10^{-3}$ is deduced from the fit.

1.6. ARE CP AND T VIOLATIONS EQUIVALENT ?

If one assumes that CPT is conserved, then the time reversal symmetry T should be violated with the same strength as CP violation. The comparison of the transition $K^o(t=0) \to \overline{K^o}(t)$ and $\overline{K^o}(t=0) \to K^o(t)$ tests the T symmetry:

$$A_T(t) = \frac{\mathcal{P}(\overline{K^o} \to K^o) - \mathcal{P}(K^o \to \overline{K^o})}{\mathcal{P}(\overline{K^o} \to K^o) + \mathcal{P}(K^o \to \overline{K^o})} \tag{41}$$

This measurement requires the knowledge of the nature of the initial kaon (K^o or $\overline{K^o}$). As already said, CPLEAR is able to get this information by tagging the neutral kaon with the charge of the accompanying $K^{\pm}$. The

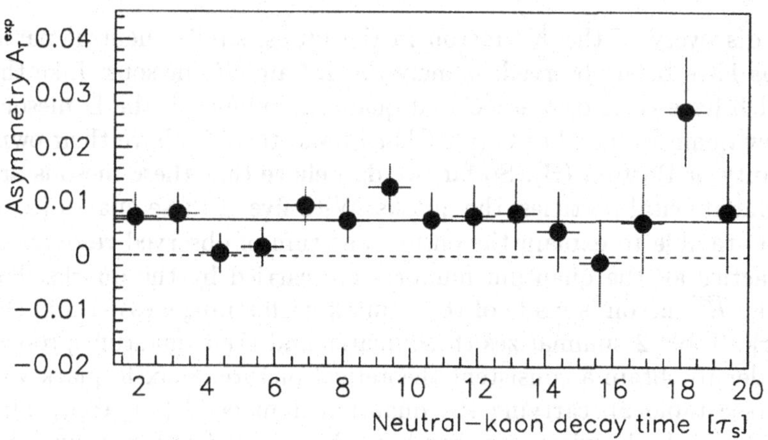

Figure 4. The asymmetry A_T showing a first evidence for the violation of the Time Reversal symmetry using semileptonic K decays. The solid line is a straight line fit to the data.

asymmetry $A_T(t)$ has been measured by CPLEAR [14] using the semileptonic decays. Defining $\bar{R}^{\pm}(t) \equiv \overline{K^o}(t=0) \to \ell^{\pm}\nu\pi^{\mp}(t)$ and $R^{\pm}(t) \equiv K^o(t=0) \to \ell^{\pm}\bar{\nu}\pi^{\mp}(t)$, they obtain

$$\begin{aligned} A_T(t) &= \frac{\bar{R}^+(t) - R^-(t)}{\bar{R}^+(t) + R^-(t)} \\ &= 4Re(\epsilon_K) - \frac{4Im(x)\sin(\Delta m_K t)e^{-\frac{\Gamma_S+\Gamma_L}{2}t}}{e^{-\Gamma_S t} + e^{-\Gamma_L t} - 2\cos(\Delta m_K t)e^{-\frac{\Gamma_S+\Gamma_L}{2}t}} \end{aligned}$$

$$= \quad (6.6 \pm 1.3_{\text{stat}} \pm 1.0_{\text{syst}}) \times 10^{-3} \tag{42}$$

Here $x = \mathcal{A}(\Delta S = -\Delta Q)/\mathcal{A}(\Delta S = \Delta Q)$ quantifies a possible violation of the $\Delta S = \Delta Q$ rule. First evidence for T violation is therefore observed at the 4.0 σ level(Fig. 4). The measured asymmetry is in agreement with its expected value: $A_T \simeq 2\delta(\ell) = (6.54 \pm 0.24) \times 10^{-3}$ [4]. Using $A_T(t)$ and the asymmetry

$$A_2(t) = \frac{[\bar{R}^-(t) + \bar{R}^+(t)] - [R^-(t) + R^+(t)]}{[\bar{R}^-(t) + \bar{R}^+(t)] + [R^-(t) + R^+(t)]} \tag{43}$$

which allows a better accuracy on $Im(x)$, CPLEAR also obtains $Im(x) = (1.2 \pm 1.9) \times 10^{-3}$ improving significantly the world average [14].

2. The B^o-$\overline{B^o}$ system

Since the discovery of the K^o meson in the fifties, similar neutral particle systems have been observed; namely the D^o and B^o mesons. Like the kaons, which have their own associated quantum number S, the D mesons carry a new quantum number C (for Charm) and the B^o's have their own, called Beauty or Bottom (B). So far, we do believe that these mesons are made of fundamental particles; the quarks. With five of these quarks (u, d, s, c, b) we are able to explain the entire spectrum of observed resonances and in practice all the quantum numbers are carried by the quarks. For example the $\overline{K^o}$ meson is made of the s quark with strangeness -1 and the $\bar{d}$ antiquark. Table 2 summarizes these quarks and their quantum properties. In order to obtain a consistent theoretical picture, a sixth quark was expected (see table 2), carrying the quantum number T (for Top). This quark, which is the heaviest one, has been discovered few years ago [15]. All these quantum numbers U, D, S, C, B and T are commonly called the flavors of the quarks. It is natural within this scheme to find other systems such as the K^o-$\overline{K^o}$ one and therefore it is justified to wander whether the D^o-$\overline{D^o}$ system (made of the c and u quarks) or the B^o-$\overline{B^o}$ system (made of the b and d quarks) have a similar behaviour. To be more precise, is CP violation a general property of the theory or is it simply an accident proper to the kaons? In the following we will describe the general formalism for any X^o-$\overline{X^o}$ system and then discuss the possible consequences of CP violation in the most general framework.

2.1. THE B^O- $\overline{B^O}$ MIXING FORMALISM.

To be practical, let us describe the formalism using the B^o and $\overline{B^o}$ mesons. However, it is obvious that it can be transposed to any other system com-

Quarks	Spin	Electric charge	Strangeness S	Charm C	Beauty B	Top T	Strong isospin I_3^S	Weak isospin I_3^W
d	1/2	−1/3	0	0	0	0	−1/2	−1/2
u	1/2	2/3	0	0	0	0	1/2	1/2
s	1/2	−1/3	−1	0	0	0	0	−1/2
c	1/2	2/3	0	1	0	0	0	1/2
b	1/2	−1/3	0	0	−1	0	0	−1/2
t	1/2	2/3	0	0	0	1	0	1/2

TABLE 2. The list of the "discovered" quarks with the associated quantum numbers.

posed of a neutral particle and its antiparticle. We have shown in the previous chapter that the K^o and $\overline{K^o}$ are not the physical state. Therefore, let us consider the arbitrary state :

$$\psi \equiv a\left|B^o\right\rangle + b\left|\overline{B^o}\right\rangle \tag{44}$$

This state has to satisfy the Schrödinger equation

$$i\frac{d\psi}{dt} = \mathcal{H}\psi \quad \equiv \quad i\frac{d}{dt}\begin{pmatrix} a \\ b \end{pmatrix} = \mathcal{H}\begin{pmatrix} a \\ b \end{pmatrix} \tag{45}$$

where we have used $\hbar = 1$ and

$$\mathcal{H} = M - i\frac{\Gamma}{2} \equiv \begin{pmatrix} M_{11} & M_{12} \\ M_{21} & M_{22} \end{pmatrix} - \frac{i}{2}\begin{pmatrix} \Gamma_{11} & \Gamma_{12} \\ \Gamma_{21} & \Gamma_{22} \end{pmatrix} \tag{46}$$

– <u>M is the mass matrix.</u> It is due to processes involving intermediate virtual states (the so-called dispersive part)

$$M_{ij} = m_B\,\delta_{ij} + \left\langle i\left|\mathcal{T}^{(\Delta B=2)}\right|j\right\rangle + \sum_v \frac{\left\langle i\left|\,\mathcal{T}^{(\Delta B=1)}\right|v\right\rangle\,\left\langle v\left|\,\mathcal{T}^{(\Delta B=1)}\right|j\right\rangle}{m_B - E_v} \tag{47}$$

where v is an intermediate virtual state and $\mathcal{T}$ denotes the transition matrix which connects two states with different quantum number B.

- M_{11} and M_{22} are due to the mass of the constituent quarks and to their binding energy.

   - M_{12} and M_{21} are generated by the transitions $B^o \to \overline{B^o}$ and $\overline{B^o} \to B^o$ involving virtual processes.

– Γ is the disintegration matrix. It is due to processes involving intermediate real states (the so-called absorbtive part)

$$\Gamma_{ij} = 2\pi \sum_f C_f \left\langle i \left| \mathcal{T}^{(\Delta B=1)} \right| f \right\rangle \left\langle f \left| \mathcal{T}^{(\Delta B=1)} \right| j \right\rangle \tag{48}$$

where C_f is a "phase space factor" and f is a real state.

   - Γ_{11} and Γ_{22} are due to $B^o \to f$ and $\overline{B^o} \to \bar{f}$ decays.
   - Γ_{12} and Γ_{21} are generated by the transitions $B^o \to f \to \overline{B^o}$ and $\overline{B^o} \to f \to B^o$ where f is a real state common to both B^o and $\overline{B^o}$.

One can easily verify that even though $H \neq H^\dagger$, $M = M^\dagger$ and $\Gamma = \Gamma^\dagger$. Let us now examine the consequences of the conservation of CPT for M and Γ. We will use the convention $CP\left|B^o\right\rangle = \left|\overline{B^o}\right\rangle$. However, one has to keep in mind that this is arbitrary and that the states $\left|B^o\right\rangle$ and $\left|\overline{B^o}\right\rangle$ are **only defined to the extend of a relative phase.**

$$\begin{aligned} \langle B^o|\mathcal{H}|f\rangle &= \left\langle B^o \left|(CPT)^\dagger \mathcal{H}(CPT)\right| f \right\rangle \\ &= \eta_{CP} \left\langle \overline{B^o} \left| T^\dagger \mathcal{H}\, T \right| \bar{f} \right\rangle \\ &= \eta_{CP} \left\langle \overline{B^o}|\mathcal{H}|\bar{f}\right\rangle^* \equiv \eta_{CP} \left\langle \bar{f}|\mathcal{H}|\overline{B^o}\right\rangle \end{aligned} \tag{49}$$

where η_{CP} is the phase factor coming from the transformation $CP|f\rangle = \eta_{CP}\left|\bar{f}\right\rangle$.

- For $|f\rangle = |B^o\rangle$, one has

$$\langle B^o|\mathcal{H}|B^o\rangle = \left\langle \overline{B^o}|\mathcal{H}|\overline{B^o}\right\rangle^* = \left\langle \overline{B^o}|\mathcal{H}|\overline{B^o}\right\rangle \Rightarrow \begin{cases} M_{11} = M_{22} \\ \Gamma_{11} = \Gamma_{22} \end{cases} \tag{50}$$

with M_{11}, M_{22}, Γ_{11} and Γ_{22} real.

- For $\left|f\right\rangle = \left|\overline{B^o}\right\rangle$, one derives

$$\left\langle B^o|\mathcal{H}|\overline{B^o}\right\rangle = \left\langle \overline{B^o}|\mathcal{H}|B^o\right\rangle^* \Rightarrow \begin{cases} M_{21} = M_{12}^* \\ \Gamma_{21} = \Gamma_{12}^* \end{cases} \tag{51}$$

The reader can verify that when the conservation of the CP symmetry is required, one obtains:

$$\begin{aligned} &M_{11} = M_{22} \quad \text{and} \quad \Gamma_{11} = \Gamma_{22} \\ &M_{12} = M_{21} \quad \text{and} \quad \Gamma_{12} = \Gamma_{21} \text{ with } M_{12} \text{ and } \Gamma_{12} \text{ real.} \end{aligned} \tag{52}$$

2.2. THE PHYSICAL STATES.

In order to obtain the mass eigenstates, the matrix $\mathcal{H}$ has to be diagonalized. Let us write the eigenstates $|B_S\rangle$ and $|B_L\rangle$

$$|B_S\rangle \;=\; \frac{p\,|B^o\rangle + q\,\left|\overline{B^o}\right\rangle}{\sqrt{|p|^2+|q|^2}} \quad , \quad |B_L\rangle \;=\; \frac{u\,|B^o\rangle + v\,\left|\overline{B^o}\right\rangle}{\sqrt{|u|^2+|v|^2}} \tag{53}$$

where p, q, u and v are complex coefficients. We have used the notations B_S and B_L by analogy to the kaons. The eigenvalues λ_S and λ_L are obtained by solving the equation

$$(\mathcal{H} - \lambda) = 0 \tag{54}$$

Imposing CPT invariance ($H_{11} = H_{22}$ and $H_{21} = H_{12}^*$), one gets

$$\lambda_S = M_S - \frac{i\Gamma_S}{2} = H_{11} + \sqrt{H_{12}\,H_{12}^*} \ , \ \lambda_L = M_L - \frac{i\Gamma_L}{2} = H_{11} - \sqrt{H_{12}H_{12}^*} \tag{55}$$

The eigenstates $|B_S\rangle$ and $|B_L\rangle$ have well defined masses and widths

$$\begin{aligned} M_S &= M_{11} + \mathrm{Re}\sqrt{\left(M_{12} - \frac{i}{2}\,\Gamma_{12}\right)\left(M_{12}^* - \frac{i}{2}\,\Gamma_{12}^*\right)} \\ \Gamma_S &= \Gamma_{11} - 2\mathrm{Im}\sqrt{\left(M_{12} - \frac{i}{2}\,\Gamma_{12}\right)\left(M_{12}^* - \frac{i}{2}\,\Gamma_{12}^*\right)} \\ M_L &= M_{11} - \mathrm{Re}\sqrt{\left(M_{12} - \frac{i}{2}\,\Gamma_{12}\right)\left(M_{12}^* - \frac{i}{2}\,\Gamma_{12}^*\right)} \\ \Gamma_L &= \Gamma_{11} + 2\mathrm{Im}\sqrt{\left(M_{12} - \frac{i}{2}\,\Gamma_{12}\right)\left(M_{12}^* - \frac{i}{2}\,\Gamma_{12}^*\right)} \end{aligned} \tag{56}$$

and thus

$$\begin{aligned} \Delta m &= M_L - M_S = -2\mathrm{Re}\sqrt{\left(M_{12} - \frac{i}{2}\,\Gamma_{12}\right)\left(M_{12}^* - \frac{i}{2}\,\Gamma_{12}^*\right)} \\ \Delta\Gamma &= \Gamma_L - \Gamma_S = 4\mathrm{Im}\sqrt{\left(M_{12} - \frac{i}{2}\,\Gamma_{12}\right)\left(M_{12}^* - \frac{i}{2}\,\Gamma_{12}^*\right)} \end{aligned} \tag{57}$$

Identifying the eigenstates, one deduces

$$\frac{q}{p} = -\frac{v}{u} = \sqrt{\frac{M_{12}^* - \frac{i}{2}\,\Gamma_{12}^*}{M_{12} - \frac{i}{2}\,\Gamma_{12}}} \tag{58}$$

Let us see now what happens when CP conservation is required (i.e. M_{12} and Γ_{12} are real). One gets $\frac{p}{q} = -\frac{u}{v} = 1$, and thus

$$\left.\begin{aligned} |B_S\rangle &= \frac{|B^o\rangle + |\overline{B^o}\rangle}{\sqrt{2}} \equiv |B_+\rangle \\ |B_L\rangle &= \frac{|B^o\rangle - |\overline{B^o}\rangle}{\sqrt{2}} \equiv |B_-\rangle \end{aligned}\right\} \text{ CP eigenstates} \tag{59}$$

An ideal mixing is obtained. The parameter x which quantifies the B^o-$\overline{B^o}$ oscillation is

$$x = \frac{\Delta m}{\left(\frac{\Gamma_S+\Gamma_L}{2}\right)} = \frac{2\,M_{12}}{\Gamma\,(B^o \to X)} \tag{60}$$

where we have neglected Γ_{12} as we will justify it later in the framework of the Standard Model [16]. This quantity is measured directly by ARGUS, CLEO [17–21] for the B_d system ($x_d = 0.708 \pm 0.080$) and by the LEP and hadron collider experiments [22, 23] in the case of an admixture of B_d and B_s. Knowing the total width of B_d^o mesons, by example from the lifetime measurement ($\tau_B \simeq (1.56 \pm 0.04) \times 10^{-12} s$) [4], one gets $\Delta m_{B_d} \simeq 3.0\ 10^{-4}$ eV. The quantity Δm_{B_d} is also measured directly by observing the period of oscillation between B_d^o and $\overline{B_d^o}$, as is discussed in the next chapter. Averaging the measurements at LEP, SLC and at the Tevatron, one obtains $\Delta m_{B_d} = (0.477 \pm 0.017)\hbar\ ps^{-1}$ or $\Delta m_{B_d} \simeq 3.1\ 10^{-4}$ eV. This latter method allows a better precision on Δm_{B_d} and will be used to check the consistency of the Standard Model in the chapter 4. For now, let us only comment on the comparison of the measurements of x_d made by ARGUS and CLEO and the value of Δm_{B_d} obtained by observing $B_d^o - \overline{B_d^o}$ oscillations. For some time, the only available information on the $B_d^o - \overline{B_d^o}$ mixing was extracted by measuring the time integrated probability

$$\chi_d \equiv \mathcal{P}(B_d^o(t=0) \text{ decay as } \overline{B_d^o}) = \frac{1}{2}\frac{x_d^2}{1+x_d^2} \tag{61}$$

Experiments running at the $\Upsilon(4S)$ measure χ_d since no B_s^o meson is produced while at higher energy, the mixing probability $\chi = f_d\chi_d + f_s\chi_s$ (where $f_d(f_s)$ is the fraction of $B_d^o(B_s^o)$ produced) is measured. Since the unambiguous indication of the $B_d^o - \overline{B_d^o}$ mixing using fully reconstructed events [24], all statistically significant measurements of x_d are obtained from inclusive studies. In the initial measurements, the fraction, R, of like sign lepton pairs was used. At the $\Upsilon(4S)$, the $B_d^o - \overline{B_d^o}$ system is in a coherent quantum state with an antisymmetric wave function (see chapter 6). Therefore one has

$$R = \frac{1}{2}\frac{x_d^2}{1+x_d^2}\frac{1}{1+h} = \frac{\chi_d}{1+h} \tag{62}$$

where $h = f_\pm \tau_\pm^2 / f_o \tau_o^2$ accounts for the fraction of $\Upsilon(4S) \to B^+B^-$ decays, $(f_\pm)$, and the semileptonic branching ratio of charged B's compared to the one of the neutrals. From the charged and neutral B lifetimes assumes equal semileptonic widths. Using the measured value of $f_\pm / f_o$ and the lifetime ratio $\tau_{B^\pm}/\tau_{B^o}$ [4], one gets $h = 1.1 \pm 0.2$. A similar technique has also been developed using the decay $B_d^o \to D^{*-}\ell^+\nu$ with a partial reconstruction of the D^{*-} decay to $\overline{D^o}\pi^-$, the second B meson being tagged with a lepton. In this latter case, the factor $(1+h)$ in Eq. 62 is not needed. Lately ARGUS has complemented this method by replacing the tagging lepton with a charged kaon [17]. Finally keeping the kaon tagging, they have also replaced the partially reconstructed semileptonic B decay by a fully reconstructed charged D*. All χ_d results are shown in Fig. 5.

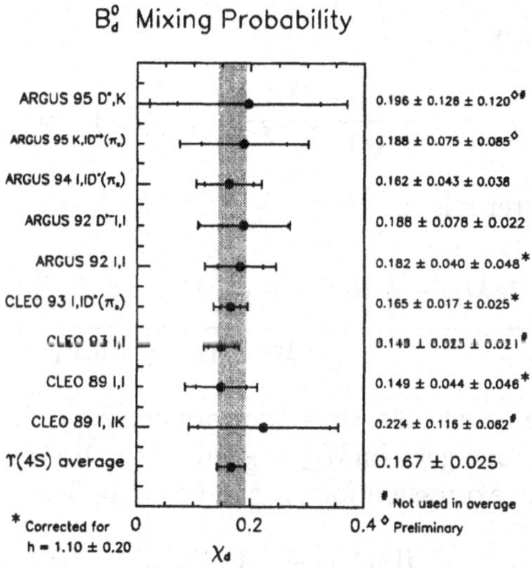

Figure 5. Summary of the measurements of χ_d at the $\Upsilon(4S)$.

Combining the results with proper care of the common systematics and keeping the most precise results when unknown statistical overlap is present, one find the world average:

$$\chi_d = 0.167 \pm 0.025 \tag{63}$$

Using Eq. 61, one gets

$$\Delta m_{\mathrm{B}_d} = 0.452 \pm 0.053\ \hbar/ps \tag{64}$$

The comparison of this value with the one derived from oscillation measurements can be viewed as a test of quantum mechanics. Indeed, should quantum mechanics fail and the two $\mathrm{B_d^0}$ mesons be produced incoherently, one would have to use

$$\mathrm{R} = \frac{1}{2}\frac{2x_d^2 + x_d^4}{(1+x_d^2)^2}\frac{1}{1+h} = \frac{2(\chi_d - \chi_d^2)}{1+h} \tag{65}$$

instead of Eq. 62 and one would derive $x_d = 0.474 \pm 0.048$ in contradiction with the measurement of Δm_{B_d} at the 4.2 σ level.
If the CP symmetry is not conserved, one would observe a deviation from the ideal mixing. Defining the complex parameter $\varepsilon = \frac{p-q}{p+q}$, the physical states read

$$|B_S\rangle = \frac{(1+\varepsilon)\,|B^o\rangle + (1-\varepsilon)\,\left|\overline{B^o}\right\rangle}{\sqrt{2\,(1+|\varepsilon|^2)}}\ ,\ |B_L\rangle = \frac{(1+\varepsilon)\,|B^o\rangle - (1-\varepsilon)\,\left|\overline{B^o}\right\rangle}{\sqrt{2\,(1+|\varepsilon|^2)}} \tag{66}$$

or equivalently:

$$|B_S\rangle = \frac{|B_+\rangle + \varepsilon\,|B_-\rangle}{\sqrt{1+|\varepsilon|^2}}\quad ,\quad |B_L\rangle = \frac{|B_-\rangle + \varepsilon\,|B_+\rangle}{\sqrt{1+|\varepsilon|^2}} \tag{67}$$

The parameter ε is :

$$\varepsilon = \frac{\mathrm{Re}M_{12} - \frac{i}{2}\,\mathrm{Re}\Gamma_{12} - \sqrt{\left(M_{12} - \frac{i}{2}\,\Gamma_{12}\right)\left(M_{12}^* - \frac{i}{2}\,\Gamma_{12}^*\right)}}{i\ \mathrm{Im}\ M_{12} + \frac{1}{2}\ \mathrm{Im}\ \Gamma_{12}} \tag{68}$$

It should be noted that, since the parameter $|\varepsilon_K|$ in the K^o system is found to be small, one gets $\mathrm{Im}M_{12} - \frac{i}{2}\mathrm{Im}\Gamma_{12} << \mathrm{Re}M_{12} - \frac{i}{2}\mathrm{Re}\Gamma_{12}$ from Eq. 68 and thus the expression for ε_K can be simplified

$$\varepsilon_K \simeq \frac{i\mathrm{Im}M_{12} + \frac{1}{2}\ \mathrm{Im}\Gamma_{12}}{2\mathrm{Re}\ M_{12} - i\ \mathrm{Re}\ \Gamma_{12}} \simeq \frac{i\mathrm{Im}M_{12} + \frac{1}{2}\ \mathrm{Im}\Gamma_{12}}{\Delta m - \frac{i}{2}\ \Delta\Gamma} \tag{69}$$

Using a particular phase convention which imposes Γ_{12} to be real, one deduces the value of the phase of ε_K from Eq. 69 and the experimental measurements of Δm_K and $\Delta\Gamma_K$ [4]

$$\phi_{\varepsilon_K} \equiv \text{phase of } \varepsilon_K \approx 43.5^o \tag{70}$$

The previous simplification cannot be done for the B^o system as $|\varepsilon_B|$ is not necessarily small.
At this stage, it is important to note that the parameter $\frac{p}{q} = \frac{1+\varepsilon}{1-\varepsilon}$ **is not an observable.** Indeed, since $|B^o\rangle$ and $\left|\overline{B^o}\right\rangle$ are only defined to a relative phase, one may change our convention by using $CP\left|B^o\right\rangle = e^{i\alpha}\left|\overline{B^o}\right\rangle$ instead of $CP\left|B^o\right\rangle = +\left|\overline{B^o}\right\rangle$. This is equivalent of using $e^{i\alpha}\left|B^o\right\rangle$ for the B^o wave function. One finds that $p/q \equiv \sqrt{\left\langle B^o|\mathcal{T}(\Delta B=2)|\overline{B^o}\right\rangle / \left\langle \overline{B^o}|\mathcal{T}(\Delta B=2)|B^o\right\rangle}$ is not invariant since $\frac{p}{q}$ becomes $e^{i\alpha}\frac{p}{q}$ and therefore only its modulus can be measured.

2.3. TIME EVOLUTION OF B^O MESONS.

Let us now examine what happens to B^o mesons once they are created. Starting with the time dependent equations of the physical states,

$$B_S(t) \;=\; e^{-(im_S+\Gamma_S/2)t}\left|B_S(0)\right\rangle \;,\; B_L(t) \;=\; e^{-(im_L+\Gamma_L/2)t}\left|B_L(0)\right\rangle \quad (71)$$

one obtains the time dependence for B^o and $\overline{B^o}$ from Eqs. 66 and 71 .

$$\begin{aligned} \left|B^o(t)\right\rangle \;=\; & \frac{1}{2}e^{-(im+\frac{\Gamma}{2})t}\left\{\left(e^{(\frac{\Delta\Gamma}{4}+i\frac{\Delta m}{2})t} + e^{-(\frac{\Delta\Gamma}{4}+i\frac{\Delta m}{2})t}\right)\left|B^o\right\rangle \right. \\ & \left. + \left(e^{(\frac{\Delta\Gamma}{4}+i\frac{\Delta m}{2})t} - e^{-(\frac{\Delta\Gamma}{4}+i\frac{\Delta m}{2})t}\right)\left|\frac{q}{p}\right| e^{2i\Phi_M}\left|\overline{B^o}\right\rangle\right\} \end{aligned} \quad (72)$$

$$\begin{aligned} \left|\overline{B^o}(t)\right\rangle \;=\; & \frac{1}{2}e^{-(im+\frac{\Gamma}{2})t}\left\{\left(e^{(\frac{\Delta\Gamma}{4}+i\frac{\Delta m}{2})t} + e^{-(\frac{\Delta\Gamma}{4}+i\frac{\Delta m}{2})t}\right)\left|\overline{B^o}\right\rangle \right. \\ & \left. + \left(e^{(\frac{\Delta\Gamma}{4}+i\frac{\Delta m}{2})t} - e^{-(\frac{\Delta\Gamma}{4}+i\frac{\Delta m}{2})t}\right)\left|\frac{p}{q}\right| e^{-2i\Phi_M}\left|B^o\right\rangle\right\} \end{aligned} \quad (73)$$

Here, the following notations have been used:

$$\Gamma = \frac{\Gamma_S+\Gamma_L}{2},\; m = \frac{m_S+m_L}{2},\; \Delta\Gamma = \Gamma_L - \Gamma_S \;\text{ et }\; \Delta m = m_L - m_S. \quad (74)$$

It is now straightforward to compute the probability of the transition $B^o(t) \to f$.

$$
\begin{aligned}
|\langle f|\mathcal{T}|B^o(t)\rangle|^2 &= \frac{e^{-\Gamma t}}{2}\left\{ \left(\mathrm{ch}\frac{\Delta\Gamma}{2}t + \cos\Delta mt\right) |\langle f|\mathcal{T}|B^o\rangle|^2 \right.\\
&+ \left(\mathrm{ch}\frac{\Delta\Gamma}{2}t - \cos\Delta mt\right)\left|\frac{q}{p}\right|^2 \left|\left\langle f|\mathcal{T}|\overline{B^o}\right\rangle\right|^2\\
&+ \left(\mathrm{sh}\frac{\Delta\Gamma}{2}t - i\sin\Delta mt\right)\left|\frac{q}{p}\right| e^{-2i\Phi_M}\langle f|\mathcal{T}|B^o\rangle \left\langle f|\mathcal{T}|\overline{B^o}\right\rangle^*\\
&+ \left.\left(\mathrm{sh}\frac{\Delta\Gamma}{2}t + i\sin\Delta mt\right)\left|\frac{q}{p}\right| e^{2i\Phi_M}\langle f|\mathcal{T}|B^o\rangle^* \left\langle f|\mathcal{T}|\overline{B^o}\right\rangle\right\}
\end{aligned}
\tag{75}
$$

For B^o_d mesons, the Standard Model predicts $\Delta\Gamma \ll \Delta m$, (as we will see later). Therefore Eq. 75 simplifies into

$$
\begin{aligned}
|\langle f|\mathcal{T}|B^o(t)\rangle|^2 &= e^{-\Gamma t}\left\{\cos^2\frac{\Delta m}{2}t\,|\langle f|\mathcal{T}|B^o\rangle|^2\right.\\
&+ \sin^2\frac{\Delta m}{2}t\left|\frac{q}{p}\right|^2\left|\left\langle f|\mathcal{T}|\overline{B^o}\right\rangle\right|^2\\
&- \frac{i}{2}\left|\frac{q}{p}\right|\sin\Delta mt e^{-2i\Phi_M}\langle f|\mathcal{T}|B^o\rangle\left\langle f|\mathcal{T}|\overline{B^o}\right\rangle^*\\
&+ \left.\frac{i}{2}\left|\frac{q}{p}\right|\sin\Delta mt e^{2i\Phi_M}\langle f|\mathcal{T}|B^o\rangle^*\left\langle f|\mathcal{T}|\overline{B^o}\right\rangle\right\}
\end{aligned}
\tag{76}
$$

When the initial meson is a $\overline{B^o}$, its time evolution is obtained in a similar way.

$$
\begin{aligned}
\left|\left\langle f|\mathcal{T}|\overline{B^o}(t)\right\rangle\right|^2 &= e^{-\Gamma t}\left\{\cos^2\frac{\Delta m}{2}t\left|\left\langle f|\mathcal{T}|\overline{B^o}\right\rangle\right|^2\right.\\
&+ \sin^2\frac{\Delta m}{2}t\left|\frac{p}{q}\right|^2|\langle f|\mathcal{T}|B^o\rangle|^2\\
&+ \frac{i}{2}\left|\frac{p}{q}\right|\sin\Delta mt e^{-2i\Phi_M}\langle f|\mathcal{T}|B^o\rangle\left\langle f|\mathcal{T}|\overline{B^o}\right\rangle^*\\
&- \left.\frac{i}{2}\left|\frac{p}{q}\right|\sin\Delta mt e^{2i\Phi_M}\langle f|\mathcal{T}|B^o\rangle^*\left\langle f|\mathcal{T}|\overline{B^o}\right\rangle\right\}
\end{aligned}
\tag{77}
$$

For a final state $\bar{f} \equiv CP\,|f\rangle$, equations identical to 76 and 77 are found and therefore the comparison of the transition probabilities $|\langle f|\mathcal{T}|B^o(t)\rangle|^2$ and $\left|\left\langle \bar{f}|\mathcal{T}|\overline{B^o}(t)\right\rangle\right|^2$ or $\left|\left\langle f|\mathcal{T}|\overline{B^o}(t)\right\rangle\right|^2$ and $|\langle \bar{f}|\mathcal{T}|B^o(t)\rangle|^2$ may exibit CP violating effects.

2.4. THE DIFFERENT EFFECTS REVEALING CP NON CONSERVATION

In the following, we will describe in broad sense the mechanisms leading to CP violation without invoquing any particular theory. The Standard Model predictions will be discussed in details in chapter 5.

CP violating effects can be generated by three possible mechanisms and might be observed by studying two generic classes of final states.

2.4.1. *Flavor Specific Final States.*
Any final state f which enables to determine unambiguiously the nature (B^o or $\overline{B^o}$) of its mother meson is called flavor specific. The simplest examples are the semileptonic final states.

$$\begin{array}{llll} B^o & \rightarrow \ell^+\nu X \equiv f & , \quad \overline{B^o} & \rightarrow \ell^-\bar{\nu} X \equiv \bar{f} \\ \overline{B^o} & \not\rightarrow \ell^+\nu X \equiv f & , \quad B^o & \not\rightarrow \ell^-\bar{\nu} X \equiv \bar{f} \end{array} \tag{78}$$

The charge of the lepton allows one to deduce the flavor of the B^o meson. Assuming that the flavor of the B^o meson is known at $t = 0$, there exists two possibilities.
a) The Final State f can only be produced by the Flavor of the Initial B^o.
This is equivalent of having $B^o(t) = B^o(0)$. With this requirement, one obtains:

$$\begin{array}{lll} \mathcal{P}r\left(B^o(t) \rightarrow f\right) \equiv |\langle f|\mathcal{T}|B^o(t)\rangle|^2 & = & e^{-\Gamma t}\cos^2\frac{\Delta m}{2}t\ |\langle f|\mathcal{T}|B^o\rangle|^2 \\ \mathcal{P}\cdot\left(\overline{B^o}(t) \rightarrow \bar{f}\right) \equiv \left|\left\langle \bar{f}|\mathcal{T}|\overline{B^o}(t)\right\rangle\right|^2 & = & e^{-\Gamma t}\cos^2\frac{\Delta m}{2}t\ \left|\left\langle \bar{f}|\mathcal{T}|\overline{B^o}\right\rangle\right|^2 \end{array} \tag{79}$$

If $\mathcal{P}r\left(B^o(t) \rightarrow f\right) \not\equiv \mathcal{P}r\left(\overline{B^o}(t) \rightarrow \bar{f}\right)$, then $|\langle f|\mathcal{T}|B^o\rangle|^2 \not\equiv \left|\left\langle \bar{f}|\mathcal{T}|\overline{B^o}\right\rangle\right|^2$. Should this be observed, one would have demonstrated that the interaction responsible of the disintegration of the B meson is at the origin of CP violation independently of the B^o-$\overline{B^o}$ mixing. This is called **DIRECT CP VIOLATION**. Note that charged mesons could be used as well.
b) The Final State cannot be Produced by the Flavor of the Initial B^o.
It is equivalent of imposing the B^o- $\overline{B^o}$ mixing.

$$\begin{array}{lll} \mathcal{P}r\left(B^o(t) \rightarrow \bar{f}\right) & = & e^{-\Gamma t}\sin^2\frac{\Delta m}{2}t\left|\frac{q}{p}\right|^2\ \left|\left\langle \bar{f}|\mathcal{T}|\overline{B^o}\right\rangle\right|^2 \\ \mathcal{P}r\left(\overline{B^o}(t) \rightarrow f\right) & = & e^{-\Gamma t}\sin^2\frac{\Delta m}{2}t\left|\frac{p}{q}\right|^2\ |\langle f|\mathcal{T}|B^o\rangle|^2 \end{array} \tag{80}$$

If one is certain that $\left|\left\langle \bar{f}|\mathcal{T}|\overline{B^o}\right\rangle\right| = |\langle f|\mathcal{T}|B^o\rangle|$, for example because no direct CP violating effect was observed for the final state that is used, then one may search for the asymmetry:

$$\mathcal{A}s = \frac{\mathcal{P}r\left(\overline{B^o}(t) \rightarrow f\right) - \mathcal{P}r\left(B^o(t) \rightarrow \bar{f}\right)}{\mathcal{P}r\left(\overline{B^o}(t) \rightarrow f\right) + \mathcal{P}r\left(B^o(t) \rightarrow \bar{f}\right)} = \frac{\left|\frac{p}{q}\right|^2 - \left|\frac{q}{p}\right|^2}{\left|\frac{p}{q}\right|^2 + \left|\frac{q}{p}\right|^2} \simeq \frac{4\ \mathrm{Re}\varepsilon_B}{1 + |\varepsilon_B|^2} \tag{81}$$

This asymmetry is solely due to the B^o-$\overline{B^o}$ mixing. It reveals that

$$B^o \rightarrow \overline{B^o} \quad \not\equiv \quad \overline{B^o} \rightarrow B^o \tag{82}$$

and is therefore a measurement of T violation.

2.4.2. *Non Flavor Specific Final State.*
All final states reachable by both B^o and $\overline{B^o}$ belong to this class.

$$\begin{array}{lll} B^o \rightarrow f & , & \overline{B^o} \rightarrow \bar{f} \\ \overline{B^o} \rightarrow f & , & B^o \rightarrow \bar{f} \end{array} \tag{83}$$

This condition is fulfilled by any final state which is a CP eigenstate,

$$CP\,|f_{CP}\rangle = \eta_{CP}\,|f_{CP}\rangle \tag{84}$$

where η_{CP} is the CP parity of the final state ($\eta_{CP} = \pm 1$). Should there exist a CP violating phase in the decay, one would have:

$$e^{-i\Phi_D}\,\langle f_{CP}|\mathcal{T}|B^o\rangle \rightarrow \eta_{CP}\; e^{i\Phi_D}\left\langle f_{CP}|\mathcal{T}|\overline{B^o}\right\rangle \tag{85}$$

where Φ_D is the phase coming from the disintegration process. From Eqs. 76 and 77 and using the prediction of the Standard Model $\left|\frac{p}{q}\right| \simeq 1$, one has:

$$\begin{array}{rcl} \mathcal{P}r\left(B^o(t) \rightarrow f_{CP}\right) & = & |\langle f_{CP}|\mathcal{T}|B^o\rangle|^2 \times \\ & & e^{-\Gamma t}\left\{1 - \eta_{CP}\,\sin 2\left(\Phi_M + \Phi_D\right)\sin \Delta m t\right\} \\ \mathcal{P}r\left(\overline{B^o}(t) \rightarrow f_{CP}\right) & = & \left|\left\langle f_{CP}|\mathcal{T}|\overline{B^o}\right\rangle\right|^2 \times \\ & & e^{-\Gamma t}\left\{1 + \eta_{CP}\,\sin 2\left(\Phi_M + \Phi_D\right)\sin \Delta m t\right\} \end{array} \tag{86}$$

where we have assumed

$$|\langle f_{CP}|\mathcal{T}|B^o\rangle| = \left|\left\langle f_{CP}|\mathcal{T}|\overline{B^o}\right\rangle\right|. \tag{87}$$

Hence, the time dependent asymmetry $\mathcal{A}s(t)$ reads

$$\begin{array}{rcl} \mathcal{A}s(t) & = & \dfrac{\mathcal{P}r\left(\overline{B^o}(t) \rightarrow f_{CP}\right) - \mathcal{P}r\left(B^o(t) \rightarrow f_{CP}\right)}{\mathcal{P}r\left(\overline{B^o}(t) \rightarrow f_{CP}\right) + \mathcal{P}r\left(B^o(t) \rightarrow f_{CP}\right)} \\ & = & \eta_{CP}\,\sin 2\left(\Phi_M + \Phi_D\right)\,\sin \Delta m t \end{array} \tag{88}$$

After time integration, one obtains:

$$\begin{aligned} \mathcal{A}s &= \frac{\int_0^\infty \mathcal{P}r\left(\overline{B^o}(t) \to f_{CP}\right) dt - \int_0^\infty \mathcal{P}r\left(B^o(t) \to f_{CP}\right) dt}{\int_0^\infty \mathcal{P}r\left(\overline{B^o}(t) \to f_{CP}\right) dt + \int_0^\infty \mathcal{P}r\left(B^o(t) \to f_{CP}\right) dt} \\ &= \eta_{CP}\ \frac{x}{1+x^2}\ \sin 2\Phi \end{aligned} \tag{89}$$

with $\Phi = \Phi_M + \Phi_D$ and $x = \frac{\Delta m}{\Gamma}$. For $B^o_d(\bar{b}\, d)$, $x_d \simeq 0.7$ and thus

$$\mathcal{A}s \simeq 0.47\ \eta_{CP}\ \sin 2\Phi. \tag{90}$$

Note that Eq. 87 is not always correct and thus this simple picture may be complicated by the presence of direct CP violation as we will discuss in chapter 5.

2.5. CONCLUSION

In summary, there exists three possibilities for measuring CP violation.

1. From the decay of B mesons.

$$B \to f \quad \not\equiv \quad \overline{B} \to \bar{f} \tag{91}$$

This measurement

- is time independent
- does not require the B^o-$\overline{B^o}$ mixing ($\overset{(-)}{B^o}$ and $B^\pm$ may be used).

Example:

$$\Gamma\left(B^+ \to K^+ \rho^o\right) \quad \not\equiv \quad \Gamma\left(B^- \to K^- \rho^o\right) \tag{92}$$

2. From the B^o-$\overline{B^o}$ mixing.

$$B^o \to \overline{B^o} \quad \not\equiv \quad \overline{B^o} \to B^o \tag{93}$$

This measurement

- is time dependent. However it is possible to do a time integrated measurement.
- requires the B^o-$\overline{B^o}$ mixing.

Example:

$$B^o\overline{B^o} \to \ell^+\ell^+ + X \quad \not\equiv \quad B^o\overline{B^o} \to \ell^-\ell^- + X \tag{94}$$

3. From the interplay between the decay and the mixing.
   This measurement

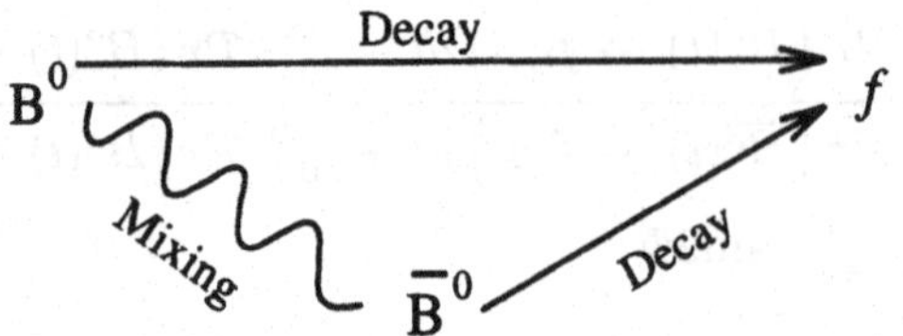

– is time dependent. It is possible in some cases to make a time integrated measurement.

– requires the B^o-$\overline{B^o}$ mixing.

Example:

$$B^o \rightarrow J/\psi\ K_S \quad \not\equiv \quad \overline{B^o} \rightarrow J/\psi\ K_S \tag{95}$$

3. CP violation in the Standard Model.

We have seen in the previous chapter how CP violation may be observed. We will now study how this phenomenon is generated [25] in the Standard Model [16] and estimate the magnitude of the expected effects [26]. However, before discussing these issues, it is worth-while underlying the importance of CP violation in cosmology. Here, CP violation, as well as the baryon number violation and C violation in an expanding universe (with deviations from thermal equilibrium), is an essential ingredient to generate the Baryon Asymmetry of the Universe (BAU) [27]. For a while, it was thought that the BAU could only be generated at the scale of the Grand Unification [28] ($\sim 10^{15\text{-}16}$ GeV). In 1985 it was realized [29] that the electroweak phase transition could have dramatic consequences on any baryon asymmetry generated at higher temperature, and could even be at the origin of the observed number of baryon to photon ratio ($n_b/n_\gamma \sim 10^{-9} - 10^{-10}$) through anomalous electroweak baryon-number violation [30]. Clearly the possibility to generate CP non-conservation at the electroweak scale, makes the study of the "Standard" scenario, for which we have definite predictions, extremely exciting. However, after some controversy it is fair to say that as of today, it seems very difficult to produce large enough BAU at the electroweak scale [31], at least in the minimal $SU(2) \times U(1)$ Standard Model with one scalar Higgs doublet [32]. Therefore, one is tempted to predict other sources of CP violation at larger scales and thus a comprehensive study of this phenomenon is of prime importance.

3.1. THE CABIBBO-KOBAYASHI-MASKAWA (CKM) MATRIX

Non conservation of the CP symmetry has been introduced in the SM in 1973 [25] by requiring three massless families of quark doublets before the spontaneous $SU(2)\times U(1)$ symmetry breaking by the Higgs mechanism [33]. At that time and contrary to our present knowledge where three families of fermions [4] are identified, there was "evidence" for only two. The three generations of fermions are:

$$\begin{pmatrix} u \\ \\ d \end{pmatrix} \quad \begin{pmatrix} c \\ \\ s \end{pmatrix} \quad \begin{pmatrix} t \\ \\ b \end{pmatrix}$$
$$\begin{pmatrix} \nu_e \\ \\ e^- \end{pmatrix} \quad \begin{pmatrix} \nu_\mu \\ \\ \mu^- \end{pmatrix} \quad \begin{pmatrix} \nu_\tau \\ \\ \tau^- \end{pmatrix} \tag{96}$$

Since the recent discovery of the t quark [15], the only particle not observed directly remains the τ neutrino. The measured t quark mass ($m_t = (173.2 \pm 5.2)$ GeV/c^2 [4]) is consistent with the SM prediction derived from the measured mass ratio of the Z and W bosons through higher order corrections [34]. This constitute one of the remarkable success of the SM. Although we do not derive here the SM and refer the reader to other lectures [35], it is appropriate to discuss in some detail the fermion mass generation mechanism. After the phase transition at the electroweak scale, the remaining neutral Higgs particle allows one to generate the mass of each fermion through its Yukawa type coupling. The interaction Lagrangian is

$$\mathcal{L}_{Yukawa} = -\left(\overline{\mathbf{u}}'_L \ \mathbf{m} \ \mathbf{u}'_R \ + \ \overline{\mathbf{d}}'_L \ \widetilde{\mathbf{m}} \ \mathbf{d}'_R \ + \ h.c.\right)\left(1 + \frac{\Phi_o}{v}\right). \tag{97}$$

Here Φ_o stands for the scalar Higgs field and v is its value in the new vacuum. The quark fields $\mathbf{u}'_{L,R}$ and $\mathbf{d}'_{L,R}$ are the 3-component vectors in flavour space for the up- and down-type quarks respectively,

$$\mathbf{u}'_{L,R} = \left(\frac{1 \mp \gamma_5}{2}\right)\begin{pmatrix} u' \\ c' \\ t' \end{pmatrix}, \qquad \mathbf{d}'_{L,R} = \left(\frac{1 \mp \gamma_5}{2}\right)\begin{pmatrix} d' \\ s' \\ b' \end{pmatrix}, \tag{98}$$

and $\mathbf{m}$, $\widetilde{\mathbf{m}}$ are 3×3 mass matrices of arbitrary complex numbers, the elements of which are $m_{ij} = -\frac{v}{\sqrt{2}} Y_{ij}$ and $\widetilde{m}_{kl} = -\frac{v}{\sqrt{2}} \widetilde{Y}_{kl}$ where Y_{ij} and $\widetilde{Y}_{kl}$ are the Yukawa coupling constants with $i, j \equiv u', c'$ or t' and $k, l \equiv d', s'$ or b'. Since $\mathbf{m}$ and $\widetilde{\mathbf{m}}$ are not diagonal, one needs to define the physical fermion fields $\mathbf{u}_{L,R} = V_{L,R}\mathbf{u}'_{L,R}$ and $\mathbf{d}_{L,R} = \widetilde{V}_{L,R}\mathbf{d}'_{L,R}$ where $V_{L,R}$ and $\widetilde{V}_{L,R}$

are unitary matrices. Writing Eq. 97 in terms of these new fields diagonalizes the mass matrices with $\mathbf{m}_D = V_L \mathbf{m} V_R^\dagger$ and $\widetilde{\mathbf{m}}_D = \widetilde{V}_L \widetilde{\mathbf{m}} \widetilde{V}_R^\dagger$. It can be verified easily that the coupling of the physical quarks to the neutral Z preserves the observed absence of flavour-changing neutral currents (FCNC), while their coupling to the charged $W^\pm$ introduces the mixing between families. The charged-current couplings are:

$$\mathcal{L}_W = \frac{g}{\sqrt{2}} \left\{ \overline{\mathbf{u}}_L \gamma^\mu W_\mu^+ \mathbf{V} \mathbf{d}_L \; + \; \overline{\mathbf{d}}_L \gamma^\mu W_\mu^- \mathbf{V}^\dagger \mathbf{u}_L \right\}, \tag{99}$$

where $\mathbf{V} \equiv V_L \widetilde{V}_L^\dagger$ is a unitary 3×3 matrix called the quark-mixing or Cabibbo-Kobayashi-Maskawa (CKM) matrix [25,36]:

$$\mathbf{V} \; = \; \begin{bmatrix} V_{ud} & V_{us} & V_{ub} \\ V_{cd} & V_{cs} & V_{cb} \\ V_{td} & V_{ts} & V_{tb} \end{bmatrix} . \tag{100}$$

We have briefly sketched how the CKM matrix is generated to outline its interconnection with the generation of the mass of the fermions. Unfortunately at present, there exists no obvious way to deduce the values of the elements of $\mathbf{V}$ from this mechanism and we are left (in the three generation case) with ten arbitrary parameters accounting for the mass and the mixing of the quarks. Several attempts to connect these parameters to the quark masses have been proposed in the framework of particular models [37], none of which seems to be satisfactory when confronting the data. It is therefore very important to measure the matrix elements V_{ij} in order to probe the predicted relations and obtain some hints on the nature of the quark-mass matrices $\mathbf{m}$ and $\widetilde{\mathbf{m}}$. Let us now see how the CKM matrix can be constructed and what are its properties.

3.1.1. *The Unitarity Condition.*

This matrix has to be unitary and therefore it is possible to construct it from the product of rotation matrices. In practical, V can be obtained using three complex rotation matrices.

$$\mathcal{R}_{12}(\theta_{12},\ \delta_{12}) = \begin{pmatrix} \cos\theta_{12} & \sin\theta_{12}\ e^{i\delta_{12}} & 0 \\ -\sin\theta_{12}\ e^{-i\delta_{12}} & \cos\theta_{12} & 0 \\ 0 & 0 & 1 \end{pmatrix}$$

$$\mathcal{R}_{23}(\theta_{23},\ \delta_{23}) = \begin{pmatrix} 1 & 0 & 0 \\ 0 & \cos\theta_{23} & \sin\theta_{23}\ e^{i\delta_{23}} \\ 0 & -\sin\theta_{23}\ e^{-i\delta_{23}} & \cos\theta_{23} \end{pmatrix} \qquad (101)$$

$$\mathcal{R}_{13}(\theta_{13},\ \delta_{13}) = \begin{pmatrix} \cos\theta_{13} & 0 & \sin\theta_{13}\ e^{i\delta_{13}} \\ 0 & 1 & 0 \\ -\sin\theta_{13}\ e^{-i\delta_{13}} & 0 & \cos\theta_{13} \end{pmatrix}$$

Thus $V = \mathcal{R}_{ij}(\theta_{ij},\ \delta_{ij}) \times \mathcal{R}_{k\ell}(\theta_{k\ell},\ \delta_{k\ell}) \times \mathcal{R}_{mn}(\theta_{mn}, \delta_{mn})$ with $i \neq j$, $k \neq \ell$ and $m \neq n$. The most general expression of **V** can be derived from $n(n-1)/2$ mixing angles and $(n-1)(n-2)/2$ phases where n is the number of fermion families; several equivalent representations can be found in the literature. It is worthwhile noting that with more than two families, the elements of **V** can be complex numbers which allow the possibility for generating CP violation through the interference of two diagrams involving different matrix elements. So far, with three families, the three mixing angles and the single phase are fundamental parameters of the theory. Obviously, should there be experimental evidence for more fermion families, the same procedure could be extended using more "elementary" rotation matrices. For example with four families, one would have used 6 such matrices with 3 arbitrary phases.

Since there exists many different parametrizations, let us discuss how many of them could be derived with 3 families from the above rules.

- As already seen, only one phase should be used.
- Furthermore, $k\ell \neq ij$ and $k\ell \neq mn$ since $\mathcal{R}_{k\ell} \times \mathcal{R}_{k\ell} = \mathcal{R}_{k\ell}$

With these two constraints, one gets 36 possibilities. The parametrization generally used [4] is

$$V = \mathcal{R}_{23}(\theta_{23},\ 0) \times \mathcal{R}_{13}(\theta_{13},\ -\delta) \times \mathcal{R}_{12}(\theta_{12},\ 0) \qquad (102)$$

$$V = \begin{bmatrix} C_{12}\ C_{13} & S_{12}C_{13} & S_{13}e^{-i\delta} \\ -S_{12}C_{23} - C_{12}S_{23}S_{13}e^{i\delta} & C_{12}\ C_{23} - S_{12}S_{23}S_{13}e^{i\delta} & S_{23}C_{13} \\ S_{12}S_{23} - C_{12}C_{23}S_{13}e^{i\delta} & -C_{12}S_{23} - S_{12}C_{23}S_{13}e^{i\delta} & C_{23}C_{13} \end{bmatrix} \qquad (103)$$

where $C_{ij} = \cos\theta_{ij}$ and $S_{ij} = \sin\theta_{ij}$. An approximated parametrization proposed by Wolfenstein [38] is also often used.

$$V = \begin{bmatrix} 1-\frac{\lambda^2}{2} & \lambda & A\lambda^3(\rho - i\eta) \\ -\lambda & 1-\frac{\lambda^2}{2} & A\lambda^2 \\ A\lambda^3(1-\rho-i\eta) & -A\lambda^2 & 1 \end{bmatrix} + \mathcal{O}\left(\lambda^4\right) \quad (104)$$

with $\lambda = \sin\theta_{12}$, $A = \frac{\sin\theta_{23}}{\lambda^2}$, $\rho = \frac{\sin\theta_{13}}{A\lambda^3}\cos\delta$ and $\eta = \frac{\sin\theta_{13}}{A\lambda^3}\sin\delta$. The angle θ_{12} is the Cabibbo angle ($\lambda = 0.2205\pm\ 0.0018$). It should be noted that the approximation is at the order λ^3. This accurary is good enough at present but when experiments will be able to reach very high sensitivities on the parameters above, it will be necessary to use an approximation at a higher order. This can be simply done [39] by writing V_{td} as $A\lambda^3(1-\bar{\rho}-i\bar{\eta})$ where $\bar{\rho} = \rho(1-\lambda^2/2)$ and $\bar{\eta} = \eta(1-\lambda^2/2)$ and adding the imaginary parts $-iA^2\lambda^5\eta$ and $-iA\lambda^4\eta$ to the expressions for V_{cb} and V_{ts} respectively. In the following, we will kept Eq. 104 as is since the redefinitions above are not necessary at present. The numerical values of the matrix elements are [4].

$$\begin{pmatrix} 0.9753\ \pm 0.0007 & 0.2205 \pm 0.0018 & 0.0026 \text{ to } 0.0038 \\ 0.2205 \pm 0.0018 & 0.9745 \pm 0.0008 & 0.0395 \pm 0.0020 \\ 0.0062 \text{ to } 0.0104 & 0.0395 \pm 0.0020 & 0.9993 \pm 0.0001 \end{pmatrix} \quad (105)$$

In order to get further informations on the matrix elements, one can use the unitarity of this matrix.

$$VV^\dagger = 1 \Rightarrow V_{\alpha j}\, V^*_{\beta j} + V_{\alpha k} V^*_{\beta k} + V_{\alpha \ell} V^*_{\beta \ell} = 0 \quad (106)$$

$$V^\dagger V = 1 \Rightarrow V^*_{\alpha j}\, V_{\alpha k} + V^*_{\beta j} V_{\beta k} + V_{\gamma j} V_{\gamma k} = 0 \quad (107)$$

where the grec and latin indices stand for up and down quarks respectively. Six independent equations can be extracted from Eq. 106 and Eq. 107:

$$\begin{array}{llllllll} 1) & V_{cd}V^*_{ud} & + & V_{cs}V^*_{us} & + & V_{cb}V^*_{ub} & = & 0 \\ 2) & V_{cd}V^*_{td} & + & V_{cs}V^*_{ts} & + & V_{cb}V^*_{tb} & = & 0 \\ 3) & V_{ud}V^*_{td} & + & V_{us}V^*_{ts} & + & V_{ub}V^*_{tb} & = & 0 \\ 4) & V^*_{us}V_{ud} & + & V^*_{cs}V_{cd} & + & V^*_{ts}V_{td} & = & 0 \\ 5) & V^*_{ub}V_{us} & + & V^*_{cb}V_{cs} & + & V^*_{tb}V_{ts} & = & 0 \\ 6) & V^*_{ub}V_{ud} & + & V^*_{cb}V_{cd} & + & V^*_{tb}V_{td} & = & 0 \end{array} \quad (108)$$

These equations can be represented by six triangles in the complex plane (ρ, η) [40] defined with the Wolfenstein parametrization. One can verify with the numerical values that equations 1, 2, 4 and 5 in 108 are describing four almost flat triangles and therefore little informations can be extracted from them. The third and sixth equations (which are almost equivalent) are more interesting since their corresponding triangles have same order sides, the measurements of which require the study of B meson decays. One generally chooses to represent the last equations (shown in figure 6) which can be simplified by keeping the matrix elements which are different from ~ 1.

$$V_{td} + \lambda \cdot V_{cb}^* + V_{ub}^* \simeq 0 \tag{109}$$

As will be seen later, it is precisely the angles of this triangle (often called the unitarity triangle) which govern the CP violating effects in the B meson system.

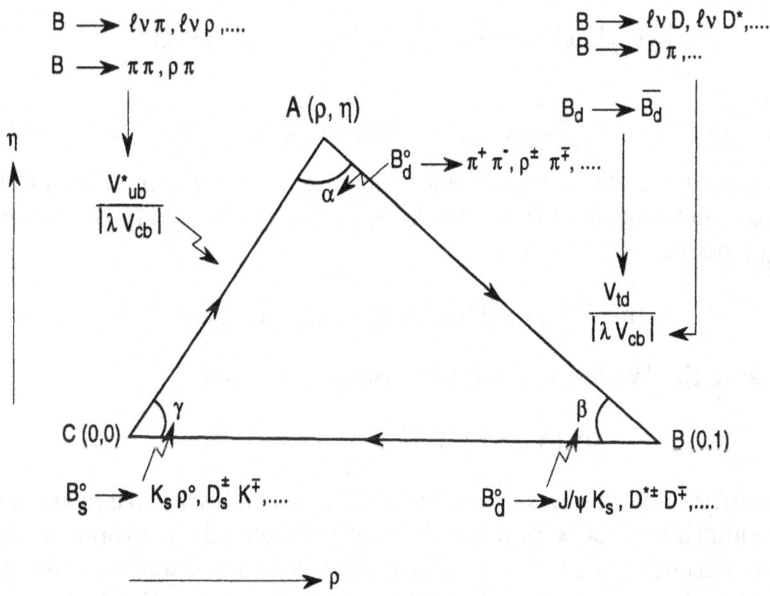

Figure 6. The unitarity triangle. We have also shown in this figure some B decay modes which allow one to measure the sides and the angles.

3.1.2. *The Invariants of the CKM Matrix.*

Since one could use 36 different parametrizations for V_{CKM}, one may ask what are the quantities that are invariant?

They are of two types :

– The magnitudes of the coupling between the quarks, $|V_{ij}|$.
– The imaginary part of $\lambda_{\alpha\beta jk}$ where

$$\mathrm{Im}\,(\lambda_{\alpha\beta \mathrm{jk}}) = \mathrm{Im}\left(\mathrm{V}_{\alpha \mathrm{j}}\mathrm{V}^*_{\alpha \mathrm{k}}\mathrm{V}_{\beta \mathrm{k}}\mathrm{V}^*_{\beta \mathrm{j}}\right) \quad \text{with} \quad \alpha \neq \beta \text{ and } \mathrm{j} \neq \mathrm{k} \tag{110}$$

The first type of invariant is obvious, therefore let us examine the second one. The subscripts α, β, j and k occur always twice in Eq. 110, (once with V and once with V^*). As a consequence, each times that one changes the phase of a quark field and absorbes it in V, one has to do the same thing with V^*. The net result is that nothing is changed. One can in fact generalize this type of invariant to $\mathrm{Im}\,(\lambda_{\alpha\beta \mathrm{jkmn}\,...}) = \mathrm{Im}(\mathrm{V}_{\alpha \mathrm{j}}\mathrm{V}^*_{\alpha \mathrm{k}}\mathrm{V}_{\beta \mathrm{m}}\mathrm{V}^*_{\beta \mathrm{n}}\mathrm{V}_{\mathrm{nk}}\mathrm{V}^*_{\mathrm{mj}}...)$ as long as the rule noted above holds.
An important property of $\mathrm{Im}\,(\lambda_{\alpha\beta \mathrm{jk}})$ has to be underlined. The absolute value of this quantity is always the same whatever α, β, j and k are. In order to verify this, let us multiply Eq. 106 by $V^*_{\alpha k}V_{\beta k}$. One obtains:

$$V_{\alpha j}V^*_{\beta j}V^*_{\alpha k}V_{\beta k} + V_{\alpha \ell}V^*_{\beta \ell}V^*_{\alpha k}V_{\beta k} = \; -\,|V^*_{\alpha k}V_{\beta k}|^2 \tag{111}$$

and thus

$$\mathrm{Im}\,(\lambda_{\alpha\beta \mathrm{jk}}) = \; -\mathrm{Im}\,(\lambda_{\alpha\beta \ell \mathrm{k}}) \tag{112}$$

One can repeat this demonstration for the other subscripts α, β and k. Using the parametrization of the Particle Data Group (PDG) [4], the quantity $\mathrm{Im}\,(\lambda_{\alpha\beta \mathrm{jk}})$ reads:

$$\mathrm{Im}\,(\lambda_{\alpha\beta \mathrm{jk}}) = \mathrm{C}_{12}\mathrm{C}_{23}\mathrm{C}^2_{13}\;\mathrm{S}_{12}\mathrm{S}_{23}\mathrm{S}_{13}\;\sin\delta \tag{113}$$

While, using the Wolfenstein parametrization, one gets

$$\mathrm{Im}\,(\lambda_{\alpha\beta \mathrm{jk}}) = \mathrm{A}^2\lambda^6\;\eta \simeq \mathcal{O}(10^{-4}) \tag{114}$$

This quantity is twice the surface area of the six unitarity triangles. Thus any CP violating effect within the Standard Model will be proportional to this area. Although all CP violation effects have the same absolute amplitude, the observed size of the effects may vary dramatically from one channel to the other. Indeed, for a particular mode, the absolute amplitude in Eq. 114 has to be scaled by the width of the system which is studied. Thus one can choose a system for which the effects of CP violation will be the largest. This comparison is made qualitatively in the table 3.

One sees that the largest effects are expected in the B system. It should however be noted that it is also possible, in principle, to select a particular decay mode which has a small partial width (say of the order of $A^2\lambda^6$) and for which CP violation may show up with a very large effect. To summarize,

	Dominant quark decays	Resonance width	CP violating effect
K	$s \longrightarrow u$	$\propto \lambda^2 \times$ phase space	$\propto A^2\lambda^4\eta$
D	$c \longrightarrow s$	$\propto 1 \times$ phase space	$\propto A^2\lambda^6\eta$
B	$b \longrightarrow c$	$\propto A^2\lambda^4 \times$ phase space	$\propto \lambda^2\eta$

TABLE 3. Relative magnitude of the CP violating effect for different $q_1\overline{q_2}$ systems.

one has two options: either using decay modes which have large branching fractions but produce small CP violating effects or rare decays which could exibit large asymmetries.

4. The constraints of the unitarity triangle

We have seen in the previous chapter that with three families, the unitarity of the CKM matrix leads to the triangle in figure 6. It is of prime importance for experimentalists to know what are the expected sizes of the three angles in order to estimate their measurement feasibility. In the following we will use present experimental results to constrain the region of the plane (ρ, η) in which the apex of the unitarity triangle is located.

4.1. CONSTRAINTS DUE TO CP VIOLATION IN THE KAON SYSTEM.

We have shown in the Chapter I that the observed CP violation in the K^o system is due to the K^o-$\overline{K^o}$ mixing. The diagrams contributing to the mixing are shown in the figure 7. The parameter ε_K, discussed in chapter I,

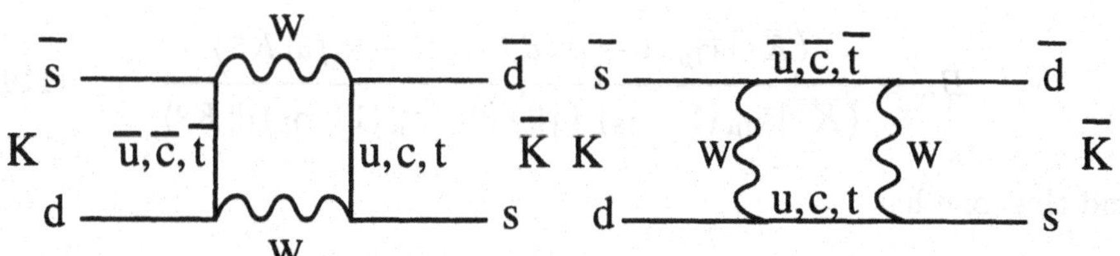

Figure 7. Feynman diagrams responsible of the K^o-$\overline{K^o}$ mixing.

is derived in the framework of the Standard Model [41],

$$
\begin{aligned}
|\varepsilon_K| &= \frac{G_F^2 m_K m_W^2 f_K^2 B_K}{6\pi^2 \sqrt{2}\, \Delta m_K} A^2 \lambda^5 \eta \times \\
&\left\{ [\eta_3 f_3(y_t) - \eta_1] y_c \lambda + \eta_2 y_t f_2(y_t) A^2 \lambda^5 (1-\rho) \right\}
\end{aligned} \qquad (115)
$$

where λ, A, ρ, and η are the four parameters of the CKM matrix in the Wolfenstein parametrization. In addition, $G_F, m_K, \Delta m_K, m_W$, and f_K are the Fermi constant, the kaon mass, the mass difference $m_{K_L} - m_{K_S}$, the W boson mass and the kaon decay constant respectively. Finally, $\eta_1 = 1.32 \pm 0.22$, $\eta_2 = 0.57 \pm 0.01$ and $\eta_3 = 0.47 \pm 0.04$ are the constants accounting for the QCD correction factors computed at the Next to Leading Order [42] and the functions $f_2(y_i)$ and $f_3(y_i)$ are [43] :

$$
\begin{aligned}
f_2(y_t) &= 1 - \frac{3}{4} \frac{y_t(1+y_t)}{(1-y_t)^2} \left[1 + \frac{2y_t}{(1-y_t^2)} \ln(y_t) \right] \\
f_3(y_t) &= \ln\left(\frac{y_t}{y_c}\right) - \frac{3}{4} \frac{y_t}{1-y_t} \left[1 + \frac{y_t}{1-y_t} \ln(y_t) \right]
\end{aligned} \qquad (116)
$$

with $y_i = \frac{m_i^2}{m_W^2}$, $i = c, t$.

Although most of the parameters in Eq. 115 are reasonably well known [4],

$$
\begin{gathered}
G_F = 1.16639(1)\ 10^{-5} \mathrm{GeV}^{-2}, \ m_W = 80.41 \pm 0.10\ \mathrm{GeV}/c^2 \\
m_K \simeq 0.497672(31)\ \mathrm{GeV}/c^2, \ \Delta m_K = (3.489 \pm 0.009)\ 10^{-15} \mathrm{GeV}/c^2 \\
\lambda = 0.2205 \pm 0.0018, \ f_K = 0.160 \pm 0.003\ \mathrm{GeV}
\end{gathered} \qquad (117)
$$

a few others have non negligible errors,

$$A \simeq 0.79 \pm 0.04, \quad m_t = 173.8 \pm 5.2\ \mathrm{GeV}/c^2, \quad 0.71 < \hat{B}_K < 0.99 \text{ [44]} \qquad (118)$$

The parameter B_K (called the "Bag"factor) is a factor accounting for the non pertubative corrections to the vaccuum insertion approximation :

$$B_K \equiv \frac{\left\langle \overline{K^o} \left| \bar{s}\gamma_\mu (1-\gamma_5) d\ \bar{s}\ \gamma_\mu (1-\gamma_5) d \right| K^o \right\rangle}{\left\langle \overline{K^o} \left| \bar{s}\gamma_\mu (1-\gamma_5) d \right| 0 \right\rangle \langle 0 | \bar{s}\ \gamma_\mu (1-\gamma_5) d | K^o \rangle} \qquad (119)$$

and thus one has

$$\left\langle \overline{K^o} \left| \bar{s}\gamma_\mu (1-\gamma_5) d\ \bar{s}\ \gamma_\mu (1-\gamma_5) d \right| K^o \right\rangle = \frac{8}{3}\ B_K\ f_K^2\ m_K^2 \qquad (120)$$

The value of B_K is determined from QCD sum rules [45] and lattice calculations [46]. It should be noted that the value of B_K is scale dependent and

therefore one usually quotes the renormalization group invariant quantity $\widehat{B}_K$ [47].
Using Eq. 115, one can extract the value of η as function of ρ. To estimate the sensitivity on the determination of η versus ρ due to the uncertainties of the parameters in Eq. 118, let us rewrite Eq. 115 as

$$|\varepsilon_K| = CB_K V_{cb}^2 \eta \left\{ [\eta_3 f_3(y_t) - \eta_1] y_c + \eta_2 y_t f_2(y_t) V_{cb}^2 (1-\rho) \right\} \quad (121)$$

where $C = \frac{G_F^2 m_W^2 \lambda^2 m_K f_K^2}{6\pi^2 \sqrt{2} \Delta m_K} = (1.86 \pm 0.04) \times 10^3$. Using the equation above, one can constrain the position of the apex of the unitary triangle in the ρ-η plane.

$$\left.\begin{aligned} |V_{cb}| &= 0.0395 \pm 0.0020 \\ \widehat{B}_K &= 0.85 \pm 0.14 \\ m_t &= 173.8 \pm 5.2 \text{ GeV}/c^2 \end{aligned}\right\} \Rightarrow \ \eta = f(1-\rho) \times [1 \pm \underbrace{0.14}_{V_{cb}} \pm \underbrace{0.16}_{B_K} \pm \underbrace{0.05}_{m_t}] \quad (122)$$

As one can see in the equation above, the main sources of error are due to V_{cb} and B_K, and despite the good precision on those parameters improvements on both of them are still desirable. For illustration, we show in the figure 8 the region of (ρ, η) allowed by the measurement of ε_K (dashed lines). This figure shows also the other constraints which will be discussed in the following and thus gives the presently allowed region of (ρ, η) when one combines them all.

4.2. CONSTRAINTS FROM $B_Q^o\,(\bar{B}Q) - \overline{B_Q^o}(B\bar{Q})$ MIXING: $|V_{TD}|$ MEASUREMENT.

The Schrödinger equation which describes the time evolution of the B^o and $\overline{B^o}$ system leads to the mixing of the two states with a period $T_{\rm osc} = 2\pi/\Delta m$ (see Eqs. 76 and 77). The probability to observe the decay of an initial B_d^o as a $\overline{B_d^o}$ at time t is

$$\mathcal{P}[B_d^o(t=0) \to \overline{B_d^o}(t)] = \frac{e^{-\Gamma t}}{2} \left(\text{ch} \frac{\Delta\Gamma_{B_d}}{2} t - \cos \Delta m_{B_d} t \right) \quad (123)$$

where $\Delta m_{B_d} = m_L - m_S$ and $\Delta\Gamma_{B_d} = \Gamma_S - \Gamma_L$. The subscripts S and L refer to the states with Short and Long lifetimes. In general, $\Delta\Gamma_B$ is suppressed relative to Δm_B by a factor $(m_b/m_t)^2$ and thus one approximates Eq. 124 with

$$\mathcal{P}(B_d^o(t=0) \to \overline{B_d^o}(t)) = \frac{e^{-\Gamma t}}{2} (1 - \cos \Delta m_{B_d} t) \quad (124)$$

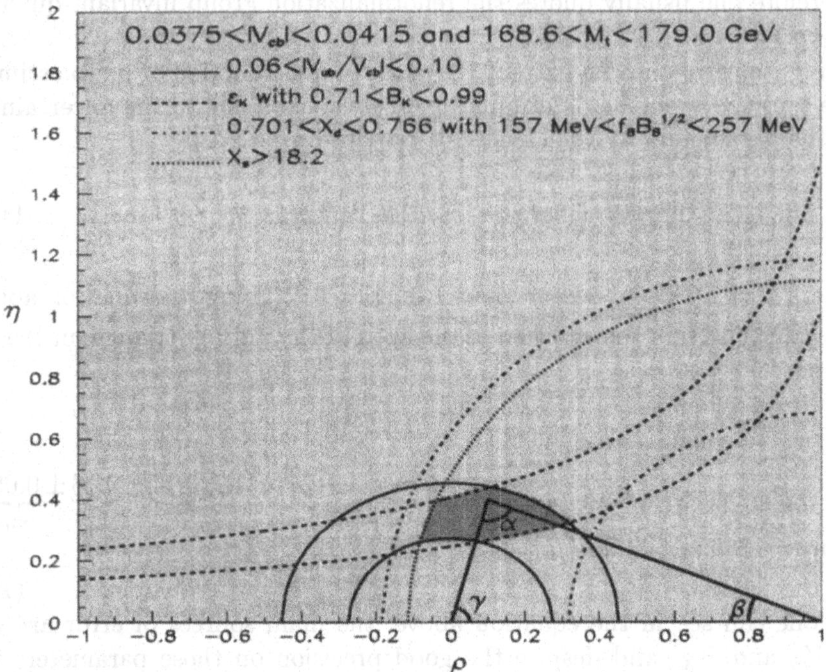

Figure 8. Present experimental constraints of the unitarity triangle in the complex plane (ρ, η).

This approximation is well justified in the B_d system since the magnitude of $\Delta\Gamma_{B_d}$ is expected to be small. However it should be noted that, in the B_s system, the above statement is not correct as one expects $\Delta\Gamma_{B_s}/\Gamma_{B_s} \approx 0.15$ [48] is likely.
The transition $B^o_d \to \overline{B^o_d}$ has been observed some 10 years ago with the ARGUS detectors [24]. The parameter x_d quantifying this mixing is calculated in the Standard Model [49] by computing the diagrams similar to those of figure 7 where the quark s is replaced by the b quark. One gets

$$x_d = \frac{\Delta m_{B_d}}{\Gamma} = \frac{G_F^2}{6\pi^2}\,\tau_B m_{B_d}\, m_t^2\, f_2\,(y_t)\;f_B^2 B_B\,\eta_{QCD}\,|V_{tb}^* V_{td}|^2 \qquad (125)$$

where τ_B, η_{QCD}, f_B and B_B are respectively the B meson lifetime, a QCD correction factor [50] ($\eta_{QCD} = .55$), the B decay constant, and the "Bag" factor for the B^o_d system. Using the Wolfenstein parametrization, $|V_{td}V_{tb}^*|^2 = A^2\lambda^6\left[(1-\rho)^2+\eta^2\right]$ and thus the measurement of Δm_{B_d} constrains the apex of the unitarity triangle to lie on a circle, the center of which is at (1,0) in the plane (ρ, η). Several experiments have contributed to the evaluation of Δm_{B_d}, using two types of method:

- The time integrated method which provides the mixing measurements.
- The time dependent technique which allows one to observe the B^o-$\overline{B^o}$oscillations.

We have discussed in Chapter 2 the former type of measurements. However, the study of the time dependent oscillations allows a better accuracy on Δm_B. In the following we will describe in some detail how this measurement is made. The analyses carried out at LEP, SLC and at the Tevatron, make intensive use of vertex detectors since the mean flight length of B mesons is small ($c\tau \approx 470\ \mu$m). Besides the knowledge of the flavor of the B^o meson at its production and decay time, the B^o decay length, ℓ, and its momentum, p, have to be measured since $t = \ell m_{B^o}/pc$. The proper time resolution is

$$\left(\frac{\sigma_t}{\tau}\right)^2 = \left(\frac{\sigma_\ell}{<\ell>}\right)^2 + \left(\frac{t}{\tau}\right)^2 \left(\frac{\sigma_p}{p}\right)^2 \tag{126}$$

In general, $\sigma_\ell/ <\ell>$ is in the 10-15% range whereas σ_p/p may vary from 10% to 20% depending on the final state which is used and the momentum evaluation method. It should be noted that this latter term dominates at long decay times. It induces a damping of the observed oscillation amplitude by a factor $e^{-(\sigma_t x_q/\tau)^2/2}$ making the measurement of x_s in the B_s^o system considerably more difficult than that of x_d as we will discuss later.
Three generic methods are used to identify the flavor of the B^o meson at its production and decay time:

- Jet charge measurement
- Inclusive lepton
- Inclusive D meson or "exclusive" B final state

The first two methods are used to tag the charge of the b quark at $t = 0$ while the last two are generally used to identify the B^o flavor at decay time. Besides the resolution of the decay length and B^o_d momentum, the precision on Δm_{B_d} obviously depends on the efficiency of the tagging but also on the wrong tag probability ω. Indeed one should add the dilution term $(1 - 2\omega)$ in front ot the cosine term in Eq. 124 to account for imperfect tagging. Although there are differences from one experiment to the other, we give in Table 4 the typical values obtained for the tagging efficiency and the fraction of wrong tags.

4.2.1. *Measurement of* Δm_{B_d}.

Some examples of analyses involving the techniques discussed above are shown in the Fig. 9 and 10 whereas more details can be found in Ref. [51]. The results obtained by the experiments are summarized in Fig. 11.
Combining all the results with proper care of common systematics, one find $\Delta m_{B_d} = (0.477 \pm 0.017)\hbar/ps$ [51] or equivalently $x_d = 0.744 \pm 0.033$

Method	Efficiency	Wrong tags
Lepton	$\sim 10\%$	$\sim 10\%$
Jet charge	$\sim 90\%$	$\sim 35\%$

TABLE 4. Typical tagging efficiencies and fraction of wrong tags.

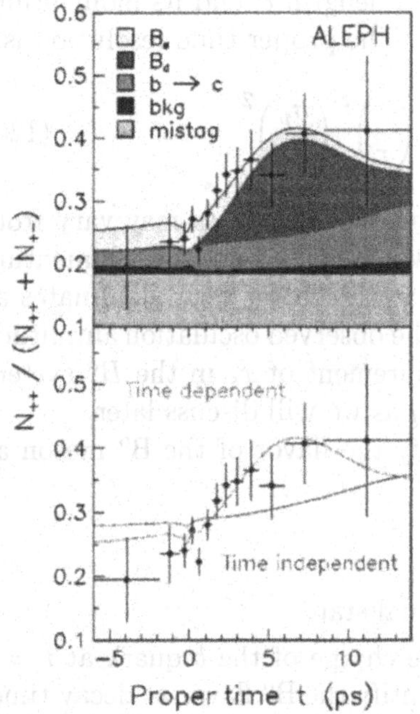

Figure 9. Time dependence of the fraction of pairs of like sign leptons from ALEPH. The solid line is a fit for Δm_{B_d}.

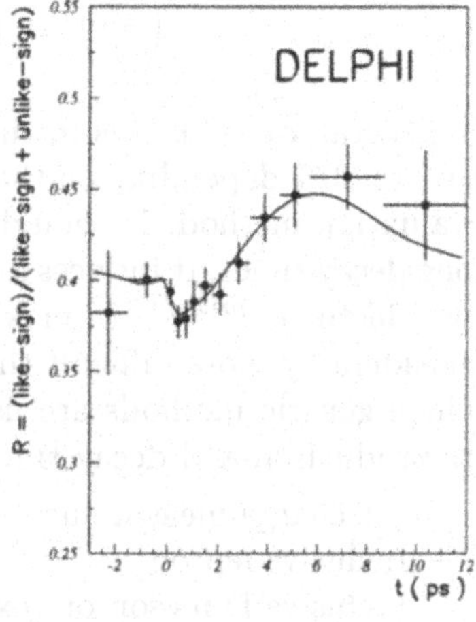

Figure 10. Time dependence of the fraction of lepton and Jet charge with the same sign from DELPHI. The solid line is a fit for Δm_{B_d}.

where we have used $\tau_{B^0_d} = (1.56 \pm 0.04)\ ps$ [4]. It is interesting to note that this value is in good agreement with the one obtained at the $\Upsilon(4S)$. Thus, being confident that quantum mechanics holds for the B^o-$\overline{B^o}$ system, we are entitled to combine all the measurements in Fig. 11 and get the world average :

$$\Delta m_{B_d} = (0.471 \pm 0.016)\ \hbar/ps \text{ or } x_d = 0.735 \pm 0.031 \qquad (127)$$

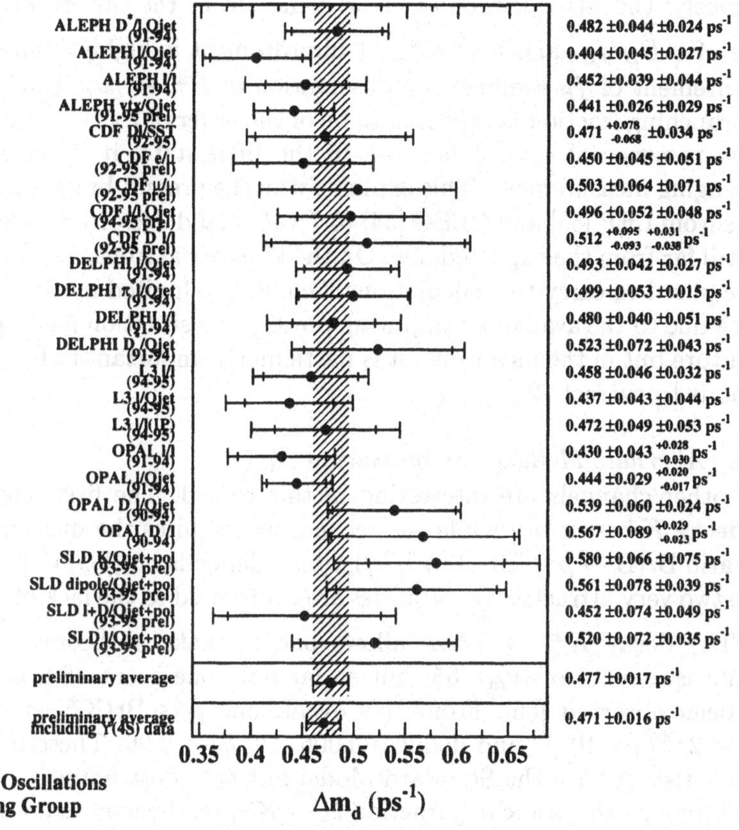

Figure 11. Summary of Δm_{B_d} measurements.

From Eq. 125 and assuming $V_{tb}^* = 1$, we derive

$$|V_{td}| = (8.3 \pm \underbrace{2.0}_{\sqrt{B_{B^o}} f_{B^o}} \pm \underbrace{0.3}_{m_t} \pm \underbrace{0.2}_{B_d^o}) \times 10^{-3} \tag{128}$$

where we have used $\sqrt{\widehat{B}_{B^o}} f_{B^o} = 207 \pm 50$ MeV, a conservative range compatible with lattice [44] and QCD sum rule [52] calculations. The measurement of $|V_{td}|$ constrains the apex of the unitarity triangle to lie in a circle (see Fig. 8) with center (1,0) and radius

$$\sqrt{(1-\rho)^2 + \eta^2} = 0.95 \pm 0.24 \tag{129}$$

Obviously, the main source of error comes from the uncertainty on the value of $\sqrt{\widehat{B}_{B^o}} f_{B^o}$ and a significant improvement is highly desirable. The measurement of f_B requires the observation of $B^{\pm} \to \tau\nu_\tau$. Unfortunately the branching fraction is expected to be of the order of 7×10^{-5} and because of the presence of several neutrini in the final state, it is an extremely challenging measurement. This explains why the present branching fraction limits from LEP [53] and CLEO [54] (5.7×10^{-4} and 2.2×10^{-3} respectively) are still far from the expected one. On the theoretical front, approximations are required to carry the calculations. One is also limited by the size of the lattice due to the available computing power. The situation may improve in the future but in the meantime, it is particularly important to find different methods to extract $|V_{td}|$.

4.2.2. *Alternate Methods for measuring $|V_{td}|$*

Two other channels are interesting in this regard. The first one involves the decay $K^{\pm} \to \pi^{\pm}\nu\nu$ while the second one requires the measurement of the ratio $\mathrm{Br}(\mathrm{B} \to \rho\gamma)/\mathrm{Br}(\mathrm{B} \to \mathrm{K}^*\gamma)$ which depends on $|V_{td}/V_{ts}|$. However, these two very attractive possibilities are confronted to the lack of statistics.

- The decay $K^{\pm} \to \pi^{\pm}\nu\nu$ allows one to make a theoretically clean measurement of $|V_{td}|$ [55] but so far only one event of this type has been observed [56]. From this event, one gets $\mathrm{Br}(\mathrm{K}^{\pm} \to \pi^{\pm}\nu\nu) = (4.2^{+9.7}_{-3.5}) \times 10^{-10}$ and deduces $0.006 < |V_{td}| < 0.06$. These bounds are consistent with the Standard Model but not constraining.
- Although the branching fraction $B \to K^*\gamma$ has been measured [$\mathrm{Br}(\mathrm{B} \to \rho\gamma) = (4.2 \pm 0.8 \pm 0.6) \times 10^{-5}$] [57], the $B \to \rho\gamma$ transitions have not been seen yet and only an upper limit is devived; $\mathrm{Br}(\mathrm{B} \to \rho\gamma)/\mathrm{Br}(\mathrm{B} \to \mathrm{K}^*\gamma) < 0.19$ [57]. This later limit translates to $|V_{td}/V_{ts}| < 0.56$ depending on the size of SU(3) breaking.

4.2.3. *Study of $B^0_s - \overline{B^0_s}$ Oscillation*

The main source of uncertainty on the extraction of $|V_{td}|$ from $B^o_d - \overline{B^o_d}$ oscillations is the large theoretical uncertainty due to $\sqrt{B_{B_d}} f_{B_d}$. Reducing this uncertainty is possible by measuring Δm_{B_s}, the mass difference between the two B^o_s mass eigenstates. Indeed since $V_{ts} \simeq V_{cb}$ from unitarity, the ratio

$$\frac{\Delta m_{\mathrm{B}_s}}{\Delta m_{\mathrm{B}_d}} = \frac{m_{\mathrm{B}^o_s}}{m_{\mathrm{B}^o_d}} \underbrace{\frac{B_{\mathrm{B}^o_s} f^2_{\mathrm{B}^o_s}}{B_{\mathrm{B}^o_d} f^2_{\mathrm{B}^o_d}}}_{\xi^2_s} \left|\frac{V_{ts}}{V_{td}}\right|^2 \tag{130}$$

would allow to extract V_{td} with a smaller error because the ratio ξ_s, slightly different from unity due to SU(3) breaking, is better known: $\xi_s = 1.14\pm0.08$, a range compatible with QCD sum rules [58] and QCD on lattice [44]. Using the Wolfenstein parametrization, one writes

$$\frac{\Delta m_{\mathrm{B}_s}}{\Delta m_{\mathrm{B}_d}} = \xi_s^2 \frac{1}{\lambda^2} \frac{1}{(1-\rho)^2+\eta^2} \tag{131}$$

Therefore a measurement of Δm_{B_s} allows a more precise contraint on ρ and η. Inversely the present ρ, η region in Fig. 8 implies $7.2 < \Delta m_{\mathrm{B}_s} < 20.5$ possibly leading to an extremely difficult measurement.
The ingredients used in the analyses searching for $B_s^0 - \overline{B_s^0}$ oscillations are essentially the same as for $\mathrm{B_d^0}-\overline{\mathrm{B_d^0}}$. The highest sensitivities are obtained by using dilepton events or lepton-Jet charge correlations. A maximum likelihood technique is used to fit the time dependence of the like-sign fraction or the asymmetry: (unlike-sign - like-sign) / (unlike-sign + like-sign). However these analyses suffer from a large contamination of non $\mathrm{B_s^0}$ decays. Typically the fraction of events with a $\mathrm{B_s^0}$ meson is estimated to be of the order of 10% or lower. Other methods have been tried using final states which allows one to enrich significantly the sample with $\mathrm{B_s^0}$ (for example by using the final state $\ell\nu \mathrm{D}_s$ with the subsequent decays $\mathrm{D}_s \to \phi\pi$ and K*K). However, despite the very large fraction of candidate events with a $\mathrm{B_s^0}$ meson ($\sim$ 75%), such an analysis suffers from the limited statistics. As more data and more D_s final states are added this method becomes competitive. To enrich the lepton-Jet charge sample with $\mathrm{B_s^0}$ meson, one can also require a fragmentation kaon in the lepton side. The identified kaon must come from the primary vertex and raises the fraction of $\mathrm{B_s^0}$ to about 35%.
However and despite intensive efforts, none of the analyses has established $B_s^o - \overline{B_s^o}$ oscillations [59]. The combination of the various measurements has been presented at the ICHEP98 [59]: From Fig. 12, one gets

$$\Delta m_{\mathrm{B}_s} > 12.4\ \hbar/ps \text{ or } x_s > 18.2 \tag{132}$$

The lower limit on x_s is obtained using $\tau_{\mathrm{B_s^0}} = (1.54 \pm 0.07) ps$ [4] and by allowing for a one σ fluctuation. Taking the average value of Δm_{B_d}, one gets

$$\begin{aligned} \frac{\Delta m_{\mathrm{B}_s}}{\Delta m_{\mathrm{B}_d}} &> 25.5 \text{ at } 95\% \text{ C.L.} \\ \left|\frac{V_{ts}}{V_{td}}\right| &> 4.0 \text{ at } 95\% \text{ C.L.} \end{aligned} \tag{133}$$

Using Eq. 131, this limit can be translated into

$$\sqrt{(1-\rho)^2+\eta^2} < 1.12 \tag{134}$$

By comparing the above result to Eq. 129, one sees that the limit on Δm_{B_s} is now competitive and provides a useful bound (Fig. 8).

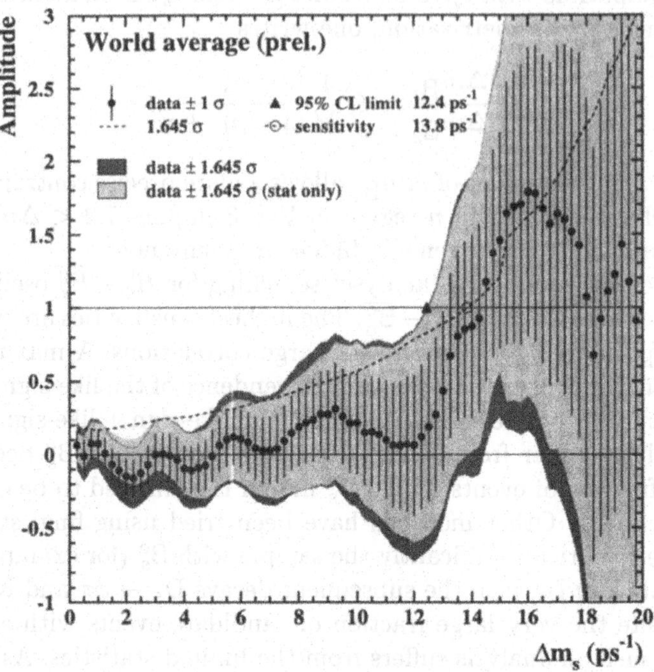

Figure 12. The limit on Δm_{B_s} at the 95% CL as presented at the ICHEP98.

4.3. MEASUREMENTS OF V_{CB}

The value of V_{cb} is an important input for constraining the unitarity triangle. In figure 6 it is used as the normalization to the sides of the triangle. It is therefore desirable to determine it with the smallest possible error. To explain how this is done, let us first describe in very general terms how B meson decays proceed and discuss the main difficulties for extracting useful information from the measured branching fractions.
The B meson decay can be described at first order as a weak decay of the b quark (Fig. 13a) in the same way as the decay of the muon. Thus, the partonic decay width for $b \to c$ transition is

$$\Gamma(\bar{b} \to \bar{c} f_1 \overline{f_2}) = \underbrace{\frac{G_F^2 m_b^5}{192\pi^3}}_{\Gamma_0} |V_{cb}^* V_{f_1 f_2}|^2 \times I\left(\frac{m_c}{m_b}, \frac{m_{f_1}}{m_b}, \frac{m_{f_2}}{m_b}\right) \tag{135}$$

where Γ_0 is the muon decay width when m_b is replaced by m_μ and the

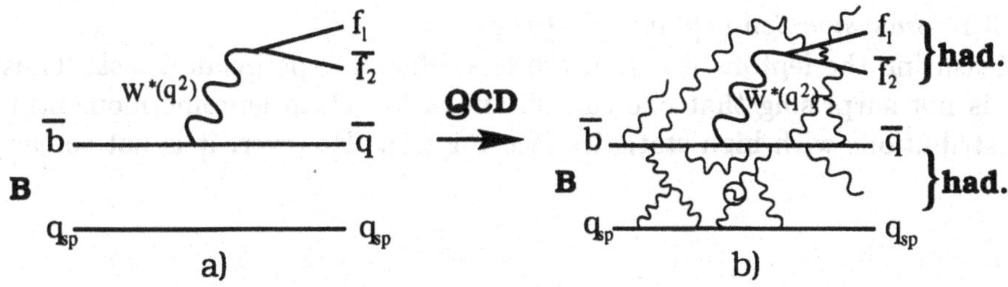

Figure 13. Dominant Feynman diagram describing B meson decays. Figure (a) shows the simple 4-fermion interaction at the parton level while figure (b) visualizes the full complexity of the problem when QCD is turned on to account for hadronization.

function $I(y_1, y_2, y_3)$ is a phase space factor accounting for the mass of the final state fermions. Neglecting the mass of f_1 and f_2, one gets

$$I(y,0,0) = 1 - 8y + 8y^2 - y^4 - 12y^2 \log y \tag{136}$$

The main difficulty is to translate $\Gamma(\bar{b} \to \bar{c} f_1 \overline{f_2})$ into $\Gamma(\bar{b} \to X_c Y...)$ by introducing the gluonic radiative corrections, the bound state effects of the initial hadron and the hadronization of the final state. Two avenues have been pursued

- Development of phenomenological models
- Simplification of QCD to allow calculations $\Rightarrow$ development of Heavy Quark Effective Theory (HQET) [60]

By looking at Fig. 13b, it is obvious that an important simplification can be obtained by considering semileptonic final states where $f_1 \overline{f_2}$ is replaced by $\ell\nu$. In the latter decays all gluonic interactions are restricted to the b quark, the spectator quark q_{sp} and the produced final quark. Looking at Fig. 13b, one guess that there are 3 relevant energy scales in the B meson decay:

- The electroweak scale $\frac{1}{m_W}$
- The QCD perturbative scale $\frac{1}{m_b}$ accounting for gluon bremstrahlung and vertex corrections.
- The QCD non-perturbative scale $\frac{1}{\Lambda_{QCD}}$ accounting for hadronic state formation.

In order to simplify further the picture, we need to find decay schemes where the hadronization is not important or can be calculated. Such schemes exist and are encountered by either considering **inclusive** semileptonic decays or **exclusive** semileptonic decays to heavy hadrons.

4.3.1. *Inclusive Semileptonic B Decays*

Measuring the leptons is one of the less difficult experimental task. Thus, it is not surprising that one can obtain rather clean leptonic momentum distributions with high statistics (see Fig 14). However, it is not so easy

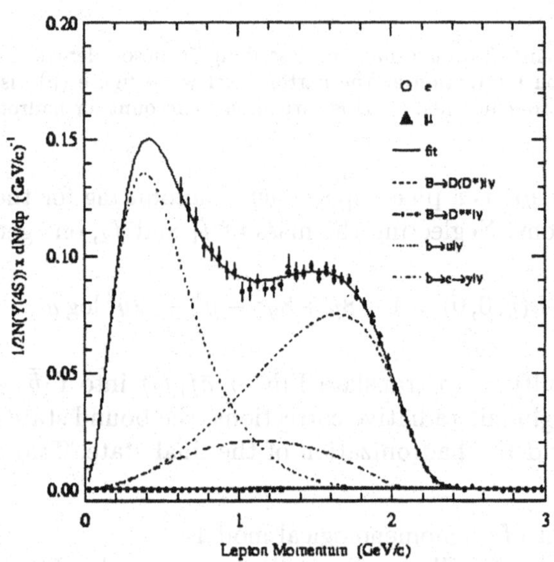

Figure 14. The lepton momentum spectrum in semileptonic B decays measured by CLEO: The solid line is a fit of the data including the direct ($\bar{b} \to \bar{c}\ell^+\nu_\ell$; dashed line) and the cascade ($\bar{b} \to \bar{c} \to y\ell^-\overline{\nu_\ell}$; dash-dotted line) leptons using a particular model [62].

to separate the part coming from the direct leptons ($\bar{b} \to \bar{c}\ell^+\nu_\ell$) from the part due to cascades ($\bar{b} \to \bar{c} \to y\ell^-\overline{\nu_\ell}$) and it is only since recently that the contribution from these 2 sources have been measured (Fig. 15) [61].

Inclusive measurement are very useful in the sense that most of the hadronization effects become unimportant if one integrates over all possible final states as the transition amplitude is then unity. However one effect which remains is the interaction of the quarks in the initial state, i.e. the bound state effects due to the potential and the spin-spin interactions. Let us first discuss how one handles these effects in a particular phenomenological approach: the ACCMM model [63]. Here non perturbative effects in the initial state are accounted for by introducing the Fermi motion of the b quark in

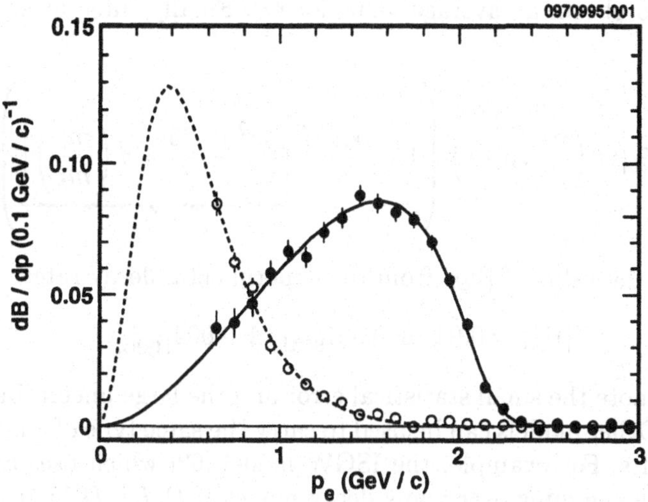

Figure 15. The electron momentum spectrum in semileptonic B decays measured by CLEO: $B \to Xe^+\nu_e$ (filled circles) and $b \to \bar{c} \to ye^-\overline{\nu_e}$ (open circles).

the B meson. Hence

$$m_B^2 - m_b^2 = 2m_B\sqrt{p^2 + m_{sp}^2} + m_{sp}^2 \tag{137}$$

where the momentum distribution function is

$$\Phi(|p|) = \frac{4}{\sqrt{\pi}p_F^3}e^{-(|p|^2/p_F^2)} \tag{138}$$

In this model, the b quark has no definite mass and the free parameters are p_F and m_{sp}. The authors have calculated the differential decay width as a function of the lepton energy.

$$\begin{aligned}\frac{d\Gamma(m_b,x)}{dx} &= 2\Gamma_0|V_{cb}|^2\frac{x^2(x_m-x)^2}{(1-x)^3}\left[(1-x)(3-2x)-(1-x_m)(3-x)\right] \\ &\times\left[1-\frac{\alpha_s(\mu)}{\pi}G(x,\epsilon)\right]\end{aligned} \tag{139}$$

where $x = 2E_\ell/m_b$, $\epsilon = m_c/m_b$ and $x_m = 1-\epsilon^2$. The function $G(x,\epsilon)$ includes the QCD perturbative corrections. With a likelihood method, one extract p_F and m_{sp} from the leptonic energy spectrum using Eq. 139. This allows one to obtain an average value for m_b. Finally, integrating Eq. 139, one obtains

$$\Gamma \simeq \Gamma_0 |V_{cb}|^2 I\left(\frac{m_c}{m_b}, 0, 0)\right) \left(1 - \frac{\alpha_s(\mu)}{\pi} \underbrace{\left[\frac{2\pi^2}{3} - \frac{25}{6} + \delta\left(\frac{m_c^2}{m_b^2}\right)\right]}_{\sim 1.54}\right) \tag{140}$$

and derives the value of $|V_{cb}|$ from the experimental decay rate:

$$|V_{cb}| = 0.041 \pm 0.001_{\text{stat.}} \pm 0.004_{\text{theo.}} \tag{141}$$

One should note the small statistical error and the large uncertainty due to the model. Other models can be used to carry the same type of analysis with similar results. For example, the ISGW model [62] which uses a different approach (i.e summing over many decay modes $\ell\nu D, D^*, D^{**}$) gives $|V_{cb}| = 0.040 \pm 0.001_{\text{stat.}} \pm 0.004_{\text{theo.}}$

It is possible to avoid model dependence by applying a new method using the heavy quark expansion [64]. However, it assumes the hypothesis of quark–hadron duality. The assumption of duality in B decays is that the decay rates are calculable in QCD after an "averaging" procedure has been applied [65]. In semileptonic decays, it is the integration over the lepton and neutrino phase space that provides the proper averaging over the invariant hadronic mass of the final state. This is so-called global duality in opposition to local duality used for inclusive nonleptonic decays where the mass of the observed hadron is fixed, and it is only the fact that one sums over many final hadronic states that provides any averaging. Therefore global duality is a weaker assumption than local duality.

Using the optical theorem, the inclusive decay width of a B meson is written as

$$\Gamma(B \to X) = \frac{1}{m_B} \, \text{Im} \, \langle B | \, \mathbf{T} \, | B \rangle \,, \tag{142}$$

where the transition operator $\mathbf{T}$ is given by

$$\mathbf{T} = i \int d^4x \, T\{\, \mathcal{L}_{\text{eff}}(x), \mathcal{L}_{\text{eff}}(0) \,\} \,, \tag{143}$$

and $\mathcal{L}_{\text{eff}}$ is the effective weak Lagrangian. The semileptonic inclusive width can be written [64]

$$\begin{aligned} \Gamma(B \to X_c \ell^+ \nu) \quad &= \quad \frac{G_F^2 m_b^5}{192\pi^3} |V_{cb}|^2 \times \\ &\left\{ c_3(y) \, \langle \bar{b}b \rangle_H + c_5(y) \, \frac{\langle \bar{b} \, g_s \sigma_{\mu\nu} G^{\mu\nu} b \rangle_H}{m_b^2} + \ldots \right\} \end{aligned} \tag{144}$$

where the c_n are calculable short-distance coefficients and are functions of $(m_c/m_b)^2$

$$\begin{aligned} c_3(y) &= (1-8y+8y^3-y^4-12y^2\log y)\times[1+\mathcal{O}(\alpha_s)]\,, \\ c_5(y) &= -(1-y)^4 \end{aligned} \tag{145}$$

and

$$\begin{aligned} \langle \bar{b}b\rangle &= 1+\frac{\lambda_1+3\lambda_2}{2m_b^2}+\mathcal{O}(1/m_b^3)\,, \\ \frac{\langle \bar{b}\, g_s\sigma_{\mu\nu}G^{\mu\nu}b\rangle}{m_b^2} &= \frac{6\lambda_2}{m_b^2}+\mathcal{O}(1/m_b^3)\,. \end{aligned} \tag{146}$$

Here λ_1 and λ_2 parameterize the matrix elements of the heavy-quark kinetic energy and chromomagnetic interaction inside the B meson, respectively. The parameters are independent of the heavy quark mass and are therefore the same in the D meson system (to the extend that the c quark is heavy enough). The equations 144, 145 and 146 show still a dependence on m_b which can be evaded by expanding the B meson mass in powers of m_b [66, 67].

$$\begin{aligned} m_B &= m_b+\bar{\Lambda}-\frac{\lambda_1+3\lambda_2}{2m_b}+\ldots \\ m_B^* &= m_b+\bar{\Lambda}-\frac{\lambda_1-\lambda_2}{2m_b}+\ldots \end{aligned} \tag{147}$$

where $\bar{\Lambda}\sim\Lambda_{QCD}$ represents the energy contribution of the light degrees of freedom. Introducing the spin-averaged masses $\overline{m}_B=\frac{1}{4}(m_B+3m_{B^*})$ and $\overline{m}_D=\frac{1}{4}(m_D+3m_{D^*})$, one obtains

$$\begin{aligned} m_b-m_c &= (\overline{m}_B-\overline{m}_D)\left(1+\frac{(-\lambda_1)}{2\overline{m}_B\overline{m}_D}+\ldots\right)\,, \\ m_{B^*}^2-m_B^2 &= 4\lambda_2+\ldots \end{aligned} \tag{148}$$

Using Eqs. 145 and 146, on the one hand, and doing the substitution of Eqs. 147 and 148 on the other hand, one rewrites Eq. 144 [68]

$$\begin{aligned} \Gamma(B\to X_c\ell^+\nu) &= \frac{G_F^2|V_{cb}|^2m_B^5}{192\pi^3}\times 0.369\times \\ &\quad \left[1-1.54\,\frac{\alpha_s}{\pi}-1.65\,\frac{\bar{\Lambda}}{m_B}\left(1-0.87\,\frac{\alpha_s}{\pi}\right)\right. \\ &\quad \left.-0.95\,\frac{\bar{\Lambda}^2}{m_B^2}-3.18\,\frac{\lambda_1}{m_B^2}+0.02\,\frac{\lambda_2}{m_B^2}\right] \\ &\equiv |V_{cb}|^2\gamma_c \end{aligned} \tag{149}$$

From the second relation in 148, it follows that $\lambda_2 \approx 0.12$ GeV^2. The parameter $\bar{\Lambda}$ and the kinetic-energy parameter λ_1, on the other hand, are given in terms of a quark mass and a difference of quark masses respectively and cannot be determined from hadron spectroscopy. Extracting these parameters from data is possible but will not be discussed here (see for example [68]). Various model calculations have been used to obtain $\bar{\Lambda}$ and λ_1 as well as QCD sum rules and QCD on lattice evaluations. The range of predictions obtained from a variety of methods is 0.2 GeV $< \bar{\Lambda} <$ 0.7 GeV and 0.1 GeV$^2 < -\lambda_1 <$ 0.6 GeV^2 [69].
Using Eq. 149, one extracts $|V_{cb}|$ from the measurement of the semileptonic branching fraction and B meson lifetime, τ_B:

$$|V_{cb}| = \xi_{\rm th} \left(\frac{{\rm Br}(B \to X_c \ell^+ \nu)}{10.5\%} \right)^{1/2} \left(\frac{1.6 \text{ ps}}{\tau_B} \right)^{1/2}, \tag{150}$$

where $\xi_{\rm th} = 0.0403 \pm 0.0019$ if all theoretical errors are summed quadratically. The main sources of theoretical uncertaintiess are due to higher-order perturbative corrections, the dependence on the mass difference $m_b - m_c$, the dependence on the b-quark mass, and unknown $1/m_b^3$ corrections. Comparing the above result to Eq. 141, one should notice the significant improvement of the theoretical uncertainty.

4.3.2. *Exclusive semileptonic B Decays*
The exclusive semileptonic B decays $B \to D\ell^+\nu_\ell$ and $B \to D^*\ell^+\nu_\ell$ (Fig. 16) offer an other interesting way to measure $|V_{cb}|$.

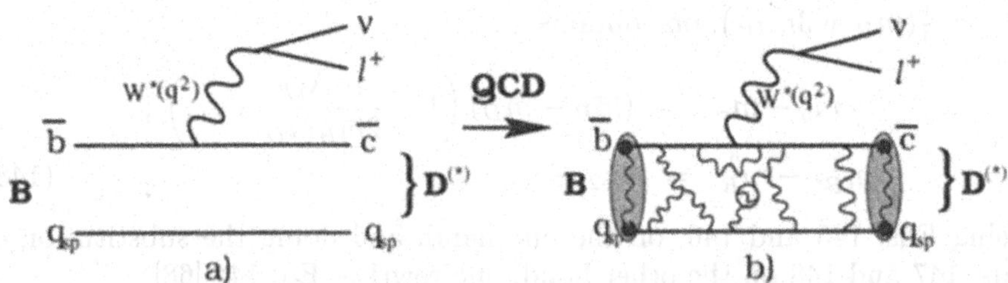

Figure 16. Feynman diagram describing $B \to D^{(*)}\ell^+\nu$ decays. Figure (a) shows the simple 4-fermion interaction at the parton level while figure (b) visualizes QCD interaction.

In such decays, one can factorize the transition with a leptonic and a hadronic current.

$$\begin{aligned} &\langle D\ell^+\nu|\, \bar{c}\gamma^\mu(1-\gamma^5) b\, \bar{\nu}\gamma_\mu(1-\gamma^5)\ell\, |B\rangle = \\ &\langle D|\, \bar{c}\gamma^\mu(1-\gamma^5) b\, |B\rangle \times \langle \ell^+\nu|\, \bar{\nu}\gamma_\mu(1-\gamma^5)\ell\, |0\rangle. \end{aligned} \tag{151}$$

While the leptonic part involves only weak currents, one needs in general to introduce several form factors [70] to write the hadronic part to account for the hadronization. Two form factors are required for a pseudoscalar to pseudoscalar transition in which only a vector current is allowed:

$$
\begin{aligned}
< P(p')|V_\mu|B(p) > &= \left[(p'+p)_\mu - \frac{m_B^2 - m_P^2}{q^2} q_\mu\right] F_1(q^2) \\
&+ \frac{m_B^2 - m_P^2}{q^2} q_\mu F_0(q^2),
\end{aligned} \tag{152}
$$

where P and V_μ denote a pseudoscalar meson and a vector current respectively, and $q = p - p'$. Note that often a pole dependence of the type $F(q^2) = F(0)/[1 - q^2/(m_b^2 + m_c^2)]$ is used for the form factors and that $F_1(0) = F_0(0)$.
In a pseudoscalar to vector transition four form factors are necessary since both a vector and axial vector current are present:

$$
\begin{aligned}
\langle V(\epsilon, p')|(V_\mu - A_\mu)|B(p)\rangle &= \frac{2}{m_B + m_V} i\epsilon_{\mu\nu\alpha\beta}\epsilon^{\nu *} p^\alpha p'^\beta V(q^2) \\
&-(m_B + m_V)[\epsilon_\mu - \frac{\epsilon^*.q}{q^2} q_\mu] A_1(q^2) \\
&+ \frac{\epsilon^*.q}{m_B + m_V}[(p+p')_\mu - \frac{m_B^2 - m_V^2}{q^2} q_\mu] A_2(q^2) \\
&-\epsilon^*.q \frac{2m_V}{q^2} q_\mu A_0(q^2)
\end{aligned} \tag{153}
$$

The form factor A_0 is related to A_1 and A_2 with:

$$
A_0(0) = \frac{m_B + m_V}{2m_V} A_1(0) - \frac{m_B - m_V}{2m_V} A_2(0). \tag{154}
$$

In principle all these form factors could be calculated using QCD. However we are dealing here with non pertubative effects and thus this evaluation is difficult. To do so, one either introduces model dependence or uses QCD sum rules or QCD on lattice calculations. All these methods have their advantages and disavantages which turn out giving somewhat large uncertainties. It is therefore highly desirable to find a framework where it is possible to derive easily these factors from first principles. Such a framework exists and is HQET. It is not the purpose of these lectures to make an explicit derivation of the theory and we leave it to the interested readers to consult the literature [71]. Instead we just give below a brief and intuitive description of the main features of the theory.
As we have said in the introduction on the measurement of V_{cb}, the B meson decay is characterized by 3 scales, the electroweak scale $1/m_W$, the b

quark scale $1/m_b$, and the QCD scale $1/\Lambda_{QCD}$. The B meson system can be viewed as a heavy quark surrounded by a valence light quark and a "cloud" of gluons and light sea quarks or antiquarks strongly interacting with it at the scale $1/\Lambda_{QCD}$. This process determines the size of the hadron. The soft gluons exchanged between the heavy quark and the light constituents can only resolve distances much larger than the heavy quark scale and hence as $m_Q \to \infty$, the light quarks only see the color field of the heavy quark irrespective of its flavor or spin quantum numbers. This resulting flavor and spin symmetry provides useful relations between B, D, B^* and D^*, or the heavy baryons Λ_b and Λ_c up to $1/m_Q$ and $\alpha_s(m_Q)$ corrections which account for the fact that the mass of the b and c quarks is not infinite.
In the semileptonic decay $B \to D\ell^+\nu$, the transition $B \to D$ is governed by the velocity of the incoming heavy B meson (v) and the outgoing heavy D meson (v'). According to what we have said above, when both velocity are equal the light quarks will not "notice" any change (flavor and spin symmetry). On the other hand, when $v \neq v'$, the light constituents suddenly find themselves interacting with a moving color source. Soft gluons have to be exchanged to rearrange them so as to form a D meson moving at velocity v'. This reorganization leads to a form-factor suppression as the velocities become more and more different. It is the reflection of the fact that harder and harder gluon exchanges are required or equivalently that the probability for an elastic transition decreases. It follows that, in the limit $m_b \to \infty$, the form factor can only depend on the Lorentz boost $w = v{\cdot}v'$ (or the momentum transfer $q^2 = m_B^2 + m_D^2 - 2m_B m_D w$) connecting the rest frames of the initial- and final-state mesons. Thus, in this limit a dimensionless probability amplitude $\xi(v \cdot v')$ describes the transition. It is sometimes called the "Isgur–Wise function" [72]. The transition process can be written as

$$\frac{1}{\sqrt{m_D m_B}} \langle \bar{D}(v') | \,\bar{c}_{v'} \gamma^\mu b_v \,| \overline{\mathrm{B}}(v) \rangle = \xi(v \cdot v')\,(v + v')^\mu . \qquad (155)$$

Here b_v and $c_{v'}$ are the velocity-dependent heavy-quark fields of the HQET. According to our description above, the function $\xi(v \cdot v') = 1$ for $v \cdot v' = 1$. This is the translation of the fact that "nothing is changed" (i.e. the transition amplitude $B \to \bar{D}$ is unity due to current conservation).
In Eq. 152, we described the transition $B \to D$ in terms of the form factors $F_1(q^2)$ and $F_0(q^2)$. It follows that, in the heavy-quark limit, these form factors are no longer independent, but they must obey the relation

$$F_1(q^2) = \left[1 - \frac{q^2}{(m_B + m_D)^2}\right]^{-1} F_0(q^2) = \frac{m_B + m_D}{2\sqrt{m_B m_D}}\,\xi(v \cdot v'), \qquad (156)$$

A similar simplification takes place for the form factors related to the $B \to D^*$ transition and one gets

$$\begin{aligned} V(q^2) = A_0(q^2) = A_2(q^2) &= \left[1 - \tfrac{q^2}{(m_B+m_{D^*})^2}\right]^{-1} A_1(q^2) \\ &= \frac{m_B + m_{D^*}}{2\sqrt{m_B m_{D^*}}}\,\xi(v\cdot v') . \end{aligned} \tag{157}$$

One can now write the differential width for the decay $B \to \bar{D}\ell^+\nu_\ell$ and $B \to \bar{D}^*\ell^+\nu_\ell$ [73] using the relations in Eq. 156 and 157.

$$\frac{d\Gamma(B\to\bar{D}\,\ell^+\,\nu)}{dw} = \frac{G_F^2}{48\pi^3}|V_{cb}|^2\,(m_B+m_D)^2\,m_D^3\,(w^2-1)^{3/2}\,\mathcal{G}^2(w)\,, \tag{158}$$

$$\begin{aligned} \frac{d\Gamma(B\to\bar{D}^*\ell^+\,\nu)}{dw} &= \frac{G_F^2}{48\pi^3}|V_{cb}|^2\,(m_B-m_{D^*})^2\,m_{D^*}^3\,\sqrt{w^2-1}\,(w+1)^2 \\ &\times\left(1+\frac{4w}{w+1}\,\frac{m_B^2-2w\,m_B m_{D^*}+m_{D^*}^2}{(m_B-m_{D^*})^2}\right)\mathcal{F}^2(w) \end{aligned} \tag{159}$$

where the form factors $\mathcal{G}(w)$ and $\mathcal{F}(w)$ coincides with $\xi(w)$ up to small symmetry-breaking corrections [74] and [75–77] .

$$\begin{aligned} &\mathcal{G}(w) = \mathcal{G}(1)\,\left[1-\widehat{\rho}_g^2\,(w-1)+\widehat{c}_g\,(w-1)^2+\ldots\right] \text{ with} \\ &\mathcal{G}(1) = [1+\delta_{QED}][1+g(\alpha_s)][1+\delta'_{1/m}+\delta'_{1/m^2}] = 0.99\pm0.07\ \ [74] \end{aligned} \tag{160}$$

and

$$\begin{aligned} &\mathcal{F}(w) = \mathcal{F}(1)\,\left[1-\widehat{\rho}^2\,(w-1)+\widehat{c}\,(w-1)^2+\ldots\right] \text{ with} \\ &\mathcal{F}(1) = [\underbrace{1+\delta_{QED}}_{\approx 1.007\ [75]}][\underbrace{1+f(\alpha_s)}_{0.960\pm0.007\ [76]}][\underbrace{1+\delta_{1/m^2}}_{0.945\pm0.025\ [77]}] = 0.913\pm0.027 \end{aligned} \tag{161}$$

There is no $(1/m)$ corrections [78] in Eq. 161 and thus one obtains a better resolution for the extraction of $|V_{cb}|$ using $B \to \bar{D}^*\ell^+\nu_\ell$ data. Fig. 17 shows an example of the measurement of $|V_{cb}|$. It can be seen from Eq. 161 that as long as the value of $w-1$ is small, the w dependence can be fitted as a function of $\widehat{\rho}^2$. Combining the results from ARGUS, CLEO and the LEP experiments one finds:

$$\begin{aligned} V_{cb} &= 0.0386\pm0.0015_{\text{stat.}}\pm0.0012_{\text{th.}} \\ \text{with } \widehat{\rho}^2 &= 0.69\pm0.08 \end{aligned} \tag{162}$$

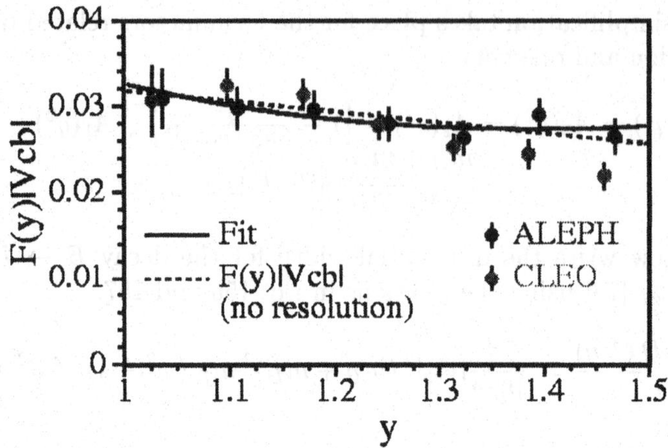

Figure 17. The determination of $\mathcal{F}(1) \times |V_{cb}|$ using $B \to D^{(*)}\ell^+\nu$ decays from ALEPH and CLEO data. The solid line shows the $y(\equiv w)$ dependent fit.

The analysis of exclusive semileptonic decays gives results compatible with the inclusive analysis although with different systematic errors. Here the precision is still limited by the statistics and therefore improvements are to be expected in the near future. Combining both techniques, one gets the present results on $|V_{cb}|$:

$$|V_{cb}| = 0.0395 \pm 0.0020 \tag{163}$$

4.4. MEASUREMENT OF V_{UB}

The Feynman diagram describing the semileptonic charmless B decays is shown in the figure 18. Accordingly, one is tempted to do a naive transposition of the techniques involved to measure V_{cb} to extract the value of V_{ub}. This translates either in studying the inclusive semileptonic spectrum for final states where no charm is produced or measuring exclusive charmless decay rates such as for example $B \to \pi(\rho)\ell^+\nu_\ell$. Unfortunately, the parallel with V_{cb} is limited by a couple of serious difficulties. In the inclusive decays, one faces an experimental problem; namely how to warranty that the observed decay do not contain a charmed quark. In the exclusive decays, no simplification of the form factors is possible since we are now dealing with heavy to light quark transitions for which HQET cannot be applied.

4.4.1. *Inclusive charmless Decays*

The first evidence for $b \to u$ transitions has been established using the observed lepton momentum spectrum [79] (Fig. 14). With the vertical scale

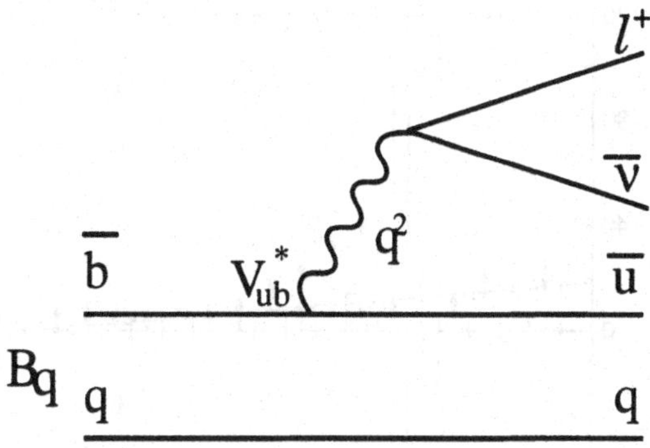

Figure 18. Dominant Feynman diagram responsible for the charmless semileptonic B decays.

of that figure, the distribution of $\bar{b} \to \bar{u}\ell^+\nu$ decays would run along the absissa since $|V_{ub}| << |V_{cb}|$ making the identification of $\bar{b} \to \bar{u}$ transition by vetoing charm a difficult experimental task. Therefore one is constrained to use a part of the phase space where no $\bar{b} \to \bar{c}$ are present. Thanks to the mass of the c quark, such a region is found by requiring

$$E_\ell \geq \left.\frac{m_B^2 - m_D^2}{2m_B}\right|_{\text{B frame}} \approx 2.45 \text{ GeV}|_{\Upsilon(4S)\text{ frame}} \tag{164}$$

As one can see in Fig. 19, an excess of lepton is observed in that region and $|V_{ub}|$ can be extracted in principle using

$$|V_{ub}|^2 = \frac{\Delta Bu|_{\text{measured}}}{\tau_B} \times \underbrace{\frac{1}{\gamma_u f(p)}}_{\text{theory}} \tag{165}$$

where ΔBu is the observed branching fraction of leptons due to $\bar{b} \to \bar{u}$ transitions. The term γ_u is similar to the one in Eq. 149 and is the theoretical total rate for charmless semileptonic decays divided by $|V_{ub}|^2$ ($\gamma_u = \Gamma_{\text{th.}}(B \to X_u\ell^+\nu)/|V_{ub}|^2$). Finally, $f(p)$ is the theoretical fraction of leptons expected in the observed region. However because a very small fraction of the spectrum is measured there is a large uncertainty on $f(p)$. Indeed, if one decides to use a model, a very strong model dependence is introduced which is difficult to estimate. Nevertheless such analyses have been carried out using for example the ACCMM model and one obtains $|V_{ub}| = (3.4 \pm 0.3_{\text{exp.}} \pm ?_{\text{th.}}) \times 10^{-3}$. Other models can be used to get an idea of the theoretical uncertainty. From the range of the results, one estimates

$$2.9 \times 10^{-3} < |V_{ub}| < 3.8 \times 10^{-3}. \tag{166}$$

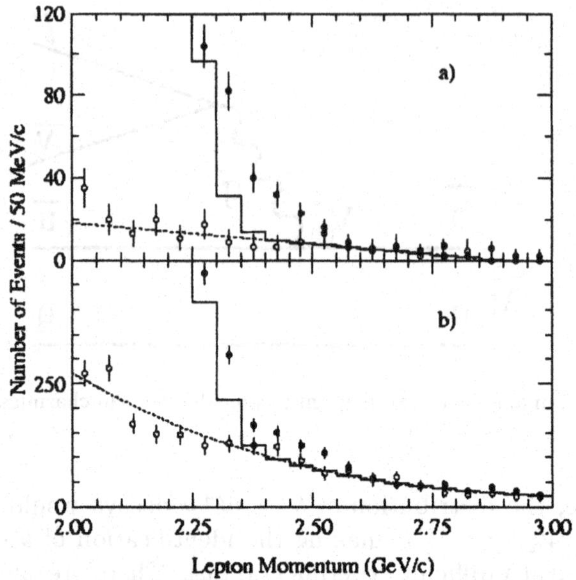

Figure 19. CLEO data on the end of the semileptonic spectrum. The excess of events indicates $\bar{b} \to \bar{u}$ transition. Different cuts are applied in the two histograms.

One might think that we should do much better if the OPE approach is used. However, here, one faces a problem related to the breakdown of OPE close to the end point of the lepton momentum spectrum. A somewhat simple way of picturing the problem it that for very high lepton momentum, the recoiling hadronic mass is very small and therefore one is dominated by a few exclusive resonant states such as $\pi \ell \nu$ and the hadron parton duality does not apply anymore. It was proposed recently to use the recoiling hadronic mass to circumvent this problem. In $\bar{b} \to \bar{c}\ell^+\nu$ decays, the average hadronic invariant mass is larger than m_D since at the least a D meson has to be formed. In contrast, the average hadronic invariant mass for $\bar{b} \to \bar{u}\ell^+\nu$ transitions is expected to be significantly lower. Using this variable one may then identify this latter decays and thus relax the momentum cut on the lepton. In practice, such analyses are very difficult to perfom but attempt have been carried out recently at LEP and results compatible to the one in

Eq. 166 are obtained.

$$
\begin{array}{lll}
|V_{ub}| = & (4.16 \pm 0.70_{\text{stat.}} \pm 0.64_{\text{syst.}} \pm 0.31_{\text{model}}) \times 10^{-3} & \text{ALEPH [80]} \\
|V_{ub}| = & (4.08 \pm 0.47_{\text{stat.}} \pm 0.69_{\text{syst.}} \pm 0.35_{\text{model}}) \times 10^{-3} & \text{DELPHI [81]} \\
|V_{ub}| = & (6.0 \pm 0.9_{\text{stat.}} \pm 1.9_{\text{syst.}} \pm 0.6_{\text{model}}) \times 10^{-3} & \text{L3 [82]}
\end{array}
\tag{167}
$$

One should note that there is still a model dependence since in these analyses, one needs to estimate the fraction of $\bar{b} \to \bar{u}\ell^+\nu$ transitions kept after applying the cuts from the Monte Carlo. However this model dependence is somewhat reduced compared to case where only the end spectrum is used.

4.4.2. *Exclusive charmless Decays*

Exclusive semileptonic decays have been observed since 1996 by the CLEO collaboration [83] (Fig. 20) and the branching fractions are:

$$
\begin{array}{lll}
\text{Br}(B^o \to \pi^-\ell^+\nu) & = & (1.8 \pm 0.4_{\text{stat.}} \pm 0.3_{\text{syst.}} \pm 0.2_{\text{mod.}}) \times 10^{-4} \\
\text{Br}(B^o \to \rho^-\ell^+\nu) & = & (2.8 \pm 0.4_{\text{stat.}} \pm 0.4_{\text{syst.}} \pm 0.6_{\text{mod.}}) \times 10^{-4}
\end{array}
\tag{168}
$$

Now the difficulty is to translate these numbers in measurements of $|V_{ub}|$. One cannot use HQET in the same way as for the extraction of $|V_{cb}|$ since the transition is from a heavy quark to a light one. Therefore one has to use values of the form factors in Eqs. 156 and 157 either obtained from models [70, 84–86], lattice calculations [87] or light cone sum rules [88]. Finally, it is possible to use HQET indirectly and obtain the form factors by relating two heavy to light quark transitions [89] (namely, $\bar{B} \to \rho(\pi)\ell^+\nu$ and $\bar{D} \to \rho(\pi)\ell^+\nu$ or $\bar{D} \to K\ell^+\nu$ if SU(3) is assumed). Making an average of value of $|V_{ub}|$ obtained using these different theoretical inputs, one gets:

$$
|V_{ub}| = (3.2 \pm 0.30_{\text{stat.}} \pm 0.25_{\text{syst.}} \pm 0.6_{\text{model}}) \times 10^{-3} \tag{169}
$$

Combining the inclusive and exclusive measurements, we derive the somewhat conservative limits

$$
0.27 < \frac{|V_{ub}|}{\lambda |V_{cb}|} = \sqrt{\rho^2 + \eta^2} < 0.45 \tag{170}
$$

which define the circle centered at (0,0) in the (ρ, η) plane and add a useful constraint on the position of the apex of the unitarity triangle as shown in Fig. 8.

4.5. SUMMARY

From figure 8, one can define the region in which the apex of the unitarity triangle has to lie. This area can be translated in terms of $\sin 2\beta$ versus

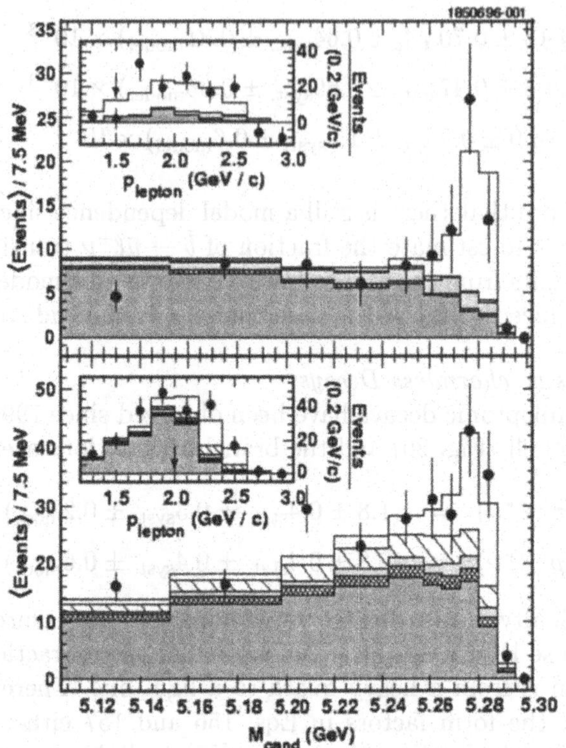

Figure 20. The mass distribution of reconstructed $\pi \ell \nu$ (top) and $\rho \ell \nu$ (bottom) events by CLEO. The dark area represents the estimated background. The excess at the B mass allows CLEO to estimate the branching fractions.

$\sin 2\alpha$ or $\sin 2\beta$ versus $\sin 2\gamma$ as shown in the figure 21 from which one deduces:

$$-0.6 \ \leq \sin 2\alpha \leq 0.9 \ , \ \ 0.45 \leq \sin 2\beta \leq 0.82 \ , \ \ -0.60 \ \leq \sin 2\gamma \leq 1 \tag{171}$$

One may also attempt to do a global fit using all these constraints but it is not obvious how to treat the theoretical uncertainties due to the averaging over several models as these errors are not gaussian. For more detail see Ref. [1].

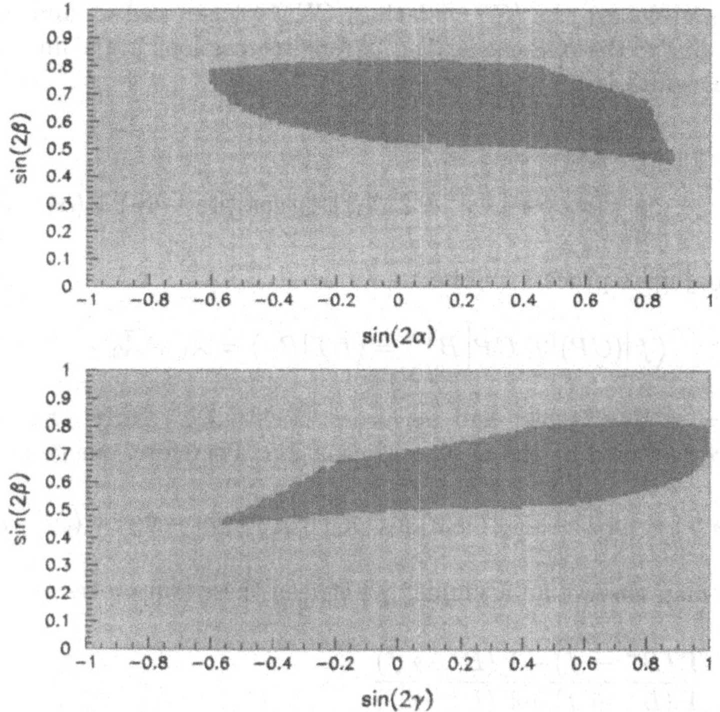

Figure 21. The allowed region in the ($\sin 2\alpha$,$\sin 2\beta$) plane derived from the constraints discussed in this chapter.

5. The CP violating effects in the Standard Model.

In chapter two we have divided the CP violating effects into three different categories. In the following, we will be more specific and discuss what is really measured from the observation of a CP asymmetry. We will see in particular whether and how the phase of the CKM matrix can be extracted and highlight the methods allowing a clean theoretical interpretation of the effects.

5.1. DIRECT CP VIOLATION.

In the Standard Model, the interference of several decay amplitudes involving different CKM complex elements may result into CP violating effects [90]. Let us illustrate this by assuming that two amplitudes $\mathcal{A}_1$ and $\mathcal{A}_2$ contribute to the decay $B^+ \to f$ where f is a particular final state. The total amplitude reads

$$\mathcal{A}\left(B^+ \to f\right) \equiv \langle f|\mathcal{T}|B^+\rangle = \mathcal{A}_1 + \mathcal{A}_2 \equiv e^{i\Phi_1} e^{i\alpha_1} \left|\mathcal{A}_1\right| + e^{i\Phi_2} e^{i\alpha_2} \left|\mathcal{A}_2\right| \quad (172)$$

where Φ_1 and Φ_2 are the (CP violating) CKM phases and α_1 and α_2 are the phases due to the (CP conserving) strong interactions in the final state. The partial width is

$$\begin{aligned}\Gamma\left(B^+ \to f\right) &= \left|\mathcal{A}_1 + \mathcal{A}_2\right|^2 \\ &= \left|\mathcal{A}_1\right|^2 + \left|\mathcal{A}_2\right|^2 + 2\left|\mathcal{A}_1\right|\left|\mathcal{A}_2\right|\cos\left[\left(\Phi_1 - \Phi_2\right) + \left(\alpha_1 - \alpha_2\right)\right]\end{aligned} \tag{173}$$

The CP conjugate process will be

$$\left\langle f \left| (CP)^{\dagger} \mathcal{T}\, CP \right| B^+ \right\rangle = \langle \bar{f} | \mathcal{T} | B^- \rangle = \bar{\mathcal{A}}_1 + \bar{\mathcal{A}}_2 \tag{174}$$

with $\bar{\mathcal{A}}_1 = e^{-i\Phi_1} e^{i\alpha_1} \left|\mathcal{A}_1\right|$ and $\bar{\mathcal{A}}_2 = e^{-i\Phi_2} e^{i\alpha_2} \left|\mathcal{A}_2\right|$. **Only the weak phases** are modified by the CP transformation. Therefore, one gets:

$$\Gamma\left(B^- \to \bar{f}\right) = \left|\mathcal{A}_1\right|^2 + \left|\mathcal{A}_2\right|^2 + 2\left|\mathcal{A}_1\right|\left|\mathcal{A}_2\right| \cos\left[-\left(\Phi_1 - \Phi_2\right) + \left(\alpha_1 - \alpha_2\right)\right] \tag{175}$$

Thus, one may observe an asymmetry between these two processes.

$$\begin{aligned}A &= \frac{\Gamma\left(B^- \to \bar{f}\right) - \Gamma\left(B^+ \to f\right)}{\Gamma\left(B^- \to \bar{f}\right) + \Gamma\left(B^+ \to f\right)} \\ &= \frac{2\left|\mathcal{A}_1\right|\left|\mathcal{A}_2\right| \sin\left(\Phi_1 - \Phi_2\right) \sin\left(\alpha_1 - \alpha_2\right)}{\left|\mathcal{A}_1\right|^2 + \left|\mathcal{A}_2\right|^2 + 2\left|\mathcal{A}_1\right|\left|\mathcal{A}_2\right| \cos\left(\Phi_1 - \Phi_2\right) \cos\left(\alpha_1 - \alpha_2\right)}\end{aligned} \tag{176}$$

This asymmetry is significantly different from 0 if the three following conditions are satisfied:

* The magnitudes of the amplitudes are of the same order, $\left|\mathcal{A}_1\right| \simeq \left|\mathcal{A}_2\right|$,
* The amplitudes have different CKM phases, $\Phi_1 - \Phi_2 \neq 0$,
* The amplitudes have different strong phases, $\alpha_1 - \alpha_2 \neq 0$.

Let us illustrate how these conditions may be fulfilled with an example. Several Feynman diagrams are possible for the reaction $B^+ \to K^+ \rho^o$ as shown in figure 22. The CKM matrix elements contributing in the graphs 22a,b,c are $V_{ub}^* V_{us}$. Therefore one can sum up these amplitudes into a single one since no CP violation can be generated through their interference. In fact, among these amplitudes, the one corresponding to the diagram 22a is the dominant one [70]. The diagrams in figure 22d have different CKM matrix elements. The exchange of the c and t quarks involves the products $V_{cb}^* V_{cs}$ and $V_{tb}^* V_{ts}$ respectively. It is also possible to exchange the u quark, however we will neglect this contribution here since the $V_{ub}^* V_{us}$ term is about 20 times smaller than the two others. Let us estimate in somewhat more details whether the interference of the tree diagrams (Fig. 22a,b) and the "penguins" (Fig. 22d) could lead to a sizeable CP violating effect.

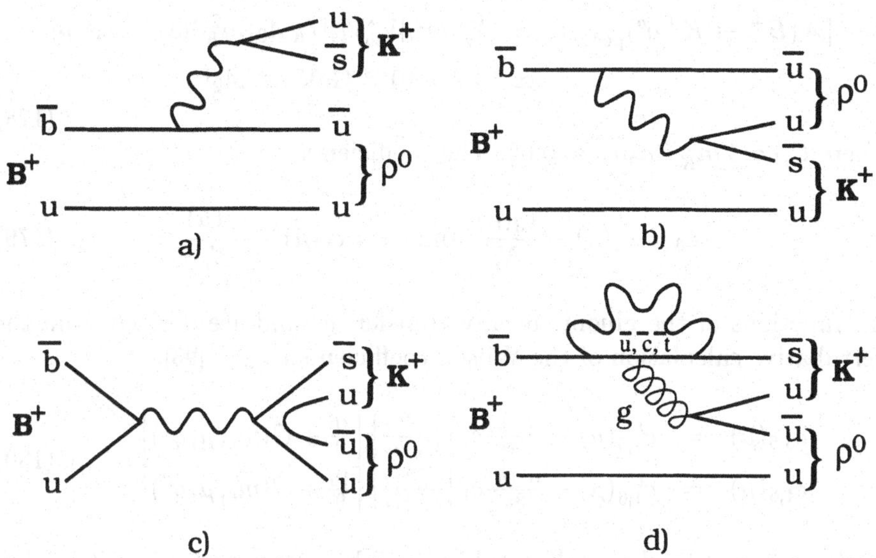

Figure 22. Feynman diagrams for the reaction $B^+ \to K^+\rho^o$. The graphs a, b, c and d are often called *External Spectator* , *Internal Spectator* , *Annihilation* and *Penguin* diagrams, respectively.

5.1.1. *Estimation of the Amplitudes.*

The diagrams in the figure 22a and b are calculated using the Operator Product Expansion with the factorization hypothesis [70] (for simplicity we neglect the annihilation diagram).

$$
\begin{aligned}
|\mathcal{A}\,(B^+ \to K^+\rho^o)_{\text{spect.}}| &\simeq \frac{G_F}{2} m_B |V_{ub}^* V_{us}| f_K A_0(m_K^2) \left(a_1 + \frac{f_\rho F_1(m_\rho^2)}{f_K A_0(m_K^2)} a_2 \right) \\
&\simeq 1.3 \times 10^{-8} \text{ GeV } \equiv |\mathcal{A}_1|
\end{aligned}
\tag{177}
$$

where $f_K(= 0.16$ GeV) and $f_\rho(\approx 0.21$ GeV) are the kaon and ρ decay constants and F_1 and A_0 the form factors for the $B \to K$ and $B \to \rho$ transitions, respectively. The scale dependent factors $a_1 = c_1(\mu) + c_2(\mu)/N_c$ and $a_2 = c_2(\mu) + c_1(\mu)/N_c$ (N_c is the number of color) are introduced by the OPE and are written in terms of the Wilson coefficients $c_{1,2}$. These coefficients include the gluonic radiative corrections calculated pertubatively at the next to the leading order [91]. To obtain the value of $|\mathcal{A}_1|$ above we have used $|V_{ub}| = 0.0035$, $F_1 = A_0 = 0.5$, $a_1 = 1.0$ and $a_2 = 0.2$.

The penguin diagrams are more difficult and uncertain to evaluate [92].

$$|\mathcal{A}(B^+ \to K^+\rho^o)_{\text{peng.}}| \simeq \frac{G_F}{2} m_B^2 |V_{tb}^* V_{ts}| f_K A_0(m_K^2) |a_4 - a_6 C_P| \\ \simeq 4.7 \times 10^{-9}\ \text{GeV} \equiv |\mathcal{A}_2| \tag{178}$$

where $C_P \simeq 2m_K^2/m_b m_s \approx 0.65$. The coefficients

$$a_4 = c_4(\mu) + \frac{c_3(\mu)}{N_c} \text{ and } a_6 = c_6(\mu) + \frac{c_5(\mu)}{N_c} \tag{179}$$

are functions of the gluonic energy transfer, q^2 and are derived from the pertubative calculation of the Wilson coefficients $c_{i=3-6}$ [93].

$$\begin{aligned} c_{3,5}(\mu) &= c_{3,5}^t(\mu) - \frac{\alpha_s(\mu)}{24\pi} c_1 \left|\frac{V_{cb}^* V_{cs}}{V_{tb}^* V_{ts}}\right| \left[\frac{10}{9} + G(m_c, \mu, q^2)\right] \\ c_{4,6}(\mu) &= c_{4,6}^t(\mu) + \frac{\alpha_s(\mu)}{8\pi} c_1 \left|\frac{V_{cb}^* V_{cs}}{V_{tb}^* V_{ts}}\right| \left[\frac{10}{9} + G(m_c, \mu, q^2)\right] \end{aligned} \tag{180}$$

In the expressions above, the doubly Cabibbo suppressed u-loop term has been neglected. The c_i^t are the contributions from the t-loop diagram and

$$G(m_c, \mu, q^2) = 4 \int_0^1 x(1-x) \ln \frac{m_c^2 - x(1-x)q^2}{\mu^2} dx \tag{181}$$

For $\mu = m_b$ and $< q^2 >= m_b^2/2$, one finds [93]

$$c_3^t = 0.017\ ,\ c_4^t = -0.037\ ,\ c_5^t = 0.010\ ,\ c_6^t = -0.045 \tag{182}$$

However, these values suffer from uncertainties due to the renormalization scheme and the lack of knowledge on the proper value of q^2 to be used. Let us consider two boundary conditions [93, 94].

- In the dispersive mode ($4\ m_c^2 \gg q^2$), the function $G(m_c, \mu, q^2)$ is real and one has:

$$\begin{aligned} c_{3,5}(\mu) &= c_{3,5}^t(\mu) - \frac{\alpha_s(\mu)}{24\pi} c_1 \left[\frac{2}{3} \ln \frac{m_c^2}{\mu^2}\right] \\ c_{4,6}(\mu) &= c_{4,6}^t(\mu) + \frac{\alpha_s(\mu)}{8\pi} c_1 \left[\frac{2}{3} \ln \frac{m_c^2}{\mu^2}\right] \end{aligned} \tag{183}$$

- In the absorptive mode ($4\ m_c^2 \ll q^2$), the function $G(m_c, \mu, q^2)$ is pure imaginary and thus the strong phase would be maximal. However q^2 is bounded by m_b^2 and therefore the strong phase is smaller. For example using $\mu = m_b$ and $< q^2 >= m_b^2/2$

$$a_4 - a_6 C_P \simeq -0.0062 - i\ 0.0066 \simeq -0.009 e^{i\frac{\pi}{4}} \tag{184}$$

According to our estimations in Eqs. 177 and 178, the requirement for 2 amplitudes of the same order is met since one has $3\,|\mathcal{A}_2| \simeq |\mathcal{A}_1|$. However, one should note that the branching fraction is not large:

$$
\begin{aligned}
\frac{\Gamma\left(B^+ \to K^+\rho^o\right)}{\Gamma_{\text{tot}}} &\simeq \frac{\tau_B}{16\pi m_B}|\mathcal{A}_1 + \mathcal{A}_2|^2 \\
&\simeq \frac{\tau_B G_F^2}{64\pi} m_B^3 f_K^2 A_0^2(m_K^2) \times \\
&\quad \left| V_{ub}^* V_{us}\left(a_1 + \frac{f_\rho F_1(m_K^2)}{f_K A_0(m_K^2)} a_2\right) - V_{tb}^* V_{ts}(a_4 - a_6 C_P)\right|^2 \\
&\simeq 9 \times 10^{-7}
\end{aligned}
\tag{185}
$$

5.1.2. *Evaluation of the Strong Phase Difference* $[\sin(\alpha_1 - \alpha_2)]$.
In general, one does not expect strong interaction in the final state for the tree diagrams [95]. Assuming $\alpha_1 = 0$, the strong phase α_2 has to be different from 0 in order to generate CP violating effect. This requires a non negligible contribution of the absorptive part in the "penguin" diagrams ($q^2 \gg 4\ m_c^2$). From the estimation in Eq. 184, one finds :

$$\sin\Delta\alpha = -\sin\alpha_2 \approx -0.7 \tag{186}$$

At this point, it is important to say some words of caution. The estimates which have been made here rely heavily on the factorization Ansatz. This Ansatz is in general verified for tree diagrams involving $b \to c$ transitions but is by no means verified for $b \to u$ transitions. In our particular example, the penguin diagrams are an essential ingredient for which factorization is not on a solid ground, although present observation do not show significant inconsistencies. Finally it should be noted that even if there is no re-interaction of the quarks in the final state, there could be some at the hadron level. For example, one could have the final state $K^+\rho^o$ from the re-interaction of the intermediate state $D_S^+ D^{*o}$.

$$B^+ \to D_S^+ \overline{D^{*o}} \to K^+\rho^o \tag{187}$$

For all the reasons above, it is very difficult to deduce a reliable value for $\Delta\alpha$ and thus it will be difficult to extract the value of the CKM phases.

5.1.3. *Evaluation of the Weak Phase Difference* $[\sin(\Phi_1 - \Phi_2)]$.
From the figures 22a,b and 22d (with t (c) quark exchange), one sees that the CKM matrix elements contributing to the amplitudes are $V_{ub}^* V_{us}$ and $V_{tb}^* V_{ts}$ ($V_{cb}^*\ V_{cs}$) respectively. Since $V_{tb}^* V_{ts} \simeq -V_{cb}^* V_{cs}$ and is essentially real, the weak phase of the interference term is

$$\sin(\Phi_1 - \Phi_2) = \frac{\mathrm{Im}\left(\mathrm{V_{ub}^* V_{us} V_{ts}^* V_{tb}}\right)}{|\mathrm{V_{ub}^* V_{us} V_{ts}^* V_{tb}}|} \simeq \frac{\eta}{\sqrt{\rho^2 + \eta^2}} = \sin\gamma \tag{188}$$

where γ is one of the angles of the unitarity triangle. From the figure 8, one deduces that the value of $\sin\gamma$ can be as large as 1. Hence using Eqs. 177, 178, 186 and 188, one finds

$$\frac{\Gamma\left(B^- \to K^-\rho^o\right) - \Gamma\left(B^+ \to K^+\rho^o\right)}{\Gamma\left(B^- \to K^-\rho^o\right) + \Gamma\left(B^+ \to K^+\rho^o\right)} \simeq 0.35 \tag{189}$$

The calculated effect is large. However the uncertainty on this prediction is large as well and it is unfortunately possible that the asymmetry be significantly smaller. In general, the trend using the factorization Ansatz is that to get a strong phase, penguin diagrams are required. To enhance the effect, one needs to suppress the real part of the amplitude making its contribution small if the branching fraction is large. The net result is that one expects large asymmetries for modes which have small branching fractions. In any case, the observation of any asymmetry is very important as it will demonstrate that direct **CP violation** exists assessing the CKM scenario. Unfortunately, it would be very difficult to extract a reliable value for $\sin\gamma$ due to the uncertainties on $\sin(\alpha_1 - \alpha_2)$, $|\mathcal{A}_1|$ and $|\mathcal{A}_2|$. Other decay channels of the same type could be used as it is shown in the table 5.

Mode	Theoretical Br × 10^{-5}	Expected Asymmetry
$B^\pm \to K^\pm\,\pi^o$	$\sim \mathcal{O}(1)$	$0 \to 10\%$
$K^{*\pm}\,\pi^o$	$\sim \mathcal{O}(1)$	$0 \to 10\%$
$\pi^\pm\,\rho^o$	$\sim \mathcal{O}(0.1)$	$0 \to 30\%$
$K^\pm K^{*o}$	$\sim \mathcal{O}(0.01)$	0%

TABLE 5. The theoretical branching fractions and asymmetries for some particular B decay modes [94,96].

The interpretation of a CP non-conserving effect in terms of the angles of the unitarity triangle may be better achieved by studying the charged B decays $B^\pm \to D^o_{1,2}X^\pm$ [97], where $X^\pm$ is any state with the flavor of a $K^\pm$, and $D^o_{1,2}$ denote the CP eigenstates $(D^o \pm \overline{D^o})/\sqrt{2}$. Let's consider the decay $B^+ \to D^o_1K^+$, where the CP-even eigenstate D^o_1 is identified by one of its CP-even decay products*. There are two different amplitudes contributing to this process, which correspond to the transitions $B^+ \to$

*We neglect D^o–$\overline{D^o}$ mixing and CP violation in D decays.

D^oK^+ and $B^+ \to \overline{D^o}K^+$ (Fig. 23). At the quark level, they are associated with the decays $\overline{b} \to \overline{u}c\overline{s}$ and $\overline{b} \to \overline{c}u\overline{s}$, which have different CKM factors $[V_{ub}^* V_{cs} \approx A\lambda^3(\rho+i\eta)$ and $V_{cb}^* V_{us} \approx A\lambda^3$ respectively$]$. These two amplitudes can be separately identified by looking for events where the D^o ($\overline{D^o}$) decays semileptonically; the flavor states D^o and $\overline{D^o}$ are distinguished by the charge of the final lepton.

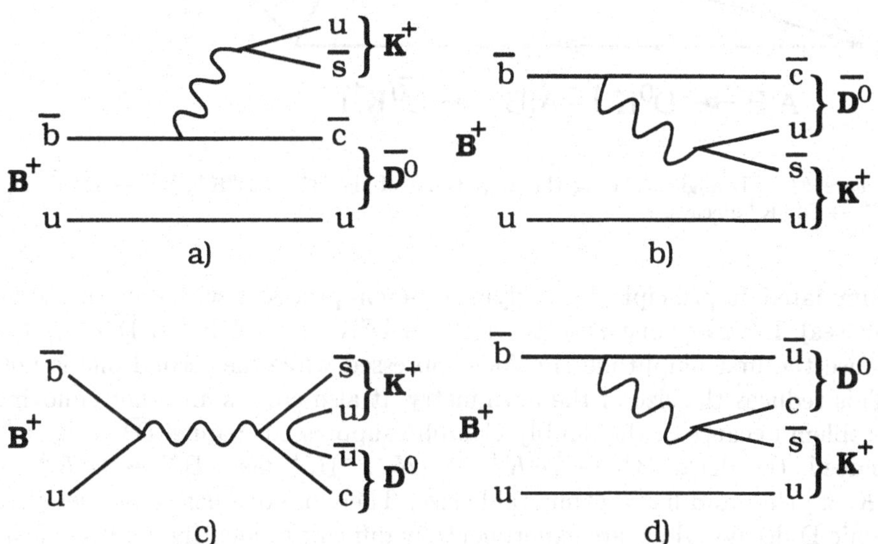

Figure 23. Feynman diagrams contributing to the $B^+ \to D^oK^+$ and $B^+ \to \overline{D^o}K^+$ decays.

In the standard CKM parametrization the amplitudes of the three processes can be written in the form

$$
\begin{aligned}
A_D \equiv A(B^+ \to D^oK^+) &= |A_D|\, e^{i\gamma}\, e^{i\delta_D}, \\
A_{\overline{D}} \equiv A(B^+ \to \overline{D^o}K^+) &= |A_{\overline{D}}|\, e^{i\delta_{\overline{D}}}, \\
A_{D_1} = A(B^+ \to D_1^oK^+) &= \tfrac{1}{\sqrt{2}}\,(A_D + A_{\overline{D}}),
\end{aligned}
\tag{190}
$$

where δ_D and $\delta_{\overline{D}}$ are the corresponding final-state-interaction phases, and γ is one of the weak angles of the unitarity triangle. The amplitudes of the charge-conjugated processes $B^- \to D^oK^-$, $B^- \to \overline{D^o}K^-$ and $B^- \to D_1^oK^-$ ($\bar{A}_D$, $\bar{A}_{\overline{D}}$ and $\bar{A}_{D_1}$, respectively) are obtained from the A amplitudes by simply changing the sign of the CKM phase γ. Note that $|\bar{A}_D| = |A_{\overline{D}}|$, $|\bar{A}_{\overline{D}}| = |A_D|$, but $|\bar{A}_{D_1}| \neq |A_{D_1}|$ if $\gamma \neq 0$.

From the two triangle relations relating the B^+ (B^-) decays into D^o, $\overline{D^o}$ and $D_{1,2}^o$ (Fig. 24), one could obtain $\sin\gamma$ up to a four-fold ambiguity [97] (the ambiguity, which is due to the presence of final-state interactions, could be

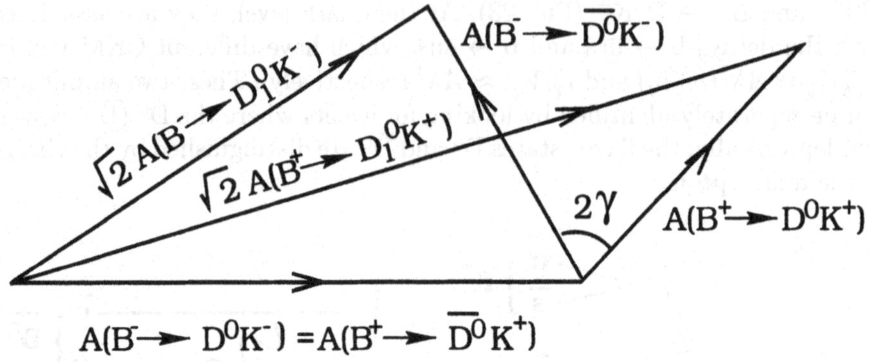

Figure 24. Triangles relating the amplitudes of the $B^+ \to D^oK^+$, $B^+ \to \overline{D^o}K^+$ and $B^+ \to D^o_{1,2}K^+$ decays.

eliminated, in principle, by studying different processes with the same weak phases). However, one expects $|A(B^+ \to D^oK^+)| \ll |A(B^+ \to \overline{D^o}K^+)|$, because the first amplitude is color suppressed while the second one is not. This reduces the size of the asymmetry. It also creates an other ennoying problem because of the doubly Cabibbo suppressed channel $\overline{D^o} \to K^-\pi^+$. Indeed, the decay $B^+ \to \overline{D^o}K^+ \to (K^-\pi^+)K^+$ fakes $B^+ \to D^oK^+ \to (K^-\pi^+)K^+$ and has a comparable size. Therefore one has to use semileptonic D decays which are experimentally difficult to identify because of the presence of neutrino and the large contamination from semileptonic B decays. One may try to take advantage of doubly Cabibbo suppressed modes and modify the analysis to overcome the problem [98]. The self-tagging decays $B^o_d \to D^o_{1,2}K^{*o}$ could perhaps provide a better way to extract γ [99]. Here the beauty flavor is deduced from the decay $K^{*o} \to K^+\pi^-$. The analysis proceeds as in the $B^\pm$ case, but this time the two interfering amplitudes ($B^o_d \to D^oK^{*o}$ and $B^o_d \to \overline{D^o}K^{*o}$) are likely to have comparable magnitudes with branching fractions of the order of 10^{-5}.

Another place for studying direct CP violation are decays of bottom baryons [100,101], and a somewhat similar analysis may be performed [101]. CP violation can show up as a rate asymmetry and in various decay parameters. For instance, the decay $1/2 \to 1/2 + 0$ (like $\Lambda_b \to \Lambda\overline{D^o}$) involves, in addition to the partial decay rate Γ, the three decay parameters α, β, γ, that characterize the angular distribution in the rest frame of the decaying particle. If barred quantities describe the charge-conjugated process, then CP invariance requires $\bar{\Gamma} = \Gamma$, $\bar{\alpha} = -\alpha$, $\bar{\beta} = -\beta$ and $\bar{\gamma} = \gamma$. Neither tagging nor time-dependences are required to observe CP violation with modes of baryons, in contrast to B^0 modes that involve mixing. One has to bear in mind that part of the mentioned observables need spin-polarization of

the initial state in order to show the effect. Sizeable CP effects are expected in modes involving a D^o, which is seen in a final state that can also be fed from a $\overline{D^o}$; six related processes, such as $\Lambda_b \to \Lambda D^o$, $\Lambda_b \to \Lambda\overline{D^o}$, $\Lambda_b \to \Lambda D^o_{1,2}$ and their charge conjugated counterparts, could be used to extract the angle γ. In [101], it is predicted that $Br(\Lambda_b \to \Lambda D^o) \sim 10^{-5}$, thus $Br(\Lambda_b \to \Lambda D^o_{1,2}) \sim 10^{-7}$. Under favorable circumstances CP asymmetries are estimated to occur at the few tens of percent level.
In flavor non-specific decays of the neutral B mesons (i.e. those final states which are common to B^o and $\overline{B^o}$ decays), one needs to know the flavor of the original B meson in order to study CP-violating rate asymmetries. However, there are some signals of CP violation which do not require flavor identification. These involve CP-odd energy asymmetries in the sum of all B^o and $\overline{B^o}$ events [102]. For instance, in the decay $B^o_d \to K_S\pi^+\pi^-$, the distribution in the $\pi^+\pi^-$ energies (the Dalitz-plot distribution) need not be symmetric under exchange of π^+ and π^-. There will in general be an energy asymmetry of the form [102]

$$\begin{aligned} \Gamma(B^o_d \to K_S\pi^+\pi^-) &= a + b(E_+ - E_-) \\ \Gamma(\overline{B^o_d} \to K_S\pi^+\pi^-) &= \bar{a} + \bar{b}(E_+ - E_-), \end{aligned} \tag{191}$$

where a, b, $\bar{a}$ and $\bar{b}$ are symmetric functions of the energies. CP invariance requires $\bar{a} = a$, $\bar{b} = -b$, so that there is no energy asymmetry left if one sums together all $K_S\pi^+\pi^-$ events (provided there is an equal number of B^o_d and $\overline{B^o_d}$ in the initial state). Therefore, the measurement of a net energy asymmetry in the sum of all B^o_d and $\overline{B^o_d}$ events would be a signal of CP violation. Such an effect could originate, in principle, from both indirect and direct CP violation. As discussed in the previous section, however, the mixing-induced asymmetry is expected to be very small in the Standard Model; thus, this Dalitz-plot asymmetry should be a direct CP-violation effect, requiring two interfering amplitudes with different weak and strong phases. The decays $B^o_d \to K_S\pi^+\pi^-$ and $B^o_s \to \phi\pi^+\pi^-$ seem to be the most promising ones, since they proceed through the $b \to u\bar{u}s$ mechanism, where the penguin amplitude is relatively enhanced with respect to the direct tree contribution.

5.2. CP VIOLATION DUE TO THE B^O-$\overline{B^O}$ MIXING.

The B^o-$\overline{B^o}$ mixing may also exibit an asymmetry in the transition $B^o \to \overline{B^o}$ and $\overline{B^o} \to B^o$. If $B^o\overline{B^o}$ pairs are produced, this asymmetry can be observed, by example, by studying the difference between the number of same sign positive leptons ($\ell^+\ell^+$) and same sign negative leptons ($\ell^-\ell^-$). We have derived this asymmetry earlier.

$$\mathcal{A}s = \frac{\mathcal{P}r\left(\overline{B^o} \to B^o\right) - \mathcal{P}r\left(B^o \to \overline{B^o}\right)}{\mathcal{P}r\left(\overline{B^o} \to B^o\right) + \mathcal{P}r\left(B^o \to \overline{B^o}\right)} = \frac{\left|\frac{p}{q}\right|^2 - \left|\frac{q}{p}\right|^2}{\left|\frac{p}{q}\right|^2 + \left|\frac{q}{p}\right|^2} \simeq \frac{4\mathrm{Re}\epsilon_B}{1+|\epsilon_B|^2} \qquad (192)$$

with $\frac{p}{q} = \sqrt{\frac{M_{12}-\frac{i}{2}\Gamma_{12}}{M_{12}^*-\frac{i}{2}\Gamma_{12}^*}}$. The values of M_{12} and Γ_{12} are obtained by computing the magnitude of the box diagram in the figure 25.

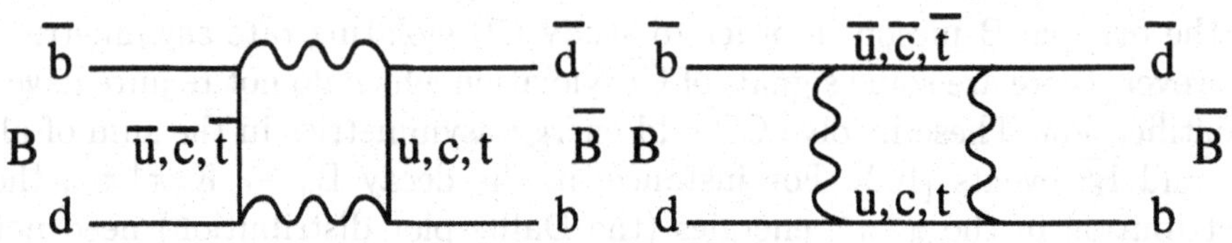

Figure 25. Feynman diagrams responsible for the $B^o - \overline{B^o}$ mixing

The dispersive part of these diagrams gives M_{12} while the absorptive part allows one to compute Γ_{12} [103].

$$M_{12} \simeq \frac{G_F^2 B_B f_B^2 m_B m_t^2}{12\ \pi^2}\ (V_{td}V_{tb}^*)^2 \times G\left(\frac{m_t^2}{m_W^2}\right) \qquad (193)$$

where $G(x)$ is a known function [26]

$$G(x) = 1 - \frac{3}{4} \times \frac{x+x^2}{(1-x)^2} - \frac{3}{2} \times \frac{x^2}{(1-x)^3} \ln x \qquad (194)$$

and

$$\begin{aligned} \Gamma_{12} \ \simeq \ & \frac{G_F^2 B_B f_B^2 m_B m_b^2}{8\pi} \left[(V_{ub}^* V_{ud} + V_{cb}^* V_{cd})^2 \right. \\ & \left. -\frac{8}{3}\frac{(m_c^2 - m_u^2)}{m_b^2} \times V_{cb}^* V_{cd} \left(V_{ub}^* V_{ud} + V_{cb}^* V_{cd}\right)\right] \end{aligned} \qquad (195)$$

Using the unitarity relations of the CKM matrix, one finds

$$\frac{\Gamma_{12}}{M_{12}} \simeq \frac{3\pi}{2}\frac{m_b^2}{m_t^2}\left[1 + \frac{8}{3}\frac{m_c^2}{m_b^2}\frac{V_{cb}^* V_{cd}}{V_{td}V_{tb}^*}\right] \times \frac{1}{G\left(\frac{m_t^2}{m_W^2}\right)} \ll 1 \qquad (196)$$

And thus

$$\begin{aligned} \left|\frac{p}{q}\right| &\simeq 1+\frac{1}{2}\,\mathrm{Im}\,\frac{\Gamma_{12}}{\mathrm{M}_{12}} \simeq 1+2\pi\,\frac{\mathrm{m}_c^2}{\mathrm{m}_t^2}\,\frac{1}{\mathrm{G}\left(\frac{\mathrm{m}_t^2}{\mathrm{m}_W^2}\right)}\,\frac{\mathrm{Im}\,(\mathrm{V}_{cb}^*\mathrm{V}_{cd}\mathrm{V}_{td}^*\mathrm{V}_{tb})}{|\mathrm{V}_{td}\mathrm{V}_{tb}^*|^2} \\ &\simeq 1-10^{-3}\sin\beta \end{aligned} \tag{197}$$

Taking into account the various uncertainties, one obtains the following limit

$$\mathcal{A}s \lesssim 2\;10^{-3} \tag{198}$$

Although such a small asymmetry is very difficult to observe, it is very important to make this measurement since any sizeable asymmetry (say larger than 1%) would be a clear indication of new physics.
Finally, note that in the case of the B_s mesons, the imaginary term in Eq. 197 would read $\mathrm{Im}\,(V_{cb}^* V_{cs} V_{ts}^* V_{tb})$. Since the CKM elements involved are almost real, the asymmetry is expected to be extremely small ($\mathcal{A}s \lesssim 10^{-4}$) for that system.

5.3. CP VIOLATION INVOLVING THE $B^O - \overline{B^O}$ MIXING AND THE DISINTEGRATION.

Let us see what happens when the final state f is not a CP eigenstate but could nevertheless be produced from both B^o and $\overline{B^o}$ [104] using Eq. 76. Let us write the direct transition amplitudes

$$\begin{aligned} \left\langle f|\mathcal{T}|\overline{B^o}\right\rangle &= \sum_{j=1}^{n} e^{i\Phi_j}\, e^{i\alpha_j}\mathcal{A}_j \;\equiv\; e^{i\beta}\mathcal{A} \\ \langle f|\mathcal{T}|B^o\rangle &= \sum_{j=1}^{n} e^{-i\Phi'_j}\, e^{i\alpha'_j}\mathcal{A}'_j \;\equiv\; e^{-i\beta'}\mathcal{A}' \end{aligned} \tag{199}$$

where

- Φ_j and Φ'_j are the weak phases from the CKM matrix such than $\Phi_i \neq \Phi_j$ for $i \neq j$.
- α_j and α'_j are the phases due to the strong interactions in the final state.
- $\mathcal{A}_j$ and $\mathcal{A}'_j$ are the magnitudes of the amplitudes corresponding to different Feynman diagrams. However when several diagrams involve the same CKM matrix elements, we will consider their sum as a single amplitude since no CP violating effect can be generated via their interference.

We note that in the case of a final state which is a CP eigenstate, one has $\Phi'_j = \Phi_j$, $\mathcal{A}'_j = \mathcal{A}_j$ and $\alpha'_j = \alpha_j$ or $\alpha'_j = \alpha_j + \pi$ depending on the CP parity

(± 1). By replacing the transition amplitudes in 76 with the expressions 199, one obtains:

$$\begin{aligned} \mathcal{P}r\left(\overset{(-)}{B^o}\rightarrow f\right) &= \frac{\mathcal{A}^2+\mathcal{A}'^2}{2}\, e^{-\frac{t}{\tau}}\left\{1 \overset{(+)}{-} \mathcal{R}\cos x\frac{t}{\tau}\right. \\ &\left. \overset{(+)}{-} \mathcal{D}\sin\left(2\Phi_M+\beta+\beta'\right)\times\ \sin x\frac{t}{\tau}\right\} \end{aligned} \tag{200}$$

where $\mathcal{R}=\frac{\rho^2-1}{\rho^2+1}$, $\mathcal{D}=\frac{2\rho}{\rho^2+1}=\sqrt{1-\mathcal{R}^2}$ with $\rho=\frac{\mathcal{A}}{\mathcal{A}'}$. We have used the approximation $\left|\frac{p}{q}\right|=1$ as discussed earlier. In Eq. 200, $x=\Delta m/\Gamma$ and τ is the B^o meson lifetime.

The transition amplitude $\overset{(-)}{B}\rightarrow\bar{f}$, where $\bar{f}$ is the CP conjugate final state, are obtained from the Eqs. 199

$$\begin{aligned} \langle\bar{f}|\mathcal{T}|B^o\rangle &= \sum_{j=1}^{n} e^{-i\Phi_j}e^{i\alpha_j}\mathcal{A}_j \equiv e^{-i\bar{\beta}}\bar{\mathcal{A}} \\ \left\langle\bar{f}|\mathcal{T}|\overline{B^o}\right\rangle &= \sum_{j=1}^{n} e^{i\Phi'_j}e^{i\alpha'_j}\mathcal{A}'_j \equiv e^{i\bar{\beta}'}\bar{\mathcal{A}}' \end{aligned} \tag{201}$$

Only the phases coming from the CKM matrix have the opposite signs. The probability of the transition $\overset{(-)}{B^o}\rightarrow\bar{f}$ is

$$\begin{aligned} \mathcal{P}r\left(\overset{(-)}{B^o}\rightarrow\bar{f}\right) &= \frac{\bar{\mathcal{A}}^2+\bar{\mathcal{A}}'^2}{2} e^{-\frac{t}{\tau}}\left\{1 \overset{(-)}{+} \bar{\mathcal{R}}\cos x\frac{t}{\tau}\right. \\ &\left. \overset{(+)}{-} \bar{\mathcal{D}}\sin\left(2\Phi_M+\bar{\beta}+\bar{\beta}'\right)\sin x\frac{t}{\tau}\right\} \end{aligned} \tag{202}$$

where $\bar{\mathcal{R}}=\frac{\bar{\rho}^2-1}{\bar{\rho}^2+1}$, $\bar{\mathcal{D}}=\frac{2\bar{\rho}}{\bar{\rho}^2+1}$ with $\bar{\rho}=\frac{\bar{\mathcal{A}}}{\bar{\mathcal{A}}'}$. If CP is conserved, the following equalities hold:

$$\begin{aligned} \bar{\mathcal{A}} = \mathcal{A} \quad &, \quad \bar{\mathcal{A}}' = \mathcal{A}' \\ S\equiv\sin\left(2\Phi_M+\beta+\beta'\right) &= -\bar{S}\equiv-\sin\left(2\Phi_M+\bar{\beta}+\bar{\beta}'\right) \end{aligned} \tag{203}$$

From the total number of events in each of the four distributions 200 and 202 and a global fit of their time dependence, it is possible to extract the quantities, ρ ,$\bar{\rho}$, S ,$\bar{S}$,$\mathcal{A}^2+\mathcal{A}'^2$ et $\bar{\mathcal{A}}^2+\bar{\mathcal{A}}'^2$. The inequalities $\rho\neq\bar{\rho}$ or/and $\mathcal{A}^2+\mathcal{A}'^2\neq\bar{\mathcal{A}}^2+\bar{\mathcal{A}}'^2$ imply that $\mathcal{A}\neq\bar{\mathcal{A}}$ and $\mathcal{A}'\neq\bar{\mathcal{A}}'$. This observation would be an evidence for **direct CP** violation meaning that several amplitudes contribute in Eqs. 199 and 201. In such circumstances, it will be difficult to separate the CKM phases from the strong phases as only the quantities S and $\bar{S}$ are measured. We will see later how one could distinguish the various contributions for some particular final states. If only one amplitude dominates in the transitions $B \to f$ and $\overline{B} \to f$, then the interpretation of the measurements is much more easy. The hypothesis of a single amplitude implies the equalities $\rho = \bar{\rho}$ and $\mathcal{A}^2 + \mathcal{A}'^2 = \bar{\mathcal{A}}^2 + \bar{\mathcal{A}}'^2$ which can be verified experimentally. However it is important to underline that the observation of these equalities is not necessarily sufficient to prove that $\mathcal{A} = \bar{\mathcal{A}}$. This can be seen with the Eqs. 199 and 201 by assuming two contributing amplitudes.

$$\begin{aligned}
\mathcal{A}^2 &= \mathcal{A}_1^2 + \mathcal{A}_2^2 + 2\mathcal{A}_1\mathcal{A}_2 \cos\left[(\Phi_1 - \Phi_2) + (\alpha_1 - \alpha_2)\right] \\
\mathcal{A}'^2 &= \mathcal{A}_1'^2 + \mathcal{A}_2'^2 + 2\mathcal{A}_1'\mathcal{A}_2' \cos\left[(\Phi_1' - \Phi_2') + (\alpha_1' - \alpha_2')\right] \\
\bar{\mathcal{A}}^2 &= \mathcal{A}_1^2 + \mathcal{A}_2^2 + 2\mathcal{A}_1\mathcal{A}_2 \cos\left[-(\Phi_1 - \Phi_2) + (\alpha_1 - \alpha_2)\right] \\
\bar{\mathcal{A}}'^2 &= \mathcal{A}_1'^2 + \mathcal{A}_2'^2 + 2\mathcal{A}_1'\mathcal{A}_2' \cos\left[-(\Phi_1' - \Phi_2') + (\alpha_1' - \alpha'_2)\right]
\end{aligned} \tag{204}$$

One notes that $\mathcal{A} = \bar{\mathcal{A}}$ if $\sin(\Phi_1 - \Phi_2)\sin(\alpha_1 - \alpha_2) = 0$ and $\mathcal{A}' = \bar{\mathcal{A}}'$ if $\sin(\Phi_1' - \Phi_2')\sin(\alpha_1' - \alpha_2') = 0$. This eventuality depends on the different contributing diagrams. For example, if spectator diagrams and "penguins" have magnitudes of the same order, the difference $\sin(\alpha_1 - \alpha_2)$ is in general expected to be non zero as already discussed in this chapter.
Let us now assume that one amplitude dominates. Thus, one obtains:

$$\begin{aligned}
\mathcal{A} &= \bar{\mathcal{A}} \quad , \quad \mathcal{A}' = \bar{\mathcal{A}}' \\
\beta &= \Phi_f + \alpha \quad , \quad \beta' = \Phi'_f - \alpha' \\
\bar{\beta} &= \Phi_f - \alpha \quad , \quad \bar{\beta}' = \Phi'_f + \alpha'
\end{aligned} \tag{205}$$

and

$$S = \sin\left(2\Phi_M + \Phi_f + \Phi'_f + \Delta\alpha\right) \quad , \quad \bar{S} = \sin\left(2\Phi_M + \Phi_f + \Phi'_f - \Delta\alpha\right) \tag{206}$$

where Φ_f and Φ'_f are the CKM phases in the amplitudes which dominate in the transition $\overline{B^o} \to f$ and $B^o \to f$ respectively and $\Delta\alpha = \alpha - \alpha'$ is the strong phase difference. From Eqs. 206, one can extract the CKM phases.

$$\sin^2\left(2\Phi_M + \Phi_f + \Phi'_f\right) = \frac{1}{2}\left[1 + S\bar{S} \pm \sqrt{(1 - S^2)(1 - \bar{S}^2)}\,\right] \tag{207}$$

There is a twofold discrete ambiguity for $\sin^2\left(2\Phi_M + \Phi_f + \Phi'_f\right)$. The other solution corresponds to $\cos^2\Delta\alpha$.
Let us see with an example what is the quantity which is measured. We will use the transition $\overset{(-)}{B^o} \rightarrow f \equiv \rho^+\pi^-$ for which the Feynman diagrams are shown in the figure 26.

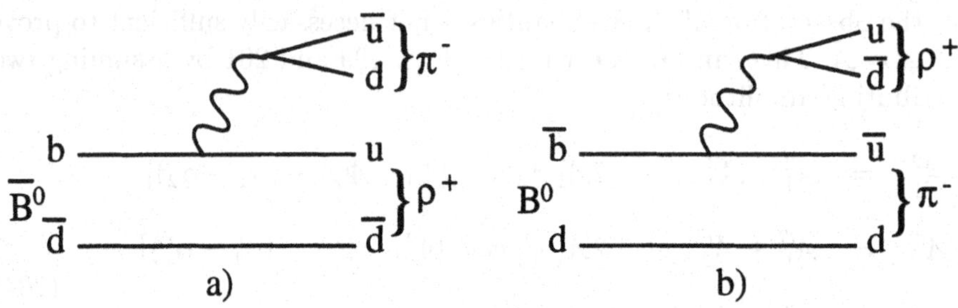

Figure 26. The dominant Feynman diagrams for the transition $\overset{(-)}{B^o} \rightarrow \rho^+\pi^-$.

The figure 27 shows the time dependence of the transition probabilities (Eq. 200 and 202). For our example, one has $e^{i\Phi_f} = \frac{V_{ub}V^*_{ud}}{|V_{ub}V^*_{ud}|}$, $e^{-i\Phi'_f} = \frac{V^*_{ub}V_{ud}}{|V^*_{ub}V_{ud}|}$ and $e^{i\Phi_M} = \frac{V_{td}V^*_{tb}}{|V_{td}V^*_{tb}|}$.
Finally, one gets:

$$\sin\left[2\Phi_M + \Phi_f + \Phi'_f\right] = \mathrm{Im}\left[\frac{(V_{ub}V^*_{ud}V_{td}V^*_{tb})^2}{|V_{ub}V^*_{ud}V_{td}V^*_{tb}|^2}\right] \equiv \sin 2\left(\Phi_M + \Phi_f\right) \quad (208)$$

The angle $\Phi \equiv \Phi_M + \Phi_f$ is nothing else that angle at the apex of the unitarity triangle. The ambiguity that we have discussed above can be resolved if one of the solutions in Eq. 207 is zero. In that case, it is possible to attribute this solution to $\cos\Delta\alpha$ since the dominating amplitudes are probably corresponding to the tree diagrams. A more conservative attitude is to measure CP violation in an other mode involving the same angle Φ, for example the decay mode $B^0 \rightarrow \pi^+\pi^-$. This final state is a CP eigenstate with the CP parity $\eta_{CP} = +1$. Here, the final state strong phases are identical regardless of the mother B^o or $\overline{B^o}$ since the two relevant diagrams similar to those of figures 26a and 26b are equivalent. Thus $\mathcal{A} = \mathcal{A}'$ and the transition probabilities are [105](figure 28)

$$\mathcal{P}r\left(\overset{(-)}{B^o} \rightarrow \pi^+\pi^-\right) = \mathcal{A}^2\, e^{-\frac{t}{\tau}}\{1 \overset{(+)}{-} \sin 2\Phi \sin x\frac{t}{\tau}\} \quad (209)$$

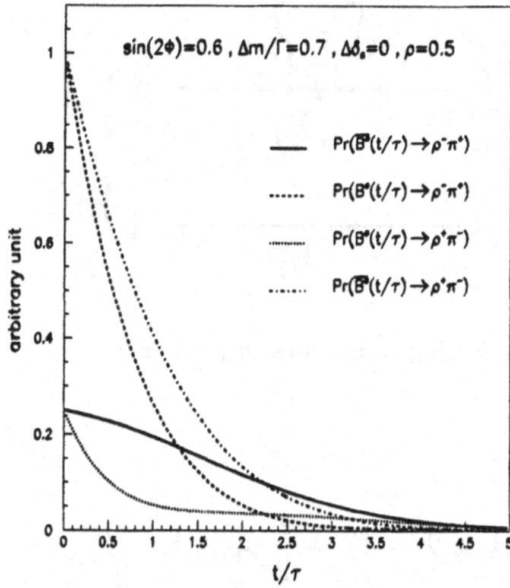

Figure 27. Transition probabilities as function of the B meson proper time for the decays $\overset{(-)}{B^o} \rightarrow \rho^\pm \pi^\mp$.

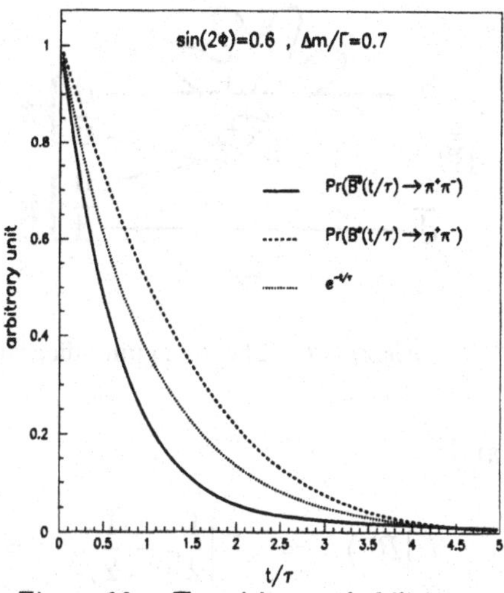

Figure 28. Transition probabilities as function of the B meson proper time for the decays $\overset{(-)}{B^o} \rightarrow \pi^+\pi^-$.

The CP violating asymmetry is:

$$\mathcal{A}s = \frac{\mathcal{P}r\left(\overline{B^o} \rightarrow \pi^+\pi^-\right) - \mathcal{P}r\left(B^o \rightarrow \pi^+\pi^-\right)}{\mathcal{P}r\left(\overline{B^o} \rightarrow \pi^+\pi^-\right) + \mathcal{P}r\left(B^o \rightarrow \pi^+\pi^-\right)} = \sin 2\Phi \sin x\frac{t}{\tau} \quad (210)$$

Knowing $x \simeq 0.7$ and $\tau \simeq 1.6\ 10^{-12}s$, it is obvious to obtain sin2Φ without any ambiguity. However for this particular example, one cannot verify whether a single amplitude dominates using the same method than for $f = \rho^+\pi^-$. Let us assume that, in addition to the two diagrams in figure 26 where ρ^+ is replaced by π^+, there exists a contribution from "penguins" (Fig. 29). The additional information can be extracted [106] from the transitions $B^\pm \rightarrow \pi^\pm\pi^o$ and $B^o \rightarrow \pi^o\pi^o$. These transitions can be written in term of two isospin amplitudes:

$$\mathcal{T} = \bar{\mathcal{A}}_{3/2}|3/2\ ,\ -1/2\rangle\ +\ \bar{\mathcal{A}}_{1/2}|1/2\ ,\ -1/2\rangle \quad (211)$$

where $\bar{\mathcal{A}}_{3/2}$ and $\bar{\mathcal{A}}_{1/2}$ are the transition amplitudes with $\Delta I = 3/2$ et $\Delta I = 1/2$. For the diagrams shown in figure 29 only $\Delta I = 1/2$ transitions are allowed. Therefore, one obtains:

$$\begin{aligned} \mathcal{T}\left|\overline{B^o}\right\rangle = \mathcal{T}\left|\frac{1}{2}\ ,+\frac{1}{2}\right\rangle &= \frac{1}{\sqrt{2}}\,\bar{\mathcal{A}}_{3/2}|2,0\rangle - \frac{1}{\sqrt{2}}\,\bar{\mathcal{A}}_{3/2}|1,0\rangle \\ &\quad +\frac{1}{\sqrt{2}}\bar{\mathcal{A}}_{1/2}|1,0\rangle - \frac{1}{\sqrt{2}}\,\bar{\mathcal{A}}_{1/2}|0,0\rangle \end{aligned} \quad (212)$$

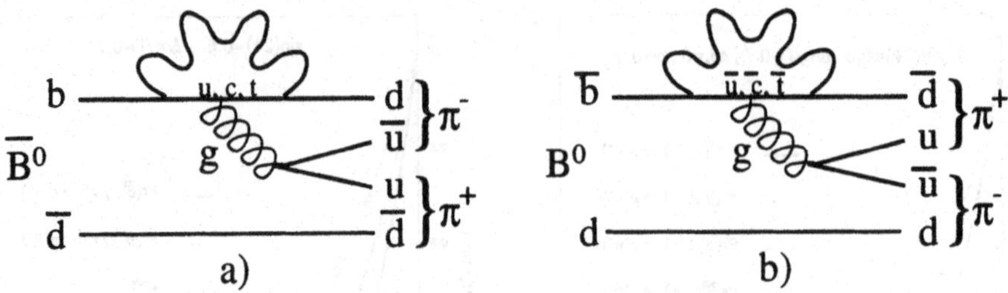

Figure 29. The "penguin" diagrams contributing to the final state $\pi^+\pi^-$.

and

$$\mathcal{T}|B^-\rangle = \mathcal{T}\left|\frac{1}{2}, -\frac{1}{2}\right\rangle = \frac{\sqrt{3}}{2}\bar{\mathcal{A}}_{3/2}|2, -1\rangle + \frac{1}{2}\bar{\mathcal{A}}_{3/2}|1, -1\rangle + \bar{\mathcal{A}}_{1/2}|1, -1\rangle \tag{213}$$

Since the $\pi^+\pi^-$ system has to be in S wave, the generalization of the Bose-Einstein symmetry requires that the isospin wave function is symmetric. Therefore, only even isospin states are allowed.

$$\begin{aligned}|\pi^+\pi^-\rangle &= \frac{1}{\sqrt{3}}|2, 0\rangle + \sqrt{\frac{2}{3}}|0, 0\rangle \\ |\pi^o\pi^o\rangle &= \sqrt{\frac{2}{3}}|2, 0\rangle - \frac{1}{\sqrt{3}}|0, 0\rangle \\ |\pi^-\pi^o\rangle &= |2, -1\rangle\end{aligned} \tag{214}$$

Finally, one gets

$$\begin{aligned}\bar{\mathcal{A}}^{+-} &= \left\langle\pi^+\pi^-|\mathcal{T}|\overline{B^o}\right\rangle = \frac{\bar{\mathcal{A}}_{3/2}}{\sqrt{6}} - \frac{1}{\sqrt{3}}\bar{\mathcal{A}}_{1/2} \\ \bar{\mathcal{A}}^{oo} &= \left\langle\pi^o\pi^o|\mathcal{T}|\overline{B^o}\right\rangle = \frac{1}{\sqrt{3}}\bar{\mathcal{A}}_{3/2} + \frac{1}{\sqrt{6}}\bar{\mathcal{A}}_{1/2} \\ \bar{\mathcal{A}}^{-o} &= \langle\pi^-\pi^o|\mathcal{T}|B^-\rangle = \frac{\sqrt{3}}{2}\bar{\mathcal{A}}_{3/2}\end{aligned} \tag{215}$$

Thus the transition $\Delta I = \frac{3}{2}$ and $\Delta I = \frac{1}{2}$ correspond to the isospin 2 and 0 final states, respectively. Denoting

$$\bar{A}_2 = \frac{\bar{\mathcal{A}}_{3/2}}{2\sqrt{3}}, \quad \bar{A}_0 = \frac{1}{\sqrt{6}}\bar{\mathcal{A}}_{1/2}, \tag{216}$$

one finds

$$\left.\begin{array}{rcl} \frac{\bar{\mathcal{A}}^{+-}}{\sqrt{2}} & = & \bar{\mathcal{A}}_2 - \bar{\mathcal{A}}_o \\ \bar{\mathcal{A}}^{oo} & = & 2\,\bar{\mathcal{A}}_2 + \bar{\mathcal{A}}_o \\ \bar{\mathcal{A}}^{-o} & = & 3\,\bar{\mathcal{A}}_2 \end{array}\right\} \Rightarrow \quad \frac{\bar{\mathcal{A}}^{+-}}{\sqrt{2}} + \bar{\mathcal{A}}^{oo} = \bar{\mathcal{A}}^{-o} \qquad (217)$$

This equation correspond to a triangle. Similar equations are obtained for the meson B^o and B^+.

$$\frac{\mathcal{A}^{+-}}{\sqrt{2}} + \mathcal{A}^{oo} = \mathcal{A}^{+o} \qquad (218)$$

Since the spectator diagrams are the only ones contributing to the disintegrations $B^\pm \to \pi^\pm\pi^o$, the two triangles described by Eqs. 217 and 218 have a common side (Fig. 30).

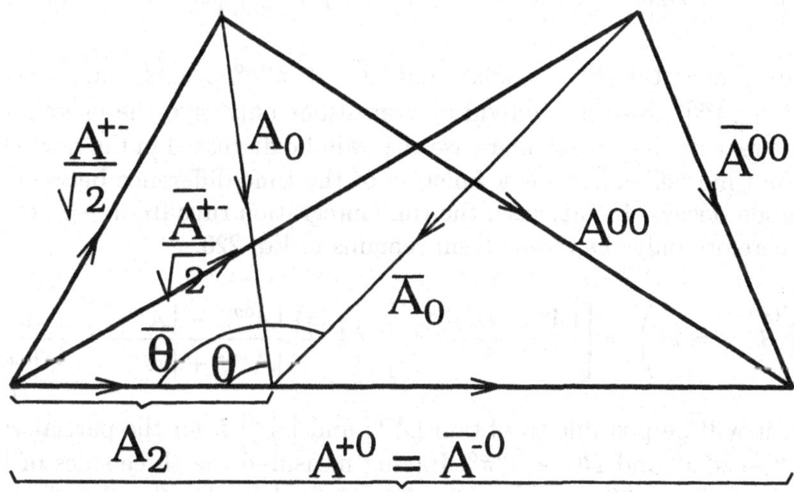

Figure 30. Triangles corresponding to the Eqs. 217 and 218.

If CP is conserved in the disintegration, these two triangles are identical. This is equivalent to say that the relative phase θ between $\mathcal{A}_o$ and $\mathcal{A}_2$ is the same than $\bar{\theta}$, the one between $\bar{\mathcal{A}}_o$ and $\bar{\mathcal{A}}_2$. The quantities $|\mathcal{A}^{+o}|$ and $|\bar{\mathcal{A}}^{-o}|$ are easy to measure since $|\mathcal{A}^{+o}|^2 \equiv \Gamma\left(B^+ \to \pi^+\pi^o\right)$ and $|\bar{\mathcal{A}}^{-o}|^2 \equiv \Gamma\left(B^- \to \pi^-\pi^o\right)$. The values of $|\mathcal{A}^{+-}|$ and $|\bar{\mathcal{A}}^{+-}|$ are more difficult to extract because of the B^o-$\overline{B^o}$ mixing. The transition probability $B \to \pi^+\pi^-$

reads

$$\mathcal{P}r\left(\overset{(-)}{B^o}\left(\frac{t}{\tau}\right)\rightarrow\pi^+\pi^-\right)=\left[\frac{|\mathcal{A}^{+-}|^2+|\bar{\mathcal{A}}^{+-}|^2}{2}\right]e^{-\frac{t}{\tau}}\times$$
$$\left\{1\overset{(-)}{+}\frac{|\mathcal{A}^{+-}|^2-|\bar{\mathcal{A}}^{+-}|^2}{|\mathcal{A}^{+-}|^2+|\bar{\mathcal{A}}^{+-}|^2}\cos x\frac{t}{\tau}\overset{(+)}{-}2\frac{\mathrm{Im}\left[\mathrm{e}^{i2\Phi}\bar{\mathcal{A}}^{+-}\mathcal{A}^{+-*}\right]}{|\mathcal{A}^{+-}|^2+|\bar{\mathcal{A}}^{+-}|^2}\sin x\frac{t}{\tau}\right\} \tag{219}$$

One extracts $|\mathcal{A}^{+-}|$ and $|\bar{\mathcal{A}}^{+-}|$ from the total number of events and the fit of the time dependence. The measurements of $|\mathcal{A}^{oo}|$ and $|\bar{\mathcal{A}}^{oo}|$ are somewhat more complicated because it is impossible to determine with a sufficient precision the $\pi^o\pi^o$ vertex (a resolution of about 100 μm is necessary). The transition probability for the neutral mode is:

$$\mathcal{P}r\left(\overset{(-)}{B^o}\left(\frac{t}{\tau}\right)\rightarrow\pi^o\pi^o\right)=\left[\frac{|\mathcal{A}^{oo}|^2+|\bar{\mathcal{A}}^{oo}|^2}{2}\right]e^{-\frac{t}{\tau}}\times$$
$$\left\{1\overset{(-)}{+}\frac{|\mathcal{A}^{oo}|^2-|\bar{\mathcal{A}}^{oo}|^2}{|\mathcal{A}^{oo}|^2+|\bar{\mathcal{A}}^{oo}|^2}\cos x\frac{t}{\tau}\overset{(+)}{-}2\frac{\mathrm{Im}\left[\mathrm{e}^{i2\Phi}\bar{\mathcal{A}}^{oo}\mathcal{A}^{oo*}\right]}{|\mathcal{A}^{oo}|^2+|\bar{\mathcal{A}}^{oo}|^2}\sin x\frac{t}{\tau}\right\} \tag{220}$$

Summing over the $B^o\rightarrow\pi^o\pi^o$ and $\overline{B^o}\rightarrow\pi^o\pi^o$ events, one measures $|\mathcal{A}^{oo}|^2+|\bar{\mathcal{A}}^{oo}|^2$, but the individual transitions only give the overall term in the brackets. In a particular case that will be discussed in the next chapter, the time dependence is a function of the time difference between two B^o meson decays. In this case, the time integration runs from $-\infty$ to $+\infty$ and therefore only the cosine term remains in Eq. 220.

$$\mathcal{P}r\left(\overset{(-)}{B^o}\rightarrow\pi^o\pi^o\right)=\left[\frac{|\mathcal{A}^{oo}|^2+|\bar{\mathcal{A}}^{oo}|^2}{2}\right]\left\{1\overset{(-)}{+}\frac{|\mathcal{A}^{oo}|^2-|\bar{\mathcal{A}}^{oo}|^2}{|\mathcal{A}^{oo}|^2+|\bar{\mathcal{A}}^{oo}|^2}\times\frac{1}{1+x^2}\right\} \tag{221}$$

Thus, it will be possible to obtain $|\mathcal{A}^{oo}|$ and $|\bar{\mathcal{A}}^{oo}|$ from the partial width for $B^o\rightarrow\pi^o\pi^o$ and $\overline{B^o}\rightarrow\pi^o\pi^o$. Having measured the three sides of both triangles in figure 30, one extracts $|\mathcal{A}_o|, |\bar{\mathcal{A}}_o|, |\mathcal{A}_2|, |\bar{\mathcal{A}}_2|, \theta$ and $\bar{\theta}$. Since the time dependence of the decay $\overset{(-)}{B^o}\rightarrow\pi^+\pi^-$ gives

$$\mathrm{Im}\left[e^{i2\Phi}\bar{\mathcal{A}}^{+-}\mathcal{A}^{+-*}\right]=2\mathrm{Im}\left[e^{i2\Phi}\left(|\bar{\mathcal{A}}_2|-|\bar{\mathcal{A}}_o|\,e^{\pm i\bar{\theta}}\right)\left(|\mathcal{A}_2|-|\mathcal{A}_o|\,e^{\pm i\theta}\right)\right], \tag{222}$$

one obtains sin2Φ with a four fold ambiguity corresponding to the various possible orientations of the triangles in the figure 30 (the apex of the triangle may be up side or down side). In this particular case, Φ represents the angle α of the unitarity triangle.

Finally, we will discuss an other category of final states which can be used for measuring the angles of unitarity triangle. This class of decays involves the final states which have a definite CP parity once the orbital momentum is known. A good example is the decay $B^o \to \psi K^{*o}$ with $K^{*o} \to K_S\pi^o$. Since the two particles of the final state are vectors, they may have the helicity $\lambda = 0$ or $\lambda = \pm 1$. One has three possibillities.

$$\begin{aligned}
&|\psi(\lambda=0)K^{*o}(\lambda=0)\rangle \Rightarrow CP\ |f\rangle = +|f\rangle \\
&|\psi(\lambda=1)K^{*o}(\lambda=1)\rangle + |\psi(\lambda=-1)K^{*o}(\lambda=-1)\rangle \Rightarrow\ CP\ |f\rangle = +|f\rangle \\
&|\psi(\lambda=1)K^{*o}(\lambda=1)\rangle - |\psi(\lambda=-1)K^{*o}(\lambda=-1)\rangle \Rightarrow CP\ |f\rangle = -|f\rangle
\end{aligned} \tag{223}$$

We have shown that for a CP eigenstate, one has

$$\mathcal{P}r\left(\overset{(-)}{B^o}\ (\frac{t}{\tau}) \to f_{\rm CP}\right) \propto e^{-\frac{t}{\tau}}\ \left\{1 \overset{(+)}{-}\ \eta_{\rm CP}\ \sin 2\Phi\ \sin x\ \frac{t}{\tau}\right\} \tag{224}$$

If one does not distinguish the various contributions, $\eta_{\rm CP}$ is not known and should be replaced by a dilution factor

$$d\ =\ \frac{\mathcal{A}_+^2 - \mathcal{A}_-^2}{\mathcal{A}_+^2 + \mathcal{A}_-^2} \tag{225}$$

Here $\mathcal{A}_+$ and $\mathcal{A}_-$ are the amplitudes corresponding to a final state with $\eta_{\rm CP}\ =\ +1$ and $\eta_{\rm CP}\ =\ -1$ respectively. Let us note that if the helicity state $\lambda = 0$ dominates, then $d = +1$ and there is no dilution. It is relatively simple to verify this by measuring the angular dependence of the K^{*o} decay. The angular distribution of K_S in the K^* rest frame has to be measured relatively to the K^* flight direction (see Fig. 31).

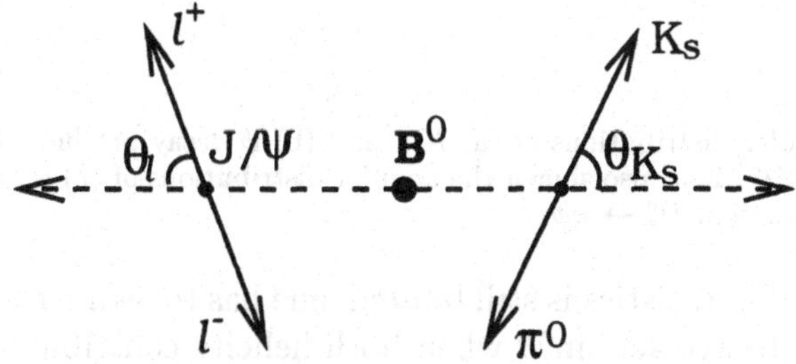

Figure 31. Definition of the angle K_s decay angle.

The expected angular distributions of the K_S depend on the helicity of the K^* and are

$$\frac{d\Gamma\left[K^*(\lambda=0)\right]}{d\cos\theta_{K_S}} \propto \cos^2\theta_{K_S} \quad , \quad \frac{d\Gamma\left[K^*(\lambda=\pm 1)\right]}{d\cos\theta_{K_S}} \propto \sin^2\theta_{K_S}\,. \tag{226}$$

The experimental results seem to favor the $\cos^2\theta_{K_S}$ distribution [107] (Fig. 32).

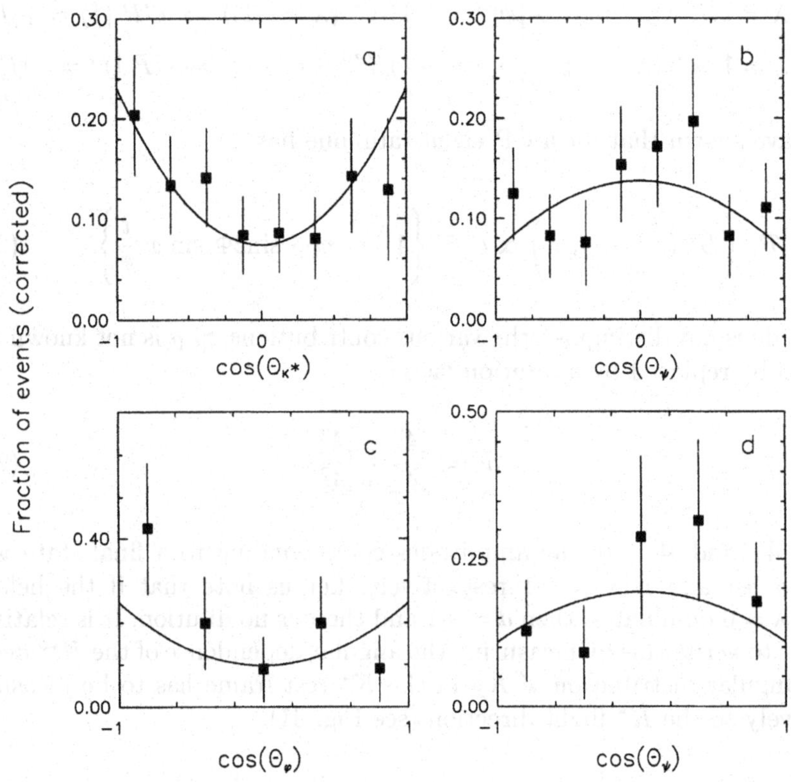

Figure 32. Angular distributions of (a) K^{*o} and (b) ψ decay products in the reaction $B^o \to \psi K^{*o}$ by CDF. It is also shown the angular distributions of (c) ϕ and (d) ψ decay products in the reaction $B_s^o \to \psi\phi$

However since the statistics is still limited, one has to wander whether there is a possibility to extract $\sin 2\theta$ when both helicity contribute. It is in fact possible to do so by studying the correlations between the angles of the K^* and ψ decay products. One can separate the contributions with $\eta_{\rm CP} = +1$

and $\eta_{\rm CP} = -1$ and extract sin2Φ at the cost of a worse sensitivity [108]. The figure 33 shows that even when $\eta_{\rm CP} = +1$ and $\eta_{\rm CP} = -1$ amplitudes have the same magnitude, sin2Φ is still measured but the error is about three times larger than for the pure $\eta_{CP} = +1$ case.

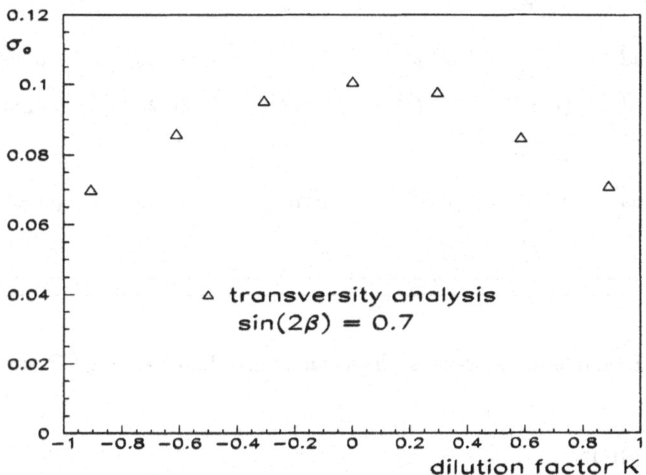

Figure 33. The error on sin 2Φ as a function of the fraction of $\eta_{\rm CP} = +1$ and $\eta_{\rm CP} = -1$ amplitudes by doing a angular dependent analysis of the data.

We have seen in this section that there exists a large varity of decay modes common to both B^o and $\overline{B^o}$ allowing one to get valuable information on the angles of the unitarity triangle by studying asymmetries in the decay of B^o and $\overline{B^o}$. These asymmetries are directly proportional to the angles α, β and γ defined in the figure 6 and may exibit large effects. Many different modes allow one to measure these quantities leading to a comprehensive test of the theory on the one hand and to a good measurement sensitivity when several channels measuring the same angle are added up. We have discussed three types of final states:

- Class 1: The CP eigenstates (ex. $\pi^+\pi^-$).
- Class 2: The final states which may be reached both from the B^o and $\overline{B^o}$ mesons but that are not CP eigenstates (ex: $\rho^\pm\pi^\mp$).
- Class 3: Final states that are C eigenstates but which are not CP eigenstates (ex: $J/\psi K_S$).

Some interesting final states are summarized in the tables 6 and 7.

Quark decays	Class 1	Class 2	Class 3	Measured angle	Penguin contribution
$\bar{b} \to \bar{c} + c\bar{s}$	ψK_s		$\psi(K_s\pi^o)_{K^*}$	sin 2β	negligible
	$\psi' K_s$		$\psi K_s \pi^o$		
	$\chi_{c_1} K_s$				
$\bar{b} \to \bar{c} + u\bar{d}$		$\overset{(-)}{D^o}\ \rho^o$		sin 2β	none
$\bar{b} \to \bar{c} + c\bar{d}$	D^+D^-	$D^{*\pm}D^{\mp}$	$D^{*\pm}D^{*\mp}$	sin 2β	small
	$D^o\overline{D^o}$				
$\bar{b} \to \bar{u} + u\bar{d}$	$\pi^+\pi^-$	$\rho^\pm\pi^\mp$	$\rho^+\rho^-$	sin 2α	possible
	$\rho^o\pi^o$	$a_1^\pm\pi^\mp$			(< 20%)

TABLE 6. Examples of $\overset{(-)}{B_d^o}$ decays which can be used for observing CP violation.

5.4. CONCLUSION

We have shown in this chapter that many different methods can be used to study CP violation with B mesons in the framwork of the Standard Model. Therefore one can conclude that the B system will be the unavoidable tool for testing the consistency of the theory in this sector and any measured inconsistency would give the first hint for new physics beyond the Standard Model.

6. Experimental Aspects and Prospects for studying CP Violation

We have seen in these lectures how CP violation is generated within the Standard Model and studied it consequences in B decays. We will now discuss the experimental feasibility of the proposed measurements. The aim of this chapter is by no means to carry an extensive and detailed description of how the measurements will be done but rather give an overview of the requirements, the methods used, the difficulties and the expected sensitivities.
There are presently four major efforts directed toward studying direct CP violation in the "strange" sector [109–112] which are summarized in the table 8. The primary goal of the experiments using the kaon decays is to measure ϵ'/ϵ. However, the last project in this table uses the $\Xi + \Lambda$ system

Quark decays	Class 1	Class 2	Class 3	Measured angle	Penguin contribution
$\bar{b} \to \bar{c} + c\bar{s}$			$\psi\Phi$	$2\frac{\lvert V_{us}^* V_{ub}\rvert}{\lvert V_{ud}^* V_{cb}\rvert}$	negligible
			$\psi'\Phi$	$\times \sin\gamma$	
			$\chi_{c_1}\Phi$		
$\bar{b} \to \bar{c} + c\bar{d}$	ψK_s		$\psi(K_s\pi^o)_{K^*}$	$2\frac{\lvert V_{us}^* V_{ub}\rvert}{\lvert V_{ud}^* V_{cb}\rvert}$	small
	$\psi' K_s$		$\psi K_s\pi^o$	$\times \sin\gamma$	
	$\chi_{c_1} K_s$				
$\bar{b} \to \bar{c} + u\bar{s}$		$D_s^{\pm}K^{\mp}$	$D_s^{*\pm}K^{*\mp}$	$\sin\gamma$	none
		$D_s^{*\pm}K^{\mp}$			
		$D_s^{\pm}K^{*\mp}$			
$\bar{b} \to \bar{u} + u\bar{d}$	$\rho^o K_s$			$\sin 2\gamma$	probably large
	K^+K^-				
	ωK_s				

TABLE 7. Examples of $\overset{(-)}{B_s^o}$ decays which can be used for observing CP violation.

System	Experiment/location	Facility	status
	NA48/CERN	K beams	in operation
K	KTeV/Fermilab	K Beams	in operation
	KLOE/Frascati	ϕ Factory	start in 1998
$\Xi + \Lambda$	E871/Fermilab	p beam	in operation

TABLE 8. The projects studying CP violation in the "strange" sector.

and searches for an asymmetry in the angular distributions of their decay products for the particles versus the antiparticles. We will not discuss these techniques here but rather concentrate on the projects dedicated to the b quark sector.

There are also four major area of activities directed toward studying CP violation in the "beauty" sector [113–122] (Table 9). In fact very prelim-

Experiment/location	Facility	status
CLEO/Cornell	Υ(4S)Factory	in operation
BELLE/KEK	Υ(4S)Factory	start in 1999
BABAR/SLAC	Υ(4S)Factory	start in 1999
HERA-B	p beam	start in 1999
CDF/Fermilab	2 TeV pp collider	start in 2000
D0/Fermilab	2 TeV pp collider	start in 2000
BTeV/Fermilab	2 TeV pp collider	in study
ATLAS/CERN	14 TeV pp collider	start in 2005
CMS/CERN	14 TeV pp collider	start in 2005
LHCB/CERN	14 TeV pp collider	start in 2005

TABLE 9. The projects studying CP violation in the "beauty" sector.

inary results have already been obtained using the $b\bar{b}$ pairs produced at LEP/CERN and at the Tevatron/Fermilab but these measurements are limited by the statistics.

6.1. REQUIRED NUMBER OF B MESONS

Let us now study the implications of the sensitivity which is required for studying CP violation using the three mechanism descibed in the previous chapter.

6.1.1. *Direct CP Violation.*

To get evidence for CP violation, one has to compare the number of $B \to f$ and $\overline{B} \to \bar{f}$ transitions. The total number of events required for observing an asymmetry $\mathcal{A}s$ with S standard deviation is:

$$N = \frac{S^2}{Br\,\mathcal{A}s^2\,\varepsilon}(1 + R_{B/S}) \tag{227}$$

where Br is the branching fraction of the final state f, ε is its detection efficiency and $R_{B/S}$ is the background to signal ratio. If one wants to be sensitive to an asymmetry of 1%, about 10^{10} B mesons are necessary assuming no background and a detection efficiency of $\sim$ 30% for a final state having a branching fraction of 3×10^{-5} (the typical order of magnitude for the decays of interest). However, the expected asymmetries here are very uncertain and one may hope for larger effets (up to 10%) eventually reducing the required number of B mesons by one to two orders of magnitude.

6.1.2. *CP Violation due to B^o-$\overline{B^o}$ Mixing.*
The experimental technique for observing such an effect is the simplest one. One needs to compare the number of like sign lepton with positive charge to the one with negative charge. The asymmetry is defined by

$$\mathcal{A}s = \frac{N(\ell^+\ell^+) - N(\ell^-\ell^-)}{N(\ell^+\ell^+) + N(\ell^-\ell^-)} = \frac{4\mathrm{Re}\epsilon_B}{1 + |\epsilon_B|^2} \tag{228}$$

The number of events necessary for observing this asymmetry with a sensitivity of S standard deviation is

$$N = \frac{S^2}{\mathcal{A}s^2 \; Br^2 \; \varepsilon^2 \; f_0 \; \chi} \tag{229}$$

where Br is the semileptonic branching fraction $(e + \mu)$, ε is the detection efficiency, f_0 is the fraction of $B^o(\overline{B^o})$ mesons produced and χ is the $B^o - \overline{B^o}$ mixing probability. Using the values $Br = 20\%$, $\varepsilon = 50\%$, $f_0 = 50\%$ and $\chi = 18\%$, about 10^{10} $b\bar{b}$ pairs are required in order to measure an asymmetry of 10^{-3} at the 3 standard deviation level. Clearly reaching this level of sensitivity is extremely difficult and requires a very good understanding of the systematics. Nevertheless it is important to try to make this measurement as the observation of any effect at the percent level would require the presence of new physics to explain it. It should be noted that this type of study can also be done with single leptons since the mixing probability is well known but the requirement in terms of events are similar [123].

6.1.3. *CP Violation due to the Interplay of B Decay and B^o-$\overline{B^o}$ Mixing.*
The measurement of this class of asymmetry requires the reconstruction of exclusive final states which can be produced in B^o and $\overline{B^o}$ decays. In general the branching fractions of such modes are small. It is therefore necessary to have a very large trigger/reconstruction efficiency. The table 10 gives a few examples for some interesting final states with their expected reconstruction efficiency. For illustration, the figure 34 shows the reconstructed invariant mass of $J/\psi\ K_S$ with the CDF detector at the Tevatron. This is the largest existing sample of such decays. It can be noticed that there is some (although not large) background for this particular final state at a hadron collider (while there is essentially none at e^+e^- colliders). The

Modes	Total efficiency	Useddecay modes
$\psi\ K_S$	$\sim$ 5%	$\psi \to \ell^+\ell^-$ $K_S \to \pi^+\pi^-,\ \pi^o\pi^o$
$\pi^+\pi^-$	$\sim$ 60%	
$\rho^\pm\pi^\mp$	$\sim$ 20%	$\rho^\pm \to \pi^\pm\pi^o$

TABLE 10. Reconstruction efficiency estimated for some interesting final states with the BaBar detector.

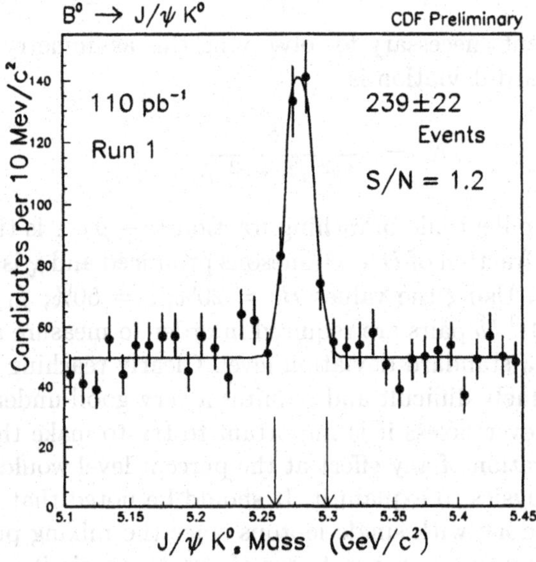

Figure 34. The reconstructed invariant mass for the final state $J/\psi\ K_S$ at CDF.

number of B^o mesons required to observe an asymmetry in the decay rates $B^o \to f_{CP}$ vs $\overline{B^o} \to f_{CP}$ with S standard deviation can be written as:

$$N_B = S^2 \frac{1+4x^2}{2x^2} \frac{1+\eta_z \frac{B}{S}}{d^2 \sin^2(2\phi)} \frac{e^{\sigma_t^2 x^2/2}}{\mathrm{Br}\ \epsilon_f\ \epsilon_{\mathrm{tag}}\ (1-2\omega)^2} \tag{230}$$

where

- $x = \frac{\Delta m}{\Gamma} \simeq 0.73$ for B_d^o mesons

- η_z is a rejection factor due a different time dependent behaviour of the background with respect to the signal (in general $\eta_z \approx 0.1$)
- $\frac{B}{S}$ is the background over signal ratio (~ 0 for ψK_S)
- d is an eventual dilution factor depending of the final state (=1 for CP eigenstates)
- ϕ is one of the angle of the unitarity triangle depending on the studied mode
- σ_t is the resolution on the flight time in proper time units (in general ≤ 0.3)
- Br is the branching fraction of the studied mode ($\text{Br} \simeq 4 \times 10^{-4}$ for ψK_S)
- ϵ_f and ϵ_{tag} are the final state reconstruction and tagging efficiencies respectively ($\epsilon_f \approx 5\%$ and $\epsilon_{\text{tag}} \leq 0.3$)
- ω is the fraction of wrong tagged B mesons (≥ 0.1)

By using the typical numerical values of the parameters in the equation above and taking $\sigma(\sin 2\beta) = 0.4$ for the decay $B \to \psi K_S$, one may observed an asymmetry at 3 standard deviation with 5×10^7 produced B^o.

6.2. WHICH FACILITY TO CHOOSE?

Obviously from the dicussion above, the first requirement is the production of a large number of B mesons. We show in figure 35 the $b\bar{b}$ cross section at various existing or proposed accelerator.
As it can be seen, cross sections may vary by several order of magnitude from one facility to the other. Still it is not so straightforward to decide which is the most convenient source of B mesons as in order to make a choice, one needs also to consider the important following criterium:

- The cleanliness of the environment in which the B mesons are produced. In other words, what are the B reconstruction and tagging efficiencies and what is the signal over background ratio?

From Table 9, one identify three main options presently pursued.

- e^+e^- colliders at the Υ(4S)energy.
- Fixed target experiment using a high energy proton beam.
- pp colliders at very high energies.

Table 11 summarizes the main characteristics of these options, including LEP for comparison. As can be seen from this table, some numbers require very challenging technical developments in order to have an operational experiment and require comments.

- <u>*B Factory:*</u> Here the main limitation is coming from the luminosity of the collider. The highest luminosity so far achieved is $7 \times 10^{32}\text{cm}^{-2}\text{s}^{-1}$ with CESR at Cornell. One would need a factor 5 to 10 improvement

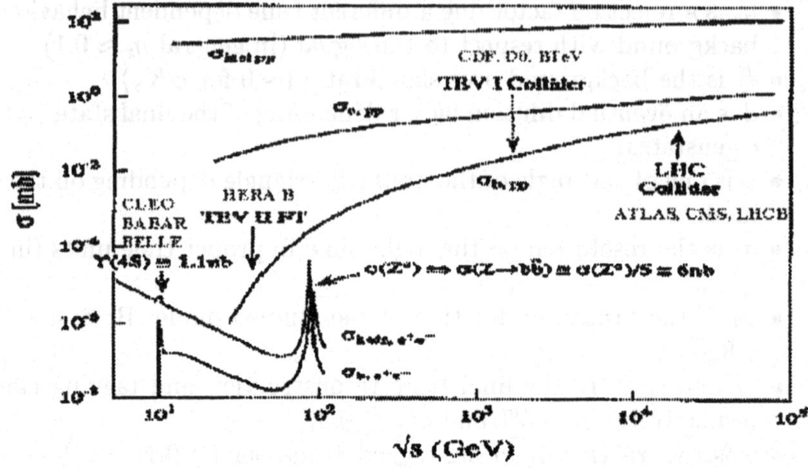

Figure 35. The cross section for the production of $b\bar{b}$ pairs at various existing or proposed accelerator.

Facility	$b\bar{b}$/year	$b\bar{b}$/All	Radiation	π^0	Input rate
Υ(4S)Factory	3×10^7 to 10^8	0.25	0.2 Mrad/y	yes	~ 100 Hz
LEP I	4×10^5	0.2	negligible	yes	~ 0.2 Hz
HERA-B	10^9	5×10^{-6}	Mrad/y	no	$\sim 10^7$ Hz
LHC	4×10^{10}	2×10^{-3}	1-10 Mrad/y	no	$\sim 2 \times 10^6$ Hz

TABLE 11. Some of the main characteristics at several planned or existing facilities.

and this is what is planned at the asymmetric facilities in construction at KEK and SLAC. These colliders have two rings with collision at a single point. No technical show stopper have been identified so far. Apart from the luminosity the main difficulty will be to keep the background from the accelerator at a level manageable by the detectors. The detector by themselves do not represent a big challenge as most of the necessary techniques are already in operation at other facilities.

– <u>*LEP collider:*</u> This would be probably the best place to study CP violation, thanks to the relatively large $b\bar{b}$ cross section ($\sim$ 6 nb at the Z pole), the long B flight lenght ($\sim$3 mm) and the clean environnement. However LEP in now running at higher energy and there is no plan to get back to the Z pole. Furthermore, the luminosity achieved so far is way too low and increasing it would be a complicated and costly project.

– *HERA-B:* The basic idea is to use the halo of the intense HERA proton beam around which several wires are suspended and serve as targets. The main difficulty resides in the small $b\bar{b}$ cross section ($\sim$ 10 nb) relative to the large inelastic hadronic cross section ($\sim$ 50 mb). Consequently, one needs to develop a triggering system being able to reduce the input rate of 10^7 Hz down to a manageable level while keeping a good efficiency for $b\bar{b}$ events. This requires the developments of detectors standing very high rates and radiation doses. The reconstruction becomes also a major challenge because of the superimposed interactions ($\sim$ 5) and the large boost which keeps all tracks in a small solid angle. As a consequence of the above difficuties, only a few modes containing a J/ψ and no neutrals can be studied.
– *LHC:* Experiments at LHC (and earlier at the Tevatron) are confronted essentially to the same problems than HERA-B, although in a somewhat less hostile environnement. The trigger problem is easier because of the much more favorable ratio $b\bar{b}$ events versus inelastic interation. The solid angle being larger, the track density is lower and because of the very large $b\bar{b}$ cross section, one has the option to operate at a lower luminosity improving further the latter difficulty and reducing the radiation damages. Because of the less difficult conditions, more modes are reacheable althouth the use of neutrals is still very challenging.

In general terms all these different options require essentially the same basic detector components and thus a generic detectors can be sketched. Each detector needs a silicon vertex detector, a tracker which can be made of silicon detector and/or wire chambers, a particle identification system based on de/dx, Time of Flight or Cherenkov counters or imaging devices and electromagnetic calorimetry using crystals or charge collection. For illustration and personal bias, I will concentrate in the following on the $\Upsilon(4S)$ option (which have some peculiarities) and mention the differences at other facilities as required. However this choice does not preclude the ability of the other options to achieve our common goal: performing a comprehensive study of the CP violation phenomenon.

6.3. THE E^+E^- COLLIDERS AT THE Υ(4S)

Up to now, it is fair to say that the most extensive study of the B meson decays has been performed at e^+e^- colliders running at a center of mass energy of about 10.6 GeV corresponding to the mass of the Υ(4S) resonance. It is therefore natural to ask ourselves whether the studies (involving many exclusive decays) which were discussed in previous chapters can be done at such colliders. First, let's clarify what we mean by a B Factory at the Υ(4S).

- B "Factory" is an expression commonly used to describe any accelerator which is able to produce a large number of B mesons ($\geq 10^7 B$ per year).
- $\Upsilon(4S)$ is a resonance made of a $b\bar{b}$ quark pair, the mass of which is about 10.58 GeV and which decays into a pair of $B\overline{B}$ meson ($\sim$ 50% B^+B^- and $\sim$ 50% $B^o\overline{B^o}$). The quantum numbers of this resonance are $J^{CP} = 1^{--}$.

What are the advantages of a B Factory at the $\Upsilon(4S)$?

- Although not huge, the cross section is acceptable ($\sigma[e^+e^- \to \Upsilon(4S)] \simeq 1.1\, nb$) (Fig. 36).

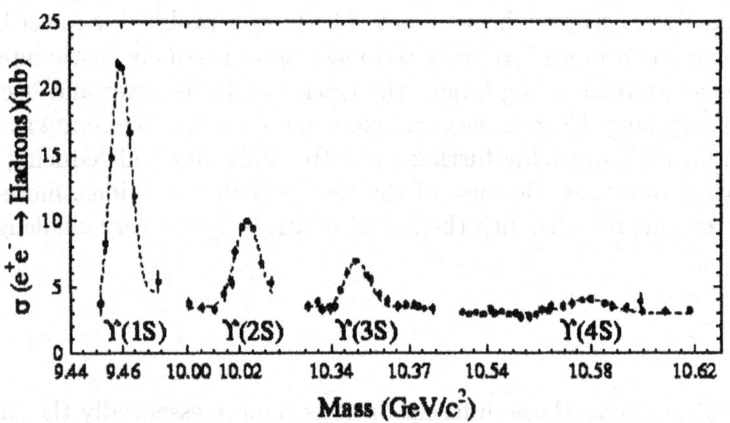

Figure 36. Hadronic cross section in the threshold region for $B\overline{B}$ production.

- It is the cleanest source of $B\overline{B}$ pairs: No other particule is produced allowing one to get a good tagging efficiency as we will see later. The multiplicity of the event is small (10 charged and 10 photons are produce on average) and the tracks are spread over the full solid angles. Therefore the reconstruction efficiency is large with a rather small background contribution. The background is essentially due to $q\bar{q}$ continuum events ($e^+e^- \to u\bar{u}$, $d\bar{d}$, $s\bar{s}$, $c\bar{c}$) and can be reduced by using the fact that the B mesons are monoenergetic ($P_B \simeq 300\ MeV$). The measure of the B mass is sensibly improved by using the energy of the beam as the energy of the B mesons and thus it is possible to get a resolution of 2 MeV/c^2 instead of 20 MeV/c^2.

6.3.1. *Coherent $B\overline{B}$ Pair Production.*

Since the $\Upsilon(4S)$ decays exclusively into a $B\overline{B}$ pair, the $B\overline{B}$ system is in a coherent quantum state [124]. This is a consequence of the generalization of the Bose-Einstein statistics. Thus the wave functions of the $B\overline{B}$ pair is antisymmetric and therefore one should consider the $B\overline{B}$ system as a whole instead of the individual B mesons. The wave function

$$\frac{\left|B\,(t_1)\ \overline{B}\,(t_2)\,\right\rangle - \left|\overline{B}\,(t_1)\ B\,(t_2)\,\right\rangle}{\sqrt{2}} \tag{231}$$

should be used in all the previous calculations involving time evolution. The main consequence is that the neutral B^o and $\overline{B^o}$ will evolve coherently until one of them decays. It is only at that time that the nature of the second meson (B^o or $\overline{B^o}$) will be defined. This second B will have the flavor opposite to the one of the first B. Let us see what happens to equations 76 and 77 when this wave function is used.

$$\begin{aligned}
\mathcal{P}r\left(\overline{B^o}\,(t_1)_{\rm tag}\,,\ B\,(t_2) \to f\right) = e^{-\Gamma(t_1+t_2)} \Big\{ & \cos^2\frac{\Delta m}{2}\ \Delta t\ |\langle f|\mathcal{T}|B^o\rangle|^2 \\
& + \sin^2\frac{\Delta m}{2}\ \Delta t\ \left|\frac{q}{p}\right|^2 \left|\left\langle f|\mathcal{T}|\overline{B^o}\right\rangle\right| \\
& - \frac{i}{2}\sin\Delta m\ \Delta t\ \left|\frac{q}{p}\right|\ e^{-2i\Phi_M}\ \langle f|\mathcal{T}|B^o\rangle\ \left\langle f|\mathcal{T}|\overline{B^o}\right\rangle^* \\
& + \frac{i}{2}\sin\Delta m\ \Delta t\ \left|\frac{q}{p}\right|\ e^{2i\Phi_M}\ \langle f|\mathcal{T}|B^o\rangle^*\ \left\langle f|\mathcal{T}|\overline{B^o}\right\rangle \Big\}
\end{aligned} \tag{232}$$

where $\Delta t = t_2 - t_1$ and $\overline{B^o}\,(t_1)_{\rm tag}$ means that the $\overline{B^o}$ meson has been identified whithout any ambiguity at time t_1. Similarly, if the meson B^o

was identified at time t_1, one gets.

$$\mathcal{P}r\left(B^o\left(t_1\right)_{\text{tag}},\, B\left(t_2\right) \to f\right) = e^{-\Gamma(t_1+t_2)} \times \left\{\cos^2\frac{\Delta m}{2}\,\Delta t\,\left|\left\langle f|\mathcal{T}|\overline{B^o}\right\rangle\right|^2\right.$$

$$+\sin^2\frac{\Delta m}{2}\,\Delta t\,\left|\frac{p}{q}\right|^2\,|\langle f|\mathcal{T}|B^o\rangle|^2$$

$$+\frac{i}{2}\sin\Delta m\,\Delta t\,\left|\frac{p}{q}\right|\,e^{-2i\Phi_M}\,\langle f|\mathcal{T}|B^o\rangle\,\left\langle f|\mathcal{T}|\overline{B^o}\right\rangle^*$$

$$\left.-\frac{i}{2}\sin\Delta m\,\Delta t\,\left|\frac{p}{q}\right|\,e^{2i\Phi_M}\,\langle f|\mathcal{T}|B^o\rangle^*\,\left\langle f|\mathcal{T}|\overline{B^o}\right\rangle\right\} \tag{233}$$

The reference time, t_0, is not the $\Upsilon(4S)$ decay time anymore, but the time at which one of the B mesons is identified. The consequence of this is somewhat unpleasant since the terms with $\sin\Delta m\Delta t$ can have a positive or negative value. Therefore if Δt is not measured and one integrates the time dependence from $-\infty$ to $+\infty$, the $\sin\Delta m\Delta t$ term vanishes. No CP violating effect can be observed for B decays involving the interplay of B decays and B^o-$\overline{B^o}$ mixing. This fact is easily seen (Fig. 37) for example in $B \to \pi^+\pi^-$ for which the transition probabilities are:

$$\mathcal{P}r\left(\overset{(-)}{B^o}\left(t_1\right)_{\text{tag}},\, B\left(t_2\right) \to \pi^+\pi^-\right) \propto e^{-\frac{|\Delta t|}{\tau}}\left[1 \overset{(-)}{+} \sin 2\Phi\,\sin x\,\frac{\Delta t}{\tau}\right] \tag{234}$$

It is therefore mandatory to measure Δt if one wants to use this class of events.

Is there any alternative? In principle, one could use a $B\overline{B}$ system for which the wave function is symmetric. It can be verified that in such case, the time dependence is $t_1 + t_2$ instead of $t_2 - t_1$. The time integrated asymmetry will then be

$$\frac{2x}{\left(1+x^2\right)^2}\sin 2\Phi \simeq 0.6\sin 2\Phi \tag{235}$$

The question is how to produce a $B\overline{B}$ system having a symmetric wave function? The answer is to produce a $B^o\overline{B^{o*}}$ system with $J^{CP} = 1^{--}$. Since the B^{o*} meson decay to $B^o\gamma$ and since the photon has $C_\gamma = -1$, the remaining $B^o\overline{B^o}$ system will have $C = +1$. Unfortunately the cross section at the threshold for $B\overline{B^*} + \overline{B}B^*$ is about 6 times smaller than for the $\Upsilon(4S)$ [125] and therefore the total production rate will be suppressed. So Δt has to be extracted by measuring the flight distance of the B mesons using

$$L = \beta\,\gamma\,c\,t \tag{236}$$

where β is the particle velocity $\left(\beta = \frac{P}{E}\right)$, $\gamma = \frac{E}{M}$ and c is the speed of light. Should the B meson move, it would be possible to measure its flight dis-

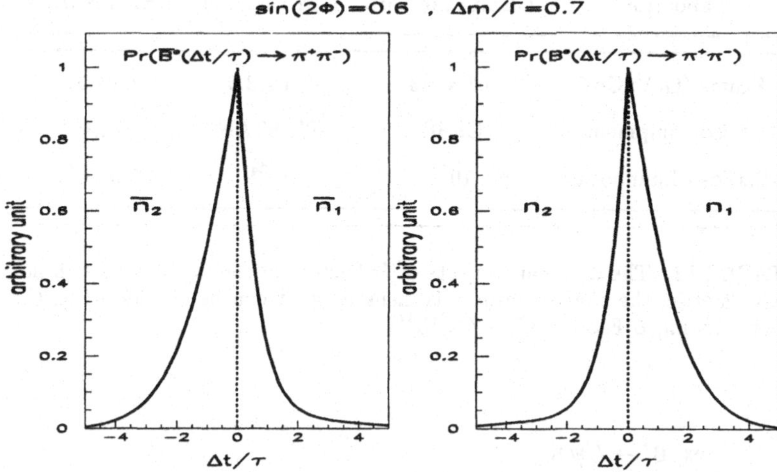

Figure 37. The expected Δt dependent decay distributions for B^o and $\overline{B^o}$ mesons. The asymmetry in B^o and $\overline{B^o}$ decay would vanish if one integrates over Δt ($n_1+n_2 = \bar{n}_1+\bar{n}_2$). However an integrated asymmetry would be possible to measure if one is able to distinguish positive and negative Δt ($\mathcal{A}s = \frac{(n_1+\bar{n}_2)-(n_2+\bar{n}_1)}{n_1+\bar{n}_2+n_2+\bar{n}_1}$).

tance and thus the time difference between the two decays. But as we said previously, the B mesons are produced almost at rest ($P_B \simeq 300$ MeV/c) in the $\Upsilon(4S)$ center of mass and therefore the average distance between both B decay vertices is about 50 μm. This distance is of the same order than the resolution which is possible to obtain with the present technology. It has therefore been suggested to create the $\Upsilon(4S)$ with a boost by using heteroenergetic beams energies [126, 127]. In that case, the B mesons are produced with a boost along the direction of the high energy beam. Many studies for the feasibility of such a machine has been carried out in several laboratories [128] and two of them are being constructed (see Table 12). However no asymmetric e^+e^- collider has been yet built and its construction poses several technical problems. In particular, the beam intensities have to be about 10 times larger than the ones currently achieved. Nevertheless, the accelerator physicists are confident that it can be achieved and that luminosities larger than 2 10^{33} $cm^{-2}s^{-1}$ can be obtained.

6.3.2. *Boosting the B Mesons.*

Let us examine now what happens when the $\Upsilon(4S)$ resonance is produced with a boost $\beta\gamma$. The figure 38 shows conceptually how a $B\overline{B}$ event might look like.

The distance Δz can be written in term of t_1 and t_2.

Laboratory	CESR (Cornell)	KEKB(KEK)	PEPII(SLAC)
Beams (GeVxGeV)	5.3 x 5.3	8.0 x 3.5	9.0 x 3.1
Used equipment	CESR	TRISTAN	PEP
Planned Luminosity	10^{33}	10^{34}	3×10^{33}

TABLE 12. The different projects of B Factory in the world Although not asymmetric, the Cornell project is included as it can be considered as the first existing B Factory ($L > 7 \times 10^{32}$).

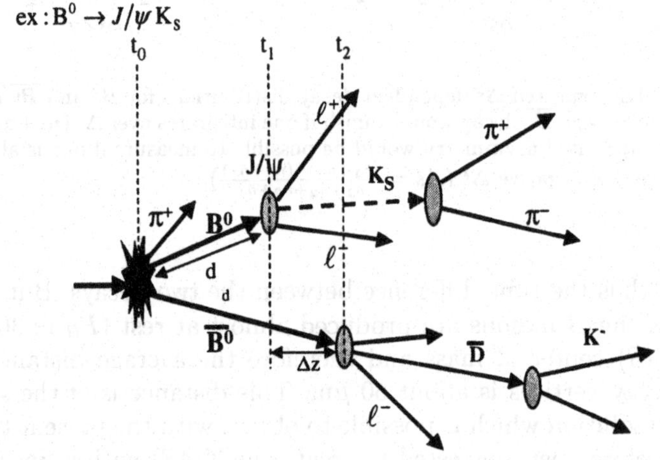

Figure 38. A schematic view of an interaction with asymmetric beam energies. The shaded particles may be used for tagging purpose (the π^+ coming from the prmary vertex is not present at $\Upsilon(4S)$ energy).

$$\Delta z = \beta\gamma\, c\tau \left[\frac{(t_2 - t_1)}{\tau}\right] + \gamma\, \beta_{cm} c\tau \cos\theta_B^* \left[\frac{t_2 + t_1}{\tau}\right] \tag{237}$$

Here, $\beta\gamma$ is the boost of the Υ(4S) ($\beta\gamma \simeq \frac{E}{10.6} - \frac{2.6}{E}$ where E is the energy of the most energetic beam), $c\tau$ is the average flight distance of a B meson and β_{cm} is its velocity in the Υ(4S) center of mass ($\beta_{cm} \simeq 0.07$). Finally, θ_B^* is the angle at which the B meson is produced in the Υ(4S) rest frame with respect to the beam direction. For most studies the last term in Eq. 237 can be neglected. Assuming a high energy beam of 9 GeV, $\beta\gamma \simeq 0.56$ and thus $\beta\gamma c\tau \approx 250\ \mu$m. What is the minimal value of $\beta\gamma$ which is necessary in order to observe a time dependent asymmetry? This obviously depends of the experimental resolution on Δz (we shall see later that $\sigma(\Delta z) \approx 100\ \mu m$ typically). All studies have shown that the CP asymmetry can safely be measured with the high energy beam in the range $9 \gtrsim E_H \gtrsim 8$ GeV.

6.4. MEASURING THE CP ASYMMETRIES

We will now discuss the sensitivities which may be achieved, the so-called "CP Reach" for the 3 classes of asymmetry. In order to be quantitative, one needs to define the number of $b\bar{b}$ events which is expected in a year of operation. The numbers given in the table 11 are based on a "standard year" of 10^7 seconds. For the B Factories, we will use 3×10^7 $b\bar{b}$ corresponding to a luminosity of 3×10^{33} cm^{-2} s^{-1}.

6.4.1. *Direct CP Violation.*

The main issue here is to reconstruct the largest possible sample of final states susceptible of exhibiting CP violation. In general, these decays are charmless rare decays. Some of them have already been observed [129–131].

$$\begin{aligned}
\mathrm{Br}(B^+ \to K^o\pi^+) &= (2.3^{+1.1}_{-1.0} \pm 0.3 \pm 0.2) \times 10^{-5} \\
\mathrm{Br}(B^+ \to \omega K^+) &= (1.5^{+0.7}_{-0.6} \pm 0.2) \times 10^{-5} \\
\mathrm{Br}(B^+ \to \eta' K^+) &= (6.5^{+1.5}_{-1.4} \pm 0.9) \times 10^{-5} \qquad (238) \\
\mathrm{Br}(B^o \to K^+\pi^-) &= (1.5^{+0.5}_{-0.4} \pm 0.1 \pm 0.1) \times 10^{-5} \\
\mathrm{Br}(B^o \to \eta' K^o) &= (4.7^{+2.7}_{-2.0} \pm 0.9) \times 10^{-5}
\end{aligned}$$

In general a good momentum resolution and an excellent particle identification are necessary in order to separate kaons and pions. This is mandatory for the mode $B \to K^+\pi$ for which the background due to $B \to \pi^+\pi$ might be large. The figure 39 shows how difficult it is to separate $K\pi$ and $\pi\pi$ final states with kinematical cuts only.

There are many rare two body modes which are candidates for searching Direct CP violation. However all these modes have a branching fraction at the 10^{-5} level or significantly lower in some case. Unfortunately, it is generally the one with very low branching fraction which are the most promising candidates for a large ($\sim 10\%$) asymmetry. We give in the table 13 the estimated sensitivities on the asymmetries (δA) for some typical final states at an asymmetric B factory with 1 year of data taking at a luminosity of 3×10^{33} cm^{-2} s^{-1}. In the previous chapter, we had estimated that the asymmetries due to direct CP violation would in general be less than 10% and more likely around 1%. It would therefore be very difficult to observe

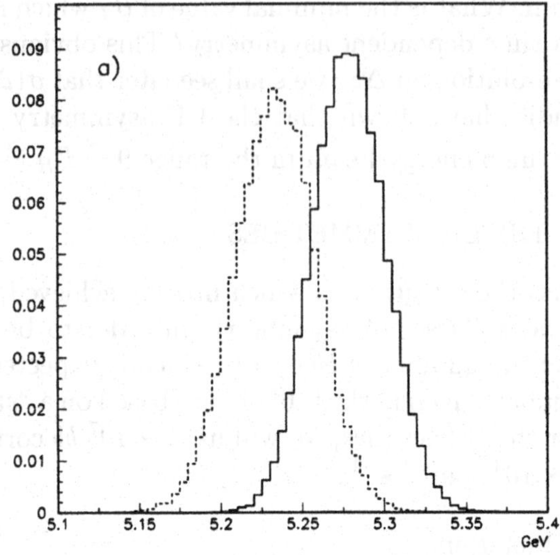

Figure 39. The invariant mass for the $K^+\pi^-$ and $\pi^+\pi^-$ pairs produced in B meson decays without any particle identification, i.e. all particles are assumed to be pions.

Mode	Theoretical Br	Asymmetry sensitivity
$B^\pm \to K^\pm\pi^o$	$\sim 1 \times 10^{-5}$	$\sim$ 10%
$\eta' K^\pm$	$\sim 6 \times 10^{-5}$	$\sim$ 7%
$\eta\pi^\pm$	$\sim 5 \times 10^{-6}$	$\sim$ 23%
$K^\pm K^{*o}$	$\sim 1 \times 10^{-7}$	$\sim$ 40%

TABLE 13. Estimated experimental sensitivity for the observation of direct CP violation using some particular decay modes.

an effect for these interesting decay modes. Nevertheless it is extremely important to search for any effect since it is probable that a large asymmetry will point to physics beyond the Standard Model. Although more B mesons would be produced at hadron facilities, the measurement is likely not to be easier as one would need to keep these decays at the trigger level. Furthermore only the charged modes could be reconstructed and one would need to identify the kaons and the pions, a difficult task in the environnement of hadron accelerators. If not, the asymmetry may be diluted making the measurement even more difficult.

6.4.2. *Mixing Induced CP Violation.*

The search for this type of CP violation can be carried out by using leptons as discussed earlier in this chapter. If an asymmetry is observed, the parameter which will be extracted is $\mathrm{Re}(\epsilon_B)$ (see Eq. 228). A search for an asymmetry has already been made some time ago by CLEO [132] using dileptons. No asymmetry was observed ($\mathcal{A}s = 0.031 \pm 0.096 \pm 0.032$) and a limit on $\mathrm{Re}(\epsilon_{B_d})$ was obtained $\mathrm{Re}(\epsilon_{B_d}) < 0.045$ at 90% confidence level. A similar search has been made by CDF [133] using a sample of pairs of high p_T muons with the same sign. In the same way as for CLEO, the comparison of the number of positive pairs to the negative could reveal CP violation induced by the $B^o\overline{B^o}$ mixing. However, the measured asymmetry is now a combination of ϵ_{B_d} due to B_d^o mesons and ϵ_{B_s} induced by B_s^o mesons:

$$\begin{aligned}
\mathcal{A}s &= \frac{N(\ell^+\ell^+) - N(\ell^-\ell^-)}{N(\ell^+\ell^+) + N(\ell^-\ell^-)} \\
&= \frac{8(1-\chi)}{D}\left\{ f_d\chi_d \frac{Re(\epsilon_{B_d})}{1+|\epsilon_{B_d}|^2} + f_s\chi_s\frac{Re(\epsilon_{B_s})}{1+|\epsilon_{B_s}|^2}\right\} \qquad (239) \\
&= (2.76 \pm 6.32_{\mathrm{stat}} \pm 3.28_{\mathrm{syst}}) \times 10^{-2}
\end{aligned}$$

where $D = 2\chi(1-\chi) + f_{\mathrm{seq}}[\chi^2(1-\chi)^2]$, f_{seq} being the fraction of selected muons from the cascade $b \to c \to \mu$ estimated with the Monte Carlo and $f_{d(s)}$ and $\chi_{d(s)}$ the fraction of produced $B_{d(s)}$ and their mixing probability respectively. No asymmetry is observed and therefore CDF sets limits on these parameters as shown in Fig. 40.

Others measurements have been carried out at LEP by DELPHI and OPAL, and their results are

$$\begin{aligned}
\mathrm{DELPHI:}\quad \mathrm{Re}(\epsilon_{B_d}) &= (-12 \pm 15 \pm 5) \times 10^{-3} \\
\mathrm{Re}(\epsilon_{B_d}) &= (-6 \pm 11 \pm 3) \times 10^{-3} \text{ if CPT assumed} \\
\mathrm{OPAL:}\quad \mathrm{Re}(\epsilon_{B_d}) &= (6 \pm 10 \pm 6) \times 10^{-3} \\
\mathrm{Re}(\epsilon_{B_d}) &= (2 \pm 7 \pm 3) \times 10^{-3} \text{ if CPT assumed}
\end{aligned} \qquad (240)$$

Here again no effects is observed and the limits which can be derived are in the same range than the one above. We have seen that in the Standard Model, the expected value for $\mathrm{Re}(\epsilon_{B_d})$ and $\mathrm{Re}(\epsilon_{B_s})$ are at the 10^{-3} and 10^{-4} level [136] and therefore present data are not sensitive enough. However it is important to pursue these measurements as the observation of an effect, even at the percent level, would reveal the presence of new Physics. At

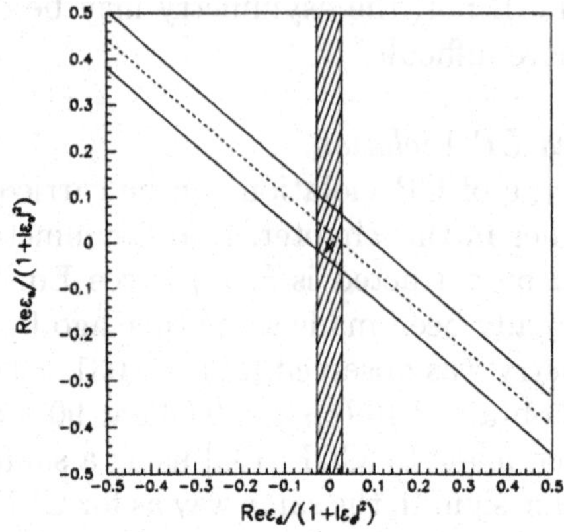

Figure 40. Measurements of $\mathrm{Re}(\epsilon_{B_d})$ versus $\mathrm{Re}(\epsilon_{B_s})$ from CLEO and CDF.

future facilities, the large number of events would allow to get the statistical error at the few 10^{-3} level (see for example [1]). The main difficulty resides in the systematics and a "perfect" understanding of the detector will be required.

6.4.3. *CP Violation Induced by the Interplay of Decay and Mixing.* Besides reconstructing the relevant final state, two essential ingredients are necessary to display a time dependent asymmetry.

– Identifying the flavor of the B meson at t=0.

– Measuring its decay time from its flight length.

a) The Tagging: The initial flavor of a B meson can be determined from the charge of its companion b quark as $b\bar{b}$ pair are produced. The tagging is a method which determines this charge. Several techniques have been developed but we will concentrate here on two of them which are likely to be the most useful ones: **tagging using leptons and kaons.**

– The Leptonic Tagging.
The decay $b \to \ell\nu c$ allows one to deduce the charge of the b quark by means of the lepton charge as can be seen from Fig. 41a and c. Since the semileptonic branching fraction $[Br(b \to e\nu_e c) + Br(b \to \mu\nu_\mu c)]$ is about 20% these decays are a good tagging tool. There exists two sources of background leading to wrong tagging. The first one is purely experimental and is caused by false lepton identification while the second source has a physical origin. It is due to the semileptonic decays

of the c quark as shown in the figure 41b. In cascade c decays, the lep-

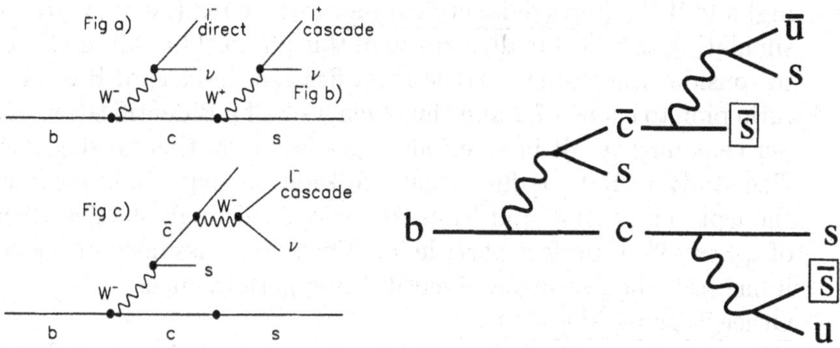

Figure 41. The different source of production of leptons from b quark decays.

Figure 42. The different source of production of s quarks.

tons have a sign opposite to the direct leptons, leading to an erroneous tagging. Fortunately the momentum distribution of the direct leptons is different from the one generated in the cascade $b \to c \to \ell$ (see the figure 14). The simpliest method is to use a cut $P^*_\ell > 1.4$ GeV/c (P^*_ℓ is the momentum of the lepton in the $\Upsilon(4S)$ center of mass). About 50% of the direct leptons are kept with a purity of $\mathcal{P}$=96%. The effective tagging efficiency would then be $\epsilon_{\text{eff}} = \epsilon_{\text{tag}}(2\mathcal{P}-1)^2 \approx 9\%$. However one could take also benefit of the low energy (or cascade) leptons as they also allow to get the b charge by flipping their sign. To maximize ϵ_{eff}, an efficient option is to use Neural Networks. A detailed study has been carried out in Ref. [1] and one obtains $\epsilon_{\text{eff}} \approx 23\%$ if perfect lepton identification is assumed and still $\epsilon_{\text{eff}} \approx 14\%$ if one has a pessimistic lepton identification.

- <u>The Kaon Tagging.</u>
  The direct cascade $b \to c \to s$ shown in figure 42 could be used as a tagging method. Indeed in this direct chain, the charge of the $s(\bar{s})$ quark is the same than the charge of the $b(\bar{b})$ quark, and therefore if the s quark produces a charged kaon, its charge should have the same sign than the charge of the initial $b(\bar{b})$ quark. Unfortunately there are other possibilities to produce $s(\bar{s})$ quark in B decays and some of them lead to an incorrect tagging as shown in the figure 42. In this figure, the wrong sign s quark have been encircled. The most ennoying "wrong" strange quark in Fig 42 comes from the $\bar{c}$ quark in the upper vertex. This contribution represents about 10% to 15% of the b decays. In

order to identify such "spurious" decays, it is useful to observe all the charged and neutral kaons in the event. But this can only be effective if the detector acceptance is very good (i.e. few charged particles missing) and if the particle identification is very good (i.e. very little pions simulating kaons). Finally, on top of this physical background, one has to consider the wrong particle identification. In general B decays, the ratio pion to kaon is 7:1 and thus even a small misidentification of a few per cent may result in a serious degradation on this tagging method. The study in Ref. [1] finds that the kaon tagging would complement the leptonic method and leads to a combined total effective efficiency of $\epsilon_{\text{eff}} \approx 36\%$ if perfect particle identification is assumed or $\epsilon_{\text{eff}} \approx 23\%$ if one uses the pessimistic identification performance.

- Other Tagging Methods.
  Other tagging methods have been also developed. Amongst those, it is worthwhile indicating the Jet Charge, and the accompanying π. The first one has been used at the LEP for measuring the $B\overline{B}$ oscillation and consist of determining the charge of the b quark in a jet by constructing a weighted charge, for example [137]:

$$Q = \frac{\sum_i |\vec{p}_i \cdot \vec{e}_T|^{\kappa} q_i}{\sum_i |\vec{p}_i \cdot \vec{e}_T|^{\kappa}} \tag{241}$$

  where $\vec{e}_T$ is the Thrust axis, $\vec{p}_i$ and q_i are the momentum and the sign of the charged particle i and κ is a power factor which has an optimal value arround 0.5. The second method takes advantage of the fact that the sign of the most energic pion accompanying a B meson carry some information about the sign of the b quark [138]. This is true either because the B originated from the decay of a higher excited state such as $B^{**+} \rightarrow B^{(*)o}\pi^+$ ($B^{**-} \rightarrow \overline{B}^{(*)o}\pi^-$) or because the charged pion came "first" in the fragmentation process. In the example $b + (\bar{d}d) + (\bar{u}u) + ... \rightarrow \overline{B^o} + \pi^- + ...$, the "first" pion has the same charge sign as the b quark and tends to be the most energic amongst the fragmentation pions.

To summarize, we show in the table 14 the expected effective tagging efficiencies at various facilities.

b) The Δz measurement: Once a B meson has been reconstructed and its initial flavor is tagged, the next step is to measure the distance Δz between the reconstructed B meson and the vertex of the tagging B (or the distance d to the primary vertex). One of the advantages of the facilities where energic B mesons are produced is that it is easier to measure their flight distance provided one has a good vertex detector. Indeed at LEP, HERA-B, Tevatron and LHC, $\sigma_d / < d >$ is rather "comfortable" ($\approx$ 5 to 20). It is

Facility	Tagging methods	$\epsilon_{\text{eff}} = \epsilon_{\text{tag}}(2\mathcal{P} - 1)^2$
Υ(4S)	ℓ, K	0.36
HERA-B	ℓ, K, Jet Charge	0.09
Tevatron	ℓ, K, Jet Charge, accom. π	0.10

TABLE 14. The expected effective tagging efficiency at different facilities.

somewhat more difficult at asymmetric B Factories because of the small boost of the B mesons and of the lower energy of their decay products. Because of this latter point, it is necessary to minimized the amount of material in front of the first detection layer. We show in the figure 43 the expected resolution on Δz at BaBar. It can in general be parametrized by a narrow ($\sigma \approx 80$ μm) and a wider ($\sigma \approx 180$ μm) gaussian.

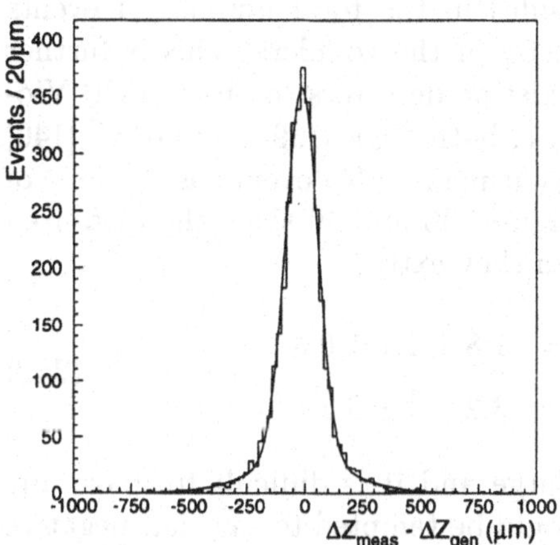

Figure 43. The resolution on the distance Δz from the simulation of a vertex detector.

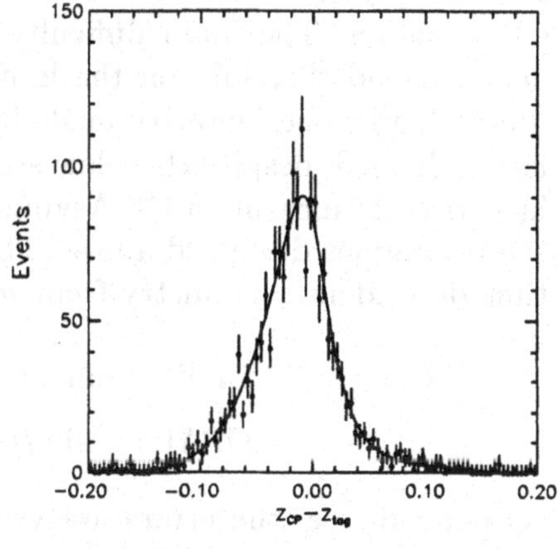

Figure 44. Decay probabilities for the reaction $B^o \to \eta' K_S$ as a function of the proper time using the full detector simulation of the BaBar detector at PEPII. The value $\sin 2\beta = 0.7$ was assumed here.

The Sensitivity to CP Asymmetry:

Using the ingredient described above, one can measure the time dependent decay rate for a particular mode. For illustration we show such a distribution simulated for the decay $B^o \to \eta' K_S$ in figure 44. The conclusion is

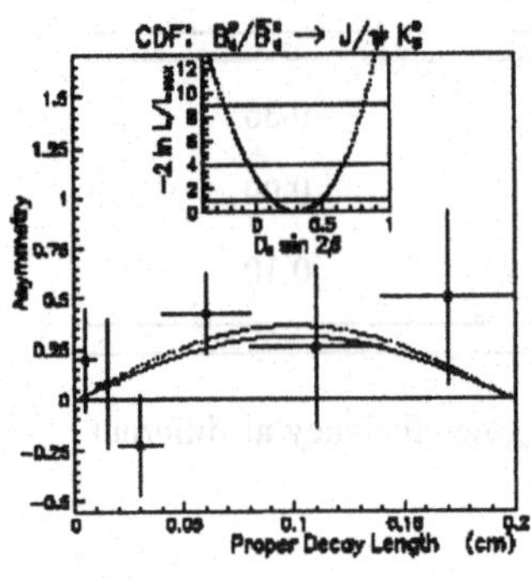

Figure 45. The time dependent asymmetry measured at the CDF detector with $J/\psi K_S$ events.

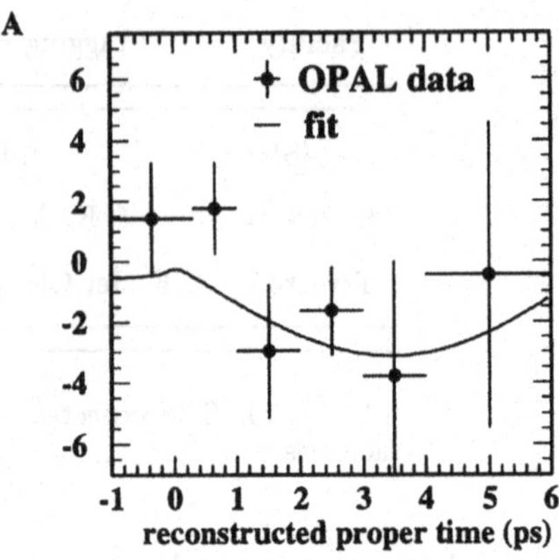

Figure 46. The time dependent asymmetry measured at the OPAL detector with $J/\psi K_S$ events.

that the present detector technologies allows in principle to observe such CP violation. The main difficulty resides in the total number of events detected and therefore on the luminosity of the colliders. This is further illustrated by the tentative made by existing detectors to observe CP Violation. Indeed, despite their low statistic, both CDF [139] and OPAL [140] have tried to measure a CP Asymmetry using $J/\psi K_S$ events using some of the techniques discussed above. The figures 45 and 46 show the measured time dependent asymmetry from which they extract

$$\begin{aligned} \text{CDF}: \quad \sin 2\beta &= 1.8 \pm 1.1 \pm 0.3 \\ \text{OPAL}: \quad \sin 2\beta &= 3.2^{+1.8}_{-2.0} \pm 0.5 \end{aligned} \tag{242}$$

As one can see, the errors are very large and it is difficult to make any definitive conclusion, althought one would be tempted to say that negative value of $\sin 2\beta$ may be excluded.

Let us rather discuss what are the prospects for observing CP Violation in the B system with the next generation of experiments. We show in the table 15, some of the decay modes which can be used for measuring the angles of the unitarity triangle.

We give in Tables 16 and 17 the expected errors on sin2α and sin2β if some of the channels allowing to measure the same angle are summed up.

The errors which are obtained are visualized in the figure 47 where we have shown as well the anticipated improvement on $|V_{ub}|$ and $|V_{td}|$. It is

Modes	Branching fraction	Total efficiency	Background/ Signal	Signal events $30fb^{-1}$	Measured quantity
$J/\psi K_S$	4×10^{-4}	5.0×10^{-2}	$<6\times10^{-2}$	~ 660	$\sin 2\beta$
$J/\psi K_L$	4×10^{-4}	5.0×10^{-2}	~ 0.6	~ 650	$\sin 2\beta$
$\chi_{c1}K_L$	5×10^{-4}	9×10^{-3}	~ 0.8	~ 140	$\sin 2\beta$
$D^{\pm}D^{\mp}$	4.5×10^{-4}	$\sim 10^{-2}$	2.8	~ 140	$\sin 2\beta$
$\pi^+\pi^-$	1×10^{-5}	0.63	0.9	~ 200	$\sin 2\alpha$
$\rho^{\pm}\pi^{\mp}$	5.4×10^{-5}	0.35	3	~ 600	$\sin 2\alpha$
$a_1^{\pm}\pi^{\mp}$	6×10^{-5}	0.2	2.2	~ 200	$\sin 2\alpha$
$D^{*\pm}D^{\pm}$	4.8×10^{-4}	$\sim 6\times10^{-3}$	1.0	~ 80	$\sin 2\beta$
$\psi K_s\pi^o$	1.3×10^{-3}	1.3×10^{-2}	0.18	~ 51	$\sin 2\beta$
$D^{*\pm}D^{*\mp}$	9.7×10^{-4}	2×10^{-3}	~ 0.2	~ 60	$\sin 2\beta$
$\rho^+\rho^-$	5×10^{-5}	0.25	7	~ 420	$\sin 2\alpha$

TABLE 15. The expected number of reconstructed events with BaBar which can be used for measuring the angles β and α assuming an integrated luminosity of $30fb^{-1}$ at an PEPII. The efficiencies include the branching fraction of the resonance decays which are used to reconstruct them but not the tagging.

Modes	$\pi^+\pi^-$	$\rho^{\pm}\pi^{\mp}$ $a_1^{\pm}\pi^{\mp}$	$\rho^+\rho^-$	All
d	1	1	1	
$\sigma(\sin 2\alpha)$	0.26	0.20	0.28	0.14

TABLE 16. The expected errors on $\sin 2\alpha$ with $30fb^{-1}$ at an asymmetric B Factory using a selected sample of modes.

interesting to note that assuming the plausible scenario where $\sin 2\beta = 0.6$, a rather modest integrated luminosity of the order of 10 fb^{-1}(about 4 months of running at an asymmetric B factory with a peak luminosity

Modes	$J/\psi K_S$ $J/\psi K_L$ $D^{\pm}D^{\mp}$ $\chi_{c1}K_L$	$D^{*\pm}D^{\mp}$	$J/\psi K_S\pi^o$ $D^{*\pm}D^{*\mp}$	All
d	1	1	1	
$\sigma(\sin 2\beta)$	0.09	0.5	0.35	0.085

TABLE 17. The expected errors on $\sin 2\beta$ with $30 fb^{-1}$ at an asymmetric B Factory using a selected sample of modes.

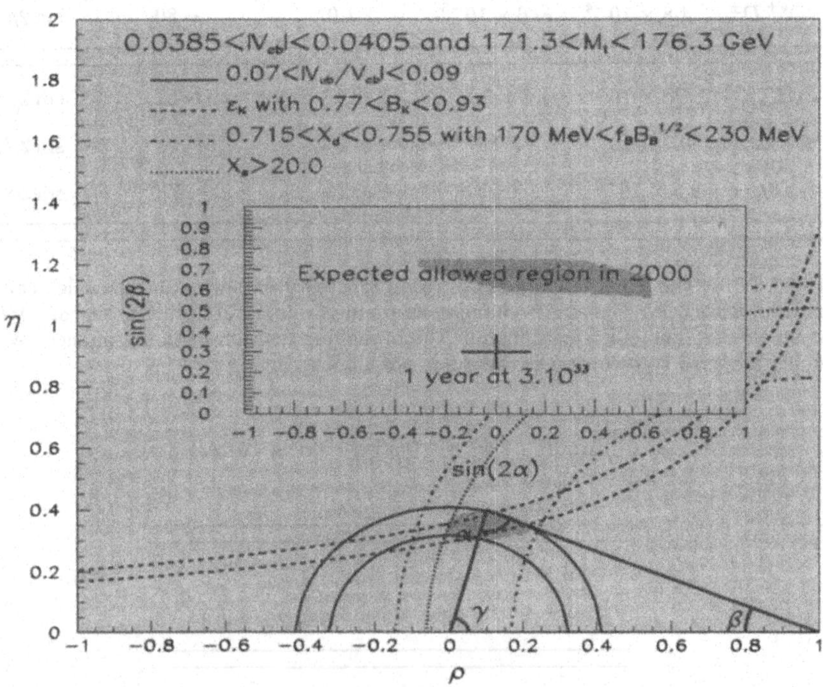

Figure 47. Anticipated experimental constraints for the apex of the unitarity triangle in about 3 years from now. The cross indicates the errors which may be achieved at an asymmetric B Factory assuming an integrated luminosity of $100 fb^{-1}$. Clearly, should these values be measured, it would be an evidence for the inconsistency of the Standard Model.

of $3 \times 10^{33} cm^{-1}s^{-1}$) would be sufficient to observe CP Violation at the 3σ level using only the $J/\psi K_S$ mode. Using a more conservative value for $\sin 2\beta$ (≥ 0.4), an integrated luminosity of 12 fb^{-1} would be required if one uses all the decay channels studied in the table 17. For comparison, we give in the table 18, the expected sensitivities at hadrons facilities.

Facility	$\delta(\sin 2\alpha)$ with $\pi^+\pi^-$	$\delta(\sin 2\beta)$ with $J/\psi K_S$	x_s sensitivity
HERA-B	-	0.18(0.11*)	20
CDF/D0	0.14(0.10**)	0.14(0.10**)	20
ATLAS/CMS	0.07/0.05	0.03(0.05**)	30

TABLE 18. The expected sensitivities for the experiments at hadron facilities (* depending on cross section, ** depending on trigger efficiency).

Amongst the three angles of the unitarity triangle, the angle γ is the most difficult to measure at an asymmetric B factory. A promising method involves the decays of the B_s^o meson into $D_s^{\mp}K^{(*)\pm}$ or $D_s^{(*)\mp}K^{\pm}$ [141]. Unfortunately the energy at the $\Upsilon(4S)$ is below the threshold of $B_s^o\overline{B_s^o}$ production and one needs to run at the $\Upsilon(5S)$ where the cross section is about $0.1nb$ [$\sim$ 10 times smaller than the B_d^o production rate at the $\Upsilon(4S)$]. Therefore this method may only be usable at hadron facilities. At e^+e^- B Factories, one is forced to study the decays $B_d^o \to D^oK_S^{(*)}$ or $B^{\pm} \to D^oK^{\pm}$ following the method discussed in chapter 5. Because the branching fractions for these decay modes are expected to be of the order of 10^{-5} and since one needs to observe the D^o decay into CP eigenstates to extract γ, the statistics for the interesting events will be rather limited even with an integrated luminosity of 300 fb^{-1}. Studies carried out by the BABAR collaboration [1] show that a sensitivity at the 10-20° level may be achieved depending on the values of γ and the strong phase. A similar study was carried out by the BELLE collaboration [142] with somewhat more optimistic conclusions. A maybe more promising method has been proposed using triangular relation between tree and penguin amplitudes in $B \to K\pi$ [143] and $B \to \pi\pi$ decays [144] assuming SU(3) symmetry. However, here also, uncertainties on rescattering effects and electroweak penguins diagrams [145] may dilute the asymmetry and/or prevent a clean extraction of γ.

7. Conclusion

The study of CP violation in the sector of the b quark is extremely promising. It should allow one to test the consistency of the Standard Model and

verify that the non-conservation of the CP symmetry is a fundamental property of the theory. We have seen that, amongst the various possibilities, the most interesting mechanism is the one involving the interplay of the B decay and the B^o-$\overline{B^o}$ mixing. Many interesting classes of events can be used. Although the simplest one is the B decay into CP eigenstate, other decay modes may also be used with a slightly more complicated analysis. Using all the decay channels discussed in these lectures, it would be possible to carry a comprehensive study and test the consistency of the Standard Model in the sector of CP violation at an asymmetric e^+e^- B Factory provided an integrated luminosity of $100fb^{-1}$ is accumulated. Hadron machines also offer the possibility to study this phenomenon in a complementary way and with higher precision for some particular modes.
In summary, we will be at the beginning of the next millenium in the position to explore in detail a so far poorly studied sector of the theory with possibly exciting surprises. Indeed, should the consistency tests fail, one would have obtained the first and exciting evidence for new Physics, possibly opening up new avenues toward the understanding of our universe.

Acknowledgements

It is a real pleasure to thank warmly Barbara and Tom Ferbel for the spendid organization of such a succesful, high quality and pleasant school.

References

1. D. Boutigny et al., SLAC Report 504, October 1998, Ed. P.F. Harrison and H.R. Quinn.
2. K. Nishijima, *Fundamental Particles*, W.A. Benjamin Inc., New York, 1963; J.J. Sakurai, *Invariance Principle and Elementary Particle*, Princeton University Press, 1964; E.P. Wigner, *Symmetries and Reflections*, Bloomington, Indiana University Press, 1967; L.H. Ryder, Elementary Particles and Symmetries, Gordon and Breach Science Publishers, London, 1975.
3. W. von Witsch, A. Richter and P. von Brentano, Phys. Rev. Lett. **19** (1967) 524.
4. C. Caso et al., Particle Data Group, "Review of Particle Physics", Eur. Phys. Journ. C **3** (1998) 1.
5. C.S. Wu et al., Phys. Rev. **105** (1957) 1413.
6. M. Gell-Mann and A.Païs, Phys. Rev. **97** (1955) 1387.
7. J.H. Christenson, J.W. Cronin, V.L. Fitch and R. Turlay, Phys. Rev. Lett. **13** (1964) 138.
8. L. Wolfenstein, Phys. Rev. Lett. **13** (1964) 562.
9. H. Burkhardt et al., Phys. Lett. B **206** (1988) 163.
10. L.K. Gibbons et al., Phys. Rev. Lett. **70** (1993) 1203.
11. P.K. Kabir, The CP Puzzle, Academic Press Inc., London, 1968; L.H. Ryder, Réference 1.
12. B. Schwingenheuer et al., E773 Collaboration, Phys. Rev. Lett. **74** (1995) 4376.
13. P. Kokkaas, CPLEAR Collaboration, presented at the XXIX ICHEP98, Vancouver, Canada, July23-29, 1998.

14. A. Angelopoulos et al., CPLEAR Collaboration, Preprint CERN/EP-98-153, October 1998.
15. F. Abe at al., CDF Collaboration, Phys. Rev. Lett. **74** (1995) 2626 and S. Abachi et al., D0 Collaboration, Phys. Rev. Lett. **74** (1995) 2632.
16. S. Weinberg, Phys. Rev. Lett. **19** (1967) 1264; A. Salam, Elementary Particle Physics, ed. by N. Svartholm (Almquist and Wiksells, Stockholm 1968), p. 367.
17. ARGUS Collaboration, Contributed paper 0148 to the EPS-HEP Conference, Brussels, July 1995.
18. H. Albrecht et al., ARGUS Collaboration, Phys. Lett. B **324** (1994) 249.
19. H. Albrecht et al., ARGUS Collaboration, Z. Phys. C **55** (1992) 357.
20. J. Bartelt et al., CLEO Collaboration, Phys. Rev. Lett. **71** (1993) 1680.
21. M. Artuso et al., CLEO Collaboration, Phys. Rev. Lett. **62** (1989) 2233.
22. D.Decamp et al., ALEPH Collaboration, Phys. Lett. B **284** (1992) 177 and contribution to the 1994 winter conferences of La Thuile and Moriond; P. Abreu et al., DELPHI Collaboration, Phys. Lett. B **322** (1994) 459, Phys. Lett. B **322** (1994) 488; M. Acciarri et al., L3 Collaboration, Phys. Lett. B **335** (1994) 542; P. Akers et al., OPAL Collaboration, Z. Phys. C **60** (1993) 199.
23. C. Albajar et al., UA1 Collaboration, Phys. Lett. B **262** (1991) 171; F. Abe et al., CDF Collaboration, Preprint FERMILAB-CONF-95-223-E and 230-E, Jul. 1995; S. Abachi et al., D0 Collaboration, Preprint FERMILAB-CONF-95-209-E, Jul. 1995.
24. H. Albrecht et al., ARGUS Collaboration, Phys. Lett. B **192** (1987) 245.
25. M. Kobayashi and T. Maskawa, Prog. Theor. Phys. **42** (1973) 652.
26. see also for example I.I. Bigi, V.A. Khoze, N.G. Uraltsev and A.I. Sanda, Advanced Series on Directions in High Energy Physics - Vol.3 p.175, CP Violation, ed. C.Jarlskog, published by World Scientific, Singapore, 1989.
27. A.D. Sakharov, Pis'ma Zh. Ekps. Teor. Fiz. **5** (1967) 32.
28. For a review of Cosmological Consequences of GUTs see: D.V. Nanopoulos, Prog. in Particle and Nuclear physics **6** (1980) 23.
29. V.A. Kuzmin, V.A. Rubakov and M.E. Shaposhnikov, Phys. Lett. B **155** (1985) 36.
30. M.E. Shaposhnikov, JETP Lett. **44** (1986) 465, Nucl. Phys. B **287** (1987) 757, Nucl. Phys. B **299** (1988) 797;
A.I Bochkarev, S.Y Khlebnikov and M.E. Shaposhnikov, Nucl. Phys. B **329** (1990) 490;
L.McLerran, Phys. Rev. Lett. **62** (1989) 1075;
L.McLerran, M.E. Shaposhnikov, N. Turok and M. Voloshin, Phys. Lett. B **256** (1991) 451;
N. Turok and P. Zadrozny, Nucl. Phys. B **358** (1991) 471;
M. Dine, P. Huet, R. Singleton and L. Susskind, Phys. Lett. B **257** (1991) 351;
A. Cohen, D.B Kaplan and A.E. Nelson, Nucl. Phys. B **349** (1991) 727, Phys. Lett. B **263** (1991) 86 and preprint UCSD-PTH-91-20 (1991).
31. M.B. Gavela, P. Hernandez, J. Orloff and O. Pene, Mod. Phys. Lett. A **9** (1994) 795.
32. M. Dine, R.G. Leigh, P. Huet, A. Linde and D. Linde, Phys. Rev. D **46** (1992) 550.
33. P. Higgs, Phys. Rev. Lett. **13** (1964) 508.
34. B.W. Lynn and R.G. Stuart, Nucl. Phys. B **253** (1985) 216; Nucl. Phys. B 253 (1985) 216; Physics at LEP, Edited by J. Ellis and R. Peccei, CERN 86-02, Vol. 1, 1986.
35. see for example M.J. Herrero in this book.
36. N. Cabibbo, Phys. Rev. Lett. **10** (1963) 531.
37. H. Fritsch, Phys. Lett. B **70** (1977) 436, Phys. Lett. B **73** (1978) 317, Nucl. Phys. B **155** (1979) 189;
H. Georgi and C. Jarlskog, Phys. Lett. B **86** (1979) 297;
B. Stech, Phys. Lett. B **130** (1983) 189;
M.V. Barnhill, Phys. Rev. D **36** (1987) 192;
G. Giudice, Mod. Phys. Lett. A **7** (1992) 2429;

S. Dimopoulos, L.J. Hall and S. Raby, Phys. Rev. Lett. **68** (1992) 1984.
38. L. Wolfenstein, Phys. Rev. Lett. **51** (1983) 1945.
39. A.J. Buras, M.E. Lautenbacher and G. Ostermaier, Phys. Rev. D **50** (1994) 3433.
40. L.-L. Chau and W.-Y. Keung, Phys. Rev. Lett. **53** (1984) 1802; J.D. Bjorken, Fermilab Preprint, 1988 (unpublished); C. Jarlskog and R. Stora, Phys. Lett. B **208** (1988) 268; J.L. Rosner, A.I. Sanda and M.P. Schmidt, in proceedings of the Workshop on High Sensitivity Beauty Physics at Fermilab, Fermilab, Nov. 11-14, 1987, ed. by A.J. Slaughter, N. Lockyer and M.P. Schmidt, p.165 .
41. J. Ellis, Mary K. Gaillard and D.V. Nanopoulos, Nucl. Phys. B **109** (1976) 213.
42. S. Herrlich and U. Nierste, Phys. Rev. D **52** (1995) 6505.
43. T. Inami and C.S. Lim, Prog. Theor. Phys. **65** (1981) 297.
44. S.R. Sharpe, plenary talk at the 29th International Conference on High Energy Physics, Vancouver, Canada, July 23-29, 1998; hep-lat/9811006v3, Dec. 1998.
45. see for example S. Narison, Phys. Lett. B **351** (1995) 369 and N. Bilic, C.A. Dominguez and G. Guberina, Z. Phys. C **39** (1988) 351.
46. see for example S. Aoki et al., Phys. Rev. Lett. **80** (1998) 5271.
47. The value of $\widehat{B}_K$ which we are using here is the one obtained by lattice QCD with the quenched approximation. The unquenched approximation gives $\widehat{B}_K = 0.94 \pm 0.15$ [44].
48. R. Aleksan et al., Phys. Lett. B **316** (1993) 567; M. Beneke, G. Buchalla and I. Dunietz, Phys. Rev. D **54** (1996) 4419.
49. J.S. Hagelin,, Nucl. Phys. B **193** (1981) 132.
50. A.J. Buras, M. Jamin and P.H. Weisz, Nucl. Phys. B **347** (1990) 491.
51. V. Andreev, presented at the 29th International Conference on High Energy Physics, Vancouver, Canada, July 23-29, 1998, see also the LEP Working Group preprint LEPBOSC 98/3, Sept. 1998 and the references therein.
52. C.A. Dominguez, *proceedings of the 3rd Workshop on the Tau-Charm Factory*, Marbella, Spain, 1-6 June 1993, Ed. J. Kirkby and R. Kirkby, Editions Frontières; see also Ref. [58] and [71].
53. M. Acciarri et al., L3 Collaboration, Phys. Lett. B **396** (1997) 327.
54. M. Artuso et al., CLEO Collaboration, Phys. Rev. Lett. **75** (1995) 785.
55. G. Buchalla and A.J. Buras, Phys. Lett. B **333** (1994) 221.
56. S. Adler et al., E787 Collaboration, Phys. Rev. Lett. **79** (1997) 2204.
57. S. Adler et al., CLEO Collaboration, Preprint CLEO CONF 96-05 and ICHEP96 PA05-093, contibuted paper at the 28th International Conference on High Energy Physics, 25-31 July 1996, Warsaw, Poland.
58. S. Narison, Phys. Lett. B **322** (1994) 247; S. Narison and A. Pivovarov, Phys. Lett. B **327** (1994) 341; E. Bagan et al., Phys. Lett. B **278** (1992) 457; P. Ball, Nucl. Phys. B **421** (1994) 593; T. Huang and C.W. Luo, Phys. Rev. D **53** (1996) 5042.
59. F. Parodi, presented at the 29th International Conference on High Energy Physics, Vancouver, Canada, July 23-29, 1998.
60. E.V Shuryak, Phys. Lett. B **93** (1980) 134, Nucl. Phys. B **198** (1982) 83; J.E. Paschalis and G.J Gounaris, Nucl. Phys. B **222** (1983) 473; F. Close, G.J Gounaris and J.E. Paschalis, Phys. Lett. B **149** (1984) 209; S. Nussinov and W. Wetzel, Phys. Rev. D **36** (1987) 130; M.B. Voloshin and M.A. Shifman, Yad. Fiz. **45** (1987) 463, Sov. J. Nucl. Phys. **45** (1987) 292, Yad. Fiz. **47** (1988) 801, Sov. J. Nucl. Phys. **47** (1988) 511; E. Eichten and B. Hill, Phys. Lett. B **234** (1990) 511, Phys. Lett. B **243** (1990) 427; H. Georgi, Phys. Lett. B **240** (1990) 447.
61. B. Barish et al., CLEO Collaboration, Phys. Rev. Lett. **76** (1996) 1570.
62. N. Isgur, D. Scora, B. Grinstein and M.B. Wise, Phys. Rev. D **39** (1989) 799.
63. G. Altarelli, N. Cabibbo, G. Corbo, L. Maiani and G. Martinelli, Nucl. Phys. B **208** (1982) 365.
64. I. I. Bigi, N. G. Uraltsev and A. I. Vainshtein, Phys. Lett. B **293** (1992) 430; [, Phys. Lett. B **297** (E) 477(1993)]; B. Blok and M. Shifman, Nucl. Phys. B **399** (1993)

441 and 459.
65. E.C. Poggio, H.R. Quinn and S. Weinberg, Phys. Rev. D **13** (1976) 1958.
66. M. Luke, Phys. Lett. B **252** (1990) 447.
67. A.F. Falk and M. Neubert, Phys. Rev. D **47** (1993) 2965.
68. A. F. Falk, M. Luke and M. J. Savage, Phys. Rev. D **53** (1996) 2491;, Phys. Rev. D **53** (1996) 6316.
69. M. Neubert, Nucl. Phys. B *Proc. Suppl.* **59** (1997) 101.
70. M. Wirbel, B. Stech and M. Bauer, Z. Phys. C **29** (1985) 637; M. Bauer, B. Stech and M. Wirbel, Z. Phys. C **34** (1987) 103; see also M. Bauer, and M. Wirbel, Z. Phys. C **42** (1989) 671.
71. see for example M. Neubert, Phys. Rev. **245** (1994) 259.
72. N. Isgur and M.B. Wise, Phys. Lett. B **232** (1989) 113;, Phys. Lett. B **237** (1990) 527.
73. M. Neubert, Phys. Lett. B **264** (1991) 455.
74. M. Neubert, Phys. Rev. D **46** (1992) 3914; Z. Ligeti, Y. Nir and M. Neubert, Phys. Rev. D **49** (1994) 1302.
75. S.J Brodsky, G.P. Lepage and P.B. Mackenzie, Phys. Rev. D **28** (1983) 228; G.P. Lepage and P.B. Mackenzie, Phys. Rev. D **48** (1993) 2250.
76. A. Czarnecki, Phys. Rev. Lett. **76** (1996) 4124; A. Czarnecki and K. Melnikov, Nucl. Phys. B **505** (1997) 65.
77. A Falk and M. Neubert, Phys. Rev. D **47** (1996) 2965 and 2982; T. Mannel, Phys. Rev. D **50** (1994) 428; M. Shifman, N.G. Uraltsev and A.I. Vainshtein, Phys. Rev. D **51** (1995) 2217; M. Neubert, Phys. Lett. B **338** (1994) 84; C.G. Boyd and I.Z. Rothstein, Phys. Lett. B **395** (1997) 96.
78. M.E. Luke, Phys. Lett. B **252** (1990) 447.
79. H. Albrecht et al, ARGUS Collaboration, DESY 90-121 Oct. 1990; R. Fulton et al, CLEO Collaboration, Phys. Rev. Lett. **64** (1990) 16; J. Bartelt et al., CLEO Collaboration, Phys. Rev. Lett. **71** (1993) 4111.
80. ALEPH Collaboration, preprint CERN-EP98-067, May 1998.
81. DELPHI Collaboration, contribution-241 at the at the 29th International Conference on High Energy Physics, Vancouver, Canada, July 23-29, 1998.
82. L3 Collaboration, preprint CERN-EP98-097, June 1998.
83. J. Alexander et al., CLEO Collaboration, Phys. Rev. Lett. **77** (1996) 5000; C.P. Jessop et al., preprint CLEO CONF 98-18 and contribution-855 at the at the 29th International Conference on High Energy Physics, Vancouver, Canada, July 23-29, 1998.
84. N. Isgur, D. Scora, B. Grinstein and M.B. Wise, Phys. Rev. D **52** (1995) 2783.
85. M. Beyer and D. Melikhov, Phys. Rev. D **53** (1996) 2160; hep-ph/9807223 (1998).
86. J.G. Körner and G.A. Schuler, Z. Phys. C **38** (1988) 511.
87. L. Del Debbio, Phys. Lett. B **416** (1998) 392.
88. P. Ball et al., hep-ph/9805422 (1998).
89. Z. Ligeti and M.B. Wise, Phys. Rev. D **53** (1996) 4937; Z. Ligeti, I.W. Stewart and M.B. Wise, Phys. Lett. B **420** (1998) 359.
90. M. Bander, D. Silverman and A. Soni, Phys. Rev. 43 (1979) 242.
91. G. Altarelli, G. Curci, G. Martinelli and S. Petrarca, Nucl. Phys. B **187** (1981) 461; A.J. Buras and P.H. Weisz, Nucl. Phys. B **333** (1990) 66.
92. M.B. Gavela et al., Phys. Lett. B **154** (1985) 425.
93. A.J. Buras, M. Jamin, M.E. Lautenbacher and P.H. Weisz, Nucl. Phys. B **370** (1992) 69; R. Fleisher, Z. Phys. C **62** (1994) 81, Z. Phys. C **58** (1993) 483.
94. J.M. Gérard and W.S. Hou, Phys.Lett. B 253 (1991) 478.
95. J.D. Bjorken, 18th Annual SLAC Summer Institute on Particle Physics, Stanford, Ca July 16-27, 1990, SLAC-PUB-5389, 1990.
96. L.-L. Chau, *B mesons, a Beautiful Source of New Physics*, CP Violation, Advanced Series on Directions in High Energy Physics Vol.3, World Scientific, Singapore, 1989; M. Bauer et al., in Reference [70]

97. M. Gronau and D. Wyler, Phys. Lett. B **265** (1991) 172.
98. D. Atwood, I. Dunietz and A. Soni, Phys. Rev. Lett. **78** (1997) 3257.
99. I. Dunietz, Phys. Lett. B **270** (1991) 75.
100. J.D. Bjorken, Nucl. Phys. B (Proc. Suppl.) **11** (1989) 325; I.I. Bigi and B. Stech, in "Proceedings of the Workshop on High Sensitivity Beauty physics" at Fermilab, Nov. 1987, ed. A.J. Slaughter, N. Lockyer and M. Schmidt, p. 239.
101. I. Dunietz, Z. Phys. C **56** (1992) 129.
102. G. Burdman and J.F. Donoghue, Phys. Rev. D **45** (1992) 187.
103. J. Ellis, M.K. Gaillard and D.V. Nanopoulos, Nucl. Phys. B109 (1976) 213.
104. R. Aleksan, I. Dunietz, B. Kayser, F. Le Diberder, Nucl. Phys. B361 (1991) 141.
105. I. Dunietz and J.L. Rosner, Phys. Rev. D34 (1986) 1404.
106. M. Gronau and D. London, Preprint DESY 90-106, Sept.1990.
107. M. Danilov, presented at the ECFA workshop on Physics at a B Factory, DESY, October 29th, 1992. at the 4th International Symposium on Heavy Flavour Physics, Orsay, 24-29 june 1991.
108. B. Kayser, M. Kuroda, R.D. Peccei and A.I. Sanda, Phys. Lett. B237 (1990) 508; I. Dunietz, H. Quinn, A. Snyder, W. Toky and H.J. Lipkin, SLAC-PUB-5270, November 1990.
109. G.D. Barr et al., Proposal CERN-SPSC-89-39, Oct. 1989 and CERN-SPSC-90-22, CERN-SPSC/P253, July 1990.
110. K. Arisaka et al., KTeV design Report, Fermilab Report FN-580, Jan. 1992.
111. A. Alosio et al., Proposal LNF-93-002-IR, Jan. 1993.
112. J. Antos et al., Fermilab Proposal P871, March 1994 and A. Chan et al, Fermilab Proposal E-871/97, Dec. 1997.
113. D. Andrews et al., CLEO Collaboration, Nucl. Inst. and Meth. **211** (1983) 47; A.H. Wolf for CLEO Collaboration, Nucl. Inst. and Meth. A **408** (1998) 58.
114. M.T. Cheng et al., BELLE Collaboration, BELLE-TDR-3-95, March 1995.
115. D. Boutigny et al., BABAR Collaboration, SLAC-Report-457, March 1995.
116. P. Krizan et al., HERA-B Collaboration, Nucl. Inst. and Meth. A **351** (1994) 111;
117. F. Abe et al., CDF Collaboration, Nucl. Inst. and Meth. A **271** (1988) 387; see also FERMILAB-CONF-95-239-E, June 1995.
118. S. Abachi et al., D0 Collaboration, Nucl. Inst. and Meth. A **338** (1994) 185; see also FERMILAB-PUB-96-357-E, Oct. 1996.
119. P.A. Kasper *for the collaboration*, BTeV Collaboration, Nucl. Inst. and Meth. A **408** (1998) 146.
120. W.W. Armstrong et al., ATLAS Collaboration, Technical Proposal, CERN-LHCC-94-43, Dec. 1994.
121. M. Della Negra et al., CMS Collaboration, Letter of Intent, CERN-LHCC-92-3, Oct. 1992.
122. S. Amato et al., LHCb Collaboration, Technical Proposal, CERN-LHCC-98-4, Feb. 1998.
123. H. Yamamoto, Preprint HUTP-97/A011, March 1997, hep-ph/9703336.
124. K.J. Foley et al. Proceedings of the Workshop on Experiments, Detectors and Experimental Areas for the Supercollider, July 7 – 17, 1987, Berkeley, California, ed. by R.Donaldson and M.G.D. Gilchriese (World Scientific, Singapore, 1988), p. 701-727.
125. Cleo-II Collaboration, D.S. Akerib et al, Preprint CLNS 91/1089, July 1991.
126. P. Oddone, Proceedings of the UCLA Workshop: Linear Collider $B\overline{B}$ Factory Conceptual Design, ed. by D.Stock (1987) p. 243; H. Nesemann, W. Schmidt-Parzefall and F. Willeke, " The use of PETRA as a B-Factory", DESY 1988; G. Feldman et al., preprint SLAC-PUB-4838, CNLS 89/884, LBL-26790 January 1989.
127. R. Aleksan, J. Bartelt, P.R. Burchat and A. Seiden, Phys. Rev. D39 (1989) 1283.
128. Proposal for an Electron Positron Collider for Heavy Flavour particle Physics and

Synchrotron Radiation, PSI-PR-88-09, July 1988; The Physics Program of a High-Luminosity Asymmetric B Factory at SLAC, SLAC-353, LBL-27856, CALT-68-1588, October 1989; Feasibility Study for a B-Meson Factory in the CERN ISR Tunnel, Ed. by T.Nakada, CERN 90-02, PSI PR-90-08, March 1990; CESR B Factory : Physics Program, CLNS 91-1043, January 1991; Wokshop on Physics and Detector Issues for a High-Luminosity Asymmetric B Factory, SLAC-373, LBL-30097, CALT-68-1697, February 1991; Physics and Detector of an Asymmetric B Factory at KEK, KEK Report 90-23, March 1991; HELENA a Beauty Factory in Hamburg, preprint DESY 92-041, March 1992.

129. R. Godang et al., CLEO Collaboration, Phys. Rev. Lett. **80** (1998) 3456.
130. T. Bergfeld et al., CLEO Collaboration, Phys. Rev. Lett. **81** (1998) 272.
131. B. Behrens et al., CLEO Collaboration, Phys. Rev. Lett. **80** (1998) 3710.
132. J. Bartelt et al., CLEO Collaboration, Phys. Rev. Lett. **71** (1993) 1680.
133. F. Abe et al., CDF Collaboration, Phys. Rev. D **55** (1997) 2546.
134. A. Ackerstaff et al., DELPHI Collaboration, preprint DELPHI 97-98 CONF-80, July 1997, contibution number 449 to the HEP'97 conference, Jerusalem, August 19-26, 1997.
135. A. Ackerstaff et al., OPAL Collaboration, Z. Phys. C **76** (1997) 401.
136. I.I Bigi, V.A. Khoze, N.G. Uraltsev and A.I. Sanda, *Advanced Series on Directions in High Energy Physics*, CP Violation, Vol. 3, p. 175, Ed. C. Jarlskog, World Scientific, Singapore 1989.
137. D. Decamp et al., ALEPH Collaboration, Phys. Lett. B **259** (1991) 377 R. Akers et al., OPAL Collaboration, Z. Phys. C **67** (1995) 365.
138. M. Gronau and J. Rosner, Preprint TECHNION-PH-93-37, Oct. 1993.
139. F. Abe et al., CDF Collaboration, Preprint FERMILAB-PUB-98/189-e, Jun. 1998.
140. K. Ackerstaff et al., OPAL Collaboration, Eur. Phys. Journ. C **5** (1998) 379.
141. R. Aleksan, I. Dunietz and B. Kayser, Z. Phys. C **54** (1992) 653.
142. BELLE Progress Report, B Factory, KEK Progress Report 96-1, 1996.
143. R. Fleischer, Phys. Lett. B **365** (1996) 399.
144. M. Gronau, J.L. Rosner and D. London, Phys. Rev. Lett. **73** (1994) 21; O.F. Hernandez et al., Phys. Lett. B **333** (1994) 500; M. Gronau et al., Phys. Rev. D **50** (1994) 4529.
145. N.G. Deshpande X.-G. He, Phys. Rev. Lett. **74** (1996) 26; [E: *ibid.* p.4099].

DIFFRACTIVE HARD SCATTERING

G. INGELMAN
DESY, Hamburg, Germany
Uppsala University, Sweden

Abstract. Diffraction is an old subject which has received much interest in recent years due to the advent of diffractive hard scattering. We discuss some theoretical models and experimental results that have shown new striking effects, *e.g.* rapidity gaps in jet and W production and in deep inelastic scattering. Many aspects can be described through the exchange of a pomeron with a parton content, but the pomeron concept is nevertheless problematic. New ideas, *e.g.* based on soft colour interactions, have been introduced to resolve these problems and provide a unified description of diffractive and non-diffractive events. This is part of the general unsolved problem of non-perturbative QCD and confinement.

1. Introduction

Ideas on diffraction have been developed over a long time. Quite old (*'old-old'*) is the Regge approach with a pomeron mediating elastic and diffractive interactions [1]. Being from pre-QCD times, Regge phenomenology only considers soft interactions described in terms of hadrons. In a modern QCD-based language one would like to understand diffraction on the parton level. This was the starting point of the by now *'old new'* idea [2] that one should probe the structure of the pomeron through a hard scattering in diffractive events. By introducing a hard scale one should resolve partons in the pomeron and also make calculations possible through perturbative QCD (pQCD). This opened the new branch of *diffractive hard scattering* with models and the discovery by UA8 [3] as discussed in section 3.

The models are based on a factorization between the new concepts of a 'pomeron flux' (in the proton) and a 'pomeron structure function' in terms of parton density functions. These ideas may be interpreted as the pomeron being analogous to a hadron (maybe a glueball?) as discussed in section 2.

T. Ferbel (ed.), Techniques and Concepts of High Energy Physics X, 597–624.

The discovery of rapidity gap events in deep inelastic scattering (DIS) at HERA was a great surprise to most people, although it had been predicted as a natural consequence of the diffractive hard scattering idea [2]. The pointlike probe in DIS makes it an ideal way to measure the parton structure of diffraction. This is discussed in section 4 together with diffractive production od jets and W's at the Tevatron. Although pomeron-based models may work phenomenologically, there are conceptual and theoretical problems as discussed in section 5.

These problems are related to the general unsolved problem of non-perturbative QCD (non-pQCD). Diffraction is one important aspect of this, others are hadronization in high energy collisions and the confinement of quarks and gluons. In recent years there has been an increased interest for these problems and efforts are made based on new ideas and methods as discussed in section 6. The hard scale in diffractive hard scattering only solves part of the problem by making the upper part of the diagrams in Fig. 1 calculable in perturbation theory. However, the soft, lower part of the interaction occurs over a large space-time as illustrated in Fig. 1c and must be treated with some novel non-pQCD methods.

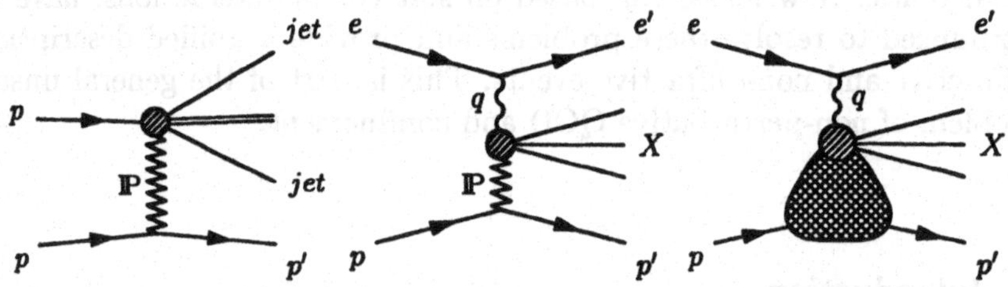

Figure 1. **(a,b) Diffractive hard scattering, in $p\bar{p}$ and DIS ep, in the pomeron approach. (c) Diffractive DIS illustrating the long space-time scale for the soft interaction at the proton 'vertex'.**

One such *'new-new'* idea is the *soft colour interaction* (SCI) model [4], which is an explicit attempt to describe non-pQCD interactions in a Monte Carlo event generation model. Although it is quite simple, it is able to describe data on different diffractive and non-diffractive interactions as discussed in section 6. However, a better theoretical basis for this kind of models is certainly needed. In addition, the rapidity gaps between high-$p_\perp$ jets observed at the Tevatron are still a challenge to understand (section 7). In conclusion (section 8), although substantial progress has been made recently, diffractive scattering is still a basically unsolved problem which provides challenges for the future.

2. Rapidity gaps and the pomeron concept

The dynamics of hadron-hadron interactions are largely not understood. Only the very small fraction of the cross section related to hard (large momentum transfer) interactions can be understood from first principles using pertubation theory, *e.g.* jet production in QCD or γ^*, W, Z production in electroweak theory. The large cross section ($\mathcal{O}(mb)$) processes, on the other hand, are given by non-pQCD for which proper theory is lacking and only phenomenological models are available. These processes are classified in terms of their final states as illustrated in Fig. 2.

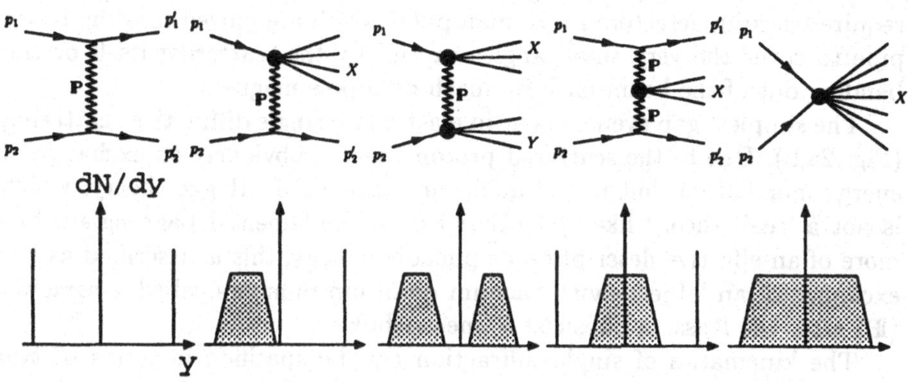

Figure 2. Rapidity distribution, dN/dy, of final state hadrons in hadron-hadron collisions with interpretations in terms of pomeron exchange: (a) elastic, (b) single diffractive, (c) double diffractive, (d) double pomeron exchange and (e) totally inelastic interactions.

The distribution of final state hadrons is then usually expressed in terms of the rapidity variables

$$\text{rapidity } y = \frac{1}{2}\ln\frac{E+p_z}{E-p_z} \approx -\ln\tan\frac{\theta}{2} = \eta \text{ pseudorapidity} \tag{1}$$

where the approximation becomes exact for massless particles and the polar angle θ is with respect to the z-axis along the beam. In a totally inelastic interaction (Fig. 2e) the hadrons are distributed with a flat rapidity plateau. This corresponds to longitudinal phase space where the transverse momenta are limited to a few hundred MeV, but longitudinal momenta cover the available phase space. This is in accordance with hadronization models, *e.g.* the Lund string model [5], where longitudinal momenta are given by a scaling fragmentation function and transverse momenta are strongly suppressed above the scale of soft interactions. The probability to have events with a gap, *i.e.* a region without particles, due to statistical fluctuations in such a rapidity distribution decreases exponentially with the size of the gap.

Experimentally one observes a much higher rate of gaps. *Diffraction* is nowadays often defined as events with *large rapidity gaps which are not exponentially suppressed* [6]. This is, however, a wider definition than that previously often used in terms of a leading proton taking a large fraction (*e.g.* $x_F \gtrsim 0.9$) of the beam proton momentum which enforces a rapidity gap simply by kinematical constraints. However, a gap can be anywhere in the event and therefore allow a forward system of higher mass than a single proton. The definition chosen reflects what the experiments actually observe. The leading protons go down the beam pipe and their detection require tracking detectors in 'Roman pots' which are moved into the beam pipe to cover the very small angles caused by the scattering itself or the bending out of the beam path by machine dipole magnets.

The simplest gap events occur in elastic and single diffractive scattering (Fig. 2a,b). Due to the scattered proton there is obviously an exchange of energy-momentum, but not of quantum numbers. In Regge theory, which is not a 'real' theory like QED based on a fundamental Lagrangian, but more of an effective description or phenomenology, this is described as the exchange of an 'object' with vacuum quantum numbers called a *pomeron* ($\mathbb{P}$) after the Russian physicist Pomeranchuk.

The kinematics of single diffraction can be specified in terms of two variables, *e.g.* the momentum fraction $x_p = p_f/p_i$ of the final proton relative to the initial one and the momentum transfer $t = (p_i - p_f)^2$. The pomeron then takes the momentum fraction $x_{\mathbb{P}} = 1 - x_p$ and has a negative mass-squared $m_{\mathbb{P}}^2 = t < 0$ meaning that it is a virtual exchanged object. The other proton produces a hadronic system X of mass $M_X^2 = x_{\mathbb{P}} s$, *i.e.* the invariant mass-squared of the 'pomeron-proton collision'. The cross section for single diffraction (SD) is experimentally found to be well described by

$$\frac{d\sigma_{SD}}{dt\,dx} \simeq \frac{1}{x_{\mathbb{P}}}\left\{a_1 e^{b_1 t} + a_2 e^{b_2 t} + \ldots\right\} \simeq \frac{a}{M_X^2}|F(t)|^2 \tag{2}$$

where the exponential damping in t can be interpreted in terms of a proton form factor $F(t)$ giving the probability that the proton stays intact after the momentum 'kick' t. With $x_{\mathbb{P}} < 0.1$ the maximum M_X reachable at ISR, $Sp\bar{p}S$, Tevatron and LHC are 20, 170, 570 GeV and 4.4 TeV, respectively. However, the rate of large M_X events is suppressed due to the dominantly small pomeron momentum fraction. This is the reason why it took until 1985 to demonstrate that the rapidity distribution of hadrons in the X-system shows longitudinal phase space [7]. Therefore, the pomeron-proton collision is similar to an ordinary hadron-proton interaction. This ruled out 'fireball models' giving a spherically symmetric final state having a

Gaussian rapidity distribution [1]. Thus, the hadronic final state provides information on the interaction dynamics producing it.

The Regge formalism relates the differential cross sections for different processes. This is achieved through the factorization of the different vertices such that the same kind of vertex in different processes is given by the same expression. The exchange of other than vacuum quantum numbers are described as, *e.g.*, meson exchanges. Since the exchanged object is not a real state, but virtual with a negative mass-squared, it is actually a representation of a whole set of states (*e.g.* mesons) with essentially the same quantum numbers. The spin versus the mass-squared of such a set gives a linear relation which can be extrapolated to $m^2 = t < 0$ and provides the trajectory $\alpha(t)$ for the exchange. This provides the essential energy dependence $\sigma \sim s^{2\alpha(t)-2}$ of the cross section. The pomeron trajectory $\alpha_{I\!P}(t) = 1 + \epsilon + \alpha' t \simeq 1.08 + 0.25t$ has the largest value at $t = 0$ (intercept) which leads to the dominant contribution to the hadron-hadron cross section. Contrary to the π and ρ trajectories, which have well known integer spin states at the pole positions $t = m^2_{meson}$, there are no real states on the pomeron trajectory. However, a recently found spin-2 glueball candidate with mass 1926 ± 12 MeV [8] fits well on the pomeron trajectory.

This would be in accord with the suggestion that the pomeron is some gluonic system [9] which may be interpreted as a virtual glueball [10]. In a modern QCD-based language it is natural to consider a *pomeron-hadron analogy* where the pomeron is a hadron-like object with a quark and gluon content. Pomeron-hadron interactions would then resemble hadron-hadron collisions and give final state hadrons in longitudinal phase space, just as observed.

There was, however, another view in terms of a *pomeron-photon analogy* [11] where the pomeron is considered to have an effective pointlike coupling to quarks. Single diffractive scattering would then be similar to deep inelastic scattering and the exchanged pomeron scatters a quark out of the proton, leading to a longitudinal phase space after hadronization. This fits well with the experimental evidence for pomeron single-quark interactions [12].

3. Idea and discovery of diffractive hard scattering

To explore the diffractive interaction further, we [2] introduced in 1984 the new idea that one should use a hard scattering process to probe the pomeron interaction at the parton level. In retrospect this seems obvious and simple, but at that time it was quite radical and was criticised. The idea was launched before the observations of longitudinal event structure in diffraction, the glueball candidate on the pomeron trajectory and the pomeron single-quark interactions discussed above. Furthermore, diffraction was at that time a side issue in particle physics that was ignored by most people.

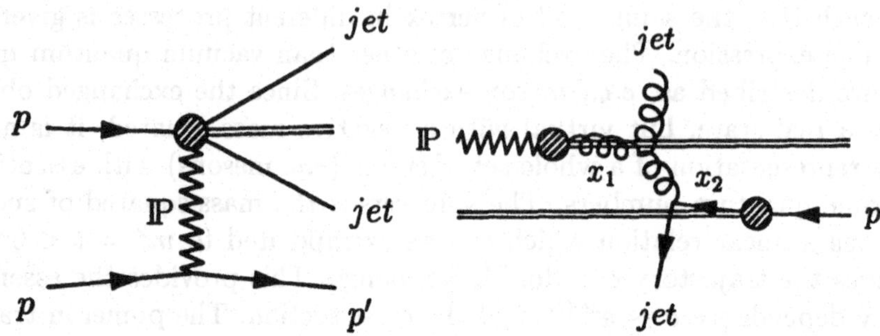

Figure 3. Single diffractive scattering with pomeron exchange giving a pomeron-proton interaction with a hard parton level subprocess producing jets in the X-system.

Based on the pomeron factorization hypothesis, the diffractive hard scattering process was considered [2] in terms of an exchanged pomeron and a pomeron-particle interaction were a hard scattering process on the parton level may take place as illustrated in Fig. 3. The diffractive hard scattering cross section can then be expressed as the product of the inclusive single diffractive cross section and the ratio of the pomeron-proton cross sections for producing jets and anything, *i.e.*

$$\frac{d\sigma_{jj}}{dtdM_X^2} = \frac{d\sigma_{SD}}{dtdM_X^2}\frac{\sigma(I\!Pp \to jj)}{\sigma(I\!Pp \to X)} \tag{3}$$

Here, $d\sigma_{SD}$ can be taken as the parametrization of data in eq. (2) and the total pomeron-proton cross section $\sigma(I\!Pp \to X)$ can be extracted from data using the Regge formalism resulting in a value of order 1 mb. Together these parts of eq. (3) can be seen as an expression for a *pomeron flux* $f_{I\!P/p}(x_{I\!P},t)$ in the beam proton. The cross section for pomeron-proton to jets, $\sigma(I\!Pp \to jj)$, is assumed to be given by pQCD as

$$\sigma(I\!Pp \to jj) = \int dx_1\, dx_2\, d\hat{t} \sum_{ij} f_{i/I\!P}(x_1,Q^2) f_{j/p}(x_2,Q^2) \frac{d\hat{\sigma}}{d\hat{t}} \tag{4}$$

where a parton density function $f_{i/I\!P}$ for the pomeron is introduced in analogy with those for ordinary hadrons. The pomeron parton density functions were basically unknown, but assuming the pomeron to be gluon dominated it was resonable to try $xg(x) = ax(1-x)$ or $xg(x) = b(1-x)^5$ for the cases of only two gluons or of many gluons similar to the proton. Similarly, if the pomeron were essentially a $q\bar{q}$ system one would guess $xq(x) = cx(1-x)$. The normalisation constants a, b, c can be chosen to saturate the momentum sum rule $\int_0^1 dx \sum_i x f_{i/\mathbb{P}}(x) = 1$, which seems like a reasonable assumption to get started.

This formalism allows numerical estimates for diffractive hard scattering cross sections. Diffractive jet cross sections at the CERN S$p\bar{p}$S collider energy were found [2] to be large enough to be observable. Furthermore, turning the formalism into a Monte Carlo (MC) program (precursor to POMPYT [13] described below) to simulate complete events, demonstrated a clearly observable event signature: a leading proton ($x_F \gtrsim 0.9$) separated by a large rapidity gap from a central hadronic system with high-$p_\perp$ jets.

Based on these predictions, the UA8 experiment was approved and constructed. It had Roman pots in the beam pipes to measure the momentum of leading (anti)protons and used the UA2 central detector to observe jets. The striking event signature were observed in 1987 [3] signalling the discovery of the diffractive hard scattering phenomenon, which was investigated further with more data [14, 15].

The observed jets showed the characteristic properties of QCD jets as quantified in the Monte Carlo, *e.g.* jet $E_\perp$ and angular distributions and energy profiles. The longitudinal momentum of the jets gives information on the momentum fraction (x_1 in Fig. 3b) of the parton in the pomeron; a change in the shape of the x_1-distribution shifts the parton-parton cms with respect to the X cms and thereby the momentum distribution of the jets [2]. Comparison of data and the Monte Carlo shows a clear preference for a hard parton distribution [14]. Using a quark or gluon distribution $xf(x) \sim x(1-x)$ gives a resonable description of the observed x_F-distribution of the jets, although giving too little in the tail at large x_F. This is more clearly seen, if instead of considering individual jets, one takes both jets in each event and plot the longitudinal momentum of this pair, Fig. 4. The excess at large x_F can be described by having 30% of the pomeron structure function in terms of a *super-hard* component with partons taking the entire pomeron momentum, *i.e.* $xf(x) \sim \delta(1-x)$. The δ-function can be seen as a representation of some more physical distribution which is very hard, *e.g.* $xf(x) \sim 1/(1-x)$.

With the UA8 data alone, one cannot distinguish between gluons or quarks in the pomeron. The UA1 experiment has given some evidence for diffractive bottom production [16]. This may be interpreted with a gluon-dominated pomeron such that the $gg \to b\bar{b}$ subprocess can be at work, but no firm conclusion can be made given the normalization uncertainty in the model and the experimental errors [17].

UA8 have recently provided the absolute cross section for diffractive jet production [15]. This shows that, although the Monte Carlo model repro-

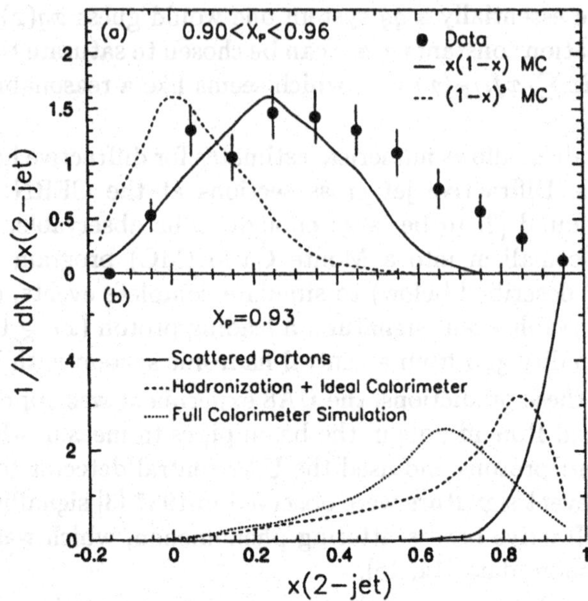

Figure 4. **(a) Distribution of scaled longitudinal momentum x_F for the two-jet system in diffractive events defined by a leading proton ($0.90 < x_p < 0.96$). Data from UA8 compared to the Monte Carlo model using the indicated hard and soft parton densities in the pomeron. (b) Monte Carlo results at the parton level, after hadronization and after detector simulation assuming 'super-hard' partons in the pomeron ($xf(x) \sim \delta(1-x)$). From [14].**

duces the shapes of various distributions, it overestimates the absolute cross section; $\sigma(data)/\sigma(model) = 0.30 \pm 0.10$ or 0.56 ± 0.19 for the model with the pomeron as a gluonic or a $q\bar{q}$ state, respectively. This have raised questions concerning the normalization of the pomeron flux and the pomeron structure function, as will be discussed in section 5.

In summary, diffractive hard scattering has been discovered by UA8 and the main features can be interpreted in terms of an exchanged pomeron with a parton structure.

4. Rapidity gap events at HERA and the Tevatron

The above model for diffractive hard scattering can be naturally extended to other kinds of particle collisions, $a + p \rightarrow p + X$ where a can not only be any hadron but also a lepton or a photon. Based on the pomeron factorization hypothesis [2, 18] the cross section is $d\sigma(a + p \rightarrow p + X) = f_{I\!P/p}(x_{I\!P}, t)\, d\sigma(a + I\!P \rightarrow X)$. The pomeron flux can be taken as a simple parametrization of data in terms of exponentials as above, or obtained from

Regge phenomenology in the form [19]

$$f_{\mathbb{P}/p}(x_{\mathbb{P}},t) = \frac{9\beta_0^2}{4\pi^2}\left(\frac{1}{x_{\mathbb{P}}}\right)^{2\alpha_{\mathbb{P}}(t)-1}[F_1(t)]^2 \qquad (5)$$

with parameters obtained from data on hadronic inclusive diffractive scattering. Here, $\beta = 3.24\ GeV^2$ is the mentioned effective pomeron-quark coupling and $F_1(t) = (4m_p^2 - At)/(4m_p^2 - t) \cdot (1 - t/B)^{-2}$ is a proton form factor with m_p the proton mass and parameters $A = 2.8$, $B = 0.7$. The pomeron trajectory is $\alpha_{\mathbb{P}}(t) \simeq 1.08 + 0.25t$.

For the hard scattering cross section $d\sigma(a + \mathbb{P} \to X)$ one should use the relevant convolution of parton densities and parton cross sections, *e.g.* eq. (4) for hadron-pomeron collisions. In order to simulate complete events this formalism has been included in the Monte Carlo program POMPYT [13] based on the Lund Monte Carlo PYTHIA [20]. In particular, there are options for different pomeron flux factors and parton densities. Moreover, POMPYT also contains pion exchange processes where a pion, with a flux factor and parton densities, replaces the pomeron as an example of other possible Regge exchanges.

4.1. DIFFRACTIVE DIS AT HERA

As suggested already in [2], one should probe the pomeron structure with deep inelastic scattering, *e.g.* at HERA. The advantage would be to have a clean process with a well understood point-like probe with high resolving power Q^2. The experimental signature should be clear; a quasi-elastically scattered proton (going down the beam pipe) well separated by a rapidity gap from the remaining hadronic system. The kinematics is then described by the diffractive variables $x_{\mathbb{P}}$ (or $x_p = 1 - x_{\mathbb{P}}$) and t, as above, and the standard DIS variables $Q^2 = -q^2 = -(p_e - p_{e'})^2$ and Bjorken $x = Q^2/2P \cdot q$ (where $P, p_e, p_{e'}, q$ are the four-momenta of the initial proton, initial electron, scattered electron and exchanged photon, respectively).

The cross section for diffractive DIS can then be written [21]

$$\frac{d\sigma(ep \to epX)}{dx dQ^2 dx_{\mathbb{P}} dt} = \frac{4\pi\alpha^2}{xQ^4}\left(1 - y + \frac{y^2}{2}\right) F_2^D(x, Q^2; x_{\mathbb{P}}, t) \qquad (6)$$

where the normal proton structure function F_2 has been replaced by a corresponding diffractive one, F_2^D, with $x_{\mathbb{P}}$ and t specifying the diffractive conditions. Only the dominating electromagnetic interaction is here considered and $R = \sigma_L/\sigma_T$ is neglected for simplicity. If pomeron factorization holds, then F_2^D can be factorized into a pomeron flux and a pomeron structure function, *i.e.* $F_2^D(x, Q^2; x_{\mathbb{P}}, t) = f_{\mathbb{P}/p}(x_{\mathbb{P}}, t)\, F_2^{\mathbb{P}}(z, Q^2)$ where the pomeron structure function $F_2^{\mathbb{P}}(z,Q^2) = \sum_f e_f^2 \left(zq_f(z,Q^2) + z\bar{q}_f(z,Q^2)\right)$ is given by the densities of (anti)quarks of flavour f and with a fraction $z = x/x_{\mathbb{P}}$ of the pomeron momentum. Since the photon does not couple directly to gluons, they will only enter indirectly through $g \to q\bar{q}$ as described by QCD evolution or the photon-gluon fusion process.

Although diffractive DIS had been predicted in this way [2, 21, 22], it was a big surprise to many when it was observed first by ZEUS [23] and then by H1 [24]. Since leading proton detectors were not available at that time, it was the large rapidity gap that was the characteristic observable, *i.e.* no particle or energy depositions in the forward part of the detector as shown in Fig. 5a. Leading protons have later been clearly observed [25], but the efficiency is low so the dominant diffractive data samples are still defined in terms of rapidity gaps. A simple observable to characterize the effect is η_{max} giving, in each event, the maximum pseudo-rapidity where an energy deposition is observed. Fig. 5b shows the distribution of this quantity.

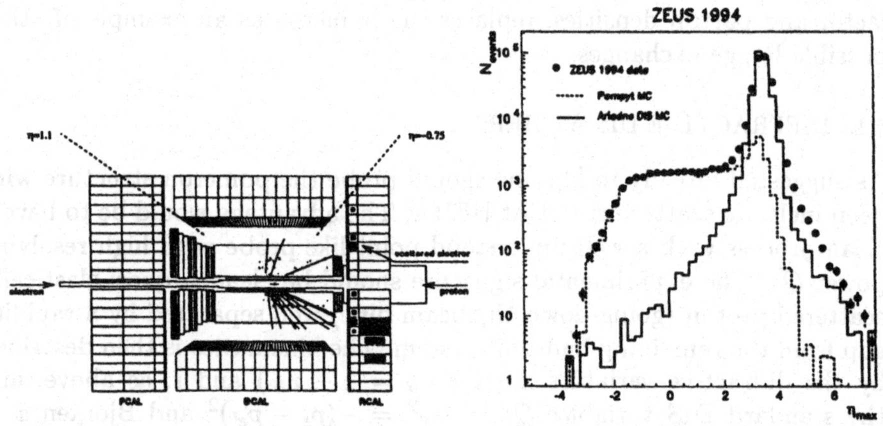

Figure 5. **(a) The ZEUS detector with a rapidity gap event, *i.e.* having no tracks or energy depositions in the forward (proton beam direction) part of the detector. (b) Distribution of η_{max}, the maximum pseudorapidity of observed tracks/energy, in ZEUS data compared to Monte Carlo simulations using ARIADNE [26] (full histogram) for ordinary DIS and POMPYT [13] (dashed) for DIS on a pomeron with a hard quark density ($zq(z) \sim z(1-z)$), with normalisation adjusted such that the sum fits the data. Courtesy of the ZEUS collaboration.**

Although the bulk of the data with η_{max} in the forward region is well described by ordinary DIS Monte Carlo events, there is a large excess with a smaller η_{max} corresponding to the central region or even in the electron hemisphere. This excess is well described by POMPYT as deep inelastic scattering on an exchanged pomeron with a hard quark density. The gap events have the same Q^2 dependence as normal DIS and are therefore not some higher twist correction. Their overall rate is about 10% of all events, so it is not a rare phenomenon.

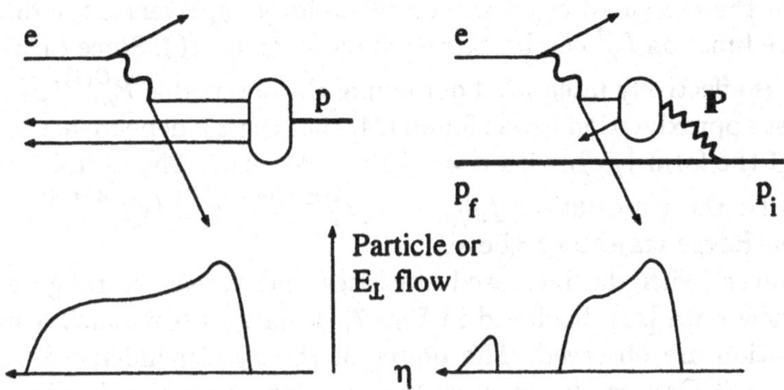

Figure 6. Illustration of (a) normal DIS and (b) diffractive DIS at HERA and the resulting particle or transverse energy flow in laboratory pseudorapidity.

In normal DIS, a quark is scattered from the proton leaving a colour charged remnant (diquark in the simplest case). This gives rise to a colour field (*e.g.* a string) between the separated colour charges, such that the hadronization gives particles in the whole intermediate phase space region as illustrated in Fig. 6a. The gap events correspond to the scattering on a colour singlet object, Fig. 6b, which gives no colour field between the hard scattering system and the proton remnant system. Therefore, no hadrons are produced in the region between them, *i.e.* a rapidity gap appears. The size of the gap is basically a kinematic effect. The larger fraction of the proton beam momentum that is carried by the forward going colour singlet proton remnant system, the smaller fraction remains for other particles which therefore emerge at smaller rapidity. The forward going system Y must have a small invariant mass in order to escape undetected in the beam pipe. It is mostly a proton with a large fraction of the beam momentum and only a very small angular deflection.

Since the Y system is not observed the t variable is not measured, but is usually negligibly small (*c.f.* the proton form factor above). However, with the invariant definitions

$$x_{I\!P} = \frac{q\cdot(P-p_Y)}{q\cdot P} = \frac{Q^2+M_X^2-t}{Q^2+W^2-m_p^2} \approx \frac{x(Q^2+M_X^2)}{Q^2} \quad (7)$$

$$z=\beta = x/x_{I\!P} = \frac{-q^2}{2q\cdot(P-p_Y)} = \frac{Q^2}{Q^2+M_X^2-t} \approx \frac{Q^2}{Q^2+M_X^2} \quad (8)$$

$x_{I\!P}$ can be reconstructed from the DIS variables and the X-system. Likewise, z (or β) can be measured and corresponds to Bjorken-x for DIS on the pomeron and can therefore be interpreted as the momentum fraction of the parton in the pomeron.

From the measured cross section of rapidity gap events, the diffractive structure function F_2^D can be extracted based on eq. (6). Since t is not measured it is effectively integrated out giving the observable $F_2^{D(3)}(x_{I\!P}, \beta, Q^2)$. To a first approximation it was found [24] that the $x_{I\!P}$ dependence factorises and is of the form $1/x_{I\!P}^n$ with $n = 1.19 \pm 0.06 \pm 0.07$. This is in basic agreement with the expectation $f_{I\!P/p} \sim 1/x_{I\!P}^{2\alpha_{I\!P}(t)-1} \simeq 1/x_{I\!P}^{1.16+0.5t}$ from the pomeron Regge trajectory above.

However, with the increased statistics and kinematic range available in the new data [27] displayed in Fig. 7, deviations from such a universal factorisation are observed. The power of the $x_{I\!P}$-dependence is found to depend on β. One way to interpret this is to introduce a subleading reggeon (IR) exchange with expected trajectory $\alpha_{I\!R}(t) \simeq 0.55 + 0.9t$ and quantum numbers of the ρ, ω, a or f meson [27]. Fits to the data (Fig. 7) show that although the pomeron still dominates, the meson exchange contribution is important at larger $x_{I\!P}$ and causes F_2^D to decrease slower (or $x_{I\!P}F_2^{D(3)}$ to even increase). The fit gives the intercepts $\alpha_{I\!R}(0) = 0.50 \pm 0.19$, in agreement with the expectation, and $\alpha_{I\!P}(0) = 1.20 \pm 0.04$ which is, however, significantly larger than 1.08 obtained from soft hadronic cross sections.

There is no evidence for a β or Q^2 dependence in these intercepts and one can therefore integrate over $x_{I\!P}$ (using data and the fitted dependence), resulting in the measurement of $F_2^D(\beta, Q^2)$ shown in Fig. 8. Following the above framework, this quantity can be interpreted as the structure function of the exchanged colour singlet object, which is mainly the pomeron. The fact that F_2^D is essentially scale independent, *i.e.* almost constant with Q^2, shows that the scattering occurs on point charges. The small Q^2 dependence present is actually compatible with being logarithmic as in normal QCD evolution, although the rise with lnQ^2 persists up to large values of β in contrast to the proton structure function. There is only a weak dependence on β such that the partons are quite hard and there is no strong decrease at large momentum fraction which is characteristic for ordinary hadrons.

These features are in accordance with a substantial gluon component in the structure of the diffractive exchange, as confirmed by a quantitative QCD analysis [27]. Standard next-to-leading order DGLAP evolution [28] gives a good fit of $F_2^D(\beta, Q^2)$ as demonstrated in Fig. 8. The fitted momentum distributions of quarks and gluons in the pomeron are shown in Fig. 9. Clearly, the gluon dominates and carries 80–90% of the pomeron momentum depending on Q^2. At low Q^2 the gluon distribution may even be peaked at large momentum fractions, *c.f.* the superhard component ob-

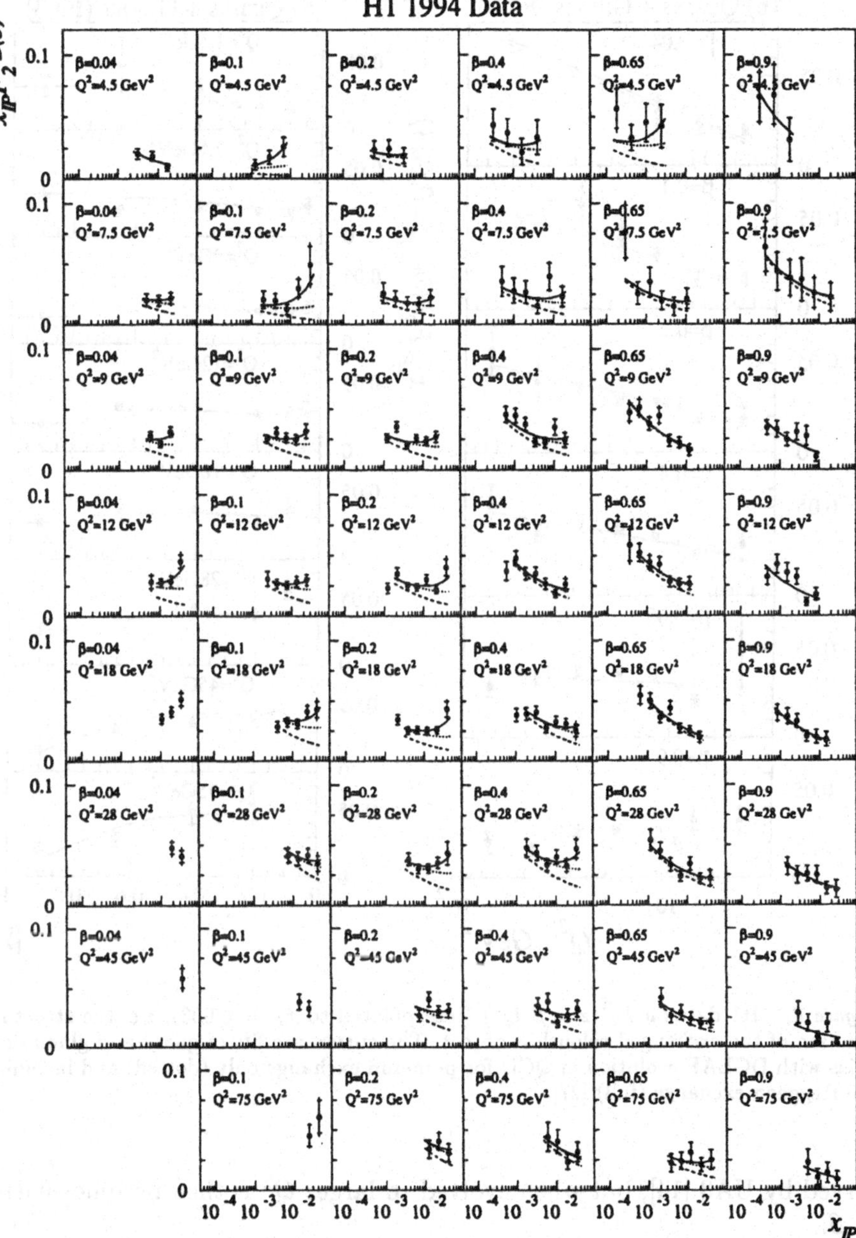

Figure 7. The diffractive structure function $x_{IP}F_2^{D(3)}$ for bins in β and Q^2. H1 data with Regge fits for pomeron and one reggeon exchange with their interference (full curves), only pomeron exchange (dashed curves) and pomeron exchange plus the interference (dotted curves) [27].

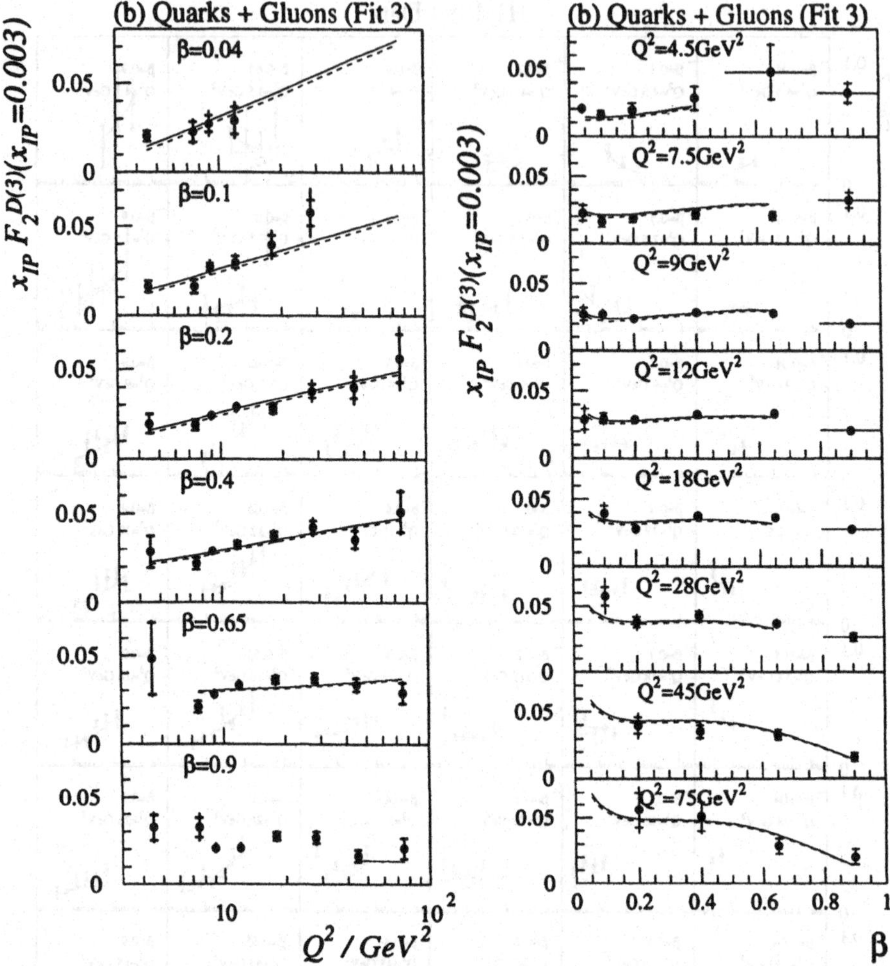

Figure 8. **H1 data on $F_2^D(z = \beta, Q^2)$ (interpolated to $x_{I\!P} = 0.003$), *i.e.* the structure function of the exchanged colourless object. The curves are fits of quark and gluon densities with DGLAP evolution in QCD for pomeron exchange only (dotted) and including the Reggeon exchange (full) [27].**

served by UA8 [14], but when evolved to larger Q^2 it then becomes flatter in β.

The general conclusion from these HERA data is therefore that the concept of an exchanged pomeron with a parton density seems appropriate. Moreover, Monte Carlo models, like POMPYT [13] and RAPGAP [29] (which is also based on the above pomeron formalism), can give a good description of the observed rapidity gap events.

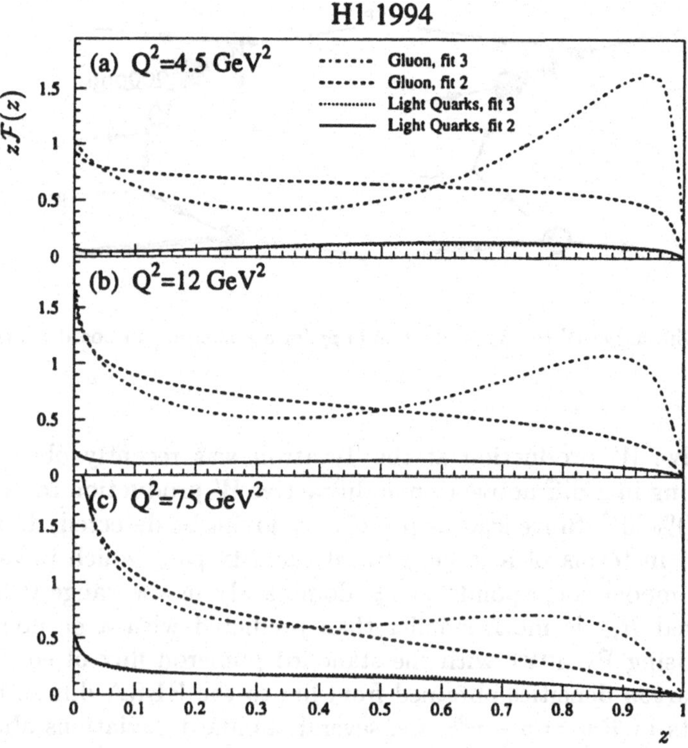

Figure 9. Momentum-weighted distributions in fractional momenta z of gluons and light quarks in the exchanged colorless object; obtained at different Q^2 from a NLO DGLAP fit to $F_2^D(\beta, Q^2)$ in Fig. 8. From [27].

4.2. DIFFRACTIVE W AND JETS AT THE TEVATRON

Based on the POMPYT model, predictions were also made [30] for diffractive W and Z production at the Tevatron $p\bar{p}$ collider, which provides sufficient energy in the pomeron-proton subsystem. With partons in the pomeron this occurs through the subprocesses $q\bar{q} \to W$ and $gq \to qW$ as illustrated in Fig. 10. The latter requires an extra QCD vertex $g \to q\bar{q}$ and is therefore suppressed by a factor α_s. Thus, a gluon-dominated pomeron leads to a smaller diffractive W cross section than a $q\bar{q}$-dominated pomeron. However, in both cases the cross sections were found to be large enough to be observable and the decay products of the W (Z) often emerge in a central region covered by the detectors. Moreover, a measurement of these decay products, ideally muons from Z decay, allows a reconstruction of the x-shape of the partons in the pomeron [30].

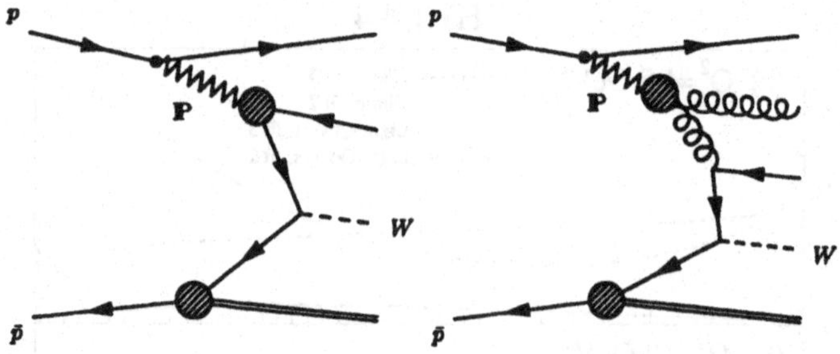

Figure 10. Diffractive W (or Z) production in $p\bar{p}$ for a pomeron composed of (a) $q\bar{q}$ and (b) gluons.

Diffractive W production at the Tevatron was recently observed by CDF resulting in a diffractive to non-diffractive W production ratio $R_W = (1.15 \pm 0.55)\%$ [31]. Since leading protons could not be detected, diffraction was defined in terms of a large forward rapidity gap, which in terms of a pomeron model corresponds to $x_{I\!P}$ dominantly in the range 0.01–0.05. The observed R_W is much smaller than predicted with a $q\bar{q}$ dominated pomeron. Using POMPYT with the standard pomeron flux of eq. (5) and pomeron parton densities obtained from fits to the HERA diffractive DIS data, results in $R_W = 5 - 6\%$, *i.e.* several standard deviations above the measured value!

Diffractive hard scattering has also been observed at the Tevatron in terms of rapidity gap events with two high-$p_\perp$ jets (dijets) as in UA8. The detailed definitions of gaps and jets differ somewhat between CDF and D0, but the results are similar. The ratio of diffractive to non-diffractive dijet events found at $\sqrt{s} = 1800\,GeV$ by CDF is $R_{jj} = (0.75 \pm 0.05 \pm 0.09)\%$ [32] and by D0 $R_{jj} = (0.76 \pm 0.04 \pm 0.07)\%$ [33]. D0 has also obtained the ratio $R_{jj} = (1.11 \pm 0.11 \pm 0.20)\%$ at the lower cms energy $\sqrt{s} = 630\,GeV$. These rates are significantly lower than those obtained with the standard pomeron model with parton densities that fit the diffractive HERA data.

The inability to describe the data on hard diffraction from both HERA and the Tevatron with the same pomeron model raises questions on the universality of the model, *e.g.* concerning the pomeron flux and structure function. This is examined in Fig. 11 in terms of the momentum sum of the partons and the amount of gluons needed to fit the data. The region acceptable to HERA data is compatible with a saturated momentum sum rule, but in disagreement with the internally consistent $p\bar{p}$ collider data.

CDF has also very recently observed events with a central dijet system and rapidity gaps on both sides. On one side a high-x_F antiproton is actually detected. This can be interpreted as double pomeron exchange (*c.f.*

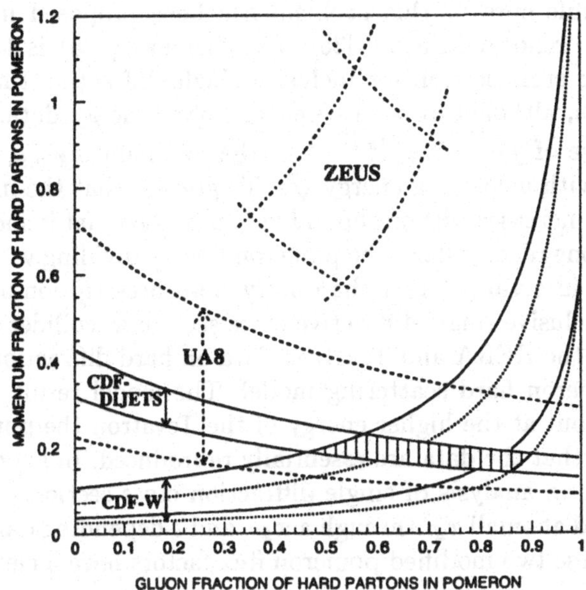

Figure 11. Total momentum fraction versus gluon fraction of hard partons in the pomeron evaluated by comparing measured diffractive rates with Monte Carlo results based on the standard pomeron flux showing a mismatch between results from $p\bar{p}$ and *ep*. From [32].

Fig. 2d), one from each of the quasi-elastically scattered proton and antiproton, where the two pomerons interact to produce the jets. The diffractive hard scattering model then contains a convolution of two pomeron flux factors and two pomeron parton densities with a QCD parton level cross section. The observed ratio of two-gap jet events to the single-gap jet events is found by CDF to be $(0.26 \pm 0.05 \pm 0.05)\%$ [34]. An important observation is also that the $E_{\perp}$-spectrum of the jets in these two-gap events have the same *shape* as in single-gap and no-gap events. This hints at the same underlying hard scattering dynamics which does not change with the soft processes that cause gaps or no-gaps. It is not yet clear whether this feature appears naturally in the pomeron model. However, the double pomeron exchange model, with pomeron flux and parton densities based on diffractive HERA data, seems to overestimate the rate of two-gap jet events [34].

5. Pomeron problems

The inability to describe both HERA and $p\bar{p}$ collider data on hard diffraction is a problem for the pomeron model. It shows that the 'standard' pomeron flux factor and pomeron parton densities cannot be used universally. A possible cure to this problem has been proposed in terms of a pomeron flux 'renormalization' [35]. The flux in eq. (5) is found to give a much larger cross section for inclusive single diffraction than measured at $p\bar{p}$ colliders, although it works well for lower energy data. This is due to the increase of $f_{I\!P} \sim 1/x_{I\!P}^{2\alpha_{I\!P}(t)-1}$ as the minimum $x_{I\!P\,min} = M^2_{X\,min}/s$ gets smaller with increasing energy $\sqrt{s}$. To prevent that the integral of the pomeron flux increases without bound, it is proposed that it should saturate at unity, *i.e.* one renormalizes the pomeron flux by dividing with its integral whenever the integral is larger than unity. This prescription not only gives the correct inclusive single diffractive cross section at collider energies, but it also makes the HERA and Tevatron data on hard diffraction compatible with the pomeron hard scattering model. The model result for HERA is not affected, but at the higher energy of the Tevatron the pomeron flux is reduced such that the data are essentially reproduced. In another proposal [36] based on an analysis of single diffraction cross sections, the pomeron flux is reduced at small $x_{I\!P}$ through a $x_{I\!P}$- and t-depending damping factor. Neither of these two modified pomeron flux factors have a clear theoretical basis.

A difference between diffraction in ep and $p\bar{p}$ is the possibility for coherent pomeron interactions in the latter [37]. In the incoherent interaction only one parton from the pomeron participates and any others are spectators. However, in the pomeron-proton interaction with $I\!P = gg$ both gluons may take part in the hard interaction giving a coherent interaction. For example, in the $I\!Pp$ hard scattering subprocess $gg \to q\bar{q}$, the second gluon from the pomeron may couple to the gluon from the proton. Such diagrams cancel when summing over all final states for the inclusive hard scattering cross section (the factorization theorem). For gap events, however, the sum is not over all final states and the cancellation fails leading to factorization breaking and these coherent interactions where the whole pomeron momentum goes into the hard scattering system. With momentum fraction x of the first gluon and $1-x$ of the second, a factor $1/(1-x)$ arises from the propagator of the second, soft gluon in the pomeron. This may motivate a super-hard component in the pomeron with effective structure function $1/(1-x) \approx \delta(1-x)$ as in the UA8 data discussed above. This coherent interaction cannot occur in the same way in DIS since the pomeron interacts with a particle without coloured constituents. This difference between ep and $p\bar{p}$ means that there should be no complete universality of parton densities in the pomeron.

Although modified pomeron models may describe the rapidity gap events reasonably well, there is no satisfactory understanding of the pomeron and its interaction mechanisms. On the contrary, there are conceptual and theoretical problems with this framework. The pomeron is not a real state, but can only be a virtual exchanged spacelike object. The concept of a structure function is then not well defined and, in particular, it is unclear whether a momentum sum rule should apply. In fact, the factorisation into a pomeron flux and a pomeron structure function cannot be uniquely defined since only the product is an observable quantity [38].

It may be incorrect to consider the pomeron as being 'emitted' by the proton, having QCD evolution as a separate entity and being 'decoupled' from the proton during and after the hard scattering. Since the pomeron-proton interaction is soft, its time scale is long compared to the short space-time scale of the hard interaction. It may therefore be natural to expect soft interactions between the pomeron system and the proton both before and after the snapshot of the high-Q^2 probe (as illustrated in Fig. 1c). The pomeron can then not be considered as decoupled from the proton and, in particular, is not a separate part of the QCD evolution in the proton.

Large efforts have been made to understand the pomeron as two-gluon system or gluon ladder in pQCD. By going to the soft limit one may then hope to gain understanding of non-pQCD. Perhaps one could establish a connection between pQCD in the small-x limit and Regge phenomenology. More explicitly, attempts have been made to connect the Regge pomeron with gluon ladders in pQCD. For example, the analogy between the Regge triple pomeron diagram for single diffractive scattering has been connected with the gluon ladder fan diagram in pQCD to estimate the pomeron gluon density [39]. The fan diagrams are described by the GLR equation [40] which gives a novel QCD evolution with non-linear effects due to gluon recombination $gg \to g$. This reduces the gluon density at small-x (screening); an effect that could be substantial in the pomeron [21].

Diffractive DIS has been considered in terms of models based on two-gluon exchange in pQCD, see *e.g.* [41]. The basic idea is to take two gluons in a colour singlet state from the proton and couple them to the $q\bar{q}$ system from the virtual photon. With higher orders included the diagrams and calculations become quite involved. Nevertheless, these formalisms can be made to describe the main features of the diffractive DIS data. Although this illustrates the possibilities of the pQCD approach to the pomeron, one is still forced to include non-perturbative modelling to connect the two gluons in a soft vertex to the proton. Thus, even if one can gain understanding by working as far as possible in pQCD, one cannot escape the fundamental problem of understanding non-pQCD.

6. Non-perturbative QCD and soft colour interactions

The main problem in understanding diffractive interactions is related to our poor theoretical knowledge about non-pQCD. The Regge approach with a pomeron can apparently be made to work phenomenologically, but has problems as discussed above. Therefore, new models have recently been constructed without using the pomeron concept or Regge phenomenology. Instead, they are based on new ideas on soft colour interactions that give colour rearrangements which affect the hadronization and thereby the final state. These models have first been developed for diffractive DIS which is a simpler and cleaner process than diffraction in $p\bar{p}$ collisions.

One model [42] to understand diffractive DIS at HERA exploits the dominance of the photon-gluon fusion process $\gamma^* g \to q\bar{q}$ at small-x. The $q\bar{q}$ pair is produced in a colour octet state, but it is here assumed that soft interactions with the proton colour field randomizes the colour. The $q\bar{q}$ pair would then be in an octet or singlet state with probability 8/9 and 1/9, respectively. When in a singlet state, the $q\bar{q}$ pair hadronizes independently of the proton remnant, which should result in a lack of particles in between. From the photon-gluon fusion matrix element one then obtains the diffractive structure function

$$F_2^D(x, Q^2, \xi) \simeq \frac{1}{9} \cdot \frac{\alpha_s}{2\pi} \sum_q e_q^2 g(\xi) \cdot \beta\{[\beta^2 + (1-\beta)^2] \ln \frac{Q^2}{m_g^2 \beta^2} - 2 + 6\beta(1-\beta)\} \tag{9}$$

where 1/9 is the colour singlet probability. The next factor, including the density $g(\xi)$ of gluons with momentum fraction ξ, corresponds to a pomeron flux factor. The β-dependent factor corresponds to the pomeron structure function $F_2^D(\beta, Q^2)$ above, with $\beta = x/\xi$ as usual. Thus, there is an effective factorisation which is similar to pomeron models. The gluon mass parameter m_g regulates the divergence in the QCD matrix element and is chosen so as to saturate the DIS cross section at small-x with the photon-gluon fusion process. The model reproduces main features of the gap events, such as their overall rate and Q^2 dependence. However, it is simple and does not take into account higher order parton emissions and hadronization. Therefore, it cannot give as detailed predictions as the Monte Carlo models above.

In the same general spirit another model was developed independently using a Monte Carlo event generator approach [4, 43]. The starting point is the normal DIS parton interactions, with pQCD corrections in terms of matrix elements and parton showers in the initial and final state. The basic new idea is that there may be additional soft colour interactions (SCI) between the partons at a scale below the cut-off Q_0^2 for the perturbative treatment. Obviously, interactions will not disappear below this cut-off, the question is rather how to describe them properly. The proposed SCI mechanism can be viewed as the perturbatively produced quarks and gluons interacting softly with the colour medium of the proton as they propagate through it. This should be a natural part of the process in which 'bare' perturbative partons are 'dressed' into non-perturbative ones and the formation of the confining colour flux tube in between them. These soft interactions cannot change the momenta of the partons significantly, but may change their colour and thereby affect the colour structure of the event. This corresponds to a modified topology of the string in the Lund model approach, as illustrated in Fig. 12, such that another final state will arise after hadronization.

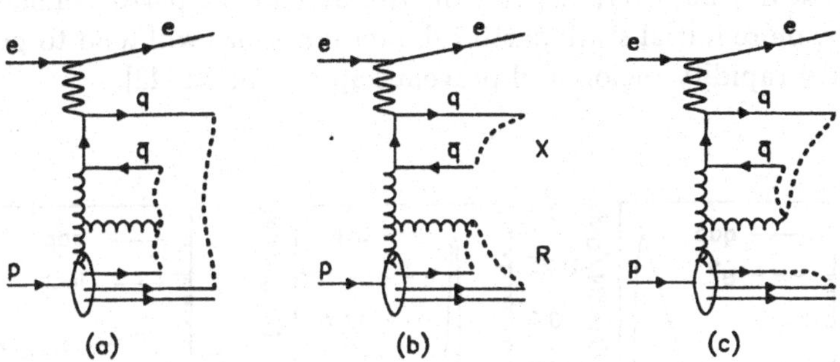

Figure 12. Gluon-induced DIS event with examples of colour string connection (dashed lines) of partons in (a) conventional Lund model based on the colour order in pQCD, and (b,c) after soft colour interactions.

Lacking a proper understanding of non-perturbative QCD processes a simple model was constructed to describe and simulate soft colour interactions. The hard parton level interactions are treated in the normal way using the LEPTO Monte Carlo [44] based on the standard electroweak cross section together with pQCD matrix elements and parton showers. The perturbative parts of the model are kept unchanged, since these hard processes cannot be altered by softer non-pQCD ones. Thus, the set of partons, including the quarks in the proton remnant, are generated as in conventional DIS. The SCI model is added by giving each pair of these colour charged partons the possibility to make a soft interaction, changing only the colour and not the momentum. This may be viewed as soft non-perturbative gluon exchange. Being a non-perturbative process, the exchange probability cannot be calculated and is therefore described by a phenomenological parameter R. The number of soft exchanges will vary event-by-event and change the colour topology such that, in some cases, colour singlet subsystems arise separated in rapidity as shown in Fig. 12bc. Here, (b) can be seen as a switch of anticolour between the antiquark and the diquark and (c) as a switch of colour between the two quarks. Colour exchange between the perturbatively produced partons and the partons in the proton remnant are of particular importance for the gap formation.

Both gap and no-gap events arise in this model. The rate and main properties of the gap events are qualitatively reproduced [43], *e.g.* the η_{max} distribution in Fig. 5b and the diffractive structure function $F_2^{D(3)}$. The gap rate depends on the parameter R, but the dependence is not strong giving a stable model with $R \simeq 0.2$–0.5. This colour exchange probability is the only new parameter in the model. Other parameters belong to the conventional DIS model [44] and have their usual values. The rate and size of gaps do, however, depend on the amount of parton emission. In particular, more initial state parton shower emissions will tend to populate the forward rapidity region and prevent gap formation [43].

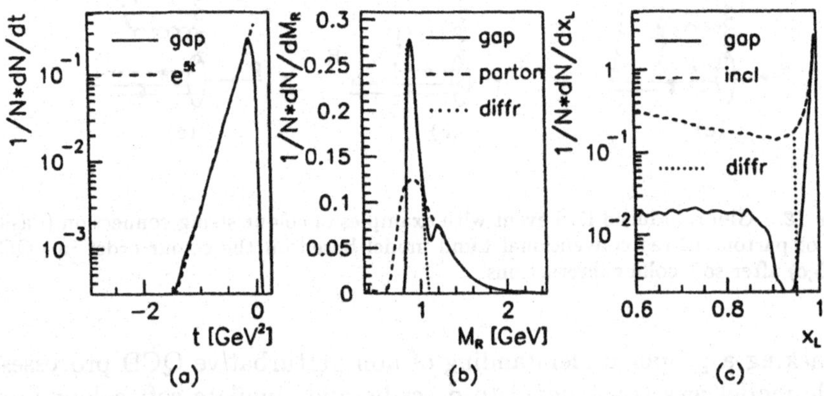

Figure 13. (a) Squared momentum transfer t from initial proton to remnant system R for 'gap' events compared with the exponential slope $1/\sigma_i^2 = 5\,GeV^2$. (b) Invariant mass M_R of the forward remnant system for 'gap' events at the hadron level (solid line) and parton level (dashed line) compared with 'diffractive' events at the hadron level (dotted line). (c) Longitudinal momentum fraction x_L for final protons in 'gap' events (solid line), all events (dashed line) and 'diffractive' events (dotted line). From [43]

The gap events show properties characteristic of diffraction as demonstrated in Fig. 13. The exponential t-dependence arises in the model from the gaussian intrinsic transverse momentum (Fermi motion) of the interacting parton which is balanced by the proton remnant system, *i.e.* $exp(-k_\perp^2/\sigma_i^2)$ with $\sigma_i \simeq 0.4 GeV$ and $t \simeq -k_\perp^2$. The forward system (Fig. 13b) is dominantly a single proton, as in diffractive scattering, but there is also a tail corresponding to proton dissociation. The longitudinal momentum spectrum of protons in Fig. 13c shows a clear peak at large fractional momentum x_L. Defining events having a leading proton with $x_L > 0.95$ as 'diffractive', one observes in Fig. 13bc that most of these events fulfill the gap requirement.

One may ask whether this kind of soft colour interaction model is essentially a model for the pomeron. This is not the case as long as no pomeron or Regge dynamics is introduced. The behaviour of the data on $F_2^D(\beta, Q^2)$ in Fig. 8 is in the SCI model understood as normal pQCD evolution in the proton. The rise with lnQ^2 also at larger β is simply the normal behaviour at the small momentum fraction $x = \beta x_{I\!P}$ of the parton in the proton. Here, $x_{I\!P}$ is only an extra variable related to the gap size or M_X (eq. (7)) which does not require a pomeron interpretation. The flat β-dependence (Fig. 8b) of $x_{I\!P} F_2^D = \frac{x}{\beta} F_2^D$ is due to the factor x compensating the well-known increase at small-x of the proton structure function F_2.

This Monte Carlo model gives a general description of DIS, with and without gaps. In fact, it can give a fair account for such 'orthogonal' observables as rapidity gaps and the large forward $E_\perp$ flow [43]. Diffractive events are in this model defined through the topology of the final state, in terms of rapidity gaps or leading protons just as in experiments. There is no particular theoretical mechanism or description in a separate model, like pomeron exchange, that defines what is labelled as diffraction. This provides a smooth transition between diffractive gap events and non-diffractive no-gap events [45]. In addition, leading neutrons are also obtained in fair agreement with recent experimental measurements [25]. In a conventional Regge-based approach, pomeron exchange would be used to get diffraction, pion exchange added to get leading neutrons and still other exchanges added to get a smooth transition to normal DIS. The SCI model indicates that a simpler theoretical description can be obtained.

The same SCI model can also be applied to $p\bar{p}$ collisions, by introducing it in the PYTHIA Monte Carlo [20]. This leads to gap events in hard scattering interactions as illustrated for W production in Fig. 14. It is amazing that the same SCI model, normalized to the diffractive HERA data, re-

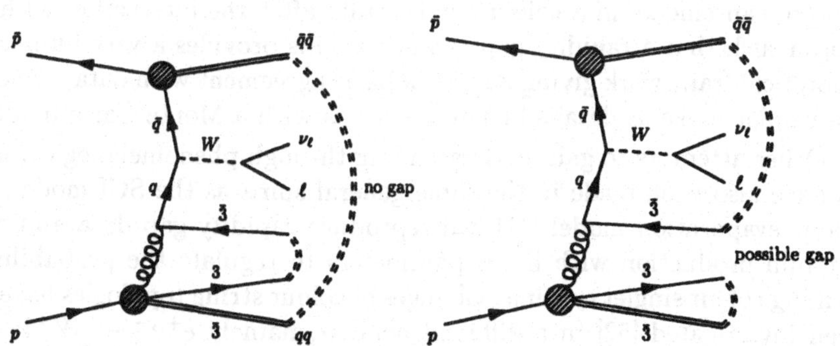

Figure 14. W production in $p\bar{p}$ with examples of colour string connections (dashed lines) of partons in (a) the conventional PYTHIA model and (b) after soft colour interactions.

produces the above discussed rates of diffractive W's and diffractive jet production observed at the Tevatron [46].

The soft colour interactions do not only lead to rapidity gaps, but also to other striking effects. They have been found [47] to reproduce the observed rate of high-$p_\perp$ charmonium and bottomonium at the Tevatron, which are factors of 10 larger than predictions based on conventional pQCD. The SCI model included in PYTHIA accomplish this through the standard pQCD parton level processes of heavy quark pair production. The most important contribution comes from a high-$p_\perp$ gluon which splits in a $Q\bar{Q}$ pair, *e.g.* the next-to-leading process $gg \to gQ\bar{Q}$, where the colour octet charge of the $Q\bar{Q}$ can be turned into a singlet through SCI. The $Q\bar{Q}$ pairs with mass below the threshold for open heavy flavour production, are then mapped onto the various quarkonium states using spin statistics. The results [47] are in good agreement with the data, both in terms of absolute normalization and the shapes. Also details like the rates of different quarkonium states and the fraction of J/ψ produced directly or from decays are reproduced quite well.

This simple model for soft colour interactions is quite successful to describe a lot of data, both for diffractive and non-diffractive events. Of course, it is only a very simple model and far from a theory, but it may lead to a proper description. A very recent step in this direction is the use an area law for string dynamics [48].

The SCI model has similarities with other attempts to understand soft dynamics. Soft interactions of a colour charge moving through a colour medium has been considered and argued to give rise to large K-factors in Drell-Yan processes and synchrotron radiation of soft photons [49]. A semi-classical approach to describe the interaction of a $q\bar{q}$ pair with a background colour field of a proton has been developed into a model for diffraction in DIS [50]. The $q\bar{q}$, which is here a fluctuation of the exchanged virtual photon, can emerge in a colour singlet state after the interaction with the proton such that a rapidity gap can arise. This provides a very interesting theoretical framework giving results in basic agreement with data, although one cannot make as detailed comparisons as with a Monte Carlo model.

Other attempts to gain understanding through phenomenological models have also been made in the same general spirit as the SCI model. The colour evaporation model [51] can reproduce rapidity gap data and charmonium production with fitted parameters to regulate the probability of forming colour singlet systems. Changes of colour string topologies have also been investigated [52] in a different context, namely $e^+e^- \to W^+W^- \to q_1\bar{q}_2q_3\bar{q}_4$. This gives two strings that may interact and cause colour reconnections resulting in a different string topology affecting the W mass reconstruction and Bose-Einstein effects.

In conclusion, there has been an increased interest in recent years to explore non-pQCD through various theoretical attempts and phenomenological soft interaction models.

7. Rapidity gaps between jets

The diffractive events discussed so far always had a rapidity gap adjacent to a leading proton or small mass system. The momentum transfer between the initial proton and this very forward system is always very small (exponential t-distribution) as characteristic of soft processes. This applies whether the high-mass X-system contains hard scattering or not. In $p\bar{p}$ collisions at the Tevatron one has discovered a new kind of rapidity gaps, namely where the gap is in the central region and between two jets with high $p_{\perp}$, *i.e.* 'jet-gap-jet' events.

In a sample of $p\bar{p}$ events at $\sqrt{s} = 1800\,GeV$ having two jets with transverse energy $E_{\perp}^{jet} > 20\,GeV$, pseudorapidity $1.8 < |\eta_{jet}| < 3.5$ and $\eta_{jet1}\eta_{jet2} < 0$, CDF finds [53] that a fraction $R_{jgj} = (1.13 \pm 0.12 \pm 0.11)\%$ has a rapidity gap within $|\eta| < 1$ between the jets. At $\sqrt{s} = 630\,GeV$ the CDF result is $R_{jgj} = (2.7 \pm 0.7 \pm 0.6)\%$ with $E_{\perp}^{jet} > 8\,GeV$ which corresponds to approximately the same momentum fraction x of the interacting partons at the two cms energies. D0 finds [54] very similar results in terms of 'colour singlet fractions' $f_s = (0.94 \pm 0.04 \pm 0.12)\%$ for $E_{\perp}^{jet} > 30\,GeV$ at $\sqrt{s} = 1800\,GeV$ and $f_s = (1.85 \pm 0.09 \pm 0.37)\%$ for $E_{\perp}^{jet} > 12\,GeV$ at $\sqrt{s} = 630\,GeV$. Although the CDF and D0 event selections and analyses differ, the resulting relative rates of jet-gap-jet events are quite similar. They are definitely larger at the lower energy. In D0 the ratios tend to increase with increasing $E_{\perp}^{jet}$ and rapidity separation between the jets, but the CDF data shows no significant such effect.

The jet-gap-jet events can be interpreted in terms of colour singlet exchange. However, the momentum transfer $|t| \sim E_{\perp jet}^{2} > 100\,GeV^{2}$ is very large in contrast to the small t in ordinary diffraction. An interpretation in terms of the Regge pomeron is therefore not possible, but attempts have been made using pQCD models of two-gluon exchange. Such models seems at first to give energy and $E_{\perp}^{jet}$ dependences that are not consistent with the data, but recent developments indicate that this need not be the case [55]. The salient features of the data can, on the other hand, be interpreted in terms of the colour evaporation model [56]. A problem with both these approaches is however, that they do not take proper account of higher order pQCD parton emissions, multiple parton-parton scattering and hadronization. These are well known problems for the understanding of the 'underlying event' in hadron-hadron collisions and must be investigated with detailed Monte Carlo models. For example, the perturbative radiation in a high-$p_\perp$ scattering must be included since it cannot be screened by soft interactions. The proposed models attempts to describe all these effects through a 'gap survival probability' [6]. However, a real understanding of gap between jets is still lacking.

8. Conclusions

Diffractive hard scattering has in recent years been established as a field of its own with many developments in both theory and experiment. Rapidity gap events have been observed with various hard scattering processes; high-$p_\perp$ jet and W production, and deep inelastic scattering.

The model with a pomeron having a parton structure is quite successful in describing data, in particular for diffractive DIS at HERA where parton densities in the pomeron have been extracted. However, the pomeron model has some problems. The pomeron flux and/or the pomeron parton densities are not universal to all kinds of interactions, or they are more complicated with, *e.g.*, a flux renormalization. Even if such modified pomeron models can be made to describe data both from ep and $p\bar{p}$, there are conceptual problems with the pomeron. In particular, it is doubtful whether the pomeron can be viewed as a separate entity which is decoupled from the proton during the long space-time scale of the soft interaction.

The general problem is soft interactions in non-perturbative QCD. Perhaps Regge theory is the proper soft limit of QCD, but it may also exist more fruitful roads towards a theory for soft interactions. This has generated an increased interest to explore new theoretical approaches and phenomenological models.

A new trend is to consider the interactions of partons with a colour background field. The hard pQCD processes should then be treated as usual, but soft interactions are added which change the colour topology resulting in a different final state after hadronization. In the Monte Carlo model for soft colour interactions this gives a unified description with a smooth transition between diffractive and non-diffractive events. The different event classes can then be defined as in experiments, *e.g.* in terms of rapidity gaps or leading protons. This model and others in a similar general spirit can describe the salient features of many different kinds of experimental data.

Nevertheless, there are many unsolved problems that are challenging to solve. In particular, the events with a rapidity gap between two high-$p_\perp$ jets are poorly understood. Progress in the field of diffractive hard scattering will contribute to the ultimate goal: to understand non-perturbative QCD.

Acknowledgments: I am grateful to Tom Ferbel and all the participants for a most enjoyable school.

References

1. K. Goulianos, Phys. Rep. 101 (1983) 169
2. G. Ingelman, P.E. Schlein, Phys. Lett. B152 (1985) 256
3. UA8 collaboration, R. Bonino et al., Phys. Lett. B211 (1988) 239
4. A. Edin, G. Ingelman, J. Rathsman, Phys. Lett. B366 (1996) 371
5. B. Andersson, G. Gustafson, G. Ingelman, T. Sjöstrand, Phys. Rep. 97 (1983) 31
6. J.D. Bjorken, Phys. Rev. D47 (1993) 101; SLAC-PUB-6463, SLAC-PUB-6477 (1994)
7. A.M. Smith et al., Phys. Lett. B167 (1986) 248
   D. Bernard et al. UA4 collaboration, Phys. Lett. B166 (1986) 459
8. WA91 collaboration, Phys. Lett. B324 (1994)509; B353 (1995) 589
9. F.E. Low, Phys. Rev. D12 (1975) 163
   S. Nussinov, Phys. Rev. Lett. 34 (1975) 1286; Phys. Rev. D14 (1976) 246
10. Yu.A. Simonov, Phys. Lett. B249 (1990) 514; Nucl. Phys. B (Proc. Suppl.) 23 (1991) 283
11. A. Donnachie, P.V. Landshoff, Nucl. Phys. B244 (1984) 322
12. A.M. Smith et al., Phys. Lett. B163 (1985) 267
13. P. Pruni, G. Ingelman, Phys. Lett. B311 (1993) 317
    P. Bruni, A. Edin, G. Ingelman, POMPYT 2.6, http://www3.tsl.uu.se/thep/pompyt/
14. UA8 collaboration, A. Brandt et al., Phys. Lett. B297 (1992) 417
15. UA8 collaboration, A. Brandt et al., Phys. Lett. B421 (1998) 395
16. K. Eggert, UA1, in 'Elastic and diffractive scattering', Ed. K. Goulianos, Editions Frontieres 1988, p. 1
17. P. Bruni, G. Ingelman, DESY-93-187, in proc. International Europhysics Conference on High Energy Physics, Marseille 1993, Eds. J. Carr, M. Perrottet, Editions Frontieres 1988, p. 595
18. E.L. Berger, J.C. Collins, D.E. Soper, G. Sterman, Nucl. Phys. B286 (1987) 704
19. A. Donnachie, P.V. Landshoff, Phys. Lett. B191 (1987) 309; Nucl. Phys. B303 (1988) 634
20. T. Sjöstrand, Comp. Phys. Commun. 82 (1994) 74
21. G. Ingelman, K. Janson-Prytz, *in* proc. Physics at HERA, Eds. W. Buchmüller, G. Ingelman, DESY 1991, p. 233
    G. Ingelman, K. Prytz, Z. Phys. C58 (1993) 285
22. A. Donnachie, P.V. Landshoff, *in* proc. Physics at HERA, Ed. R.D. Peccei, DESY 1987, p. 351
    K.H. Streng, *ibid* p. 365
    P. Bruni, G. Ingelman, A. Solano, *in* proc. Physics at HERA, Eds. W. Buchmüller, G. Ingelman, DESY 1991, p. 353
23. ZEUS collaboration, M. Derrick et al., Phys. Lett. B315 (1993) 481; Z. Phys. C68 (1995) 569
24. H1 collaboration, Nucl. Phys. B429 (1994) 477; Phys. Lett. B348 (1995) 681
25. C. Adloff et al., H1 collaboration, DESY-98-169
26. L. Lönnblad, ARIADNE version 4, Comp. Phys. Comm. 71 (1992) 15
27. H1 collaboration, C. Adloff et al., Z. Phys. C76 (1997) 613
28. G. Altarelli, G. Parisi, Nucl. Phys. B126 (1977) 298
    V.N. Gribov, L.N. Lipatov, Sov. J. Nucl. Phys. 15 (1972) 438
    Yu. Dokshitzer, Sov. Phys. JETP 46 (1977) 641
29. H. Jung, Comp. Phys. Commun. 86 (1995) 147
30. P. Bruni, G. Ingelman, Phys. Lett. B311 (1993) 317
31. F. Abe et al., CDF collaboration, Phys. Rev. Lett. 78 (1997) 2698
32. F. Abe et al., CDF collaboration, Phys. Rev. Lett. 79 (1997) 2636
33. D0 collaboration, submitted to the XXIX International Conferrence on High Energy Physics, Vancouver 1998
34. M. Albrow, CDF Collaboration, FERMILAB-CONF-98/138-E *in* Proc. LISHEP 98 workshop on Diffractive Physics, Rio de Janeiro 1998

35. K. Goulinaos, Phys. Lett. B358 (1995) 379; hep-ph/9805496
36. S. Erhan, P.E. Schlein, Phys. Lett. B427 (1998) 389
37. J.C. Collins, L. Frankfurt, M. Strikman, Phys. Lett. B307 (1993) 161
38. P.V. Landshoff, *in* Workshop on DIS and QCD, Paris 1995, Eds. J.F. Laporte, Y. Sirois, p. 371
39. J. Bartels, G. Ingelman, Phys. Lett. B235 (1990) 175
40. L.V. Gribov, E.M. Levin, M.G. Ryskin, Phys. Rep. 100 (1983) 1
41. N.N. Nikolaev, B.G. Zakharov, Z. Phys. C53 (1992) 331
M. Wüsthoff, Phys. Rev. D56 (1997) 4311
J. Bartels, J. Ellis, H. Kowalski, M. Wusthoff, CERN-TH-98-67, hep-ph/9803497
42. W. Buchmüller, A. Hebecker, Phys. Lett. B355 (1995) 573
43. A. Edin, G. Ingelman, J. Rathsman, Z. Phys. C75 (1997) 57
44. G. Ingelman, A. Edin, J. Rathsman, Comput. Phys. Commun. 101 (1997) 108
http://www3.tsl.uu.se/thep/lepto/
45. A. Edin, G. Ingelman, J. Rathsman, in proc. 'Future physics at HERA', Eds. G. Ingelman, A. De Roeck, A. Klanner, DESY 96-235, p. 580
46. R. Enberg, G. Ingelman, N. Timneanu, in preparation
47. A. Edin, G. Ingelman, J. Rathsman, Phys. Rev. D56 (1997) 7317
48. J. Rathsman, SLAC-PUB-8034, hep-ph/9812423
49. O. Nachtmann, A. Reiter, Z. Phys. C24 (1984) 283
G.W. Botz, P. Heberl, O. Nachtmann, Z. Phys. C67 (1995) 143
50. W. Buchmüller, A. Hebecker, Nucl. Phys. B476 (1996) 203
51. J.F. Amundson, O.J.P. Eboli, E.M. Gregores, F. Halzen, Phys. Lett. B372 (1996) 127; Phys.Lett.B390 (1997) 323
52. G. Gustafson, U. Pettersson, P. Zerwas, Phys. Lett. B209 (1988) 90
T. Sjöstrand, V. Khoze, Z. Phys. C62 (1994) 281
G. Gustafson, J. Häkkinen, Z. Phys. C64 (1994) 659
53. F. Abe et al., CDF collaboration, Phys. Rev. Lett. 80 (1998) 1156; *ibid.* 81 (1998) 5278
54. B. Abbott et al., D0 collaboration, Phys. Lett. B440 (1998) 189
55. R. Oeckl, D. Zeppenfeld, Phys.Rev.D58:014003,1998
56. O.J.P. Eboli, E.M. Gregores, F. Halzen, Phys.Rev.D58:114005,1998

Participants at the ASI (scanning from left to right): Iris Abt, Andrea Valassi, Arthur Maghakian (companion), Tejinder Virdee, Brian Earl, David Whittum, Wander Baldini, Martin Braeuer, David Calvet, James Stirling, Cristina Corloganu, Pascal Zini, Breese Quinn, Alex Oh, James Weatherall, Alessandro Buffini, Rene Janicek, Nick Robertson, Jim Graham, Roberta Arcidiacono, Ivan Mikulec, Dmitry Vavilov, Nicholas Savvas, Artur Vaitaitis, Kate Frame, Roy Aleksan, Thomas Nunnemann, Tom Fahland, Doug Bryman (barely visible), Robin Erbacher, Tom Ferbel, Rifat Hossein, Claudia Rossi, Luca Casagrande, Vitaliano Ciulli, Larry Nodulman (barely visible), Phil Clark, Niccolo Moggi, Felicitas Pauss, Amanda Weinstein, Rocky Kolb, Jonathan Hays, Vittorio Palmieri, Deborah Sciarrino, Edmund Clay, Gunnar Ingelman, Malcolm John, Sergey Likhoded, Arnold Pompos, Joao Guimaraes de Costa, Ahmet Ayan, Scott Metzler, Alexey Ershov, Aysel Kayis, Max Sang, Jane Tinslay, Andrei Solodsky, Martijn Mulders, Paul Harrison (companion), Nirmalya Parua, Phil Strother, Bjoern Schwenninger, Wolfgang Wagner, Jeff McDonald (barely visible), Volker Buescher, Raphael Granier de Cassagnac, Paul Bergbush, Maria-Jose Herrero, Felix Rosenbaum, David Eatough, Theresa Champion, Sophie Hoorelbeke, Paul Ngan, Borut Erzen, Christina Mesropian, Kenzo Nakamura, and Matthew Chalmers (Missing: Harpreet Singh).

INDEX